Lecture Notes in Computer Science 8957

Commenced Publication in 1973
Founding and Former Series Editors:
Gerhard Goos, Juris Hartmanis, and Jan van Leeuwen

More information about this series at http://www.springer.com/series/7410

Dongdai Lin · Moti Yung
Jianying Zhou (Eds.)

Information Security and Cryptology

10th International Conference, Inscrypt 2014
Beijing, China, December 13–15, 2014
Revised Selected Papers

 Springer

Editors
Dongdai Lin
SKLOIS, Institute of Engineering
Chinese Academy of Sciences
Beijing 100093
China

Jianying Zhou
Infocomm Security Department
Institute for Infocomm Research
Singapore
Singapore

Moti Yung
Computer Science Department
Columbia University
New York, NY 10027
USA

ISSN 0302-9743   ISSN 1611-3349   (electronic)
Lecture Notes in Computer Science
ISBN 978-3-319-16744-2   ISBN 978-3-319-16745-9   (eBook)
DOI 10.1007/978-3-319-16745-9

Library of Congress Control Number: 2015934900

LNCS Sublibrary: SL4 – Security and Cryptology

Printed on acid-free paper

Springer International Publishing AG Switzerland is part of Springer Science+Business Media
(www.springer.com)

Preface

This volume contains the papers presented at Inscrypt 2014: The 10th China International Conference on Information Security and Cryptology held during December 13–15, 2014 in Beijing, China. Since its inauguration in 2005, Inscrypt has become a well-recognized annual international forum for security researchers and cryptographers to exchange ideas.

The conference received 93 submissions. Each submission was reviewed by 2–4 Program Committee members. The Program Committee decided to accept 29 papers. The overall acceptance rate was, therefore, about 31%. The program also included six invited talks.

Inscrypt 2014 was held in cooperation with the International Association for Cryptologic Research (IACR), and was co-organized by the State Key Laboratory of Information Security (SKLOIS) of the Chinese Academy of Sciences (CAS), and the Chinese Association for Cryptologic Research (CACR). Inscrypt 2014 was also partly supported by the Institute of Information Engineering (IIE) of the Chinese Academy of Sciences, the Natural Science Foundation of China (NSFC), and the Priority Strategic Program NICT of Chinese Academy of Sciences. The conference could not have been a success without the support of these organizations, and we sincerely thank them for their continued assistance and help.

We would also like to thank the authors who submitted their papers to Inscrypt 2014, and the conference attendees for their interest and support that made the conference possible. We thank the Organizing Committee for their time and efforts that allowed us to focus on selecting papers. We thank the Program Committee members and the external reviewers for their hard work in reviewing the submissions; the conference would not have been possible without their expert reviews. Last but not least, we thank the EasyChair system and its operators, for making the entire process of the conference convenient.

December 2014

Dongdai Lin
Moti Yung
Jianying Zhou

Inscrypt 2014

10th China International Conference on Information Security and Cryptology

Beijing, China
December 13–15, 2014

Sponsored and organized by

State Key Laboratory of Information Security
(Chinese Academy of Sciences)
Chinese Association for Cryptologic Research

in cooperation with

International Association for Cryptologic Research

General Chair

Xiaoyun Wang   Tsinghua University, China

Program Co-chairs

Dongdai Lin   SKLOIS, Institute of Information Engineering,
                 Chinese Academy of Sciences, China
Moti Yung   Google Inc. and Columbia University, USA
Jianying Zhou   Institute for Infocomm Research, Singapore

Organization Co-chairs

Yanping Yu   Chinese Association for Cryptologic Research,
                 China
Chuankun Wu   SKLOIS, Institute of Information Engineering,
                 Chinese Academy of Sciences, China
Yongbin Zhou   SKLOIS, Institute of Information Engineering,
                 Chinese Academy of Sciences, China

Publicity Chair

Yu Chen   SKLOIS, Institute of Information Engineering,
           Chinese Academy of Sciences, China

Technical Program Committee

Long, Yu
Lu, Yang
Lu, Yao
Lucks, Stefan
Ma, Hui
Mao, Xianping
Mikhalev, Vasily
Ohkubo, Miyako
Omote, Kazumasa
Perret, Ludovic
Petit, Christophe
Pitropakis, Nikolaos
Rao, Vanishree
Rizomiliotis, Panagiotis
Roy, Sankardas
Ruan, Ou
Samanthula, Bharath Kumar
Sarkar, Santanu
Shirase, Masaaki
Snook, Michael
Sui, Han
Susil, Petr
Tanaka, Satoru
Tang, Fei
Tang, Qiang
Tao, Chengdong
Tischhauser, Elmar

Tsoulkas, Vasilis
Venkateswarlu, Ayineedi
Vouyioukas, Demosthenes
Vrakas, Nikos
Wang, Daniel
Wang, Ding
Wang, Gaoli
Wang, Liangliang
Wang, Yanfeng
Wenger, Erich
Xu, Hong
Yan, Fei
Yang, Rupeng
Yasuda, Takanori
Yuan, Wei
Zhang, Bingsheng
Zhang, Huang
Zhang, Lei
Zhang, Liqiang
Zhang, Tao
Zhang, Yuanyuan
Zhang, Yuexin
Zhang, Zongyang
Zhao, Chang'An
Zhao, Fangming
Zomlot, Loai

Contents

Privacy and Anonymity

An Efficient Privacy-Preserving E-coupon System

Weiwei Liu[✉], Yi Mu, and Guomin Yang

School of Computer Science and Software Engineering,
University of Wollongong, Wollongong, NSW 2522, Australia
wl265@uowmail.edu.au, {ymu,gyang}@uow.edu.au

Abstract. Previous work on electronic coupon (e-coupon) systems mainly focused on security properties such as unforgeability, double-redemption detection, and anonymity/unlinkability. However, achieving both traceability against dishonest users and anonymity for honest users without involving any third party is an open problem that has not been solved by the previous work. Another desirable feature of an e-coupon system that has not been studied in the literature is user privacy, which means the shop cannot identify the good (among all the choices specified in the coupon) that has been chosen by the customer during the redemption process. In this paper, we present a novel e-coupon system that can achieve all these desirable properties. We define the formal security models for these new security requirements, and show that our new e-coupon system is proven secure in the proposed models.

Keywords: Unforgeability · Anonymity · Double-redemption detection · Traceability

1 Introduction

An electronic coupon (or e-coupon) can be used by a user to obtain an electronic good or service from a service provider, which is usually a coupon issuer. E-coupon systems are similar but different from electronic cash (or e-cash) systems [2,4,9,15]. One major difference is that an e-coupon system only involves two parties: the user and the coupon issuer, while in an e-cash system there is a third party which is the bank. An e-coupon system has less algorithms/protocols compared with e-cash. The coupon issuer can issue a coupon to a user through a coupon issuing algorithm; then the user can redeem the coupon at a later time to obtain the good/service specified in the coupon. Similar to e-cash systems, e-coupon systems are very useful in e-commerce, especially when the shops don't want to have the bank involved in the transactions.

There are a number of e-coupon systems proposed in the literature. Chen et al. [10] presented a privacy-preserving e-coupon system, in which the users are allowed to redeem a single e-coupon for a pre-determined number of times. In order to reduce the cost for issuing and redeeming coupons, Nguyen [16]

© Springer International Publishing Switzerland 2015
D. Lin et al. (Eds.): Inscrypt 2014, LNCS 8957, pp. 3–15, 2015.
DOI: 10.1007/978-3-319-16745-9_1

later presented another more efficient e-coupon system which has constant communication and computation costs. In [7], Canard et al. also proposed another interesting multi-coupon system which allows a user to transfer some value of a multi-coupon to another user.

Although user anonymity has been considered in all the above e-coupon systems, none of them has considered traceability against dishonest users. That is, if a dishonest user redeems a multi-coupon more than the pre-determined number of times, it is desirable to allow the coupon issuer to trace the identity of the user. On the other hand, for honest users, we should ensure that their identities will remain anonymous to the coupon issuer.

Another desirable feature of an e-coupon system is user privacy. Different from user anonymity, here we are concerning the privacy of the goods/services chosen by the users during the redemption process. In general, an e-coupon can contain a number of options that have the same value and a user can choose any one of them. If the shop can know the good/service chosen by the user in the redemption process, then it is possible for the shop to link two redemptions performed by the same customer (e.g., a customer may prefer to redeem the same item among several transactions). Therefore, it is also desirable to keep the buying behavior of a user secret from the coupon issuer. It is worth noting that user privacy is possible in the electronic world since we are considering electronic goods, so there is no physical reduction of the goods from the shop's 'warehouse', while in the physical world, the shop can always trace the number of each good to find out the item redeemed by the user.

In this paper, we propose a new e-coupon system, which can achieve all the desirable properties mentioned above, namely unforgeability, anonymity for honest users, double redemption detection, traceability against dishonest users, and user privacy. It is worth noticing that different from a fair e-cash system [12,14,18], the traceability in our e-coupon system is performed by the merchant (i.e., coupon issuer) rather than the bank, which makes the task more challenging. In order to achieve unforgeability, anonymity for honest users, double redemption detection and traceability against dishonest users, we design a new variant of blind signature which allows the signer (which is essentially the coupon issuer) to issue a signature (coupon) on a message without seeing its content. However, different from conventional blind signature schemes [1,5,6,8,13], our scheme involves an extra dynamic challenge-response verification in the verification phase to ensure that if a coupon is redeemed more than once, the identity of the coupon holder can be calculated. In order to achieve user privacy, we employ a mechanism that is similar to oblivious transfer in the redemption protocol. In addition, we present a multi-coupon extension of our basic scheme to allow multi-use of an e-coupon.

The rest of the paper is organized as follows. We provide some definitions and security assumptions in Sect. 2. The formal security model for our e-coupon system is presented in Sect. 3. We then present our construction in Sect. 4 and prove its security in Sect. 5. We propose an extension of our system in Sect. 6 and the paper is concluded in Sect. 7.

2 Definition and Assumptions

2.1 Definition of an E-coupon System

An e-coupon system consists of two participants, namely, a user and a coupon issuer, which is also a service provider or shop. Our e-coupon system consists of the following algorithms.

1. **ParamGen:** On input a security parameter κ, the parameter generation algorithm outputs the public parameters.

$$params \leftarrow \textbf{ParamGen}(1^\kappa).$$

2. **KeyGen:** On input the public parameter $params$, the key generation algorithm outputs a key pair for a user or a service provider.

$$(pk, sk) \leftarrow \textbf{KeyGen}(params).$$

3. **Issue:** The issue algorithm is an interactive protocol between the service provider $\mathcal{S}$ and a user $\mathcal{U}$,

$$C \leftarrow \textbf{Issue}(\mathcal{S}(pk_\mathcal{S}, sk_\mathcal{S}), \mathcal{U}(pk_\mathcal{U}, params)).$$

The output is an e-coupon for the user.

4. **Redeem:** The redeem algorithm is an interactive protocol between a user $\mathcal{U}$ and the service provider $\mathcal{S}$, taking as input an e-coupon C, the public key $pk_\mathcal{S}$ of the service provider, the public parameters $params$, a challenge c from the service provider $\mathcal{S}$ and a corresponding response R from the user $\mathcal{U}$. The output of this algorithm for the service provider is accept or reject.

$$\text{'}accept\text{'} \text{ or } \text{'}reject\text{'} \leftarrow \textbf{Redeem}_\mathcal{S}(pk_\mathcal{S}, params, C, c, R).$$

The output of this algorithm for the user is the item $item_i$ of his choice or a failure symbol.

$$\text{'}\bot\text{'} \text{ or } item_i \leftarrow \textbf{Redeem}_\mathcal{U}(pk_\mathcal{S}, params, C, c, R).$$

5. **Reveal:** The reveal algorithm is executed by the service provider, taking a sequence of the challenge-response pairs and the corresponding redeemed coupon $\{(c_1, R_1), (c_2, R_2), C\}$ and the public parameters $params$ as input, outputs the identity $ID_\mathcal{U}$ of the corresponding user or a failure symbol '$\bot$'.

$$ID_\mathcal{U} \text{ or } \text{'}\bot\text{'} \leftarrow \textbf{Reveal}((c_1, R_1), (c_2, R_2), C, params).$$

2.2 Complexity Assumptions

Definition 1 *Decisional Diffie-Hellman (DDH) Assumption: Given a cyclic group G of order q, the DDH assumption states that, given $g, g^a, g^b, Z \in G$ for some unknown $a, b \in \mathbb{Z}_q$ and a random generator g, decide whether $Z = g^{ab}$.*

Definition 2 *Knowledge of Exponent Assumption [3]: Given a cyclic group G of order q, for any adversary $\mathcal{A}$ that takes input q, g, g^a and returns (C, Y) with $Y = C^a$, there exists an "extractor" $\bar{\mathcal{A}}$, which given the same inputs as $\mathcal{A}$ returns c such that $g^c = C$.*

Definition 3 *One More Discrete Logarithm (OMDL) Assumption: Given a cyclic group G of order q and g is a generator of G, Let $DLog_g(\cdot)$ be the discrete logarithm oracle that takes an element in G_q and returns the discrete logarithm in base g. Let $C(\cdot)$ be a challenge oracle which takes no input and returns a random element in G_q. Let $W_1, W_2, \ldots, W_t$ denote the challenges returned by $C(\cdot)$, we say an OMDL adversary $\mathcal{A}$ wins if $\mathcal{A}$ can output a sequence of $w_1, w_2, \ldots, w_t \in \mathcal{Z}_q$ satisfying $g^{w_i} = W_i$ and the number of queries make by $\mathcal{A}$ to the discrete logarithm oracle $DLog(\cdot)$ is less than t.*

3   Security Model

We formalize four security requirements for our e-coupon system, that is unforgeability, user anonymity, double-redemption detection, and user privacy.

3.1   Unforgeability

Unforgeability requires that an adversary $\mathcal{A}$ (could be a malicious user) cannot forge a new valid coupon that can be redeemed successfully with an honest service provider $\mathcal{S}$.

The adversarial game for unforgeability between an adversary $\mathcal{A}$ and a simulator $\mathcal{B}$ is defined as follows:

1. **ParamGen:** The simulator $\mathcal{B}$ runs algorithm **ParamGen** to generate public parameters $params$.
2. **KeyGen:** The simulator $\mathcal{B}$ generates two key pairs $(pk_\mathcal{S}, sk_\mathcal{S})$ and $(pk_\mathcal{U}, sk_\mathcal{U})$, $\mathcal{B}$ sends $pk_\mathcal{S}$ and $(pk_\mathcal{U}, sk_\mathcal{U})$ to $\mathcal{A}$ and keeps $sk_\mathcal{S}$ secret.
3. **Issue queries:** Assume $\mathcal{A}$ makes q_s issue queries to the issuing oracle $\mathcal{I}(\cdot)$, for the i-th query, $1 \leq i \leq q_s$, $\mathcal{A}$ runs the issue protocol with $\mathcal{B}$ in an interactive manner, after each query, $\mathcal{A}$ obtains a coupon

   $$C_i \leftarrow \text{Issue}(\mathcal{S}(pk_\mathcal{S}, sk_\mathcal{S}), \mathcal{U}(pk_\mathcal{U}, params)).$$

4. **Challenge:** Finally, $\mathcal{A}$ outputs a new coupon C^*. We say $\mathcal{A}$ wins the game if this coupon has not appeared in any issue query but can be redeemed successfully by $\mathcal{A}$, i.e.,
   - 'accept' $\leftarrow$ **Redeem**$_\mathcal{S}(pk_\mathcal{S}, params, C^*, c, R)$.
   - $C^* \neq C_i$, for $1 \leq i \leq q_s$.

Define the advantage of a adversary $\mathcal{A}$ in winning the unforgeability game as

$$\mathbf{Adv}_\mathcal{A}^{unf}(\kappa) = \Pr[\mathcal{A} \text{ wins the game}]$$

Definition 1. *An e-coupon system is said to be unforgeable if $\mathbf{Adv}_\mathcal{A}^{unf}(\kappa)$ is negligible for any PPT adversary $\mathcal{A}$.*

3.2   Anonymity

Anonymity requires that if one user follows the protocol honestly, even a malicious service provider cannot link one redeemed coupon to the identity of the user. The adversarial game between $\mathcal{A}$ and simulator $\mathcal{B}$ for anonymity is defined as follows:

1. **ParamGen:** The simulator $\mathcal{B}$ runs algorithm **ParamGen** to generate public parameters $params$.
2. **KeyGen:** The simulator $\mathcal{B}$ generates key pairs for a service provider $(pk_{\mathcal{S}}, sk_{\mathcal{S}})$ and two users $\mathcal{U}_0$ $(pk_{\mathcal{U}_0}, sk_{\mathcal{U}_0})$ and $\mathcal{U}_1$ $(pk_{\mathcal{U}_1}, sk_{\mathcal{U}_1})$ respectively, $\mathcal{B}$ sends $(pk_{\mathcal{S}}, sk_{\mathcal{S}}, pk_{\mathcal{U}_0}, pk_{\mathcal{U}_1})$ to $\mathcal{A}$.
3. **Issue queries:** Assume $\mathcal{A}$ runs **Issue** algorithm q times with $\mathcal{U}_0$ and $\mathcal{U}_1$ respectively. Let $C^0 = \{C_{\mathcal{U}_0}^1, C_{\mathcal{U}_0}^2, \ldots, C_{\mathcal{U}_0}^q\}$ and $C^1 = \{C_{\mathcal{U}_1}^1, C_{\mathcal{U}_1}^2, \ldots, C_{\mathcal{U}_1}^q\}$ be the coupon set obtained by $\mathcal{U}_0$ and $\mathcal{U}_1$.
4. **Challenge:** $\mathcal{A}$ outputs an index $1 \leq i \leq q$. $\mathcal{B}$ flips a coin to decide a value $b^* \in \{0, 1\}$, and returns $C_{\mathcal{U}_{b^*}}^i$ to $\mathcal{A}$. $\mathcal{A}$ makes a guess b' of the value b^*.

We say $\mathcal{A}$ wins the game if $b' = b^*$. Define the advantage of a adversary $\mathcal{A}$ in winning the game as

$$\mathbf{Adv}_{\mathcal{A}}^{Ano}(\kappa) = \Pr[\mathcal{A} \text{ wins the game}] - \frac{1}{2}$$

Definition 2. *An e-coupon system is said to provide anonymity if $\mathbf{Adv}_{\mathcal{A}}^{Ano}(k)$ is negligible for any PPT adversary $\mathcal{A}$.*

3.3   Double-Redemption Detection

Detection of double-redemption is a major concern for any digital coupon system. An e-coupon system is said to provide double-redemption detection if one user cannot redeem one coupon twice with the same service provider without being caught. In our e-coupon system, if one coupon is redeemed twice, the service provider can find a polynomial time algorithm to trace the identity of the user with overwhelming probability. The adversarial game for double-redemption detection is defined as follows:

1. **ParamGen:** The simulator $\mathcal{B}$ runs algorithm **ParamGen** to generate public parameters $params$.
2. **KeyGen:** The simulator $\mathcal{B}$ generates two key pairs $(pk_{\mathcal{U}}, sk_{\mathcal{U}})$ and $(pk_{\mathcal{S}}, sk_{\mathcal{S}})$, $\mathcal{B}$ sends $(pk_{\mathcal{U}}, sk_{\mathcal{U}})$ and $pk_{\mathcal{S}}$ to $\mathcal{A}$.
3. **Issue queries:** Assume $\mathcal{A}$ makes q_d coupon issuing queries. $\mathcal{S}$ runs the **Issue** algorithm with $\mathcal{A}$ to issue a sequence of coupons $\{C_1, C_2, \ldots, C_{q_d}\}$ for $\mathcal{A}$.
4. **Redeem queries:** $\mathcal{A}$ runs the redeem protocol with $\mathcal{S}$ with any coupon of his choice.
5. **Challenge:** $\mathcal{A}$ outputs two pairs (C^*, c_1^*, R_1^*) and (C^*, c_2^*, R_2^*). We say $\mathcal{A}$ wins the game if
   - $(C^*, c_1^*, R_1^*) \neq (C^*, c_2^*, R_2^*)$.

– $\mathbf{Redeem}_\mathcal{S}(pk_\mathcal{S}, params, C^*, c_1^*, R_1^*) = 1$ and $\mathbf{Redeem}_\mathcal{S}(pk_\mathcal{S}, params,$
   $C^*, c_2^*, R_2^*) = 1$.
– '$\perp$'$\leftarrow$$\mathbf{Reveal}((c_1^*, R_1^*), (c_2^*, R_2^*), C^*, params)$

Define the advantage of $\mathcal{A}$ in winning the adversarial game above as

$$\mathbf{Adv}_\mathcal{A}^{drd}(\kappa) = \Pr[\mathcal{A} \text{ wins the game}]$$

Definition 3. *An e-coupon system is said to provide double-redemption detection if $\mathbf{Adv}_\mathcal{A}^{drd}(\kappa)$ is negligible for any PPT adversary $\mathcal{A}$.*

3.4   User Privacy

We formalize a new security property which has not been considered in previous e-coupon systems. When a valid user redeems an e-coupon with the service provider, it is desirable that the service provider cannot make a connection between the coupon from the user and the service that is redeemed by the user if the coupon can be used to redeem an item from a list of options. The adversarial game for user privacy is defined as follows.

1. **ParamGen:** The simulator $\mathcal{B}$ runs algorithm **ParamGen** to generate public parameters *params*.
2. **KeyGen:** The simulator $\mathcal{B}$ generates two key pairs $(pk_\mathcal{U}, sk_\mathcal{U})$ and $(pk_\mathcal{S}, sk_\mathcal{S})$, $\mathcal{B}$ sends $(pk_\mathcal{S}, sk_\mathcal{S})$ and $pk_\mathcal{U}$ to $\mathcal{A}$.
3. **Issue queries:** $\mathcal{A}$ runs the **Issue** algorithm with $\mathcal{B}$ to generate a set of coupons $C = \{C_1, C_2, \ldots, C_{q_R}\}$.
4. **Guess:** $\mathcal{A}$ outputs an index $1 \leq i \leq q_R$. $\mathcal{B}$ then redeems C_i with $\mathcal{A}$ to choose an item m_{b^*} from $\{m_1, m_2, \ldots, m_n\}$, which is the set of items that can be redeemed by $\mathcal{B}$. Finally, $\mathcal{A}$ makes a guess $b' \in [1, n]$ for b^*.

We say $\mathcal{A}$ wins the game if $b' = b^*$. Define the success probability of the adversary $\mathcal{A}$ in making a successful guess about the service that the user choose as

$$\mathbf{Adv}_\mathcal{A}^{up}(\kappa) = \Pr[\mathcal{A} \text{wins the game}] - \frac{1}{n}$$

Definition 4. *An e-coupon system is said to provide user privacy if $\mathbf{Adv}_\mathcal{A}^{up}(\kappa)$ is negligible for any PPT adversary $\mathcal{A}$.*

4   Construction of Our E-coupon System

We denote in our system the service provider by $\mathcal{S}$ and a user by $\mathcal{U}$. Denote $\{m_1, m_2, \ldots, m_n\}$ the set of items that can be redeemed. The detail description of our e-coupon system is as follows.

1. **ParamGen:** On input a security parameter $\kappa \in \mathbb{N}$, generates the system parameters $paras = (G, g, p, q, H_1, H_2)$, where G_q is the subgroup of $\mathbb{Z}_p$ with prime order q and g is a generator of G_q, where $p = 2q + 1$ is also prime. $H_1 : \{0,1\}^* \to G_q$ and $H_2 : G_q \to \{0,1\}^\kappa$ are two collision-resistant hash functions.

2. **KeyGen:** On input a security parameter $\kappa \in \mathbb{N}$ and the public parameter $params$, randomly choose $x, y \in_R \mathbb{Z}_q^*$ and calculate g^x, g^y and output the private and public key pairs $(sk_{\mathcal{U}} = x, pk_{\mathcal{U}} = g^x)$ and $(sk_{\mathcal{S}} = y, pk_{\mathcal{S}} = g^y)$ for the user and service provider respectively.

3. **Issue:** The issue protocol is performed through interactive communications between the service provider $\mathcal{S}$ and a user $\mathcal{U}$. The result of the issue protocol is that $\mathcal{S}$ generates a valid coupon for a user $\mathcal{U}$.
   - On receiving a request from $\mathcal{U}$, $\mathcal{S}$ chooses $k \in_R \mathbb{Z}_q^*$ and computes $\delta_1 \leftarrow pk_{\mathcal{U}}^k$ and $\delta_2 \leftarrow g^k$ sends (δ_1, δ_2) to $\mathcal{U}$.
   - After receiving (δ_1, δ_2) from $\mathcal{S}$, $\mathcal{U}$ checks whether $\delta_1 = \delta_2^{sk_{\mathcal{U}}}$. If the verification fails, $\mathcal{U}$ stops; otherwise, $\mathcal{U}$ chooses $x_1 \in \mathbb{Z}_p^*$ and computes $\alpha \leftarrow (g^{xy})^{x_1}$, $\beta \leftarrow (g^x)^{x_1}$ and $\lambda = g^{x_1}$, $m \leftarrow H_1(\alpha, \beta, \lambda)$. $\mathcal{U}$ chooses two different random number a, b and computes $r \leftarrow m\beta^a \delta_1^{bx_1}$ and $m' \leftarrow H_1(m, r)/b$, $\mathcal{U}$ sends m' to $\mathcal{S}$.
   - $\mathcal{S}$ computes the signature $s' = m'y + k$ on the blind message m' and sends s' to $\mathcal{U}$.
   - $\mathcal{U}$ verifies if $g^{s'} \equiv Y^{m'} \delta_2 \mod p$, if the equation holds, $\mathcal{U}$ removes the blind factor b by calculating $s = s'b + a$ and stores $(\alpha, \beta, \lambda, r, s)$; otherwise, abort.

4. **Redeem:** The redeem protocol is performed as follows:
   - After receiving a redeem request from the user, $\mathcal{S}$ generates a challenge $c = H_1(ID_{\mathcal{S}}||Date||Time)$ and sends c to $\mathcal{U}$.
   - After receiving c, $\mathcal{U}$ computes $R = x_1 + cx_1x$ and choose $\sigma_i \in \{1, 2, \ldots, n\}$ and a random number $a_i \in \mathbb{Z}_q^*$, $w_{\sigma_i} = H_1(\sigma_i)$ and $A = w_{\sigma_i} g^{a_i}$. $\mathcal{U}$ sends $(c, R, \alpha, \beta, \lambda, r, s)$ and A to $\mathcal{S}$.
   - $\mathcal{S}$ checks if $H_1(\alpha, \beta, \lambda) = \beta^{-s} \alpha^{H_1(H_1(\alpha, \beta, \lambda), r)} r$ and $g^R = \lambda \beta^c$. If the equation not holds, aborts; otherwise, $\mathcal{S}$ computes $D = A^y$, $w_i = H_1(i)$ and $c_i = m_i \oplus H_2(w_i^y)$, $i = 1, 2, \ldots, n$. $\mathcal{S}$ sends D and $c_1, c_2, \ldots, c_n$ to $\mathcal{U}$.
   - $\mathcal{U}$ computes $K = D/Y^{a_i}$ and recover $m_{\sigma_i} = c_{\sigma_i} \oplus H_2(K)$.

5. **Reveal:** Assume the coupon $C = (\alpha, \beta, \lambda, r, s)$ is redeemed twice, the $\mathcal{S}$ could get two challenge-response pairs (R_1, c_1) and (R_2, c_2) on C such that $R_1 = x_1 + c_1x_1x$ and $R_2 = x_1 + c_2x_1x$. It is obvious that $\mathcal{S}$ could calculate x and x_1, thus the identity of $\mathcal{U}$ is traced by $\mathcal{S}$.

6. **Correctness:** The correctness check for validity of the coupon is as follows:

$$\beta^{-s} \alpha^{H_1(H_1(\alpha, \beta, \lambda), r)} r$$
$$= (pk_{\mathcal{U}}^{x_1})^{-s} (g^{xyx_1})^r r$$
$$= (pk_{\mathcal{U}}^{x_1})^{-H_1(H_1(\alpha, \beta, \lambda), r)y - kb - a} (pk_{\mathcal{U}}^{x_1})^{H_1(H_1(\alpha, \beta, \lambda), r)y} m(pk_{\mathcal{U}}^{x_1})^a (pk_{\mathcal{U}}^{x_1})^{kb}$$
$$= m$$
$$= H_1(\alpha, \beta, \lambda)$$

The correctness check for a user $\mathcal{U}$ to recover the correct message is as follows:

$$c_{\sigma_i} \oplus H_2(K)$$
$$= m_{\sigma_i} \oplus H_2(w_{\sigma_i}^y) \oplus H_2(A^y/Y^{a_i})$$
$$= m_{\sigma_i} \oplus H_2(w_{\sigma_i}^y) \oplus H_2((w_{\sigma_i} g^{a_i})^y/Y^{a_i})$$
$$= m_{\sigma_i} \oplus H_2(w_{\sigma_i}^y) \oplus H_2((w_{\sigma_i})^y)$$
$$= m_{\sigma_i}$$

5   Security Analysis

5.1   Unforgeability

Theorem 1. *The proposed e-coupon system is unforgeable.*

Proof. The security proof is by contradiction. We will prove that if there exists a PPT adversary $\mathcal{A}$ that can forge a coupon, then there exists another algorithm $\mathcal{B}$ that can break the OMDL assumption with a non-negligible probability. Suppose there exists a polynomial time forge adversary $\mathcal{A}$ which can break the unforgeability of our system with a non-negligible probability ϵ. $\mathcal{B}$ is the simulator in our proof and has access to two types of oracles. The first is discrete logarithm oracle $DLog_{G_{q,g}}(\cdot)$ which takes $P_i \in G_q$ as input and returns $p_i \in \mathbb{Z}_q$ such that $g^{p_i} = P_i$. The second is a challenge oracle $C(\cdot)$ which takes noting as input, but for each time it is invoked it returns a challenge $P \in G_q$. Besides, $\mathcal{B}$ maintains an H-table to record all the hash queries and the corresponding answers. Assume $\mathcal{A}$ makes q_h hash queries and q_s coupon issuing queries, the simulation is as follows:

1. **ParamGen:** $\mathcal{B}$ runs algorithm **ParamGen** to generate public parameters (G, p, q, g, H_1, H_2).
2. **KeyGen:** $\mathcal{B}$ runs **KeyGen** to generate a key pair $(sk_{\mathcal{U}}, pk_{\mathcal{U}})$. $\mathcal{B}$ queries the challenge oracle $C(\cdot)$ and sets the response P_0 as the public key of the shop $pk_{\mathcal{S}} = P_0$. $\mathcal{B}$ sends $(p, g, pk_{\mathcal{S}})$ and $(sk_{\mathcal{U}}, pk_{\mathcal{U}})$ to $\mathcal{A}$.
3. **Hash queries:** For each hash query with an input message m, $\mathcal{B}$ first checks the H-table:
   - If there exists a pair (m, h) in the H-table, where m refers to the message queried before, $\mathcal{B}$ returns h as the answer to $\mathcal{A}$.
   - Otherwise, $\mathcal{B}$ chooses a random $h \in \mathbb{Z}_q$, sends h to $\mathcal{A}$ as the answer for the hash query, and adds (m, h) into the H-table.
4. **Issue queries:** Upon receiving an issuing query, $\mathcal{B}$ make a query to the challenge oracle $C(\cdot)$ and obtains a challenge P_i. $\mathcal{B}$ then sets $(\delta_1, \delta_2) = (P_i^{sk_{\mathcal{U}}}, P_i)$ and sends (δ_1, δ_2) to the adversary. After receiving a message m_i, $\mathcal{B}$ sends $P_i P_0^{m_i}$ to the discrete logarithm oracle $DLog(\cdot)$ and gets a response z_i, and sends z_i to $\mathcal{A}$. Since $z_i = DLog_{G_{q,g}}(P_i P_0^{m_i}) = DLog_{G_{q,g}}(P_i) + m_i DLog_{G_{q,g}}(P_0)$. In $\mathcal{A}$'s view, $\mathcal{B}$ simulates the signer perfectly.

5. **Challenge:** Suppose $\mathcal{A}$ can successfully forge a new coupon $C^* = (\alpha^*, \beta^*, \lambda^*, r^*, s^*)$ where $s^* = ep_0 + r'$, and C^* can pass the redemption protocol. According to the Forking lemma [17] by rewinding $\mathcal{A}$ to the step where $H_1(m^*, r^*) = e$ is determined and providing a new hash value for $H_1(m^*, r^*) = \hat{e}$. $\mathcal{B}$ can generate another valid coupon $\hat{C}^* = (\alpha^*, \beta^*, \lambda^*, r^*, \hat{s}^*)$ where $\hat{s}^* = e'p_0 + r'$. Then $\mathcal{B}$ can compute

$$p_0 = DLog_{G_{q,g}}(P_0) = \frac{s^* - \hat{s}^*}{e - \hat{e}}.$$

Once $\mathcal{B}$ obtains p_0, for each challenge P_i from the challenge oracle $C(\cdot)$, it can calculate $p_i = z_i - m_i p_0$ for each P_i. Therefore, $\mathcal{B}$ can successfully solve the OMDL problem.

5.2 Anonymity

Theorem 2. *The proposed e-coupon system provides anonymity.*

Proof. Anonymity of the user requires the service provider cannot link a redeemed coupon to a honest user. The proof is by contradiction, suppose that there exists a PPT adversary $\mathcal{A}$ which can break anonymity of our e-coupon system with a non-negligible probability ϵ, then we can build an algorithm $\mathcal{B}$ that use $\mathcal{A}$ to solve the DDH problem with a non-negligible probability. Let (g, g^a, g^b, g^z) be an instance of the DDH problem, the purpose of $\mathcal{B}$ is to decide whether $q^z = q^{ab}$. The simulation is as follows:

1. **ParamGen:** $\mathcal{B}$ runs algorithm **ParamGen** to generate public parameters (G, p, q, g, H_1, H_2).
2. **KeyGen:** $\mathcal{B}$ choose two random number $s^*, r_0 \in \mathbb{Z}_p^*$ and computes $S^* = g^{s^*}$, $\mathcal{S}$ sets key pair of the service provider as $(pk_\mathcal{S}, sk_\mathcal{S}) = (S^*, s^*)$ and the public keys of two valid users $\mathcal{U}_0$ and $\mathcal{U}_1$ as $pk_{\mathcal{U}_0} = g^{r_0}$ and $pk_{\mathcal{U}_1} = g^b$ respectively. $\mathcal{B}$ sends $(pk_\mathcal{S}, sk_\mathcal{S})$ and $pk_{\mathcal{U}_0}$, $pk_{\mathcal{U}_1}$ to $\mathcal{A}$.
3. **Issuing queries:** $\mathcal{B}$ performs **Issuing queries** with $\mathcal{A}$ as follows.
   (a) For $\mathcal{U}_0$, $\mathcal{B}$ knows the private key of $\mathcal{U}_0$. Thus $\mathcal{B}$ just follows the **Issue** protocol to obtains a set of coupons $\{C_{\mathcal{U}_0}^1, C_{\mathcal{U}_0}^2, \ldots, C_{\mathcal{U}_0}^{q_c}\}$;
   (b) For $\mathcal{U}_1$, $\mathcal{B}$ simulates the queries as follows:
      – Upon receiving a pair $(\delta_1, \delta_2) = (pk_{\mathcal{U}_1}^{k_i}, g^{k_i})$ from $\mathcal{A}$. $\mathcal{B}$ executes the extractor defined in the KEA assumption to extract the value k_i. If $\mathcal{A}$ misbehaves to generate a fake pair (δ_1', δ_2'). The extractor will return a failure symbol '$\perp$' and thus $\mathcal{B}$ stops this query. Otherwise, $\mathcal{B}$ chooses a random number r_i and sets $\alpha_i = g^{(z)s^*r_i}, \beta_i = g^{(z)r_i}, \lambda_i = g^{(a)r_i}$ and computes $m_i = H_1(\alpha_i, \beta_i, \lambda_i)$.
      – $\mathcal{B}$ chooses two random number $a_i, b_i \in \mathbb{Z}_q$ and computes $r = m_i g^{zr_i a_i} \cdot g^{zk_i b_i}$ and $m' = \frac{H_1(m_i, r)}{b}$. $\mathcal{B}$ sends m' to $\mathcal{A}$.
      – On receiving an $\bar{s}$ from $\mathcal{A}$, $\mathcal{B}$ calculates $s = \bar{s}b_i + a_i$, and stores $(\alpha_i, \beta_i, \lambda_i, r, s)$.

Let $C^0 = \{C^1_{\mathcal{U}_0}, C^2_{\mathcal{U}_0}, \ldots, C^{q_c}_{\mathcal{U}_0}\}$ and $C^1 = \{C^1_{\mathcal{U}_1}, C^2_{\mathcal{U}_1}, \ldots, C^{q_c}_{\mathcal{U}_1}\}$ be the q coupons generated for $\mathcal{U}_0$ and $\mathcal{U}_1$ respectively in this phase.

4. **Challenge:** After receiving the index i, $\mathcal{B}$ flips a coin to decide a value $b^* \in \{0,1\}$ and returns $C^i_{\mathcal{U}_{b^*}}$ to $\mathcal{A}$. $\mathcal{A}$ finally returns b'. $\mathcal{B}$ outputs '1' if $b' = b^*$. Otherwise, $\mathcal{B}$ outputs '0'.

We finish the simulation for the e-coupon system. Assume a PPT $\mathcal{A}$ have a non-negligible probability ϵ in breaking anonymity of our scheme. Then the probability of $\mathcal{B}$ to solve the DDH problem $\mathcal{A}^{DDH}_{\mathcal{B}}(\kappa)$ can be calculated as follows:

$$
\begin{aligned}
\mathcal{A}^{DDH}_{\mathcal{B}}(\kappa) \\
&= \Pr[\mathcal{A} \text{ wins}|g^z = g^{ab}] - \Pr[\mathcal{A} \text{ wins}|g^z = g^r] \\
&= \Pr[b^* = b'|g^z = g^{ab}] - \Pr[b^* = b'|g^z = g^r] \\
&= \frac{1}{2} + \epsilon - (\Pr[b^* = b'|g^z = g^r \wedge b^* = 0]\Pr[b^* = 0] + \Pr[b^* = b'|g^z = g^r \wedge b^* = 1] \\
&\quad \cdot \Pr[b^* = 1]) \\
&= \frac{1}{2} + \epsilon - \frac{1}{2}(\frac{1}{2} + \epsilon + \frac{1}{2}) \\
&= \frac{1}{2}\epsilon
\end{aligned}
$$

which is non-negligible. Thus, we reach a contradiction.

5.3   Double-Spend Detection

Theorem 3. *The proposed e-coupon system provides double-redemption detection.*

Proof. According to our e-coupon system, if $\mathcal{U}$ has double-redeemed a coupon $(\alpha, \beta, \lambda, r, s)$, then $\mathcal{B}$ obtains two different challenge-response pairs (c_1, R_1) and (c_2, R_2) on the coupon where $R_1 = x_1 + c_1 x_1 x$ and $R_2 = x_1 + c_2 x_1 x$, therefore, the secret key x of $\mathcal{U}$ can be easily calculated as follows:

$$
x = \frac{R_2 - R_1}{c_2 R_1 - c_1 R_2}
$$

Thus, the public key of the user could be obtained by the service provider by further calculating $y = g^x \mod p$.

5.4   User Privacy

Theorem 4. *The proposed e-coupon system provides unconditionally user privacy.*

Proof. User privacy of our e-coupon system can be prove by following the receiver's privacy in the oblivious transfer scheme proposed in [11]. For any $A = w_{\sigma_i} g^{a_i}$, there exists w_l and a'_l such that $l \neq \sigma_i$, but $A = w_l g^{a'_l}$. Thus in the service provider's view, A could be a masked value of any index. Thus, the user's choices are unconditionally secure.

6   An Extension of Our E-coupon System

In order to enable the user to redeem one coupon for multiple times, we propose an extension of our e-coupon system. In the extensional system, the user is able to redeems a single coupon for k times. However, if the user misuse the coupon for more than k times then the service provider can trace the identity of the user. The extension scheme is as follows:

1. **ParamGen** and **KeyGen:** Same as above.
2. **Issue:** The issue protocol is performed through interactive communications between the service provider $\mathcal{S}$ and a user $\mathcal{U}$. The result of the issue protocol is that $\mathcal{S}$ generates a valid coupon for a user $\mathcal{U}$.
   - On receiving a request from $\mathcal{U}$, $\mathcal{S}$ chooses $k \in_R \mathbb{Z}_q^*$ and computes $\delta_1 \leftarrow pk_{\mathcal{U}}^k$ and $\delta_2 \leftarrow g^k$ sends (δ_1, δ_2) to $\mathcal{U}$
   - After receiving (δ_1, δ_2) from $\mathcal{S}$, $\mathcal{U}$ checks whether $\delta_1 = \delta_2^{sk_{\mathcal{U}}}$. If the verification fails, $\mathcal{U}$ stops; otherwise, $\mathcal{U}$ chooses $x_1, a_1, a_2, \ldots, a_k \in \mathbb{Z}_q^*$ and computes $\alpha \leftarrow (g^{xy})^{x_1}$, $\beta \leftarrow (g^x)^{x_1}$, $\lambda = g^{x_1}$ and $A_1 = g^{a_1}, A_2 = g^{a_2}, \ldots, A_k = g^{a_k}$, $m \leftarrow H_1(\alpha, \beta, \lambda, A_1, \ldots, A_k)$. $\mathcal{U}$ chooses two different random number a, b and computes $r \leftarrow m\beta^a \delta_1^{bx_1}$ and $m' \leftarrow H_1(m, r)/b$, $\mathcal{U}$ sends m' to $\mathcal{S}$.
   - $\mathcal{S}$ computes the signature $s' = m'y + k$ on the blind message m' and sends s' to $\mathcal{U}$.
   - $\mathcal{U}$ verifies if $g^{s'} \equiv Y^{m''} \delta_2 \mod p$, if the equation holds, $\mathcal{U}$ removes the blind factor b by calculating $s = s'b + a$; otherwise, abort. $\mathcal{U}$ stores $(\alpha, \beta, \lambda, A_1, A_2, \ldots, A_k, r, s)$.
3. **Redeem:** The redeem protocol is performed as follows:
   - After receiving a redeem request from the user, $\mathcal{S}$ generates a challenge $c_i = H_1(ID_{\mathcal{S}}\|Date\|Time)$ and sends c_i to $\mathcal{U}$.
   - After receiving the challenge c_i from the service provider and $\mathcal{U}$ computes $R_i = x_1 + a_1c_i + a_2c_i^2 + \ldots + a_kc_i^k$ and choose $\sigma_i \in \{1, 2, \ldots, n\}$ and a random number $b_i \in \mathbb{Z}_q^*$, $\mathcal{U}$ computes $w_{\sigma_i} = H_1(\sigma_i)$ and $A = w_{\sigma_i}g^{b_i}$. $\mathcal{U}$ sends $(c, R, \alpha, \beta, \lambda, r, s, A_1, \ldots, A_k)$ and A to $\mathcal{S}$.
   - $\mathcal{S}$ checks if $H_1(\alpha, \beta, \lambda, A_1, \ldots, A_k) = \beta^{-s}\alpha^{H_1(H_1(\alpha,\beta,\lambda,A_1,\ldots,A_k),r)}r$ and $g^{R_i} = \lambda A_1^{c_i} A_2^{c_i^2} \ldots A_k^{c_i^k}$. If the equation does not hold, abort; otherwise, $\mathcal{S}$ computes $D = A^y$, $w_i = H_1(i)$ and $c_i = m_i \oplus H_2(w_i^y)$, $i = 1, 2, \ldots, n$. $\mathcal{S}$ sends D and $c_1, c_2, \ldots, c_n$ to $\mathcal{U}$.
   - $\mathcal{U}$ computes $K = D/Y^{b_i}$ and recover $m_{\sigma_i} = c_{\sigma_i} \oplus H_2(K)$.
4. **Reveal:** If the user redeem one single coupon for $k+1$ times, then the service provider can get the following equations:

$$R_1 = x_1 + a_1c_1 + a_2c_1^2 + \ldots + a_kc_1^k$$
$$R_2 = x_1 + a_1c_2 + a_2c_2^2 + \ldots + a_kc_2^k$$
$$\ldots$$
$$R_{k+1} = x_1 + a_1c_{k+1} + a_2c_{k+1}^2 + \ldots + a_kc_{k+1}^k$$

It is obvious that $\mathcal{S}$ can obtain the value x_1 from the $k+1$ equations above. Once $\mathcal{S}$ calculates x_1, he makes an exhaustive search in his database to determine the public key of the user such that $pk_{\mathcal{U}}^{x_1} = \beta$. In this way, the identity of the user is exposed.

7   Conclusion

In this paper, we proposed a practical e-coupon system which enables the coupon issuer to trace the identity of misbehaving users, while maintaining the anonymity for the honest users. We achieved this without requiring any third party in the system. In addition, we formalized the notion of user privacy during the coupon redemption process and proved that our new e-coupon system also satisfied this property.

Acknowledgments. We are grateful to the anonymous reviewers for their helpful comments on this paper and we would like to thank Dr. Fuchun Guo for his valuable discussion.

References

1. Abe, M., Okamoto, T.: Provably secure partially blind signatures. In: Bellare, M. (ed.) CRYPTO 2000. LNCS, vol. 1880, pp. 271–286. Springer, Heidelberg (2000)
2. Au, M.H., Wu, Q., Susilo, W., Mu, Y.: Compact e-cash from bounded accumulator. In: Abe, M. (ed.) CT-RSA 2007. LNCS, vol. 4377, pp. 178–195. Springer, Heidelberg (2006)
3. Bellare, M., Palacio, A.: The knowledge-of-exponent assumptions and 3-round zero-knowledge protocols. In: Franklin, M. (ed.) CRYPTO 2004. LNCS, vol. 3152, pp. 273–289. Springer, Heidelberg (2004)
4. Brands, S.: Untraceable off-line cash in wallets with observers. In: Stinson, D.R. (ed.) CRYPTO 1993. LNCS, vol. 773, pp. 302–318. Springer, Heidelberg (1994)
5. Camenisch, J.L., Koprowski, M., Warinschi, B.: Efficient blind signatures without random oracles. In: Blundo, C., Cimato, S. (eds.) SCN 2004. LNCS, vol. 3352, pp. 134–148. Springer, Heidelberg (2005)
6. Camenisch, J.L., Piveteau, J.-M., Stadler, M.A.: Blind signatures based on the discrete logarithm problem. In: De Santis, A. (ed.) EUROCRYPT 1994. LNCS, vol. 950, pp. 428–432. Springer, Heidelberg (1995)
7. Canard, S., Gouget, A., Hufschmitt, E.: A handy multi-coupon system. In: Zhou, J., Yung, M., Bao, F. (eds.) ACNS 2006. LNCS, vol. 3989, pp. 66–81. Springer, Heidelberg (2006)
8. Chaum, D.: Blind signatures for untraceable payments. In: CRYPTO, pp. 199–203 (1982)
9. Chaum, D., Fiat, A., Naor, M.: Untraceable electronic cash. In: Goldwasser, S. (ed.) CRYPTO 1988. LNCS, vol. 403, pp. 319–327. Springer, Heidelberg (1990)
10. Chen, L., Enzmann, M., Sadeghi, A.-R., Schneider, M., Steiner, M.: A privacy-protecting coupon system. In: S. Patrick, A., Yung, M. (eds.) FC 2005. LNCS, vol. 3570, pp. 93–108. Springer, Heidelberg (2005)

11. Chu, C.-K., Tzeng, W.-G.: Efficient k-out-of-n oblivious transfer schemes with adaptive and non-adaptive queries. In: Vaudenay, S. (ed.) PKC 2005. LNCS, vol. 3386, pp. 172–183. Springer, Heidelberg (2005)
12. Frankel, Y., Tsiounis, Y., Yung, M.: Fair off-line e-cash made easy. In: Ohta, K., Pei, D. (eds.) ASIACRYPT 1998. LNCS, vol. 1514, pp. 257–270. Springer, Heidelberg (1998)
13. Juels, A., Luby, M., Ostrovsky, R.: Security of blind digital signatures. In: Kaliski Jr., B.S. (ed.) CRYPTO 1997. LNCS, vol. 1294, pp. 150–164. Springer, Heidelberg (1997)
14. Mu, Y., Nguyen, K.Q., Varadharajan, V.: A fair electronic cash scheme. In: Kou, W., Yesha, Y., Tan, C.J.K. (eds.) ISEC 2001. LNCS, vol. 2040, pp. 20–32. Springer, Heidelberg (2001)
15. Nguyen, K.Q., MU, Y., Varadharajan, V.: A new digital cash scheme based on blind Nyberg-Rueppel digital signature. In: Okamoto, E. (ed.) ISW 1997. LNCS, vol. 1396, pp. 313–320. Springer, Heidelberg (1998)
16. Nguyen, L.: Privacy-protecting coupon system revisited. In: Di Crescenzo, G., Rubin, A. (eds.) FC 2006. LNCS, vol. 4107, pp. 266–280. Springer, Heidelberg (2006)
17. Pointcheval, D., Stern, J.: Security proofs for signature schemes. In: Maurer, U.M. (ed.) EUROCRYPT 1996. LNCS, vol. 1070, pp. 387–398. Springer, Heidelberg (1996)
18. Stadler, M.A., Piveteau, J.-M., Camenisch, J.L.: Fair blind signatures. In: Guillou, L.C., Quisquater, J.-J. (eds.) EUROCRYPT 1995. LNCS, vol. 921, pp. 209–219. Springer, Heidelberg (1995)

Spatial Bloom Filters: Enabling Privacy in Location-Aware Applications

Paolo Palmieri[1]([⊠]), Luca Calderoni[2], and Dario Maio[2]

[1] Parallel and Distributed Systems Group, Delft University of Technology,
Mekelweg 4, 2628CD Delft, The Netherlands
p.palmieri@tudelft.nl
[2] Department of Computer Science and Engineering, Università di Bologna,
via Sacchi 3, 47521 Cesena, Italy
{luca.calderoni,dario.maio}@unibo.it

Abstract. The wide availability of inexpensive positioning systems made it possible to embed them into smartphones and other personal devices. This marked the beginning of location-aware applications, where users request personalized services based on their geographic position. The location of a user is, however, highly sensitive information: the user's privacy can be preserved if only the minimum amount of information needed to provide the service is disclosed at any time. While some applications, such as navigation systems, are based on the users' movements and therefore require constant tracking, others only require knowledge of the user's position in relation to a set of points or areas of interest. In this paper we focus on the latter kind of services, where location information is essentially used to determine membership in one or more geographic sets. We address this problem using Bloom Filters (BF), a compact data structure for representing sets. In particular, we present an extension of the original Bloom filter idea: the Spatial Bloom Filter (SBF). SBF's are designed to manage spatial and geographical information in a space efficient way, and are well-suited for enabling privacy in location-aware applications. We show this by providing two multi-party protocols for privacy-preserving computation of location information, based on the known homomorphic properties of public key encryption schemes. The protocols keep the user's exact position private, but allow the provider of the service to learn when the user is close to specific points of interest, or inside predefined areas. At the same time, the points and areas of interest remain oblivious to the user.

Keywords: Location privacy · Bloom filters · Secure multi-party computation

1 Introduction

In recent years, location-aware applications spread widely. These applications usually require the user to disclose her exact position, in order to receive content

---

L. Calderoni—Part of this research work was accomplished while visiting the Parallel and Distributed Systems group of Delft University of Technology (The Netherlands).

© Springer International Publishing Switzerland 2015
D. Lin et al. (Eds.): Inscrypt 2014, LNCS 8957, pp. 16–36, 2015.
DOI: 10.1007/978-3-319-16745-9_2

and information relevant to the user's location. Examples of such location-aware services are local advertising, traffic or weather information, or suggestions about points of interest (PoI) in the user's surroundings [5]. In general, the ability to detect movements and exact position of the users in outdoor environments has become widespread since the introduction of smartphones equipped with positioning systems such as a GPS receiver.

The ability to track a user's position raises however deep privacy concerns, due to the sensitive nature of location information. In fact, a number of potentially sensitive professional and personal information about an individual can be inferred knowing only her presence at specific places and times [1,4]. Even anonymized position data sets (not containing name, phone number or other obvious references to the person) do not prevent precise identification of the user: in fact, just four mobility traces may be enough to identify her [11]. The more users disclose their data, the more providers are able to profile them in an accurate way. This is for instance the case discussed by Wicker in [21], where a marketing company database model is used in conjunction with anonymous mobile phone location traces. Smartphones and location-aware services are now an integral part of our everyday life, but it is reasonable to predict that in the coming years users will demand privacy safeguards for their information with respect to the involved service provider [17] and more specifically for location information [10,22]. The real challenge is therefore how to protect user's privacy without losing the ability to deliver services based on her location [15].

A common application scenario of location-based services requires the service provider to learn when the user is close to some sensitive or interesting locations. This is the case, for instance, of "around-me" applications or security and military systems [5]. If we add to this scenario the requirement of the user position to stay private, the problem becomes an interesting and fundamental research question. In literature a similar problem, known as *private proximity testing* has been studied: Alice can test if she is close to Bob without either party revealing any other information about their location [12]. Narayanan et al. proposed a solution based on location tags (features of the physical environment) and relying on Facebook for the exchange of public keys [12]. His protocol was later improved in efficiency by Saldamli et al. [16]. Location tags and proximity tests are also used in [8], as a way of providing local authentication, while [23] presents a secure handshake for communication between the two actors in proximity. The security of the basic proximity testing protocol has been further improved in [13]. In [19], Tonicelli et al. propose a solution for proximity testing based on pre-distributed data, secure in the Universal Composability framework. Finally, the problem of checking the proximity in a specific time is addressed in [18].

In this paper we do not focus on proximity testing, but on a broader and more general problem: testing in a private manner whether a user is within one of a set of areas of arbitrary size and shape. By solving this problem and applying an intelligent conformation of areas, we can also solve the proximity testing problem (for one or multiple points simultaneously), and we are actually able to identify with some precision the distance of the user from the point of interest. Given the conceptual similarity of our problem with membership testing in sets, we base our solution on a novel modification of Bloom Filters (BF). Bloom filters are a compact data structure that allows to compute whether an element is a member of the set the filter has been built upon, without knowledge of the set itself [2]. Bloom filters have already been used in privacy-preservation protocols, and they are particularly suited to be used in conjunction with the homomorphic properties of certain public key encryption schemes [9].

1.1 Contribution

In this paper we propose a modification of Bloom filters aimed at managing location information, and we present two private positioning protocols for privacy-preserving location-aware applications.

The novel variant of Bloom filters we introduce, which we call *Spatial Bloom Filter* (SBF), is specifically designed to deal with location information. Similarly to the classic Bloom filters, SBFs are also well suited to be used in privacy preserving applications, and we show this by presenting two protocols for private positioning. The first protocol is based on a two-party setting, where communication happens directly between the user of a location-based service and the service provider. A more complex scenario is defined in the second protocol, that involves a three-party setting in which the service provider outsources to a third party the communication with the user. In both settings we do not assume any trust between the different parties involved. The protocols allow secure computation of location-aware information, while keeping the position of the user private. The only information disclosed to the provider is the user's vicinity to specific points of interest or his presence within predefined areas. At the same time, the areas of interest are not disclosed to the user. Therefore, unlike other works on location privacy, which is usually discussed from the end-user point of view, we work here in the secure multi-party computation, where both parties have an interest in keeping their information private. Military and government applications are just the most immediate examples of when location privacy represents a key-problem for the provider as well.

In the paper, we discuss the security and the computational cost of the proposed schemes, as well the probabilistic and storage properties of the SBF.

2 Preliminaries

We introduce in the following some useful notions and definitions, that will be used later in the paper.

2.1 Bloom Filters

A Bloom Filter (BF) is a data structure that represents a set of elements in a space-efficient manner. A BF generated for a specific set allows membership queries on the originating set without knowledge of the set itself. The BF always determines positively if an element is in the set, while elements outside the set are generally determined negatively, but with a probabilistic false positive error.

Definition 1. *We define a* Bloom filter $B(S)$ *representing a set* $S = \{a_1, \ldots, a_n\}$ *as the set*

$$B(S) = \bigcup_{a \in S, h \in H} h(a), \tag{1}$$

where $H = \{h_1, \ldots, h_k\}$ *is a set of* k *hash functions such that each* $h_i \in H :$ $\{0,1\}^* \to \{1, \ldots, m\}$, *that is, the hash functions take binary strings as input and output a random number uniformly chosen in* $\{1, \ldots, m\}$.

A Bloom filter $B(S)$ can be represented as a binary vector b composed of m bits, where the i-th bit

$$b[i] = \begin{cases} 1 & \text{if } i \in B(S) \\ 0 & \text{if } i \notin B(S) \end{cases}. \tag{2}$$

A detailed discussion about Bloom filters construction and verification as reported in literature, along with an explanation of false positive probability and optimal number of hash functions is proposed in Appendix A.

2.2 Cryptographic Primitives

In part of our construction we use the homomorphic properties of encryption schemes. In general, a cipher has homomorphic properties when it is possible to perform certain computations on a ciphertext without decrypting it and, therefore, without knowledge of the decryption key. In particular, we say an encryption scheme is *additively homomorphic* when a specific operation $\boxplus$ applied on two ciphertexts $(Enc(p_1), Enc(p_2))$ decrypts to the sum of their corresponding plaintexts $(p_1 + p_2)$:

$$Dec(Enc(p_1) \boxplus Enc(p_2)) = p_1 + p_2. \tag{3}$$

There is additive homomorphism also when an operation on a ciphertext and a plaintext results in the sum of the two plaintexts. We have instead *multiplicative homomorphism* between an encrypted plaintext and a plaintext when an operation $\boxdot$ results into the multiplication of the two plaintexts:

$$Dec(Enc(p_1) \boxdot p_2) = p_1 \cdot p_2. \tag{4}$$

An example of encryption scheme that is both additively and multiplicatively homomorphic is the Paillier cryptosystem [14]. In this case, the product of two ciphertexts will decrypt to the sum of their corresponding plaintexts (additive property), while an encrypted plaintext raised to the power of another plaintext will decrypt to the product of the two plaintexts (multiplicative property).

Private Hadamard Product. The Hadamard (or entrywise) product of two vectors, one binary (owned by Alice) and one composed of natural numbers (owned by Bob), is performed in a privacy-preserving manner by Algorithm 1. The algorithm is private with respect to the input vectors, and only reveals the product vector to Alice. The security of the algorithm is based on the encryption of Alice's vector using a public key encryption scheme that is multiplicative homomorphic for operation $\boxdot$.

Algorithm 1. Private Hadamard product of an encrypted binary vector for a cleartext vector of natural numbers

Input Alice: $\mathbf{X} = (\mathbf{x}_1, \ldots, \mathbf{x}_n)$, $\mathbf{X} \in \{0,1\}^n$.
Input Bob: $\mathbf{Y} = (\mathbf{y}_1, \ldots, \mathbf{y}_n)$, $\mathbf{Y} \in \mathbb{N}^n$.
Output Alice: $\mathbf{X} \cdot \mathbf{Y}$.

1 Alice generates a public and private key pair using a multiplicative homomorphic encryption scheme, and sends the public key to Bob.
2 Alice sends to Bob the ciphertext vector $\mathbf{E} = (Enc(\mathbf{x}_1), \ldots, Enc(\mathbf{x}_n))$.
3 Bob computes the vector $\mathbf{C} = (Enc(\mathbf{x}_1) \boxdot \mathbf{y}_1, \ldots, Enc(\mathbf{x}_n) \boxdot \mathbf{y}_n)$ and sends the result to Alice.
4 Alice uses her secret key to decrypt $\mathbf{C}$ and obtains $\mathbf{D} = Dec(\mathbf{C}) = \mathbf{X} \cdot \mathbf{Y}$.

A more conservative version of the algorithm requires Bob to multiply a randomly chosen prime number p, larger than any $\mathbf{y} \in \mathbf{Y}$, to each value in the vector, before performing the homomorphic multiplication. Alice can then obtain $\mathbf{X} \cdot \mathbf{Y}$ by calculating p using any greatest common divisor algorithm.

In general, we assume that the parties participating in the proposed construction do not deviate from the protocol, but gather all available information in order to try to learn private information of other parties. We are, therefore, in the semi-honest setting.

Security Model. We assume the parties are *honest-but-curious*, that is, the parties will follow the protocol but try to learn additional information about other parties private data.

3   Spatial Representation

The construction we present in this paper is based on a novel variant of BFs aimed at managing location information. Since BFs are constructed over finite sets of elements, we need to represent location information – that is, a geographical position – as an element that is part of the finite and discrete set of all possible positions. Therefore, instead of considering a location as a point, we divide Earth's surface into a set of distinct areas, and we identify a position as the corresponding element in this set. Considering that we can set the dimension of such areas to an arbitrarily small size, there is no loss in the precision of

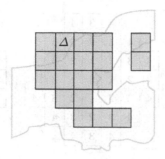

Fig. 1. A sample area covered by an arbitrary grid.

the location information. In particular, we do not use this approach in order to obfuscate or partially hide an exact position: on the contrary, we are interested in retaining a precision as high as the one allowed by the location sensor used in the specific application. A detailed discussion concerning spatial model designed for the presented method is provided in Appendix B.

For the purpose of this work we choose to consider the grid defined by longitude and latitude values with a precision of three decimal point places. This grid divides Earth's surface in a number of regions. We define the set of all regions as follows.

Definition 2. *We define $\mathcal{E}$ as the set of all regions in which Earth's surface is divided by the grid defined by the circles (called parallels) of latitude distant multiples of $0.001°$ from the equator and the arcs (called meridians) of longitude distant multiples of $0.001°$ from the Prime Meridian.*

The sides (in meters) of a region of side 0.001 degrees in terms of longitude and latitude vary depending on its position on the globe. Appendix B contains some reference values.

3.1 Areas and Points of Interest (AoI & PoI)

The purpose of this paper is to present a method able to preserve both user's and provider's privacy in location-aware applications. We imagine a scenario in which the provider of such an application wants to be notified of the presence of the user in one of a predefined set of areas of interest (AoI). The areas of interest are selected by the provider, and each is composed of an arbitrary number of regions in $\mathcal{E}$, defined above. An area may, for instance, represent a sensitive or interesting location for the purposes of the application. A number of concentric areas around a point of interest (PoI) can be used to detect the user's vicinity to the PoI. In the following we present two ways in which the set-based location information described above can be used to achieve this goal. Both approaches are used in the following of the paper as strategies to select the areas of interest by the service provider, but we stress here that our construction is independent

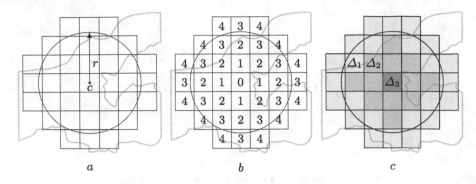

Fig. 2. An example of the area coverage algorithm applied to a point of interest of interest. After defining the grid (a), the Manhattan distance from the center region is computed (b). Finally, each region is assigned to the right set (c). In this case, the maximum distance (σ) is 4, so we assign the regions belonging to the distance classes 4 and 3 to Δ_1, those belonging to the classes 2 and 1 to Δ_2 and the sole region belonging to the 0 class to the set Δ_3.

of the strategy used, and therefore can accommodate any other set selection mechanism.

The first strategy to define a set of areas of interest follows naturally from the idea of detecting the presence/absence of a person in a given area. In order to do that, the provider of the service defines an area by selecting a subset of $\mathcal{E}$ (Fig. 1). The regions in the area need not to be contiguous, and there is no limitations in shape or size of the area. The set containing all of these regions is defined as Δ (we get back on this subject in Sect. 4).

A second approach is instead to monitor the user by detecting his proximity to the area's center as he approaches it. We achieve this goal without knowing the user's exact location by defining several concentric areas around the point of interest to be monitored. In the example shown in Fig. 2 we use three areas for this purpose, but this parameter can take any value deemed useful.

Let c be the center of the area (having coordinates lng_c, lat_c) and let r be the range we are interested to monitor users around the center itself. First of all we choose a region such that it is the element of $\mathcal{E}$ that contains the point c. Then a number of adjacent elements (all belonging to $\mathcal{E}$) are added in order to form a grid, until the circle of center c and radius r is completely included in the grid, as shown in Fig. 2a. Now let us label each region with its distance from the center region, using the standard *Manhattan distance*. Assume that σ is the maximum distance value in the generated grid; we need to discuss two cases. If $(\sigma + 1) \mod 3 = 0$, we assign to the set Δ_3 each region labeled from 0 to $q - 1$, where $q = (\sigma + 1)/3$. Similarly, we fill the set Δ_2 with each square labeled from q to $2q - 1$ and the set Δ_1 with each square labeled from $2q$ to σ. If 3 does not divide $\sigma + 1$ exactly (i.e. $(\sigma + 1) \mod 3 \neq 0$) some rounding is required; we could for instance assign the first remaining class to Δ_1 and the second optionally remaining class to Δ_2. In that case, given $q = \lfloor (\sigma + 1)/3 \rfloor$, the procedure can be formalized assigning each region labeled from 0 to $q-1$ to the set Δ_3, each region labeled from q to $2q$ to the set Δ_2 and each region labeled from $2q+1$ to σ to the set Δ_1.

4   Spatial Bloom Filter

After defining a spatial representation $\mathcal{E}$ of Earth's surface and providing a way to identify geographical areas (and points) as elements of a subset of $\mathcal{E}$, we can use a set-based data structure like the Bloom filter to encode this information. However, the original definition of BF proves to be quite inefficient for this task, as it would be possible to encode only one area for each BF.

In the following we define a novel data structure called *Spatial Bloom Filter*. A spatial Bloom filter can be used, likewise the original BF, to perform membership queries on the originating set of elements without knowledge of the set itself. Contrary to the BF, however, a spatial Bloom filter can be constructed over multiple sets, and querying a spatial Bloom filter for an element returns the identifier of the specific set among all the originating sets in which the element is contained, minus a false positive probability (of assigning the element to the wrong set). Similarly to a classical BF, there is also a false positive probability that querying a SBF with an element outside the originating sets returns a positive result (wrongly assigning the element to one of the originating sets).

An important property of SBF is that the probability of false positives, that is, the probability that an element is wrongly recognized as belonging to a specific originating set, depends on the order in which the sets have been encoded in the filter: a false positive can occur either when an element outside the originating sets is recognized as being part of one, or when an element that is part of an originating set is recognized as being belonging to a different one (sets are disjoint). The latter case, however, can only happen if the wrongly recognized set has been encoded later than the actual originating set.

This fundamental property allows to define an order of priority for the different originating sets, thus reducing the error probability for elements (areas) deemed more important. Considering the strategies described in the previous section for selecting areas of interests, this property is particularly useful when using SBFs to store location information. In the example presented in Sect. 3.1, for instance, we used a set of three different areas $S = \{\Delta_1, \Delta_2, \Delta_3\}$. Assuming the provider would prefer a more accurate monitoring of the area's central region, we assigned the highest label value (3) to the inner area. In the following we generally consider the sets as already ordered by priority, meaning that set Δ_2 is considered as having higher priority than Δ_1.

Definition 3. *Let $S = \{\Delta_1, \Delta_2, \dots, \Delta_s\}$ be a set of areas of interest such that $\Delta_i \subseteq \mathcal{E}$ and S is a partition of the union set $\bar{S} = \bigcup_{\Delta_i \in S} \Delta_i$. Let O be the strict total order over S for which $\Delta_i < \Delta_j$ for $i < j$. Let also $H = \{h_1, \dots, h_k\}$ be a set of k hash functions such that each $h_i \in H : \{0,1\}^* \rightarrow \{1, \dots, m\}$, that is, each hash function in H takes binary strings as input and outputs a random* *number uniformly chosen in* $\{1, \ldots, m\}$. *We define the* Spatial Bloom Filter *(SBF) over* (S, O) *as the set of pairs*

$$B^{\#}(S, O) = \bigcup_{i \in I} \langle i, \max L_i \rangle, \tag{5}$$

where I *is the set of all values output by hash functions in* H *for elements of* $\bar{S}$

$$I = \bigcup_{\delta \in \bar{S}, h \in H} h(\delta), \tag{6}$$

and L_i *is the set of labels* l *such that:*

$$L_i = \{l \mid \exists \delta \in \Delta_l, \exists h \in H : h(\delta) = i\}. \tag{7}$$

A spatial Bloom filter $B^{\#}(S, O)$ can be represented as a vector $b^{\#}$ composed of m values, where the i-th value

$$b^{\#}[i] = \begin{cases} l & \text{if } \langle i, l \rangle \in B^{\#}(S, O) \\ 0 & \text{if } \langle i, l \rangle \notin B^{\#}(S, O) \end{cases}. \tag{8}$$

In the following, when referring to a SBF, we refer to its vector representation $b^{\#}$.

A SBF is built as follows. Initially all values in $b^{\#}$ are set to 0. Then, for each element $\delta \in \Delta_1$ and for each $h \in H$ we calculate $h(\delta) = i$, and set the i-th value of $b^{\#}$ to 1 (that is, to the label of Δ_1). We do the same for the elements belonging to the set Δ_2, setting $b^{\#}[i]$ to 2. We proceed incrementally until all sets in S have been encoded in $b^{\#}$. We observe that, following Definition 3, should a collision occur, the label with higher value is the one stored at the end of the process. Thus, values in the filter corresponding the elements in Δ_s will never be overwritten. This procedure is formalized in Algorithm 2 and depicted in Fig. 3.

The verification process shall check whether an element δ_u is contained in a set $\Delta_i \in S$. Hence we verify whether $\delta_u \in \Delta_i$ if

$$\exists h \in H : b^{\#}[h(\delta_u)] = i \quad \text{and} \quad \forall h \in H, b^{\#}[h(\delta_u)] \geq i. \tag{9}$$

The procedure is described in Algorithm 3.

In practice, if any value of $b^{\#}$ in a position that corresponds to the output of one of the hash functions for δ_u is 0, then $\delta_u \notin \bar{S}$. If all the hashes map to elements of value i, then $\delta_u \in \Delta_i$ minus a false positive probability which is discussed in the following. The same applies if at least one hash maps to an element of value i and the remaining hashes map to elements of value $> i$. In fact, since when a collision occurs the highest value is stored, a lower value could be overwritten.

Similarly to the case of the original Bloom filter (Sect. 2.1), a false positive probability p exists when determining whether an element belongs to the set $\bar{S}$ or not. In the case of a spatial Bloom filter $B^{\#}(S, O)$, however, the probability p can be split into several probabilities p_i, each one subset-specific. Specifically,

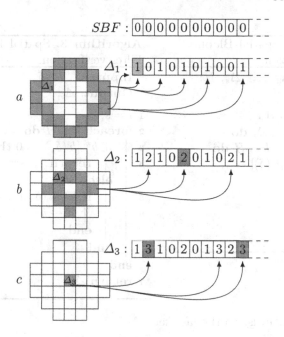

Fig. 3. Areas Δ_1, Δ_2 and Δ_3 are used to construct a SBF. Three hash functions are used to map each element into the filter. Only the first ten elements of the SBF are shown. In a, two elements belonging to Δ_1 are processed by the hash functions, resulting in six 1 value elements to be written into the SBF. The first element collides as highlighted. This kind of collision is the same that may occur in a classic Bloom Filter. After each element in Δ_1 is processed, the algorithm processes elements in Δ_2 (b) and finally in Δ_3 (c). Note that the collisions in b and c are different from the previous one and are SBF specific. Areas marked with a greater label are assumed to be more important from the provider point of view and overwrite elements of lower value on collision.

p_i is the probability that an element δ is wrongly recognized as belonging to the set Δ_i, while either $\delta \notin \bar{S}$ or $\delta \in \Delta_j$, with $j < i$. For instance, a false positive assigned to the set Δ_s occurs if each hash collides with a value s in $b^{\#}$. Thus we can denote this probability as follows:

$$p_s \approx \left(1 - e^{-\frac{k|\Delta_s|}{m}}\right)^k. \qquad (10)$$

The above probability follows from the false probability of classical Bloom filters (15), as explained in Appendix A. Similarly, we can compute the probability to wrongly assign an element to the set Δ_{s-1} considering all of the possible collisions with elements belonging to Δ_s and Δ_{s-1}, excluding those deriving from collisions with elements belonging to Δ_s entirely. Hence

$$p_{s-1} \approx \left(1 - e^{-\frac{k|\Delta_s \cup \Delta_{s-1}|}{m}}\right)^k - p_s. \qquad (11)$$

Algorithm 2. Spatial Bloom Filter construction.	**Algorithm 3.** Spatial Bloom Filter verification.
Input: $\Delta_1, \Delta_2, \ldots, \Delta_s, H$; **Output:** $b^{\#}$;	**Input:** $b^{\#}, H, \delta_u, s$; **Output:** Δ_i;

<table>
<tr><td>

1 **for** $i \leftarrow 1$ **to** s **do**
2   **foreach** $\delta \in \Delta_i$ **do**
3   **foreach** $h \in H$ **do**
4 | $b^{\#}[h(\delta)] \leftarrow i$;
     end
   end
  end
5 **return** $b^{\#}$;

</td><td>

1 $i = s$;
2 **foreach** $h \in H$ **do**
3   **if** $b^{\#}[h(\delta_u)] = 0$ **then**
4 | **return** $false$;
   else
5   **if** $b^{\#}[h(\delta_u)] < i$ **then**
6 | $i \leftarrow b^{\#}[h(\delta_u)]$;
     end
   end
  end
7 **return** Δ_i;

</td></tr>
</table>

We can proceed likewise to the last set:

$$p_1 \approx \left(1 - e^{-\frac{k|\bar{S}|}{m}}\right)^k - p_s - p_{s-1} - \cdots - p_2. \tag{12}$$

It follows that $p_1 + p_2 + \cdots + p_s = p$, where p is the same false positive probability provided in (15) if $|\bar{S}| = n$.

In the following we assume that the possibility of false positives among sets (that is, having elements in $\bar{S}$ assigned to the wrong set) is deemed as generally acceptable when using a SBF.

Let us finally note that a SBF bears some resemblance to a bloomier filter [6,7], a variant of the classical Bloom filter used for storing binary functions instead of sets. We could in fact define the originating sets through a function, and build the corresponding bloomier filter. However, in the case of a spatial Bloom filter we have an error probability between different Δ's, but we know exactly whether a $\delta \in S$ or not. A bloomier filter, instead, would behave in the opposite way: the function always outputs the correct Δ, but there exists a probability that a $\delta \notin S$ will be wrongly recognized as belonging to one Δ. Considering location-aware applications, we deem an error in positioning over two contiguous areas of interest as acceptable, while mistakenly recognizing a position outside the areas of interests (even by far) as inside as much more problematic. Therefore, we believe that the proposed spatial Bloom filters are better suited to be used in the location-aware context, while bloomier filters might still be useful in specific application scenarios.

5 Private Positioning Protocols for Spatial Bloom Filters

A major feature of SBFs is that they allow private computation of location based information. We show this by providing two protocols based on spatial

Bloom filters that address the problem of location privacy in a location-aware application. In general, a location-aware application is any service that is based on (partial) knowledge of the geographic position of the user. In this work, however, we focus on applications in which the service provider has an interest in learning when the user is within an area (or close to a point) of interest.

The protocols we present are designed for a secure multi-party computation setting, where the user and the service provider are mutually distrusting, and therefore do not want to disclose private information to the other party. In the case of the user, private information is his exact location. The service provider, instead, does not want to disclose the monitored areas. We address this problem by providing a scheme that allows the provider of a service to detect when the user is within an area of interest, without requiring the user to reveal his exact position to the provider. At the same time, the privacy of the provider is also guaranteed with respect to the areas of interest. The privacy benefits for the user are double: first and foremost, the relative location is only revealed when the user is within predetermined areas, and remains private otherwise. Secondly, even when presence in an area is detected, only this generic information is learned by the provider, and not the actual position. Following the area coverage mechanism proposed in Sect. 3.1, for instance, the provider learns the distance from the central area to a certain extent, while the direction from which the user approaches it stays private. Dividing the area around the point of interest in a different manner may reveal instead the direction but conceal the distance within the area range.

In the following we discuss two different settings: in the first setting the user communicates directly with the service provider, who computed beforehand a spatial Bloom filter relative to the areas he is interested in monitoring. In the second setting, instead, the service provider computes the SBF, but communication with the user is handled by a third party, to which the provider outsources the task. In both setting, no trust is implied among the parties, including the third party, and we assume the parties do not collude with each other. We work in the honest-but-curious setting, as defined in the preliminaries (Sect. 2).

5.1   Two-Party Scenario

In the two-party scenario the communication happens between the service provider *Paul* and the user *Ursula*. We assume the user has access to a positioning system that allows her to determine her geographic position. Ursula is interested in using a location-aware service provided by Paul, but she does not want to disclose her exact position. Paul, on the other hand, wants to learn if Ursula is close to some points of interest or is within an area of interest, but he does not want to share with her these locations. Since the two parties are mutually distrusting, this is a secure multi-party computation problem.

We propose Protocol 1, that addresses the problem securely by disclosing only the identifier i of the area Δ_i in which the user is. Intuitively, the protocol works as follows. Paul creates a SBF for the points and areas of interest as described in the previous sections. He encrypts the filter (by encrypting each value therein) with an encryption scheme that allows the private Hadamard product defined in Algorithm 1, and sends it to Ursula. Ursula creates a SBF for the set composed only of her position in the grid. The filter is binary, since 0's and 1's are the only possible values in a filter with only one point of interest. Then Ursula computes the entrywise homomorphic product of the received SBF with the one she just computed: this way, only the values of the encrypted filter corresponding to a 1 in her filter are preserved, while the others take value 0. Then she shuffles the values in the resulting encrypted filter and sends the randomly ordered filter back to Paul.

Protocol 1. Two-party private positioning protocol between service provider Paul and user Ursula.

Before any communication, the provider selects the areas of interest $\Delta_1, \ldots, \Delta_s \subset \mathcal{E}$. Then, he selects the desired false positive probability p, and determines k and m according to (16) and (17) respectively. Finally, following the notation of Definition 3, the provider computes the spatial Bloom filter $b^{\#}$ over $\bar{S}$ using Algorithm 2.

1 The service provider Paul generates a public and private key pair using a multiplicative homomorphic encryption scheme, and sends the public key to the user Ursula.

2 Paul sends to Ursula the encryption of the precomputed SBF $Enc\left(b^{\#}\right)$, the set of k hash functions H, the value m and the conventional grid $\mathcal{E}$.

3 At regular time intervals, or when required by the specific application, Ursula determines her geographic position and selects the corresponding grid region $e_u \in \mathcal{E}$. Then, following Algorithm 2 and using the values and functions shared by Paul, she builds a spatial Bloom filter $b_u^{\#}$ over $\{e_u\}$ and counts the number z of values equal to 1 therein.

4 Ursula computes $e^{\#} = Enc\left(b^{\#}\right) \boxdot b_u^{\#}$ using the homomorphic properties of the encryption scheme (Algorithm 1). Then she applies a random permutation to the values in the filter, and sends z and the result to Paul.

5 Paul decrypts $e^{\#}$ and counts all non-zero values. If the resulting number is $< z$, Ursula's position is outside of the areas on which the SBF was built. Otherwise, the value i, corresponding to area Δ_i identifying Ursula's position (minus error probability p_i), is the smallest non-zero value in $Dec\left(e^{\#}\right)$.

Security Definition. In a two-party setting implementing Protocol 1, the computation is achieved privately if at the end of the protocol execution Paul learns only $i \in \{1, \ldots, s\}$, and Ursula learns nothing.

In the following we analyze the security of the protocol with respect to the above definition. In order to quantify the information learned by Paul during the protocol execution, we introduce an arbitrarily small security parameter ϵ. Then, we prove that the probability of Paul learning useful information is upper-bounded by the chosen ϵ.

Security Analysis. As stated in the security definition, a successful execution of Protocol 11 should guarantee three conditions: correctness of the result for Paul, privacy for Ursula's position and privacy of the areas encoded in the filter by Paul. We discuss the three conditions in the following.

The protocol ends correctly if the number of non-zero values read in the decrypted $e^{\#}$ by Paul is $< z$ in case Ursula is outside the areas of interests; in case Ursula is within an area, the protocol ends correctly if the number of non-zero values is equal to z, and the area is identified by the smallest non-zero value, minus error probability p_i. The former case is always true, for the properties of Definition 3, as explained in Sect. 4. In the latter case, the false positive probability p_i for each area i is determined by Paul according to (12) during filter creation. It is therefore Paul himself who decides the correctness bounds of the protocol.

The second condition (Ursula's privacy) is respected if Paul learns only in which (predefined) area the user is, and not her exact position at the end of the protocol. If the user is outside the areas of interest, the provider should learn nothing. Ursula encodes her position in $b_u^{\#}$ at step 3 of the protocol, and sends the encrypted filter $e^{\#} = Enc\left(b^{\#}\right) \boxdot b_u^{\#}$ back to Paul after performing a random permutation on the order of its values. The homomorphic properties of a public key encryption scheme guarantee that Paul can only learn a number of values from $b^{\#}$ that corresponds to non-zero values in $b_u^{\#}$ [9]. At the same time, the random permutation prevents him from understanding to which position in $b^{\#}$ each of these values corresponds to, therefore making it impossible to reconstruct Ursula's filter based on the order of elements. If the number of non-zero values is z, and all take the value i corresponding to an area of interest, Paul only learns the area of interest. In case, instead, some values are $> i$ for some of the positions on the grid within the area of interest, then Paul learns the area of interest Δ_i and a pattern of values. The same applies in case Ursula is outside of any area of interest, but the decryption of $e^{\#}$ reveals a number of non-zero values $w < z$. In the following we focus on the latter scenario, as a potential attack exploiting the pattern information could reveal the user's position even when she is outside the areas of interests. In fact, if the pattern is unique for a position on the grid, Paul may be able to learn Ursula's position by performing an exhaustive search on all the possible positions on the grid: given the irreversibility of (spatial) Bloom filters, the complexity of the attack is linear to the number of such positions. We prevent this attack by having each pattern shared by at least a possible positions: in which case we achieve a-anonimity for the user's position even in case of an exhaustive search. We define an arbitrarily small security parameter ϵ, and we consider the privacy condition to be met if the probability of Paul learning Ursula's position is $\frac{1}{a} < \epsilon$. For each number $w \in \{1, \ldots, z\}$ of non-zero values obtained by Paul, we can estimate the value of a based on the number of possible positions in $\mathcal{E}$ and the number of areas of interest s. In particular, we calculate the number of possible patterns for a given w as the combinations with repetitions of length w, $\dbinom{s + w - 1}{w}$. Based on this, we can estimate the average value $\bar{a}$ for the different a's of all possible combination with repetitions to be

$$\bar{a} = \frac{|\mathcal{E}|}{\sum_{w=1}^{k} \binom{s+w-1}{w} + 1}, \tag{13}$$

if we assume a linear distribution of the values $\{1, \ldots, s\}$ over the filter. The security condition is hence met if $\frac{1}{a} < \epsilon$ for all a's relative to any possible w. We note, from the formula above, that this mostly depends on the number of areas of interest s and, on a lesser extent, on the number of hashes k (since $z \leq k$). These two values can therefore be tuned in order to achieve the desired security parameter ϵ, as both values are selected before the creation of the filter. Considering the order of magnitude of $|\mathcal{E}|$, which is 10^{12}, an appropriately built filter can satisfy a security parameter $\epsilon = 10^{-6}$ for most values of k and s. Thanks to the fine grained nature of the grid, even geographically limited settings which restricts the area of potential positions of the user can achieve reasonable security margins ($\epsilon \approx 10^{-3}$): in fact, small areas of a few square kilometers already include several millions possible positions (Sect. 3 and Appendix B).

Finally, the privacy of the service provider, that is, the secrecy of the areas encoded in the filter, is ensured by the encryption of the filter itself. Ursula, in fact, never learns the cleartext of the filter, as she is able to perform the multiplication of step 4 in the encrypted domain thanks to the homomorphic properties of the public key encryption scheme.

Computation and Communication Analysis. The computational complexity for the insertion and the verification of a single element in a SBF are linear in the number k of hash functions used for the filter. The private Hadamard product has instead a computational cost linear to the length of the filter m.

Since we intend this primitive to be used in concrete scenarios, in the following we provide an evaluation of actual communication costs, and number of computational operations to be performed during the execution of the protocol (Table 1). While being a generally compact data structure, a SBF built over a significantly large number of sets can consume a sizeable amount of memory. While m bits are required for storing a classical Bloom filter b, a SBF needs more bits due to the labels relative to the subsets Δ_i. More precisely, in order to store $b^{\#}$, $(\lfloor \log_2 s \rfloor + 1)\, m$ bits are needed. Depending on the number of areas and the desired error probability, a SBF could require a storage space (and communication cost when transmitted) not suitable for constrained scenarios, as in the case of mobile devices. For instance, consider hash functions with a 16 bit digest (i.e. $m = 2^{16}$) and an area of interest divided into six sub areas. Since $s = 6$, a SBF built on these functions needs $(\lfloor \log_2 6 \rfloor + 1)\, 2^{16}$ bits, resulting in approximately 24 KB data structure. For this reason, we introduce in the next section a protocol involving a third party which offloads user's bandwidth consumption.

5.2   Three-Party Scenario

In the three-party scenario the communication does not happen directly between the service provider and the user. The service provider is responsible for creating

Table 1. Computation and communication load for stakeholders.

	User	Provider	Third party
Comp. (2-p)	1 SBF-insertion, 1 Private Hadamard Product	1 decryption, 1 match count	
Comp. (3-p)	k hashes	1 decryption, 1 match count	1 SBF-completion, 1 Private Hadamard Product
Comm. (2-p)	$\mathcal{O}(m)$ $(\lfloor \log_2 s \rfloor + 1)\, m$	$\mathcal{O}(m)$ $(\lfloor \log_2 s \rfloor + 1)\, m$	
Comm. (3-p)	$\mathcal{O}(\log_2 m)$ $k\,(\lfloor \log_2 m \rfloor + 1)$	$\mathcal{O}(m)$ $(\lfloor \log_2 s \rfloor + 1)\, m$	$\mathcal{O}(m)$ $k\,(\lfloor \log_2 m \rfloor + 1) + (\lfloor \log_2 s \rfloor + 1)\, m$

and managing the filter, but the verification of user values and therefore all direct communication with the user is outsourced to a third party, whom we call *Olga*. We introduce the third party in order to decrease the computation and communication burden imposed on the user Ursula. In fact, while it is reasonable to assume that the service provider has adequate resources in terms of computational power and bandwidth to manage filters of big size, the same assumption can not be made for the user, who might be constrained to the limited resources of a mobile device such as a smartphone. Therefore, we offload all onerous tasks to the provider and the third party, who is also assumed to be communication and computationally capable.

Security Definition. In a three-party setting implementing Protocol 2, assuming that no information other than the one implied by the protocol is shared between the parties (parties do not collude), the computation is achieved privately if at the end of the protocol execution Paul learns only $i \in \{0, \dots, s\}$, while Olga and Ursula learn nothing.

Security Analysis. The security of the three-party protocol follows that of the two-party protocol above. The introduction of the third party means however that the user sends her unencoded hash values to the third party, who performs the private Hadamard product. This exposes the user to an attack on the spatial Bloom filter by the third party. While Bloom filters have proved to be irreversible, an exhaustive search may reveal to Olga the input used to produce the received hash outputs. This attack, however, assumes knowledge of $\mathcal{E}$ by Olga. The conventional grid $\mathcal{E}$ represents in fact the coding scheme (or ordering) of the elements on the geographical grid: that is, which value is to be given as input to the hash functions for each position. Since this information is not required by Olga for the execution of the protocol, the user and the provider can agree on an encoding scheme (which can simply be a random ordering of the geographical grid elements) unknown to the third party, thus preventing her from running a search attack. We note that the same goal can also be achieved by using keyed hash functions, which would however require a key exchange between the two parties.

A second threat to which the user is exposed is due to the deterministic nature of the hash results for the same input. In fact, the third party may easily know if the user is revisiting the same grid position twice by comparing the hash digests. In settings in which this is considered unacceptable, a temporal-based variation of the above encoding of the geographical grid can be used.

Protocol 2. Three-party private positioning protocol among provider Paul, third party Olga and user Ursula.

Before any communication, the provider selects the areas of interest and creates the corresponding spatial Bloom filter similarly to Protocol 1.

1 The service provider Paul generates a public and private key pair using a multiplicative homomorphic encryption scheme, and sends the public key to the third party Olga.

2 Paul sends to Olga the encryption of the precomputed spatial Bloom filter $Enc\left(b^{\#}\right)$ and the value m. Then, Paul sends to the user Ursula the set of k hash functions H and the conventional grid $\mathcal{E}$.

3 At regular time intervals, or when required by the specific application, Ursula determines her geographic position and selects the corresponding grid region $e_u \in \mathcal{E}$. Then, she computes the values $\{v_1, \ldots, v_k\}$ where $v_i = h_i\left(e_u\right)$, and sends them to Olga.

4 Olga receives the values from Ursula and builds $b_o^{\#}$, by assigning $b_o^{\#}\left[v_i\right] = 1$ for every $v_i \in \{v_1, \ldots, v_k\}$. Then, she calculates z as the number of 1's in $b_o^{\#}$.

5 Olga computes $e^{\#} = Enc\left(b^{\#}\right) \boxdot b_o^{\#}$ using the homomorphic properties of the encryption scheme (Algorithm 1). Then she applies a random permutation to the values in the filter, and sends z and the result to Paul.

6 Paul decrypts $e^{\#}$ and counts all non-zero values. If the resulting number is $< z$, Ursula's position is outside of the areas on which the SBF was built. Otherwise, the value i, corresponding to Ursula's area Δ_i (minus error probability p_i), is the smallest non-zero value in $Dec\left(e^{\#}\right)$.

---

6   Conclusions

In this paper we present a novel privacy-preserving primitive, the spatial Bloom filter. Based on the classical Bloom filter, the SBF extends it by allowing multiple different sets to be encoded in a single filter. Spatial Bloom filters are particularly suited to store location information represented in a set-based format. We provide a spatial representation system for geographic areas, which allows us to encode or query positioning information (such as the one produced by GPS devices) into an SBF. In this paper we propose two protocols for privacy preservation in location-based services based on spatial Bloom filters. We imagine a scenario in which the provider of the service is interested in detecting the presence of a user within predetermined areas of interest, or his proximity to points of interest. Thanks to the properties of SBF, the provider can build a filter over a number of geographic areas, and the user can query the filter to determine whether his current location lies within those areas without knowing the areas themselves, thus preserving the privacy of the provider. By using the homomorphic properties of a public key encryption scheme, we can also guarantee the user's privacy, by allowing the provider to only learn in which (predefined) area the user is, and not his exact position. The provider learns nothing in case the user is outside the predefined set of areas.

Acknowledgments. The authors would like to acknowledge Marco Miani for the code used in producing Fig. 4.

A Bloom Filters Properties

The bloom filter is built as follows. Initially all bits are set to 0. Then, for each element $a \in S$ and for each $h \in H$ we calculate $h(a) = i$, and set the corresponding i-th bit of b to 1. Thus, m bits are needed in order to store b.

We test an element a_u against b to determine membership in S, that is, we verify whether $a_u \in S$ if

$$\forall h \in H, b[h(a_u)] = 1. \tag{14}$$

If any bit in b that corresponds to a value output by one of the hash functions for a_u is 0, then $a_u \notin S$. If, instead, all the hashes map to bits of value 1, then $a_u \in S$ minus a false positive probability p determined by the number n of elements in S, the number k of hash functions in H and the maximum possible value m output by the hash functions (equal to the binary length of b) as follows:

$$p = \left(1 - \left(1 - \frac{1}{m}\right)^{kn}\right)^{k} \approx \left(1 - e^{-\frac{kn}{m}}\right)^{k}. \tag{15}$$

This small false positive probability is due to the potential collision of hashes evaluated on different inputs, resulting into all bits associated to an element outside the originating set having value 1. As such, it is determined largely by k: if k is sufficiently small for given m and n, the resulting b is sufficiently sparse and collisions are infrequent. If we consider the approximation in (15), we can calculate the optimal number of hashes k as

$$opt(k) = \frac{m}{n} \ln 2, \tag{16}$$

from which we can infer

$$m = \left\lceil -\frac{n \ln p}{(\ln 2)^2} \right\rceil. \tag{17}$$

However, the number of hashes also determines the number of bits read for membership queries, the number of bits written for adding elements to the filter, and the computational cost of calculating the hashes themselves. Therefore, in constrained settings, we may choose to use a less than optimal k, according to performance reasons, if the resulting p is considered sufficiently low for the specific application domain.

B   More on Spatial Representation

The most natural spatial representation for Earth is the standard geographic coordinate system. In the geographic coordinate system every location on Earth can be specified by using a set of values, called coordinates. Standard coordinates are *latitude*, *longitude* and *elevation*. For the purposes of this work we focus on longitude and latitude only, as the combination of these two components is enough to determine the position of any point on the planet (excluding elevation or depth). The whole Earth is divided with 180 parallels and 360 meridians; the plotted grid resulting on the surface is known as the *graticule* (Fig. 4).

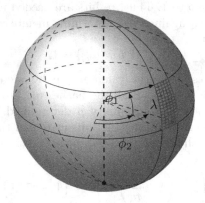

Fig. 4. An example of the planet's surface and the grid plotted on it. ϕ_1 and ϕ_2 are longitude values while λ is a latitude value.

Longitude (`lng`) and latitude (`lat`) can be stored and represented according to several formats. In the following we use the *decimal degrees plus/minus* format, where latitude is positive if it is north of the equator (negative otherwise), and longitude is positive if it is east of the prime meridian (negative otherwise); for instance, 31.456764° (lat) and −85.887734° (lng) are two possible values.

Using a fixed precision in longitude and latitude (that is, choosing a fixed number of decimal points for their values) allows us to easily divide the planet's surface into a discrete grid. Since meridians get closer as they converge the poles, as can be seen in Fig. 4, the portions of the Earth's surface defined by such a grid have varying areas depending on their position (Table 2). While the construction proposed in the following is not dependent on the size or shape of the regions, for simplicity in the discussion it is reasonable to approximate such portions to rectangles and assume they have the same area.

Table 2. Some reference values of accuracy using three decimal places for coordinate representation.

0.001 degrees	lng	lat	
Equator	111.32 m	~ 111.00 m	
23th parallel N/S	102.47 m	~ 111.00 m	Cuba
45th parallel N/S	78.71 m	~ 111.00 m	Italy
67th parallel N/S	43.50 m	~ 111.00 m	Alaska

In actual applications, the precision in decimal points for longitude and latitude should reflect the expected error of the device or sensor used for learning the location information. The precision and accuracy of mobile devices in determining their geographic position were proved to vary considerably depending on the context (urban areas, rural areas, etc.) [20].

In a detailed experiment on the accuracy of GPS sensors installed on mobile devices, Blum et al. show that the location is reported with a precision varying from 10 to 60 meters, depending on the device orientation and type, and, in cities, on the surrounding buildings [3]. Hence, when designing a system based on mobile devices it would reasonable to consider regions with sides tens of meters long.

References

1. Avoine, G., Calderoni, L., Delvaux, J., Maio, D., Palmieri, P.: Passengers information in public transport and privacy: can anonymous tickets prevent tracking? Int. J. Inf. Manag. **34**(5), 682–688 (2014)
2. Bloom, B.H.: Space/time trade-offs in hash coding with allowable errors. Commun. ACM **13**(7), 422–426 (1970)
3. Blum, J.R., Greencorn, D.G., Cooperstock, J.R.: Smartphone sensor reliability for augmented reality applications. In: Zheng, K., Li, M., Jiang, H. (eds.) MobiQuitous 2012. LNICST, vol. 120, pp. 127–138. Springer, Heidelberg (2013)
4. Blumberg, A.J., Eckersly, P.: On locational privacy, and how to avoid losing it forever, April 2009. https://www.eff.org/wp/locational-privacy
5. Calderoni, L., Maio, D., Palmieri, P.: Location-aware mobile services for a smart city: design, implementation and deployment. JTAER **7**(3), 74–87 (2012)
6. Charles, D., Chellapilla, K.: Bloomier filters: a second look. In: Halperin, D., Mehlhorn, K. (eds.) ESA 2008. LNCS, vol. 5193, pp. 259–270. Springer, Heidelberg (2008)
7. Chazelle, B., Kilian, J., Rubinfeld, R., Tal, A.: The bloomier filter: an efficient data structure for static support lookup tables. In: SODA, pp. 30–39. SIAM (2004)

8. Jiazhu, D., Zhilong, L.: A location authentication scheme based on proximity test of location tags. In: ICINS 2013, pp. 1–6 (2013)
9. Kikuchi, H., Sakuma, J.: Bloom filter bootstrap: Privacy-preserving estimation of the size of an intersection. JIP **22**(2), 388–400 (2014)
10. Kulik, L.: Privacy for real-time location-based services. SIGSPATIAL Spec. **1**(2), 9–14 (2009)
11. de Montjoye, Y.-A., Hidalgo, C.A., Verleysen, M., Blondel, V.D.: Unique in the crowd: the privacy bounds of human mobility. Sci. Rep. **3**(1376) (2013). doi:10.1038/srep01376
12. Narayanan, A., Thiagarajan, N., Lakhani, M., Hamburg, M., Boneh, D.: Location privacy via private proximity testing. In: NDSS. The Internet Society (2011)
13. Nielsen, J.D., Pagter, J.I., Stausholm, M.B.: Location privacy via actively secure private proximity testing. In: PerCom Workshops, pp. 381–386. IEEE (2012)
14. Paillier, P.: Public-key cryptosystems based on composite degree residuosity classes. In: Stern, J. (ed.) EUROCRYPT 1999. LNCS, vol. 1592, p. 223. Springer, Heidelberg (1999)
15. Pan, X., Meng, X.: Preserving location privacy without exact locations in mobile services. Front. Comput. Sci. **7**(3), 317–340 (2013)
16. Saldamli, G., Chow, R., Jin, H., Knijnenburg, B.P.: Private proximity testing with an untrusted server. In: WISEC, pp. 113–118. ACM (2013)
17. Shu, X., Yao, D.D.: Data leak detection as a service. In: Keromytis, A.D., Di Pietro, R. (eds.) SecureComm 2012. LNICST, vol. 106, pp. 222–240. Springer, Heidelberg (2013)
18. Sun, J., Zhang, R., Zhang, Y.: Privacy-preserving spatiotemporal matching. In: INFOCOM, pp. 800–808. IEEE (2013)
19. Tonicelli, R., David, B.M., de Morais Alves, V.: Universally composable private proximity testing. In: Boyen, X., Chen, X. (eds.) ProvSec 2011. LNCS, vol. 6980, pp. 222–239. Springer, Heidelberg (2011)
20. von Watzdorf, S., Michahelles, F.: Accuracy of positioning data on smartphones. In: LocWeb, p. 2. ACM (2010)
21. Wicker, S.B.: The loss of location privacy in the cellular age. Commun. ACM **55**(8), 60–68 (2012)
22. Zakhary, S., Radenkovic, M., Benslimane, A.: The quest for location-privacy in opportunistic mobile social networks. In: IWCMC, pp. 667–673. IEEE (2013)
23. Zheng, Y., Li, M., Lou, W., Hou, Y.T.: SHARP: private proximity test and secure handshake with cheat-proof location tags. In: Foresti, S., Yung, M., Martinelli, F. (eds.) ESORICS 2012. LNCS, vol. 7459, pp. 361–378. Springer, Heidelberg (2012)

Security of Direct Anonymous Authentication Using TPM 2.0 Signature

A Possible Implementation Flaw

Tao Zhang$^{(\boxtimes)}$ and Sherman S.M. Chow

Department of Information Engineering,
Chinese University of Hong Kong, Sha Tin, N.T., Hong Kong
{zt112,sherman}@ie.cuhk.edu.hk

Abstract. Direct Anonymous Attestation (DAA) is a digital signature scheme designed for anonymous authentication. A major application of DAA is privacy-preserving remote authentication of a trusted platform module (TPM). The private key used by DAA is stored within the TPM. The resource of TPM is limited, thus TPM devices usually implement only necessary secret-related algorithms and only store sensitive data. Recently, in CCS 2013, Chen and Li proposed the notion of TPM 2.0 signature, which implements a simple yet generic algorithm taking the private key as an input, for a wide range of higher applications such as DAA and others (e.g., Schnorr's signature, U-Prove). However, the reuse of the same TPM algorithm and private key for multiple purposes may introduce vulnerability, even within the same context of DAA. In particular, there are two situations in which the DAA scheme uses the same signature scheme and private key, namely, signing or authentication, and joining the system (for proving the knowledge of the private key to the issuer of the DAA credential). In this paper, we analyzed the current security model of DAA schemes with this in mind, identified the weakness and the corresponding implementation flaw which leads to insecurity, and suggested a fix. Our study provides more comprehensive security analysis for DAA which suggests a prudent practice of DAA implementation.

Keywords: Accountable privacy · Direct anonymous authentication · Trusted platform module · DAA · TPM 2.0

1 Introduction

Direct anonymous authentication (DAA) are originally designed for privacy-preserving remote authentication of a trusted platform module (TPM). A DAA scheme involves three kinds of entities: an issuer, signers, and verifiers. The issuer,

---

This work is supported by grant 439713 from Research Grants Council (RGC), Hong Kong, and grants (4055018, 4930034) from Chinese University of Hong Kong. Sherman Chow is supported by the Early Career Award from RGC, Hong Kong. The authors would like to thank Liqun Chen for inspiration of this research.

© Springer International Publishing Switzerland 2015
D. Lin et al. (Eds.): Inscrypt 2014, LNCS 8957, pp. 37–48, 2015.
DOI: 10.1007/978-3-319-16745-9_3

after successfully authenticated the signer via some means outside of the DAA system, issues a credential to each signer. The signer can prove membership to a verifier anonymously with a DAA signature generated from the credential. The signer can choose to make a certain set of DAA signatures linkable, by specifying the same "base name" during the creation of these signatures. This functionality is known as user-controlled-linkability. The verifiers are in charge of verifying the membership of the signers by verifying the DAA signatures the signers created. The verifiers cannot learn the identity of each signer, beyond what could be inferred from linkability.

TPM [1] is a secure multipurpose hardware chip which provides support for various cryptographic functions. TPM devices implement a secure environment to store and operate on the sensitive data. The computation power and storage space of TPM devices are usually more expensive than common (untrusted) devices such as desktop computers. Thus, the secure storage space is limited, and the algorithms implemented in a TPM device are fixed and only provide minimum functionalities. TPM devices only store sensitive data, or more often, a private key for the cryptographic function such as decryption of the sensitive data or authentication. Moreover, TPM devices only process the operations which need direct access to the sensitive data.

DAA scheme was originally designed for TPM authentication. It splits the signer role into two parts, namely, a Host (e.g., a desktop computer), and the TPM device embedded in the Host. The TPM device, as the principle signer, is in charge of the sensitive data related operations, while the Host, as the assistant signer, is in charge of the non-sensitive operations. The Host has more computation and storage resources, which complements TPM's deficiency in the resources.

From the cost perspective, it is desirable for the TPM device to support only one signature scheme, and the algorithm implementation within the TPM device should also be fixed. Chen and Li [2] proposed a new signature scheme, called TPM 2.0 signature, which provides a single implementation of signing functionality with the goal of supporting more than one higher-level signing-related applications. The scheme splits the traditional signing algorithm Sign() into two parts, namely, a commitment algorithm Commit() and a signature generating algorithm GenSig(). With this new design, higher applications can just call these two algorithms according to the specification of a specific scheme when the private key embedded in the TPM is expected as an input.This is a nice and neat concept which can potentially help the widespread application of TPM 2.0 in DAA, regular signature (e.g., Schnorr's signature), or anonymous credential systems (e.g., U-Prove) [2].

Related Work. DAA can be seen as a group signature without the feature that the signatures can be opened and linked to the signer by the group manager. Brickell *et al.* [3] introduced the concept and proposed the first concrete DAA construction. Since then, DAA has drawn a lot of attentions [4–11] from both industry and academia. Brickell *et al.* [8,12] proposed the first pairing-based DAA constructed with Camenisch-Lysyanskaya (CL) signature. Chen [5] reduced the resource requirement on TPM for DAA scheme under Strong Diffie-Hellman

(SDH) assumption. Brickell and Li [11] further reduced the resource requirement. Brickell *et al.* [13] proposed a DAA scheme under LRSW assumption supporting batch proof. Researchers have also worked on security analysis of DAA [7,14]. For example, Leung *et al.* considered a possible privacy flaw in DAA [6].

Our Contribution. The nature of DAA scheme requires multiple uses of the same private key. To obtain a DAA credential, a non-interactive proof of the signing key should be produced by the TPM. At the same time, the same signing key will of course be used in the actual Sign algorithm. Both are inevitable. Many DAA schemes [3,5,11,13] employ this design. An implementer may expect the non-interactive proof produced by the TPM to follow a certain fixed algorithm implementation available in the TPM, say, via the Commit() and GenSig() oracle as advocated by the work of Chen and Li [2], On the other hand, the issuer is normally considered to be trusted, so is its interaction with the TPM via the Join algorithm. It may explain why existing security models did not consider the usage of the private key by the adversary via this interface.

In this paper, we point out that, when a DAA scheme is not properly implemented, in the sense that all different computations involving the same private key really follow just a single implementation (either using the same hash function within the TPM, or relying on the external hash value supplied to the TPM), it is possible that an adversary may take advantage of the Join() operation to make a forgery. The rest of the paper exploits this vulnerability with several existing DAA designs as illustrations, and proposes a new security model for DAA schemes.

2 Preliminaries

This section reviews the recently proposed TPM 2.0 signature scheme and the definition of DAA's algorithms suite.

2.1 TPM 2.0 Signature

TPM 2.0 signature (or simply TPM signature) is basically a normal signature scheme. Unlike the usual signature schemes, Sign algorithm is split into Commit and GenSig. When used in TPM 2.0 device, the signing algorithm(s) should have a fixed implementation for cost reason. This separation makes the application more flexible since one may realize more than one applications such as different DAA constructions or anonymous credentials as long as they share some similarities in the usage of the private key embedded in the TPM.

TPM 2.0 signature is a tuple of four algorithms (Setup, KeyGen, Sign, Verify), where Sign is actually a protocol between the TPM and the Host.

Setup(1^λ): Let $\mathbb{G}_1$ be a cyclic group. Randomly choose $h_1 \xleftarrow{\$} \mathbb{G}_1$. Choose the collision resistant hash functions $\mathbb{H}_3 : \{0,1\}^* \rightarrow \mathbb{G}_1$, $\mathbb{H}_4 : \{0,1\}^* \rightarrow \mathbb{Z}_p$, $\mathbb{H}_5 : \{0,1\}^* \rightarrow \mathbb{Z}_p$. Output the system parameter gpk $= (\mathbb{G}_1, h_1, \mathbb{H}_3, \mathbb{H}_4, \mathbb{H}_5)$.

KeyGen(gpk): Randomly choose $f \leftarrow \mathbb{Z}_p$ and compute $F = h_1^f$. Output (tpk, tsk)$= (F, f)$.

Sign(gpk, tsk, m):

– Commit(str, P_1): Given $P_1 \in \mathbb{G}_1$ and str $\in \{0,1\}^*$:
  1. If str $= \perp$, set $J = 1_{\mathbb{G}_1}$, otherwise, $J = \mathbb{H}_3(\text{str})$.
  2. Randomly choose $r \leftarrow \mathbb{Z}_p$ and compute $R_1 = J^r$, $R_2 = P_1^r$, $K = J^f$.
  3. Output (K, R_1, R_2).
– GenSig(c_h, m): Given c_h and m as input, where $c_h = \mathbb{H}_4(R_1 \| R_2)$ in general cases, and it is computed outside the TPM:
  1. Compute $c = \mathbb{H}_5(c_h \| m)$, $s = r + c \cdot f$, where r is from Commit and will be deleted after this operation.
  2. Output (c, s).

The signature on m is $\sigma = (P_1, J, K, R_1, R_2, c, s)$. It is a signature of knowledge

$$\text{SPK}\{(f) : K_1 = P_1^f \wedge K = J^f\}(m).$$
Verify(m, σ, K_1): Parse $\sigma = (P_1, J, K, R_1, R_2, c, s)$:

1. If $P_1 = 1_{\mathbb{G}_1}$ and $P_2 = 1_{\mathbb{G}_1}$, reject and return 0.
2. Verify if $c = \mathbb{H}_5(\mathbb{H}_4(R_1 \| R_2) \| m)$. Reject and return 0 if the equation does not hold.
3. Verify if $R_1 = J^s \cdot K^{-c}$ and $R_2 = P_1^s \cdot K_1^{-c}$. Reject and return 0 if either of the equations does not hold.
4. Return 1 if all the above succeed.

2.2 Direct Anonymous Attestation (DAA)

A DAA scheme involves three entities: an issuer, signers, and verifiers. The signer role is split into a Host and the TPM embedded in it. A DAA scheme has five polynomial-time algorithms (Setup, Join, Sign, Verify, Link).

– Setup. An issuer runs this probabilistic polynomial-time algorithm to setup the system. On input of a security parameter, Setup output a pair (isk, param), where isk is the secret key of the issuer, param is the global public parameters for the system including issuer's public key, a description of the DAA credential space, a description of a finite message space, and a description of a finite DAA signature space.
– Join. An issuer and a signer jointly run this probabilistic polynomial-time algorithm to authorize a signer (a Host-TPM pair). On the TPM part, TPM runs Join to produce a pair (tsk, comm) where tsk is the TPM secret key and comm is a commitment on tsk associated with the issuer. TPM sends comm to the issuer. On the issuer part, the issuer produces a DAA credential cred with comm and isk for certifying tsk. This credential cred is given to both TPM and the corresponding Host.
– Sign. TPM and the corresponding Host jointly run this probabilistic algorithm to generate an anonymous signature with authentication for a specific verifier. On input the TPM secret key tsk, the DAA credential cred, a basename bsn from the target verifier (or a special symbol $\perp$), a message m, and a nonce n_V from the target verifier, Sign outputs a signature σ on m and n_V under (tsk, cred) associated with bsn. The basename bsn is used for user-controlled linkability.

- Verify. A verifier $\mathcal{V}$ runs this deterministic polynomial-time algorithm to verify the authenticity of a signer. On input a message m, a basename bsn, a signature σ, and a rogue list RogueList containing signers with revoked credential, Verify returns accept or reject.
- Link. A verifier $\mathcal{V}$ runs this deterministic polynomial-time algorithm to determine whether two signatures are linked. On input two DAA signatures σ_0 and σ_1, Link returns linked, unlinked, or $\bot$. Link outputs $\bot$ only if either σ_0 or σ_1 is rejected by Verify with an empty rogue list (i.e., the rogue list does not influence the result of Link). That means the corresponding message and the basename should also be part of the inputs.

Chen and Li [2] instantiated the DAA schemes in [11, 15] with TPM 2.0 signatures described in Sect. 2.1. The DAA schemes in [11, 15] use different algorithms to issue credentials to the signers, but both use a Schnorr type signature in Join and Sign operations. The use of Schnorr type signature in Join and Sign is compatible with TPM 2.0. The high-level idea to apply existing DAA schemes to TPM 2.0 is described below.

- Join. A TPM device invokes GenSig to create a signature of knowledge on its private key. An issuer verifies the signature and issues a credential associated with the TPM's private key.
- Sign. A Host randomizes the credential. The TPM takes in a message along with the randomized credential, and invokes GenSig to create a signature on the message and the credential with the TPM's private key.

3 Review of the Current Security Model

This section presents the current security model for DAA schemes (e.g., [5, 11, 15]), and briefly states the possible security problem of the model.

3.1 Current Security Model

Correctness. If both the signer and verifier are honest, the signatures generated by the signer will be accepted by the verifier and can be linked together with overwhelming probability. This means that the DAA algorithms must meet the following consistency requirement. If

$$(\text{isk}, \text{gpk}) \leftarrow \text{Setup}(1^\lambda)$$
$$(\text{tsk}, \text{cred}) \leftarrow \text{Join}(\text{isk}, \text{gpk})$$
$$(m_b, \sigma_b) \leftarrow \text{Sign}(m_b, \text{bsn}, \text{tsk}, \text{cred}, \text{gpk})|_{b=0,1}$$

then we must have

$$1 \leftarrow \text{Verify}(m_b, \text{bsn}, \sigma_b, \text{gpk}, \text{RogueList})|_{b=0,1}$$
$$1 \leftarrow \text{Link}(\sigma_0, \sigma_1, \text{gpk})|_{\text{bsn} \neq \bot}$$

User-Controlled-Anonymity. The notion of user-controlled-anonymity is defined via a game played by a PPT challenger $\mathcal{C}$ and a PPT adversary $\mathcal{A}$ as below:

- *Initial:* $\mathcal{A}$ runs Setup, obtains (isk, gpk), and publishes gpk. $\mathcal{C}$ verifies the validity of gpk.
- *Phase 1:* $\mathcal{A}$ makes the following queries to $\mathcal{C}$:
  - Signing query: $\mathcal{A}$ submits a signer's identity ID, a basename bsn (either $\perp$ or a data string) and a message m of his choice to $\mathcal{C}$, who runs Sign to get a signature σ and responds with σ.
  - Joining query: $\mathcal{A}$ submits a signer's identity ID of his choice to $\mathcal{C}$, who runs Join with $\mathcal{A}$ to create tsk and to obtain cred from $\mathcal{A}$. $\mathcal{C}$ verifies the validity of cred and keeps tsk secret.
  - Corrupt query: $\mathcal{A}$ submits a signer's identity ID of his choice to $\mathcal{C}$, who responds with the value tsk of the signer.
- *Challenge:* At the end of *Phase 1*, $\mathcal{A}$ outputs two signers' identities ID_0 and ID_1, a message m and a basename bsn of his choice to $\mathcal{C}$. $\mathcal{A}$ must not have made any Corrupt query on either ID_0 or ID_1, and not have made the Sign query with the same bsn if bsn $\neq \perp$ with either ID_0 or ID_1. To make the challenge, $\mathcal{C}$ chooses a bit $b \in \{0,1\}$ uniformly at random, signs m associated with bsn under $(tsk_b, cred_b)$ to get a signature σ and returns it to $\mathcal{A}$.
- *Phase 2:* Same as *Phase 1*, but it is not allowed to corrupt signer ID_0 or ID_1, or to make any Signing query with bsn if bsn $\neq \perp$ with either ID_0 or ID_1.
- *Response:* $\mathcal{A}$ returns a bit b'. We say that the adversary wins the game if $b' = b$.

Definition 1 (User-Controlled-Anonymity). *The advantage of the adversary $\mathcal{A}$ in user-controlled-anonymity game is $Adv[\mathcal{A}_{DAA}^{anon}] = |\Pr[b' = b] - \frac{1}{2}|$. A DAA scheme is user-controlled-anonymous if for any PPT adversary $\mathcal{A}$, $Adv[\mathcal{A}_{DAA}^{anon}]$ is negligible.*

User-Controlled-Traceability. The notion of User-Controlled-Traceability is defined via a game played by a $\mathcal{C}$ and an $\mathcal{A}$ as below:

- *Initial:* There are two initial cases.
  - *Case 1:* $\mathcal{C}$ runs Setup and gives the resulting gpk to $\mathcal{A}$, and $\mathcal{C}$ keeps isk secret.
  - *Case 2:* $\mathcal{C}$ receives gpk from $\mathcal{A}$ and does not know the value of isk.
- *Query:* $\mathcal{A}$ makes the following queries to $\mathcal{C}$:
  - Signing query: The same as in the game of user-controlled-anonymity.
  - Semi-signing query: $\mathcal{A}$ submits a signer identity ID along with the data transmitted from the host to the TPM in Sign of his choice to $\mathcal{C}$, who acts as the TPM in Sign and responds with the data transmitted from the TPM to the host in Sign.
  - Joining query: There are three cases of this query. The first two are used with *Initial Case 1*, while the last one is used with *Initial Case 2*. We assume that $\mathcal{A}$ does not use a single ID for more than one join query.
    * Join Case 1a: $\mathcal{A}$ submits a signer's identity ID of his choice to $\mathcal{C}$, who runs Join to create tsk and cred for the signer, and finally $\mathcal{C}$ sends cred to $\mathcal{A}$ and keeps tsk secret.

∗ Join Case 1b: $\mathcal{A}$ submits a signer's identity ID with a tsk of his choice to $\mathcal{C}$, who runs Join to create cred for the signer and puts the given tsk on the rogue list RogueList. $\mathcal{C}$ responds $\mathcal{A}$ with cred.
∗ Join Case 2: $\mathcal{A}$ submits a signer's identity ID of his choice to $\mathcal{C}$, who runs Join with $\mathcal{A}$ to create tsk and to obtain cred from $\mathcal{A}$. $\mathcal{C}$ verifies the validity of cred and keeps tsk secret.
• Corrupt query: This is the same as in the game of user-controlled-anonymity, except that at the end $\mathcal{C}$ puts the revealed tsk on the rogue list RogueList.
− Forge: $\mathcal{A}$ returns a signer's identity ID, a message m, a signature σ on m, and the associated basename bsn. We say that the $\mathcal{A}$ wins the game if either of the following two situations is true:
  1. With Initial Case 1 where $\mathcal{A}$ does not have access to isk,
    (a) Verify$(m, \text{bsn}, \sigma) = 1$, but σ is neither a response of the existing Signing queries nor a response of the existing Semi-signing queries.
    (b) In the case of bsn $\neq \bot$, there exists another signature σ' associated with the same identity and bsn, and Link$(\sigma, \sigma') = 0$.
  2. With Initial Case 2 where $\mathcal{A}$ knows isk, the same as in (a), in the condition that the secret key tsk used to create σ was generated in Join Case 2 where $\mathcal{A}$ does not have access to tsk.

Definition 2 (User-Controlled-Traceability). *The advantage of the adversary $\mathcal{A}$ in user-controlled-traceability game is $Adv[\mathcal{A}_{DAA}^{trace}] = \Pr[\mathcal{A} \ wins]$. A DAA scheme is user-controlled-traceable if for any PPT adversary $\mathcal{A}$, $Adv[\mathcal{A}_{DAA}^{trace}]$ is negligible.*

3.2 Security Concerns

TPM 2.0 signature provides two signing-related interface (APIs), a commitment API Commit() and a signing API GenSig(). The current security model of DAA does not cover all the information an adversary can get from a DAA scheme. Generally speaking, the more APIs a system provide, the more potential attack vectors one may exploit. Xi *et al.* [16] and Acar *et al.* [17] both have studied the impact on security by TPM API Commit(). Here, we located another vulnerability brought in by the multiple uses of GenSig for different purposes. This vulnerability is not as obvious as the one in [16,17], but is also vital for the security of DAA schemes.

In more details, DAA with TPM devices typically employ one signature scheme for all usage due to the resource limitation of TPM devices. Thus, both Join() and Sign() uses GenSig() API. For example, the DAA scheme of Brickell and Li [11] used the same signing key for Sign() (for the actual signing) and Join() (for proving its knowledge). It is possible for an adversary to exploit this for creating a forgery. Section 4.2 describes a possible breach and gives an attack when existing DAA schemes are not properly implemented.

4   Our Analysis

We explain why the same signing algorithm may be used in the actual Sign algorithm and (perhaps surprisingly) the Join algorithm. We then describe the necessary modification on the current security model of DAA schemes. The second part of this section analyzes the current security model, explains the vulnerability in the current security model, and explains what could go wrong with this vulnerability in some existing DAA schemes.

4.1   Revision of Security Model

For DAA, the user secret key should be created by the TPM. Obviously, this secret key should be certified by the issuer in the Join algorithm. Yet, for the security proof of the whole DAA scheme, it is common to expect the TPM to produce a proof of knowledge of this user secret key during the Join algorithm. On the other hand, as argued in the introduction, the TPM usually provides a single fixed interface for every operation involving this secret key. In other words, DAA schemes use the same "signature scheme" (since a non-interactive proof-of-knowledge of a signing key can be considered as a signature scheme) and the same signing key for both Join and Sign operations.

We consider the situation that the target TPM's Host or an issuer are corrupted by the adversary. Now, Join operation acts as an oracle which returns signature of a certain type to the adversary. The adversary does not have control on the choice of the messages for these signatures, however, with the corrupted issuer, the adversary can meddle with part of the input of signatures. Hence, the adversary is able to produce a forgery with this join oracle provided by the corrupted issuer.

To make the security model more comprehensive, the following oracle needs to be added into the security model of DAA schemes. This oracle complements the current security model for user-controlled-traceability in Sect. 3.1.

Definition 3 (Join Oracle 2). *Join oracle 2 outputs the transcript of* TPM. *An adversary inputs a nonce* n_{Issuer} *as in* Join *operation, Join oracle 2 invokes the* GenSig *oracle of the* TPM *oracles to generate the transcript of* TPM *in* Join *operation.*

In the security model, the adversary has the access to Join oracle 2 besides all the other oracles stated in Sect. 3.1.

4.2   Potential Vulnerability

Now we describe attacks on a possible mis-implementation of the schemes in [11] (SDH-DAA) and [15] (LRSW-DAA). We keep the notations used in the original paper, yet we omit the encryption scheme employed in Join as it is not essential in the forgery. For the attack, we assume that the hash functions $\mathbb{H}_2$ and $\mathbb{H}_5$ are instantiated with the same hash function $\mathbb{H}^*$, and the arrangement of the inputs are concatenated in a specific order (for both of the original schemes). This assumption is strong but may hold when TPM devices implement only one signature scheme at a time to reduce the cost. For example, as described in Chen and Li's work [2], there is only one implementation of GenSig algorithm. To instantiate the DAA schemes in [11,15] with TPM 2.0 signature, Join and Sign have to invoke the same API, GenSig, with a fixed hash algorithm implemented.

While it may be pretty obvious to cryptographers that the reuse of hash functions for different purposes (even within the same scheme) is bad to security, it may not be apparent to the developers who actually implement the scheme, especially under the cost constraints of TPM which only affords a single implementation, and the TPM 2.0 signature idea advocated by Chen and Li which allows multiple higher applications to use the same "basic" signature scheme implemented within the TPM.

We stress that our attacks below are not applicable to the original DAA schemes as described in the original papers [11,15], if they do not utilize a single implementation via TPM 2.0 signature. Yet, we have reasons to believe that it may be the case since only the only available way to access the private key is via the same GenSig API. Moreover, we believe that it is the job of cryptographers to propose a security model which is as comprehensible (in terms of possible vulnerabilities) as possible, and point out what could go wrong if cryptographic schemes are not properly implemented.

Attack on SDH-DAA Scheme. An adversary $\mathcal{A}$ can generate a signature for a verifier $\mathcal{V}$ with basename bsn in a variant of utilizing TPM 2.0 signature to instantiate the SDH-based DAA scheme [11] in this way:

1. $\mathcal{A}$ starts a Sign operation with $\mathcal{V}$ and receives a nonce $n_{\mathcal{V}}$ from $\mathcal{V}$.
2. $\mathcal{A}$ makes a sign query on an arbitrary message for $\mathcal{V}$, and extract $(B = \mathbb{H}_3(\mathsf{bsn}), K = B^f, \mathsf{bsn})$ from the output. We assume that the signature query for $\mathcal{V}$ with basename bsn has been made prior to the attack, thus $\mathcal{A}$ can obtain the value $K = B^f$ without the knowledge of f. If this has not happened, a signing query can always be made on a random message. The adversary only concerns about the tuple (B, K, bsn).
3. $\mathcal{A}$ performs the following calculation as the Join operation of TPM and Host in the real Sign operation:
   (a) $\mathcal{A}$ chooses $r_f^*, r_x^*, r_a^*, r_b^* \xleftarrow{\$} \mathbb{Z}_p$.
   (b) $\mathcal{A}$ computes $b = a \cdot x \mod p$ and $T = A \cdot h_2^a$.
   (c) $\mathcal{A}$ computes $R_1^* = B^{r_f^*}$ and $R_2^* = e(h_1^{r_f^*} \cdot T^{-r_x^*} \cdot h_2^{r_b^*}, g_2) \cdot e(h_2, w)^{r_a^*}$.
   (d) $\mathcal{A}$ computes $c_h = \mathbb{H}_4(\mathsf{gpk}, B, K, T, R_1^*, R_2^*, n_{\mathcal{V}})$.
4. $\mathcal{A}$ starts Join oracle 2 with the victim TPM, and sends c_h as the nonce n_{Issuer} as in Join algorithm.
   (a) TPM chooses $r_f \xleftarrow{\$} \mathbb{Z}_p$, and computes $R = h_1^{r_f}$.
   (b) TPM computes $c = \mathbb{H}^*(\mathsf{gpk}||c_h||F||R)$.
   (c) TPM computes $s_f = r_f + c \cdot f$ and outputs $\sigma = (s_f, c)$.

Note that in the preceding steps, $\mathcal{A}$ plays Issuer's role. However, we only need a TPM signature on the nonce input by $\mathcal{A}$. Join oracle 2 ends after outputting the TPM signature on n_{Issuer} to $\mathcal{A}$.

5. $\mathcal{A}$ computes $\hat{R} = h_1^{s_f} \cdot F^{-c}$ and outputs a DAA signature (s_f, c) on $F||\hat{R}$ for $\mathcal{V}$. Obviously, the signature (s_f, c) can be verified for message $F||\hat{R}$ in our variant of the SDH-DAA scheme. And $\mathcal{A}$ can easily compute (s_x^*, s_a^*, s_b^*) with the signature (s_f, c) and finally output the forgery.

Attack on LRSW-DAA Scheme. An adversary $\mathcal{A}$ can generate a signature for a verifier $\mathcal{V}$ with basename bsn in a variant of utilizing TPM 2.0 signature to instantiate the LRSW-based DAA scheme [15] in this way:

1. $\mathcal{A}$ starts a Sign operation with $\mathcal{V}$ and receives a nonce $n_{\mathcal{V}}$ from $\mathcal{V}$.
2. $\mathcal{A}$ makes a sign query on an arbitrary message for $\mathcal{V}$, and then extract $(R, S, T, W, J, K = [\mathsf{sk}_T]J, \mathsf{bsn})$ from the output. We assume that the signature query for $\mathcal{V}$ with basename bsn has been made prior to the attack, thus $\mathcal{A}$ can obtain the value $K = [\mathsf{sk}_T]J$ without the knowledge of sk_T. If this has not happened, a signing query can always be made on a random message. The adversary only concerns about the tuple $(R, S, T, W, J, K, \mathsf{bsn})$.
3. $\mathcal{A}$ perform the following operation:
   (a) $\mathcal{A}$ chooses $l^* \xleftarrow{\$} \mathbb{Z}_p$ and re-randomize the credential $(R^* = [l^*]R, S^* = [l^*]S, T^* = [l^*]T, W^* = [l^*]W)$.
   (b) $\mathcal{A}$ chooses $r^* \xleftarrow{\$} \mathbb{Z}_p$ and computes $R_1^* = [r^*]J, R_2^* = [r^*]S^*$.
   (c) $\mathcal{A}$ computes $\mathsf{str}^* = J||K||\mathsf{bsn}||R_1^*||R_2^*$.
   (d) $\mathcal{A}$ computes $c^* = \mathbb{H}_4(R^*||S^*||T^*||W^*||n_{\mathcal{V}})$.
   (e) $\mathcal{A}$ computes $c_h = c||\mathsf{str}^*$.
4. $\mathcal{A}$ starts Join oracle 2 with the victim TPM, and sends c_h as the nonce n_{Issuer} as in Join algorithm.
   (a) TPM chooses $u \xleftarrow{\$} \mathbb{Z}_p$ and computes $Q_2 = [\mathsf{sk}_T]P_1, U = [u]P_1$.
   (b) TPM computes $\mathsf{str} = X||Y||n_{\mathsf{Issuer}} = X||Y||c_h$.
   (c) TPM computes $v = \mathbb{H}^*(P_1||Q_2||U||\mathsf{str})$ and $w = u + v \cdot \mathsf{sk}_T \mod p$.
   Note that in the preceding two steps, $\mathcal{A}$ plays Issuer's role. However, we only need a TPM signature on the nonce input by $\mathcal{A}$. Join oracle 2 ends after outputting the TPM signature (w, v) on n_{Issuer} to $\mathcal{A}$.
5. $\mathcal{A}$ outputs a DAA signature (w, v) on $P_1||Q_2||U$ for $\mathcal{V}$. Obviously, the signature (w, v) can be verified for message $P_1||Q_2||U$ in the variant of the LRSW-DAA scheme.

5   Conclusion

In this paper, we proposed a new security model for DAA schemes. We analyzed the current security model of DAA schemes, and found a possible vulnerability in the current model. This possible vulnerability affects the DAA schemes which use the same signing algorithm and the same signing key for different usages.

We conducted an attack on potential mis-implementations of DAA to show the consequence of the vulnerability. As a concrete (yet well-known) advice, it is dangerous to reuse the same hash function for more than one purpose, even within the same context. We leave it as an interesting (but maybe not necessarily a pure-research type) question to strike a balance between security of the higher applications and the generality of the single building block implementation for those higher applications.

References

1. Sumrall, N., Novoa, M.: Trusted computing group (TCG) and the TPM 1.2 specification. In: Intel Developer Forum 2003, vol. 32 (2003)
2. Chen, L., Li, J.: Flexible and scalable digital signatures in TPM 2.0. In: CCS 2013, pp. 37–48. ACM (2013)
3. Brickell, E., Camenisch, J., Chen, L.: Direct anonymous attestation. In: CCS 2004, pp. 132–145. ACM (2004)
4. Brickell, E., Li, J.: Enhanced privacy ID: a direct anonymous attestation scheme with enhanced revocation capabilities. In: Proceedings of the 2007 ACM Workshop on Privacy in Electronic Society, pp. 21–30. ACM (2007)
5. Chen, L.: A DAA scheme requiring less TPM resources. In: Bao, F., Yung, M., Lin, D., Jing, J. (eds.) Inscrypt 2009. LNCS, vol. 6151, pp. 350–365. Springer, Heidelberg (2010)
6. Leung, A., Chen, L., Mitchell, C.J.: On a possible privacy flaw in direct anonymous attestation (DAA). In: Lipp, P., Sadeghi, A.-R., Koch, K.-M. (eds.) Trust 2008. LNCS, vol. 4968, pp. 179–190. Springer, Heidelberg (2008)
7. Rudolph, C.: Covert identity information in direct anonymous attestation (DAA). In: Venter, H., Eloff, M., Labuschagne, L., Eloff, J., von Solms, R. (eds.) New Approaches for Security, Privacy and Trust in Complex Environments, pp. 443–448. Springer, New York (2007)
8. Brickell, E., Chen, L., Li, J.: Simplified security notions of direct anonymous attestation and a concrete scheme from pairings. Int. J. Inf. Secur. **8**(5), 315–330 (2009)
9. Chen, L., Morrissey, P., Smart, N.P.: On proofs of security for DAA schemes. In: Baek, J., Bao, F., Chen, K., Lai, X. (eds.) ProvSec 2008. LNCS, vol. 5324, pp. 156–175. Springer, Heidelberg (2008)
10. Chen, X., Feng, D.: Direct anonymous attestation for next generation TPM. J. Comput. **3**(12), 43–50 (2008)
11. Brickell, E., Li, J.: A pairing-based DAA scheme further reducing TPM resources. In: Acquisti, A., Smith, S.W., Sadeghi, A.-R. (eds.) TRUST 2010. LNCS, vol. 6101, pp. 181–195. Springer, Heidelberg (2010)
12. Brickell, E., Chen, L., Li, J.: A new direct anonymous attestation scheme from bilinear maps. In: Lipp, P., Sadeghi, A.-R., Koch, K.-M. (eds.) Trust 2008. LNCS, vol. 4968, pp. 166–178. Springer, Heidelberg (2008)
13. Brickell, E., Chen, L., Li, J.: A (corrected) DAA scheme using batch proof and verification. In: Chen, L., Yung, M., Zhu, L. (eds.) INTRUST 2011. LNCS, vol. 7222, pp. 304–337. Springer, Heidelberg (2012)
14. Backes, M., Maffei, M., Unruh, D.: Zero-knowledge in the applied pi-calculus and automated verification of the direct anonymous attestation protocol. In: IEEE SP 2008, pp. 202–215. IEEE (2008)

15. Chen, L., Page, D., Smart, N.P.: On the design and implementation of an efficient DAA scheme. In: Gollmann, D., Lanet, J.-L., Iguchi-Cartigny, J. (eds.) CARDIS 2010. LNCS, vol. 6035, pp. 223–237. Springer, Heidelberg (2010)
16. Xi, L., Yang, K., Zhang, Z., Feng, D.: DAA-related APIs in TPM 2.0 revisited. In: Holz, T., Ioannidis, S. (eds.) Trust 2014. LNCS, vol. 8564, pp. 1–18. Springer, Heidelberg (2014)
17. Acar, T., Nguyen, L., Zaverucha, G.: A TPM Diffie-Hellman oracle. Technical Report MSR-TR-2013-105, Microsoft Research (2013) Also available at Cryptology ePrint Archive 2013/667

Multiparty and Outsource
Computation

Revocation in Publicly Verifiable Outsourced Computation

James Alderman$^{(\boxtimes)}$, Christian Janson, Carlos Cid, and Jason Crampton

Information Security Group, Royal Holloway, University of London,
Egham, Surrey TW20 0EX, UK
{James.Alderman.2011,Christian.Janson.2012}@live.rhul.ac.uk
{Carlos.Cid,Jason.Crampton}@rhul.ac.uk

Abstract. The combination of software-as-a-service and the increasing use of mobile devices gives rise to a considerable difference in computational power between servers and clients. Thus, there is a desire for clients to outsource the evaluation of complex functions to an external server. Servers providing such a service may be rewarded per computation, and as such have an incentive to cheat by returning garbage rather than devoting resources and time to compute a valid result.

In this work, we introduce the notion of Revocable Publicly Verifiable Computation (RPVC), where a cheating server is revoked and may not perform future computations (thus incurring a financial penalty). We introduce a Key Distribution Center (KDC) to efficiently handle the generation and distribution of the keys required to support RPVC. The KDC is an authority over entities in the system and enables revocation. We also introduce a notion of blind verification such that results are verifiable (and hence servers can be rewarded or punished) without learning the value. We present a rigorous definitional framework, define a number of new security models and present a construction of such a scheme built upon Key-Policy Attribute-based Encryption.

Keywords: Publicly Verifiable Outsourced Computation · Key Distribution Center · Key-Policy Attribute-Based Encryption · Revocation

1 Introduction

It is increasingly common for mobile devices to be used as general computing devices. There is also a trend towards cloud computing and enormous volumes of

J. Alderman acknowledges support from BAE Systems Advanced Technology Centre under a CASE Award.

C. Cid—This research was partially sponsored by US Army Research laboratory and the UK Ministry of Defence under Agreement Number W911NF-06-3-0001. The views and conclusions contained in this document are those of the authors and should not be interpreted as representing the official policies, either expressed or implied, of the US Army Research Laboratory, the U.S. Government, the UK Ministry of Defence, or the UK Government. The US and UK Governments are authorized to reproduce and distribute reprints for Government purposes notwithstanding any copyright notation hereon.

© Springer International Publishing Switzerland 2015
D. Lin et al. (Eds.): Inscrypt 2014, LNCS 8957, pp. 51–71, 2015.
DOI: 10.1007/978-3-319-16745-9_4

data ("big data") which means that computations may require considerable computing resources. In short, there is a growing discrepancy between the computing resources of end-user devices and the resources required to perform complex computations on large datasets. This discrepancy, coupled with the increasing use of software-as-a-service, means there is a requirement for a client device to be able to delegate a computation to a server.

Consider, for example, a company that operates a "bring your own device" policy, enabling employees to use personal smartphones and tablets for work. Due to resource limitations, it may not be possible for these devices to perform complex computations locally. Instead, a computation is outsourced over some network to a more powerful server (possibly outside the company, offering software-as-a-service, and hence untrusted) and the result of the computation is returned to the client device. Another example arises in the context of battlefield communications where each member of a squadron of soldiers is deployed with a reasonably light-weight computing device. The soldiers gather data from their surroundings and send it to regional servers for analysis before receiving tactical commands based on results. Those servers may not be fully trusted e.g. if the soldiers are part of a coalition network. Thus a soldier must have an assurance that the command has been computed correctly. A final example could consider sensor networks where lightweight sensors transmit readings to a more powerful base station to compute statistics that can be verified by an experimenter.

In simple terms, given a function F to be computed by a server S, the client sends input x to S, who should return $F(x)$ to the client. However, there may be an incentive for the server (or an imposter) to cheat and return an invalid result $y \neq F(x)$ to the client. The server may wish to convince a client of an incorrect result, or (particularly if servers are rewarded per computation performed) the server may be too busy or may not wish to devote resources to perform the computation. Thus, the client wishes to have some assurance that the result y returned by the server is, in fact, $F(x)$. This problem, known as *Verifiable Outsourced Computation* (VC), has attracted a lot of attention in the community recently. In practical scenarios, it may well be desirable that cheating servers are prevented from performing future computations, as they are deemed completely untrustworthy. Thus, future clients need not waste resources delegating to a 'bad' server, and servers are disincentivised from cheating in the first place as they will incur a significant (financial) penalty from not receiving future work. Many current schemes have an expensive pre-processing stage run by the client. However, it is likely that many different clients will be interested in outsourcing computations, and that functions of interest to each client will substantially overlap, as in the "bring your own device" scenario above. It is also conceivable that the number of servers offering to perform such computations will be relatively low (limited to a reasonably small number of trusted companies with plentiful resources). Thus, it is easy to envisage a situation in which many computationally limited clients wish to outsource computations of the same (potentially large) set of functions to a set of untrusted servers. Current VC schemes do not support this kind of scenario particularly well.

Our main contribution, then, is to introduce the new notion of *Revocable Publicly Verifiable Computation* (RPVC). We also propose the introduction of a Key Distribution Center (KDC) to perform the computationally intensive parts of VC and manage keys for all clients, and we simplify the way in which the computation of multiple functions is managed. We enable the revocation of misbehaving servers (those detected as cheating) such that they cannot perform further computations until recertified by the KDC, as well as "blind verification", a form of output privacy, such that the verifier learns whether the result is valid but not the value of the output. Thus the verifier may reward or punish servers appropriately without learning function outputs. We give a rigorous definitional framework for RPVC, that we believe more accurately reflects real environments. This new framework both removes redundancy and facilitates additional functionality, leading to several new security notions.

In the next section, we briefly review related work. In Sect. 3, we define our framework and the relevant security models. In Sect. 4, we provide an overview, technical details and a concrete instantiation of our framework using Attribute-based Encryption as well as full security proofs. Additional background details can be found in the Appendix.

Notation. In the remainder of this paper we use the following notation. If A is a probabilistic algorithm we write $y \leftarrow A(\cdot)$ for the action of running A on given inputs and assigning the result to an output y. We denote the empty string by ϵ and use PPT to denote probabilistic polynomial-time. We say that $\mathrm{negl}(\cdot)$ is a negligible function on its input and κ denotes the security parameter. We denote by $\mathcal{F}$ the family of Boolean functions closed under complement – that is, if F belongs to $\mathcal{F}$ then $\overline{F}$, where $\overline{F}(x) = F(x) \oplus 1$, also belongs to $\mathcal{F}$. We denote the domain of F by $\mathrm{Dom}(F)$ and the range by $\mathrm{Ran}(F)$. By $\mathcal{M}$ we denote a message space and the notation $\mathcal{A}^{\mathcal{O}}$ is used to denote the adversary $\mathcal{A}$ being provided with oracle access. Finally, $[n]$ denotes the set $\{1, \ldots, n\}$.

2 Verifiable Computation Schemes and Related Work

The concept of non-interactive verifiable computation was introduced by Gennaro et al. [5] and may be seen as a protocol between two polynomial-time parties: a *client*, C, and a *server*, S. A successful run of the protocol results in the provably correct computation of $F(x)$ by the server for an input x supplied by the client. More specifically, a VC scheme comprises the following steps [5]:

1. KeyGen (*Run once*):C computes evaluation information EK_F that is given to S to enable it to compute F;
2. ProbGen (*Run multiple times*):C sends the encoded input $\sigma_{F,x}$ to S;
3. Compute (*Run multiple times*):S computes $F(x)$ using EK_F and $\sigma_{F,x}$ and returns an encoding of the output $\theta_{F(x)}$ to C;
4. Verify (*Run multiple times*):C checks whether $\theta_{F(x)}$ encodes $F(x)$.

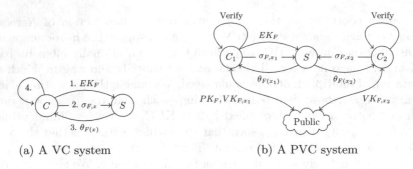

(a) A VC system   (b) A PVC system

Fig. 1. The operation of verifiable computation schemes

The operation of a VC scheme is illustrated in Fig. 1a. KeyGen may be computationally expensive but the remaining operations should be efficient for the client. The cost of setup is amortized over multiple computations of F.

In prior work, Gennaro et al. [5] gave a construction using Garbled Circuits [11], which provides a "one-time" Verifiable Outsourced Computation allowing a client to outsource the evaluation of a function on a single input. However, the construction is insecure if the circuit is reused on a different input and thus this cost cannot be amortized. Moreover, the cost of generating a new garbled circuit is approximately equal to the cost of evaluating the function itself. The authors therefore suggested using fully homomorphic encryption [6] to re-randomise the circuit to allow multiple executions. In independent and concurrent work, Carter et al. [3] introduce a third party to generate garbled circuits for such schemes but require this entity to be online throughout and model the system as a secure multi-party computation between the client, server and third-party. Some works [4,7] consider the multi-client case where functions are computed over joint input from multiple clients. Parno et al. [10] introduced *Publicly Verifiable Computation* (PVC), where a single client C_1 computes EK_F, as well as publishing information PK_F that enables other clients to encode inputs (so only one client has to run the expensive pre-processing stage). A client submits an input x and may publish $VK_{F,x}$ to allow other clients to verify the output. The operation of a PVC scheme is illustrated in Fig. 1b. It uses the same four algorithms as VC but KeyGen and ProbGen now output public values that other clients may use to encode inputs and verify outputs. Parno et al. gave an instantiation of PVC using Key-Policy Attribute-based Encryption (KP-ABE) for a class of Boolean functions. Further details are available in Appendix A.

3   Revocable Publicly Verifiable Computation

We now describe our new notion of PVC, which we call *Revocable Publicly Verifiable Computation* (RPVC). We assume there is a Key Distribution Center (KDC) and many clients which make use of multiple untrusted or semi-trusted servers to perform complex computations. Multiple servers may be certified, by the KDC, to compute the same function F. As we briefly explained in the introduction, there appear to be good reasons for adopting an architecture of this nature and several scenarios in which such an architecture would be appropriate. The increasing popularity of relatively lightweight mobile computing devices in the workplace means that complex computations may best be performed by more powerful servers run by the organization or in the cloud and we would wish to have some guarantee that those servers are certified to perform certain functions. It is essential that we can verify the results of the computation. If cloud services are competing on price to provide "computation-as-a-service" then it is important that a server cannot obtain an unfair advantage by simply not bothering to compute $F(x)$ and returning garbage instead. It is also important that a server who is not certified cannot return a result without being detected.

Key Distribution Center. Existing frameworks assume that a client or clients run the expensive phases of a VC scheme and that a single server performs the outsourced computation. We believe that this is undesirable for a number of reasons, irrespective of whether the client is sufficiently powerful. First, in a real-world system, we may wish to outsource the setup phase to a trusted third party. In this setting, the third party would operate rather similarly to a certificate authority, providing a trust service to facilitate other operations of an organization (in this case outsourced computation, rather than authentication). Second, we may wish to enforce an access control policy limiting the functions each client can outsource; an internal trusted entity would operate both as a facilitator of outsourced computation and as a policy enforcement point (We will examine the integration of RPVC and access control in future work.)

We consider the KDC to be a separate entity to illustrate separation of duty between the clients that request computations, and the KDC that is authoritative on the system and users. The KDC could be authoritative over many sets of clients (e.g. at an organizational level as opposed to a work group level), and we minimise its workload to key generation and revocation only. It may be tempting to suggest that the KDC, as a trusted entity, performs all computations itself. However we believe that this is not a practical solution in many real world scenarios, e.g. the KDC could be an authority within the organization responsible for user authorization that wishes to enable workers to securely use cloud-based software-as-a-service. As an entity within organization boundaries, performing all computations would negate the benefits gained from outsourcing computations to externally available servers.

System Architecture. In this paper we consider two system architectures, which we call the Standard Model and the Manager Model. The *standard model* is a natural extension of the PVC architecture with the addition of a KDC (as shown in Fig. 2a). The entities comprise a set of clients, a set of servers and a KDC. The KDC initializes the system and generates keys to enable verifiable computation. Clients submit computation requests to a particular server and publish some verification information. Any party can verify the correctness of a server's output. If the output is incorrect, the client may report the server to the

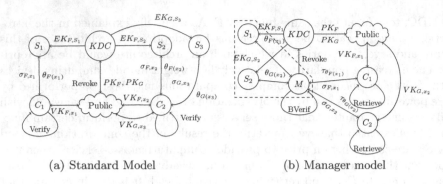

(a) Standard Model   (b) Manager model

Fig. 2. The operation of RPVC

KDC for revocation which will prevent the server from performing any further computations. The *manager model*, in contrast, employs an additional Manager entity who "owns" a pool of computation servers (as shown in Fig. 2b). Clients submit jobs to the manager, who will select a server from the pool based on workload scheduling, available resources or by a bidding process if servers are to be rewarded per computation. A plausible scenario is that servers enlist with a manager to "sell" the use of spare resources, whilst clients subscribe to utilise these through the manager. Results are returned to the manager who should be able to verify the server's work. The manager forwards correct results to the client whilst a misbehaving server may be reported to the KDC for revocation, and the job assigned to another server. In some situations we may not desire external entities to access the result, yet there remain legitimate reasons for the manager to perform verification. Thus we introduce "blind verification", as hinted by Parno et al. [10], such that the manager (or other entity) may verify the validity of the computation without learning the output, while the client holds an extra key that enables the output to be retrieved.

3.1 Formal Definition

Definition 1. A *Revocable Publicly Verifiable Outsourced Computation Scheme (RPVC)* comprises the following algorithms:

- $(PP, MK) \leftarrow \mathsf{Setup}(1^\kappa)$: Run by the KDC to establish public parameters PP and a master secret key MK.
- $PK_F \leftarrow \mathsf{FnInit}(F, MK, PP)$: Run by the KDC to generate a public delegation key, PK_F, for a function F.
- $SK_S \leftarrow \mathsf{Register}(S, MK, PP)$: Run by the KDC to generate a personalised signing key SK_S for a computation server S.
- $EK_{F,S} \leftarrow \mathsf{Certify}(S, F, MK, PP)$: Run by the KDC to generate a certificate in the form of an evaluation key $EK_{F,S}$ for a function F and server S.
- $(\sigma_{F,x}, VK_{F,x}, RK_{F,x}) \leftarrow \mathsf{ProbGen}(x, PK_F, PP)$: ProbGen is run by a client to delegate the computation of $F(x)$ to a server. The output value $RK_{F,x}$ is used to enable output retrieval after the blind verification step.

- $\theta_{F(x)} \leftarrow$ Compute$(\sigma_{F,x}, EK_{F,S}, SK_S, PP)$: Run by a server S holding an evaluation key $EK_{F,S}$, SK_S and an encoded input $\sigma_{F,x}$ of x, to output an encoding, $\theta_{F(x)}$, of $F(x)$, including an identifier of S.
- $(\tilde{y}, \tau_{\theta_{F(x)}}) \leftarrow$ Verify$(\theta_{F(x)}, VK_{F,x}, RK_{F,x}, PP)$:
  Verification comprises:
  - $(RT_{F,x}, \tau_{\theta_{F(x)}}) \leftarrow$ BVerif$(\theta_{F(x)}, VK_{F,x}, PP)$: Run by any verifying party (standard model), or by the manager (manager model), in possession of $VK_{F,x}$ and an encoded output, $\theta_{F(x)}$. This outputs a token $\tau_{\theta_{F(x)}} = ($accept, $S)$ if the output is valid, or $\tau_{\theta_{F(x)}} = ($reject, $S)$ if S misbehaved. It also outputs a retrieval token $RT_{F,x}$ which is an encoding of the actual output value.
  - $\tilde{y} \leftarrow$ Retrieve$(\tau_{\theta_{F(x)}}, RT_{F,x}, VK_{F,x}, RK_{F,x}, PP)$: Run by a verifier holding $RK_{F,x}$ to retrieve the actual result $\tilde{y}$ which is either $F(x)$ or $\bot$.[1]
- $\{EK_{F,S'}\}$ or $\bot \leftarrow$ Revoke$(\tau_{\theta_{F(x)}}, MK, PP)$: Run by the KDC if a misbehaving server is reported i.e. that Verify returned $\tau_{\theta_{F(x)}} = ($reject, $S)$ (if $\tau_{\theta_{F(x)}} = ($accept, $S)$ then this algorithm outputs $\bot$). It revokes all evaluation keys $EK_{\cdot,S}$ of the server S thereby preventing S from performing any further evaluations. Updated evaluation keys $EK_{\cdot,S'}$ are issued to all servers.[2]

Although not stated, the KDC may update the public parameters PP during any algorithm. A RPVC scheme is *correct* if the verification algorithm almost certainly outputs accept when run on a valid verification key and an encoded output, where the encoded output is honestly produced by a computation server given a validly generated encoded input and evaluation key. That is, if all algorithms are run honestly then the result should almost certainly be accepted.

3.2 Security Models

We now formalize several notions of security as a series of cryptographic games. The adversary against a particular function F is modelled as a PPT algorithm $\mathcal{A}$ run by a challenger with input parameters chosen to represent the knowledge of a real attacker as well the security parameter κ and a parameter $q_t > 1$ denoting the number of queries the adversary makes to the Revoke oracle before the challenge is generated. The adversary algorithm may maintain state and be multi-stage and we overload the notation by calling each of these adversary algorithms $\mathcal{A}$. The notation $\mathcal{A}^{\mathcal{O}}$ denotes the adversary $\mathcal{A}$ being provided with oracle access to the following functions: FnInit$(\cdot, MK, PP)$, Register$(\cdot, MK, PP)$, Certify$(\cdot, \cdot, \cdot, MK, PP)$ and Revoke$(\cdot, \cdot, \cdot, MK, PP)$.[3] For each game, we define the *advantage* and *security* of $\mathcal{A}$ as:

---

[1] Note that if a server is not given $RK_{F,x}$ then it too cannot learn the output.

[2] In some instantiations, it may not be necessary to issue entirely new evaluation keys to each entity. In Sect. 4, we only need to issue a partially updated key for example.

[3] We do not need to provide a Verify oracle since this is a publicly verifiable scheme and $\mathcal{A}$ is given verification keys (thus we also avoid the rejection problem).

Definition 2. The *advantage* of a PPT adversary $\mathcal{A}$ making a polynomial number of queries q (including q_t Revoke queries) is defined as follows, where $\mathbf{X} \in \{sSS\text{-}PubVerif, sSS\text{-}Revocation, sSS\text{-}VindictiveM\}$:

- $Adv_{\mathcal{A}}^{\mathbf{X}}(\mathcal{RPVC}, F, 1^\kappa, q) = \Pr[\mathbf{Exp}_{\mathcal{A}}^{\mathbf{X}}[\mathcal{RPVC}, F, q_t, 1^\kappa] = 1]$
- $Adv_{\mathcal{A}}^{VindictiveS}(\mathcal{RPVC}, F, 1^\kappa, q) = \Pr[\mathbf{Exp}_{\mathcal{A}}^{VindictiveS}[\mathcal{RPVC}, F, 1^\kappa] = 1]$
- $Adv_{\mathcal{A}}^{BVerif}(\mathcal{RPVC}, F, 1^\kappa, q) \quad = \quad \Pr[\mathbf{Exp}_{\mathcal{A}}^{BVerif}[\mathcal{RPVC}, F, 1^\kappa] \quad = \quad 1] \ - \ \max_{y \in \mathsf{Ran}(F)} (\Pr_{x \in \mathsf{Dom}(F)}[F(x) = y])$.

A RPVC is *secure against Game* $\mathbf{X}$, *VindictiveS or BVerif* for a function F, if for all PPT adversaries $\mathcal{A}$, $Adv_{\mathcal{A}}^{\mathbf{X}, VindictiveS, BVerif}(\mathcal{RPVC}, F, 1^\kappa, q) \leq \mathrm{negl}(\kappa)$.

Public Verifiability. In Game 1 we extend the Public Verifiability game of Parno et al. [10] to formalize that multiple colluding servers should be unable to convince *any* verifying party of an incorrect output (i.e. that Verify returns accept on an encoded output not representing the true output of the computation). We define a selective, semi-static notion[4] such that the adversary must select its challenge input before seeing the public parameters and must declare a list of entities that must be revoked at the challenge time before receiving oracle access.

The adversary first selects an input value to be outsourced. The challenger initializes a list of currently revoked entities Q_{Rev} and a time parameter t before running Setup and FnInit to create a public delegation key for the function F (lines 2–5). The adversary is given the generated public parameters and must output a list $\overline{R}$ of servers to be revoked when the challenge is created. It is then given oracle access to the above functions which simulate all values known to a real server as well as those learnt through corrupting entities. The challenger responds to Certify and Revoke queries as detailed in Oracle Queries 1 and 2 respectively. It must ensure that Q_{Rev} is kept up-to-date by adding or removing the queried entity, and in the case of revocation must increment the time parameter. It also ensures that issued keys will not lead to a trivial win.

Once the adversary has finished this query phase (and in particular, due to the parameterisation of the adversary, after exactly q_t Revoke queries), the challenger must check that the queries made by the adversary has indeed left the list of revoked entities to be at least that selected beforehand by the adversary. If there is a server that the adversary included on $\overline{R}$ but is not currently revoked, then the adversary loses the game. Otherwise, the challenger generates the challenge by running ProbGen on $x^\star$. The adversary is given the resulting encoded input and oracle access again, and wins the game if it creates an encoded output that verifies correctly yet does not encode the correct value $F(x^\star)$.

---

[4] This is due to the selective IND-sHRSS game that we base the construction upon. Since this is used in a black-box manner however, a stronger primitive may allow this game to be improved accordingly.

Game 1. $\mathrm{Exp}_{\mathcal{A}}^{sSS\text{-}PubVerif}$ $[\mathcal{RPVC}, F, q_t, 1^\kappa]$:

1: $x^\star \leftarrow \mathcal{A}(1^\kappa)$;
2: $Q_{\mathrm{Rev}} = \epsilon$;
3: $t = 1$;
4: $(PP, MK) \leftarrow \mathsf{Setup}(1^\kappa)$;
5: $PK_F \leftarrow \mathsf{FnInit}(F, MK, PP)$;
6: $\bar{R} \leftarrow \mathcal{A}(PK_F, PP)$;
7: $\mathcal{A}^{\mathcal{O}}(PK_F, PP)$;
8: if $(\bar{R} \not\subseteq Q_{\mathrm{Rev}})$ return 0;
9: $(\sigma_{F,x^\star}, VK_{F,x^\star}, RK_{F,x^\star}) \leftarrow \mathsf{ProbGen}(x^\star, PK_F, PP)$;
10: $\theta^\star \leftarrow \mathcal{A}^{\mathcal{O}}(\{\sigma_{F,x^\star}, VK_{F,x^\star}, RK_{F,x^\star}\}, EK_{F,\mathcal{A}}, SK_{\mathcal{A}}, PK_F, PP)$;
11: if $((((\tilde{y}, \tau_{\theta^\star}) \leftarrow \mathsf{Verify}(\theta^\star, VK_{F,x^\star}, RK_{F,x^\star}, PP))$
    and $((\tilde{y}, \tau_{\theta^\star}) \neq (\perp, (\mathrm{reject}, \cdot)))$ and $(\tilde{y} \neq F(x^\star))))$
12:       return 1;
13: else return 0;

Oracle Query 1. $\mathcal{O}^{\mathsf{Certify}}(S, F', MK, PP)$:

1: if $((F' = F$ and $S \notin \bar{R})$ or $(t = q_t$ and $\bar{R} \not\subseteq Q_{\mathrm{Rev}} \setminus S))$ return $\perp$;
2: $Q_{\mathrm{Rev}} = Q_{\mathrm{Rev}} \setminus S$;
3: return $\mathsf{Certify}(S, F', MK, PP)$;

Oracle Query 2. $\mathcal{O}^{\mathsf{Revoke}}(\tau_{\theta_{F'(x)}}, MK, PP)$:

1: $t = t + 1$;
2: if $(\tau_{\theta_{F'(x)}} = (\mathrm{accept}, \cdot))$ return $\perp$;
3: if $(t = q_t$ and $\bar{R} \not\subseteq Q_{\mathrm{Rev}} \cup S)$ return $\perp$;
4: $Q_{\mathrm{Rev}} = Q_{\mathrm{Rev}} \cup S$;
5: return $\mathsf{Revoke}(\tau_{\theta_{F'(x)}}, MK, PP)$;

Revocation. The notion of revocation requires that any subsequent computations by a server detected as misbehaving should be rejected (even if the result is correct). Thus a misbehaving server may be completely removed from the system and will be punished by not receiving rewards for future work.

The selective, semi-static notion of Revocation given in Game 2 proceeds exactly as the sSS-PubVerif game except that the adversary wins if it outputs *any* result (even a correct encoding of $F(x^\star)$) that is accepted as a valid response from any server that was revoked at the time of the challenge. This game also uses the Certify and Revoke oracles specified in Oracle Queries 1 and 2 respectively.

Vindictive Server. This notion is motivated by the manager model where the client does not a priori know the identities of servers selected from the pool. Since an invalid result can lead to revocation, this reveals a new threat model (particularly if servers are rewarded per computation). A malicious server may return incorrect results but attribute them to an alternate server ID such that an (honest) server is revoked, thus reducing the size of the server pool and increasing

Game 2. Exp$_{\mathcal{A}}^{sSS\text{-}Revocation}$ $[\mathcal{RPVC}, F, q_t, 1^{\kappa}]$:

1: $x^{\star} \leftarrow \mathcal{A}(1^{\kappa})$;
2: $Q_{\text{Rev}} = \epsilon$;
3: $t = 1$;
4: $(PP, MK) \leftarrow \text{Setup}(1^{\kappa})$;
5: $PK_F \leftarrow \text{FnInit}(F, MK, PP)$;
6: $\overline{R} \leftarrow \mathcal{A}(PK_F, PP)$;
7: $\mathcal{A}^{\mathcal{O}}(PK_F, PP)$;
8: **if** $(\overline{R} \not\subseteq Q_{\text{Rev}})$ **return** 0;
9: $(\sigma_{F,x^{\star}}, VK_{F,x^{\star}}, RK_{F,x^{\star}}) \leftarrow \text{ProbGen}(x^{\star}, PK_F, PP)$;
10: $\theta^{\star} \leftarrow \mathcal{A}^{\mathcal{O}}(\sigma_{x^{\star}}, VK_{F,x^{\star}}, RK_{F,x^{\star}}, PK_F, PP)$;
11: **if** $(((\tilde{y}, (\text{accept}, S)) \leftarrow \text{Verify}(\theta^{\star}, VK_{F,x^{\star}}, RK_{F,x^{\star}}, PP))$
     and $(S \in \overline{R})$ **then**
12:     **return** 1
13: **else**
14:     **return** 0

Game 3. Exp$_{\mathcal{A}}^{VindictiveS}$ $[\mathcal{RPVC}, F, 1^{\kappa}]$:

1: $Q_{\text{Reg}} = \epsilon$;
2: $(PP, MK) \leftarrow \text{Setup}(1^{\kappa})$;
3: $PK_F \leftarrow \text{FnInit}(F, MK, PP)$;
4: $x^{\star} \leftarrow \mathcal{A}^{\mathcal{O}}(PK_F, PP)$;
5: $(\sigma_{F,x^{\star}}, VK_{F,x^{\star}}, RK_{F,x^{\star}}) \leftarrow \text{ProbGen}(x^{\star}, PK_F, PP)$;
6: $\tilde{S} \leftarrow \mathcal{A}^{\mathcal{O},\text{Register2}}(\sigma_{F,x^{\star}}, VK_{F,x^{\star}}, RK_{F,x^{\star}}, PK_F, PP)$ subject to (1);
7: $\theta^{\star} \leftarrow \mathcal{A}^{\mathcal{O},\text{Compute},\text{Register2}}(\sigma_{F,x^{\star}}, VK_{F,x^{\star}}, RK_{F,x^{\star}}, PK_F, PP)$ subject to (2);
8: **if** $((\tilde{y}, \tau_{\theta^{\star}}) \leftarrow \text{Verify}(\theta^{\star}, VK_{F,x^{\star}}, RK_{F,x^{\star}}, PP))$
     and $((\tilde{y}, \tau_{\theta^{\star}}) = (\bot, (\text{reject}, \tilde{S})))$ **and** $(\bot \nleftarrow \text{Revoke}(\tau_{\theta^{\star}}, MK, PP)))$ **then**
9:     **return** 1
10: **else**
11:     **return** 0

---

the future reward for the malicious server. In Game 3, the challenger maintains a list of registered entities Q_{Reg}. The game proceeds similarly to the previous notions, except that, on lines 6 and 7, the adversary selects a target server ID, $\tilde{S}$, he wishes to be revoked and generates an encoded output that will cause this. He is given oracle access subject to the following constraints to avoid trivial wins:

(1) No query of the form $\mathcal{O}^{\text{Register}}(\tilde{S}, MK, PP)$ was made;
(2) As above and no query $\mathcal{O}^{\text{Compute}}(\sigma_{F,x_i^{\star}}, EK_{F,\tilde{S}}, SK_{\tilde{S}}, PP)$ was made.

In addition, he is provided with an oracle, Register2, which performs the Register algorithm but *does not* return the resulting key SK_S (it may however update the public parameters to reflect the additional registered entity). The adversary may query *any* identity to Register2 (including $\tilde{S}$). We also modify the standard Register oracle such that if an identity has been previously queried to the Register2 oracle, it generates the same parameters (and vice versa). The adversary wins if the KDC believes $\tilde{S}$ returned $\tilde{y}$ and revokes $\tilde{S}$.

Game 4. $\text{Exp}_{\mathcal{A}}^{sSS\text{-}VindictiveM}[\mathcal{RPVC}, F, q_t, 1^\kappa]$:

1: $x^* \leftarrow \mathcal{A}(1^\kappa)$;
2: $Q_{\text{Rev}} = \epsilon$;
3: $t = 1$
4: $(PP, MK) \leftarrow \text{Setup}(1^\kappa)$;
5: $PK_F \leftarrow \text{FnInit}(F, MK, PP)$;
6: $\bar{R} \leftarrow \mathcal{A}(PK_F, PP)$
7: $\mathcal{A}^{\mathcal{O}}(PK_F, PP)$;
8: **if** $((\bar{R} \not\subseteq Q_{\text{Rev}})$ **or** $(\bar{R} = \mathcal{U}_{\text{ID}}))$ **return** 0;
9: $S \xleftarrow{\$} \mathcal{U}_{\text{ID}} \setminus \bar{R}$;
10: $SK_S \leftarrow \text{Register}(S, MK, PP)$;
11: $EK_{F,S} \leftarrow \text{Certify}(S, F, MK, PP)$;
12: $(\sigma_{F,x^*}, VK_{F,x^*}, RK_{F,x^*}) \leftarrow \text{ProbGen}(x^*, PK_F, PP)$;
13: $\theta_{F(x^*)} \leftarrow \text{Compute}(\sigma_{F,x^*}, EK_{F,S}, SK_S, PP)$;
14: $(RT_{F,x^*}, \tau_{\theta_{F(x^*)}}) \leftarrow \mathcal{A}^{\mathcal{O}}(\sigma_{F,x^*}, \theta_{F(x^*)}, VK_{F,x^*}, PK_F, PP)$;
15: **if** $(\tilde{y} \leftarrow \text{Retrieve}(\tau_{\theta_{F(x^*)}}, RT_{F,x^*}, VK_{F,x^*}, RK_{F,x^*}, PP))$
     and $(\tilde{y} \neq F(x^*))$ **and** $(\tilde{y} \neq \perp)$ **then**
16:     **return** 1
17: **else**
18:     **return** 0

Vindictive Manager. This is a natural extension of the Public Verifiability notion to the manager model where a vindictive manager may attempt to provide a client with an incorrect answer. We remark that instantiations may vary depending on the level of trust given to the manager: a completely trusted manager may simply return the result to a client, whilst an untrusted manager may have to provide the full output from the server. Here we consider a semi-trusted manager where the clients would still like a final, efficient check.

The security model is presented in Game 4. First, the adversary selects its challenge input x^*, and the challenger initializes a list of revoked entities Q_{Rev} and a time paramter t. It also sets up the system and gives the public parameters to the adversary, who must select a list $\bar{R}$ of servers to be revoked at the challenge time. We require that $\bar{R}$ is not the full set of all servers in the system, as one non-revoked identity is required to generate the challenge. The adversary then gets oracle access (using the Certify and Revoke oracles specified in Oracle Queries 1 and 2 respectively). If, after finishing this query phase (and in particular after q_t Revoke queries), the list of revoked entities does not include $\bar{R}$ then the adversary loses the game. Otherwise, a server S is chosen at random from the set of all server identities $\mathcal{U}_{\text{ID}}$ excluding $\bar{R}$ (as these must be revoked at the challenge time). This server is used to generate the challenge. If not already done, the challenger registers and certifies S for F, and runs ProbGen on the challenge input, before finally running Compute to generate an encoded output $\theta_{F(x^*)}$. The adversary is then given the encoded input, verification key and $\theta_{F(x^*)}$, as well as oracle access, and must output a retrieval token RT_{F,x^*} and an acceptance token $\tau_{\theta_{F(x^*)}}$. The challenger runs Retrieve on $RT_{F,x}$ to get an output value $\tilde{y}$, and the adversary wins if the challenger accepts this output and $\tilde{y} \neq F(x^*)$.

Game 5. Exp$_{\mathcal{A}}^{BVerif}[\mathcal{RPVC}, F, 1^\kappa]$:

1: $(PP, MK) \leftarrow$ Setup(1^κ);
2: $PK_F \leftarrow$ FnInit(F, MK, PP);
3: $x \xleftarrow{\$}$ Dom(F);
4: $S \xleftarrow{\$} \mathcal{U}_{\mathrm{ID}}$;
5: $SK_S \leftarrow$ Register(S, MK, PP);
6: $EK_{F,S} \leftarrow$ Certify(S, F, MK, PP);
7: $(\sigma_{F,x}, VK_{F,x}, RK_{F,x}) \leftarrow$ ProbGen(x, PK_F, PP);
8: $\theta_{F(x)} \leftarrow$ Compute$(\sigma_{F,x}, EK_{F,S}, SK_S, PP)$;
9: $\hat{y} \leftarrow \mathcal{A}^\mathcal{O}(\theta_{F(x)}, VK_{F,x}, PK_F, PP)$;
10: if $(\hat{y} = F(x))$ then
11:     return 1
12: else
13:     return 0

Blind Verification. With this notion we aim to show that a verifier that does not hold the retrieval token $RT_{F,x}$ chosen in ProbGen cannot learn the value of $F(x)$ given the encoded output. This property was hinted at by Parno et al. [10] but was not formalized. The game begins as usual with the challenger initializing the system. The challenger then selects an input at random from the domain of F, and a random server S. It registers and certifies S, runs ProbGen for the chosen input and runs Compute to generate an output $\theta_{F(x)}$. This is given to the adversary along with the verification key and oracle access, and the adversary wins if it can guess the value of $F(x)$ without seeing the retrieval key. Clearly, the adversary can trivially make a guess for $F(x)$ based on a priori knowledge of the distribution of F over all possible inputs. Unless F is balanced (i.e. outputs 1 exactly half the time), the adversary could gain an advantage. Thus, we define security by subtracting the most likely guess for $F(x)$.

4   Construction

We now provide an instantiation of a RPVC scheme. Our construction is based on that used by Parno et al. [10] (summarised in Appendix A) which uses Key-Policy Attribute-based Encryption (KP-ABE) in a black-box manner to outsource the computation of a Boolean function.[5] Notice that to achieve the outsourced evaluation of functions with n bit outputs, it is possible to evaluate n different functions, each of which applies a mask to output the single bit in position i.

Recall that if $\bot$ is returned by the server then the verifier is unable to determine whether $F(x) = 0$ *or* whether the server misbehaved. To avoid this issue, we follow Parno *et al.* and restrict the family of functions $\mathcal{F}$ to be the set of Boolean

---

[5] Following Parno et al. we restrict our attention to Boolean functions, and in particular the complexity class NC^1 which includes all circuits of depth $\mathcal{O}(\log n)$. Thus functions we can outsource can be built from common operations such as AND, OR, NOT, equality and comparison operators, arithmetic operators and regular expressions.

functions closed under complement. That is, if $F \in \mathcal{F}$ then $\overline{F}(x) = F(x) \oplus 1$ also belongs to $\mathcal{F}$. Then, the client encrypts two random messages m_0 and m_1. The server is required to return the decryption of those ciphertexts. Thus, a well-formed response $\theta_{F(x)}$, comprising recovered plaintexts (d_b, d_{1-b}), satisfies the following, where $RK_{F,x} = b$:

$$(d_b, d_{1-b}) = \begin{cases} (m_b, \perp), & \text{if } F(x) = 1; \\ (\perp, m_{1-b}), & \text{if } F(x) = 0. \end{cases} \tag{1}$$

4.1 Technical Details

We require an *indirectly revocable KP-ABE scheme* comprising the algorithms ABE.Setup, ABE.KeyGen, ABE.KeyUpdate, ABE.Encrypt and ABE.Decrypt. We also use a signature scheme with algorithms Sig.KeyGen, Sig.Sign and Sig.Verify, and a one-way function g. Let $\mathcal{U}$ be the universe of attributes acceptable by the ABE scheme, and let $\mathcal{U} = \mathcal{U}_{\text{attr}} \cup \mathcal{U}_{\text{ID}} \cup \mathcal{U}_{\text{time}} \cup \mathcal{U}_{\mathcal{F}}$ where: attributes in $\mathcal{U}_{\text{attr}}$ form characteristic tuples for input data, as detailed in Appendix A; $\mathcal{U}_{\text{ID}}$ comprises attributes representing entity identifiers; $\mathcal{U}_{\text{time}}$ comprises attributes representing time periods issued by the time source $\mathbb{T}$; and finally $\mathcal{U}_{\mathcal{F}}$ comprises attributes that represent functions in $\mathcal{F}$. Define a bijective mapping between functions $F \in \mathcal{F}$ and attributes $f \in \mathcal{U}_{\mathcal{F}}$. Then the policy $F \wedge f$ denotes adding a conjunctive clause requiring the presence of the label f to the expression of the function F, and $(x \cup f)$ denotes adding the function attribute to the attribute set representing the input data x. This will prevent servers using alternate evaluation keys for a given input and hence we are able to certify servers to compute multiple functions.

Parno et al. [10] considered two models of publicly verifiable computation. In single function PVC, the function to be computed is embedded in the public parameters, whilst in multi-function PVC delegation keys for multiple functions can be generated and a single encoded input can be used to input the same data to multiple functions. To achieve this latter notion, Parno et al. required the somewhat complex primitive of KP-ABE with Outsourcing [9]. In this work, we take a different approach. We believe that in practical environments it is unrealistic to expect a server to compute just a single function, and we also believe that it is a reasonable cost expectation to prepare an encoded input per computation, and that the input data to different functions may well differ. Thus, whereas Parno et al. use complex primitives to allow an encoded input to be used for computations of different functions on the same data, we use the simple trick of adding a conjunctive clause to the functions requiring the presence of the appropriate function label in the input data – that is, the function F is encoded in a decryption key for the policy $F \wedge f$ where f is the attribute representation of F in $\mathcal{U}_{\mathcal{F}}$; the complement function $\overline{F}$ is encoded as a key for $\overline{F} \wedge f$; and we encode the input data x to the function F as $x \cup f$. Thus, the client must perform the ProbGen stage per computation as the function label in the input data will differ, but servers can be certified for multiple functions and may not use a key for one function to compute on data intended for another (since the function label required by the conjunctive clause in the key will not be present in the input data). As a result, and unlike the single function notion of Parno et al., we are able to provide the adversary with oracle access in our security games.

The scheme of Parno et al. required a one-key IND-CPA notion of security for the underlying KP-ABE scheme. This is a more relaxed notion than considered in the vast majority of the ABE literature (where the adversary is given a KeyGen oracle and the scheme must prevent collusion between holders of different decryption keys). Parno et al. could use this property due to their restricted system model where the client is certified for only a single function per set of public parameters (so the client must set up a new ABE environment per function). In our setting, we must be able to certify servers for multiple functions and hence the KDC must be able to issue multiple keys and we require the more standard, multi-key notion of security usually considered for ABE schemes.

4.2   Instantiation

Informally the scheme operates as follows.

1. RPVC.Setup establishes public parameters and a master secret key by calling the ABE.Setup algorithm twice. This algorithm also initializes a time source[6] $\mathbb{T}$, a list of revoked servers, and a two-dimensional array of registered servers L_{Reg} – the array is indexed in the first dimension by server identities and the first dimension will store signature verification keys while the second will store a list of functions that server is authorized to compute.
2. RPVC.FnInit simply outputs the public parameters.
3. RPVC.Register creates a public-private key pair by calling the signature KeyGen algorithm. This is run by the KDC (or the manager in the manager model) and updates L_{Reg} to store the verification key for S.
4. RPVC.Certify creates the key $EK_{F,S}$ that will be used by a server S to compute F by calling the ABE.KeyGen and ABE.KeyUpdate algorithms twice – once with a "policy" for F and once with the complement $\overline{F}$. It also updates L_{Reg} to include F. Note that since we have a form of multi-function PVC, we must prevent a server certified to perform two different functions, F and G (that differ on their output) from using the key for G to retrieve the plaintext and claiming it as a result for F. To prevent this, we add an additional attribute to the input set in ProbGen encoding the function the input should applied to, and add a conjunctive clause for such an attribute to the key policies. Thus an input set intended for F (including the F attribute) will only satisfy a key issued for F (comprising the F conjunctive clause), and a key for G will not be satisfied as G is not in the input set.
5. RPVC.ProbGen creates a problem instance $\sigma_{F,x} = (c_b, c_{1-b})$ by encrypting two randomly chosen messages under an attribute set corresponding to x, and a verification key $VK_{F,x}$ by applying a one-way function g to the messages.

---

[6] $\mathbb{T}$ could be a counter that is maintained in the public parameters or a networked clock.

The ciphertexts and verification tokens are ordered randomly according to $RK_{F,x} = b$ for a random bit b, such that the positioning of an element does not imply whether it relates to F or to $\overline{F}$.

6. RPVC.Compute is run by a server S. Given an input $\sigma_{F,x} = (c_b, c_{1-b})$ it returns $(m_0, \perp)$ if $F(x) = 1$ or $(\perp, m_1)$ if $F(x) = 0$ (ordered according to $RK_{F,x}$ chosen in RPVC.ProbGen) and a signature on the output.

7. RPVC.Verify either accepts the output $\theta_{F(x)} = (d_b, d_{1-b})$ or rejects it. This algorithm verifies the signature on the output and confirms the output is correct by applying g and comparing with $VK_{F,x}$. In RPVC.BVerif the verifier can compare pairwise between the components of $\theta_{F(x)}$ and $VK_{F,x}$ to determine correctness but as they are unaware of the value of $RK_{F,x}$, they do not know the order of these elements and hence whether the correct output corresponds to F or $\overline{F}$ being satisfied i.e. if $F(x) = 1$ or 0 respectively. The verifier outputs an accept or reject token as well as the output value $RT_{F,x} \in \{d_b, d_{1-b}, \perp\}$ where $RK_{F,x} = b$. Parno et al. [10] gave a one line remark that permuting the key pairs and ciphertexts given out in ProbGen could give output privacy. We believe that doing so would require four decryptions in the Compute stage to ensure the correct keys have been used (since an incorrect key,associated with different public parameters, but for a satisfying attribute set will return an incorrect, random plaintext which is indistinguishable from a valid, random message). Since our construction fixes the order of the key pairs, we do not have this issue and only require two decryptions. In RPVC.Retrieve a verifier that has knowledge of $RK_{F,x}$ can check whether the output from BVerif matches m_0 or m_1.

8. RPVC.Revoke is run by the KDC and redistributes fresh keys to all non-revoked servers. This algorithm first refreshes the time source $\mathbb{T}$ (e.g. increments $\mathbb{T}$ if it is a counter). It then updates L_{Reg} and L_{Rev}, and updates $EK_{F,S}$ using the results of two calls to the ABE.KeyUpdate algorithm.

More formally, our scheme is defined by Algorithms 1–9.

Algorithm 1. $(PP, MK) \leftarrow$ RPVC.Setup(1^{κ})

1: Let $\mathcal{U} = \mathcal{U}_{\text{attr}} \cup \mathcal{U}_{\text{ID}} \cup \mathcal{U}_{\text{time}} \cup \mathcal{U}_{\mathcal{F}}$
2: $(MPK_{\text{ABE}}^0, MSK_{\text{ABE}}^0) \leftarrow$ ABE.Setup($1^{\kappa}, \mathcal{U}$)
3: $(MPK_{\text{ABE}}^1, MPK_{\text{ABE}}^1) \leftarrow$ ABE.Setup($1^{\kappa}, \mathcal{U}$)
4: **for** $S \in \mathcal{U}_{\text{ID}}$ **do**
5:     $L_{\text{Reg}}[S][0] = \epsilon$
6:     $L_{\text{Reg}}[S][1] = \{\epsilon\}$
7: $L_{\text{Rev}} = \epsilon$
8: Initialise $\mathbb{T}$
9: $PP = (MPK_{\text{ABE}}^0, MPK_{\text{ABE}}^1, L_{\text{Reg}}, \mathbb{T})$
10: $MK = (MSK_{\text{ABE}}^0, MSK_{\text{ABE}}^1, L_{\text{Rev}})$

Algorithm 2. $PK_F \leftarrow$ RPVC.FnInit(F, MK, PP)

1: Set $PK_F = PP$

Algorithm 3. $SK_S \leftarrow$ RPVC.Register(S, MK, PP)

1: $(SK_{\text{Sig}}, VK_{\text{Sig}}) \leftarrow$ Sig.KeyGen(1^κ)
2: $SK_S = SK_{\text{Sig}}$
3: $L_{\text{Reg}}[S][0] = VK_{\text{Sig}}$

Algorithm 4. $EK_{F,S} \leftarrow$ RPVC.Certify(S, F, MK, PP)

1: $L_{\text{Reg}}[S][1] = L_{\text{Reg}}[S][1] \cup F$
2: $L_{\text{Rev}} = L_{\text{Rev}} \setminus S$
3: $t \leftarrow \mathbb{T}$
4: $SK^0_{\text{ABE}} \leftarrow$ ABE.KeyGen$(S, F \wedge f, MSK^0_{\text{ABE}}, MPK^0_{\text{ABE}})$
5: $SK^1_{\text{ABE}} \leftarrow$ ABE.KeyGen$(S, \overline{F} \wedge f, MSK^1_{\text{ABE}}, MPK^1_{\text{ABE}})$
6: $UK^0_{L_{\text{Rev}},t} \leftarrow$ ABE.KeyUpdate$(L_{\text{Rev}}, t, MSK^0_{\text{ABE}}, MPK^0_{\text{ABE}})$
7: $UK^1_{L_{\text{Rev}},t} \leftarrow$ ABE.KeyUpdate$(L_{\text{Rev}}, t, MSK^1_{\text{ABE}}, MPK^1_{\text{ABE}})$
8: $EK_{F,S} = (SK^0_{\text{ABE}}, SK^1_{\text{ABE}}, UK^0_{L_{\text{Rev}},t}, UK^1_{L_{\text{Rev}},t})$

Algorithm 5. $(\sigma_{F,x}, VK_{F,x}, RK_{F,x}) \leftarrow$ RPVC.ProbGen(x, PK_F, PP)

1: $t \leftarrow \mathbb{T}$
2: $(m_0, m_1) \xleftarrow{\$} \mathcal{M} \times \mathcal{M}$
3: $b \xleftarrow{\$} \{0, 1\}$
4: $c_b \leftarrow$ ABE.Encrypt$(m_b, (x \cup f), t, MPK^0_{\text{ABE}})$
5: $c_{1-b} \leftarrow$ ABE.Encrypt$(m_{1-b}, (x \cup f), t, MPK^1_{\text{ABE}})$
6: Output: $\sigma_{F,x} = (c_b, c_{1-b})$, $VK_{F,x} = (g(m_b), g(m_{1-b}), L_{\text{Reg}})$ and $RK_{F,x} = b$

Algorithm 6. $\theta_{F(x)} \leftarrow$ RPVC.Compute$(\sigma_{F,x}, EK_{F,S}, SK_S, PP)$

1: Input: $EK_{F,S} = (SK^0_{\text{ABE}}, SK^1_{\text{ABE}}, UK^0_{L_{\text{Rev}},t}, UK^1_{L_{\text{Rev}},t})$ and $\sigma_{F,x} = (c_b, c_{1-b})$
2: Parse $\sigma_{F,x}$ as (c, c')
3: $d_b \leftarrow$ ABE.Decrypt$(c, SK^0_{\text{ABE}}, MPK^0_{\text{ABE}}, UK^0_{L_{\text{Rev}},t})$
4: $d_{1-b} \leftarrow$ ABE.Decrypt$(c', SK^1_{\text{ABE}}, MPK^1_{\text{ABE}}, UK^1_{L_{\text{Rev}},t})$
5: $\gamma \leftarrow$ Sig.Sign$((d_b, d_{1-b}, S), SK_S)$
6: Output: $\theta_{F(x)} = (d_b, d_{1-b}, S, \gamma)$

Algorithm 7. $(RT_{F,x}, \tau_{\theta_{F(x)}}) \leftarrow$ RPVC.BVerif$(\theta_{F(x)}, VK_{F,x}, PP)$

1: Input: $VK_{F,x} = (g(m_b), g(m_{1-b}), L_{\text{Reg}})$ and $\theta_{F(x)} = (d_b, d_{1-b}, S, \gamma)$
2: **if** $F \in L_{\text{Reg}}[S][1]$ **then**
3:    **if** accept $\leftarrow$ Sig.Verify$((d_b, d_{1-b}, S), \gamma, L_{\text{Reg}}[S][0])$ **then**
4:       **if** $g(m_b) = g(d_b)$ **then** Output $(RT_{F,x} = d_b, \tau_{\theta_{F(x)}} = (\text{accept}, S))$
5:       **else if** $g(m_{1-b}) = g(d_{1-b})$ **then** Output $(RT_{F,x} = d_{1-b}, \tau_{\theta_{F(x)}} = (\text{accept}, S))$
6:       **else**
            Output $(RT_{F,x} = \perp, \tau_{\theta_{F(x)}} = (\text{reject}, S))$
7: Output $(RT_{F,x} = \perp, \tau_{\theta_{F(x)}} = (\text{reject}, \perp))$

Algorithm 8. $\tilde{y} \leftarrow$ RPVC.Retrieve$(\tau_{\theta_{F(x)}}, RT_{F,x}, VK_{F,x}, RK_{F,x}, PP)$

1: Input: $VK_{F,x} = (g(m_b), g(m_{1-b}), L_{\text{Reg}})$, $\theta_{F(x)} = (d_b, d_{1-b}, S, \gamma)$, $RK_{F,x} = b$, and $(RT_{F,x}, \tau_{\theta_{F(x)}})$ where $RT_{F,x} \in \{d_b, d_{1-b}, \perp\}$
2: **if** $(\tau_{\theta_{F(x)}} = (\text{accept}, S)$ **and** $g(RT_{F,x}) = g(m_0))$ **then** Output $\tilde{y} = 1$
3: **else if** $(\tau_{\theta_{F(x)}} = (\text{accept}, S)$ **and** $g(RT_{F,x}) = g(m_1))$ **then** Output $\tilde{y} = 0$
4: **else** Output $\tilde{y} = \perp$

Algorithm 9. $\{EK_{F,S'}\}$ or $\bot \leftarrow$ RPVC.Revoke$(\tau_{\theta_{F(x)}}, MK, PP)$

1: **if** $\tau_{\theta_{F(x)}} = (\text{reject}, S)$ **then**
2:    $L_{\text{Reg}}[S][1] = \{\epsilon\}$
3:    $L_{\text{Rev}} = L_{\text{Rev}} \cup S$
4:    Refresh $\mathbb{T}$
5:    $t \leftarrow \mathbb{T}$
6:    $UK^0_{L_{\text{Rev}},t} \leftarrow$ ABE.KeyUpdate$(L_{\text{Rev}}, t, MSK^0_{\text{ABE}}, MPK^0_{\text{ABE}})$
7:    $UK^1_{L_{\text{Rev}},t} \leftarrow$ ABE.KeyUpdate$(L_{\text{Rev}}, t, MSK^1_{\text{ABE}}, MPK^1_{\text{ABE}})$
8:    **for all** $S \in \mathcal{U}_{\text{ID}}$ **do**
9:       Parse $EK_{F,S}$ as $(SK^0_{\text{ABE}}, SK^1_{\text{ABE}}, UK^0_{L_{\text{Rev}},t-1}, UK^1_{L_{\text{Rev}},t-1})$
10:       Update and send $EK_{F,S} = (SK^0_{\text{ABE}}, SK^1_{\text{ABE}}, UK^0_{L_{\text{Rev}},t}, UK^1_{L_{\text{Rev}},t})$
11: **else**
12:    output $\bot$

Theorem 1. *Given a revocable KP-ABE scheme secure in the sense of indistinguishability against selective-target with semi-static query attack (IND-sHRSS) [2] for a class of Boolean functions $\mathcal{F}$ closed under complement, an EUF-CMA secure signature scheme and a one-way function g. Let $\mathcal{RPVC}$ be the RPVC scheme defined in Algorithms 1–9. Then $\mathcal{RPVC}$ is secure in the sense of selective semi-static Public Verifiability, selective semi-static Revocation, Vindictive Servers, Blind Verification and selective semi-static Vindictive Managers.*

Lemma 1. *The $\mathcal{RPVC}$ construction defined by Algorithms 1–9 is secure against Vindictive Servers (Game 3) under the same assumptions as in Theorem 1.*

Proof. Let $\mathcal{A}_{VC}$ be an adversary with non-negligible advantage against the Vindictive Servers game (Game 3) when instantiated by Algorithms 1–9. We show that an adversary $\mathcal{A}_{Sig}$ with non-negligble advantage δ in the EUF-CMA signature game can be constructed using $\mathcal{A}_{VC}$. $\mathcal{A}_{Sig}$ interacts with the challenger $\mathcal{C}$ in the EUF-CMA security game and acts as the challenger for $\mathcal{A}_{VC}$ in the security game for Vindictive Servers for a function F as follows. The basic idea is that $\mathcal{A}_{Sig}$ can create a VC instance and play the Vindictive Servers game with $\mathcal{A}_{VC}$ by executing Algorithms 1–9 himself. $\mathcal{A}_{Sig}$ will guess a server identity that he thinks the adversary will select to vindictively revoke. The signature signing key that would be generated during the Register algorithm for this server will be implicitly set to be the signing key in the EUF-CMA game and any Compute oracle queries for this identity will be forwarded to the challenger to compute. Then, assuming that $\mathcal{A}_{Sig}$ guessed the correct server identity, $\mathcal{A}_{VC}$ will output a forged signature that $\mathcal{A}_{Sig}$ may output as its guess in the EUF-CMA game.

1. $\mathcal{C}$ initializes $Q = \epsilon$ to be an empty list of messages queried to the Sig.Sign oracle and runs Sig.KeyGen(1^κ) to generate a challenge signing key $\overline{SK}$ and verification key $\overline{VK}$. $\mathcal{C}$ sends $\overline{VK}$ to $\mathcal{A}_{Sig}$.
2. $\mathcal{A}_{Sig}$ chooses a function F on which to instantiate $\mathcal{A}_{VC}$.
3. $\mathcal{A}_{Sig}$ initializes the revocation list $Q_{\text{Reg}} = \epsilon$. Furthermore, it chooses a server identity from $\mathcal{U}_{\text{ID}} \setminus \mathcal{A}_{VC}$ which will be denoted by $\overline{S}$.

4. $\mathcal{A}_{Sig}$ runs RPVC.Setup(1^κ) and RPVC.FnInit(F, MK, PP), as specified in Algorithms 1 and 2 and passes PK_F and PP to the VC adversary $\mathcal{A}_{VC}$.

5. $\mathcal{A}_{VC}$ may now perform oracle queries to RPVC.FnInit RPVC.Register RPVC.Certify and RPVC.Revoke which $\mathcal{A}_{Sig}$ handles by running Algorithms 2, 3, 4 and 9 respectively.

6. Eventually, $\mathcal{A}_{VC}$ finishes querying and declares the challenge input $x^\star$.

7. $\mathcal{A}_{Sig}$ runs RPVC.ProbGen on the challenge $x^\star$ as specified in Algorithm 5.

8. $\mathcal{A}_{VC}$ is given the values of $PK_F, PP, \sigma_{F,x^\star}, VK_{F,x^\star}$ and $RK_{F,x^\star}$. It is also given oracle access to the following functions. $\mathcal{A}_{Sig}$ simulates these oracles and maintains a state of the generated parameters for each query.
   - FnInit($\cdot, MK, PP$): $\mathcal{A}_{Sig}$ runs this step as per Algorithm 2.
   - Register($\cdot, MK, PP$): If, for a queried server S, $S = \bar{S}$ then return $\bot$. Otherwise, $\mathcal{A}_{Sig}$ makes queries to $\mathcal{O}^{\text{Register}}(S, MK, PP)$. If S has not been registered before and therefore does not appear on the registration list Q_{Reg} then the oracle returns a signing key SK_S for S and adds the pair (S, SK_S) to Q_{Reg}. Otherwise, the stored signing key is returned.
   - Certify($\cdot, \cdot, MK, PP$): $\mathcal{A}_{Sig}$ honestly runs Algorithm 4.
   - Revoke($\cdot, MK, PP$): $\mathcal{A}_{Sig}$ operates as in Algorithm 9.
   - Register2($\cdot, MK, PP$): $\mathcal{A}_{Sig}$ responds in the same way as for standard Register queries above, but *always* returns $\bot$ and not a signing key.

   $\mathcal{A}_{VC}$ eventually outputs a target server identity $\tilde{S}$.

9. If $\tilde{S} \neq \bar{S}$ then $\mathcal{A}_{Sig}$ outputs $\bot$ and stops. Else, $\mathcal{A}_{VC}$ continues with oracle access as in Step 8 as well as a Compute oracle. $\mathcal{A}_{VC}$ submits queries of the form $\mathcal{O}^{\text{Compute}}(\sigma_{F,x}, EK_{F,S}, SK_S, PP)$ for its choice of server S and $\sigma_{F,x}$ If $S \neq \bar{S}$ then $\mathcal{A}_{Sig}$ simply follows Algorithm 6 using the decryption and signing keys generated during the oracle queries. Otherwise, $S = \bar{S}$ and $\mathcal{A}_{Sig}$ does not have access to the signing key $SK_{\bar{S}}$. Thus, he runs the ABE.Decrypt operations correctly to generate plaintexts d_0 and d_1, and submits $m = (d_0, d_1, \bar{S})$ as a Sig.Sign oracle query to $\mathcal{C}$. $\mathcal{C}$ adds m to the list Q and returns $\gamma \leftarrow$ Sig.Sign($m, \overline{SK}$), which $\mathcal{A}_{Sig}$ uses to return $\theta_{F(x)} = (d_0, d_1, \bar{S}, \gamma)$.

10. $\mathcal{A}_{VC}$ finally outputs $\theta^\star$ which appears to be an invalid result computed by $\tilde{S}$. Thus, Verify will output a reject token for $\tilde{S}$ and accept $\leftarrow$ Sig.Verify($(d_0, d_1, \tilde{S}), \gamma, \overline{VK}$). Thus, γ is a valid signature under key $\overline{SK}$.

11. $\mathcal{A}_{Sig}$ outputs $m^\star = (d_0, d_1, \tilde{S})$ and $\gamma^\star = \gamma$ to $\mathcal{C}$.

Note that due to Constraint 2 in Game 3, $\mathcal{A}_{VC}$ is not allowed to have made a query for $\mathcal{O}^{\text{Compute}}(\sigma_{x^\star}, EK_{F,\tilde{S}}, SK_{\tilde{S}}, PP)$ and thus the forgery $(m^\star, \gamma^\star)$ output by $\mathcal{A}_{Sig}$ will satisfy the requirement in that $m^\star \notin Q$. We argue that, assuming $\bar{S} = \tilde{S}$ (i.e. $\mathcal{A}_{Sig}$ correctly guessed the challenge identity) then $\mathcal{A}_{Sig}$ succeeds with the same non-negligible advantage δ as $\mathcal{A}_{VC}$. We assume that $n = |\mathcal{U}_{\text{ID}}|$ is polynomial (else the KDC could not efficiently search the list L_{Reg}). The probability that $\mathcal{A}_{Sig}$ correctly guesses $\bar{S} = \tilde{S}$ is $\frac{1}{n}$ and

$$Adv_{\mathcal{A}_{Sig}} \geq \frac{1}{n} Adv_{\mathcal{A}_{VC}} \geq \frac{\delta}{n} \geq negl(\kappa)$$

We conclude that if $\mathcal{A}_{VC}$ has a non-negligible advantage in the Vindictive Servers game then $\mathcal{A}_{Sig}$ has the same advantage in the EUF-CMA game, but since the signature scheme is assumed EUF-CMA secure, $\mathcal{A}_{VC}$ may not exist. $\square$

5 Conclusion

We have introduced the new notion of RPVC and provided a rigorous framework that we believe to be more realistic than the purely theory oriented models of prior work, especially when the KDC is an entity responsible for user authorization within a organization. We believe our model more accurately reflects practical environments and the necessary interaction between entities for PVC. Each server may provide services for many different functions and for many different clients. The first model of Parno et al. [10] considered evaluations of a single function, while their second allowed for multiple functions but required a more exotic type of ABE scheme. This allowed a single ProbGen stage to encode input for any function, whilst in our model, we also allow multiple functions but use a simpler ABE scheme that also permits the revocation functionality. We require ProbGen to be run for each unique $F(x)$ to be outsourced which we believe to be reasonable. Additionally, in our model, any clients may submit multiple requests to any available servers, whereas prior work considered just one server.

We have shown that by using a revocable KP-ABE scheme we can revoke misbehaving servers such that they receive a penalty for cheating and that, by permuting elements within messages, we achieve output privacy (as hinted at by Parno et al. although seemingly with two fewer decryptions than their brief description implies). We have shown that this blind verification could be used when a manager runs a pool of servers and rewards correct work – he needs to verify but is not entitled to learn the result. We have extended previous notions of security to fit our new definitional framework, introduced new models to capture additional threats (e.g. vindictive servers using revocation to remove competing servers), and provided a provably secure construction.

We believe that this work is a useful step towards making PVC practical and provides a natural set of baseline definitions from which to add future functionality. For example, in future work we will introduce an access control framework (using our scheme as a black box construction) to restrict the set of functions that clients may outsource, or to restrict (using the blind verification property) the set of verifiers that may learn the output. In this scenario, the KDC entity may, in addition to certifying servers and registering clients, determine access rights for such entities. The full version of this paper is available online [1].

A PVC Using KP-ABE

Parno et al. [10] provide a instantiation using *Key-policy Attribute-based Encryption*[7] (KP-ABE) [8], for Boolean functions. Define a universe $\mathcal{U}$ of n attributes and associate $V \subseteq \mathcal{U}$ with a binary n-tuple (the *characteristic tuple* of V) where the ith place is 1 if and only if the ith attribute is in V. Thus, there is a natural one-to-one correspondence between n-tuples and attribute sets; we write

---

[7] If input privacy is required then a predicate encryption scheme could be used in place of the KP-ABE scheme.

A_x to denote the set associated with x. A function $F \colon \{0,1\}^n \to \{0,1\}$ is monotonic if $x \leqslant y$ implies $F(x) \leqslant F(y)$, where $x = (x_1, \ldots, x_n)$ is less than or equal to $y = (y_1, \ldots, y_n)$ if and only if $\forall i, x_i \leqslant y_i$. For a monotonic F, the set $\mathbb{A}_F = \{x \in \{0,1\}^n : F(x) = 1\}$ defines a monotonic access structure. Informally, for a Boolean function F, the client generates a private key $SK_{\mathbb{A}_F}$ using the KeyGen algorithm.

Given an input x, a client encrypts a random message m "with" A_x using the Encrypt algorithm and publishes $VK_{F,x} = g(m)$ where g is a suitable one-way function (e.g. a pre-image resistant hash function). The server decrypts the message using the Decrypt algorithm, which will either return m (when $F(x) = 1$) or $\bot$.

The server returns m to the client. Any client can test whether the value returned by the server is equal to $g(m)$. Note, however, that a "rational" malicious server will always return $\bot$, since returning any other value will (with high probability) result in the verification algorithm returning a reject decision. Thus, it is necessary to have the server compute both F and its "complement" (and for both outputs to be verified).

Note that, to compute the private key $SK_{\mathbb{A}_F}$, it is necessary to identify all minimal elements x of $\{0,1\}^n$ such that $F(x) = 1$. There may be exponentially many such x. Thus, the initial phase is indeed computationally expensive for the client. Note also that the client may generate different private keys to enable the evaluation of different functions.

References

1. Alderman, J., Janson, C., Cid, C., Crampton, J.: Revocation in publicly verifiable outsourced computation. Cryptology ePrint Archive, Report 2014/640 (2014). http://eprint.iacr.org/
2. Attrapadung, N., Imai, H.: Attribute-based encryption supporting direct/indirect revocation modes. In: Parker, M.G. (ed.) Cryptography and Coding 2009. LNCS, vol. 5921, pp. 278–300. Springer, Heidelberg (2009)
3. Carter, H., Lever, C., Traynor, P.: Whitewash: outsourcing garbled circuit generation for mobile devices. In: Payne, Jr., C.N., Hahn, A., Butler, K.R.B., Sherr, M. (eds.) Proceedings of the 30th Annual Computer Security Applications Conference, ACSAC 2014, pp. 266–275. ACM (2014)
4. Choi, S.G., Katz, J., Kumaresan, R., Cid, C.: Multi-client non-interactive verifiable computation. In: Sahai, A. (ed.) TCC 2013. LNCS, vol. 7785, pp. 499–518. Springer, Heidelberg (2013)
5. Gennaro, R., Gentry, C., Parno, B.: Non-interactive verifiable computing: outsourcing computation to untrusted workers. In: Rabin, T. (ed.) CRYPTO 2010. LNCS, vol. 6223, pp. 465–482. Springer, Heidelberg (2010)
6. Gentry, C.: Fully homomorphic encryption using ideal lattices. In: Mitzenmacher, M. (ed.) STOC, pp. 169–178. ACM (2009)
7. Goldwasser, S., Gordon, S.D., Goyal, V., Jain, A., Katz, J., Liu, F.-H., Sahai, A., Shi, E., Zhou, H.-S.: Multi-input functional encryption. In: Nguyen, P.Q., Oswald, E. (eds.) EUROCRYPT 2014. LNCS, vol. 8441, pp. 578–602. Springer, Heidelberg (2014)

8. Goyal, V., Pandey, O., Sahai, A., Waters, B.: Attribute-based encryption for fine-grained access control of encrypted data. In: Juels, A., Wright, R.N., di Vimercati, S.D.C. (eds.) Proceedings of the 13th ACM Conference on Computer and Communications Security, CCS 2006, pp. 89–98. ACM (2006)
9. Green, M., Hohenberger, S., Waters, B.: Outsourcing the decryption of ABE ciphertexts. In: 2011 Proceedings of the 20th USENIX Security Symposium, San Francisco, CA, USA, August 8–12. USENIX Association (2011)
10. Parno, B., Raykova, M., Vaikuntanathan, V.: How to delegate and verify in public: verifiable computation from attribute-based encryption. In: Cramer, R. (ed.) TCC 2012. LNCS, vol. 7194, pp. 422–439. Springer, Heidelberg (2012)
11. Yao, A.C.-C.: How to generate and exchange secrets (extended abstract). In: FOCS, pp. 162–167. IEEE Computer Society (1986)

Private Aggregation
with Custom Collusion Tolerance

Constantinos Patsakis[1]([⊠]), Michael Clear[2], and Paul Laird[2]

[1] Department of Informatics, University of Piraeus, Piraeus, Greece
kpatsak@unipi.gr
[2] Distributed Systems Group, School of Computer Science and Statistics,
Trinity College, Dublin, Ireland
{clearm,lairdp}@scss.tcd.ie

Abstract. While multiparty computations are becoming more and more efficient, their performance has not yet reached the required level for wide adoption. Nevertheless, many applications need this functionality, while others need it for simpler computations; operations such as multiplication or addition might be sufficient. In this work we extend the well-known multiparty computation protocol (MPC) for summation of Kurswave *et al.* More precisely, we introduce two extensions of the protocol one which bases its security on the Decisional Diffie-Hellman hypothesis and does not use pairings, and one that significantly reduces the pairings of the original. Both protocols are proven secure in the semi-honest model. Like the original, the protocols are entirely broadcast-based and self-bootstrapping, but provide a significant performance boost, allowing them to be adopted by devices with low processing power and can also be extended naturally to achieve t-privacy in the malicious model, while remaining practical. Finally, the protocols can further improve their performance if users decide to decrease their collusion tolerance.

Keywords: Multiparty computation · Private aggregation · Cryptographic protocols

1 Introduction

The core of many applications needs input from other entities to generate the expected output. Nevertheless, the point where other entities on the Internet could be trusted has long been passed. This lack of trust makes users unwilling in many cases to provide the necessary feedback, unless they have credible guarantees that it will not be disclosed. This has generated an increased interest among the research community towards secure multiparty computation (MPC). The role of MPC is to allow users to evaluate a given function with their inputs, without disclosing any information to any other entity. Additionally, some MPC protocols provide additional measures to counter malicious acts, *e.g.* detecting users who try to disturb the execution of the protocol or leak information.

© Springer International Publishing Switzerland 2015
D. Lin et al. (Eds.): Inscrypt 2014, LNCS 8957, pp. 72–89, 2015.
DOI: 10.1007/978-3-319-16745-9_5

Consider a scenario where a user might have to calculate a function in collaboration with other entities that cannot be trusted. Hence, user input to the function should be hidden in such a way that other users cannot disclose his input, while preserving the ability for the output of the function to be calculated correctly and efficiently. More formally, we have n entities that want to calculate function $f(x_1, x_2, ...x_n)$ without disclosing $x_i, i \in \{1, 2, ..., n\}$ to any other entity. This is in fact a special case of MPC. We refer to such functions f as *(public) aggregation functions*. Note that the output of f may be learned by everyone, not merely the n participants.

While MPC has been around for a few decades, it is only recently that efficient protocols have started to appear. Throughout this period, a wide range of protocols emerged with a view to providing solutions to simplified problems or problems with relaxed security requirements. The reason is that many applications do not need the evaluation of complex functions, or the intended users of these applications do not have any incentive to break the correctness of the protocol, they just want their own input to remain secret. A typical example can be seen in the case of smart meters in a smart grid. Each smart meter has to send its expected power consumption to the power plant, so that the latter balances the demand without wasting any resources. Clearly, if these values are disclosed, one could deduce many things regarding the daily habits and preferences of individuals [14, 16, 18, 22]. Additionally, the power plant needs only to calculate the overall consumption, the sum of these values. Other such applications include collective decision making, sensor aggregation in smart cities or anonymous statistics. Such applications introduce requirements that state of the art in MPC protocols cannot meet. Such requirements include scalability (number of users), low computational cost (for constrained devices like sensors) and low bandwidth consumption.

The focus of this work is on extending the protocol of Kursawe *et al.* [15]. In general, the protocol has several nice features, for instance, it is **self-bootstrapping**. Practically, this means that there is no trusted third party to provide users with secret shares. These shares are generated only by the users and are known only to them. It also implies that no setup phase is needed involving point-to-point contact among the participants. The protocol is also efficient enough to perform the calculations much faster than standard MPC. Finally, it is **broadcast-based**; no one-to-one communication takes place among the participants. This decreases significantly the computational effort of each node, while simultaneously diminishing the communication overhead.

The self-bootstrapping quality means that no trusted dealer is needed for setup. Protocols which are not self-bootstrapping may allow one execution of the protocol without requiring a dealer, such as the protocol of Yang et al. [23], based on the ElGamal encryption scheme, but cannot be used for multiple executions of the protocol (in [23] this would invalidate the security of ElGamal). Alternatively, they may allow multiple rounds of aggregation to be executed, as in the protocol of Shi et al. [21] for time-series data, but require a dealer at the outset. Naturally, it is possible to bootstrap [21] using a round of [23] to simulate a dealer, but the ability to run a single protocol for multiple rounds of aggregation without a separate set-up phase or dealer is desirable for reasons of simplicity, security (a security proof with a single underlying security assumption), and lower communication cost (fewer communication rounds).

Protocols are considered to be "broadcast-based" where no point-to-point channels are assumed to exist between pairwise parties. Secure point-to-point channels are a strong assumption, which is not always appropriate, particularly where nodes are resource-constrained and communicate wirelessly. In a broadcast-based protocol, the security of the protocol must be unaffected by message interceptions; the content of any message from one party to another is available in the clear to all other parties. The required architecture typically involves each party broadcasting two types of protocol message: (1) a public key, and (2) an input for a given "round of aggregation" (i.e. an independent aggregate sum, referred to in this work as simply an *aggregation*). The first type of protocol message, namely a public key, is broadcast in the first stage of the protocol. Each subsequent stage corresponds to a different *aggregation*. In practice, the broadcast channel might be implemented in a number of alternative ways, such as a public registry where each party stores its messages, communication through an aggregator that relays authenticated protocol messages between parties. Note that the messages sent by each party are assumed to be authenticated.

Originally, the Kursawe, Danezis and Kohlweiss [15] protocol, which we refer to here as KDK, was designed to be used for private aggregation in the smart grid as a faster alternative to standard MPC. Nevertheless, due to its heavy reliance on pairings, the protocol cannot be used from devices with low processing capabilities. Notably, as we experimentally demonstrate, for large scale implementations, the measurements cannot be sent in less than a second, something that is going to be needed soon for the smart grid. It has to be highlighted that according to the UK's smart metering equipment technical specifications v2 [10], the current needed rate is one measurement per 10 s. Nevertheless, our experiments show that for 1000 users the calculations on a common laptop take roughly 4 s, clearly indicating that it cannot face current needs in urban environments where we have thousands of users.

1.1   Related Work

The concept of secure multiparty computation was introduced by Yao in [24] with the introduction of the famous millionaires problem. In this problem, two millionaires want to compare their assets in order to verify which one is the richest, but without revealing the actual value of their assets. Some important foundational advances were made soon after such as [3,6,12,25], however it took almost two decades for real-world implementations to became practical [2,5,17].

Secure multiparty computation can be broadly divided into schemes which allow arbitrary computations to be performed without leaking information, and specialized protocols for the evaluation of one particular function, such as summation, while keeping individual inputs private.

Schemes for specific arithmetic operations trade generality for greater efficiency in calculating their particular operation. Many also allow further trade-offs, such as collusion-resistance against efficiency.

Clifton et al. in [7] proposed a very efficient and elegant scheme for summing the shares of n entities anonymously. Their scheme involves communication among parties in a ring, and their is a trade-off between the number of passes needed and the collusion tolerance supported. However, the protocol relies on point-to-point channels. Another significant disadvantage of the approach is that to achieve collusion tolerance of $n - 2$, a total of $n\lfloor\frac{n}{2}\rfloor$ messages must be sent serially, with each node sending and receiving $\lfloor\frac{n}{2}\rfloor$ messages. This is not practical where latency is an issue or where large numbers of nodes are involved.

Protocols involving self-canceling blinding variables alleviate this problem, allowing a constant number of communication rounds for an arbitrary number of participants. Yang et al. use two rounds of communication, with the first establishing common key material [23], but the key material cannot be used for more than one round of aggregation. Shi et al. allow aggregation to be performed with a single round of communication [21], however the protocol relies upon the existence of a secret share of zero, which must be supplied in advance by a trusted dealer or created using another protocol. For protocols based on secret sharing which support an additive homomorphism, either a trusted dealer is needed to distribute the shares, or point-to-point channels are needed. Examples of the latter include BGW [3] and more recently SPDZ [9], whereas the former is exemplified by [21].

The KDK protocol is entirely broadcast-based and self-bootstrapping. Kursawe et al. present both a *single-aggregation* and a *multi aggregation* variant of this protocol; the latter supports an unbounded number of independent aggregations from the same set of public keys published by the parties. The latter relies heavily on bilinear pairings, and as such, computing a round of aggregation is computationally expensive. In this work, we present an alternative multi-aggregation protocol that relies on the standard Decisional Diffie-Hellman (DDH) assumption, and does not need pairings. As such, it significantly outperforms multi-aggregation KDK. However, our protocol only supports a bounded number of aggregations from the same public keys (as opposed to an unbounded number in the case of multi-aggregation KDK), and there is a connection between the number of rounds supported and the collusion tolerance guaranteed.

1.2   Overview of Our Protocols

At the heart of single-aggregation KDK, described formally in Sect. 2.2, is the following fixed $n \times n$ matrix A, known to all n parties where:

$$A = \begin{bmatrix} 0 & -1 & -1 & -1 & -1 \\ 1 & 0 & -1 & -1 & -1 \\ 1 & 1 & 0 & \cdots & \vdots \\ \vdots & \vdots & \vdots & \ddots & \vdots \\ 1 & 1 & 1 & \cdots & 0 \end{bmatrix}$$

Suppose party P_i's public key is $u_i = g^{x_i}$ where g is a public generator of a group $\mathbb{G}$ of order p, and $x_i \in \mathbb{Z}_p$ is party P_i's private key. At first, each party P_i broadcasts her public key. Subsequently, each party P_i computes a "blinded" version of her private input m_i by first computing $w_i \leftarrow \prod_{j\in\{1,...,n\}} u_j^{A_{i,j}}$ where $A_{i,j}$ denotes the j-th column of the i-th row of A, and then broadcasting $w_i^{x_i} \cdot g^{m_i}$. This will be discussed in more detail later in the paper.

Observe that A is a skew-symmetric matrix, that is $-A = A^T$, but clearly any such matrix can be used instead of the proposed one, as any such matrix fulfills the basic principle which is $\sum x_i y_i = 0$, where y_i is the sum of the elements other than element x_i, as originally set in [13]. So far this does not give us any advantages over the standard protocol of Kursawe et al.

As previously discussed, users could agree upon a skew-symmetric matrix to generate the coefficients $A_{i,j}$. However, if they would like to compute one more summary, then they could not use the previous values. The reason is that an adversary could deduce important information from that. For instance, if they had originally published $g^{x_i y_i + m_{i_1}}$ and then $g^{x_i y_i + m_{i_2}}$, where $y_i = \sum_{j=1} A_{i,j} x_j$, given that m_1 and m_2 are small, the value $m_1 - m_2$ could easily be calculated. It becomes apparent that a new matrix should be generated to protect users' privacy. This way, users could co-operate in the generation of the skew-symmetric matrix, defining their collusion thresholds, that is how many users should collude to recover their input. Undeniably, by lowering the threshold users are making certain compromises in their privacy. Nevertheless, this is speeding up the calculations and decreasing the communication overhead. The compromise, depending on the nature of the network, the trust of the users to each other, and the importance of protected information can imply a good balance in the user's benefit.

Thus, we argue that in order to minimize the bandwidth overhead and communication between users, instead of agreeing on a matrix A, users on the initialization of the scheme agree on a random seed that is used to generate a series of matrices A_ρ. However, we must allow each A_ρ's independent entries to be uniformly random over $\mathbb{Z}_p$ instead of over $\{-1, 1\}$ to avoid a straightforward linear algebra attack. Instead of using the public keys of the parties (i.e. the g^{x_i}) for a single aggregation, we observe that we can re-use them for multiple aggregations provided that we use a different skew-symmetric matrix A in each round, where the independent entries of each A are uniformly random over $\mathbb{Z}_p$. We use a hash function to derive the matrix A for each round.

Nevertheless, since the reuse of g^{x_i} in many aggregations might enable collusion attacks, the number of matrices that can be generated is bounded by $\frac{n-t}{2}$, where t is the minimum threshold. More details on the latter bound are given in Sect. 3. The advantage of allowing users to reuse their published values is that it decreases the computational and communication cost. Rather than going through the first step of the algorithm again; generating, publishing g^{x_i} and downloading the output of the other users, users recompute locally the new instance of matrix A and compute the new values of g^{y_i}. It turns out that Kursawe et al. propose a variant of their protocol that similarly supports multiple aggregations, but they rely on bilinear pairings to achieve this. However, their multi-aggregation protocol supports an unbounded number of aggregations, as opposed to $\frac{n-t}{2}$ aggregations for our basic protocol, as described above. As a result we can consider two multi-aggregation protocols:

- **Basic protocol with a bounded number of aggregations**: This is our protocol that supports $\frac{n-t}{2}$ independent aggregations with the same public key information and does not use any pairings. Arithmetic is performed over a finite cyclic group $\mathbb{G}$. We abbreviate the basic protocol as "BP".
- **Hybrid protocol with unbounded number of aggregations**: This protocol combines our basic protocol (BP) with ideas from multi-aggregation KDK to achieve greater efficiency than multi-aggregation KDK while still allowing an unbounded number of aggregations with the same public-key information. Arithmetic is performed over a "larger" finite cyclic group $\mathbb{G}_T$. We abbreviate the hybrid protocol as "HP".

BP can be viewed as supporting a finite *batch* of aggregations of size $b = \frac{n-t}{2}$, where $t \in [n]$ is the collusion tolerance. HP uses BP as a building block along with multi-aggregation KDK (MA-KDK). Every aggregation in MA-KDK requires $n - 1$ pairings, which are considerably expensive, as demonstrated later in Sect. 5. The main idea in HP is to amortize this cost over a whole batch of aggregations; the larger the batch size, the greater the savings. But since the batch size $b = \frac{n-t}{2}$ grows with n for any fixed collusion threshold, the savings over MA-KDK also grow with n.

2 Preliminaries

2.1 Notation and Definitions

Let k be an integer. We denote the contiguous set of integers $\{1, \ldots, k\}$ by $[k]$. Let X and Y be distributions. The notation $X \underset{C}{\approx} Y$ denotes the fact that both distributions are computationally indistinguishable to any probabilistic polynomial time (PPT) algorithm.

In order to show that the proposed protocol provides the necessary privacy to the participants, we have to show that it provides privacy against collusions of up to t users when executed with at most $\max(1, \lfloor \frac{n-t}{2} \rfloor)$ rounds. Intuitively, suppose $n - 2$ users collude, then it should not be possible for the colluding users to learn anything about the 2 honest users' inputs beyond their sum. If $n - 1$ users collude, then we expect them to learn the honest party's input. So for the case of $n - 1 \leq t \leq n$, there is no privacy requirement, and thus these trivial cases are easily handled in meeting our security definition below.

We adopt the standard simulation-based definition of security in the semi-honest model with static adversaries. We base our definition below on Definition 2.1 in [1]. Here we consider only computational security, and relax the more standard definition to deterministic functionalities with a single output, since this paper is concerned with aggregation. Note that this definition is general enough to accommodate multi-aggregation aggregation as provided by our protocol.

Let $\boldsymbol{m} \in (\{0,1\}^*)^n$ be a vector of the inputs from each party and let π be a protocol. We define $\mathsf{OUTPUT}^\pi(m_1, \ldots, m_n)$ as the final aggregated result computed with protocol π from the input vector $\boldsymbol{m}$. Furthermore, we define the view of a party P_i in the execution of protocol π with input vector $\boldsymbol{m}$ as

$$\mathsf{VIEW}_i^\pi(\boldsymbol{m}) = (m_i, r_i, \mu_i^{(1)}, \ldots, \mu_i^{(\ell)})$$

where m_i is party $P_i's$ input, r_i is its random coins and $\mu_i^{(1)}, \ldots, \mu_i^{(\ell)}$ are the ℓ protocol messages it received during the protocol execution. Similarly, the combined view of a set of $I \subseteq \{1, \ldots, n\}$ parties is denoted by $\mathsf{VIEW}_I^\pi(\boldsymbol{x})$.

Definition 1 (t-privacy of n-party protocols for deterministic aggregation functionalities). *Let $f : (\{0,1\}^*)^n \to (\{0,1\}^*)$ be a deterministic n-ary functionality and let π be a protocol. We say that π t-privately computes f if for every $\boldsymbol{m} \in (\{0,1\}^*)^n$ where $|m_1| = \ldots = |m_n|$,*

$$\mathsf{OUTPUT}^\pi(m_1, \ldots, m_n) = f(m_1, \ldots, m_n) \tag{1}$$

and there exists a PPT algorithm $\mathcal{S}$ such that for every $I \subset [n]$ with $|I| \leq t$, and every $\boldsymbol{m} \in (\{0,1\})^n$ where $|m_1| = \ldots = |m_n|$, it holds that:

$$\{\mathsf{VIEW}_I^\pi(\boldsymbol{m})\} \underset{C}{\approx} \{\mathcal{S}(I, \boldsymbol{m}_I, f(\boldsymbol{m}))\}. \tag{2}$$

2.2 Single-Aggregation KDK

Kursawe, Danezis and Kohlweiss (KDK) [15] present a specialized multiparty computation (MPC) protocol for private summation, which is shown to be secure in the semi-honest model under the Decisional Diffie-Hellman (DDH) assumption. We refer to this protocol as KDK. In their protocol, n parties $P_1, \ldots, P_n$ can compute a joint sum of their inputs $m_1, \ldots, m_n \in \{0, \ldots, \beta\}$ for some positive integer β. An overview of their protocol follows.

Let p be a prime. The "public parameters" used in the protocol consist of a description of a cyclic group $\mathbb{G}$ of order p together with a generator g of $\mathbb{G}$. It is assumed that DDH is intractable in $\mathbb{G}$. These public parameters $\mathsf{PP} = (\mathbb{G}, g, p)$ are known to all parties P_i. The group operation of $\mathbb{G}$ is written multiplicatively.

1. **Setup:** Party P_i generates a secret key $x_i \in \mathbb{Z}_p$ and computes her public key $u_i = g^{x_i} \in \mathbb{G}$. She broadcasts u_i.
2. **Main Round:**
   - Party P_i chooses her input $m_i \in \{0, \ldots, \beta\}$.
   - Compute $w \leftarrow \prod_{j \in 1}^{i-1} u_j^{-1} \cdot \prod_{j \in i+1}^n u_j \in \mathbb{G}$.
   - Broadcast $v_i \leftarrow w^{x_i} \cdot g^{m_i} \in \mathbb{G}$.
3. **Output:** The protocol produces an output in $\{0, \ldots, n\beta\}$, namely the sum of the user inputs. To compute the sum σ:
   - Compute $z \leftarrow \prod_{j=1}^n v_j$.

- Use Pollard's Lambda algorithm to compute the discrete log $\sigma \in \{0, \ldots, n\beta\}$ of z with respect to g in G. The time complexity of Pollard's lambda algorithm is $\sqrt{n\beta}$.
- Output σ.

It can be easily observed that $\prod_{j=1}^{n} v_j = g^{\sum_{j=1}^{n} m_j}$.

2.3 Multi-aggregation KDK (MA-KDK)

If the protocol must be run a number of times, it would be desirable to avoid re-running the "Setup" phase above which involves each party generating and broadcasting a new public key; in practice, a verification step for these keys may also be needed. To re-use the published keys $u_i, \ldots, u_n$ for more than a single round of aggregation, Kursawe et al. propose an extension of their protocol that facilitates multiple aggregations. In fact, their multi-aggregation protocol accommodates an unbounded number of aggregations. They make use of bilinear pairings to achieve this. More details on bilinear pairings are provided in Sect. 5 when we address practical issues.

Let $\mathbb{G}_1$, $\mathbb{G}_2$ and $\mathbb{G}_T$ be cyclic groups of prime order p. Let $e : \mathbb{G}_1 \times \mathbb{G}_2 \to \mathbb{G}_T$ be a cryptographic bilinear pairing. Furthermore, the Bilinear Decisional Diffie Hellman (BDDH) assumption is expected to hold with respect to $\mathbb{G}_1$, $\mathbb{G}_2$, $\mathbb{G}_T$ and e. Let $H_2 : \mathbb{Z} \to \mathbb{G}_2$ be a hash function. The main changes to KDK to support multiple rounds are as follows (optimizations are discussed later):

- The public parameters include generators $P \in \mathbb{G}_1$, $Q \in \mathbb{G}_2$ and $g = e(P, Q) \in \mathbb{G}_T$.
- The public keys are generated as $U_i \leftarrow x_i P \in \mathbb{G}_1$ for all $1 \leq i \leq n$.
- In aggregation k, party P_i computes
  - $Q_k \leftarrow H_2(k) \in \mathbb{G}_2$ (i.e. for a good choice of H_2, we have $Q_k = rQ$ for some uniformly random r, which is intractable to find).
  - $w \leftarrow \prod_{j=1}^{i-1} e(U_j, Q_k)^{-1} \cdot \prod_{j \in i+1}^{n} e(U_j, Q_k) \in \mathbb{G}_T$.

The rest of the protocol remains unchanged except that the computations are performed in $\mathbb{G}_T$, and P_i may choose a different input value in every round. Naturally, the output of the protocol is then $(\sigma_1, \ldots, \sigma_\ell) \in \{0, \ldots, n\beta\}^\ell$ if ℓ rounds are executed.

3 Protocol for Bounded Number of Aggregations

We start with a description of our basic MPC protocol, i.e. BP, that extends KDK to achieve a bounded number of aggregations from the same public key information.

3.1   Main Protocol Description

Our protocol builds on the work in [15] to add support for multiple aggregation rounds using the same public keys generated by all parties in the initial stage. Moreover, instead of relying on additional assumptions to achieve this (as is the case in [15] where bilinear groups are employed), our construction provides security in the semi-honest model for $\max(1, \lfloor \frac{n-t}{2} \rfloor)$ rounds where t is the collusion tolerance. This is optimal for our techniques.

Let $A \in \mathbb{Z}_p^{k \times k}$ be a skew-symmetric matrix with uniformly random entries in $\mathbb{Z}_p$. Each row of A represents a quadratic polynomial over $\mathbb{Z}_p$ in k unknowns. There are $\frac{k(k-1)}{2}$ possible monomials. Thus, A can be transformed into a coefficient matrix $B \in \mathbb{Z}_p^{k \times k(k-1)/2}$. We write this as $B = \mathsf{coeff}(A)$. It follows that $\mathsf{rank}(B) = k - 1$. Therefore, no linear combination of $k - 1$ equations yields 0.

Let $H : \{0,1\}^* \to \mathbb{Z}_p$ be a hash function. We define a function $\chi : \mathbb{Z}_p \times \mathbb{Z} \to \mathbb{Z}_p^{n \times n}$ that takes a random seed and a round number, and outputs a pseudorandom skew-symmetric matrix over $\mathbb{Z}_p$. Let $s \in \{0,1\}^*$ be a seed. To compute the cells of the skew-symmetric matrix $A^{(i)}$ for round i, we set $A_{j,k}^{(i)} \leftarrow H(s \parallel i \parallel j \parallel k)$ and $A_{k,j}^{(i)} \leftarrow -A_{j,k}^{(i)}$ for every j, k satisfying $1 \leq j < k \leq n$. The remaining entries of $A^{(i)}$ are set to zero. By construction, $A^{(i)}$ is skew-symmetric. Furthermore, $\mathsf{rank}(\mathsf{coeff}(A^{(i)})) = n - 1$.

Suppose there are n parties and the desired collusion tolerance is $T \leq n - 2$. Then the protocol can accommodate $\ell \leq \frac{\lfloor n-T \rfloor}{2}$ independent aggregations. Let $\beta > 0$ denote the size of the message space i.e. every party chooses her input for a given round of aggregation from the set $\{0, \ldots, \beta\}$. Therefore, the sums are bounded from above by $n\beta$.

A public seed s is deterministically derived from the users' public keys generated in the first stage of the protocol. Alternatively, s may be pre-agreed or collaboratively generated. In the security proof, it is assumed to be unique for a given protocol execution.

The "public parameters" used in the protocol consist of a description of a cyclic group $\mathbb{G}$ of order p together with a generator g of $\mathbb{G}$. It is assumed that the Decisional Diffie-Hellman (DDH) problem is intractable in $\mathbb{G}$. These public parameters $\mathsf{PP} = (\mathbb{G}, g, p)$ are known to all parties P_i.

The protocol proceeds in the following stages:

1. **Setup:** Party P_i generates a secret key $x_i \in \mathbb{Z}_p$ and computes her public key $u_i = g^{x_i} \in \mathbb{G}$. She broadcasts u_i.
2. **Aggregation r:** For every $r \in \{1, \ldots, \ell\}$:
   - Party P_i chooses her input $m_i^{(r)} \in \{0, \ldots, \beta\}$.
   - Compute $A^{(r)} \leftarrow \chi(s, i)$.
   - Compute $w \leftarrow \prod_{j \in 1}^{n} u_j^{A_{i,j}^{(r)}} \in \mathbb{G}$.
   - Broadcast $v_i^{(r)} \leftarrow w^{x_i} \cdot g^{m_i^{(r)}} \in \mathbb{G}$.
3. **Output:** The protocol produces an output of ℓ elements, namely the sum of the inputs in each round. To compute the sum σ_r for round r:

- Compute $z \leftarrow \prod_{j=1}^{n} v_j^{(r)}$.
- Use Pollard's Lambda algorithm to compute the discrete log $\sigma_r \in \{0, \dots, n\beta\}$ of z with respect to g in G. The time complexity of Pollard's lambda algorithm is $\sqrt{n\beta}$.
- The final output is $(\sigma_1, \dots, \sigma_\ell)$.

It can be easily observed that for any $1 \leq r \leq \ell$, $\prod_{j=1}^{n} v_j^{(r)} = g^{\sum_{j=1}^{n} m_j^{(r)}}$.

Theorem 1. *Under the DDH assumption, our multi-aggregation protocol is computationally t-private for all $t \leq n$ with at most* $\mathsf{max}(1, \lfloor (n - t)/2 \rfloor)$ *rounds in the random oracle model.*

The proof of Theorem 1 is given in Appendix A.

3.2 Security Against Malicious Adversaries

Our protocol, like KDK, can be extended to provide t-privacy in the presence of malicious adversaries. However in the malicious setting, we relax the correctness requirement given by Eq. 1 in Definition 1. Requiring both properties in the malicious setting is captured by the stronger notion of t-security defined in [1], and although our protocol can also be adapted to meet this definition, we focus only on privacy in this work.

First we consider why the basic version of our protocol as previously described is not t-private in the malicious model. It turns out there is quite a simple attack that can be mounted by an active adversary. A dishonest party P_i can set her public key u_i to depend on one or more other party's public keys. For example, P_i might set $u_i \leftarrow u_j^{c_j}$ for a carefully chosen $c_j \in \mathbb{Z}_p$ such that some monomial $x_j x_k$ "cancels out" in the exponent, and as a result, P_i can learn more information about the honest users' inputs. This is a very straightforward attack to mount by an active adversary.

Fortunately this can be circumvented in a standard manner by requiring that P_i proves in zero knowledge that she knows the private key corresponding to u_i; that is, the discrete log of u_i in $\mathbb{G}$ with respect to g. However, an interactive zero-knowledge proof would impede the attractive broadcast nature of the protocol. Hence as suggested in [15], we can employ a non-interactive zero-knowledge (NIZK) argument system for proving knowledge of discrete logs. We can do this in the random oracle model by applying the Fiat-Shamir heuristic [11] to the well-known and practical Schnorr Protocol [20] (an honest-verifier zero-knowledge protocol for proving knowledge of discrete logs). The latter is computationally inexpensive, involving only one exponentiation and one multiplication in $\mathbb{G}$ for the prover, and two exponentiations and one multiplication for the verifier. We defer to Appendix B a discussion on how to prove this extended protocol t-private in the malicious model. For a discussion on how to yield the stronger notion of t-security, we refer the interested reader to Sect. 4.1 in [15].

4 Hybrid Protocol

The hybrid protocol (HP) presented in this section exploits our MPC protocol (BP) from the last section together with multi-aggregation KDK (MA-KDK) to create a protocol that supports an unbounded number of aggregations from the same public key information, but with improved performance over MA-KDK. The only price we way is a reduction in the collusion tolerance t. However, as n grows, the collusion tolerance need not be lowered by much to enjoy performance gains over MA-KDK. In practice, however, a collusion tolerance of $n \leq n/2$ or $n \leq \frac{3n}{4}$ is usually adequate. We introduce a new parameter b, referred to as the "batch size". This parameter represents the number of aggregations that can be performed with a single execution of BP (with parameters n and t). The maximum value of this parameter is $b = \lfloor (n - t)/2 \rfloor$.

Recall the description of MA-KDK from Sect. 2.3. Let $\mathbb{G}_1$, $\mathbb{G}_2$ and $\mathbb{G}_T$ be cyclic groups of prime order p. Let $e : \mathbb{G}_1 \times \mathbb{G}_2 \to \mathbb{G}_T$ be a cryptographic bilinear pairing. Furthermore, the Bilinear Decisional Diffie Hellman (BDDH) assumption is expected to hold with respect to $\mathbb{G}_1$, $\mathbb{G}_2$, $\mathbb{G}_T$ and e. Let $H_2 : \mathbb{Z} \to \mathbb{G}_2$ be a hash function. Our protocol HP is given as follows:

- The public parameters include generators $P \in \mathbb{G}_1$.
- The public keys are generated as $U_i \leftarrow x_i P \in \mathbb{G}_1$ for all $1 \leq i \leq n$.
- In aggregation k, party P_i performs the following steps:
  1. If $k \equiv 0 \mod b$:
     (a) $Q \leftarrow H_2(k \mod b) \in \mathbb{G}_2$ (i.e. for a good choice of H_2, we have $Q = rQ$ for some uniformly random r, which is intractable to find).
     (b) Compute $u_j \leftarrow e(U_j, Q) \in \mathbb{G}_T$ for $j \in [n] \setminus \{i\}$:
     (c) Compute $g \leftarrow e(P, Q) \in \mathbb{G}_T$.
     (d) Set $\pi \leftarrow$ BPInstance$((\mathbb{G}_T, g), n, t, \{u_1, \ldots, u_n\})$ where BPInstance instantiates the BP protocol with the group $\mathbb{G}_T$ and generator $g \in \mathbb{G}_T$, and parameters n and t, along with public keys $u_1, \ldots, u_n \in \mathbb{G}_T$.
  2. Set $\rho \leftarrow k \mod b$.
  3. Run aggregation round ρ of protocol π for party P_i.

The security proof of the HP protocol is similar with the proof of the BP protocol. More precisely, the only change that should be introduced is the re-keying of the matrix generation so that the new seed is derived from each time a pairing is performed.

We will see in the next section how HP considerably improves upon the performance of MA-KDK.

5 Performance

5.1 Computation of an Aggregation Round

We first turn our attention to comparing the cost of an aggregation round in BP vs. MA-KDK. Later, we take a look at HP.

To begin the comparison, we compare the necessary group operations that a party P_i must perform in a given aggregation round. MA-KDK requires $n - 1$ pairings, $n - 1$ multiplications in $\mathbb{G}_T$ and an exponentiation in $\mathbb{G}_T$. Note the omission of the inversions in $\mathbb{G}_T$ for $1 \leq j < i$. The reason for this is that in the *Setup* phase, party P_i can compute $U_j \leftarrow -U_j$ for $1 \leq j < i$ where $U_j \in \mathbb{G}_1$ is P_j's public key. Thus by bilinearity of e, no inversions are needed in $\mathbb{G}_T$.

On the other hand, BP needs n exponentiations and n multiplications in group $\mathbb{G}$, Derivation of the *per-round information* for KDK involves computing $Q_k \leftarrow H_2(k) \in \mathbb{G}_2$ whereas our protocol involves computing $A^{(k)} \leftarrow \chi(s, k)$. However, the latter can be optimized since only a single row of the matrix $A^{(k)}$ is needed by party P_i. Recall that χ uses a hash function $H : \mathbb{Z}_p \times \mathbb{Z} \times \mathbb{Z} \times \mathbb{Z} \rightarrow \mathbb{Z}_p$ to generate $A_{i,j}^{(k)}$; that is, $A_{i,j}^{(k)} \leftarrow H(s, k, i, j)$. However, derivation of the *per-round informa-tion* in both protocols is negligible relative to the cost of the group operations.

At present, all known efficient realizations of bilinear pairings are based on elliptic curves. Therefore, in order to implement multi-aggregation KDK, we had to instantiate $\mathbb{G}_1$ and $\mathbb{G}_2$ with elliptic curve groups. There is far less freedom when choosing an elliptic curve when pairings are involved, since the chosen curve must satisfy additional properties. Notwithstanding, to provide a fair performance comparison between both protocols, the same curve was used for both protocols.

Consider an elliptic curve E over $\mathbb{F}_q$ for prime q whose order is $\#E(\mathbb{F}_q) = p$. For our protocol, the group $\mathbb{G}$ may be instantiated by the additive group formed by $E(\mathbb{F}_q)$. For multi-aggregation KDK, we have opted to use the Modified Tate Pairing to instantiate e since efficient implementations exist in libraries such as MIRACL.

Now the embedding degree k of E is the smallest positive integer such that $p \mid q^k - 1$. Concretely, the Tate pairing takes two points on $E(\mathbb{F}_{q^k})[p]$ and outputs an element of $\mathbb{F}_{q^k}^*$ (more precisely, an element of a multiplicative subgroup of order p of $\mathbb{F}_{q^k}^*$), where $E(\mathbb{F}_{q^k})[p]$ denotes the set of points on $E(\mathbb{F}_{q^k})$ of order p i.e. the set of points A with $pA = \mathcal{O}$, where $\mathcal{O}$ is the additive identity (*point at infinity*). Basically, $\mathbb{G}_1$ and $\mathbb{G}_2$ must be two distinct subgroups of $E(\mathbb{F}_{q^k})$ of order p. In fact, we can set $\mathbb{G}_1$ to $E(\mathbb{F}_q)$ to make the pairing calculation faster. Furthermore, certain pairing-friendly curves E allow us to choose $\mathbb{G}_2$ such that it is isomorphic to a sub-group of $E'(\mathbb{F}_{q^{k/d}})$ where E' is related to E (known as the "twisted curve"); this means arithmetic operations can be carried out in the smaller field $\mathbb{F}_{q^{k/d}}$ where d is the "twist degree".

The curve chosen for our implementation is a member of the pairing-friendly BN family from [19] with a 254-bit prime q, embedding degree $k = 12$, and "twist degree" $d = 6$. As a consequence of the latter, arithmetic operations in $\mathbb{G}_2$ can be carried out in $\mathbb{F}_{q^2}$ instead of $\mathbb{F}_{q^{12}}$. In addition, $\mathbb{G}_T$ is the group generated by $g = e(P, Q) \in \mathbb{F}_{q^k}^*$, and thus its arithmetic operations are carried out in the "big" field $\mathbb{F}_{q^k}$. This has implications for message recovery, since Pollard's lambda algorithm is much slower.

We implemented both protocols in C++ using the MIRACL C/C++ library version 5[1] using the BN curve as described above. The code was compiled with

---

[1] On Github, commit https://github.com/CertiVox/MIRACL/commit/6d7bb13285e7 962ccfa110b4149fa8a63db2ed52.

g++ with the compiler flags "-m64 -O2" as recommended in the MIRACL documentation, and was executed on a machine with an Intel Core i5-3340M CPU (2.7 GHz) and 4 GB of RAM, and running 64-bit Debian GNU/Linux 3.2.41. For each protocol. we measured the time taken to compute a single round per participant (recall that a round involves preparing the value v_i for some party P_i). We ran this 100 times for different values of n. Note that on each run a random index $i \in \{1, \ldots, n\}$ was chosen, and the round was executed for P_i. Our results are shown in Table 1. As expected, the cost of a round is roughly linear in n. Moreover, the difference in times between KDK and our protocol is significant; on average our protocol outperforms KDK by a factor of ≈ 437 based on the times in Table 1. For even a moderate number of users such as $n = 100$, it is clear that multi-aggregation KDK is not suited to time-sensitive applications. This is more pronounced for resource-constrained devices such as wireless sensors.

We also implemented the hybrid protocol (HP) as described in Sect. 4. HP uses an instantiation of BP with the group $\mathbb{G}_T$, and thus the exponentiations involved in a round of aggregation are more costly. However, the dominant cost is that of the $n - 1$ pairings. Unlike MA-KDK, these pairings need only be carried out once per "batch" of aggregations. Table 1 reports the per-party time to compute a round of aggregation as $\tau = \tau_{\mathsf{BP}} + \tau_P / b$ where τ_{BP} is the time taken to compute a round of aggregation in BP (instantiated with $\mathbb{G}_T$), τ_P is the time taken to compute the $n - 1$ pairings, and b is the batch size. In other words, τ represents the amortized time. The batch size is $b = 1 - Tn/2$ for some collusion tolerance $T \in [0, 1]$. For our experiments, we set $T = 1/2$ and thus our results for HP pertain to a batch size of $b = n/4$.

Table 1. Mean time in ms (over 100 runs) for a party to compute a round of aggregation (standard deviation in parenthesis). Note that for HP, the threshold tolerance is $T = 1/2$, so the batch size is $b = (1 - T)n/2 = n/4$.

	n=10	n=100	n=1000
Multi-aggregation KDK	47.78 (0.25)	480.71 (2.34)	4795.33 (6.69)
Our Protocol (BP)	0.94 (0.053)	1.33 (0.01)	5.33 (0.07)
Hybrid Protocol (HP)(col. tolerance 1/2)	34.72 (0.31)	53.79 (0.2)	324.23 (1.34)

5.2 Recovery of the Sum with Pollard's Lambda Algorithm

Now we turn our attention to the aggregation phase of our BP protocol. In any given round, this entails multiplying all elements v_i to calculate $z \leftarrow \prod v_i$, then finding a discrete logarithm with respect to a generator g to recover the sum $\sigma = \sum m_i$ of the parties' inputs in the round. For this purpose, Pollard's Lambda algorithm is employed. In our protocol, the v_i belong to $\mathbb{G}$ whereas in the multi-aggregation KDK or the HP, they belong to $\mathbb{G}_T$. Recall that our implementation with elliptic curves of BP instantiates $\mathbb{G}$ as $E(\mathbb{F}_q)$ whereas $\mathbb{G}_T$ is instantiated as a subgroup of $\mathbb{F}_{q^k}^*$. As such, this phase is more expensive for multi-aggregation

KDK because the field operations take place in a "bigger" field. Pollard's Lambda algorithm dominates recovery of the sum. Its time complexity is $O(\sqrt{M})$ where M denotes the size of the message space. In this case, $M = n\beta$ since each party chooses her message in $\{0, \ldots, \beta\}$.

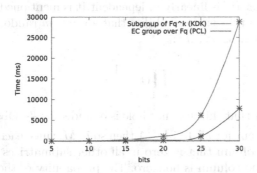

Fig. 1. Average time to find discrete logs in $E(\mathbb{F}_q)$ (Our Protocol, denoted PCL in the figure) and $\mathbb{F}_{q^k}$ (multi-aggregation KDK) for different value ranges (upper bound in bits).

In order to compare multi-aggregation KDK to our protocol in this phase, we measured the time taken to compute Pollard's Lambda algorithm for different message space sizes. Moreover, values were randomly generated in the set $m \xleftarrow{\$} \{2^{b-2}, \ldots, 2^b\}$ for different values of b and the time taken to recover m given g^m using Pollard's Lambda algorithm was measured (the range given to the algorithm was $\{0, \ldots, 2^b\}$); here g denotes the generator of the group in question and multiplicative notation is employed arbitrarily. The measurements were performed in Sage version 5.9 on the same machine and operating system as that used for the previous experiments above. We ran the experiment 10 times each for $b \in \{5, 10, 15, 20, 25, 30\}$ for both $E(\mathbb{F}_q)$ and the group $\langle e(P, Q) \rangle \subset \mathbb{F}_{q^k}^*$ (recall that $k = 12$ for the curve we used). Our results are shown in Fig. 1. Observe that for $\approx$30-bit numbers, multi-aggregation KDK takes almost half a minute to recover the result. Hence, for large values of β, the recovery phase acts a big bottleneck in multi-aggregation KDK.

A   Proof of Theorem 1

Lemma 1. Let $m = \lfloor k/2 \rfloor$. Let $A^{(1)}, \ldots, A^{(m)}$ be skew-symmetric $k \times k$ matrices with uniformly random entries in $\mathbb{Z}_p$. Let $B^{(1)} = \mathsf{coeff}(A^{(1)})[1, \ldots, k-1], \ldots, B^{(m)} = \mathsf{coeff}(A^{(m)})[1, \ldots, k-1]$ where the notation $[1, \ldots, k-1]$ signifies the first $k - 1$ rows of the matrix. Let $M = (B^{(1)}; \ldots; B^{(m)}) \in \mathbb{Z}_p^{m(k-1) \times m(k-1)}$ be the joint matrix consisting of $k - 1$ rows from each of the m coefficient matrices. Then $\Pr[\mathsf{rank}(M) \neq m(k-1)] \leq \frac{\mathsf{poly}(k)}{p}$.

Proof. We can rearrange the rows of M such that the t-th block $M^{(t)}$ consists of the t-th rows of the coefficient matrices. In each such row, there are only $k - 1$ nonzero entries. Eliminating the zero columns results in an $m \times (k - 1)$ matrix $M^{(t)\prime}$ with independent and uniformly random elements from $\mathbb{Z}_p$. Since $m < k-1$, the probability that $M^{(t)\prime}$ is linearly independent is at least the probability that its left $m \times m$ submatrix is linearly independent.It is mentioned in [4] (the result is due to Cooper [8]) that the probability that an $m \times m$ random matrix over $\mathbb{Z}_p$ is linearly independent is at least:

$$\prod_{i=1}^{m}(1 - \frac{1}{p^i}).$$

Now the probability that this does not hold is bounded by $\frac{m}{p}$. Observe that if $M^{(t)}$ is linearly independent for all $1 \leq k - 1$ then so is M, since each submatrix $M^{(t)}$ contains a unique column that is zero in all other submatrices, provided that a submatrices's unique column is nonzero. The probability of the latter not holding is $\frac{k-1}{p^m}$. Therefore, an upper bound on the probability of M not being linearly independent is:

$$\frac{k - 1}{p^m} + \frac{m(k - 1)}{p} = \frac{\mathsf{poly}(k)}{p}.$$

$\square$

Theorem 1. *Under the DDH assumption, our multi-aggregation protocol is computationally t-private for all $t \leq n$ with at most $\mathsf{max}(1, \lfloor (n - t)/2 \rfloor)$ rounds in the random oracle model.*

Proof. Let $\ell \leq \mathsf{max}(1, \lfloor (n-t)/2 \rfloor)$ be the number of aggregations. Let $h = n-t$ be the number of honest users. If $h \leq 1$, it is trivial to construct a simulator $\mathcal{S}$ since $\mathcal{S}$ can fully learn $\boldsymbol{m}$ and then simulate all parties. Therefore, we assume that $h \geq 2$. Let $w = h(h - 1)/2$. Consider the following series of Hybrids.

Hybrid 0: This is the same as the real distribution i.e. the LHS of Eq. 2 with the exception that we "simulate" each honest party P_k using input $m_k^{(\rho)}$; therefore we have access to x_k.

For $1 \leq q \leq w$: Hybrid q involves two honest parties which we denote by P_i and P_j. Their equations share the monomial $x_i x_j$. There are $w = h(h-1)/2$ such monomials and the goal of each Hybrid q is to replace the q-th monomial with a uniformly random element.

Hybrid q: The changes between Hybrid q and Hybrid $q - 1$ involve changing the protocol messages of the honest parties P_i and P_j in all ℓ aggregations. Let $m_i^{(\rho)}$ and $m_j^{(\rho)}$ be the inputs of these honest parties in round ρ. Generate a uniformly random integer $r \in \{0, \ldots, p-1\}$ and replace all occurrences of $g^{x_i x_j}$ by g^r in the computation of the second messages in all aggregations.

Hybrid $q - 1$ and Hybrid q are computationally indistinguishable under the DDH assumption. Hybrid $q - 1$ involves the DDH instance $(g, g^{x_i}, g^{x_j}, g^{x_i x_j})$ and

Hybrid q involves the DDH instance $(g, g^{x_i}, g^{x_j}, g^r)$ where x_i, x_j and r are uniformly distributed in $\{0, \dots, p-1\}$. A non-negligible advantage distinguishing between Hybrid 0 and Hybrid 1 implies a non-negligible advantage against DDH.

Hybrid $w + 1$: (where $w = h(h-1)/2$) H is modelled as a random oracle and as such the skew-symmetric matrices contain uniformly random elements in $\mathbb{Z}_p$. In this Hybrid, we program H such that the joint coefficient matrix $M \in \mathbb{Z}_p^{\ell(n-t-1) \times (n-t)(n-t-1)/2}$ formed from the coefficient matrix in every aggregation is linearly independent. By Lemma 1, the probability of M not being linearly independent when generated as in the real world is at most $\frac{\mathsf{poly}(n-t)}{p}$. Because p is superpolynomial in the security parameter, an adversary has a negligible chance between distinguishing Hybrid $w + 1$ and Hybrid w.

Hybrid $w + 2$: Without loss of generality, assume that parties $P_1, \dots, P_h$ are the honest parties. For all $1 \le i < h$ and $1 \le \rho \le \ell$, replace the protocol message $v_i^{(\rho)}$ of party P_i in aggregation ρ with $g^{r_i^{(\rho)}} \cdot g^{m_i^{(\rho)}}$ for uniformly random $r_i^{(\rho)} \in \mathbb{Z}_p$. Furthermore, for every $1 \le \rho \le \ell$, replace the protocol message $v_h^{(\rho)}$ with $g^{-\sum_{j=1}^{h-1} r_j^{(\rho)} + m_h^{(\rho)}}$. Due to the linear independence of the coefficient matrix $M \in \mathbb{Z}_p^{\ell(n-t-1) \times (n-t)(n-t-1)/2}$, distinguishing between Hybrid $w + 2$ and Hybrid $w + 1$ is impossible.

Hybrid $w + 3$: Finally, in this Hybrid, the inputs $m_1^{(\rho)}, \dots, m_h^{(\rho)}$ are replaced by a random partition of $\sum_{k=1}^{h} m_k^{(\rho)}$, namely the values $s_1^{(\rho)}, \dots, s_h^{(\rho)}$ for every $\rho \in \{1, \dots, \ell\}$.

An adversary has a zero advantage distinguishing Hybrid $w + 3$ and Hybrid $w + 2$. To see this, suppose the adversary could distinguish the hybrids. Then it can determine that some party's input (say P_i) in some aggregation ρ is not $s_i^{(\rho)}$. But $v_i^{(\rho)} = g^{r'}$ for some uniformly random r', which provides no information about the message (whether it is $m_i^{(\rho)}$ or $s_i^{(\rho)}$). Note that $v_h^{(\rho)}$ gives no additional information since it can be derived from the information known to the adversary (recall that the sum in each aggregation is known).

Since Hybrid $w + 3$ no longer relies on the honest parties' messages, and all other information needed to construct the distribution can be derived from the simulators' inputs in Eq. 2, it follows that there exists an algorithm S that can simulate the real distribution.   □

B t-privacy in the Malicious Setting

We only give a brief overview here of how to prove t-privacy of the extended protocol described in Sect. 3.2 in the presence of malicious adversaries. Recall that the protocol uses a NIZK argument system (Setup, Prove, Verify) for statements of the form $S_i = \{(x_i) : u_i = g^{x_i}\}$. The common reference string $\sigma \leftarrow \mathsf{Setup}(1^\kappa)$ is known to all parties and consists of a description of a hash function H_{NIZK}, which is modeled in the proof as a random oracle. A party P_j rejects a public key and proof pair $(u_i, \mathfrak{p}_i)$ if $\mathsf{Verify}(\sigma, S_i, \mathfrak{p}_i) \ne 1$. As a result, we can argue that the x_i for

$i \in I$ are independent of $\{x_j\}_{j \in [n] \setminus I}$ with all but negligible probability. The main modification to the proof of Theorem 1 involves the simulation of the NIZK proofs for the honest parties, since we need to embed DDH challenges and thus do not know the exponents. Before embedding the DDH challenges, we have a series of $h = n - t$ hybrids, where in the k-th such hybrid, we invoke the zero-knowledge property of the NIZK argument system to simulate (which will involve programming the oracle H_{NIZK}) the proof string $\mathfrak{p}_k$ for honest party P_k with a computationally indistinguishable proof string $\mathfrak{p}_k'$. The remainder of the proof proceeds in the same manner as the proof of Theorem 1.

References

1. Asharov, G., Lindell, Y.: A full proof of the BGW protocol for perfectly-secure multiparty computation. Electron. Colloq. Comput. Complex. (ECCC) **18**, 36 (2011)
2. Ben-David, A., Nisan, N., Pinkas, B.: Fairplaymp: a system for secure multi-party computation. In: Proceedings of the 15th ACM Conference on Computer and Communications Security, pp. 257–266. ACM (2008)
3. Ben-Or, M., Goldwasser, S., Wigderson, A.: Completeness theorems for non-cryptographic fault-tolerant distributed computation. In Proceedings of the Twentieth Annual ACM Symposium on Theory of Computing, pp. 1–10. ACM (1988)
4. Blake, I.F., Studholme, C.: Properties of random matrices and applications. Unpublished report (2006). http://www.cs.toronto.edu/~cvs/coding
5. Bogetoft, P., Christensen, D.L., Damgård, I., Geisler, M., Jakobsen, T.P., Krøigaard, M., Nielsen, J.D., Nielsen, J.B., Nielsen, K., Pagter, J., et al.: Multiparty computation goes live. IACR Cryptology ePrint Archive 2008, p. 68 (2008)
6. Chaum, D., Crépeau, C., Damgard, I.: Multiparty unconditionally secure protocols. In: Proceedings of the Twentieth Annual ACM Symposium on Theory of Computing, pp. 11–19. ACM (1988)
7. Clifton, C., Kantarcioglu, M., Vaidya, J., Lin, X., Zhu, M.Y.: Tools for privacy preserving distributed data mining. ACM SIGKDD Explor. Newsl. **4**(2), 28–34 (2002)
8. Cooper, C.: On the rank of random matrices. Random Struct. Algorithms **16**, 2000 (2000)
9. Damgrd, I., Pastro, V., Smart, N.P., Zakarias, S.: Multiparty computation from somewhat homomorphic encryption. IACR Cryptology ePrint Archive 2011, p. 535 (2011)
10. Department of Energy and Climate Change. Smart metering equipment technical specifications: second version July 2013. https://www.gov.uk/government/consultations/smart-metering-equipment-technical-specifications-second-version
11. Fiat, A., Shamir, A.: How to prove yourself: practical solutions to identification and signature problems. In: Odlyzko, A.M. (ed.) CRYPTO 1986. LNCS, vol. 263, pp. 186–194. Springer, Heidelberg (1987)
12. Goldreich, O., Micali, S., Wigderson, A.: How to play any mental game. In: Proceedings of the Nineteenth Annual ACM Symposium on Theory of Computing, pp. 218–229. ACM (1987)
13. Hao, F., Zieliński, P.: A 2-round anonymous veto protocol. In: Christianson, B., Crispo, B., Malcolm, J.A., Roe, M. (eds.) Security Protocols. LNCS, vol. 5087, pp. 202–211. Springer, Heidelberg (2009)
14. Hart, G.W.: Nonintrusive appliance load monitoring. Proc. IEEE **80**(12), 1870–1891 (1992)

15. Kursawe, K., Danezis, G., Kohlweiss, M.: Privacy-friendly aggregation for the smart-grid. In: Fischer-Hübner, S., Hopper, N. (eds.) PETS 2011. LNCS, vol. 6794, pp. 175–191. Springer, Heidelberg (2011)
16. Laughman, C., Lee, K., Cox, R., Shaw, S., Leeb, S., Norford, L., Armstrong, P.: Power signature analysis. IEEE Power Energy Mag. $1(2)$, 56–63 (2003)
17. Malkhi, D., Nisan, N., Pinkas, B., Sella, Y.: Fairplay-secure two-party computation system. In: USENIX Security Symposium, pp. 287–302 (2004)
18. Molina-Markham, A., Shenoy, P., Fu, K., Cecchet, E., Irwin, D.: Private memoirs of a smart meter. In: Proceedings of the 2nd ACM Workshop on Embedded Sensing Systems for Energy-efficiency in Building, pp. 61–66. ACM (2010)
19. Pereira, G.C.C.F., Simplício Jr., M.A., Naehrig, M., Barreto, P.S.L.M.: A family of implementation-friendly bn elliptic curves. J. Syst. Softw. $84(8)$, 1319–1326 (2011)
20. Schnorr, C.: Efficient identification and signatures for smartcards. pp. 239–252 (1990)
21. Shi, E., Chow, R., Chan, T.H.H., Song, D., Rieffel, E.: Privacy-Preserving Aggregation of Time-Series Data. Technical report, UC Berkeley (2011)
22. Weiss, M., Helfenstein, A., Mattern, F., Staake, T.: Leveraging smart meter data to recognize home appliances. In: 2012 IEEE International Conference on Pervasive Computing and Communications (PerCom), pp. 190–197. IEEE (2012)
23. Yang, Z., Zhong, S., Wright, R.N.: Privacy-preserving classification of customer data without loss of accuracy. In: SIAM International Conference on Data Mining, pp. 1–11 (2005)
24. Yao, A.C.-C.: Protocols for secure computations. In: FOCS, vol. 82, pp. 160–164 (1982)
25. Yao, A.C.-C.: How to generate and exchange secrets. In: 27th Annual Symposium on Foundations of Computer Science, 1986, pp. 162–167. IEEE (1986)

Signature and Security Protocols

Ring Signatures of Constant Size Without Random Oracles

Fei Tang[1,2,3](✉) and Hongda Li[1,2]

[1] State Key Laboratory of Information Security, Institute of Information Engineering of Chinese Academy of Sciences, Beijing, China
[2] Data Assurance and Communication Security Research Center of Chinese Academy of Sciences, Beijing, China
[3] University of Chinese Academy Sciences, Beijing, China
tangfei127@163.com, lihongda@iie.ac.cn

Abstract. Ring signatures allow a signer to anonymously sign on behalf of a group of users, the so-called ring; the only condition is that the signer is a member of the ring. At PKC 2007, Shacham and Waters left an open problem, *"obtain a ring signature secure without random oracles and its signature size is independent of the number of signers implicated in the ring"*, which has not been solved yet. In this paper, by using a powerful tool, indistinguishability obfuscator ($i\mathcal{O}$), we construct a constant size ring signature scheme without random oracles and thus answer Shacham et al.'s open problem. Furthermore, we construct an identity-based ring signature scheme which also has constant signature size in the standard model. However, we stress that due to the low efficiency of the existing $i\mathcal{O}$ candidates, we mainly focus on the existence of the constant size ring signature schemes without random oracles, but do not care about their practicability. A shortcoming of our approach is that the ring unforgeability merely is selective but not adaptive.

Keywords: Ring signatures · Constant size · Indistinguishability obfuscation

1 Introduction

Ring signatures allow a signer to anonymously sign on behalf of a group of users, the so-called ring; the only condition is that the signer is a member of the ring. In such a signature scheme, a verifier can be convinced that someone in the ring is responsible for a valid signature, but cannot tell who is the real signer. In contrast to group signatures [12], the anonymity of the signer in a ring signature scheme cannot be revoked. Ring signatures provide an elegant way to leak authoritative secrets in an anonymous way [26], and it also can be used to implement designated verifier signatures [24].

This research is supported by the National Natural Science Foundation of China (Grant No. 60970139) and the Strategic Priority Program of Chinese Academy of Sciences (Grant No. XDA06010702).

D. Lin et al. (Eds.): Inscrypt 2014, LNCS 8957, pp. 93–108, 2015.
DOI: 10.1007/978-3-319-16745-9_6

Related works. The notion of ring signature was introduced by Rivest, Shamir, and Tauman [26]. In a ring signature scheme, the signer chooses the ring members who do not need to cooperate and may be unaware that they are included in a ring signature. Since then, ring signatures, along with the related notion of ring/ad-hoc identification schemes, have been studied extensively. We may divide these schemes into two classes according to their security model. The first class of schemes are secure in the random oracle model (e.g., [2,13,15,22,26,33]) and the second class of schemes are secure in the standard model (e.g., [1,6,11,14,29, 31,32]). Due to some negative results about the random oracles, such as [10,16], researches on cryptographic primitives that are secure in the standard model has gained much attention. The scheme of [14] is based on a strong new assumption. The scheme in [6] makes use of generic ZAPs for NP. Schacham and Waters [31] gave a construction of linear size that is secure under the computational setting of the definitions in [6]. The scheme in [11] has sub-linear signature size uses the common reference string model. Schäge and Schwenk [29] gave a CDH-based construction.

The description of the ring itself is in general linear in the number of members because it is need to specify the users included in the ring. To the best of our knowledge, most ring signature schemes known today are of linear or sub-linear size in the number of ring members, the only exception being the scheme of [15] which is secure in the random oracle model. All existing ring signature schemes (except [15]) admit signature sizes relying on the ring size. Yet, one might face a situation wherein we would like to put many members in a ring. In such situation, the size of the ring signature being constant is quite useful. Based on this consideration, Shacham and Waters [31] proposed an open problem at PKC 2007, *"obtain a ring signature secure without random oracles and its signature size is independent of the number of signers implicated in the ring"*, which has not been solved yet. The goal of this paper is to solve this open problem.

Our approach. Our constructions will make use of a powerful tool, indistinguishability obfuscator ($i\mathcal{O}$). The notion of $i\mathcal{O}$ derives from the program obfuscation which aims at making a computer program "unintelligible" while preserving its functionality. Barak et al. [4] initiated the formal study, but unfortunately, they showed that the most natural simulation-based formulation of obfuscation, i.e., "black-box obfuscation", is impossible to achieve for general programs. To circumvent this negative result, they suggested another weaker notion called indistinguishability obfuscation. A uniform indistinguishability obfuscator for a class of circuits guarantees that obfuscations of given any two equal-size circuits that compute a same function are computationally indistinguishable. A recent breakthrough by Garg et al. [17] admits the first candidate construction of an efficient $i\mathcal{O}$ for general boolean circuits. Whereafter, this powerful tool is shown to be very useful in cryptography, some successful examples including replacing random oracles [23], functional encryption [17,18], deniable encryption [30], (identity-based) multiparty key exchange [9] and so on.

Our approach makes use of Sahai and Waters' [30] technique, namely punctured program. (Please see [30] for details.) Our constructions are inspired by

Boneh and Zhandry's [9] (identity-based) multiparty key exchange schemes. In our constructions, a trusty authority is needed to create an obfuscated signing program. Then, each user is able to run this program on inputs $(M, SK_s, R = (VK_1, \ldots, VK_n), s)$, where M is the message that the user wants to sign, SK_s is his secret key, R is the ring that the user wants to sign on half of, and the index s means that his verification key in R is VK_s. The signing program will check that whether SK_s is a valid secret key corresponds to the s-th verification key in R, if it checks succeed, then a constrained pseudorandom function (PRF) [3,8,25] will act on the concatenation of M and R and output the result as the signature. Hence the signature size equals to the length of the output of the constrained PRF which is a constant with respect to the size, n, of the ring R. (See below for details.) Due to the low efficiency of the existing $i\mathcal{O}$ candidates [5,7,17], our constructions are inefficient. In addition, another shortcoming of our approach is that the ring unforgeability merely is selective but not adaptive.

Furthermore, we will show that our ring signature scheme can be easily extended to the identity-based setting, i.e., identity-based ring signatures of constant size in the standard model.

2   Building Blocks

2.1   Pseudorandom Generator

Definition 1. A function $G : \{0,1\}^\lambda \to \{0,1\}^{2\lambda}$ is a pseudorandom generator (PRG) if for any PPT adversary $\mathcal{A}$ its advantage

$$\mathbf{Adv}_{\mathcal{A},G}(\lambda) = |\Pr[\mathcal{A}(y) = 1 : y = G(s), s \xleftarrow{\$} \{0,1\}^\lambda] - \Pr[\mathcal{A}(y) = 1 : y \xleftarrow{\$} \{0,1\}^{2\lambda}]|$$

is at most negligible.

Håstad, Impagliazzo, Levin, and Luby [21] proved that there is a PRG if and only if there is a one way function (OWF).

2.2   Constrained Pseudorandom Functions

Definition 2. A puncturable family of PRFs mapping is consists of a triple of algorithms Key, Pun, and Eva, and a pair of computable functions $u(\cdot)$ and $v(\cdot)$, satisfying the following conditions:

- Functionality preserved under puncturing: For every PPT adversary $\mathcal{A}$ such that $\mathcal{A}(1^\lambda)$ outputs a set $S \subseteq \{0,1\}^{u(\lambda)}$, then for all $x \in \{0,1\}^{u(\lambda)}$ where $x \notin S$, we have:

$$\Pr[\mathsf{Eva}_K(x) = \mathsf{Eva}_{K(S)}(x) : K \leftarrow \mathsf{Key}(1^\lambda), K(S) \leftarrow \mathsf{Pun}(K, S)] = 1.$$

- Pseudorandom at punctured points: For every PPT adversary $(\mathcal{A}_1, \mathcal{A}_2)$ such that $\mathcal{A}_1(1^\lambda)$ outputs a set $S \subseteq \{0,1\}^{u(\lambda)}$ and state τ, consider an experiment where $K \leftarrow \mathsf{Key}(1^\lambda)$ and $K(S) \leftarrow \mathsf{Pun}(K, S)$. Then we have:

$$\Pr[\mathcal{A}_2(\tau, K(S), S, \mathsf{Eva}_K(S)) = 1] - \Pr[\mathcal{A}_2(\tau, K(S), S, U_{m(\lambda) \cdot |S|}) = 1] = negl(\lambda),$$

where $S = (x_1, \ldots, x_k)$, $\mathsf{Eva}_K(S)$ denotes $\mathsf{Eva}_K(x_1)||\cdots||\mathsf{Eva}_K(x_k)$ is the enumeration of the elements of S in lexicographic order, and U_ℓ denotes the uniform distribution over ℓ bits.

The GGM tree-based construction of PRFs [19] from PRG that maps $\{0,1\}^\lambda$ to $\{0,1\}^{2\lambda}$ are easily seen to yield constrained PRFs, as realized by [3,8,25]. In the GGM-based constrained PRF constructions, the length, $v(\lambda)$, of the outputs equals to λ and the length, $u(\lambda)$, of the inputs equals to the depth of the binary tree. Hence the output length is independent of the input length.

2.3   Indistinguishability Obfuscation

Definition 3. A uniform PPT machine $i\mathcal{O}$ is called an indistinguishability obfuscator for a class of circuit $\{\mathcal{C}_\lambda\}$ if it meets the following conditions:

– Functionality preservation: For all security parameters $\lambda \in \mathbb{N}$, all circuits $C \in \mathcal{C}_\lambda$, and all inputs x, we have:

$$\Pr[C'(x) = C(x) : C' \leftarrow i\mathcal{O}(\lambda, C)] = 1.$$

– Indistinguishability: For any (not necessarily uniform) PPT adversaries ($Samp$, D), there exists a negligible function $negl$ such that the following holds: if $\Pr[\forall x, C_0(x) = C_1(x) : (C_0, C_1, \tau) \leftarrow Samp(1^\lambda)] > 1 - negl(\lambda)$, then:

$$\big|\Pr[D(\tau, i\mathcal{O}(\lambda, C_0)) = 1 : (C_0, C_1, \tau) \leftarrow Samp(1^\lambda)]$$
$$-\Pr[D(\tau, i\mathcal{O}(\lambda, C_1)) = 1 : (C_0, C_1, \tau) \leftarrow Samp(1^\lambda)]\big| \le negl(\lambda).$$

Garg et al. [17] constructed such an indistinguishability obfuscator for NC^1 and all polynomial size circuits. In this paper, we make use of the version of the polynomial size circuits defined below.

Definition 4. A uniform PPT machine $i\mathcal{O}$ is called an indistinguishability obfuscator for $P/poly$ if the following holds: Let $\{\mathcal{C}_\lambda\}$ be the circuits class of size at most λ. Then $i\mathcal{O}$ is an indistinguishability obfuscator for the class $\{\mathcal{C}_\lambda\}$.

3   Definitions of Ring Signatures

In this section, we review the formal definition and security models of the ring signature schemes.

3.1   Syntax

We define a Setup algorithm, run by a trusted authority (TA), to generate public parameters and create the system of the ring signature scheme. Additionally, we refer to an ordered set $R = (VK_1, \ldots, VK_n)$ of verification keys as a ring, and let $R[i] = VK_i$. We will also freely use set notation, e.g., $VK \in R$ if there exists an index i such that $R[i] = VK$.

Definition 5. A ring signature scheme consists of the following four PPT algorithms:

- Setup algorithm takes as input a security parameter $\lambda \in \mathbb{N}$ and outputs the public parameters $\mathbb{PP}$. The public parameters contain the descriptions of the message space $\mathbb{M}$ and signature space $\mathbb{S}$. We write it $\mathbb{PP} \leftarrow \mathsf{Setup}(1^\lambda)$.[1]
- KeyGen algorithm outputs the users' signing and verification keys (SK, VK). We write it $(SK, VK) \leftarrow \mathsf{KeyGen}()$.
- Sign algorithm takes as inputs a message $M \in \mathbb{M}$ to be signed, a set of verification keys R (i.e., the ring), and an user's signing key SK_s. It is required that $VK_s \in R$ holds. The algorithm outputs a signature σ on M for the ring R. We write it $\sigma \leftarrow \mathsf{Sign}(M, SK_s, R)$.
- Vrfy algorithm takes as inputs the ring R of verification keys and a purported signature $\sigma \in \mathbb{S}$ on a message $M \in \mathbb{M}$. It outputs 1 (accept) if σ is valid. Otherwise, it outputs 0 (reject). We write it $1/0 \leftarrow \mathsf{Vrfy}(M, \sigma, R)$.

Correctness. For all security parameters $\lambda \in \mathbb{N}$, messages $M \in \mathbb{M}$, and all $\mathbb{PP} \leftarrow \mathsf{Setup}(1^\lambda), \{(SK_i, VK_i)\}_{i=1}^{n(\lambda)} \leftarrow \mathsf{KeyGen}(), \sigma \leftarrow \mathsf{Sign}(M, SK_s, R)$, if $R \subseteq \{VK\}_{i=1}^{n(\lambda)}$ and $VK_s \in R$, then we have $\Pr[\mathsf{Vrfy}(M, \sigma, R) = 1] = 1$.

3.2 Security Models

Informally, a ring signature scheme should satisfy two security properties. The first one is *unforgeability*, meaning that an adversary is able to compute a valid signature on behalf of a ring if and only if he knows a secret key corresponding to one of them. The second one is *anonymity*, meaning that a verifier is convinced that someone in the ring is responsible for the signature, but cannot tell who it is. In this paper, we follow the following two notions: unforgeability with respect to insider corruption introduced by Bender, Katz, and Morselli [6];[2] and perfect anonymity introduced by Chandran, Groth, and Sahai [11].

Ring Unforgeability. The notion of unforgeability with respect to insider corruption [6] is defined by the following game which is played by a challenger and a PPT adversary $\mathcal{A}$.

1. **Setup:** The challenger runs the Setup and KeyGen algorithms to generate the public parameters $\mathbb{PP}$ and all keys $\{(SK_i, VK_i)\}_{i=1}^{n(\lambda)}$. The adversary $\mathcal{A}$ is given $\mathbb{PP}$ and verification keys $S = \{VK_i\}_{i=1}^{n(\lambda)}$. The challenger maintains a set C to record the corrupted users. Initially, $C \leftarrow \emptyset$.

---

[1] For ease of notation on the reader, we suppress repeated $\mathbb{PP}$ arguments that are provided to all of the following algorithms. For example, we will write $(SK, VK) \leftarrow \mathsf{KeyGen}()$ instead of $(SK, VK) \leftarrow \mathsf{KeyGen}(\mathbb{PP})$.

[2] We consider a weaker version of this notion in which corruptions of honest users are allowed but adversary-chosen public keys are not allowed. This weaker notion has been used in [20,29].

2. **Signing queries:** The adversary $\mathcal{A}$ is allowed to adaptively make singing queries. A ring signing query is of the form (M, R, s). Here $M \in \mathbb{M}$ is the message to be signed, $R \subseteq S$ is a ring of verification keys, and s is an index such that $VK_s \in R$. The challenger responds with a ring signature $\sigma \leftarrow$ Sign(M, SK_s, R).
3. **Corruption queries:** The adversary $\mathcal{A}$ is allowed to adaptively make corruption queries. A corruption query is of the form $s \in [n(\lambda)]$. The challenger returns SK_s to $\mathcal{A}$ and adds VK_s into the set C.
4. **Output:** Finally, the adversary $\mathcal{A}$ outputs a tuple $(M^* \in \mathbb{M}, \sigma^* \in \mathbb{S}, R^* \in S)$. We say that $\mathcal{A}$ wins the game if (1) Vrfy$(M^*, \sigma^*, R^*) = 1$; (2) $R^* \subseteq S\backslash C$; and (3) it never made a singing query (M^*, R^*, s) for any s.

We denote the success probability of a PPT adversary $\mathcal{A}$ (taken over the random coins of the challenger and adversary) to win the above game as

$$\mathbf{Adv}_{\mathcal{A}}^{Unf} = \Pr[\mathcal{A} \text{ wins}].$$

Definition 6. We say that a ring signature scheme is unforgeable with respect to insider corruption, if for any PPT adversary $\mathcal{A}$, it cannot win the above game with non-negligible advantage.

Selective security. We consider a selective variant to the above definition where the adversary $\mathcal{A}$ is required to commit to a forgery ring/message pair (R^*, M^*)[3] before the setup phase. Then it cannot make signing query on inputs (M^*, R^*, s) for any s, and also cannot make corruption query on input s for which $VK_s \in R^*$.

Ring Anonymity. Informally, a ring signature scheme is perfectly anonymous [11], if a signature on a message $M^* \in \mathbb{M}$, a ring $R^* \subseteq S$, and a key $VK_{i_0} \in R^*$ looks exactly the same as a signature on the same message M^* and ring R^*, and a different key $VK_{i_1} \in R^*$. This means that the signer's key is hidden among all the honestly generated keys in the ring. Formally, we define the following game which is played by a challenger and an unbounded adversary $\mathcal{A}$.

1. **Setup:** The challenger runs the Setup and KeyGen algorithms to generate the public parameters $\mathbb{PP}$ and key pairs $\{(SK_i, VK_i)\}_{i=1}^{n(\lambda)}$. Then it gives the public parameters and all keys $\{(SK_i, VK_i)\}_{i=1}^{n(\lambda)}$ to the adversary $\mathcal{A}$.
2. **Challenge:** The adversary $\mathcal{A}$ submits a tuple (M^*, R^*, i_0, i_1), where $M^* \in \mathbb{M}$ is the challenge message, $R^* \subseteq \{VK_i\}_{i=1}^{n(\lambda)}$ is the challenge ring, i_0 and i_1 are two indices such that $\{VK_{i_0}, VK_{i_1}\} \subseteq R^*$, to the challenger. The challenger chooses a random $b \in \{0, 1\}$, then computes $\sigma^* \leftarrow$ Sign(M^*, SK_{i_b}, R^*) and returns σ^* to the adversary.

---

[3] In the beginning, $\mathcal{A}$ does not given the keys $S = \{VK_i\}_{i=1}^{n(\lambda)}$. In order to obtain the forgery ring R^*, we require that $\mathcal{A}$ submits a set of index $I_{R^*} = \{i_1, \dots, i_{|R^*|}\} \subseteq [n(\lambda)]$. Then after the keys $S = \{VK_i\}_{i=1}^{n(\lambda)}$ are generated, the forgery ring $R^* = \{VK_{i_1}, \dots, VK_{i_{|R^*|}}\} \subseteq S$ is also obtained.

3. **Guess:** Finally, the adversary outputs b', indicating his guess for b.

We denote the advantage of an unbounded adversary $\mathcal{A}$ (taken over the random coins of the challenger and the adversary) to win the above game as

$$\mathbf{Adv}_{\mathcal{A}}^{Ano} = |\Pr[b' = b] - \Pr[b' \neq b]|.$$

Definition 7. We say that a ring signature scheme is perfectly anonymous, if even an unbounded adversary has no non-negligible advantage to win the above game of anonymity.

4 Constant Size Ring Signature Scheme

4.1 Our Construction

The idea of our construction is the following: Each user chooses a random string x_i as his secret key. His public key is a PRG value $VK_i = \mathsf{G}(x_i)$. To allow a legal user VK_s to get a signature on behalf of a set of users $R = (VK_1, \ldots, VK_n)$, the trusty authority (TA) publishes an obfuscated signing program for constrained PRF which requires knowledge of a secret key to operate. In this way, if $VK_s \in R$, then he can get a signature that the PRF acts on a concatenation of the message and R, but anyone else will not know any of a secret key x_i such that $\exists i \subset [n]$ s.t. $VK_i = \mathsf{G}(x_i)$, will be returned back $\bot$.[4]

Let $\mathsf{Eva} : \{0,1\}^{\ell(\lambda)+2 \cdot n \cdot \lambda} \to \{0,1\}^{\lambda}$ be a constrained PRF. Let $\mathsf{G} : \{0,1\}^{\lambda} \to \{0,1\}^{2\lambda}$ be a PRG. Let $f(\cdot)$ be a OWF. The message space is $\mathbb{M} = \{0,1\}^{\ell(\lambda)}$. The signature space is $\mathbb{S} - \{0,1\}^{\lambda}$. Our ring signature scheme is as follows:

- Setup(1^{λ}): TA first chooses a random key K for the constrained PRF. Next, it creates two obfuscated programs of the signing circuit $\mathcal{P}_S$ and verification circuit $\mathcal{P}_V$, respectively. Here, we assume that these two circuits contain authentication block to check the validity of the public keys. The public parameters are $\mathbb{PP} = (i\mathcal{O}(\mathcal{P}_S), i\mathcal{O}(\mathcal{P}_V))$.
- KeyGen(): Each user chooses a random bit string $x_i \in \{0,1\}^{\lambda}$ as his secret key SK_i. Correspondingly, his verification key is $VK_i = \mathsf{G}(x_i)$.
- Sign($SK_s, M, R = (VK_1, \ldots, VK_n), s$): The holder (i.e., the real signer) of secret key SK_s with $s \in [n]$ runs the signing program $i\mathcal{O}(\mathcal{P}_S)$ on inputs (M, SK_s, R, s).
- Vrfy($M, \sigma, R = (VK_1, \ldots, VK_n)$): The verifier runs the verification program $i\mathcal{O}(\mathcal{P}_V)$ on inputs (M, σ, R).

$\mathcal{P}_S$

- **Constants:** PRF key K.
- **Input:** Message M, secret key SK_s, set $R = (VK_1, \ldots, VK_n)$, index s.
  1. Test if $\mathsf{G}(x_s) = VK_s$. Output $\mathsf{Eva}_K(M\|R)$ if true, $\bot$ if false.

---

[4] This idea is from [9] where Boneh and Zhandry constructed a non-interactive key exchange protocol.

$\mathcal{P}_V$

- **Constants:** PRF key K.
- **Inputs:** Message M, signature σ, set $R = (VK_1, \ldots, VK_n)$.
  1. Test if $f(\sigma) = f(\mathsf{Eva}_K(M\|R))$. Output 1 if true, 0 if false.

Remark 1. Correctness of the ring signature scheme is trivial by inspection. In our construction, the signer (resp. verifier) should run the signing (resp. verification) program, and hence the signer (resp. verifier) should be online or download the signing (resp. verification) program before sign messages (resp. verify signatures). In this paper, we focus on the existence of the constant size ring signature scheme without random oracles rather than its practicability.

Remark 2. The above construction restricts that the ring can contain only n verification keys, that is, it is an n-user ring signature scheme [6]. We can easily extend it to that can support flexible ring. The launching point for this issue is that by using Ramchen and Waters' [27] technique. In [27], Ramchen and Waters constructed a new constrained PRF by adapting the GGM construction. This new constrained PRF accepts variable-length inputs. (Please refer to [27] for details.) Therefore, the above n-user ring signature scheme equipped with this new PRF can form a fully-fledged ring signature scheme.

4.2   Security

Selective Unforgeability. We show that if there exists a PPT adversary $\mathcal{A}$ that can break the selective unforgeability of the above ring signature scheme, then we can construct a challenger $\mathcal{B}$ to break the security of the OWF. We describe the proof as a sequence of the following hybrid games:

- Hyb_0 : This hybrid corresponds to the honest execution of the selective unforgeability game where the adversary initially submits his target (challenge) value $T' = (M^* \in \{0,1\}^{\ell(\lambda)}, I_{R^*} \subseteq [n(\lambda)])$, then it interacts with $\mathcal{B}$:
  - **Setup:** $\mathcal{B}$ chooses a constrained PRF key K, a PRG G, and a OWF f. Then it creates the public parameters $\mathbb{PP}$. In addition, $\mathcal{B}$ chooses random $\{x_i\}_{i=1}^{n(\lambda)} \subseteq \{0,1\}^{\lambda}$ as the secret keys and computes $S = \{VK_i = \mathsf{G}(x_i)\}_{i=1}^{n(\lambda)}$ as the public keys. The adversary's challenge value T' is replaced by $T = (M^*, R^*)$. $\mathcal{B}$ maintains a set C to record the corrupted users. Initially, $C \leftarrow \emptyset$. The adversary is given $\mathbb{PP}$ and S.
  - **Queries:** There are two kinds of queries that $\mathcal{B}$ must answer: signing oracle $\mathcal{O}_{Sig}$ and corruption oracle $\mathcal{O}_{Cor}$.
    * For each query to $\mathcal{O}_{Sig}$ on inputs $(M \in \mathbb{M}, R \subseteq S, s)$, where $(M, R) \neq T$, $\mathcal{B}$ responds with a ring signature $\sigma = \mathsf{Eva}_K(M\|R)$.
    * For each query to $\mathcal{O}_{Cor}$ on input $s \in [n(\lambda)]$, where $VK_s \notin R^*$, $\mathcal{B}$ returns x_s to $\mathcal{A}$ and adds VK_s into the set C.

- **Output:** At the end the adversary outputs a forgery σ^* with respect to M^* and R^*, it succeeds if $\mathsf{Vrfy}(M^*, \sigma^*, R^*) = 1$.
- Hyb_1 : This hybrid is identical to Hyb_0 with the exception that for the challenge ring $R^* = (VK_1^*, \ldots, VK_n^*)$, $\mathcal{B}$ randomly chooses $VK_i^* \in \{0,1\}^{2\lambda}$, for $i \in [n]$, as their verification keys.
- Hyb_2 : This hybrid is identical to Hyb_1 with the exception that we let $w^* = \mathsf{Eva}_K(T)$ and replace the signing program $i\mathcal{O}(\mathcal{P}_S)$ by an obfuscation of the following circuit $\mathcal{P}_S^*$. The sizes of the circuits $\mathcal{P}_S$ and $\mathcal{P}_S^*$ are identical by appropriate padding.

$$\mathcal{P}_S^*$$

- **Constants:** PRF key $K(T)$, set $T = (M^*, VK_1^*, \ldots, VK_n^*)$, w^*.
- **Input:** Message M, secret key SK_s, set $R = (VK_1, \ldots, VK_n)$, s.
  1. Test if $\mathsf{G}(x_s) = VK_s$. Output $\perp$ if false.
  2. If $(M, R) = T$, output w^*. Otherwise, output $\mathsf{Eva}_K(m||R)$.

- Hyb_3 : This hybrid is identical to Hyb_2 with the exception that we let $z^* = f(\mathsf{Eva}_K(T))$ and replace the verification program $i\mathcal{O}(\mathcal{P}_V)$ by an obfuscation of the following circuit $\mathcal{P}_V^*$. The sizes of the circuits $\mathcal{P}_V$ and $\mathcal{P}_V^*$ are identical by appropriate padding.

$$\mathcal{P}_V^*$$

- **Constants:** PRF key $K(T)$, set $T = (m^*, VK_1^*, \ldots, VK_n^*)$, z^*.
- **Input:** Message M, signature σ, set $R = (VK_1, \ldots, VK_n)$.
  1. If $(M, R) = T$, test if $f(\sigma) = z^*$. Output 1 if true, 0 if false.
  2. Else, test if $f(\sigma) = f(\mathsf{Eva}_K(m||R))$. Output 1 if true, 0 if false.

- Hyb_4 : This hybrid is identical to Hyb_3 with the exception that $z^* = f(t)$ for t chosen uniformly at random from $\{0,1\}^\lambda$ in the verification circuit $\mathcal{P}_V^*$.

First, we argue that the advantage of any PPT adversary in forging a signature must be negligibly close in hybrids Hyb_0 and Hyb_1 since the security of the PRG. We may note that with overwhelming probability, none of values VK_i^*, for $i \in [n]$, have a preimage under PRG. Therefore, the adversary (with overwhelming probability) cannot run the signing program $i\mathcal{O}(\mathcal{P}_S^*)$ on input the target ring R^* (since there is no valid x_i^* such that $\mathsf{G}(x_i^*) = VK_i^*$) even if it obtains all the other secret keys.

Then, we argue that the advantage of any PPT adversary in forging a signature must be negligibly close in hybrids Hyb_1 and Hyb_2. We first observe that the input/output behaviors of the circuits $\mathcal{P}_S$ and $\mathcal{P}_S^*$ are identical. The only difference is that the circuit $\mathcal{P}_S$ computes $\mathsf{Eva}_K(T)$ by itself, whereas the circuit $\mathcal{P}_S^*$ is given $\mathsf{Eva}_K(T)$ as the constant w^*. Therefore, if there is a difference in advantage we can create an attacker $(Samp, D)$ to break the indistinguishability security of the $i\mathcal{O}$. $Samp$ submits two circuits $C_0 = \mathcal{P}_S$ and $C_1 = \mathcal{P}_S^*$ to the $i\mathcal{O}$ challenger. Then $Samp$ will receive an obfuscated program of C_0 or C_1, it builds the public parameters which contain the obfuscated program. If the $i\mathcal{O}$ challenger chooses C_0, then we are in Hyb_1. If $i\mathcal{O}$ challenger chooses C_1, then we are in Hyb_2. Finally, D outputs 1 if the adversary successfully forges. In conclusion, any adversary with different advantages in Hyb_1 and Hyb_2 will lead to $(Samp, D)$ as an attacker on the indistinguishability security of the $i\mathcal{O}$. Similarly, the advantage of any PPT adversary in forging a signature must be negligibly close in hybrids Hyb_2 and Hyb_3.

Next, we argue that the advantage of any PPT adversary in forging a signature must be negligibly close in hybrids Hyb_3 and Hyb_4. Otherwise, we can build an attacker $(\mathcal{A}_1, \mathcal{A}_2)$ to break the security of the constrained PRF at the punctured point T. $\mathcal{A}_1$ first obtains $T = (M^*, R^*)$ from the adversary. It submits this set to the constrained PRF challenger and receives a constrained PRF key $K(T)$ and challenge value z^*. If $z^* = f(\mathsf{Eva}_K(T))$, then we are in Hyb_3. If it was chosen uniformly at random, then we are in Hyb_4. $\mathcal{A}_2$ will output 1 if the adversary successfully forges. In conclusion, any PPT adversary with different advantages in hybrids Hyb_3 and Hyb_4 will leads to $(\mathcal{A}_1, \mathcal{A}_2)$ as an attacker to break the pseudorandomness of the constrained PRF.

Finally, we show that in the last hybrid Hyb_4, any PPT adversary cannot win the selective game with non-negligible advantage. If there is an adversary in Hyb_4, we can use it to break the security of the OWF. We build a reduction $\mathcal{B}$ that first receives set $T = (M^*, R^*)$ from the adversary and challenge instance y from the OWF challenger, it then sets $z^* = y$. If an adversary successfully forges on M^* and R^*, then by assumption he has computed a signature σ^* such that $f(\sigma^*) = z^* = y$. $\mathcal{B}$ outputs σ^* as the solution of the given OWF challenge instance. Therefore, if the OWF is secure, no PPT adversary can forge in Hyb_4 with non-negligible advantage. Since the advantage of any PPT adversary are negligibly close in each successive hybrid, this proves the selective unforgeability for the ring signature scheme.   □

Corollary 1. *If the OWF $f(\cdot)$ is injective, then the above ring signature scheme is strongly unforgeable.*

Perfect Anonymity. For any (M^*, R^*, i_0, i_1), where $M^* \in \mathbb{M}$, $R^* \subseteq \{VK_i\}_{i=1}^{n(\lambda)}$, and $i_0, i_1 \in [n]$, which are chosen by an unbounded adversary $\mathcal{A}$, both of the signatures created by the member i_0 and i_1 are $\sigma_{i_0}^* = \mathsf{Eva}_K(M^*||R^*) = \sigma_{i_1}^*$. Therefore, any member of a ring can get a same signature on a given message and a ring. The perfect anonymity follows easily from this observation.   □

5   Identity-Based Ring Signature Scheme

In an identity-based system [28], the public key of each user is his recognizable identity, e.g. email address, phone number and so on. This property avoids the necessity of certificates. In the mean time, users' secret keys are computed by a key generator center (KGC) who is in possession of a master secret key MSK. Please refer to [13] for definition and security models of the identity-based ring signature schemes.

5.1   Our Construction

It is straightforward to turn our ring signature scheme into an identity-based ring signature scheme.[5]

Let $\mathsf{Eva}_K : \{0,1\}^{\ell(\lambda)+2 \cdot n \cdot \lambda} \to \{0,1\}^\lambda$ and $\mathsf{Eva}_{K'} : \{0,1\}^{2\lambda} \to \{0,1\}^\lambda$ be two constrained PRFs. Let $G : \{0,1\}^\lambda \to \{0,1\}^{2\lambda}$ be a PRG. Let $f(\cdot)$ be a OWF. The identity space is $\mathbb{I} = \{0,1\}^{2\lambda}$. The message space is $\mathbb{M} = \{0,1\}^{\ell(\lambda)}$. The signature space is $\mathbb{S} = \{0,1\}^\lambda$. The construction of the identity-based ring signature scheme is as follows:

- Setup(1^λ): TA first chooses two random keys K and K' for the above constrained PRFs. Next, it creates two obfuscated programs of the signing circuit $\mathcal{P}_{IS}$ and verification circuit $\mathcal{P}_{IV}$, respectively. The public parameters are $\mathbb{PP} = (i\mathcal{O}(\mathcal{P}_{IS}), i\mathcal{O}(\mathcal{P}_{IV}))$, and the master secret key is $MSK = K'$ which will be given to the KGC.
- KeyGen(MSK, id): The KGC computes $SK_{\mathsf{id}} = \mathsf{Eva}_{K'}(\mathsf{id})$ as id's secret key.
- Sign($M, SK_{\mathsf{id}_s}, R = (\mathsf{id}_1, \ldots, \mathsf{id}_n), s$): The holder (i.e., the real signer) of signing key SK_{id_s} with $s \in [n]$ runs the signing program $i\mathcal{O}(\mathcal{P}_{IS})$ on inputs $(M, SK_{\mathsf{id}_s}, R, s)$.
- Vrty($M, \sigma, R = (\mathsf{id}_1, \ldots, \mathsf{id}_n)$): The verifier runs the verifying program $i\mathcal{O}(\mathcal{P}_{IV})$ on inputs (M, σ, R).

$$\mathcal{P}_{IS}$$

- **Constants:** PRF keys K and K'.
- **Inputs:** Message M, key SK_{id_s}, set $R = (\mathsf{id}_1, \ldots, \mathsf{id}_n)$, index s.
  1. Test if $G(SK_{\mathsf{id}_s}) = G(\mathsf{Eva}_{K'}(\mathsf{id}_s))$. Output $\mathsf{Eva}_K(M\|R)$ if true, $\perp$ if false.

$$\mathcal{P}_{IV}$$

- **Constants:** PRF key K.
- **Inputs:** Message M, signature σ, set $R = (\mathsf{id}_1, \ldots, \mathsf{id}_n)$.
  1. Test if $f(\sigma) = f(\mathsf{Eva}_K(M\|R))$. Output 1 if true, 0 if false.

---

[5] The idea of our identity-based ring signature scheme is from Boneh and Zhandry's [9] identity-based non-interactive key exchange scheme.

5.2 Security

Selective Unforgeability. We show that if there exists a PPT adversary $\mathcal{A}$ that can break the selective unforgeability of the above identity-based ring signature scheme, then we can construct a challenger $\mathcal{B}$ to break the security of the OWF. We describe the proof as a sequence of the following hybrid games:

– Hyb_0 : This hybrid corresponds to the honest execution of the selective unforgeability game where the adversary initially submits his target (challenge) set $T = (M^* \in \{0,1\}^{\ell(\lambda)}, R^* = (\mathsf{id}_1^*, \ldots, \mathsf{id}_n^*))$, then it interacts with $\mathcal{B}$:
  - **Setup:** $\mathcal{B}$ randomly chooses two constrained PRF keys K and K', a PRG G, and a OWF f. Then it creates the public parameters $\mathbb{PP}$. The adversary is given the public parameters $\mathbb{PP}$.
  - **Queries:** There are two kinds of queries that $\mathcal{B}$ must answer: signing oracle $\mathcal{O}_{Sig}$ and extract oracle $\mathcal{O}_{Ext}$.
    * For each query to $\mathcal{O}_{Sig}$ on inputs $(M \in \mathbb{M}, R, s)$, where $(M, R) \neq T$, $\mathcal{B}$ responds with a ring signature $\sigma = \mathsf{Eva}_K(M\|R)$.
    * For each query to $\mathcal{O}_{Ext}$ on input $\mathsf{id} \in \mathbb{I}$, where $\mathsf{id} \notin R^*$, $\mathcal{B}$ returns $SK_{\mathsf{id}} = \mathsf{Eva}_{K'}(\mathsf{id})$ to $\mathcal{A}$.
  - **Output:** At the end the adversary outputs a forgery σ^* with respect to M^* and R^*, it succeeds if $\mathsf{Vrfy}(M^*, \sigma^*, R^*) = 1$.
– Hyb_1 : This hybrid is identical to Hyb_0 with the exception that we let $w_i^* = \mathsf{G}(\mathsf{Eva}_{K'}(\mathsf{id}_i^*))$, for $\mathsf{id}_i^* \in R^*$, and replace the signing program $i\mathcal{O}(\mathcal{P}_{IS})$ by an obfuscation of the following circuit $\mathcal{P}_{IS}^*$. The sizes of the circuits $\mathcal{P}_{IS}$ and $\mathcal{P}_{IS}^*$ are identical by appropriate padding.

$$\mathcal{P}_{IS}^*$$

- **Constants:** PRF keys K and $K'(R^*)$, set R^*, values w_i^*.
- **Inputs:** Message M, key SK_{id_s}, set $R = (\mathsf{id}_1, \ldots, \mathsf{id}_n)$, index s.
  1. If $\mathsf{id}_s = \mathsf{id}_i^* \in R^*$, test if $\mathsf{G}(SK_{\mathsf{id}_s}) = w_i^*$.
  2. Otherwise, test if $\mathsf{G}(SK_{\mathsf{id}_s}) = \mathsf{G}(\mathsf{Eva}_{K'(R^*)}(\mathsf{id}_s))$.
  3. Output $\mathsf{Eva}_K(M\|R)$ if true, $\perp$ if false.

– Hyb_2 : This hybrid is identical to Hyb_1 with the exception that the values $w_i^* = t_i$ which are chosen uniformly at random from $\{0,1\}^{2\lambda}$.
– Hyb_3 : This hybrid is identical to Hyb_2 with the exception that we let $w^* = \mathsf{Eva}_K(T)$ and replace the signing program $i\mathcal{O}(\mathcal{P}_{IS}^*)$ by an obfuscation of the following circuit $\mathcal{P}_{IS}^{**}$. The sizes of the circuits $\mathcal{P}_{IS}^*$ and $\mathcal{P}_{IS}^{**}$ are identical by appropriate padding.

$$\mathcal{P}_{IS}^{**}$$

- **Constants:** PRF keys $K(T)$ and $K'(R^*)$, sets T, values w^* and w_i^*.
- **Inputs:** Message M, key SK_{id_s}, set $R = (id_1, \ldots, id_n)$, index s.
  1. If $id_s = id_i^* \in R^*$, test if $\mathsf{G}(SK_{id_s}) = w_i^*$.
  2. Otherwise, test if $\mathsf{G}(SK_{id_s}) = \mathsf{G}(\mathsf{Eva}_{K'(R^*)}(id_s))$.
  3. If $(M, R) = T$ then output w^*; else output $\mathsf{Eva}_{K(T)}(M\|R)$ if true, $\bot$ if false.

– Hyb_4 : This hybrid is identical to Hyb_3 with the exception that we let $z^* = f(\mathsf{Eva}_K(T))$ and replace the verification program $i\mathcal{O}(\mathcal{P}_{IV})$ by an obfuscation of the following circuit $\mathcal{P}_{IV}^*$. The sizes of the circuits $\mathcal{P}_{IV}$ and $\mathcal{P}_{IV}^*$ are identical by appropriate padding.

$$\mathcal{P}_{IV}^*$$

- **Constants:** PRF key $K(T)$, set T, value z^*.
- **Inputs:** Message M, signature σ, set $R = (id_1, \ldots, id_n)$.
  1. If $(M, R) = T$, test if $f(\sigma) = z^*$. Output 1 if true, 0 if false.
  2. Else, test if $f(\sigma) = f(\mathsf{Eva}_K(m\|R))$. Output 1 if true, 0 if false.

– Hyb_5 : This hybrid is identical to Hyb_4 with the exception that $z^* = f(t)$ for t chosen uniformly at random from $\{0,1\}^\lambda$ in the verification circuit $\mathcal{P}_{IV}^*$.

First, we argue that the advantage of any PPT adversary in forging a signature must be negligibly close in hybrids Hyb_0 and Hyb_1. We first observe that the input/output behaviors of the circuits $\mathcal{P}_{IS}$ and $\mathcal{P}_{IS}^*$ are identical. The only difference is that the circuit $\mathcal{P}_{IS}$ computes $\mathsf{Eva}_{K'}(id_i^*)$, for $i \in [n]$, by itself, whereas the circuit $\mathcal{P}_{IS}^*$ is given $\mathsf{Eva}_{K'}(id_i^*)$ as some constants w_i^*. Therefore, if there is a difference in advantage we can create an attacker $(Samp, D)$ to break the indistinguishability security of the $i\mathcal{O}$. $Samp$ submits two circuits $C_0 = \mathcal{P}_{IS}$ and $C_1 = \mathcal{P}_{IS}^*$ to the $i\mathcal{O}$ challenger. Then $Samp$ will receive an obfuscated program of C_0 or C_1, it builds the public parameters which contains the obfuscated program. If the $i\mathcal{O}$ challenger chooses C_0, then we are in Hyb_0. If $i\mathcal{O}$ challenger chooses C_1, then we are in Hyb_1. Finally, D outputs 1 if the adversary successfully forges. In conclusion, any adversary with different advantages in Hyb_0 and Hyb_1 will lead to $(Samp, D)$ as an attacker on the indistinguishability security of the $i\mathcal{O}$. Similarly, the advantage of any PPT adversary in forging a signature must be negligibly close in hybrids Hyb_2 and Hyb_3; and the advantage of any PPT adversary in forging a signature must be negligibly close in hybrids Hyb_3 and Hyb_4.

Then, we argue that the advantage of any PPT adversary in forging a signature must be negligibly close in hybrids Hyb_1 and Hyb_2. Otherwise, we can build an attacker $(\mathcal{A}_1, \mathcal{A}_2)$ to break the security of the constrained PRF at the punctured set R^*. $\mathcal{A}_1$ first obtains $T = (M^*, R^*)$ from the adversary. It submits the set R^* to the constrained PRF challenger and receives a constrained PRF key $K'(R^*)$ and challenge value w_i^* for $i \in [n]$. If $w_i^* = \mathsf{Eva}_{K'}(\mathsf{id}_i^*)$ for all $i \in [n]$, then we are in Hyb_1. If they were chosen uniformly at random, then we are in Hyb_2. $\mathcal{A}_2$ will output 1 if the adversary successfully forges. In conclusion, any PPT adversary with different advantages in hybrids Hyb_1 and Hyb_2 will leads to $(\mathcal{A}_1, \mathcal{A}_2)$ as an attacker to break the pseudorandomness of the constrained PRF. We may note that, in Hyb_2, with overwhelming probability, none of values id_i^*, for $i \in [n]$, have a preimage under PRG. Therefore, the adversary (with overwhelming probability) cannot run the signing program $i\mathcal{O}(\mathcal{P}_{IS}^*)$ on input the target ring R^* (since there is no valid $SK_{\mathsf{id}_i^*}$ such that $\mathsf{G}(SK_{\mathsf{id}_i^*}) = w_i^*$ even if it obtains all the other secret keys from the extract oracle. Similarly, the advantage of any PPT adversary in forging a signature must be negligibly close in hybrids Hyb_4 and Hyb_5.

Finally, we show that in the last hybrid Hyb_5, any PPT adversary cannot win the selective game with non-negligible advantage. If there is an adversary in Hyb_5, we can use it to break the security of the OWF. We build a reduction $\mathcal{B}$ that first receives set $T = (M^*, R^*)$ from the adversary and challenge instance y from the OWF challenger, it then sets $z^* = y$. If an adversary successfully forges on M^* and R^*, then by assumption he has computed a signature σ^* such that $f(\sigma^*) = z^* = y$. $\mathcal{B}$ outputs σ^* as the solution of the given OWF challenge instance. Therefore, if the OWF is secure, no PPT adversary can forge in Hyb_5 with non-negligible advantage. Since the advantage of any PPT adversary are negligibly close in each successive hybrid, this proves the selective unforgeability for the ring signature scheme. □

Perfect Anonymity. For any challenge tuple (M^*, R^*, i_0, i_1), where $M^* \in \mathbb{M}$, $R^* = (\mathsf{id}_1^*, \ldots, \mathsf{id}_n^*)$, and $i_0, i_1 \in [n]$, which is chosen by an (unbounded) adversary $\mathcal{A}$, both of the signatures created by the member i_0 and i_1 are $\sigma_{i_0}^* = \mathsf{Eva}_K(M^* \| R^*) = \sigma_{i_1}^*$. Therefore, any member of a ring can get a same signature on a given message and ring. The perfect anonymity follows easily from this observation. □

Acknowledgement. The authors would like to thank anonymous reviewers for their helpful comments and suggestions.

References

1. Au, M.H., Liu, J.K., Susilo, W., Zhou, J.: Realizing fully secure unrestricted ID-based ring signature in the standard model from HIBE. IEEE Trans. Inf. Forensics Secur. **8**(12), 1909–1922 (2013)
2. Abe, M., Ohkubo, M., Suzuki, K.: 1-out-of-n signatures from a variety of keys. In: Zheng, Y. (ed.) ASIACRYPT 2002. LNCS, vol. 2501, pp. 415–432. Springer, Heidelberg (2002)
3. Boyle, E., Goldwasser, S., Ivan, I.: Functional signatures and pseudorandom functions. Cryptology ePrint Archive, Report 2013/631 (2013)

4. Barak, B., Goldreich, O., Impagliazzo, R., Rudich, S., Sahai, A., Vadhan, S.P., Yang, K.: On the (Im)possibility of obfuscating programs. In: Kilian, J. (ed.) CRYPTO 2001. LNCS, vol. 2139, pp. 1–18. Springer, Heidelberg (2001)
5. Barak, B., Garg, S., Kalai, Y.T., Paneth, O., Sahai, A.: Protecting obfuscation against algebraic attacks. Cryptology ePrint Archive, Report 2013/631 (2013)
6. Bender, A., Katz, J., Morselli, R.: Ring signatures: stronger definitions, and constructions without random oracles. In: Halevi, S., Rabin, T. (eds.) TCC 2006. LNCS, vol. 3876, pp. 60–79. Springer, Heidelberg (2006)
7. Brakerski, Z., Rothblum, G.N.: Virtual black-box bofuscation for all circuits via generic graded encoding. Cryptology ePrint Archive, Report 2013/563 (2013)
8. Boneh, D., Waters, B.: Constrained pseudorandom functions and their applications. In: Sako, K., Sarkar, P. (eds.) ASIACRYPT 2013, Part II. LNCS, vol. 8270, pp. 280–300. Springer, Heidelberg (2013)
9. Boneh, D., Zhandry, M.: Multiparty key exchange, efficient traitor tracing, and more from indistinguishability obfuscation. Cryptology ePrint Archive, Report 2013/642 (2013). http://eprint.iacr.org
10. Canetti, R., Goldreich, O., Halevi, S.: The random oracle methodology, revisited. J. ACM (JACM) **51**(4), 557–594 (2004)
11. Chandran, N., Groth, J., Sahai, A.: Ring signatures of sub-linear size without random oracles. In: Arge, L., Cachin, C., Jurdziński, T., Tarlecki, A. (eds.) ICALP 2007. LNCS, vol. 4596, pp. 423–434. Springer, Heidelberg (2007)
12. Chaum, D., van Heyst, E.: Group signatures. In: Davies, D.W. (ed.) EUROCRYPT 1991. LNCS, vol. 547, pp. 257–265. Springer, Heidelberg (1991)
13. Chow, S.S.M., Yiu, S.-M., Hui, L.C.K.: Efficient identity based ring signature. In: Ioannidis, J., Keromytis, A.D., Yung, M. (eds.) ACNS 2005. LNCS, vol. 3531, pp. 499–512. Springer, Heidelberg (2005)
14. Chow, S.S.M., Wei, V.K., Liu, J.K., Yuen, T.H.: Ring signatures without random oracles. Proceedings of the 2006 ACM Symposium on Information, Computer and Communications Security, pp. 297–302. ACM (2006)
15. Dodis, Y., Kiayias, A., Nicolosi, A., Shoup, V.: Anonymous identification in *Ad Hoc* groups. In: Cachin, C., Camenisch, J.L. (eds.) EUROCRYPT 2004. LNCS, vol. 3027, pp. 609–626. Springer, Heidelberg (2004)
16. Dodis, Y., Oliveira, R., Pietrzak, K.: On the generic insecurity of the full domain hash. In: Shoup, V. (ed.) CRYPTO 2005. LNCS, vol. 3621, pp. 449–466. Springer, Heidelberg (2005)
17. Garg, S., Gentry, C., Halevi, S., Raykova, M., Sahai, A., Waters, B.: Candidate indistinguishability obfuscation and functional encryption for all circuits. In: FOCS 2013, pp. 40–49. IEEE (2013)
18. Goldwasser, S., Goyal, V., Jain, A., Sahai, A.: Multi-input functional encryption. Cryptology ePrint Archive, Report 2013/727 (2013). http://eprint.iacr.org
19. Goldreich, O., Goldwasser, S., Micali, S.: How to construct random functions. J. ACM (JACM) **33**(4), 792–807 (1986)
20. Herranz, J.: Some digital signature schemes with collective signers. Ph.D. thesis, Universitat Politècnica de Catalunya, Barcelona, April 2005. http://www.lix.polytechnique.fr/herranz/thesis.htm
21. Håstad, J., Impagliazzo, R., Levin, L.A., Luby, M.: A pseudorandom generator from any one-way function. SIAM J. Comput. **28**(4), 1364–1396 (1999)
22. Herranz, J., Sáez, G.: New identity-based ring signature schemes. In: López, J., Qing, S., Okamoto, E. (eds.) ICICS 2004. LNCS, vol. 3269, pp. 27–39. Springer, Heidelberg (2004)

23. Hohenberger, S., Sahai, A., Waters, B.: Replacing a random oracle: full domain hash from indistinguishability obfuscation. In: Nguyen, P.Q., Oswald, E. (eds.) EUROCRYPT 2014. LNCS, vol. 8441, pp. 201–220. Springer, Heidelberg (2014)

24. Jakobsson, M., Sako, K., Impagliazzo, R.: Designated verifier proofs and their applications. In: Maurer, U.M. (ed.) EUROCRYPT 1996. LNCS, vol. 1070, pp. 143–154. Springer, Heidelberg (1996)

25. Kiayias, A., Papadopoulos, S., Triandopoulos, N., Zacharias, T.: Delegatable pseudorandom functions and applications. In: Proceedings ACM CCS (2013)

26. Rivest, R.L., Shamir, A., Tauman, Y.: How to leak a secret. In: Boyd, C. (ed.) ASIACRYPT 2001. LNCS, vol. 2248, pp. 552–565. Springer, Heidelberg (2001)

27. Ramchen, K., Waters, B.: Fully secure and fast signing from obfuscation. Cryptology ePrint Archive, Report 2014/523 (2014). http://eprint.iacr.org

28. Shamir, A.: Identity-based cryptosystems and signature schemes. In: Blakely, G.R., Chaum, D. (eds.) CRYPTO 1984. LNCS, vol. 196, pp. 47–53. Springer, Heidelberg (1985)

29. Schäge, S., Schwenk, J.: A CDH-based ring signature scheme with short signatures and public keys. In: Sion, R. (ed.) FC 2010. LNCS, vol. 6052, pp. 129–142. Springer, Heidelberg (2010)

30. Sahai, S., Waters, B.: How to use indistinguishability obfuscation: deniable encryption, and more. IACR Cryptology ePrint Archive, 2013, p. 454 (2013)

31. Shacham, H., Waters, B.: Efficient ring signatures without random oracles. In: Okamoto, T., Wang, X. (eds.) PKC 2007. LNCS, vol. 4450, pp. 166–180. Springer, Heidelberg (2007)

32. Yuen, T.H., Liu, J.K., Au, M.H., Susilo, W., Zhou, J.: Efficient linkable and/or threshold ring signature without random oracles. Comput. J. **56**(4), 407–421 (2013)

33. Zhang, F., Kim, K.: ID-based blind signature and ring signature from pairings. In: Zheng, Y. (ed.) ASIACRYPT 2002. LNCS, vol. 2501, pp. 533–547. Springer, Heidelberg (2002)

Universally Composable Identity Based Adaptive Oblivious Transfer with Access Control

Vandana Guleria$^{(\boxtimes)}$ and Ratna Dutta

Department of Mathematics, Indian Institute of Technology Kharagpur,
Kharagpur 721302, India
vandana.math@gmail.com, ratna@maths.iitkgp.ernet.in

Abstract. We propose the *first identity based adaptive oblivious transfer protocol with access control* (IBAOT-AC) secure in *universal composable* (UC) framework. The IBAOT-AC is run between multiple senders, multiple receivers and an issuer. Each sender incorporates its identity and access policies associated with the messages in the generation of ciphertext database. Receivers whose attribute sets satisfy the access policy associated with the message and who interact with a sender generating the corresponding ciphertext can only recover the message. The scheme supports access policy in disjunctive form, thereby, realizes disjunction of attributes. The proposed scheme is UC secure in the presence of malicious adversary under q-Strong Diffie-Hellman (SDH), Decision Bilinear Diffie-Hellman (DBDH), q-Decision Bilinear Diffie-Hellman Exponent (DBDHE) and Decision Linear (DLIN) assumptions. The scheme outperforms the existing similar schemes in terms of both communication and computation.

Keywords: Oblivious transfer · Identity based encryption · Universally composable security · Access policy · Non-interactive zero-knowledge proofs · Attribute based encryption

1 Introduction

Adaptive oblivious transfer (AOT) is a two-party protocol between a sender and a receiver and is an extensively used primitive in cryptography. It has been used in many cryptographic applications including fair exchange in e-commerce and secure multi-party computation. It is useful in oblivious search of patent database, medical database etc., where database size is very large. Suppose the sender has a database of N messages and does not want to reveal the entire database to the receiver. In AOT, each message is encrypted and the generated ciphertext database is made public by the sender in initialization phase, while enabling the receiver to learn only k out of N messages of its choice sequentially in transfer phase. The $(i\text{-}1)$-th message may be obtained before deciding on the i-th index by the receiver.

The first AOT protocol was proposed by Naor and Pinkas [15] followed by several AOT constructions [6, 7] secure in *full-simulation* security framework and

© Springer International Publishing Switzerland 2015
D. Lin et al. (Eds.): Inscrypt 2014, LNCS 8957, pp. 109–129, 2015.
DOI: 10.1007/978-3-319-16745-9_7

[11,13,16] secure in universal composable (UC) framework. The UC secure protocol guarantees security even when the protocol is composed with an arbitrary set of protocols. The UC secure AOT protocols [11,13,16] do not impose any restrictions on who can access which message. Coull *et al.* [9] proposed the first AOT with access control. Since then, there is a vast literature on oblivious transfer protocols with access control [6] secure in full-simulation security framework and become insecure when composed with other protocols. To overcome this, Abe *et al.* [1] framed the first UC secure AOT with access control.

Our Contribution. We design the *first* identity based adaptive oblivious transfer with access control (IBAOT-AC) secure in UC framework. Our construction employs Water's identity based encryption (IBE) [17], ciphertext policy attribute based encryption (CP-ABE) [18] and Boneh-Boyen (BB) [4] signature. Besides, Groth-Sahai [12] proofs are used for non-interactive verification of pairing product equations. The CP-ABE [18] enables those receivers to decrypt ciphertexts whose attribute set satisfies the access policies associated with the ciphertexts. The IBE [17] allows the decryption of a ciphertext only if the receiver, whose attribute set satisfy the access policy associated with the ciphertext, interacts in transfer phase with the sender, who has generated the ciphertext in initialization phase. The malicious behavior of the sender and receiver is controlled by providing non-interactive zero-knowledge proofs and using BB [4] signature. The sender can verify whether the receiver has randomized the same ciphertext in transfer phase which was previously published by the sender in initialization phase. If the verification succeeds, then the sender is convinced that the receiver follows the protocol specifications. The sender also gives non-interactive zero knowledge (NIZK) proof [12] to convince the receiver about its honest behavior of using the same secrets in both initialization and transfer phase.

The security of the proposed protocol is analyzed in UC framework [8] that does not allow the simulator to rewind the adversary's state to previous computation state to extract the hidden secret in zero-knowledge proofs. Our IBAOT-AC is secure under Decision Bilinear Diffie-Hellman (DBDH) [17], q-Decision Bilinear Diffie-Hellman Exponent (DBDHE) [18], q-Strong Diffie-Hellman (SDH) [4] and Decision Linear (DLIN) [12] assumptions. The sender's security and the receiver's security are proved separately. The receiver's security in IBAOT-AC ensures that the sender does not learn which message is being queried and who queries the message. The sender's security is achieved by proving that a receiver– (i) engages in transfer phase only if its attribute set satisfies the access policy associated with a message, (ii) learns only one message in each transfer phase and remains oblivious to other messages.

The proposed IBAOT-AC outperforms significantly in terms of both computation and communication in contrast to [1,11,13,16], which are, to the best of our knowledge, the only existing UC secure AOT protocols. Green *et al.* [10] proposed AOT treating the index of each message as an identity. Their scheme is not IBAOT as they have used the identity concept in blinding the index of ciphertext that the receiver wants to decrypt in each transfer phase.

They further developed a generic solution for the construction of AOT from unique blind IBE. Later, Zhang *et al.* [19] designed an IBAOT together with a generic solution for the construction of IBAOT from identity based unique blind signature. The security of [10] and [19] are in the random oracle model and do not use any access control over the messages. Besides, the construction of [19] uses bilinear groups of composite order. On the contrary, our IBAOT-AC uses bilinear groups of prime order and enables the sender to put some restrictions on who can recover the messages. More interestingly, our IBAOT-AC takes into account the disjunction of attributes whereas Abe *et al.*'s [1] covers only conjunction of attributes.

2   Preliminaries

Throughout, we use ρ as the security parameter, $x \xleftarrow{\$} A$ means sample an element x uniformly at random from the set A, $y \leftarrow B$ indicates y is the output of algorithm B, $X \overset{c}{\approx} Y$ means X is computationally indistinguishable from Y and $[\ell]$ denotes $\{1, 2, \ldots, \ell\}$. A function $f(n)$ is *negligible* if $f = o(n^{-c})$ for every fixed positive constant c.

2.1   Bilinear Pairing and Mathematical Assumptions

Definition 1. *(Bilinear Pairing.) Let* $\mathbb{G}_1, \mathbb{G}_2$ *and* $\mathbb{G}_T$ *be three multiplicative cyclic groups of prime order* p *and* g_1 *and* g_2 *be generators of groups* $\mathbb{G}_1$ *and* $\mathbb{G}_2$ *respectively. Then the map* $e : \mathbb{G}_1 \times \mathbb{G}_2 \rightarrow \mathbb{G}_T$ *is bilinear if it satisfies the following conditions: (i) Bilinear* $- e(x^a, y^b) = e(x, y)^{ab} \ \forall \ x \in \mathbb{G}_1, y \in \mathbb{G}_2, a, b \in \mathbb{Z}_p$. *(ii) Non-Degenerate* $- e(x, y)$ *generates* $\mathbb{G}_T$, $\forall \ x \in \mathbb{G}_1, y \in \mathbb{G}_2, x \neq 1, y \neq 1$. *(iii) Computable* $-$ *the pairing* $e(x, y)$ *is computable efficiently* $\forall \ x \in \mathbb{G}_1, y \in \mathbb{G}_2$.

If $\mathbb{G}_1 = \mathbb{G}_2$, then e is *symmetric* bilinear pairing. Otherwise, e is *asymmetric* bilinear pairing. Throughout the paper, we use symmetric bilinear pairing.

BilinearSetup: The BilinearSetup is an algorithm which on input security parameter ρ generates params $= (p, \mathbb{G}, \mathbb{G}_T, e, g)$, where $e : \mathbb{G} \times \mathbb{G} \rightarrow \mathbb{G}_T$ is a symmetric bilinear pairing, g is a generator of group $\mathbb{G}$ and p, the order of the groups $\mathbb{G}$ and $\mathbb{G}_T$, is prime, i.e. params $\leftarrow$ BilinearSetup(1^ρ).

Definition 2. *(q-SDH [4]) The q-Strong Diffie-Hellman (SDH) assumption in* $\mathbb{G}$ *states that for all probabilistic polynomial time (PPT) algorithm* $\mathcal{A}$, *with running time in* ρ, *the advantage* $\mathsf{Adv}_{\mathbb{G}}^{q\text{-}SDH}(\mathcal{A}) = \Pr[\mathcal{A}(g, g^x, g^{x^2}, \ldots, g^{x^q}) = (c, g^{\frac{1}{x+c}})]$ *is negligible in* ρ, *where* $g \xleftarrow{\$} \mathbb{G}, x \xleftarrow{\$} \mathbb{Z}_p, c \in \mathbb{Z}_p$.

Definition 3. *(DBDH [17]) The Decision Bilinear Diffie-Hellman (DBDH) assumption in* $(\mathbb{G}, \mathbb{G}_T)$ *states that for all PPT algorithm* $\mathcal{A}$, *with running time in* ρ, *the advantage* $\mathsf{Adv}_{\mathbb{G},\mathbb{G}_T}^{DBDH}(\mathcal{A}) = \Pr[\mathcal{A}(g, g^a, g^b, g^c, e(g, g)^{abc})] - \Pr[\mathcal{A}(g, g^a, g^b, g^c, Z)]$ *is negligible in* ρ, *where* $g \xleftarrow{\$} \mathbb{G}, Z \xleftarrow{\$} \mathbb{G}_T, a, b, c \in \mathbb{Z}_p$.

Definition 4. *(q-DBDHE [17]) The q-Decision Bilinear Diffie-Hellman Exponent (DBDHE) assumption in* $(\mathbb{G}, \mathbb{G}_T)$ *states that for all PPT algorithm* $\mathcal{A}$, *with running time in* ρ, *the advantage* $\mathsf{Adv}_{\mathbb{G},\mathbb{G}_T}^{q\text{-}DBDHE}(\mathcal{A}) = \Pr[\mathcal{A}(Y, e(g,g)^{\alpha^{q+1}s})] - \Pr[\mathcal{A}(Y, Z)]$ *is negligible in* ρ, $Y = (g, g^s, g^\alpha, g^{(\alpha^2)}, \dots, g^{(\alpha^q)}, g^{(\alpha^{q+2})}, \dots, g^{(\alpha^{2q})})$, $g \xleftarrow{\$} \mathbb{G}, Z \xleftarrow{\$} \mathbb{G}_T, s, \alpha \xleftarrow{\$} \mathbb{Z}_p$.

Definition 5. *(DLIN [5]) The Decision Linear (DLIN) assumption in* $\mathbb{G}$ *states that for all PPT algorithm* $\mathcal{A}$, *with running time in* ρ, *the advantage* $\mathsf{Adv}_{\mathbb{G}}^{DLIN}(\mathcal{A})$ $= \Pr[\mathcal{A}(g, g^a, g^b, g^{ra}, g^{sb}, g^{r+s})] - \Pr[\mathcal{A}(g, g^a, g^b, g^{ra}, g^{sb}, t)]$ *is negligible in* ρ, *where* $g \xleftarrow{\$} \mathbb{G}, t \xleftarrow{\$} \mathbb{G}, a, b, r, s \in \mathbb{Z}_p$.

2.2   Linear Secret Sharing Schemes (LSSS) [2]

Definition 6. *(Access policy) Let* $\Omega = \{a_1, a_2, \dots, a_m\}$ *be the universe of attributes and* $\mathcal{P}(\Omega)$ *be the collection of all subsets of* Ω. *An access policy (structure) is a collection* $\mathbb{A}$ *of non-empty subsets of* Ω, *i.e.,* $\mathbb{A} \subseteq \mathcal{P}(\Omega)\backslash\emptyset$. *The sets in* $\mathbb{A}$ *are called the authorized sets, and the sets not in* $\mathbb{A}$ *are called the unauthorized sets.*

A secret sharing scheme $\Pi_{\mathbb{A}}$ for the access policy $\mathbb{A}$ over Ω is called *linear* (in $\mathbb{Z}_p$) if it consists of two PPT algorithms– Distribute($\mathbb{M}, \eta, s$) and Reconstruct($\mathbb{M}, \eta, w$) which are described below, where $\mathbb{M}$ is a matrix with ℓ rows and t columns, called the share generating matrix for $\Pi_{\mathbb{A}}$, $\eta : [\ell] \to I_\Omega$ is the function which maps each row of $\mathbb{M}$ to an attribute index in $\mathbb{A}$, $s \in \mathbb{Z}_p$ is the secret to be shared, $w \in \mathbb{A}$ is the set of attributes and I_Ω is the index set of Ω.

– Distribute($\mathbb{M}, \eta, s$): This algorithm upon input $(\mathbb{M}, \eta, s)$, takes $r_2, r_3, \dots, r_t \xleftarrow{\$} \mathbb{Z}_p$, sets $v = (s, r_2, r_3, \dots, r_t) \in \mathbb{Z}_p^t$ and outputs a set $\{M_i \cdot v \mid i \in [\ell]\}$ of ℓ shares, where $M_i \in \mathbb{Z}_p^t$ is the i-th row of $\mathbb{M}$. The share $\lambda_i = M_i \cdot v$ belongs to the attribute $a_{\eta(i)}$.
– Reconstruct($\mathbb{M}, \eta, w$): This algorithm takes input $(\mathbb{M}, \eta, w)$. Let $I = \{i \in [\ell] \mid a_{\eta(i)} \in w\}$.
  1. Construct each row of matrix $\mathbb{F}$ by picking i-th row of $\mathbb{M}$, $\forall a_{\eta(i)} \in w$.
  2. Find the solution vector $\overrightarrow{x} = \{x_i \in \mathbb{Z}_p \mid i \in I\}$ such that $\sum_{a_{\eta(i)} \in w} \lambda_i x_i = s$ holds by solving system of equation $\mathbb{F}^T \overrightarrow{x} = e_1$, where $\mathbb{F}^T$ is the transpose of the matrix $\mathbb{F}$ of size $\nu \times t$, ν is the length of solution vector $\overrightarrow{x}$ which is equal to the number of attributes in w, e_1 is a vector of length t with 1 at first position and 0 elsewhere and $\{\lambda_i \in \mathbb{Z}_p \mid i \in I\}$ is a valid set of shares of the secret s generated by Distribute($\mathbb{M}, \eta, s$) algorithm. The algorithm outputs $\overrightarrow{x}$.

Theorem 1. *([14]) Let* $(\mathbb{M}, \eta)$ *be a LSSS access structure realizing an access policy* $\mathbb{A}$ *over universe of attributes* Ω, *where* $\mathbb{M}$ *is the* $\ell \times t$ *share generating matrix. Let* $w \subset \Omega$. *If* $w \notin \mathbb{A}$, *then there exists a PPT algorithm that outputs a vector* $\overrightarrow{x} = (-1, x_2, x_3, \dots, x_t) \in \mathbb{Z}_p^t$ *such that* $M_i \cdot \overrightarrow{x} = 0$ *for each row* i *of* $\mathbb{M}$ *for which* $\eta(i) \in I_\Omega$.

Converting Access Policy (AP) to LSSS Matrices: In our construction, we consider the access policy as a binary access tree, where interior nodes are AND ($\wedge$) and OR($\vee$) gates and the leaf nodes correspond to attributes. Label the root node of the tree with the vector (1), a vector of length 1. Label each internal node with a vector determined by the vector assigned to its parent node recursively as discussed below. Maintain a counter c with initial value 1.

1. If the parent node is $\vee$ with a vector v, then label its children by v keeping c same.
2. If the parent node is $\wedge$ with a vector v, pad v if necessary with 0's at the end to make it of length c. Label one of its children with the vector $(v, 1)$ and other with the vector $(0, 0, \ldots, 0, -1)$, where $(0, 0, \ldots, 0)$ is zero-vector of length c. Note that $(v, 1)$ and $(0, 0, \ldots, 0, -1)$ sum to $(v, 0)$. Now increment the counter c by 1.

After labeling the entire tree, the vectors labeling the leaf nodes form the rows of LSSS matrix. If the vectors are of different length, pad the shorter ones with 0's at the end to make all the vectors of same length.

2.3 Non-Interactive Verification of Pairing Product Equation [12]

The Groth-Sahai proofs are two party protocols between a prover and a verifier for non-interactive verification of a pairing product equation

$$\prod_{q=1}^{Q} e(a_q \prod_{i=1}^{n} x_i^{\alpha_{q,i}}, b_q \prod_{i=1}^{n} y_i^{\beta_{q,i}}) = t_T, \tag{1}$$

where $a_{q=1,2,\ldots,Q} \in \mathbb{G}, b_{q=1,2,\ldots,Q} \in \mathbb{G}, \{\alpha_{q,i}, \beta_{q,i}\}_{q=1,2,\ldots,Q, i=1,2,\ldots,n} \in \mathbb{Z}_p$ and $t_T \in \mathbb{G}_T$ are the coefficients of the pairing product Eq. 1 which are given to the verifier. The prover knows secret values $x_{i=1,2,\ldots,n}, y_{i=1,2,\ldots,n} \in \mathbb{G}$ also called witnesses that satisfy the Eq. 1. The prover wants to convince the verifier in a non-interactive way that he knows x_i and y_i without revealing anything about x_i and y_i to the verifier. Let $\mathcal{W} = \{x_{i=1,2,\ldots,n}, y_{i=1,2,\ldots,n}\}$ be the set of all secret values in the pairing product Eq. 1. The set $\mathcal{W}$ is referred as witnesses of the pairing product equation. The product of two vectors is defined component wise, i.e., $(a_1, a_2, a_3)(b_1, b_2, b_3) = (a_1 b_1, a_2 b_2, a_3 b_3)$ for $(a_1, a_2, a_3), (b_1, b_2, b_3) \in \mathbb{G}^3$ for a finite order group $\mathbb{G}$.

For non-interactive verification of the pairing product Eq. 1, a trusted party upon input a security parameter ρ generates common reference string GS $=$ (params, $u_1, u_2, u_3, \mu, \mu_T$), where params $= (p, \mathbb{G}, \mathbb{G}_T, e, g) \leftarrow$ BilinearSetup(1^ρ), $u_1 = (g^a, 1, g) \in \mathbb{G}^3, u_2 = (1, g^b, g) \in \mathbb{G}^3, u_3 = u_1^{\xi_1} u_2^{\xi_2} = (g^{a\xi_1}, g^{b\xi_2}, g^{\xi_1+\xi_2}) \in \mathbb{G}^3, \xi_1, \xi_2 \xleftarrow{\$} \mathbb{Z}_p, a, b \xleftarrow{\$} \mathbb{Z}_p$ and $\mu : \mathbb{G} \to \mathbb{G}^3, \mu_T : \mathbb{G}_T \to \mathbb{G}_T^9$ are two efficiently computable embeddings such that

$$\mu(g) = (1, 1, g) \text{ and } \mu_T(t_T) = \begin{pmatrix} 1 & 1 & 1 \\ 1 & 1 & 1 \\ 1 & 1 & t_T \end{pmatrix} \forall g \in \mathbb{G}, t_T \in \mathbb{G}_T.$$

Note that $\mu_T(t_T)$ is an element of $\mathbb{G}_T^9$. For convenience, it has been written in matrix form. The product of two elements of $\mathbb{G}_T^9$ is also component wise. The trusted party publishes GS to both the parties. The prover generates commitment to all the witnesses $x_{i=1,2,\ldots,n}$ and $y_{i=1,2,\ldots,n}$ using GS. To commit $x_i \in \mathbb{G}$ and $y_i \in \mathbb{G}$, the prover picks $r_{1i}, r_{2i}, r_{3i} \xleftarrow{\$} \mathbb{Z}_p$ and $s_{1i}, s_{2i}, s_{3i} \xleftarrow{\$} \mathbb{Z}_p$, sets

$$c_i = \mathsf{Com}(x_i) = \mu(x_i) u_1^{r_{1i}} u_2^{r_{2i}} u_3^{r_{3i}}, d_i = \mathsf{Com}(y_i) = \mu(y_i) u_1^{s_{1i}} u_2^{s_{2i}} u_3^{s_{3i}}.$$

The prover generates the proof components

$$P_j = \prod_{q=1}^{Q} \left(\mu(a_q) \prod_{i=1}^{n} \mu(x_i)^{\alpha_{q,i}} \right)^{\sum_{i=1}^{n} \beta_{q,i} s_{ji}} \left(\widehat{d_q} \right)^{\sum_{i=1}^{n} \alpha_{q,i} r_{ji}}, i = 1,2,\ldots,n, j = 1,2,3,$$

using random values r_{ji}, s_{ji}, which were used for generating commitments to $x_{i=1,2,\ldots,n}, y_{i=1,2,\ldots,n}$, and gives proof $\pi = (c_1, c_2, \ldots, c_n, d_1, d_2, \ldots, d_n, P_1, P_2, P_3)$ to the verifier, where $\widehat{d_q} = \mu(b_q) \prod_{i=1}^{n} d_i^{\beta_{q,i}}$. The verifier computes

$$\widehat{c_q} = \mu(a_q) \prod_{i=1}^{n} c_i^{\alpha_{q,i}}, \widehat{d_q} = \mu(b_q) \prod_{i=1}^{n} d_i^{\beta_{q,i}},$$

using c_i, d_i, coefficients $\alpha_{q,i}, \beta_{q,i}$ and outputs VALID if the following equation holds

$$\prod_{q=1}^{Q} F(\widehat{c_q}, \widehat{d_q}) = \mu_T(t_T) \prod_{j=1}^{3} F(u_j, P_j), \tag{2}$$

where $F : \mathbb{G}^3 \times \mathbb{G}^3 \rightarrow \mathbb{G}_T^9$ is defined as

$$F((x_1, x_2, x_3), (y_1, y_2, y_3)) = \begin{pmatrix} e(x_1, y_1) & e(x_1, y_2) & e(x_1, y_3) \\ e(x_2, y_1) & e(x_2, y_2) & e(x_2, y_3) \\ e(x_3, y_1) & e(x_3, y_2) & e(x_3, y_3) \end{pmatrix}.$$

Note that the function F is also bilinear and $F((x_1, x_2, x_3), (y_1, y_2, y_3))$ is an element of $\mathbb{G}_T^9$. The product of two elements of $\mathbb{G}_T^9$ is component wise. For convenience, it has been written in matrix form.

The Eq. 2 holds if and only if Eq. 1 holds. The Eq. 1 is non-linear. If in Eq. 1 only $y_{i=1,2,\ldots,n}$ are secrets, then it is a linear equation. For a linear equation, the verifier has to verify the following equation

$$\prod_{q=1}^{Q} F\left(\mu(a_q) \prod_{i=1}^{n} \mu(x_i)^{\alpha_{q,i}}, \widehat{d_q} \right) = \mu_T(t_T) \prod_{j=1}^{3} F(u_j, P_j), \tag{3}$$

$$\text{where } P_j = \prod_{q=1}^{Q} \left(\mu(a_q) \prod_{i=1}^{n} \mu(x_i)^{\alpha_{q,i}} \right)^{\sum_{i=1}^{n} \beta_{q,i} s_{ji}}, j = 1,2,3. \tag{4}$$

Note that there are two types of settings in Groth-Sahai proofs - *perfectly sound* setting and *witness indistinguishability* setting. The common reference string $\mathsf{GS} = (u_1, u_2, u_3)$ discussed above is in perfectly sound setting, where $u_1 = (g^a, 1, g)$, $u_2 = (1, g^b, g)$, $u_3 = (g^{a\xi_1}, g^{b\xi_2}, g^{\xi_1+\xi_2})$. One who knows the *extractable trapdoor* $\mathsf{t_{ext}} = (a, b, \xi_1, \xi_2)$, can extract the secret values from their commitments. In witness indistinguishability setting, $\mathsf{GS'} = (u_1, u_2, u_3)$, where $u_1 = (g^a, 1, g)$, $u_2 = (1, g^b, g)$, $u_3 = (g^{a\xi_1}, g^{b\xi_2}, g^{\xi_1+\xi_2+1})$. One who knows the *simulation trapdoor* $\mathsf{t_{sim}} = (a, b, \xi_1, \xi_2)$, may open the commitment differently in a pairing product equation as discussed with an example given below.

Example 1. Let $\mathsf{Com}(x) = \mu(x)u_1^{\theta_1}u_2^{\theta_2}u_3^{\theta_3}$ in witness indistinguishability setting, where $\theta_1, \theta_2, \theta_3 \overset{\$}{\leftarrow} \mathbb{Z}_p$. Opening values to $\mathsf{Com}(x)$ are $(D_1 = g^{\theta_1}, D_2 = g^{\theta_2}, D_3 = g^{\theta_3})$. The simulator knowing witness x opens $\mathsf{Com}(x)$ to any value x' using $\mathsf{t_{sim}} = (a, b, \xi_1, \xi_2)$ and D_1, D_2, D_3 as follows. The simulator sets $D_1' = D_1(\frac{x'}{x})^{\xi_1}, D_2' = D_2(\frac{x'}{x})^{\xi_2}, D_3' = D_3\frac{x}{x'})$ and opens the $\mathsf{Com}(x)$ to x' by computing $\frac{xg^{\theta_1+\theta_2+\theta_3(\xi_1+\xi_2+1)}}{D_1'D_2'(D_3')^{\xi_1+\xi_2+1}}$.

In GS, $g^a, g^b, g, g^{a\xi_1}, g^{b\xi_2}, g^{\xi_1+\xi_2}$ forms a DLIN tuple, whereas in GS', $g^a, g^b, g, g^{a\xi_1}, g^{b\xi_2}, g^{\xi_1+\xi_2+1}$ is not a DLIN tuple. The commitments in both the setting are computationally indistinguishable by the following theorem.

Theorem 2. *[12] The common reference string in perfectly sound setting is computationally indistinguishable from the common reference string in witness indistinguishability setting under DLIN assumption.*

Definition 7. *(NIWI) The non-interactive witness indistinguishable (NIWI) proof states that for all PPT algorithm $\mathcal{A}$, with running time in ρ, the advantage*

$$\mathsf{Adv}_{\mathbb{G},\mathbb{G}_T}^{\mathsf{NIWI}}(\mathcal{A}) = \Pr\left[\mathcal{A}(\mathsf{GS}, \mathcal{S}, \mathcal{W}_0) = \pi\right] - \Pr\left[\mathcal{A}(\mathsf{GS}, \mathcal{S}, \mathcal{W}_1) = \pi\right]$$

is negligible in ρ under DLIN assumption, where GS is the common reference string in perfectly sound setting, $\mathcal{S}$ is a pairing product equation, $\mathcal{W}_0, \mathcal{W}_1$ are two distinct set of witnesses satisfying $\mathcal{S}$ and π is the proof for $\mathcal{S}$.

Definition 8. *(NIZK) The non-interactive zero-knowledge (NIZK) proof states that for all PPT algorithm $\mathcal{A}$, with running time in ρ, the advantage*

$$\mathsf{Adv}_{\mathbb{G},\mathbb{G}_T}^{\mathsf{NIZK}}(\mathcal{A}) = \Pr[\mathcal{A}(\mathsf{GS'}, \mathcal{S}, \mathcal{W}) = \pi_0] - \Pr[\mathcal{A}(\mathsf{GS'}, \mathcal{S}, \mathsf{t_{sim}}) = \pi_1]$$

is negligible in ρ under DLIN assumption, where GS' is the common reference string in witness indistinguishability setting, $\mathcal{S}$ is a pairing product equation, $\mathcal{W}$ is a set of witnesses satisfying $\mathcal{S}$, π_0 is the proof for $\mathcal{S}$ and π_1 is the simulated proof for $\mathcal{S}$.

The notations $\mathsf{NIWI}\left\{ (\{x_i, y_i\}_{1 \leq i \leq n}) | \prod_{q=1}^{Q} e(a_q \prod_{i=1}^{n} x_i^{\alpha_{q,i}}, b_q \prod_{i=1}^{n} y_i^{\beta_{q,i}}) = t_T \right\}$ and $\mathsf{NIZK}\left\{ (\{x_i, y_i\}_{1 \leq i \leq n}) | \prod_{q=1}^{Q} e(a_q \prod_{i=1}^{n} x_i^{\alpha_{q,i}}, b_q \prod_{i=1}^{n} y_i^{\beta_{q,i}}) = t_T \right\}$, for NIWI and NIZK proof are followed respectively in our construction. The convention is that the quantities in the parenthesis denote elements the knowledge of which are being proved to the verifier by the prover while all other parameters are known to the verifier. We have the following theorem.

Theorem 3. *[12] The Groth-Sahai proofs are composable* NIWI *and* NIZK *for satisfiability of a set of pairing product equation over a bilinear group under DLIN assumption.*

Remark 1. (Randomizing Groth-Sahai Proofs) Belenkiy *et al.* [3] proved that the Groth-Sahai proofs can be randomized such that the same statement still hold. For instance, consider a pairing equation $e(g, h) = H$, where h is secret. The prover generates commitment to h such that $F(\mu(g), \mathsf{Com}(h)) = \mu_T(H) \prod_{i=1}^{3} F(u_i, P_i)$ hold, where $\mathsf{Com}(h) = \mu(h) u_1^{r_1} u_2^{r_2} u_3^{r_3}$, $P_1 = \mu(g)^{r_1}$, $P_2 = \mu(g)^{r_2}$, $P_3 = \mu(g)^{r_3}$, $r_1, r_2, r_3 \xleftarrow{\$} \mathbb{Z}_p$. The proof $\pi = (\mathsf{Com}(h), P_1, P_2, P_3)$. To randomize the proof π, the verifier picks $s_1, s_2, s_3 \xleftarrow{\$} \mathbb{Z}_p$, sets

$$\begin{aligned} \pi' &= (\mathsf{Com}(h) u_1^{s_1} u_2^{s_2} u_3^{s_3}, P_1 \mu(g)^{s_1}, P_2 \mu(g)^{s_2}, P_3 \mu(g)^{s_3}) \\ &= (\mu(h) u_1^{r_1+s_1} u_2^{r_2+s_2} u_3^{r_3+s_3}, \mu(g)^{r_1+s_1}, \mu(g)^{r_2+s_2}, \mu(g)^{r_3+s_3}), \end{aligned}$$

such that $e(g, h) = H$ still holds. Both the proofs π and π' have the same distribution and satisfy the equation $e(g, h) = H$.

2.4   Security Model of IBAOT-AC

UC Framework: The security of the proposed IBAOT-AC is analyzed in universal composable (UC) framework assuming static corruption. The UC framework consists of a *real world* and an *ideal world*. In the real world, parties (multiple senders, multiple receivers and an issuer) and a real world adversary $\mathcal{A}$, who has the ability of corrupting the parties, interact with each other according to IBAOT-AC protocol Ψ. In the ideal world, there are dummy parties (multiple senders, multiple receivers and an issuer), an ideal world adversary $\mathcal{A}'$ and a trusted party called ideal functionality $\mathcal{F}_{\mathsf{IBAOT-AC}}$. The parties are dummy in the sense that they submit their inputs to $\mathcal{F}_{\mathsf{IBAOT-AC}}$ and receive respective outputs from $\mathcal{F}_{\mathsf{IBAOT-AC}}$ instead of performing any computation by themselves. A protocol is said to be secure in UC framework if no interactive distinguisher, called *environment machine* $\mathcal{Z}$, can distinguish the execution of the protocol Ψ in the *real world* from the execution of the protocol in the *ideal world*.

For the message in the ideal world, we follow the notation: $\langle \mathsf{type} : \mathsf{sid}, \mathsf{payload} \rangle$, where type denotes type of the message, sid is a session identity and payload is an optional parameter. The sid is provided by $\mathcal{Z}$. No two copies of $\mathcal{F}_{\mathsf{CRS}}^{\mathcal{D}}$ can have the same sid. Two parties are said to have the same sid if and only if they are participants of the same instance of a protocol. The sid is used to distinguish between different instances of the same protocol. We now describe ideal functionality $\mathcal{F}_{\mathsf{CRS}}^{\mathcal{D}}$ for the generation of common reference string (CRS) parameterized by some specific distribution $\mathcal{D}$ and ideal functionality $\mathcal{F}_{\mathsf{IBAOT-AC}}$ for IBAOT-AC protocol following [8].

CRSSetup– Upon receiving a message $\langle$CRS : sid, $P\rangle$, from a party P (either S, R or issuer), $\mathcal{F}^{\mathcal{D}}_{\mathsf{CRS}}$ first checks if there is a recorded value crs. If there is no recorded value, $\mathcal{F}^{\mathcal{D}}_{\mathsf{CRS}}$ generates crs $\xleftarrow{\$} \mathcal{D}(1^{\rho})$ and records it. Finally, $\mathcal{F}^{\mathcal{D}}_{\mathsf{CRS}}$ sends $\langle$CRS : sid, crs$\rangle$ to the party P and $\mathcal{A}'$, where sid is the session identity.

In the ideal world, parties just forward their inputs to the $\mathcal{F}_{\mathsf{IBAOT-AC}}$ and get back their respective outputs. The functionality $\mathcal{F}_{\mathsf{IBAOT-AC}}$ is as follows.

ISetup– The issuer upon receiving the message $\langle$isetup : sid$\rangle$ from $\mathcal{Z}$, passes it to $\mathcal{F}_{\mathsf{IBAOT-AC}}$. The $\mathcal{F}_{\mathsf{IBAOT-AC}}$ checks that it has not seen the message before and then forwards it to $\mathcal{A}'$. If seen, $\mathcal{F}_{\mathsf{IBAOT-AC}}$ ignores the message.

IdSkIssue– Upon receiving the message $\langle$idsk : sid, $\mathsf{ID}_S\rangle$ from S, $\mathcal{F}_{\mathsf{IBAOT-AC}}$ sends $\langle$idsk : sid, $\mathsf{ID}_S\rangle$ to the issuer. The $\mathcal{F}_{\mathsf{IBAOT-AC}}$ keeps an identity string ID_S for each sender S which is initially set to be empty. The issuer returns $\langle$idsk : sid, $b\rangle$ in response. If $b = 1$, $\mathcal{F}_{\mathsf{IBAOT-AC}}$ updates $\mathsf{ID}_S = \mathsf{ID}_S$ and sends $b = 1$ to S. Otherwise, $\mathcal{F}_{\mathsf{IBAOT-AC}}$ does nothing and simply sends $b = 0$.

DBSetup– The $\mathcal{F}_{\mathsf{IBAOT-AC}}$ upon receiving a message $\langle$dbsetup : sid, ID_S, DB_S, $N_S\rangle$ from a sender S with identity ID_S, stores DB_S, where $\mathsf{DB}_S = ((m_1, \mathsf{AP}_1), (m_2, \mathsf{AP}_2), \ldots, (m_{N_S}, \mathsf{AP}_{N_S}))$, AP_i is the access policy associated with each m_i, N_S is the size of DB_S, $i = 1, 2, \ldots, N_S$.

AttSkIssue– A receiver R upon receiving the message $\langle$attsk : sid, $w_R\rangle$ from $\mathcal{Z}$, passes it to $\mathcal{F}_{\mathsf{IBAOT-AC}}$. The $\mathcal{F}_{\mathsf{IBAOT-AC}}$ keeps an attribute set w_R for each receiver R which is initially set to be empty. Upon receiving the message $\langle$attsk : sid, $w_R\rangle$ from R, $\mathcal{F}_{\mathsf{IBAOT-AC}}$ sends $\langle$attsk : sid, $w_R\rangle$ to the issuer. The issuer returns $\langle$attsk : sid, $b\rangle$ in response. If $b = 1$, $\mathcal{F}_{\mathsf{IBAOT-AC}}$ updates $w_R = w_R$ and sends $b = 1$ to R. Otherwise, $\mathcal{F}_{\mathsf{IBAOT-AC}}$ does nothing and simply sends $b = 0$ to R.

Transfer– Upon receiving the message $\langle$transfer : sid, ID_S, $\sigma\rangle$ from R, $\mathcal{F}_{\mathsf{IBAOT-AC}}$ sends $\langle$transfer : sid, $\mathsf{ID}_S\rangle$ to S and receives $\langle$transfer : sid, ID_S, $b\rangle$ in response from S. If $b = 1$ and w_R satisfies AP_{σ}, then $\mathcal{F}_{\mathsf{IBAOT-AC}}$ returns m_{σ} to R. Otherwise, $\mathcal{F}_{\mathsf{IBAOT-AC}}$ returns $\bot$ to R.

Definition 9. *A protocol Ψ securely realizes the ideal functionality $\mathcal{F}_{\mathsf{IBAOT-AC}}$ if for any real world adversary $\mathcal{A}$, there exists an ideal world adversary $\mathcal{A}'$ such that for any environment machine $\mathcal{Z}$, $\mathsf{IDEAL}_{\mathcal{F}_{\mathsf{IBAOT-AC}}, \mathcal{A}', \mathcal{Z}} \overset{c}{\approx} \mathsf{REAL}_{\Psi, \mathcal{A}, \mathcal{Z}}$, where $\mathsf{IDEAL}_{\mathcal{F}_{\mathsf{IBAOT-AC}}, \mathcal{A}', \mathcal{Z}}$ is the output of $\mathcal{Z}$ after interacting with $\mathcal{A}'$ and dummy parties interacting with $\mathcal{F}_{\mathsf{IBAOT-AC}}$ in the ideal world and $\mathsf{REAL}_{\Psi, \mathcal{A}, \mathcal{Z}}$ is the output of $\mathcal{Z}$ after interacting with $\mathcal{A}$ and the parties running the protocol Ψ in the real world.*

3   Our Protocol

Our IBAOT-AC protocol is a tuple of the following PPT algorithms: IBAOT-AC = (CRSSetup, ISetup, IdSkIssue, DBSetup = (InitDB, DBVerify), AttSkIssue, Transfer = (RequestTra, ResponseTra, CompleteTra)). For instance, we consider

$$\text{crs} = (\text{params}, \text{GS}_S, \text{GS}_R)$$
$$\text{params} = (p, \mathbb{G}, \mathbb{G}_T, e, g)$$

Sender S	Receiver R

$(\text{pk}_{\text{DB}_S}, \text{sk}_{\text{DB}_S}, \psi_{\text{DB}_S}, \text{cDB}_S) \leftarrow \text{InitDB}$
$\text{pk}_{\text{DB}_S} = (H_1, H_2, y_1, y_2)$
$\text{sk}_{\text{DB}_S} = (x, \gamma, \widehat{h_1}, \widehat{h_2})$
$\text{cDB}_S = (\Phi_1, \Phi_2, \ldots, \Phi_{N_S})$

$$\xrightarrow{\quad \text{pk}_{\text{DB}_S}, \psi_{\text{DB}_S}, \text{cDB}_S \quad}$$

$\text{ACCEPT} \leftarrow \text{DBVerify}$

Transfer Phase

$\sigma_j \in \{1, 2, \ldots, N\}, j = 1, 2, \ldots, k$
$(\text{Req}_{\sigma_j}, \text{Pri}_{\sigma_j}) \leftarrow \text{RequestTra}$
$\text{Req}_{\sigma_j} = (V_{\sigma_j}, X_{\sigma_j}, Y_{\sigma_j}, Z_{\sigma_j}, \pi_{\sigma_j})$
$\text{Pri}_{\sigma_j} = (v_{3,\sigma_j}, t_{1,\sigma_j}, t_{2,\sigma_j}, t_{3,\sigma_j})$

$$\xleftarrow{\quad \text{Req}_{\sigma_j} \quad}$$

$\text{Res}_{\sigma_j} \leftarrow \text{ResponseTra}$
$\text{Res}_{\sigma_j} = (s_{\sigma_j}, \delta_{\sigma_j})$
$s_{\sigma_j} = e(Z_{\sigma_j}, \widehat{h_2}) \cdot e(d_1, X_{\sigma_j}) e(d_2^{-1}, Y_{\sigma_j})$

$$\xrightarrow{\quad \text{Res}_{\sigma_j} \quad}$$

$m_{\sigma_j} \leftarrow \text{CompleteTra}$

Fig. 1. Initialization phase and jth transfer phase of our IBAOT-AC protocol, $j = 1, 2, \ldots, k$.

an execution between a sender S with identity ID_S, a receiver R and an issuer. We invoke algorithm BilinearSetup described in Sect. 2.1 for the generation of bilinear groups and pairing. The identity $\text{ID}_S \in \{0,1\}^n$ and $\Omega = \{a_1, a_2, \ldots, a_m\}$ is the universe of attributes. A pictorial view of high-level description of the interaction between the sender S with identity ID_S and the receiver R is given in Fig. 1.

CRSSetup(1^ρ): This randomized algorithm on input security parameter ρ generates common reference string crs as follows. It first generates params $= (p, \mathbb{G}, \mathbb{G}_T, e, g) \leftarrow \text{BilinearSetup}(1^\rho)$, chooses $a, b, \xi_1, \xi_2, \tilde{a}, \tilde{b}, \tilde{\xi_1}, \tilde{\xi_2} \xleftarrow{\$} \mathbb{Z}_p^*$ and sets $g_1 = g^a, g_2 = g^b, \tilde{g}_1 = g^{\tilde{a}}, \tilde{g}_2 = g^{\tilde{b}}, u_1 = (g_1, 1, g), u_2 = (1, g_2, g), u_3 = u_1^{\xi_1} u_2^{\xi_2} = (g_1^{\xi_1}, g_2^{\xi_2}, g^{\xi_1 + \xi_2}), \widetilde{u_1} = (\tilde{g}_1, 1, g), \widetilde{u_2} = (1, \tilde{g}_2, g), \widetilde{u_3} = \widetilde{u_1}^{\tilde{\xi_1}} \widetilde{u_2}^{\tilde{\xi_2}} = (\tilde{g}_1^{\tilde{\xi_1}}, \tilde{g}_2^{\tilde{\xi_2}}, g^{\tilde{\xi_1} + \tilde{\xi_2}}), \text{GS}_R = (u_1, u_2, u_3), \text{GS}_S = (\widetilde{u_1}, \widetilde{u_2}, \widetilde{u_3}), \text{crs} = (\text{params}, \text{GS}_R, \text{GS}_S)$. GS_R is used for creating non-interactive witness indistinguishable (NIWI) proof by a receiver and GS_S for generating non-interactive zero-knowledge (NIZK) proof by a sender.

ISetup(params): This randomized algorithm takes as input params $= (p, \mathbb{G}, \mathbb{G}_T, e, g)$ from the issuer, selects $c, \alpha, \beta \xleftarrow{\$} \mathbb{Z}_p^*, \widehat{g_2}, f', f_1, f_2, \ldots, f_n, h_1, h_2, \ldots, h_m \xleftarrow{\$} \mathbb{G}$, sets $\widehat{g_1} = g^\alpha, g_3 = g^c, U = e(g, g)^\beta, \text{PK}_1 = (\widehat{g_1}, \widehat{g_2}, f', f_1, f_2, \ldots, f_n), \text{MSK}_1 = \widehat{g_2}^\alpha, \text{PK}_2 = (g_3, U, h_1, h_2, \ldots, h_m), \text{MSK}_2 = (c, \beta)$. The algorithm outputs two key pairs $(\text{PK}_1, \text{MSK}_1)$ and $(\text{PK}_2, \text{MSK}_2)$ to the issuer. The issuer uses key pair $(\text{PK}_1, \text{MSK}_1)$ for the generation of *identity secret key* SK_S corresponding to a unique identity ID_S of a sender S and $(\text{PK}_2, \text{MSK}_2)$ for issuing *attribute secret key* ASK_R corresponding to an attribute set w_R of a receiver R. The issuer publishes $\mathsf{PK}_1, \mathsf{PK}_2$ and keeps $\mathsf{MSK}_1, \mathsf{MSK}_2$ secret.

IdSkIssue(params, $\mathsf{PK}_1, \mathsf{MSK}_1, \mathsf{ID}_S$): The issuer upon receiving the identity $\mathsf{ID}_S \in \{0,1\}^n$ from S runs this randomized algorithm using params $= (p, \mathbb{G}, \mathbb{G}_T, e, g)$, public key $\mathsf{PK}_1 = (\widehat{g}_1, \widehat{g}_2, f', f_1, f_2, \ldots, f_n)$ and master secret key $\mathsf{MSK}_1 = \widehat{g}_2{}^\alpha$. Let $\mathcal{V}_S = \{t \mid t\text{-th bit of } \mathsf{ID}_S \text{ is } 1\} \subseteq [n]$. The algorithm first selects $t_S \overset{\$}{\leftarrow} \mathbb{Z}_p^*$, sets $d_1 = \widehat{g}_2{}^\alpha \left(f' \prod_{l \in \mathcal{V}_S} f_l\right)^{t_S}$, $d_2 = g^{t_S}$, $\mathsf{SK}_S = (d_1, d_2)$. The identity secret key SK_S is given to S through a secure communication channel. The sender S checks the correctness of the secret key SK_S by verifying the equation $e(d_1, g)e(d_2^{-1}, f' \prod_{l \in \mathcal{V}_S} f_l) = e(\widehat{g}_1, \widehat{g}_2)$. If the verification holds, S accepts the secret key SK_S. Otherwise, S aborts the execution.

InitDB($\mathsf{ID}_S, \mathsf{SK}_S, \mathsf{crs}, \mathsf{PK}_1, \mathsf{PK}_2, \mathsf{DB}_S$): This randomized algorithm upon input $(\mathsf{ID}_S, \mathsf{SK}_S, \mathsf{crs}, \mathsf{PK}_1, \mathsf{PK}_2, \mathsf{DB}_S)$ from a sender S generates database public key $\mathsf{pk}_{\mathsf{DB}_S}$, database secret key $\mathsf{sk}_{\mathsf{DB}_S}$, ciphertext database cDB_S and NIZK proof ψ_{DB_S} as follows and makes $\mathsf{pk}_{\mathsf{DB}_S}, \psi_{\mathsf{DB}_S}, \mathsf{cDB}_S$ public to all parties while keeping $\mathsf{sk}_{\mathsf{DB}_S}$ secret to itself. The database $\mathsf{DB}_S = ((m_1, \mathsf{AP}_1), (m_2, \mathsf{AP}_2), \ldots, (m_{N_S}, \mathsf{AP}_{N_S}))$, where $m_i \in \mathbb{G}_T, \mathsf{AP}_i$ is the access policy associated with each m_i, N_S is the size of the DB_S, $i = 1, 2, \ldots, N_S$, $\mathsf{SK}_S = (d_1, d_2)$, $\mathsf{crs} = (\mathsf{params}, \mathsf{GS}_R, \mathsf{GS}_S)$ and $\mathsf{ID}_S \in \{0,1\}^n$. The algorithm picks $x, \gamma \overset{\$}{\leftarrow} \mathbb{Z}_p^*, \widehat{h}_1 \overset{\$}{\leftarrow} \mathbb{G}$ and sets $\widehat{h}_2 = g^\gamma, y_1 = g^x, y_2 = \widehat{h}_1^{-x}, H_1 = e(g, \widehat{h}_1), H_2 = e(g, \widehat{h}_2)$ $\mathsf{pk}_{\mathsf{DB}_S} = (H_1, H_2, y_1, y_2)$, $\mathsf{sk}_{\mathsf{DB}_S} = (x, \gamma, \widehat{h}_1, \widehat{h}_2)$. It generates NIZK proof $\psi_{\mathsf{DB}_S} = \mathsf{NIZK}\{(d_1, d_2^{-1}, \widehat{h}_2, \widehat{g}_1{}^\gamma, g') \mid e(d_1, g)e(d_2^{-1}, f' \prod_{l \in \mathcal{V}_S} f_l)e(g', \widehat{g}_2{}^{-1}) = 1 \wedge e(g', \widehat{h}_2)e(\widehat{g}_1{}^\gamma, g^{-1}) = 1 \wedge e(g', \widehat{g}_2) = e(\widehat{g}_1, \widehat{g}_2)\}$, where $\mu : \mathbb{G} \rightarrow \mathbb{G}^3$ and NIZK are as described in Sect. 2.3, $\mathcal{V}_S$ as defined in algorithm IdSkIssue. The proof ψ_{DB_S} also consists of commitments to $\widehat{h}_1$ and $\widehat{h}_2$ generated using GS_R. The $\mathsf{Com}(\widehat{h}_1)$ and $\mathsf{Com}(\widehat{h}_2)$ generated using GS_R are used by the receiver in transfer phase as shown in algorithm RequestTra. For $i = 1$ to N_S, the algorithm generates $\Phi_i = (A_i, D_i, E_i, \mathsf{AP}_i)$ along with the description of $(\mathbb{M}_i, \eta_i)$ as follows.

1. Compute BB signature on index i as $A_i = g^{\frac{1}{x+i}}$.

2. Compute $B_i = e(A_i, \widehat{h}_1)$.

3. Encrypt B_i to generate CP-ABE ciphertext D_i under the access policy AP_i associated with message m_i as follows. Generate LSSS matrix $\mathbb{M}_i$ corresponding to access policy AP_i as described in Sect. 2.2, where $\mathbb{M}_i$ is the $n_i \times \theta_i$ matrix, n_i is the number of attributes in AP_i. The function η_i associates index of each row of $\mathbb{M}_i$ to an attribute index in AP_i. Pick $s_i, s_{i,2}, s_{i,3}, \ldots, s_{i,\theta_i} \overset{\$}{\leftarrow} \mathbb{Z}_p^*$, set $v_i = (s_i, s_{i,2}, s_{i,3}, \ldots, s_{i,\theta_i}) \in Z_p^{\theta_i}$, compute $\mathbb{M}_i \cdot v_i = (\lambda_1, \lambda_2, \ldots, \lambda_{n_i}) \in Z_p^{n_i}$ invoking Distribute algorithm described in Sect. 2.2 and set

$$D_i^{(1)} = B_i \cdot U^{s_i}, D_i^{(2)} = g^{s_i}, D_i^{(3)} = \{D_{i,\ell}^{(3)}, \ell = 1, 2, \ldots, n_i\},$$

where $D_{i,\ell}^{(3)} = g_3^{\lambda_\ell} \cdot h_{\eta_i(\ell)}^{-s_i} \ \forall a_{\eta_i(\ell)} \in \mathsf{AP}_i$, using $\mathsf{PK}_2 = (g_3, U, h_1, h_2, \ldots, h_m)$. The component $D_i = (D_i^{(1)}, D_i^{(2)}, D_i^{(3)})$ is the CP-ABE [18] of B_i along with the description of $\mathsf{AP}_i = (\mathbb{M}_i, \eta_i)$.

4. Encrypt $e(A_i, \widetilde{h_1 h_2}) \cdot m_i$ to IBE ciphertext E_i using index set $\mathcal{V}_S$, where $\mathcal{V}_S$ is as defined in algorithm IdSkIssue. Take $r_i \xleftarrow{\$} \mathbb{Z}_p$ and compute

$$E_i^{(1)} = e(A_i, \widetilde{h_1 h_2}) \cdot m_i \cdot e(\widehat{g_1}, \widehat{g_2})^{r_i}, E_i^{(2)} = g^{r_i}, \ E_i^{(3)} = (f' \prod_{l \in \mathcal{V}_S} f_l)^{r_i},$$

using $\mathsf{PK}_1 = (\widehat{g_1}, \widehat{g_2}, f', f_1, f_2, \ldots, f_n)$. The component $E_i = (E_i^{(1)}, E_i^{(2)}, E_i^{(3)})$ is the IBE [17] of $e(A_i, \widetilde{h_1 h_2}) \cdot m_i$ where the identity ID_S is embedded through $\mathcal{V}_S$ implicitly in $E_i^{(3)}$.

5. Set $\Phi_i = (A_i, D_i, E_i, \mathsf{AP}_i)$ along with the description of $(\mathbb{M}_i, \eta_i)$.

The ciphertext database is set to be $\mathsf{cDB}_S = (\Phi_1, \Phi_2, \ldots, \Phi_{N_S})$. The algorithm outputs $(\mathsf{pk}_{\mathsf{DB}_S}, \mathsf{sk}_{\mathsf{DB}_S}, \psi_{\mathsf{DB}_S}, \mathsf{cDB}_S)$ to S. The sender S publishes $\mathsf{pk}_{\mathsf{DB}_S}, \psi_{\mathsf{DB}_S}, \mathsf{cDB}_S$ to all parties and keeps $\mathsf{sk}_{\mathsf{DB}_S}$ secret to itself.

DBVerify($\mathsf{crs}, \mathsf{pk}_{\mathsf{DB}_S}, \psi_{\mathsf{DB}_S}, \mathsf{cDB}_S, \mathsf{PK}_1, \mathsf{ID}_S$): A receiver R upon receiving NIZK proof ψ_{DB_S}, ciphertext database cDB_S from S runs this algorithm. Using $\mathsf{GS}_S = (\widetilde{u_1}, \widetilde{u_2}, \widetilde{u_3})$ extracted from crs and PK_1, the algorithm first checks the correctness of proof ψ_{DB_S}, which is NIZK proof for three linear pairing product equations, with witnesses $d_1, d_2^{-1}, h_2, \widehat{g_1}^\gamma, g'$ in the similar manner as we have done for Eq. 1 in Sect. 2.3. The validity of cDB_S is checked by verifying the following pairing product equations. $e(E_i^{(2)}, f' \prod_{l \in \mathcal{V}_S} f_l) = e(g, E_i^{(3)})$, and $e(A_i, g^i \cdot y_1) = e(g, g)$, where $E_i = (E_i^{(1)}, E_i^{(2)} = g^{r_i}, E_i^{(3)} = (f' \prod_{l \in \mathcal{V}_S} f_l)^{r_i})$, $i = 1, 2, \ldots, N_S$. If the above equations hold, the algorithm outputs VALID, otherwise, INVALID.

AttSkIssue($\mathsf{params}, \mathsf{PK}_2, \mathsf{MSK}_2, w_R$): The issuer upon receiving the attribute set w_R from a receiver R runs this randomized algorithm to generate attribute secret key ASK_R as follows. The algorithm selects $t_R \xleftarrow{\$} \mathbb{Z}_p^*$, uses $\mathsf{PK}_2 = (g_3, U, h_1, h_2, \ldots, h_m), \mathsf{MSK}_2 = (c, \beta)$ and sets $K_0 = g^\beta g_3^{t_R}, K_0' = g^{t_R}, K_\ell = h_\ell^{t_R} \ \forall \ a_\ell \in w_R, \mathsf{ASK}_R = (K_0, K_0', K_\ell \ \forall \ a_\ell \in w_R)$. The algorithm outputs ASK_R to the issuer and the issuer sends it to R through a secure communication channel.

RequestTra($\mathsf{crs}, \mathsf{cDB}_S, \mathsf{ASK}_R, \sigma_j$): This randomized algorithm upon, receiving from R with attribute set w_R, input $\mathsf{crs}, \mathsf{cDB}_S = (\Phi_1, \Phi_2, \ldots, \Phi_{N_S}), \mathsf{ASK}_R$, index $\sigma_j \in [N_S]$ generates $(\mathsf{Req}_{\sigma_j}, \mathsf{Pri}_{\sigma_j})$ as follows, where $j = 1, 2, \ldots, k$.

1. Pick $\Phi_{\sigma_j} = (A_{\sigma_j}, D_{\sigma_j}, E_{\sigma_j}, \mathsf{AP}_{\sigma_j})$ from cDB_S along with the description of $(\mathbb{M}_{\sigma_j}, \eta_{\sigma_j})$.

2. Compute solution vector $\overrightarrow{x_{\sigma_j}} = \{x_\ell \in \mathbb{Z}_p \mid a_{\eta_{\sigma_j}(\ell)} \in w_R\}$ such that $\sum_{a_{\eta_{\sigma_j}(\ell)} \in w_R} \lambda_\ell x_\ell = s_{\sigma_j}$ holds using Reconstruct algorithm described in Sect. 2.2. The length of $\overrightarrow{x_{\sigma_j}}$ is equal to the number of attributes in w_R. To recover B_{σ_j} from $D_{\sigma_j} = (D_{\sigma_j}^{(1)} = B_{\sigma_j} \cdot U^{s_{\sigma_j}}, D_{\sigma_j}^{(2)} = g^{s_{\sigma_j}},$

$D_{\sigma_j}^{(3)} = \{D_{\sigma_j,\ell}^{(3)} = g_3^{\lambda_\ell} \cdot h_{\eta_{\sigma_j}(\ell)}^{-s_{\sigma_j}} \; \forall \, a_{\eta_{\sigma_j}}(\ell) \in \mathsf{AP}_{\sigma_j}\})$ using $\mathsf{ASK}_R = (K_0 = g^\beta g_3^{t_R}, K_0' = g^{t_R}, K_{\eta_{\sigma_j}(\ell)} = h_{\eta_{\sigma_j}(\ell)}^{t_R} \; \forall \, a_{\eta_{\sigma_j}}(\ell) \in w_R)$, first compute

$$B = \frac{e\left(D_{\sigma_j}^{(2)}, K_0\right)}{\displaystyle\prod_{a_{\eta_{\sigma_j}}(\ell) \in w_R} e\left(D_{\sigma_j,\ell}^{(3)}, K_0'\right)^{x_\ell} \prod_{a_{\eta_{\sigma_j}}(\ell) \in w_R} e\left(K_{\eta_{\sigma_j}(\ell)}, D_{\sigma_j}^{(2)}\right)^{x_\ell}}$$

$$= \frac{e(g^{s_{\sigma_j}}, g^\beta \cdot g_3^{t_R})}{\displaystyle\prod_{a_{\eta_{\sigma_j}}(\ell) \in w_R} e\left(g_3^{\lambda_\ell} \cdot h_{\eta_{\sigma_j}(\ell)}^{-s_{\sigma_j}}, g^{t_R}\right)^{x_\ell} \prod_{a_{\eta_{\sigma_j}}(\ell) \in w_R} e\left(h_{\eta_{\sigma_j}(\ell)}^{t_R}, g^{s_{\sigma_j}}\right)^{x_\ell}}$$

$$= e(g,g)^{s_{\sigma_j}\beta} = U^{s_{\sigma_j}} \text{ as } g_3 = g^c, \sum_{a_{\eta_{\sigma_j}}(\ell) \in w_R} \lambda_\ell x_\ell = s_{\sigma_j}.$$

$$\text{Now} \quad \frac{D_{\sigma_j}^{(1)}}{B} = \frac{B_{\sigma_j} \cdot U^{s_{\sigma_j}}}{B} = B_{\sigma_j}. \tag{5}$$

Check if $B_{\sigma_j}^{\sigma_j} \cdot e(A_{\sigma_j}, y_2) = H_1$. If the verification does not hold, abort the execution.

3. Otherwise, choose $v_{1,\sigma_j}, v_{2,\sigma_j}, v_{3,\sigma_j} \xleftarrow{\$} \mathbb{Z}_p^*$ and set $V_{\sigma_j} = B_{\sigma_j}^{\sigma_j \cdot v_{1,\sigma_j}}$, $X_{\sigma_j} = E_{\sigma_j}^{(2)} \cdot g^{v_{2,\sigma_j}} = g^{r_{\sigma_j} + v_{2,\sigma_j}}$, $Y_{\sigma_j} = E_{\sigma_j}^{(3)} \cdot (f' \prod_{l \in \mathcal{V}_S} f_l)^{v_{2,\sigma_j}} = (f' \prod_{l \in \mathcal{V}_S} f_l)^{r_{\sigma_j} + v_{2,\sigma_j}}$, $Z_{\sigma_j} = A_{\sigma_j} \cdot g^{v_{3,\sigma_j}}$, $t_{1,\sigma_j} = \Lambda_{\sigma_j}^{v_{1,\sigma_j}}$, $t_{2,\sigma_j} = \widehat{g_2}^{v_{2,\sigma_j}}$, $t_{3,\sigma_j} = (y^{\sigma_j} \cdot y_1)^{v_{3,\sigma_j}}$.

4. Generate NIWI proof π_{σ_j} using $\mathsf{GS}_R = (u_1, u_2, u_3)$ as

$$\pi_{\sigma_j} = \mathsf{NIWI}\{(g^{v_{1,\sigma_j}}, E_{\sigma_j}^{(2)}, E_{\sigma_j}^{(3)}, A_{\sigma_j}, t_{1,\sigma_j}, t_{2,\sigma_j}, t_{3,\sigma_j}, g^{\sigma_j}) \mid$$

$$V_{\sigma_j} = e(g^{v_{1,\sigma_j}}, \widehat{h_1})e(t_{1,\sigma_j}, y_2^{-1}) \wedge e(E_{\sigma_j}^{(2)}, \widehat{g_2})e(t_{2,\sigma_j}, g) = e(X_{\sigma_j}, \widehat{g_2})$$

$$\wedge \, e(E_{\sigma_j}^{(3)}, \widehat{g_2})e(t_{2,\sigma_j}, f' \prod_{l \in \mathcal{V}_S} f_l) = e(Y_{\sigma_j}, \widehat{g_2})$$

$$\wedge \, e(Z_{\sigma_j}, g^{\sigma_j} \cdot y_1) = e(g,g)e(g, t_{3,\sigma_j})\}.$$

As explained in Sect. 2.3, the proof π_{σ_j} consists of commitments to $g^{v_{1,\sigma_j}}, E_{\sigma_j}^{(2)}, E_{\sigma_j}^{(3)}, A_{\sigma_j}, t_{1,\sigma_j}, t_{2,\sigma_j}, t_{3,\sigma_j}, g^{\sigma_j}$, randomized $\mathsf{Com}(\widehat{h_1})$, randomized $\mathsf{Com}(\widehat{h_2})$ together with proof components to 4 linear pairing product equations. The $\mathsf{Com}(\widehat{h_1})$ and $\mathsf{Com}(\widehat{h_2})$ are randomized as discussed in the remark 1 using GS_R. Here $\mathsf{Com}(\widehat{h_1})$ and $\mathsf{Com}(\widehat{h_2})$ were generated by the sender in InitDB algorithm using GS_R.

5. Set $\mathsf{Req}_{\sigma_j} = (V_{\sigma_j}, X_{\sigma_j}, Y_{\sigma_j}, Z_{\sigma_j}, \pi_{\sigma_j})$ and $\mathsf{Pri}_{\sigma_j} = (v_{3,\sigma_j}, t_{1,\sigma_j}, t_{2,\sigma_j}, t_{3,\sigma_j})$. The algorithm outputs $(\mathsf{Req}_{\sigma_j}, \mathsf{Pri}_{\sigma_j})$ to R. The receiver R sends Req_{σ_j} to S and keeps Pri_{σ_j} secret to itself.

ResponseTra($\text{crs}, \text{sk}_{\text{DB}_S}, \text{Req}_{\sigma_j}, \text{SK}_S, \text{PK}_1, \text{pk}_{\text{DB}_S}$): This randomized algorithm upon receiving the input $\text{Req}_{\sigma_j} = (V_{\sigma_j}, X_{\sigma_j}, Y_{\sigma_j}, Z_{\sigma_j}, \pi_{\sigma_j})$ from S generates response Res_{σ_j} using database secret key $\text{sk}_{\text{DB}_S} = (x, \gamma, \widehat{h_1}, \widehat{h_2})$ and identity secret key $\text{SK}_S = (d_1, d_2)$ as follows.

1. The algorithm first verifies π_{σ_j}. If verification fails, abort the execution. Otherwise, set $a_1 = Z_{\sigma_j}^{\gamma}$, $s_{\sigma_j} = e(Z_{\sigma_j}, \widehat{h_2}) \cdot e(d_1, X_{\sigma_j}) e(d_2^{-1}, Y_{\sigma_j})$ using $\text{sk}_{\text{DB}_S} = (x, \gamma, \widehat{h_1}, \widehat{h_2})$ and $\text{SK}_S = (d_1, d_2)$.

2. Generate NIZK proof δ_{σ_j} by using GS_S as

$$\delta_{\sigma_j} = \text{NIZK}\{(a_1, a_2, a_3, a_4, a_5, a_6, a_7, a_8) \mid e(a_3, g) e(a_4, f' \prod_{l \in \mathcal{V}_S} f_l) e(a_5, \widehat{g_2}^{-1}) = 1$$

$$\wedge \ e(Z_{\sigma_j}, a_2) e(X_{\sigma_j}, a_3) e(Y_{\sigma_j}, a_4) e(a_1, g^{-1}) e(a_6^{-1}, a_3) e(a_7^{-1}, a_4) = 1$$

$$\wedge \ e(\widehat{g_1}, a_5) = e(\widehat{g_1}, \widehat{g_2}) \wedge e(a_6, g) = e(X_{\sigma_j}, g) \wedge e(a_7, g) = e(Y_{\sigma_j}, g)$$

$$\wedge \ e(a_5, a_2) e(a_8, g^{-1}) = 1\},$$

where $a_2 = \widehat{h_2}$, $a_3 = d_1$, $a_4 = d_2^{-1}$, $a_5 = \widehat{g_2}$, $a_6 = X_{\sigma_j}$, $a_7 = Y_{\sigma_j}$, $a_8 = \widehat{g_1}^{\gamma}$.

3. Set $\text{Res}_{\sigma_j} = (s_{\sigma_j}, \delta_{\sigma_j})$.

CompleteTra($\text{crs}, \text{PK}_1, \text{Res}_{\sigma_j}, \Phi_{\sigma_j}, \text{Pri}_{\sigma_j}$): The receiver R upon receiving $\text{Res}_{\sigma_j} = (s_{\sigma_j}, \delta_{\sigma_j})$ runs this deterministic algorithm as follows.

1. Verify δ_{σ_j} using $\text{GS}_S = (\widetilde{u_1}, \widetilde{u_2}, \widetilde{u_3})$ in the similar way as we have done for Eq. 1 in Sect. 2.3.

2. If verification fails, abort the execution. Otherwise, parse $\text{Pri}_{\sigma_j} = (v_{3,\sigma_j}, t_{1,\sigma_j}, t_{2,\sigma_j}, t_{3,\sigma_j})$, where $t_{2,\sigma_j} = \widehat{g_2}^{v_{2,\sigma_j}}$ and compute

$$\frac{E_{\sigma_j}^{(1)} \cdot e(\widehat{g_1}, t_{2,\sigma_j}) H_2^{v_{3,\sigma_j}}}{s_{\sigma_j} \cdot B_{\sigma_j}} = m_{\sigma_j}, \tag{6}$$

where $E_{\sigma_j}^{(1)} = e(A_{\sigma_j}, \widehat{h_1 h_2}) e(\widehat{g_1}, \widehat{g_2})^{r_{\sigma_j}} \cdot m_{\sigma_j}$, $t_{2,\sigma_j} = \widehat{g_2}^{v_{2,\sigma_j}}$, $B_{\sigma_j} = e(A_{\sigma_j}, \widehat{h_1})$, $H_2 = e(g, \widehat{h_2})$.

4   Security Analysis

Theorem 4. *The IBAOT-AC protocol Ψ presented in Sect. 3 securely realizes the ideal functionality $\mathcal{F}_{\text{IBAOT-AC}}$ in the $\mathcal{F}_{\text{CRS}}^{\mathcal{D}}$-hybrid model described in Sect. 2.4 assuming the hardness of DLIN, q-SDH, DBDHE and DBDH problems.*

Proof. Let $\mathcal{A}$ be a static real world adversary interacting with multiple senders, multiple receivers and an issuer. We construct an ideal world adversary $\mathcal{A}'$ interacting with $\mathcal{F}_{\text{IBAOT-AC}}$ such that no environment machine $\mathcal{Z}$ can distinguish the ideal world of ideal functionality $\mathcal{F}_{\text{IBAOT-AC}}$ with $\mathcal{A}'$ from the real world of protocol Ψ with $\mathcal{A}$. For instance, we consider the interaction between a sender S with identity ID_S, a receiver R and an issuer. We consider simultaneously the following cases: (a) when only the receiver R is honest, (b) when only the sender

S is corrupt (c) when only the receiver R is corrupt (d) when only the sender S is honest. When all the parties (the sender S, the receiver R and the issuer) are honest or all the parties are corrupt or only the issuer is honest or only the issuer is corrupt are not addressed as these are trivial cases. The collusion of the issuer with any party is restricted.

The security proof is presented using sequence of hybrid games. Let $\Pr[\text{Game } i]$ be the probability that $\mathcal{Z}$ distinguishes the transcript of Game i from that in the real execution.

(a) Simulation when the sender S and the issuer are corrupt while the receiver R is honest. The adversary $\mathcal{A}'$ simulates the honest receiver R without having the knowledge of selected index σ_j of R. The sender S and issuer are controlled by $\mathcal{A}$.

<u>Game 0</u>: This game corresponds to the real world protocol interactions in which $\mathcal{A}'$ simulates R and interacts with $\mathcal{A}$ exactly as in the real world. So, $\Pr[\text{Game } 0] = 0$.

<u>Game 1</u>: This game is the same as Game 0 except that crs is simulated by $\mathcal{A}'$ as follows. It generates $\text{params} = (p, \mathbb{G}, \mathbb{G}_T, e, g) \leftarrow \text{BilinearSetup}(1^\rho)$, chooses $a, b, \xi_1, \xi_2, \tilde{a}, \tilde{b}, \tilde{\xi}_1, \tilde{\xi}_2 \xleftarrow{\$} \mathbb{Z}_p^*$ and sets $g_1 = g^a, g_2 = g^b, \tilde{g}_1 = g^{\tilde{a}}, \tilde{g}_2 = g^{\tilde{b}}, u_1 = (g_1, 1, g), u_2 = (1, g_2, g), u_3 = u_1^{\xi_1} u_2^{\xi_2} = (g_1^{\xi_1}, g_2^{\xi_2}, g^{\xi_1 + \xi_2}), \widetilde{u_1} = (\tilde{g}_1, 1, g), \widetilde{u_2} = (1, \tilde{g}_2, g), \widetilde{u_3} = \widetilde{u_1}^{\tilde{\xi}_1} \widetilde{u_2}^{\tilde{\xi}_2} = (\tilde{g}_1^{\tilde{\xi}_1}, \tilde{g}_2^{\tilde{\xi}_2}, g^{\tilde{\xi}_1 + \tilde{\xi}_2}), \text{GS}_R = (u_1, u_2, u_3), \text{GS}_S = (\widetilde{u_1}, \widetilde{u_2}, \widetilde{u_3}), \text{crs} = (\text{params}, \text{GS}_R, \text{GS}_S), \text{trapdoors } t_{\text{ext}} = (a, b, \xi_1, \xi_2), \widetilde{t_{\text{ext}}} = (\tilde{a}, \tilde{b}, \tilde{\xi}_1, \tilde{\xi}_2)$. When the parties ask $\langle \text{CRS} : \text{sid} \rangle$, $\mathcal{A}'$ returns $\langle \text{CRS} : \text{sid}, \text{crs} \rangle$. The adversary $\mathcal{A}'$ keeps the trapdoors t_{ext} and $\widetilde{t_{\text{ext}}}$ secret to itself. The crs generated in Game 1 has the same distribution as in Game 0. Hence, $|\Pr[\text{Game } 1] - \Pr[\text{Game } 0]| = 0$.

<u>Game 2</u>: This game is the same as Game 1 except that $\mathcal{A}'$ replaces the honest request Req_{σ_j} by the simulated request $\text{Req}_{\sigma_j'}$ generated as follows.

- When $\mathcal{A}'$ receives $(\text{pk}_{\text{DB}_S}, \psi_{\text{DB}_S}, \text{cDB}_S)$ from $\mathcal{A}$, $\mathcal{A}'$ checks the validity of ψ_{DB_S} and cDB_S by running the algorithm DBVerify discussed in Sect. 3. If the verification does not hold, $\mathcal{A}'$ aborts the execution. Otherwise, $\mathcal{A}'$ parses $\text{cDB}_S = (\Phi_1, \Phi_2, \ldots, \Phi_{N_S})$ and decrypts each Φ_i using trapdoors t_{ext} and $\widetilde{t_{\text{ext}}}$ by executing the following steps.
  - Note that $\text{Com}(d_1)$ is embedded in ψ_{DB_S}. As $\text{Com}(d_1) = \mu(d_1)\widetilde{u_1}^{\hat{b}_1} \widetilde{u_2}^{\hat{b}_2} \widetilde{u_3}^{\hat{b}_3}$
    $= (\tilde{g}_1^{\hat{b}_1 + \hat{b}_3 \tilde{\xi}_1}, \tilde{g}_2^{\hat{b}_2 + \hat{b}_3 \tilde{\xi}_2}, d_1 g^{\hat{b}_1 + \hat{b}_2 + \hat{b}_3(\tilde{\xi}_1 + \tilde{\xi}_2)}), \hat{b}_1, \hat{b}_2, \hat{b}_3 \xleftarrow{\$} \mathbb{Z}_p, \mathcal{A}'$ can extract
    $d_1 = \hat{g}_2^{\alpha} \left(f' \prod_{l \in \mathcal{V}_S} f_l\right)^{ts}$ by computing $\dfrac{d_1 g^{\hat{b}_1 + \hat{b}_2 + \hat{b}_3(\tilde{\xi}_1 + \tilde{\xi}_2)}}{(\tilde{g}_1^{\hat{b}_1 + \hat{b}_3 \tilde{\xi}_1})^{\frac{1}{\tilde{a}}} (\tilde{g}_2^{\hat{b}_2 + \hat{b}_3 \tilde{\xi}_2})^{\frac{1}{\tilde{b}}}} = d_1$ as $\tilde{g}_1 =$
    $g^{\tilde{a}}, \tilde{g}_2 = g^{\tilde{b}}$ and $\mathcal{A}'$ knows $\widetilde{t_{\text{ext}}} = (\tilde{a}, \tilde{b}, \tilde{\xi}_1, \tilde{\xi}_2)$. Similarly, $\mathcal{A}'$ extracts $d_2 = g^{ts}, \hat{h}_2, \hat{g}_1^{\gamma}$ using $\widetilde{t_{\text{ext}}}$ and $\hat{h}_1$ using t_{ext}.

- From each $\Phi_i = (A_i, D_i, E_i, \mathsf{AP}_i)$ along with the description of $(\mathbb{M}_i, \eta_i)$, $\mathcal{A}'$ computes $\mathsf{val} = \dfrac{e(A_i, \widehat{h_1 h_2}) e(d_1, E_i^{(2)})}{e(d_2, E_i^{(3)})} = \dfrac{e(A_i, \widehat{h_1 h_2}) e(\widehat{g_2}^{\alpha} \left(f' \prod_{l \in \mathcal{V}_S} f_l\right)^{ts}, g^{r_i})}{e(g^{ts}, (f' \prod_{l \in \mathcal{V}_S} f_l)^{r_i}))} =$ $e(A_i, \widehat{h_1 h_2}) e(\widehat{g_2}^{\alpha}, g^{r_i}) = e(A_i, \widehat{h_1 h_2}) e(\widehat{g_1}, \widehat{g_2})^{r_i}$ as $\widehat{g_1} = g^{\alpha}$ and computes $\dfrac{E_i^{(1)}}{\mathsf{val}} = \dfrac{e(A_i, \widehat{h_1 h_2}) \cdot m_i \cdot e(\widehat{g_1}, \widehat{g_2})^{r_i}}{e(A_i, \widehat{h_1 h_2}) e(\widehat{g_1}, \widehat{g_2})^{r_i}} = m_i$.

- The adversary $\mathcal{A}'$ sets $\mathsf{DB}_S = ((m_1, \mathsf{AP}_1), (m_2, \mathsf{AP}_2), \ldots, (m_{N_S}, \mathsf{AP}_{N_S}))$ and sends $\langle \mathsf{dbsetup} : \mathsf{sid}, \mathsf{ID}_S, \mathsf{DB}_S \rangle$ to $\mathcal{F}_{\mathsf{IBAOT-AC}}$.

- The adversary $\mathcal{A}'$, upon receiving the message $\langle \mathsf{attsk} : \mathsf{sid}, w_R \rangle$ from $\mathcal{F}_{\mathsf{IBAOT-AC}}$, simulates the receiver's side of the $\mathsf{AttSkIssue}$ protocol with the real corrupted issuer handled by $\mathcal{A}$. If $\mathcal{A}'$ obtains the valid attribute secret key ASK_R, $\mathcal{A}'$ sends $\langle \mathsf{attsk} : \mathsf{sid}, b = 1 \rangle$ to $\mathcal{F}_{\mathsf{IBAOT-AC}}$, otherwise, $\langle \mathsf{attsk} : \mathsf{sid}, b = 0 \rangle$.

- The validity of ASK_R is checked as follows. The adversary $\mathcal{A}'$ picks any $\Phi_{\sigma'_j} = (A_{\sigma'_j}, D_{\sigma'_j}, E_{\sigma'_j}, \mathsf{AP}_{\sigma'_j})$ along with the description of $(\mathbb{M}_{\sigma'_j}, \eta_{\sigma'_j})$ from cDB_S for which w_R satisfies $\mathsf{AP}_{\sigma'_j}$, recovers $B_{\sigma'_j} = e(A_{\sigma'_j}, \widehat{h_1})$ as discussed in Eq. 5 given in algorithm $\mathsf{RequestTra}$ and checks the validity of equation $B_{\sigma'_j}^{\sigma'_j} e(A_{\sigma'_j}, y_2) = H_1$. If the equation does not hold, $\mathcal{A}'$ sets the bit $b = 0$, otherwise, $b = 1$.

- Upon receiving the message $\langle \mathsf{transfer} : \mathsf{sid}, \mathsf{ID}_S \rangle$ from $\mathcal{F}_{\mathsf{IBAOT-AC}}$, $\mathcal{A}'$ picks random index σ'_j of its choice for which attribute set w_R satisfies $\mathsf{AP}_{\sigma'_j}$ and generates $(\mathsf{Req}_{\sigma'_j}, \mathsf{Pri}_{\sigma'_j}) \leftarrow \mathsf{RequestTra}(\mathsf{crs}, \mathsf{cDB}_S, \mathsf{ASK}_R, \sigma'_j)$, where $\mathsf{Req}_{\sigma'_j} = \left(V_{\sigma'_j}, X_{\sigma'_j}, Y_{\sigma'_j}, Z_{\sigma'_j}, \pi_{\sigma'_j} \right)$, $\mathsf{Pri}_{\sigma'_j} = (v_{3,\sigma'_j}, t_{1,\sigma'_j}, t_{2,\sigma'_j}, t_{3,\sigma'_j})$, $V_{\sigma'_j} = B_{\sigma'_j}^{v_{1,\sigma'_j} \sigma'_j}$, $X_{\sigma'_j} = E_{\sigma'_j}^{(2)} \cdot g^{v_{2\sigma'_j}}$, $Y_{\sigma'_j} = E_{\sigma'_j}^{(3)} \cdot (f' \prod_{l \in \mathcal{V}_S} f_l)^{v_{2\sigma'_j}}$, $Z_{\sigma'_j} = A_{\sigma'_j} g^{v_{3,\sigma'_j}}$, $t_{1,\sigma'_j} = A_{\sigma'_j}^{v_{1,\sigma'_j}}$, $t_{2,\sigma'_j} = \widehat{g_2}^{v_{2,\sigma'_j}}$, $t_{3,\sigma'_j} = (g^{\sigma'_j} y_1)^{v_{3,\sigma'_j}}$, $v_{1,\sigma'_j}, v_{2,\sigma'_j}, v_{3,\sigma'_j} \xleftarrow{\$} \mathbb{Z}_p$.

The adversary $\mathcal{A}'$ replaces honest receiver R's request Req_{σ_j} by the simulated request $\mathsf{Req}_{\sigma'_j}$ and sends $\langle \mathsf{transfer} : \mathsf{sid}, \mathsf{ID}_S, \mathsf{Req}_{\sigma'_j} \rangle$ to $\mathcal{A}$ as if it is from real receiver R. The adversary $\mathcal{A}$ returns $\mathsf{Res}_{\sigma'_j} = (s_{\sigma'_j}, \delta_{\sigma'_j})$ to $\mathcal{A}'$ and $\mathcal{A}'$ checks whether $m_{\sigma'_j} = \mathsf{CompleteTra}(\mathsf{crs}, \mathsf{Res}_{\sigma'_j}, \Phi_{\sigma'_j}, \mathsf{Pri}_{\sigma'_j})$. If so, then $\mathcal{A}'$ sets $b = 1$, otherwise, $b = 0$ and returns $\langle \mathsf{transfer} : \mathsf{sid}, \mathsf{ID}_S, b \rangle$ to $\mathcal{F}_{\mathsf{IBAOT-AC}}$. As Groth-Sahai proofs are composable NIWI [12] under DLIN assumption as stated in Theorem 3, the simulated request $\mathsf{Req}_{\sigma'_j}$ is computationally indistinguishable from the honestly generated request Req_{σ_j}. Therefore, we have $|\Pr[\mathsf{Game}\ 2] - \Pr[\mathsf{Game}\ 1]| \leq \epsilon_1(\rho)$, where $\epsilon_1(\rho)$ is a negligible function.

Thus $\mathsf{Game}\ 2$ is the ideal world interaction whereas $\mathsf{Game}\ 0$ is the real world. Now $|\Pr[\mathsf{Game}\ 2] - [\mathsf{Game}\ 0]| \leq |\Pr[\mathsf{Game}\ 2] - [\mathsf{Game}\ 1]| + |\Pr[\mathsf{Game}\ 1] - [\mathsf{Game}\ 0]| \leq \epsilon_1(\rho)$, where $\epsilon_1(\rho)$ is a negligible function. Hence, $\mathsf{IDEAL}_{\mathcal{F}_{\mathsf{IBAOT-AC}}, \mathcal{A}', \mathcal{Z}} \overset{c}{\approx} \mathsf{REAL}_{\Psi, \mathcal{A}, \mathcal{Z}}$.

(b) Simulation when the sender S (b)is corrupt while the receiver R and the issuer are honest. In this case, $\mathcal{A}'$ simulates the honest receiver R and honest issuer without knowing index σ_j of R. The simulation of this case is very much similar to Case(a) except that $\mathcal{A}'$ runs the ISetup algorithm on behalf of real issuer upon receiving the message $\langle \mathsf{isetup} : \mathsf{sid} \rangle$ from $\mathcal{F}_{\mathsf{IBAOT-AC}}$. The ISetup outputs $\mathsf{PK}_1 = (\widehat{g_1}, \widehat{g_2}, f', f_1, f_2, \ldots, f_n)$, $\mathsf{MSK}_1 = \widehat{g_2}^{\alpha}$, $\mathsf{PK}_2 = (g_3, U, h_1, h_2, \ldots, h_m)$,

$\mathsf{MSK}_2 = (c, \beta)$ to $\mathcal{A}'$ and $\mathcal{A}'$ broadcasts $\mathsf{PK}_1, \mathsf{PK}_2$ to all parties keeping MSK_1, MSK_2 secret to itself. In this case, $\mathcal{A}'$ extracts m_i from ciphertext $\Phi_i = (A_i, D_i, E_i,$ $\mathsf{AP}_i)$ along with the description of $(\mathbb{M}_i, \eta_i)$ using extracted $\widehat{h_1}, \widehat{h_2}, d_1, d_2$ as in the above case. Also, $\mathcal{A}'$ simulates the issuer side of the IdSkIssue protocol upon receiving the message $\langle \mathsf{idsk} : \mathsf{sid}, \mathsf{ID}_S \rangle$ from $\mathcal{A}$. The adversary $\mathcal{A}'$ sends the message $\langle \mathsf{idsk} : \mathsf{sid}, \mathsf{ID}_S \rangle$ to $\mathcal{F}_{\mathsf{IBAOT-AC}}$. If $\mathcal{F}_{\mathsf{IBAOT-AC}}$ sends the bit $b = 1$ to $\mathcal{A}'$, $\mathcal{A}'$ issues SK_S to $\mathcal{A}$. Otherwise, $\mathcal{A}'$ sends $\perp$ to $\mathcal{A}$. Hence, $\mathsf{IDEAL}_{\mathcal{F}_{\mathsf{IBAOT-AC}}, \mathcal{A}', \mathcal{Z}} \overset{c}{\approx} \mathsf{REAL}_{\Psi, \mathcal{A}, \mathcal{Z}}$.

(c) Simulation when the sender S and the issuer are honest while the receiver R (c)is corrupt. In this case, the adversary $\mathcal{A}$ controls the corrupted receiver R and the $\mathcal{A}'$ simulates the honest sender S and issuer.

<u>Game 0</u>: This game corresponds to the real world protocol interactions in which $\mathcal{A}'$ simulates S and the issuer and interacts with $\mathcal{A}$ exactly as in the real world. So, $\Pr[\mathsf{Game}\ 0] = 0$.

<u>Game 1</u>: This game is the same as Game 0 except that crs is simulated by $\mathcal{A}'$ as follows. It generates $\mathsf{params} = (p, \mathbb{G}, \mathbb{G}_T, e, g) \leftarrow \mathsf{BilinearSetup}(1^\rho)$, chooses $a, b, \xi_1, \xi_2, \widetilde{a}, \widetilde{b}, \widetilde{\xi_1}, \widetilde{\xi_2} \xleftarrow{\$} \mathbb{Z}_p^*$ and sets $g_1 = g^a, g_2 = g^b, \widehat{g}_1 = g^{\widetilde{a}}, \widehat{g}_2 = g^{\widetilde{b}}, u_1 = (g_1, 1, g), u_2 = (1, g_2, g), u_3 = u_1^{\xi_1} u_2^{\xi_2} = (g_1^{\xi_1}, g_2^{\xi_2}, g^{\xi_1 + \xi_2}), \widetilde{u_1} = (\widehat{g}_1, 1, g), \widetilde{u_2} = (1, \widehat{g}_2, g), \widetilde{u_3} = \widetilde{u_1}^{\widetilde{\xi_1}} \widetilde{u_2}^{\widetilde{\xi_2}} (1, 1, g) = (\widehat{g}_1^{\widetilde{\xi_1}}, \widehat{g}_2^{\widetilde{\xi_2}}, g^{\widetilde{\xi_1} + \widetilde{\xi_2} + 1}), \mathsf{GS}_R = (u_1, u_2, u_3), \mathsf{GS}_S = (\widetilde{u_1}, \widetilde{u_2}, \widetilde{u_3}), \mathsf{crs} = (\mathsf{params}, \mathsf{GS}_R, \mathsf{GS}_S), \mathsf{trapdoors}\ \mathsf{t}_{\mathsf{ext}} = (a, b, \xi_1, \xi_2), \widetilde{\mathsf{t}_{\mathsf{sim}}} = (\widetilde{a}, \widetilde{b}, \widetilde{\xi_1}, \widetilde{\xi_2})$. The adversary $\mathcal{A}'$ generates GS_R in perfectly sound setting and GS_S in witness indistinguishability setting. When the parties query $\langle \mathsf{CRS} : \mathsf{sid} \rangle$, $\mathcal{A}'$ returns $\langle \mathsf{CRS} : \mathsf{sid}, \mathsf{crs} \rangle$. The adversary $\mathcal{A}'$ keeps the trapdoors $\mathsf{t}_{\mathsf{ext}}$ and $\widetilde{\mathsf{t}_{\mathsf{sim}}}$ secret to itself. The crs generated in Game 1 by $\mathcal{A}'$ and CRSSetup in actual protocol run are computationally indistinguishable by Theorem 2. Therefore, there exists a negligible function $\epsilon_1(\rho)$ such that $|\Pr[\mathsf{Game}\ 1] - \Pr[\mathsf{Game}\ 0]| \leq \epsilon_1(\rho)$.

<u>Game 2</u>: The adversary $\mathcal{A}'$ upon receiving the message $\langle \mathsf{isetup} : \mathsf{sid} \rangle$ from $\mathcal{F}_{\mathsf{IBAOT-AC}}$ runs the ISetup algorithm and generates $\mathsf{PK}_1 = (\widehat{g}_1, \widehat{g}_2, f', f_1, f_2, \ldots, f_n), \mathsf{MSK}_1 = \widehat{g}_2^\alpha, \mathsf{PK}_2 = (g_3, U, h_1, h_2, \ldots, h_m), \mathsf{MSK}_2 = (c, \beta)$. The adversary $\mathcal{A}'$ broadcasts $\mathsf{PK}_1, \mathsf{PK}_2$ to all parties keeping $\mathsf{MSK}_1, \mathsf{MSK}_2$ secret to itself. The $\mathsf{PK}_1, \mathsf{PK}_2, \mathsf{MSK}_1, \mathsf{MSK}_2$ generated in Game 2 has the same distribution as in Game 1. Hence, $|\Pr[\mathsf{Game}\ 2] - \Pr[\mathsf{Game}\ 1]| = 0$.

<u>Game 3</u>: This game is exactly the same as Game 2 except that upon receiving $\mathsf{Req}_{\sigma_j} = (V_{\sigma_j}, X_{\sigma_j}, Y_{\sigma_j}, Z_{\sigma_j}, \pi_{\sigma_j})$ from $\mathcal{A}$, $\mathcal{A}'$ checks the validity of π_{σ_j}. If invalid, $\mathcal{A}'$ aborts the execution, otherwise, $\mathcal{A}'$ uses $\mathsf{t}_{\mathsf{ext}} = (a, b, \xi_1, \xi_2)$ to extract witnesses and index from proof π_{σ_j} as follows.

– The adversary $\mathcal{A}'$ extracts first witness wit_1 from $\mathsf{Com}(g^{v_1, \sigma_j}) = \mu(g^{v_1, \sigma_j}) u_1^{\widetilde{r}_1} u_2^{\widetilde{r}_2} u_3^{\widetilde{r}_3} = (g_1^{\widetilde{r}_1 + \widetilde{r}_3 \xi_1}, g_2^{\widetilde{r}_2 + \widetilde{r}_3 \xi_2}, g^{v_1, \sigma_j} g^{\widetilde{r}_1 + \widetilde{r}_2 + \widetilde{r}_3 (\xi_1 + \xi_2)})$ as $\dfrac{g^{v_1, \sigma_j} g^{\widetilde{r}_1 + \widetilde{r}_2 + \widetilde{r}_3 (\xi_1 + \xi_2)}}{(g_1^{\widetilde{r}_1 + \widetilde{r}_3 \xi_1})^{\frac{1}{a}} (g_2^{\widetilde{r}_2 + \widetilde{r}_3 \xi_2})^{\frac{1}{b}}} = g^{v_1, \sigma_j} = \mathsf{wit}_1, \widetilde{r}_1, \widetilde{r}_2, \widetilde{r}_3 \xleftarrow{\$} \mathbb{Z}_p$. Similarly, $\mathcal{A}'$ extracts

$\text{wit}_2 = E_{\sigma_j}^{(2)}$, $\text{wit}_3 = E_{\sigma_j}^{(3)}$, $\text{wit}_4 = A_{\sigma_j}$, $\text{wit}_5 = t_{1,\sigma_j}$, $\text{wit}_6 = t_{2,\sigma_j}$, $\text{wit}_7 = t_{3,\sigma_j}$, $\text{wit}_8 = g^{\sigma_j}$, $\text{wit}_9 = \widehat{h_1}$, $\text{wit}_{10} = \widehat{h_2}$ from their respective commitments.

- The adversary $\mathcal{A}'$ checks whether $\text{wit}_4 = A_\zeta, \zeta = 1, 2, \ldots, N_S$. Suppose no matching index found, i.e., $\sigma_j \notin \{1, 2, \ldots, N_S\}$ and $\mathcal{A}$ constructs a valid proof π_{σ_j} for the ciphertext $\Phi_{\sigma_j} \notin \text{cDB}_S = (\Phi_1, \Phi_2, \ldots, \Phi_{N_S})$ in order to generate Req_{σ_j}. This eventually means that the ciphertext Φ_{σ_j} must be a correct ciphertext as the proof π_{σ_j} generated by $\mathcal{A}$ for Φ_{σ_j} is valid. This indicates that $\mathcal{A}$ generates a valid BB signature A_{σ_j} on index σ_j and outputs A_{σ_j} as a forgery contradicting the fact that the BB signature is unforgeable under chosen-message attack assuming q-SDH problem is hard [4].

The difference between Game 3 and Game 2 is negligible provided that q-SDH assumption hold. Therefore, there exists a negligible function $\epsilon_3(\rho)$ such that $|\Pr[\text{Game 3}] - \Pr[\text{Game 2}]| \leq \epsilon_3(\rho)$.

Game 4: This game is the same as Game 3. Let σ_j be the matching index extracted in Game 3. The adversary $\mathcal{A}'$ checks whether $\text{wit}_2 = E_{\sigma_j}^{(2)}$, $\text{wit}_3 = E_{\sigma_j}^{(3)}$, $\text{wit}_8 = g^{\sigma_j}$. The adversary $\mathcal{A}'$ computes $\widehat{V_{\sigma_j}} = e(t_{1,\sigma_j}, \text{wit}_9)^{\sigma_j} = e(A_{\sigma_j}^{v_{1,\sigma_j}}, \widehat{h_1})^{\sigma_j} = V_{\sigma_j}$. Note that V_{σ_j} is computed by $\mathcal{A}$ which requires to recover B_{σ_j} from the ciphertext Φ_{σ_j} using a matching attribute secret key ASK_R corresponding to attribute set w_R. Thus $\mathcal{A}$ can frame V_{σ_j} only if $\mathcal{A}$ has obtained matching ASK_R from the issuer by calling algorithm AttSkIssue on w_R. If $\mathcal{A}$ has never queried the issuer for ASK_R corresponding to w_R, then $\widehat{V_{\sigma_j}} = V_{\sigma_j}$ occurs with negligible probability. Otherwise, we can construct a solver to break the semantic security of CP-ABE of [18] under q-DBDHE assumption with black box access to $\mathcal{A}$. With extracted index σ_j, $\mathcal{A}'$ queries $\mathcal{F}_{\text{IBAOT-AC}}$ with the message $\langle \text{transfer} : \text{sid}, \text{ID}_S, \sigma_j \rangle$. The $\mathcal{F}_{\text{IBAOT-AC}}$ gives m_{σ_j} to $\mathcal{A}'$. The difference between Game 4 and Game 3 is negligible provided that DBDHE assumption hold. Therefore, there exists a negligible function $\epsilon_4(\rho)$ such that $|\Pr[\text{Game 4}] - \Pr[\text{Game 3}]| \leq \epsilon_4(\rho)$.

Game 5: This game is the same as Game 4 except that $\mathcal{A}'$ simulates the response s_{σ_j} and proof δ_{σ_j} for each transfer phase $j = 1, 2, \ldots, k$. The ciphertext $\Phi_{\sigma_j} = (A_{\sigma_j}, D_{\sigma_j}, E_{\sigma_j}, \text{AP}_{\sigma_j})$ along with the description of $(\mathbb{M}_{\sigma_j}, \eta_{\sigma_j})$. The adversary $\mathcal{A}'$ uses $\text{wit}_4 = A_{\sigma_j}$, $\text{wit}_6 = t_{2,\sigma_j} = \widehat{g_2}^{v_{2,\sigma_j}}$, $\text{wit}_9 = \widehat{h_1}, \text{wit}_{10} = \widehat{h_2}$, m_{σ_j} extracted in the previous game and $E_{\sigma_j} = (E_{\sigma_j}^{(1)} = e(A_{\sigma_j}, \widehat{h_1 h_2}) \cdot m_{\sigma_j} \cdot e(\widehat{g_1}, \widehat{g_2})^{r_{\sigma_j}}, E_{\sigma_j}^{(2)}, E_{\sigma_j}^{(3)})$ to simulate the response $\text{Res}'_{\sigma_j} = (s'_{\sigma_j}, \delta'_{\sigma_j})$ as follows. The component $s'_{\sigma_j} =$

$$\frac{E_{\sigma_j}^{(1)} e(\widehat{g_1}, \text{wit}_6) e(\frac{Z_{\sigma_j}}{\text{wit}_4}, \text{wit}_{10})}{e(\text{wit}_4, \text{wit}_9) m_{\sigma_j}} = \frac{m_{\sigma_j} e(A_{\sigma_j}, \widehat{h_1 h_2}) e(\widehat{g_1}, \widehat{g_2})^{r_{\sigma_j}} e(\widehat{g_1}, \widehat{g_2}^{v_{2,\sigma_j}}) e(g^{v_{3,\sigma_j}}, \widehat{h_2})}{e(A_{\sigma_j}, \widehat{h_1}) m_{\sigma_j}}$$

$$= e(A_{\sigma_j}, \widehat{h_2}) e(\widehat{g_1}, \widehat{g_2})^{r_{\sigma_j} + v_{2,\sigma_j}} \cdot H_2^{v_{3,\sigma_j}}.$$

The simulated s'_{σ_j} has the same distribution as honestly generated response s_{σ_j} by algorithm ResponseTra discussed in correctness of Eq. 6 in Sect. 3. The adversary $\mathcal{A}'$ also simulates δ'_{σ_j} to prove that s'_{σ_j} is correctly framed. The proof

$\delta_{\sigma_j} = \mathsf{NIZK}\{(a_1, a_2, a_3, a_4, a_5, a_6, a_7, a_8) \mid e(a_3, g)e(a_4, f' \prod_{l \in V_S} f_l)e(a_5, \widehat{g_2}^{-1}) = 1 \wedge e(Z_{\sigma_j}, a_2)e(X_{\sigma_j}, a_3)e(Y_{\sigma_j}, a_4)e(a_1, g^{-1})e(a_6^{-1}, a_3)e(a_7^{-1}, a_4) = 1 \wedge e(a_5, \widehat{g_2}) = e(\widehat{g_1}, \widehat{g_2}) \wedge e(a_6, g) = e(X_{\sigma_j}, g) \wedge e(a_7, g) = e(Y_{\sigma_j}, g) \wedge e(a_5, a_2)e(a_8, g^{-1}) = 1\}$
consists of commitments to secret values $a_1, a_2, a_3, a_4, a_5, a_6, a_7, a_8$ and proof components to 6 equations. For simulation, $\mathcal{A}'$ sets $a_1 = a_2 = a_3 = a_4 = a_5 = a_6 = a_7 = a_8 = 1$ and generate commitments to $a_1, a_2, a_3, a_4, a_5, a_6, a_7, a_8$ using GS_S. With the help of trapdoor $\widetilde{t_{\mathsf{sim}}}$, $\mathcal{A}'$ can open the commitment of $a_5 = 1$ in the first equation, $a_6 = 1$, $a_7 = 1$ in the second equation, and a_5 to $\widehat{g_1}$ in third equation, $a_6 = X_{\sigma_j}$ in the fourth equation and $a_7 = Y_{\sigma_j}$ in the fifth equation as discussed in Sect. 2.3. As Groth-Sahai proofs are composable NIZK by Theorem 3, the simulated proof δ'_{σ_j} is computationally indistinguishable from the honestly generated proof δ_{σ_j} under the DLIN assumption. Therefore, there exists a negligible function $\epsilon_5(\rho)$ such that $|\Pr[\mathsf{Game}\ 5] - \Pr[\mathsf{Game}\ 4]| \leq \epsilon_5(\rho)$.

Game 6: This game is the same as Game 5 except that the messages $m_1, m_2, \ldots, m_{N_S}$ are replaced by the random messages $\widehat{m}_1, \widehat{m}_2, \ldots, \widehat{m}_{N_S} \in \mathbb{G}$. Upon receiving the message $\langle \mathsf{attsk} : \mathsf{sid}, w_R \rangle$ from $\mathcal{A}$, $\mathcal{A}'$ simulates the issuer side of the AttSkIssue protocol. The adversary $\mathcal{A}'$ sends the message $\langle \mathsf{attsk} : \mathsf{sid}, w_R \rangle$ to $\mathcal{F}_{\mathsf{IBAOT-AC}}$. If $\mathcal{F}_{\mathsf{IBAOT-AC}}$ sends the bit $b = 1$ to $\mathcal{A}'$, $\mathcal{A}'$ issues ASK_R to $\mathcal{A}$. Otherwise, $\mathcal{A}'$ sends $\perp$ to $\mathcal{A}$. As both issuer and sender are honest, they can simulate IdSkIssue protocol upon receiving the message $\langle \mathsf{idsk} : \mathsf{sid}, \mathsf{ID}_S \rangle$ from $\mathcal{F}_{\mathsf{IBAOT-AC}}$ in order to generate identity secret key SK_S. The adversary $\mathcal{A}'$ runs the algorithm InitDB with input $(\mathsf{ID}_S, \mathsf{SK}_S, \mathsf{crs}, \mathsf{PK}_1, \mathsf{PK}_2, \mathsf{DB}'_S)$, where $\mathsf{DB}'_S = ((\widehat{m_1}, \mathsf{AP}_1), (\widehat{m_2}, \mathsf{AP}_2), \ldots, (\widehat{m_{N_S}}, \mathsf{AP}_{N_S}))$, $m_i \xleftarrow{\$} \mathbb{G}_T$, AP_i are publicly available, $i = 1, 2, \ldots, N_S$. The algorithm outputs $\mathsf{pk}_{\mathsf{DB}'_S}, \mathsf{sk}_{\mathsf{DB}'_S}, \psi_{\mathsf{DB}'_S}, \mathsf{cDB}'_S$ to $\mathcal{A}'$, where $\mathsf{pk}_{\mathsf{DB}'_S}, \mathsf{sk}_{\mathsf{DB}'_S}, \psi_{\mathsf{DB}'_S}$ have the same distribution as in the real protocol. The ciphertext database cDB'_S is the encryption of N_S random messages. In each transfer phase, the response $\mathsf{Res}_{\sigma_j} = (s_{\sigma_j}, \delta_{\sigma_j})$ is replaced by the simulated response $\mathsf{Res}'_{\sigma_j} = (s'_{\sigma_j}, \delta'_{\sigma_j})$ as in Game 5, but here the simulated response is computed on invalid statement. The only difference between Game 6 and Game 5 is in the generation of ciphertexts. In Game 5, cDB_S is encryption of perfect messages, whereas in this game cDB'_S is that of random messages. By the semantic security of IBE of [17] under the DBDH assumption, Game 5 and Game 6 are computationally indistinguishable. Therefore, $|\Pr[\mathsf{Game}\ 6] - \Pr[\mathsf{Game}\ 5]| \leq \epsilon_6(\rho)$, where $\epsilon_6(\rho)$ is a negligible function. Thus Game 6 is the ideal world interaction whereas Game 0 is the real world. Now $|\Pr[\mathsf{Game}\ 6] - [\mathsf{Game}\ 0]| \leq \sum_{l=1}^{6} |\Pr[\mathsf{Game}\ l] - [\mathsf{Game}\ (l-1)]| \leq \epsilon_7(\rho)$, where $\epsilon_7(\rho) = \epsilon_6(\rho) + \epsilon_5(\rho) + \epsilon_4(\rho) + \epsilon_3(\rho) + \epsilon_2(\rho) + \epsilon_1(\rho)$ is a negligible function. Hence, $\mathsf{IDEAL}_{\mathcal{F}_{\mathsf{IBAOT-AC}}, \mathcal{A}', \mathcal{Z}} \overset{c}{\approx} \mathsf{REAL}_{\Psi, \mathcal{A}, \mathcal{Z}}$.

(d) Simulation when the sender S is honest while the issuer and the receiver R are corrupt. In this case, $\mathcal{A}'$ simulates the honest sender S and $\mathcal{A}$ controls the activities of corrupted receiver R and the issuer. The simulation of this case is very much similar to Case(c) except that in this case, $\mathcal{A}'$ upon receiving the message $\langle \mathsf{idsk} : \mathsf{sid}, \mathsf{ID}_S \rangle$ from $\mathcal{F}_{\mathsf{IBAOT-AC}}$, simulates the sender S's side of AttSkIssue protocol with the real corrupted issuer controlled by $\mathcal{A}$. The adversary $\mathcal{A}'$ checks the validity of obtained identity secret key $\mathsf{SK}_S = (d_1, d_2)$

Table 1. Comparison Summary of computation cost in k transfer phases and initialization phase (PO stands for number of pairings, EXP for number of exponentiations, CRSG for crs generation, AP for access policy, m is the number of attributes, N and N_S are the database sizes, n_i is the number of attributes in AP_i and n_{σ_j} is the number of attributes in AP_{σ_j}).

UC Secure	Pairing PO		Exponentiation EXP			AP
Schemes	Transfer	DBSetup	Transfer	DBSetup	CRSG	
[11]	$\geq 207k$	$24N + 1$	$249k$	$20N + 13$	18	$\times$
[16]	$> 450k$	$15N + 1$	$223k$	$12N + 9$	15	$\times$
[13]	$147k$	$5N + 1$	$150k$	$17N + 5$	18	$\times$
[1]	$(100m + 199)k$	$(m + 21)N$	$(138m + 237)k$	$(2m + 21)N + 2m + 20$	$m + 26$	"$\wedge$"
Ours	$2\sum_{j=1}^{k} n_{\sigma_j} + 253k$	$5N_S + 63$	$247k$	$7N_S + 2\sum_{i=1}^{N_S} n_i + 79$	10	"$\wedge$" and "$\vee$"

Table 2. Comparison summary of communication cost in k transfer phases and initialization phase (cDB_S stands for ciphertext database, pk_{DB_S} for public key, m is the number of attributes, AC for access control, N and N_S are the database sizes and n_i is the number of attributes in AP_i).

UC Secure Schemes	Communication	Storage		Security Assumptions	AC
	Req + Res	crs-Size	(cDB_S + pk_{DB_S})		
[11]	$144k$	14	$18N + 11$	SXDH, q-LRSW, DLIN	$\times$
[16]	$93k$	23	$12N + 7$	HSDH, TDH, DLIN	$\times$
[13]	$75k$	16	$12N + 5$	q-SDH, DLIN	$\times$
[1]	$(125 + 64m)k$	$m + 28$	$16N + m + 20$	SXDH, XDLIN	$\checkmark$
Ours	$107k$	11	$6N_S + \sum_{i=1}^{N_S} n_i + 4$	q-SDH, q-DBDHE, DBDH, DLIN	$\checkmark$

by verifying the equation $e(d_1, g)e(d_2^{-1}, f'\prod_{l\in\mathcal{V}_S} f_l) = e(\widehat{g}_1, \widehat{g}_2)$. If SK_S is valid, $\mathcal{A}'$ sets the bit $b = 1$ and returns $\langle\mathsf{idsk} : \mathsf{sid}, \mathsf{ID}_S, b\rangle$ to $\mathcal{F}_{\mathsf{IBAOT-AC}}$. Otherwise, $\mathcal{A}'$ sends $b = 0$ to $\mathcal{F}_{\mathsf{IBAOT-AC}}$. Hence, $\mathsf{IDEAL}_{\mathcal{F}_{\mathsf{IBAOT-AC}},\mathcal{A}',\mathcal{Z}} \overset{c}{\approx} \mathsf{REAL}_{\Psi,\mathcal{A},\mathcal{Z}}$.   □

5   Comparison

We compare the computational and communication cost of our proposed scheme IBAOT-AC with the existing UC secure AOT [1,11,13,16]. As illustrated in Tables 1 and 2, the proposed scheme outperforms the existing schemes [1,11,13, 16]. It is clear from the Tables 1 and 2, the construction of [13] is more efficient as compared to ours, but [13] does not realizes access control. We emphasize that our scheme computes only a constant number of pairings and exponentiations while that of [1] is linear to number of attributes m in each transfer phase.

References

1. Abe, M., Camenisch, J., Dubovitskaya, M., Nishimaki, R.: Universally composable adaptive oblivious transfer (with access control) from standard assumptions. In: ACM Workshop on Digital Identity Management, pp. 1–12. ACM (2013)

2. Beimel, A.: Secure schemes for secret sharing and key distribution. Ph.D. thesis, Israel Institute of Technology, Technion, Haifa, Israel (1996)
3. Belenkiy, M., Camenisch, J., Chase, M., Kohlweiss, M., Lysyanskaya, A., Shacham, H.: Randomizable proofs and delegatable anonymous credentials. In: Halevi, S. (ed.) CRYPTO 2009. LNCS, vol. 5677, pp. 108–125. Springer, Heidelberg (2009)
4. Boneh, D., Boyen, X.: Short signatures without random oracles. In: Cachin, C., Camenisch, J.L. (eds.) EUROCRYPT 2004. LNCS, vol. 3027, pp. 56–73. Springer, Heidelberg (2004)
5. Boneh, D., Boyen, X., Shacham, H.: Short group signatures. In: Franklin, M. (ed.) CRYPTO 2004. LNCS, vol. 3152, pp. 41–55. Springer, Heidelberg (2004)
6. Camenisch, J., Dubovitskaya, M., Neven, G.: Oblivious transfer with access control. In: ACM 2009, pp. 131–140. ACM (2009)
7. Camenisch, J., Dubovitskaya, M., Neven, G., Zaverucha, G.M.: Oblivious transfer with hidden access control policies. In: Catalano, D., Fazio, N., Gennaro, R., Nicolosi, A. (eds.) PKC 2011. LNCS, vol. 6571, pp. 192–209. Springer, Heidelberg (2011)
8. Canetti, R., Lindell, Y., Ostrovsky, R., Sahai, A.: Universally composable two-party and multi-party secure computation. In: ACM 2002, pp. 494–503. ACM (2002)
9. Coull, S., Green, M., Hohenberger, S.: Controlling access to an oblivious database using stateful anonymous credentials. In: Jarecki, S., Tsudik, G. (eds.) PKC 2009. LNCS, vol. 5443, pp. 501–520. Springer, Heidelberg (2009)
10. Green, M., Hohenberger, S.: Blind identity-based encryption and simulatable oblivious transfer. In: Kurosawa, K. (ed.) ASIACRYPT 2007. LNCS, vol. 4833, pp. 265–282. Springer, Heidelberg (2007)
11. Green, M., Hohenberger, S.: Universally composable adaptive oblivious transfer. In: Pieprzyk, J. (ed.) ASIACRYPT 2008. LNCS, vol. 5350, pp. 179–197. Springer, Heidelberg (2008)
12. Groth, J., Sahai, A.: Efficient non-interactive proof systems for bilinear groups. In: Smart, N.P. (ed.) EUROCRYPT 2008. LNCS, vol. 4965, pp. 415–432. Springer, Heidelberg (2008)
13. Guleria, V., Dutta, R.: Efficient adaptive oblivious transfer in UC framework. In: Huang, X., Zhou, J. (eds.) ISPEC 2014. LNCS, vol. 8434, pp. 271–286. Springer, Heidelberg (2014)
14. Lewko, A., Waters, B.: Decentralizing attribute-based encryption. In: Paterson, K.G. (ed.) EUROCRYPT 2011. LNCS, vol. 6632, pp. 568–588. Springer, Heidelberg (2011)
15. Naor, M., Pinkas, B.: Computationally secure oblivious transfer. J. Cryptol. 18(1), 1–35 (2005)
16. Rial, A., Kohlweiss, M., Preneel, B.: Universally composable adaptive priced oblivious transfer. In: Shacham, H., Waters, B. (eds.) Pairing 2009. LNCS, vol. 5671, pp. 231–247. Springer, Heidelberg (2009)
17. Waters, B.: Efficient identity-based encryption without random oracles. In: Cramer, R. (ed.) EUROCRYPT 2005. LNCS, vol. 3494, pp. 114–127. Springer, Heidelberg (2005)
18. Waters, B.: Ciphertext-policy attribute-based encryption: an expressive, efficient, and provably secure realization. In: Catalano, D., Fazio, N., Gennaro, R., Nicolosi, A. (eds.) PKC 2011. LNCS, vol. 6571, pp. 53–70. Springer, Heidelberg (2011)
19. Zhang, F., Zhao, X., Chen, X.: ID-based adaptive oblivious transfer. In: Youm, H.Y., Yung, M. (eds.) WISA 2009. LNCS, vol. 5932, pp. 133–147. Springer, Heidelberg (2009)

Three-Round Public-Coin Bounded-Auxiliary-Input Zero-Knowledge Arguments of Knowledge

Ning Ding[1,2]([✉])

[1] NTT Secure Platform Laboratories, Tokyo, Japan
[2] Shanghai Jiao Tong University, Shanghai, China
ning.ding@lab.ntt.co.jp

Abstract. This paper investigates the exact round complexity of public-coin (bounded-auxiliary-input) zero-knowledge arguments of knowledge (ZKAOK). It is well-known that Barak's non-black-box ZK [FOCS 01], which can be adapted to a ZKAOK, is the first one achieving constant-round, public-coin and strict-polynomial-time simulation properties, and admitting a 6-round implementation shown by Ostrovsky and Visconti [ECCC 12]. This achieves the best exact round complexity for public-coin ZKAOK ever known, to the best of our knowledge. As for a specific case of bounded-auxiliary-input verifiers, i.e. the auxiliary inputs are of bounded-size, no previous works explicitly considered to improve the general result on the exact round number of public-coin ZKAOK in this case. It is also noticeable that when ignoring the argument of knowledge property, Barak *et al.* [JCSS 06] showed based on two-round public-coin universal arguments which admit a candidate construction of the two-round variant of Micali's CS-proof, there exists a two-round public-coin plain/bounded-auxiliary-input ZK argument.

So an interesting question in ZKAOK is how to improve the exact round complexity of public-coin ZKAOK in both the general and the above specific cases. This paper provides an improvement for the specific case. That is, we show that also based on two-round public-coin universal arguments, there exists a 3-round public-coin bounded-auxiliary-input ZKAOK for **NP** which admits a strict-polynomial-time non-black-box simulator and an expected-polynomial-time extractor.

Keywords: Zero knowledge · Argument of knowledge · Exact round complexity

1 Introduction

Zero-knowledge (ZK) proof systems, introduced by Goldwasser, Micali and Rackoff [19], are a fundamental notion in cryptography. Later Brassard *et al.* [10] suggested the notion of ZK argument systems, which differs from ZK proofs only in that arguments are only required to be computationally sound. Goldreich and Oren [18] refined the notion of ZK to *plain* ZK and *auxiliary-input* ZK, where

© Springer International Publishing Switzerland 2015
D. Lin et al. (Eds.): Inscrypt 2014, LNCS 8957, pp. 130–149, 2015.
DOI: 10.1007/978-3-319-16745-9_8

plain ZK requires ZK holds for all uniform PPT verifiers while auxiliary-input ZK requires it holds for all PPT verifiers with polynomial-sized auxiliary-input (for both the notions distinguishers are always defined as non-uniform polynomial-time algorithms). Since their introduction, a fundamental positive result due to Goldreich *et al.* [17] shows that every language in **NP** has a ZK proof. There are also many works constructing ZK protocols that satisfy some additional properties such as constant rounds, the proof of knowledge [6,12,19,29] and strict-polynomial-time simulation. In this paper we focus on constant-round (public-coin) ZK proofs and arguments of knowledge (ZKPOK and ZKAOK). In the following we sketch the (first) known results on the exact round complexity of ZK, ZKPOK and ZKAOK as follows.

Zero-Knowledge. For auxiliary-input ZK for **NP**, Goldreich and Kahan [16] presented a 5-round ZK proof. Feige and Shamir [13] gave a 4-round ZK (which is also a ZKAOK). The simulators of these protocols use a verifier's code in a black-box way and run in expected-polynomial-time. Hada and Tanaka [20] presented a 3-round ZK argument based on two knowledge-of-exponent assumptions.

Barak [2] presented a constant-round public-coin non-black-box ZK argument. This is the first construction achieving constant-round and strict-polynomial-time simulation properties from complexity assumptions. Ostrovsky and Visconti [27] showed a 6-round implementation for this protocol, achieving the best round complexity ever known for it. Barak's construction consists of a 3-round preamble and a WI universal argument of knowledge which uses at least 4 rounds. Since the preamble must be finished prior to the universal argument, any implementation of the protocol seems to require 6 rounds at least and thus the 6-round implementation may be optimal.

Pandey *et al.* [28] presented a 4-round (concurrent) ZK argument with strict-polynomial-time simulation from differing-input obfuscators for machines [1], which are based on differing-input obfuscators for circuits (a candidate shown in [14]), fully homomorphic encryption and SNARKs [7] that require knowledge assumptions. Thus the assumption on differing-input obfuscators for machines is quite strong. Even for differing-input obfuscation for circuits, Garg *et al.* [15] came up with an example showing some circuits cannot be differing-input obfuscated if some new assumption is true. So the 4-round protocol is not satisfactory in the consideration of assumptions.

For plain ZK for **NP**, Barak *et al.* [5] presented a 2-round public-coin ZK argument assuming the existence of two-round public-coin universal arguments, which admits a candidate construction i.e. the two-round variant of Micali's CS proof [26]. Bitansky and Panet [8] presented a 2-round ZK argument from extractable one-way functions. The simulators of the two protocols run in strict polynomial-time. By scaling the security parameters, these protocols can be made bounded-auxiliary-input ZK, i.e. the zero-knowledge property holds when auxiliary inputs to verifiers are of bounded-size.

ZKPOK and ZKAOK. For the auxiliary-input ZK, Lindell [24] presented a 5-round ZKPOK. As shown above, Feige and Shamir [13] gave a 4-round

ZKAOK. Both the two protocols are private-coin and admit black-box expected polynomial-time simulators and extractors. As well known, Barak's protocol in [2] can be made to satisfy the argument of knowledge property. For bounded-auxiliary-input ZK, Bitansky and Panet [8] presented a 3-round private-coin ZKAOK with a strict-polynomial-time simulator and an expected-polynomial-time extractor.

When focusing on the public-coin ZKAOK (and ZKPOK), we already have Barak's protocol in [2] with adaption is a 6-round public-coin ZKAOK. And as for a specific case of bounded-auxiliary-input verifiers, no previous works explicitly considered to improve the general result on the exact round number of public-coin ZKAOK in this scenario.

Some Negative Results. Goldreich and Oren [18] showed there is no 1-round plain ZK protocol and there is no 2-round (auxiliary-input) ZK protocol for any language outside **BPP**. Extending this result, Goldreich and Krawczyk [16] showed that 3-round black-box ZK proofs exist only for languages in **BPP** and Katz [22] showed that 4-round black-box ZK proofs exist only for languages whose complement is in **MA**. Barak and Lindell [4] showed black-box simulators and extractors cannot run in strict-polynomial-time if the protocols are constant-round. Barak *et al.* [5] presented some trivialities of 2-round ZK proofs from some complexity assumptions.

So an interesting question in ZKAOK is to improve the exact round complexity of public-coin ZKAOK in both the general and the above specific cases. The question can be stated as follows.

1. Can we construct a public-coin ZKAOK for **NP** which uses fewer rounds than 6?
2. Can we construct a public-coin ZKAOK for **NP** with a round number significantly smaller than 6 for the bounded-auxiliary-input case?

1.1 Our Results

This paper investigates the above question for the bounded-auxiliary-input case and the main result is a 3-round public-coin bounded-auxiliary-input ZKAOK from complexity assumptions (without using any knowledge assumption). Let Com denote a non-interactive perfectly-binding computationally-hiding commitment scheme in e.g. [9], ZAP denote the 2-round public-coin WI proof for **NP** in [11]. Let 2rUA denote a 2-round public-coin universal argument, which admits a candidate construction suggested by Micali [26], also used in [5] (that is also a variant of the four-round universal argument of knowledge in [3]). Our result is stated as follows.

Theorem 1. *Assuming* Com, ZAP, 2rUA, *there exists a 3-round public-coin bounded auxiliary-input ZKAOK for* **NP** *which admits a strict-polynomial-time non-black-box simulator and an expected-polynomial-time extractor.*

Our Techniques. We sketch the techniques used in the protocol. Our starting point is the 2-round public-coin plain zero-knowledge argument in [5]. Let $(\mathsf{2rUA_1}, \mathsf{2rUA_2})$ denote the two messages of 2rUA. In the protocol V first sends a random $10n$-bit string r and $\mathsf{2rUA_1}$ and then P computes $Z = \mathsf{Com}(0^{|\mathsf{2rUA_2}|})$ and a non-interactive WI proof using w as witness for that $x \in L$ or letting $\mathsf{2rUA_2} \leftarrow \mathsf{Com}^{-1}(Z)$, $(\mathsf{2rUA_1}, \mathsf{2rUA_2})$ is a valid proof for that there is Π of size $5n$ which outputs r in $n^{\log\log n/10}$ steps. This protocol is plain zero-knowledge and sound (under sub-exponential complexity assumptions while our construction only resorts to polynomial-time assumptions). However, it is unknown if it has an extractor.

Our first idea of providing an extractor for it is to substitute the non-interactive WI proof with a 3-round WIPOK. We choose the 3-round public-coin WIPOK due to Lapidot and Shamir [23], denoted LS, which enjoys a key property that the first two messages are independent of the witness and the public input, noted in [27]. Let $(\mathsf{LS_1}, \mathsf{LS_2}, \mathsf{LS_3})$ denote the 3 messages of LS. Thus the protocol now could be as follows. In Step 1 P sends $\mathsf{LS_1}$ to V. In Step 2, V sends $r, \mathsf{2rUA_1}, \mathsf{LS_2}$. In Step 3, P computes $Z \leftarrow \mathsf{Com}(0^{|\mathsf{2rUA_2}|})$ and $\mathsf{LS_3}$ satisfying $(\mathsf{LS_1}, \mathsf{LS_2}, \mathsf{LS_3})$ is a valid proof for that $x \in L$ or letting $\mathsf{2rUA_2} \leftarrow \mathsf{Com}^{-1}(Z)$, $(\mathsf{2rUA_1}, \mathsf{2rUA_2})$ is a valid proof for that there is Π of size $5n$ such that $\Pi(\mathsf{LS_1})$ outputs r in $n^{\log\log n/10}$ steps.

However, the substitution with LS causes a confliction that now the public input to 2rUA has $\mathsf{LS_1}$ as a part, which means LS needs to prove a statement referring to $\mathsf{LS_1}$. (This also leads to if let y denote the public input to 2rUA, $|\mathsf{LS_1}| > |y| > |\mathsf{LS_1}|$.) So this is impossible. Our further idea is to let P commit to an n-bit random seed in Step 1 and compute the messages $(\mathsf{LS_1}, \mathsf{LS_3})$ in Step 3 which are generated with pseudorandom coins from the seed. Now the message in Step 1 is of $O(n)$-bit, independent of $|\mathsf{LS_1}|$, and can still determine $(\mathsf{LS_1}, \mathsf{LS_3})$. Moreover, the seed still remains random and unknown to the verifier due to the hiding property of Com. Thus LS is still a WIAOK and the confliction above can be bypassed.

Actually, our protocol indeed adopts this as well as some additional techniques. The additional techniques, for instance, include that to employ the WI property of LS, we ask P, V to execute LS twice in parallel and then P proves to V in ZAP that one of the two executions of LS is generated honestly. With these techniques, there is an extractor that can obtain two valid transcripts of LS by rewinding the prover and then extract a witness.

1.2 Organizations

The rest of the paper is arranged as follows. We relegate the preliminaries used through this paper to Appendix A which contains the notions of commitment schemes, auxiliary-input/bounded-auxiliary-input black-box/non-black-box zero-knowledge, witness-indistinguishability, universal arguments and the LS proof system etc. In Sect. 2, we present the high-level of the protocol. In Sect. 3, we specify the protocol formally and prove the main theorem. In Sect. 4 we consider the possibility of extending the new techniques to improve the exact round complexity for the unbounded-auxiliary-input verifiers, in which we analyze the difficulty in extending the techniques and then sketch a (possibly unreasonable) idea to solve this.

2 High-Level Description

In this section we present the high-level description of the protocol. In Sect. 2.1 we show the underlying construction idea. In Sect. 2.2 we present the overview of the protocol and informally argue that it satisfies all desired properties.

2.1 Construction Idea

Let us recall the 2-round public-coin plain zero-knowledge in [5]. As sketched previously, in the protocol V first sends a random $10n$-bit string r and $2\mathsf{rUA}_1$ and then P computes $Z \leftarrow \mathsf{Com}(0^{|2\mathsf{rUA}_2|})$ and a non-interactive WI proof using w as witness for that $x \in L$ or letting $2\mathsf{rUA}_2 \leftarrow \mathsf{Com}^{-1}(Z)$, $(2\mathsf{rUA}_1, 2\mathsf{rUA}_2)$ is a valid proof for that there is Π of size $5n$ which outputs r in $n^{\log \log n/10}$ steps. With sub-exponential complexity assumptions, this protocol is sound. However, it is unknown if it has an extractor.

Our basic idea of providing an extractor for the protocol is to substitute the non-interactive WI proof with a 3-round WIPOK. To reduce the round number, we adopt the 3-round public-coin WIPOK due to Lapidot and Shamir [23], denoted LS, which enjoys a key property that the first two messages are independent of the witness and the public input, noted in [27]. Let $(\mathsf{LS}_1, \mathsf{LS}_2, \mathsf{LS}_3)$ denote the 3 messages of LS.

Thus a first modification could be as follows. In Step 1 P sends LS_1 to V. In Step 2, V sends $r, 2\mathsf{rUA}_1, \mathsf{LS}_2$. In Step 3, P computes $Z \leftarrow \mathsf{Com}(0^{|2\mathsf{rUA}_2|})$ and LS_3 satisfying $(\mathsf{LS}_1, \mathsf{LS}_2, \mathsf{LS}_3)$ is a valid proof for a similar statement as that of the original non-interactive WI proof.

At first glance if P can convince V of $x \in L$, running the extractor of LS can extract a witness for $x \in L$ or $2\mathsf{rUA}_2$. However, since now V receives LS_1 from P before sending r, the statement for $2\mathsf{rUA}$ to prove should be changed to that there is Π of size $5n$ and $\Pi(\mathsf{LS}_1)$ outputs r in $n^{\log \log n/10}$ steps. So LS_1 should be a part of the public-input to $2\mathsf{rUA}$, which means LS needs to prove a statement referring to LS_1. Also $|\mathsf{LS}_1|$ should be larger than the size of the public input to LS which should contain LS_1. So this is impossible to achieve.

We then have a second modification to bypass this impossibility. Let P_{LS} denote the randomized honest prover algorithm of LS that on input $1^{\mathrm{poly}(n)}$ outputs LS_1 where $\mathrm{poly}(n)$ denotes the length of the public input, and then on receiving $\mathsf{LS}_2 \in \{0,1\}^n$ and a public input y and witness W outputs LS_3. We use the notations $\mathsf{LS}_1 \leftarrow P_{\mathsf{LS}}(1^{|y|})$ and $\mathsf{LS}_3 \leftarrow P_{\mathsf{LS}}(1^{|y|}, \mathsf{LS}_2, y, W)$ to denote the prover's computations in the two consecutive steps.

Note that P_{LS} needs random coins in computation. It can be seen that for P_{LS}, if the coins it uses are pseudorandom LS is still a WIAOK. Our observation is that when sampling a random $u \in \{0,1\}^n$ and running a pseudorandom generator

$PRG(u)$ to provide P_{LS} pseudorandom coins, LS_1 is determined by u (and $|y|$) and LS_3 is then further determined by LS_2 and the public input y and witness W. In this scenario we say (LS_1, LS_3) is generated from (u, W) when y, LS_2 are explicitly specified in the protocol. Note that $|u| = n$ is independent of $|LS_1|$. Thus letting P commit to u in Step 1 can actually determine LS_1 while bypassing the impossibility.

For some technical considerations we let P choose two random seeds u_1, u_2 and send $Z_1 \leftarrow \mathsf{Com}(u_1, u_2, w)$ and $Z_2 \leftarrow \mathsf{Com}(0^n, 0^n)$ to V in Step 1 instead of sending LS_1 directly. When receiving $r, 2rUA_1, LS_2$ from V in Step 2, letting y denote the public input to 2rUA, P computes $Z_3 \leftarrow \mathsf{Com}(0^{|2rUA_2|})$ and $LS_1 \leftarrow P_{LS}(1^{|y|})$ and $LS_3 \leftarrow P_{LS}(1^{|y|}, LS_2, y, W)$ where P_{LS}'s coins are generated from $PRG(u_1)$. It can be seen LS_1 is actually already determined by u_1 in Step 1 and $|Z_1|$ is fixed and independent of $|LS_1|$. In extraction, by rewinding a prover to re-do Steps 2 and 3 we can gain two valid transcripts of LS, in which the two LS_1 are equal since they are determined by the seed in Z_1. Then we can run the extractor of LS to extract a witness for $x \in L$.

However, a cheating prover may not compute LS_1, LS_3 honestly. Thus to prohibit any dishonest prover behavior, we require P to send an additional proof for that LS_1 and LS_3 are generated honestly from u_1 and a witness. Concretely, we let P use ZAP to do this. Typically, to employ a WI proof, the statement to be proven by the proof is the OR of two sub-statements. Therefore, besides (LS_1, LS_2, LS_3), we also require P, V to use a same strategy to generate a more independent transcript of LS, denoted (LS'_1, LS'_2, LS'_3), which are determined by u_2 and a witness. Then P proves to V in ZAP that either (LS_1, LS_3) or (LS'_1, LS'_3) is generated as specified.

In simulation the simulator computes $Z_1 \leftarrow \mathsf{Com}(0^n, 0^n, 0^{|w|})$ and $Z_2 \leftarrow \mathsf{Com}(v_1, v_2)$ where v_1, v_2 are two random strings. Let Π denote the verifier's code and thus it is a witness for the public input to 2rUA. Then the simulator can compute $2rUA_2$ with witness Π and compute (LS_1, LS_3) from $(v_1, 2rUA_2)$ and (LS'_1, LS'_3) from $(v_2, 2rUA_2)$. Thus the protocol can be proven zero-knowledge.

2.2   Overview of The Protocol

Primitives. Let Com denote a one-message perfectly-binding computationally-hiding commitment scheme e.g. one in [9] which satisfies $|\mathsf{Com}(msg)| = O(|msg|)$. Let ZAP denote the 2-round public-coin WI proof for **NP** constructed in [11], (ZAP_1, ZAP_2) denote the two messages of ZAP. Note that ZAP is sound even if the public input is chosen adaptively after ZAP_1 is sampled. Let LS denote the 3-round WIPOK in [23], (LS_1, LS_2, LS_3) denote the 3 messages of LS. Let 2rUA denote the 2-round public-coin universal argument i.e. the 2-round variant of Micali's CS proof [26], $(2rUA_1, 2rUA_2)$ denote the 2 messages of 2rUA. Let PRG denote a pseudorandom generator constructed in e.g. [21].

Construction. Let L denote an arbitrary language in **NP** and w.l.o.g. assume all witnesses for n-bit instances are of same length $p(n) > n$. Our protocol for

---

Public input: x (statement to be proven is "$x \in L$");
Prover's auxiliary input: w, (a witness for $x \in L$).

---

1. $P \to V$: Send $Z_1 \leftarrow \mathsf{Com}(u_1, u_2, w)$, $Z_2 \leftarrow \mathsf{Com}(0^n, 0^n)$.
2. $V \to P$: Send $r \in_R \{0,1\}^{np(n)}$, $2\mathsf{rUA}_1, \mathsf{LS}_2, \mathsf{LS}_2', \mathsf{ZAP}_1$.
3. $P \to V$: Send $Z_3 \leftarrow \mathsf{Com}(0^{|2\mathsf{rUA}_2|})$, $\mathsf{LS}_1, \mathsf{LS}_1', \mathsf{LS}_3, \mathsf{LS}_3', \mathsf{ZAP}_2$.

Protocol 1. *The 3-round public-coin bounded-auxiliary-input ZKAOK.*

L is shown in Protocol 1, which follows from the construction idea above. Let x be the public input and P has a witness w for $x \in L$.

1. In Step 1, P sends $Z_1 \leftarrow \mathsf{Com}(u_1, u_2, w)$, $Z_2 \leftarrow \mathsf{Com}(0^n, 0^n)$.
2. In Step 2, V samples $r \in_R \{0,1\}^{np(n)}$, $2\mathsf{rUA}_1, \mathsf{LS}_2, \mathsf{LS}_2', \mathsf{ZAP}_1$ and sends them.
3. In Step 3, P computes $Z_3 \leftarrow \mathsf{Com}(0^{|2\mathsf{rUA}_2|})$, $(\mathsf{LS}_1, \mathsf{LS}_3)$ from (u_1, w) and $(\mathsf{LS}_1', \mathsf{LS}_3')$ from (u_2, w). $(\mathsf{LS}_1, \mathsf{LS}_2, \mathsf{LS}_3)$ and $(\mathsf{LS}_1', \mathsf{LS}_2', \mathsf{LS}_3')$ are both to prove that $x \in L$ or letting $2\mathsf{rUA}_2 \leftarrow \mathsf{Com}^{-1}(Z_3)$, $(2\mathsf{rUA}_1, 2\mathsf{rUA}_2)$ is a valid proof for that there is Π of size $5n$ and $\Pi(Z_1, Z_2)$ outputs r in $n^{\log \log n / 10}$ steps.

   Compute ZAP_2 using witness (u_1, w) such that $(\mathsf{ZAP}_1, \mathsf{ZAP}_2)$ is a valid proof for that either $(\mathsf{LS}_1, \mathsf{LS}_2, \mathsf{LS}_3)$ or $(\mathsf{LS}_1', \mathsf{LS}_2', \mathsf{LS}_3')$ is generated as specified. Send $Z_3, \mathsf{LS}_1, \mathsf{LS}_1', \mathsf{LS}_3, \mathsf{LS}_3', \mathsf{ZAP}_2$.

It can be seen that the protocol is public-coin and the completeness is satisfied. We now sketch the zero-knowledge and argument of knowledge properties and the soundness follows.

Zero-Knowledge. Let S denote the simulator, $V^* \in \{0,1\}^n$ be any PPT verifier and $x \in L$. $S(V^*, x)$ works as follows. It samples $s \in \{0,1\}^n$ and runs $\mathsf{PRG}(s)$ to provide V^* coins in simulation. Let Π denote the program that has V^*'s code, x, PRG, s hardwired and emulates $V^*(x)$'s computing while running $\mathsf{PRG}(s)$ to provide it coins. So $|\Pi|$ can be less than $5n$ and pad it to $5n$ bits.

In Step 1 S computes $Z_1 \leftarrow \mathsf{Com}(0^n, 0^n, 0^{p(n)})$, $Z_2 \leftarrow \mathsf{Com}(v_1, v_2)$ for random $v_1, v_2 \in \{0,1\}^n$. Then emulate V^* to output $(r, 2\mathsf{rUA}_1, \mathsf{LS}_2, \mathsf{LS}_2', \mathsf{ZAP}_1)$. Once V^* sends out this message of Step 2, Π is actually a program such that $\Pi(Z_1, Z_2)$ outputs r in $n^{\log \log n / 10}$ steps. In Step 3, S computes $2\mathsf{rUA}_2$ using witness Π and $Z_3 \leftarrow \mathsf{Com}(2\mathsf{rUA}_2)$. Compute $(\mathsf{LS}_1, \mathsf{LS}_3)$ from $(v_1, 2\mathsf{rUA}_2)$ and $(\mathsf{LS}_1', \mathsf{LS}_3')$ from $(v_2, 2\mathsf{rUA}_2)$. Lastly, compute ZAP_2 using witness $(v_2, 2\mathsf{rUA}_2)$.

Thus, the pseudorandomness of PRG, the hiding property of Com and the WI properties of LS and ZAP ensure the indistinguishability of simulation.

Argument of Knowledge. We show there is an extractor E such that if P' is a polynomial-sized prover that can convince V of $x \in L$ with probability ϵ, $E(P', x)$ outputs a witness for x with probability $\epsilon - \mathsf{neg}(n)$. E works by first interacting with P' of the protocol and then rewinding P' to re-do Steps 2 and

3. Then E can receive a new valid P''s message of Step 3 in some rewinding run in expected polynomial-time. Actually, we let E perform the rewinding process constant times. So totally E has constant valid transcripts of the protocol.

Due to the soundness of ZAP, in each of the constant transcripts, either $(\mathsf{LS}_1, \mathsf{LS}_3)$ or $(\mathsf{LS}_1', \mathsf{LS}_3')$ is generated as specified. Since LS_1 (or LS_1') is determined by one of u_1, u_2, v_1, v_2, which are committed in Z_1, Z_2 and thus are all fixed before the rewinding. So in the constant valid transcripts, there are at least two LS_1 (or LS_1') that are generated from a same one of u_1, u_2, v_1, v_2. Then E on the corresponding two transcripts can output a witness that is w or $2\mathsf{rUA}_2$.

Then we only need to show what E outputs is indeed w for $x \in L$. Notice that $|(Z_1, Z_2)| = O(p(n))$ and for random $r \in \{0,1\}^{np(n)}$, there exists Π of size $5n$ such that $\Pi(Z_1, Z_2)$ outputs r with negligible probability. Thus there is no polynomial-time algorithm that can output $2\mathsf{rUA}_2$ such that $(2\mathsf{rUA}_1, 2\mathsf{rUA}_2)$ is a valid proof for that there is such a Π satisfying $\Pi(Z_1, Z_2)$ outputs r with noticeable probability, due to the soundness of $2\mathsf{rUA}$. So the witness output by E must be w for $x \in L$ except for negligible probability.

3 Actual Description

In this section we formalize the protocol and prove the main theorem. To facilitate the statement we first introduce some languages underlying the protocol, which definitions are presented in Sect. 3.1, and then present the details of the protocol and the proof of the theorem in Sect. 3.2.

3.1 Underlying Languages

Now we define the following languages which are used in Protocol 1 (and the reader can refer to the protocol for the meaning of each string in these languages).

Definition 1. *We define language L_1 as follows: $(x, r, 2\mathsf{rUA}_1, Z_1, Z_2, Z_3) \in L_1$ iff letting $n \leftarrow |x|$ there is a witness for $x \in L$ or letting $2\mathsf{rUA}_2 \leftarrow \mathsf{Com}^{-1}(Z_3)$, $(2\mathsf{rUA}_1, 2\mathsf{rUA}_2)$ is a valid proof of $2\mathsf{rUA}$ for that there is a program Π of size $5n$ such that $\Pi(Z_1, Z_2)$ outputs r in $n^{\log \log n / 10}$ steps.*

Then $L_1 \in \mathbf{NP}$. Note that in Definition 1 the public input to $2\mathsf{rUA}$ is (Z_1, Z_2, r) and a witness for (Z_1, Z_2, r) is some Π. Let $P_{2\mathsf{rUA}}$ denote the prover algorithm of $2\mathsf{rUA}$ that on input the public input (Z_1, Z_2, r) and witness Π and $2\mathsf{rUA}_1$ outputs $2\mathsf{rUA}_2$. We use the notation $2\mathsf{rUA}_2 \leftarrow P_{2\mathsf{rUA}}((Z_1, Z_2, r), \Pi, 2\mathsf{rUA}_1)$ to denote this computation.

Let INS_{L_1} denote an instance of L_1 that is of form $(x, r, 2\mathsf{rUA}_1, Z_1, Z_2, Z_3)$. A witness W for $\mathsf{INS}_{L_1} \in L_1$ is either w for $x \in L$ or $2\mathsf{rUA}_2$ (and the coins in computing Z_3 that we omit for simplicity) satisfying the second requirement in Definition 1. Let LS be the proof system for L_1. We still use $\mathsf{LS}_1 \leftarrow P_{\mathsf{LS}}(1^{|\mathsf{INS}_{L_1}|})$, $\mathsf{LS}_3 \leftarrow P_{\mathsf{LS}}(1^{|\mathsf{INS}_{L_1}|}, \mathsf{LS}_2, \mathsf{INS}_{L_1}, W)$ to denote the prover's computations in the two consecutive steps of LS.

Definition 2. *We define language L_2 as follows: $(\mathsf{INS}_{L_1}, \mathsf{LS}_1, \mathsf{LS}_2, \mathsf{LS}_3) \in L_2$ where $\mathsf{INS}_{L_1} = (x, r, 2\mathsf{rUA}_1, Z_1, Z_2, Z_3)$ iff one of the following is true:*

1. *Let u be the first or second message and w be the third message in $\mathsf{Com}^{-1}(Z_1)$. Then $\mathsf{LS}_1 = P_{\mathsf{LS}}(1^{|\mathsf{INS}_{L_1}|})$ and $\mathsf{LS}_3 = P_{\mathsf{LS}}(1^{|\mathsf{INS}_{L_1}|}, \mathsf{LS}_2, \mathsf{INS}_{L_1}, w)$ where running $\mathsf{PRG}(u)$ to provide P_{LS} coins.*
2. *Let v be any of the two messages in $\mathsf{Com}^{-1}(Z_2)$ and $2\mathsf{rUA}_2$ be $\mathsf{Com}^{-1}(Z_3)$. Then $\mathsf{LS}_1 = P_{\mathsf{LS}}(1^{|\mathsf{INS}_{L_1}|})$ and $\mathsf{LS}_3 = P_{\mathsf{LS}}(1^{|\mathsf{INS}_{L_1}|}, \mathsf{LS}_2, \mathsf{INS}_{L_1}, 2\mathsf{rUA}_2)$ where running $\mathsf{PRG}(v)$ to provide P_{LS} coins.*

Then $L_2 \in \mathbf{NP}$. In Definition 2 we assume Z_1, Z_2 are commitments of more than one messages. For instance (in our protocol) $Z_1 = \mathsf{Com}(u_1, u_2, w)$ and we consider Z_1 actually consists of $\mathsf{Com}(u_1), \mathsf{Com}(u_2), \mathsf{Com}(w)$ (similarly for Z_2). Thus by saying u is the first message in $\mathsf{Com}^{-1}(Z_1)$, we mean $u = \mathsf{Com}^{-1}(\mathsf{Com}(u_1))$ and to prove it one only needs the decommitment of $\mathsf{Com}(u_1)$, denoted $\mathsf{decom}(\mathsf{Com}(u_1))$, regardless of those of $\mathsf{Com}(u_2), \mathsf{Com}(w)$. A witness for $(\mathsf{INS}_{L_1}, \mathsf{LS}_1, \mathsf{LS}_2, \mathsf{LS}_3) \in L_2$ is some (u, w) or some $(v, 2\mathsf{rUA}_2)$ (and some coins in decommitments that we omit for simplicity).

Definition 3. *We define language L_3 as follows: $(\mathsf{INS}_{L_1}, \mathsf{LS}_1, \mathsf{LS}_2, \mathsf{LS}_3, \mathsf{LS}_1', \mathsf{LS}_2', \mathsf{LS}_3') \in L_3$ iff $(\mathsf{INS}_{L_1}, \mathsf{LS}_1, \mathsf{LS}_2, \mathsf{LS}_3) \in L_2$ or $(\mathsf{INS}_{L_1}, \mathsf{LS}_1', \mathsf{LS}_2', \mathsf{LS}_3') \in L_2$.*

Then $L_3 \in \mathbf{NP}$. A witness for $(\mathsf{INS}_{L_1}, \mathsf{LS}_1, \mathsf{LS}_2, \mathsf{LS}_3, \mathsf{LS}_1', \mathsf{LS}_2', \mathsf{LS}_3') \in L_3$ is a witness either for $(\mathsf{INS}_{L_1}, \mathsf{LS}_1, \mathsf{LS}_2, \mathsf{LS}_3) \in L_2$ or for $(\mathsf{INS}_{L_1}, \mathsf{LS}_1', \mathsf{LS}_2', \mathsf{LS}_3') \in L_2$.

3.2 Detailed Specifications

Now we present the specification of Protocol 1 as follows.

1. In Step 1, P samples $u_1, u_2 \in \{0,1\}^n$, computes $Z_1 \leftarrow \mathsf{Com}(u_1, u_2, w)$, $Z_2 \leftarrow \mathsf{Com}(0^n, 0^n)$ and sends them to V.
2. In Step 2, V responds with $r \in_R \{0,1\}^{np(n)}$, $2\mathsf{rUA}_1$, LS_2, LS_2', ZAP_1.
3. In Step 3, P does the following:
   (a) Compute $Z_3 \leftarrow \mathsf{Com}(0^{|2\mathsf{rUA}_2|})$. Let $\mathsf{INS}_{L_1} \leftarrow (x, r, 2\mathsf{rUA}_1, Z_1, Z_2, Z_3)$.
   (b) Compute $\mathsf{LS}_1 \leftarrow P_{\mathsf{LS}}(1^{|\mathsf{INS}_{L_1}|})$, $\mathsf{LS}_3 \leftarrow P_{\mathsf{LS}}(1^{|\mathsf{INS}_{L_1}|}, \mathsf{LS}_2, \mathsf{INS}_{L_1}, w)$ where running $\mathsf{PRG}(u_1)$ to provide P_{LS} coins.
   (c) Compute $\mathsf{LS}_1' \leftarrow P_{\mathsf{LS}}(1^{|\mathsf{INS}_{L_1}|})$, $\mathsf{LS}_3' \leftarrow P_{\mathsf{LS}}(1^{|\mathsf{INS}_{L_1}|}, \mathsf{LS}_2', \mathsf{INS}_{L_1}, w)$ where running $\mathsf{PRG}(u_2)$ to provide P_{LS} coins.
   (d) Compute ZAP_2 for $(\mathsf{INS}_{L_1}, \mathsf{LS}_1, \mathsf{LS}_2, \mathsf{LS}_3, \mathsf{LS}_1', \mathsf{LS}_2', \mathsf{LS}_3') \in L_3$ using witness (u_1, w) corresponding to ZAP_1. (Note that (u_2, w) is also a witness, but we just let P use (u_1, w).)
   Send $(Z_3, \mathsf{LS}_1, \mathsf{LS}_1', \mathsf{LS}_3, \mathsf{LS}_3', \mathsf{ZAP}_2)$ to V, which accepts x iff $(\mathsf{ZAP}_1, \mathsf{ZAP}_2)$ is a valid proof for $(\mathsf{INS}_{L_1}, \mathsf{LS}_1, \mathsf{LS}_2, \mathsf{LS}_3, \mathsf{LS}_1', \mathsf{LS}_2', \mathsf{LS}_3') \in L_3$.

Then we restate the main theorem and present the proof as follows.

Theorem 2. *Assuming the existence of Com, ZAP, $2\mathsf{rUA}$ (which imply LS, PRG), Protocol 1 satisfies all the properties claimed in Theorem 1.*

Proof. We show the completeness, zero-knowledge and argument of knowledge properties are satisfied and the computational soundness follows from the argument of knowledge property.

Completeness. It can be seen that P can use w to finish the interaction.

Zero-Knowledge. We present a strict-polynomial-time non-black-box simulator S for any PPT $V^* \in \{0,1\}^n$ and any $x \in L$. S samples $s \in \{0,1\}^n$ and runs $\mathsf{PRG}(s)$ to provide V^* coins. Let Π denote the program that has V^*'s code and x, PRG, s hardwired and emulates $V^*(x)$'s computing while running $\mathsf{PRG}(s)$ to provide it coins. So Π's size can be less than $5n$ and pad it to $5n$ bits.

1. In Step 1 S samples $v_1, v_2 \in \{0,1\}^n$ and computes $Z_1 \leftarrow \mathsf{Com}(0^n, 0^n, 0^{p(n)})$, $Z_2 \leftarrow \mathsf{Com}(v_1, v_2)$. Then emulate V^* to output $(r, 2\mathsf{rUA}_1, \mathsf{LS}_2, \mathsf{LS}_2', \mathsf{ZAP}_1)$. Once V^* sends out this message of Step 2, Π is a program such that $\Pi(Z_1, Z_2)$ outputs r in $n^{\log \log n/10}$ steps.
2. In Step 3, S computes $(Z_3, \mathsf{LS}_1, \mathsf{LS}_3, \mathsf{LS}_1', \mathsf{LS}_3', \mathsf{ZAP}_2)$ as follows:
   (a) Compute $2\mathsf{rUA}_2 \leftarrow P_{2\mathsf{rUA}}((Z_1, Z_2, r), \Pi, 2\mathsf{rUA}_1)$ and $Z_3 \leftarrow \mathsf{Com}(2\mathsf{rUA}_2)$. Let INS_{L_1} denote $(x, r, 2\mathsf{rUA}_1, Z_1, Z_2, Z_3)$.
   (b) Compute $\mathsf{LS}_1 \leftarrow P_{\mathsf{LS}}(1^{|\mathsf{INS}_{L_1}|})$, $\mathsf{LS}_3 \leftarrow P_{\mathsf{LS}}(1^{|\mathsf{INS}_{L_1}|}, \mathsf{LS}_2, \mathsf{INS}_{L_1}, 2\mathsf{rUA}_2)$ where running $\mathsf{PRG}(v_1)$ to provide P_{LS} coins.
   (c) Compute $\mathsf{LS}_1' \leftarrow P_{\mathsf{LS}}(1^{|\mathsf{INS}_{L_1}|})$, $\mathsf{LS}_3' \leftarrow P_{\mathsf{LS}}(1^{|\mathsf{INS}_{L_1}|}, \mathsf{LS}_2', \mathsf{INS}_{L_1}, 2\mathsf{rUA}_2)$ where running $\mathsf{PRG}(v_2)$ to provide P_{LS} coins.
   (d) Compute ZAP_2 for $(\mathsf{INS}_{L_1}, \mathsf{LS}_1, \mathsf{LS}_2, \mathsf{LS}_3, \mathsf{LS}_1', \mathsf{LS}_2', \mathsf{LS}_3') \in L_3$ using witness $(v_2, 2\mathsf{rUA}_2)$ corresponding to ZAP_1.

It can be seen that S runs in polynomial-time. We now show S's output is indistinguishable from V^*'s real view interacting with $P(w)$. Let S_0 denote the interaction between V^* and $P(w)$ where V^*'s coins are from $\mathsf{PRG}(s)$. Thus the view output by S_0 is indistinguishable from V^*'s real view due to the pseudorandomness of $\mathsf{PRG}(s)$. In the following we use some hybrids to show S_0's output is indistinguishable from S's.

Hybrid 1. Let S_1 denote S_0 except that it adopts S's strategy to compute Z_2. Due to the hiding property of Com, the two outputs are indistinguishable.

Hybrid 2. Let S_2 denote S_1 except that it adopts S's strategy to compute Z_3. So similarly, S_2's output and S_1's are indistinguishable.

Hybrid 3. Let S_3 denote S_2 except that it adopts S's strategy to compute $\mathsf{LS}_1', \mathsf{LS}_3'$. That is, S_3 generates $(\mathsf{LS}_1', \mathsf{LS}_3')$ from $(v_2, 2\mathsf{rUA}_2)$, while S_2 generates them from (u_2, w). Basically, the indistinguishability of S_3's output and S_2's follows from the hiding property of Com, the pseudorandomness of PRG and the WI property of LS. We show it in detail through the following more hybrids.

1. **Hybrid 3.1.** Let $S_{3,1}$ denote S_2 except that it generates $\mathsf{LS}_1', \mathsf{LS}_3'$ from $\mathsf{PRG}(u')$ for an independently random $u' \in \{0,1\}^n$ and witness w. Thus $S_{3,1}$ differs from S_2 only in that the coins $S_{3,1}$ uses in computing $\mathsf{LS}_1', \mathsf{LS}_3'$ are $\mathsf{PRG}(u')$ and what S_2 uses is $\mathsf{PRG}(u_2)$.

Let z denote $(x, s, Z_1, Z_2, Z_3, \mathsf{decom}(\mathsf{Com}(u_1)), \mathsf{decom}(\mathsf{Com}(w)))$ generated using S_2's strategy (identically $S_{3,1}$'s strategy). Let z_u denote z except that $\mathsf{Com}(u_2)$ in Z_1 is replaced by $\mathsf{Com}(u)$ for an independent $u \in \{0,1\}^n$.

We claim for any polynomial-sized D, $(z, \mathsf{PRG}(u'))$ and $(z, \mathsf{PRG}(u_2))$ are indistinguishable. Otherwise, suppose there is a D such that $\Pr[D(z, \mathsf{PRG}(u_2)) = 1] - \Pr[D(z, \mathsf{PRG}(u')) = 1] = \epsilon$ for a noticeable ϵ. It can be first seen $\Pr[D(z_u, \mathsf{PRG}(u_2)) = 1] = \Pr[D(z, \mathsf{PRG}(u')) = 1]$ because u_2, u, u' are independently identically distributed. Thus this implies $\Pr[D(z, \mathsf{PRG}(u_2)) = 1] - \Pr[D(z_u, \mathsf{PRG}(u_2)) = 1] = \epsilon$. Then we construct a polynomial-sized D_1 that can distinguish $\mathsf{Com}(u_2)$ from $\mathsf{Com}(u)$ as follows.

D_1 samples u_2, u, u'. Send u_2, u to a challenger that responds with C which is either $\mathsf{Com}(u_2)$ or $\mathsf{Com}(u)$. Then D_1 generates z using S_2's strategy and then replaces $\mathsf{Com}(u_2)$ in Z_1 of z with C. Let z' denote the updated z. Thus z' is either still z or z_u. Lastly, call $D(z', \mathsf{PRG}(u_2))$ and output D's output. Due to D's ability, D_1 can distinguish $\mathsf{Com}(u_2)$ from $\mathsf{Com}(u)$ with probability ϵ. But this is impossible due to the hiding property of Com.

Then we claim any distinguisher D^* for $S_{3,1}$'s output and S_2's can be transformed to a distinguisher with z for $\mathsf{PRG}(u'), \mathsf{PRG}(u_2)$. That is, the distinguisher has V^*'s code hardwired and on input z and $\mathsf{PRG}(u')$ or $\mathsf{PRG}(u_2)$ sends the message of Step 1 to V^* (with coins $\mathsf{PRG}(s)$) and gains the message of Step 2 and lastly generates $\mathsf{LS}_1, \mathsf{LS}_3$ from the coins $\mathsf{PRG}(u_1)$ and w, $\mathsf{LS}'_1, \mathsf{LS}'_3$ from the coins $\mathsf{PRG}(u')$ or $\mathsf{PRG}(u_2)$ and w, and further computes ZAP_2 using $\mathsf{decom}(\mathsf{Com}(w))$ and $\mathsf{decom}(\mathsf{Com}(u_1))$ and then calls D^* with the simulated view, i.e. $(x, \mathsf{PRG}(s), Z_1, Z_2, Z_3, \mathsf{LS}_1, \mathsf{LS}'_1, \mathsf{LS}_3, \mathsf{LS}'_3, \mathsf{ZAP}_2)$, that is either $S_{3,1}$'s output or S_2's to output a decision. It is impossible. So the two outputs are indistinguishable.

2. **Hybrid 3.2.** Let $S_{3,2}$ denote $S_{3,1}$ except that it generates $\mathsf{LS}'_1, \mathsf{LS}'_3$ from truly random coins and w. Due to the pseudorandomness of $\mathsf{PRG}(u')$, their outputs are indistinguishable.

3. **Hybrid 3.3.** Let $S_{3,3}$ denote $S_{3,2}$ except that it generates $\mathsf{LS}'_1, \mathsf{LS}'_3$ from truly random coins and witness $2\mathsf{rUA}_2$. Due to the WI property of LS, their outputs are indistinguishable.

Thus S_3 differs from $S_{3,3}$ only in that the coins that S_3 uses to compute $\mathsf{LS}'_1, \mathsf{LS}'_3$ are $\mathsf{PRG}(v_2)$ and what $S_{3,3}$ uses is truly random coins. Note that they use the same witness $2\mathsf{rUA}_2$. With a similar argument in Hybrids 3.2 and 3.1 (in which now z also contains $\mathsf{decom}(Z_3)$ for computing LS'_3), the outputs of $S_{3,3}$ and S_3 are indistinguishable. Thus S_3's output and S_2's are indistinguishable.

Hybrid 4. Let S_4 denote S_3 except that it adopts S's strategy to compute ZAP_2. Due to the WI property of ZAP, S_4's output and S_3's are indistinguishable.

Hybrid 5. Let S_5 denote S_4 except that it adopts S's strategy to compute $\mathsf{LS}_1, \mathsf{LS}_3$. Thus similarly to Hybrid 3, S_5's output and S_4's are indistinguishable.

S_5 differs from S only in Z_1. So their outputs are indistinguishable due to the hiding property of Com. Therefore V^*'s real view and S's output are indistinguishable. The zero-knowledge property is satisfied for all n-size V^*. Since the security parameter n can be scaled, this protocol can be modified to be zero-knowledge for all V^* of size bounded by an a-prior fixed polynomial.

Argument of Knowledge. We show there is an extractor E such that if P' is a polynomial-sized prover that can convince V of $x \in L$ with noticeable probability ϵ, $E(P', x)$ outputs a witness for x with probability $\epsilon - \mathsf{neg}(n)$. E works as follows. First it emulates P' to send out Z_1, Z_2. In Step 2, E samples $r, 2\mathsf{rUA}_1, \mathsf{LS}_2, \mathsf{LS}_2', \mathsf{ZAP}_1$ honestly and sends them to P' and then receives P''s message of Step 3. If one of P''s messages is invalid, abort the extraction. Otherwise, perform the following process eight times: rewind P' to Step 2 in which E samples the message of Step 2 honestly and sends it to P'; if P''s response of Step 3 is invalid in the rewinding run, repeat the rewinding until its message is valid. Thus E gains eight valid P''s message of Step 3 in all rewinding runs.

Now totally E has nine valid transcripts of the protocol of which one is generated in the first run and the latter eight are generated in the rewinding processes. Due to the soundness of ZAP, in each of the eight transcripts, either $(\mathsf{INS}_{L_1}, \mathsf{LS}_1, \mathsf{LS}_2, \mathsf{LS}_3) \in L_2$ or $(\mathsf{INS}_{L_1}, \mathsf{LS}_1', \mathsf{LS}_2', \mathsf{LS}_3') \in L_2$, for which we call the one of the two transcripts of LS being proven in ZAP primary. Since LS_1 (or LS_1') of a primary transcript is determined by one of u_1, u_2, v_1, v_2, which are committed in Z_1, Z_2 and thus are all fixed before the rewinding.

In the nine valid transcripts, there are at least five in which the primary transcripts are $(\mathsf{LS}_1, \mathsf{LS}_2, \mathsf{LS}_3)$ or $(\mathsf{LS}_1', \mathsf{LS}_2', \mathsf{LS}_3')$. W.l.o.g. assume they are $(\mathsf{LS}_1, \mathsf{LS}_2, \mathsf{LS}_3)$. In these five $(\mathsf{LS}_1, \mathsf{LS}_2, \mathsf{LS}_3)$, there are at least two in which the LS_1's are generated from a same seed of u_1, u_2, v_1, v_2 and thus identical. Moreover, the LS_2's in the two transcripts are different with $1 - 2^{-n}$ probability. So E on the two transcripts can output a witness for $\mathsf{INS}_{L_1} \in L_1$, which is either w or $2\mathsf{rUA}_2$. E finally outputs it. (Note that in E's rewinding, $2\mathsf{rUA}_1$ and Z_3 may change which means the public input INS_{L_1} to LS may be different in the two transcripts. That is, there are possibly two INS_{L_1}'s in the transcripts. However, since the LS_2's in the two transcripts contain both 0 and 1 except for exponentially small probability, E can recover witnesses for the two INS_{L_1}'s. It is even possible that there are more than two transcripts that the extractor can extract a witness. So E just outputs an arbitrary one. Notice that no matter how INS_{L_1} changes in the rewinding, x is unchanged. Thus what E outputs must be a witness for $x \in L$ and some $2\mathsf{rUA}_2$.)

Let us first consider E's running-time. Consider P''s message of Step 1 is valid (otherwise E runs in polynomial-time). Let $q_1(n), q_3(n)$ denote P''s running-time in Steps 1 and 3. Let $q_2(n)$ denote E's running-time in Step 2 and $q_4(n)$ denote the running-time of the extractor of LS. Assume when fixing P''s message of Step 1, P' can send a valid message of Step 3 with probability ξ for random E's message of Step 2. Then E's total expected running-time is bounded by $q_1(n) + q_2(n) + q_3(n) + \xi \cdot (8/\xi \cdot (q_2(n) + q_3(n)) + q_4(n)) = \mathsf{poly}(n)$.

We then claim the output by E is a witness w for $x \in L$ except for negligible probability. Notice that $|(Z_1, Z_2)| = O(p(n))$ and for random $r \in \{0,1\}^{n \cdot p(n)}$, there exists Π of size $5n$ such that $\Pi(Z_1, Z_2)$ outputs r with negligible probability. Thus there is no polynomial-sized algorithm that can output $2\mathsf{rUA}_2$ such that $(2\mathsf{rUA}_1, 2\mathsf{rUA}_2)$ is a valid proof for that there is such a Π such that $\Pi(Z_1, Z_2)$ outputs r with noticeable probability, due to to the soundness of $2\mathsf{rUA}$. So is the expected polynomial-time E. So the witness output by E must be w for $x \in L$ except for negligible probability.

4 Concluding Remarks: On Extending to the Unbounded-Auxiliary-Input Case

Having constructed the 3-round public-coin bounded-auxiliary-input ZKAOK, we would like to employ the techniques in the previous sections to reduce the exact round complexity of general public-coin ZKAOK (i.e. for unbounded-auxiliary-input verifiers), for which bear in mind the best known round complexity is 6. In this section we take a glimpse at the possibility of solving this question with the techniques. However, currently we find this task extremely difficult when insisting on using known reasonable assumptions. In Sect. 4.1, we show when moving to the unbounded-auxiliary-input case and adopting some natural modifications to Protocol 1, the main difficulty is to establish the argument of knowledge property as well as the soundness, which shows bypassing the difficulty may require new assumptions. In Sect. 4.2 we sketch an idea to modify the 4-round protocol in Sect. 4.1 which requires a (possibly unreasonable) assumption to establish the argument of knowledge property. However, due to the possible unreasonability of this assumption, we would not treat the idea as a reasonable solution of constructing 4-round public-coin ZKAOK. Thus the question of improving the exact round complexity of public-coin ZKAOK is still open and will be our target in the future.

4.1 Difficulty in Extraction

Basically, we wish to extend Protocol 1 to a four-round protocol that is zero-knowledge for any $V^* \in \{0,1\}^{\text{poly}(n)}$. Thus a first idea is that let V send a hash function h in Step 1 and P computes $Z_2 \leftarrow \mathsf{Com}(0^n, 0^n, 0^{|h|})$ in Step 2 and other messages are similarly generated with those in Protocol 1 in which the underlying languages are modified correspondingly (e.g. the statement for 2rUA to prove is now that there is Π of size $< n^{\log \log n/10}$ such that $h(\Pi)$ equals the third part of $\mathsf{Com}^{-1}(Z_2)$ and $\Pi(Z_1, Z_2) = r$). In simulation the simulator S computes $Z_2 \leftarrow \mathsf{Com}(v_1, v_2, h(\Pi))$ where Π denotes the verifier's code and then generates $2\mathsf{rUA}_2$ with witness Π and $\mathsf{LS}_1, \mathsf{LS}_3, \mathsf{LS}_1', \mathsf{LS}_3', \mathsf{ZAP}_2$ with witness $(v_2, 2\mathsf{rUA}_2)$. Thus the protocol is still zero-knowledge. However, the difficulty is to present an extractor or, even worse, to prove soundness directly.

Recall the proof of the argument of knowledge property of Theorem 2, where we showed since there is no Π of size $5n$ satisfying $\Pi(Z_1, Z_2)$ outputs r which means the public input to 2rUA is false, what E outputs cannot be a valid $2\mathsf{rUA}_2$ due to the soundness of 2rUA. However, when Π can be arbitrarily polynomially long, a cheating prover may adaptively find a Π' after seeing r such that $\Pi'(Z_1, Z_2)$ can output r and $h(\Pi')$ is equal to the value committed in Z_2. So this prover may break the soundness. Also recall Barak's auxiliary-input zero-knowledge protocol in [2] for which to prove the soundness, an extractor is required for the four-round universal argument. However, we now employ the 2-round 2rUA which is unknown to admit an extractor. So currently we do not know how to establish the soundness from the underlying complexity assumptions, in particular from the collision resistance of hash functions.

4.2   Introducing New Assumptions?

Accordingly, an idea of bypassing this difficulty is possibly to strength the assumptions. The simplest way may be directly assuming the soundness. However, even with this quite strong assumption, we still cannot present an extractor. Thus we may need another assumption other than directly assuming the soundness. One way to do this is to consider to modify the protocol and then present a related assumption. A possible modificiation is to use the tree hashing scheme in [25] to hash Π instead of directly computing $h(\Pi)$ in the protocol. Let $\mathsf{Tree}_h(\Pi)$ denote the hashing tree of Π with respect to h, $\mathsf{root}_h(\Pi)$ denote the root of $\mathsf{Tree}_h(\Pi)$, $|\mathsf{root}_h|$ denote the length of the root determined by h. Then in Step 2, P computes $Z_2 \leftarrow \mathsf{Com}(0^n, 0^n, 0^{|\mathsf{root}_h|})$. In Step 3, besides the original messages, V sends a random number $\mathsf{Request}$ that requests the value of the $\mathsf{Request}^{th}$ leaf and corresponding certificate, denoted $\mathsf{LeafCert}$, in $\mathsf{Tree}_h(\Pi)$. In Step 4, P computes $Z_3 \leftarrow \mathsf{Com}(0^{|2r\mathsf{UA}_2|}, 0^{|\mathsf{LeafCert}|})$ and computes other messages basically identically as before.

In simulation the simulator S computes $Z_2 \leftarrow \mathsf{Com}(v_1, v_2, \mathsf{root}_h(\Pi))$ and in Step 4 computes $Z_3 \leftarrow \mathsf{Com}(2r\mathsf{UA}_2, \mathsf{LeafCert})$ and finishes the interaction. Let us then consider a possible extractor. By applying the extractor shown in the previous section to a prover we may extract a witness w for $x \in L$ or $(2r\mathsf{UA}_2, \mathsf{LeafCert})$. If what is extracted is $(2r\mathsf{UA}_2, \mathsf{LeafCert})$, we would like to deduce some contradiction to the collision-resistance of h. By rewinding the prover many times we may get many different $\mathsf{LeafCert}$'s. Notice that all these $\mathsf{LeafCert}$'s share a same root. Thus the leaves in them may possibly recover a program Π. However, there is still a problem that we are only ensured that each $\mathsf{LeafCert}$ is an answer to $\mathsf{Request}$ in one $\mathsf{Tree}_h(\Pi)$ in each extraction. We are not assured that all these $\mathsf{LeafCert}$'s are retrieved from a same $\mathsf{Tree}_h(\Pi)$. A cheating prover may possibly generate $\mathsf{LeafCert}$'s from different Π's in different extractions. So even if we have gained these $\mathsf{LeafCert}$'s, it is unknown how to employ them to recover a full Π. Recall again Barak's protocol [2], for which in proving soundness a full Π can be extracted in the universal argument and then we can reduce the soundness to the collision-resistance of h.

Whereas, let us pay more attention to the possibility that a cheating prover generates $\mathsf{LeafCert}$'s from different $\mathsf{Tree}_h(\Pi)$'s in different extractions. It can be seen that each witness Π should satisfy that the root of its hashing tree equals that one fixed in Z_2. So it is hard to come up with more than one Π satisfying the requirement. Then intuitively any prover cannot generate $\mathsf{LeafCert}$'s from different $\mathsf{Tree}_h(\Pi)$'s in extractions. On the contrary, it is natural that a prover generates $\mathsf{LeafCert}$'s from a same $\mathsf{Tree}_h(\Pi)$ in extractions. Thus we can introduce such an assumption that says (with overwhelming probability) there exists a Π such that $\mathsf{LeafCert}$'s in different extractions generated by an efficient prover are from the same $\mathsf{Tree}_h(\Pi)$. Then we have the following argument.

Suppose in extraction what is extracted is $(2r\mathsf{UA}_2, \mathsf{LeafCert})$. Then by performing the extractions twice we have two $\mathsf{LeafCert}$'s. Due to the new assumption, in each extraction the $\mathsf{LeafCert}$ is from a fixed $\mathsf{Tree}_h(\Pi)$ which then corresponds to a fixed Π. Since in the two extractions r is different and Π outputs r, the two Π's in the two extractions are different. Moreover, with some probability the two Request's happen to specify the position on which the two programs are different. On the occurrence of this event the two LeafCert's are different. On the other hand, they share a same root, which contradicts the collision-resistance of h. It is impossible. Thus what is extracted is actually w. The extractor works as desired.

4.3 Summary

Section 4.2 presents a possible way to construct a 4-round public-coin ZKAOK. But we would not treat it as a reasonable construction due to the newly introduced assumption. Thus how to reduce the round number 6 for general public-coin ZKAOK from reasonable assumptions is still an interesting open question.

Acknowledgments. The author shows his deep thanks to the reviewers of Inscrypt 2014 for their detailed and useful comments. This work is supported by the National Natural Science Foundation of China (Grant No. 61100209) and Doctoral Fund of Ministry of Education of China (Grant No. 20120073110094).

A Preliminaries

This section contains the notations and definitions used throughout this paper.

A.1 Basic Notions

A function $\mu(\cdot)$, where $\mu : \mathbb{N} \to [0,1]$ is called *negligible* if $\mu(n) = n^{-\omega(1)}$ (i.e., $\mu(n) < \frac{1}{p(n)}$ for all polynomial $p(\cdot)$ and large enough n's). We will sometimes use $\mathsf{neg}(n)$ to denote an unspecified negligible function. We say that two probability ensembles $\{X_n\}_{n\in\mathbb{N}}$ and $\{Y_n\}_{n\in\mathbb{N}}$ are *computationally indistinguishable* if for every polynomial-sized circuit family $\{C_n\}_{n\in\mathbb{N}}$ it holds that $|\Pr[C_n(X_n) = 1] - \Pr[C_n(Y_n) = 1]| = \mathsf{neg}(n)$. We will sometimes abuse notation and say that the two random variables X_n and Y_n are computationally indistinguishable when each of them is a part of a probability ensemble such that these ensembles $\{X_n\}_{n\in\mathbb{N}}$ and $\{Y_n\}_{n\in\mathbb{N}}$ are computationally indistinguishable. We will also sometimes drop the index n from a random variable if it can be inferred from the context. In most of these cases, n will be the security parameter.

A.2 Commitment Schemes

A commitment scheme allows a party to digitally commit to a particular string, and then to reveal this value at a later time.

Definition 4. *A non-interactive perfectly-binding computationally-hiding commitment scheme is a polynomial-time computable sequence of functions* $\{C_n\}_{n\in N}$

where $C_n : \{0,1\}^n \times \{0,1\}^{p(n)} \rightarrow \{0,1\}^{q(n)}$, and $p(\cdot), q(\cdot)$ are some polynomials, that satisfies:

Perfect Binding. For every $x \neq x' \in \{0,1\}^n$, $C_n(x, \{0,1\}^{p(n)}) \cap C_n(x', \{0,1\}^{p(n)}) = \phi$.

Computational Hiding. For every $x, x' \in \{0,1\}^n$, the random variables $C_n(x; U_n)$ and $C_n(x'; U_n)$ are computationally indistinguishable.

A non-interactive perfectly-binding computationally-hiding commitment scheme can be constructed under the assumption that one-way permutations exist [9].

A.3 Interactive Proofs and Arguments

An interactive proof [19] is a two-party protocol, where one party is called the prover and the other party is called the verifier. We use the following definition.

Definition 5. *An interactive protocol (P, V) is called an interactive proof system for a language L if the following conditions hold:*

Efficiency: *The number and total length of messages exchanged between P and V are polynomially bounded and V is a probabilistic polynomial-time machine.*

Perfect completeness: *If $x \in L$, then V will always accept x.*
Soundness: *If $x \notin L$, then the probability that V accepts x is $\mathsf{neg}(n)$.*

Let $L \in \mathbf{NP}$, an interactive argument for L [10] is the following variation on the definition of an interactive proof.

1. The soundness requirement is relaxed to quantify only over prover strategies P^* that can be implemented by a polynomial-sized circuit.
2. The system is required to have an efficient prover strategy.

A.4 Zero-Knowledge

We present the definition of zero-knowledge [19] as follows.

Definition 6 ((Auxiliary-Input) Zero-Knowledge). *Let $L = L(R)$ be some language and let (P, V) be an interactive proof or argument for L. We say (P, V) is auxiliary-input zero-knowledge if there exists a probabilistic polynomial-time algorithm, called simulator, such that for every polynomial-sized circuit V^* and every $(x, w) \in R$, the following two probability variables are computationally indistinguishable:*

1. *The view of V^* in the real execution of $(P(w), V^*)(x)$.*
2. *The output of the simulator on input (x, V^*).*

If Definition 6, if the size of V^* should be bounded by an a-priori polynomial, we call (P, V) is bounded-auxiliary-input zero-knowledge, and if V^* is a PPT machine, we call (P, V) is plain zero-knowledge.

We say that a simulator is *black-box* if the only use it makes of its input V^* is to call it as a subroutine and thus we call (P, V) black-box zero-knowledge and otherwise we call (P, V) non-black-box zero-knowledge.

A.5   Witness Indistinguishability

In a witness indistinguishable proof system [13] if both w_1 and w_2 are witnesses that $x \in L$, then it is infeasible for the verifier to distinguish whether the prover used w_1 or w_2 as auxiliary input. The formal definition is below.

Definition 7. *Let $L = L(R)$ be some language and (P, V) be a proof or argument system for L. We say that (P, V) is witness indistinguishable if for any polynomial-sized circuit V^*, any x, w_1, w_2 where $(x, w_1) \in R$ and $(x, w_2) \in R$ such that the view of V^* in the interaction with $P(x, w_1)$ is computationally indistinguishable from the view of V^* in the interaction with $P(x, w_2)$.*

A.6   Proof of Knowledge

In a proof or argument of knowledge [6,12,19,29] the prover should convince the verifier that it also knows a witness for $x \in L$. It means if the verifier is convinced with some probability p by some (possibly cheating) prover strategy, then by applying an efficient algorithm, called the knowledge extractor, to the cheating prover's strategy and private inputs, it is possible to obtain a witness for $x \in L$, with probability (almost equal to) p. The formal definition is below.

Definition 8. *Let $L = L(R)$ and let (P, V) be a proof/argument system for L. We say that (P, V) is a proof/argument of knowledge for L if there exists a probabilistic (expected) polynomial-time algorithm E (called the knowledge extractor) such that for every polynomial-sized prover strategy P^* and for every $x \in \{0, 1\}^n$, if we let p_* denote P^*'s convincing probability, then $E(P^*, x)$ outputs a witness for $x \in L$ with probability $p_* - \mathsf{neg}(n)$.*

We say that a proof/argument of knowledge has a black-box extractor if the knowledge extractor algorithm E uses its first input (i.e., P^*) as a black-box subroutine (i.e., oracle). Otherwise, we say it a non-black-box extractor.

A.7   Universal Arguments

Universal arguments, introduced by [3], are interactive arguments of knowledge for proving membership in **NEXP**. For sake of simplicity, we introduce the definition of universal arguments only for an universal language $L_\mathcal{U}$: the tuple $\langle M, x, t \rangle$ is in $L_\mathcal{U}$ if M is the verifying machine that accepts (x, w) within t steps. Clearly, every language in **NE** is linear-time reducible to $L_\mathcal{U}$ and every language in **NEXP** is polynomial-time reducible to $L_\mathcal{U}$.

Definition 9. *An universal argument system is a pair of strategies, denoted (P, V), that satisfies the following properties:*

Efficient verification: *There exists a polynomial p such that for any $y = (M, x, t)$, the total time spent by the (probabilistic) verifier strategy V, on common input y, is at most $p(|y|)$. In particular, all messages exchanged in the protocol have length smaller than $p(|y|)$.*

Completeness by a relatively-efficient prover: *For every $(y = (M, x, t), w)$ in $R_{\mathcal{U}}$, $\Pr[\langle P(w), V \rangle (M, x, t)] = 1] = 1$.*

Furthermore, there exists a polynomial p such that the total time spent by $P(w)$, on common input (M, x, t), is at most $p(T_M(x, w)) \le p(t)$.

Computational soundness: *For every polynomial-sized circuit family $\{\tilde{P}_n\}_{n \in \mathbb{N}}$, and every $(M, x, t) \in \{0, 1\}^n \setminus L_{\mathcal{U}}$, $\Pr[\langle \tilde{P}_n, V \rangle (M, x, t)] = 1] < neg(n)$.*

A weak proof of knowledge property: *For every positive polynomial p there exists a positive polynomial p' and a probabilistic polynomial-time oracle machine E such that the following holds:*

For every polynomial-sized circuit family $\{\tilde{P}_n\}_{n \in \mathbb{N}}$ and every sufficiently long $y = (M, x, t) \in \{0, 1\}^*$ if $\Pr[\langle \tilde{P}, V(M, x, t)] = 1] > \frac{1}{p(|y|)}$ then $\Pr[E^{\tilde{P}*}(y) = C \text{ s.t. } [C] \in R_{\mathcal{U}}(y)] > \frac{1}{p'(|y|)}$ (where $[C]$ denotes the function computed by the Boolean circuit C). The oracle machine E is called a (knowledge) extractor.

Note that the weaker proof of knowledge property may be considered as an auxiliary feature, which can not be mandated by the basic definition of universal arguments. [3] gave a construction of 4-round public-coin universal arguments with the weak proof of knowledge property. A candidate of 2-round public-coin constructions is the 2-round variant of Micali's CS proof [26].

A.8   The LS Proof System in [23]

Now we describe the 3-round WIPOK protocol for the **NP**-complete language graph Hamiltonicity (HC), provided by Lapidot and Shamir in [23]. This construction is special in that only the size of the public input needs to be known before the last round. The actual public input can therefore be decided during the execution of a larger protocol.

Let k be the number of vertexes of graph G. G is represented by a $k \times k$ adjacency matrix $GMatrix$ where $GMatrix[i][j] = 1$ if there exists an edge between vertexes i and j in G. A non-edge position (i, j) is a pair of vertexes that are not connected in G and for which $GMatrix[i][j] = 0$. LS consists of k parallel executions (with the same input G) of Protocol 2.

As noted by [27] LS enjoys the three properties. The first is witness indistinguishability. The second one is proof of knowledge: Getting the answer for both $b = 0$ and $b = 1$ allows the extraction of the cycle. The reason is the following. For $b = 0$ one gets the random cycle C. Then for $b = 1$ one gets the permutation mapping the random cycle in the actual cycle w that is given to P. The third

---

Public input: G (statement to be proved is "$G \in$ HC");
Prover's auxiliary input: w, (a witness for $G \in$ HC).

---

1. $P \to V$: P picks a random k-vertex cycle graph C and commits bit-by-bit to the corresponding adjacency matrix using a statistically binding commitment scheme.
2. $V \to P$: V responds with a randomly chosen bit b.
3. $P \to V$: If $b = 0$, P opens all the commitments, showing that the matrix committed in Step 1 is actually a k-vertex cycle. If $b = 1$, P sends a permutation π mapping the vertex of C in G. Then it opens the commitment of the adjacency matrix of C corresponding to the non-edges of the graph G.

Protocol 2. *The 3-round WIPOK LS in [23].*

is that the first step is independent of the witness and the public input, since it only requires the sampling of a random-cycle (k is the size of the public input and must be known in advance). The witness and the public input are used only in the last Step.

References

1. Ananth, P., Boneh, D., Garg, S., Sahai, A., Zhandry, M.: Differing-inputs obfuscation and applications. IACR Cryptology ePrint Archive 2013, 689 (2013)
2. Barak, B.: How to go beyond the black-box simulation barrier. In: FOCS, pp. 106–115 (2001)
3. Barak, B., Goldreich, O.: Universal arguments and their applications. In: IEEE Conference on Computational Complexity, pp. 194–203 (2002)
4. Barak, B., Lindell, Y.: Strict polynomial-time in simulation and extraction. In: Reif, J.H. (ed.) STOC, pp. 484–493. ACM (2002)
5. Barak, B., Lindell, Y., Vadhan, S.P.: Lower bounds for non-black-box zero knowledge. J. Comput. Syst. Sci. **72**(2), 321–391 (2006)
6. Bellare, M., Goldreich, O.: On defining proofs of knowledge. In: Brickell, E.F. (ed.) CRYPTO 1992. LNCS, vol. 740, pp. 390–420. Springer, Heidelberg (1993)
7. Bitansky, N., Canetti, R., Chiesa, A., Tromer, E.: Recursive composition and bootstrapping for snarks and proof-carrying data. In: Boneh, D., Roughgarden, T., Feigenbaum, J. (eds.) STOC, pp. 111–120. ACM (2013)
8. Bitansky, N., Canetti, R., Paneth, O.: How to construct extractable one-way functions against uniform adversaries. Cryptology ePrint Archive, Report 2013/468 (2013). http://eprint.iacr.org/
9. Blum, M.: Coin flipping by telephone. In: Gersho, A. (ed.) CRYPTO, pp. 11–15. U. C. Santa Barbara, Dept. of Elec. and Computer Eng., ECE Report No 82–04 (1981)
10. Brassard, G., Chaum, D., Crépeau, C.: Minimum disclosure proofs of knowledge. J. Comput. Syst. Sci. **37**(2), 156–189 (1988)

11. Dwork, C., Naor, M.: Zaps and their applications. SIAM J. Comput. **36**(6), 1513–1543 (2007)
12. Feige, U., Fiat, A., Shamir, A.: Zero knowledge proofs of identity. In: Aho, A.V. (ed.) STOC, pp. 210–217. ACM (1987)
13. Feige, U., Shamir, A.: Witness indistinguishable and witness hiding protocols. In: STOC, pp. 416–426. ACM (1990)
14. Garg, S., Gentry, C., Halevi, S., Raykova, M., Sahai, A., Waters, B.: Candidate indistinguishability obfuscation and functional encryption for all circuits. In: FOCS, pp. 40–49. IEEE Computer Society (2013)
15. Garg, S., Gentry, C., Halevi, S., Wichs, D.: On the implausibility of differing-inputs obfuscation and extractable witness encryption with auxiliary input. In: Garay, J.A., Gennaro, R. (eds.) CRYPTO 2014, Part I. LNCS, vol. 8616, pp. 518–535. Springer, Heidelberg (2014). http://dx.doi.org/10.1007/978-3-662-44371-2_29
16. Goldreich, O., Kahan, A.: How to construct constant-round zero-knowledge proof systems for np. J. Cryptology **9**(3), 167–190 (1996)
17. Goldreich, O., Micali, S., Wigderson, A.: Proofs that yield nothing but their validity and a methodology of cryptographic protocol design (extended abstract). In: FOCS, pp. 174–187. IEEE Computer Society (1986)
18. Goldreich, O., Oren, Y.: Definitions and properties of zero-knowledge proof systems. J. Cryptology **7**(1), 1–32 (1994)
19. Goldwasser, S., Micali, S., Rackoff, C.: The knowledge complexity of interactive proof systems. SIAM J. Comput. **18**(1), 186–208 (1989)
20. Hada, S., Tanaka, T.: On the existence of 3-round zero-knowledge protocols. In: Krawczyk, H. (ed.) CRYPTO 1998. LNCS, vol. 1462, pp. 408–423. Springer, Heidelberg (1998)
21. Håstad, J., Impagliazzo, R., Levin, L.A., Luby, M.: A pseudorandom generator from any one-way function. SIAM J. Comput. **28**(4), 1364–1396 (1999)
22. Katz, J.: Which languages have 4-round zero knowledge proofs? In: Canetti, R. (ed.) TCC 2008. LNCS, vol. 4948, pp. 73–88. Springer, Heidelberg (2008)
23. Lapidot, D., Shamir, A.: Publicly verifiable non-interactive zero-knowledge proofs. In: Menezes, A., Vanstone, S.A. (eds.) CRYPTO 1990. LNCS, vol. 537, pp. 353–365. Springer, Heidelberg (1991)
24. Lindell, Y.: A note on constant-round zero-knowledge proofs of knowledge. J. Cryptology **26**(4), 638–654 (2013)
25. Merkle, R.C.: A certified digital signature. In: Brassard, G. (ed.) CRYPTO 1989. LNCS, vol. 435, pp. 218–238. Springer, Heidelberg (1990)
26. Micali, S.: Cs proofs (extended abstracts). In: FOCS, pp. 436–453. IEEE Computer Society (1994)
27. Ostrovsky, R., Visconti, I.: Simultaneous resettability from collision resistance. Electronic Colloquium on Computational Complexity (ECCC) 19, 164 (2012). http://dblp.uni-trier.de/db/journals/eccc/eccc19.html#OstrovskyV12
28. Pandey, O., Prabhakaran, M., Sahai, A.: Obfuscation-based non-black-box simulation and four message concurrent zero knowledge for np. Cryptology ePrint Archive, Report 2013/754 (2013). http://eprint.iacr.org/
29. Tompa, M., Woll, H.: Random self-reducibility and zero knowledge interactive proofs of possession of information. In: FOCS, pp. 472–482. IEEE Computer Society (1987)

A Model-Driven Security Requirements Approach to Deduce Security Policies Based on OrBAC

Denisse Muñante Arzapalo, Vanea Chiprianov[✉],
Laurent Gallon, and Philippe Aniorté

LIUPPA, Université de Pau et des Pays de l'Adour, 64000 Pau, France
{denisseyessica.munantearzapalo,vanea.chiprianov,
laurent.gallon,philippe.aniorte}@univ-pau.fr

Abstract. Attacks on unsecured systems result in important loses. Many of the causes are related to non-conformance of system architecture and implementation to the requirements. To reduce these conformity problems, Model Driven Engineering proposes using modelling languages for defining requirements and architecture and model transformations between them. We therefore introduce a modelling language extension/ profile for defining system requirements with basic security requirement concepts. We also formalize the model transformation between this profile and a security formal verification method. We exemplify our approach on a medical case study.

Keywords: Model-driven security · Model transformation · Requirements engineering · OrBAC · i* framework

1 Introduction

Nowadays, important financial loses are caused by attacks on systems. However, many security breaches are caused by non-conformance of the system security architecture and implementation to its requirements [1]. Therefore it is necessary to ensure that the implementation of security policy mechanisms is conform to its requirement specification.

Numerous approaches to specify security requirements exist (see Sect. 2). However, if such an approach is to ensure conformity with the next phases in the security life-cycle of architecture and implementation, it needs to describe requirements in such a format so as to enable a (semi-)formal approach for translating them into architecture. An additional concern is related to the fact that such security requirements approaches usually involve security specialists. However, in practice, such an expert is not always involved in the generic requirements elicitation phase, or security specialists implied in this phase do not have sufficient knowledge about a specific security domain (e.g. access control). Therefore, such specialised security requirements approaches cannot always be used by the non-specialist actors. Nevertheless, it would be desirable to have at least fundamental security policies captured at this phase. Therefore, the requirements

© Springer International Publishing Switzerland 2015
D. Lin et al. (Eds.): Inscrypt 2014, LNCS 8957, pp. 150–169, 2015.
DOI: 10.1007/978-3-319-16745-9_9

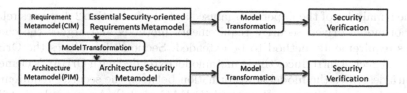

Fig. 1. Overview of the proposed approach

approach that also takes into account security should offer the actors concepts generic enough for them to be comfortable with, and also precise enough to be translatable and expanded upon at the architecture level. Additionally, it would be desirable to have preliminary verifications of the security requirements. Even if in the requirements phase the level of details is quite sparse, early conflicts may be discovered, influencing the further quality of the security architecture.

Model Driven Engineering (MDE) is an approach which uses modelling languages to describe the system at different phases of its life-cycle, and ensures translations between them through formal model transformations. Model-driven architecture (MDA) [2] uses models in the software development process and proposes three levels of abstraction, in particular, a Computation Independent Model (CIM) level presents what the system is expected to do (i.e. requirements) and a Platform Independent Model (PIM) level represents how the system fulfills its requirements and technical details (i.e. design/architecture).

To describe requirements in such a format so as to enable a (semi-)formal approach for translating them into architecture, we present in Fig. 1 an MDE approach in which requirements are modelled as CIM and the architecture is modelled as PIM. We already contributed towards the PIM level [11] - a modelling language extension based on UMLSec. In this paper we propose a language extension/profile, based on i*, for capturing security requirements at the CIM level. A model transformation can be further defined between the two levels.

The main contributions of this papers are:

- a model-driven modelling language extension/profile for capturing security requirements. It is based on i*, choice which we argue in Sect. 2, and extend it with basic security concepts inspired from the OrBAC metamodel. For this profile to be easily usable by actors which are not security specialists, we choose to model only basic, essential, generic security concepts.
- To make early evaluations and conflict detection in the requirements modelled with this profile, we define a formal model transformation towards a formal security verification method using MotOrBAC [9]. This transformation contains two steps: firstly we reduce the model written in the extended i* language in order to suppress elements which are not related to security; secondly, we deduce OrBAC security policies from this reduced model.

To sum up, we propose an MDE based approach which allows a designer non-specialist in security to model requirements which contain security concerns and to evaluate the security policies implicit in these requirements.

The remainder of the paper is organised as follows. Section 2 discusses related works of model-driven security requirements methods and argues the choices of i* as requirements method to be extended. Section 3 introduces the OrBAC model and Sect. 4 introduces the i* metamodel extended with OrBAC's elements. Section 5 describes the model transformation between the security requirements profile and the security formal method (OrBAC model). An example that illustrates our proposition is detailed in Sect. 6. Finally, Sect. 7 concludes this paper and gives future works.

2   Related Works

There are many works deal with security using various modelling languages in early stages of development systems. Abuse Frames [13], misuse cases [14] and mal-activity diagrams [15] address security concerns through negative scenarios executed by attackers. CORAS framework [20], Tropos goal-risk [21] and ISSRM [22] are based on a risk analysis process. Some goal modelling languages have been adapted to include security concepts: Secure i* [18], KAOS with anti-goal models [16], Secure Tropos [17] and GBRAM [19]. To use these referenced works, it is necessary that security specialists are part of the development team. In contrast, our approach supports developers who could not be security experts by deducing automatically security requirements from system requirements.

Some works are closed to our proposal. Mouratidis et al. [25] and Massacci et al. [23] consider social and technical dimensions to identify security requirements. Ledru et al. [24] proposes an approach called KAOS2RBAC to identify Role-Based Access Control (RBAC) [30] requirements from KAOS requirements. Graa et al. [26] generate automatically OrBAC security policies from security and functional goals through KAOS and a risk analysis method. These three requirement approaches all use concepts dedicated to security, which implies the need for security specialist to use them. In contrast to this, our approach has more generic concepts which allow actors, who are not necessarily security specialists, to use it. Hatebur et al. [27] present a method to systematically develop UMLsec [4] models from security requirements based on UML. We use i* framework to represent systems requirements in a more expressively way.

The choice of i* metamodel has been performed through an analysis of security requirement methods corresponding to model-driven and security criteria such as *Model/Standard of Development, Prototype, Security requirement, Threat, Vulnerability, Risk* ... [12]. On the other hand, we focus on OrBAC model rather than RBAC model (as in [23]) because it allows to consider more complex security rules using notions such as prohibitions, obligations, contexts, hierarchies, delegations, concrete entities, etc.

3   The Organization-Based Access Control (OrBAC)

The central entity in OrBAC [3] is the *Organization*, which can be seen as an organized group of subjects, playing some roles. OrBAC allows policy designers to define a security policy independently of the implementation thanks to the use of an abstract level and a concrete level. In OrBAC, *subject*, *action* and *object*, which corresponds to *concrete entities*, are respectively abstracted into *role*, *activity* and *view*, which corresponds to *abstract entities*.

A security policy is defined for and by an organization. This policy is a set of rules (permissions, prohibitions, obligations or dispensations). The rules only apply in a specific *context*. A context is a special condition between user, object and action that control activation of rules in the access control policy [5]. There are simple and complex contexts, for example, *working-hours* and *not in holidays* contexts are simple contexts which can be assembled to obtain "a subject has not to be in holidays and has to be in working-hours" as a complex context.

OrBAC uses predicates to define these rules. A predicate is seen as a property that a subject has or is characterized by, hence, it is an expression that can be true. Thus, the *OrBAC abstract privilege predicates* are used to define security rules as follows: *permission(org, r1, a1, v1, c1)*, *prohibition(org, r1, a1, v1, C1)* predicates are defined to indicate that *r1* is allowed/prohibited to perform *a1* on *v1* at the context *c1* for *org*. And, *obligation(org, r1, a1, v1, activationCtx, violationCtx)* compared to the *permission* and *prohibition* has two contexts: *activationCtx* expresses the condition in which the obligation is activated, and *violationCtx* expresses the condition (e.g. a deadline) in which the obligation is violated. Moreover, obligations are often associated with access control to express that some actions should be taken before, while or after resource usage. These obligations are called pre, ongoing and post obligations respectively [8].

The previous predicates are based on the others ones: (i) the *OrBAC relevant predicates* are used to indicate that an abstract entity is an relevant element for an organization, (ii) the *OrBAC abstraction predicates* are used to assign a concrete entity to an abstract entity within an organization: the *empower* predicate is used to assign subjects to roles, the *use* predicate is used to assign objects to views and the *consider* predicate is used to assign actions to activities, (iii) and the *OrBAC hold predicate* is used to associate a context to an organization.

One of the main advantages of OrBAC is that it can automatically derive concrete rules from abstract rules. OrBAC also offers the mechanisms of delegation [7] and inheritance [6] to make easier the definition of security policies.

4   The Security Requirements Profile

The specification of system requirements is conducted by adopting an approach based on RE. There are many RE methods such as *use cases*, *KAOS*, *Tropos*, *NFR*, ... and the well-known *i**. The *i* metamodel* was used to extend security concepts in [18, 23, 29]. But it is not used to derive automatically access control policies based on OrBAC. The i* framework does not support all main entities of OrBAC. In order to derive OrBAC security policies, in this section, we introduce an extension of i* metamodel with OrBAC concepts.

4.1   The I* Metamodel

An overview of the i* metamodel is show in Fig. 2. The light elements correspond to the classical i* and the shadow elements correspond to our extension. An actor is an active entity that has strategic goals and intentionality within the system, carries out activities, and produces entities to achieve goals by exercising its knowhow [28]. Actors can be roles and agents. A *role* captures an abstract characterization of the behavior of a social actor. An *agent* is an actor with concrete manifestations and can play some role.

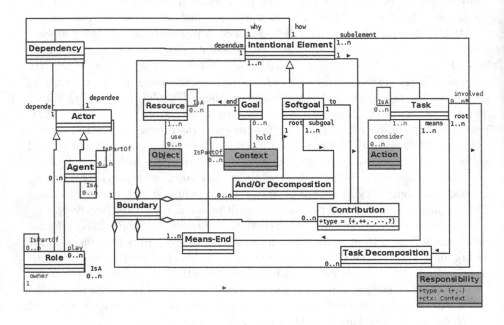

Fig. 2. The i* metamodel extended with the OrBAC concepts

Intentional elements defined by the i* framework are goals, softgoals, tasks, and resources. A *goal* represents the intentional desire of an actor. Goals are also called hard goals in contrast to *softgoals* which do not have clear criteria for deciding whether they are satisfied or not. A *task* is a set of actions which the actor needs to perform to achieve a goal. A *resource* is a physical or an informational entity used to represent assets.

The metamodel in Fig. 2 also describe the relationships between intentional elements inside the *boundary* of actors. Actors have (soft)goals and rely on other (soft)goals, tasks, and resources to achieve them. Softgoals can be decomposed into more softgoals using *AND/OR* that a goal (the *end*) can be achieved by performing alternative tasks (the *means*). Tasks can be decomposed into any other intentional elements through *task decomposition* relations. By decomposing a task into sub-elements, one can express that the sub-elements need to be satisfied or available to have the rootTask performed. Softgoals and other intentional elements can contribute either positively or negatively to the other softgoals. This is expressed by the *contribution* relations.

4.2   The I* Metamodel Extended with the OrBAC Concepts

In this paper, we present the extension of i* metamodel which is divided into OrBAC entities, OrBAC predicates and OrBAC additional mechanisms:

OrBAC Entities. In the extended i* metamodel, we add the OrBAC entities except the *organization* entity because we consider that each system model corresponds to one organization. In Fig. 2, we can use *Role*, *Task* and *Resource* entities to represent *roles*, *activities* and *views* respectively. These elements correspond to the OrBAC *abstract entities*. Moreover, for the OrBAC *concrete entities*, we can use the entity *Agent* to represent *subjects*, and we add *Objects* and *Actions* because we do not find entities related to these notions. Remember that using these abstract entities, we can associate concrete entities to the system.

OrBAC Predicates. For this paper, it is not necessary to include the *OrBAC relevant predicates* because we assume that there is only one organization for each system model, hence all the entities will belong to this "organization".

For the *OrBAC abstraction predicates*, in the i*, we can use the *play* relation between agents and roles to represent the *empower* predicate. In the same way, we add the *use* relation between objects and resources, and the *consider* relation between actions and tasks.

Related to the *OrBAC hold predicate*, contexts are attached to goals for us, i.e. a goal is the intentional desire of an actor in a particular context. To represent contexts, we add the entity *Context* and the relation *hold* to associate a context to a goal. For example, in a medical system, the goal of a physician is *to care patients* and the context attached to it is that *the patient is under the responsibility of the physician*. We also create the relation *IsPartOf* to assemble simple contexts.

On the other hand, we do not create entities to represent the *OrBAC abstract privilege predicates* used to define OrBAC security rules because we deduce them automatically (Sect. 5).

Finally, we add the new relation *responsibility* which is used to generate some security rules. In real life, people is responsible of a set of tasks according to roles that they play within an organization. This scenario is for a usual ambiance (context). However, the scenarios in unusual contexts can also be part of the organization's duty. For example, in the medical system, assistants can be in charge of managing patients' appointments in a usual context. Assistants can also be in charge of consulting of the stock of medicine in an unusual context such as urgency. We add the notion of *responsibility* to relate a role (the owner of the responsibility) to a task (task involved in the responsibility) in a specific context (Fig. 2). This notion allows to add a responsibility to a role, so-called positive responsibility (+). However, it is also possible to add negative responsibilities (-) to roles, thus a role is excluded to perform a task in a particular context. For example, physicians cannot modify medical records in an *audit* context.

OrBAC Additional Mechanisms. The delegation mechanism is beyond the scope of this paper because this mechanism is closer to an administrative work, and it is difficult to model. In contrast, we can use the relation *IsA* of *Roles* to represent the inheritance between roles. Then, we add the same relations to *resources* and *tasks* (see Fig. 2).

4.3   Correspondence Between OrBAC Concepts and Concepts of Our Extended I* Metamodel

Because the i* framework is intended to capture requirements in general, the i* models do not contain the same information than just that related to security, specifically OrBAC model. In Table 1, we show the correspondence between OrBAC concepts and concepts of our extended i* metamodel. We use the symbol (a) to distinguish OrBAC concepts added to i* metamodel, and the symbol (b) to indicate which OrBAC concepts are deduced (see third column in Table 1).

As we are interested in deducing the OrBAC security rules, we should propose a model transformation between security requirements profile (the extended i* metamodel) and the security formal method (the OrBAC model, evaluated by MotOrBAC). Notice that there are concepts of i* metamodel which are not related to OrBAC (see Table 1). Therefore, we should determine the i* concepts that be implied on our model transformation (and eliminate the other ones) to deduce OrBAC security rules.

5   The Model Transformation Between the Security Requirements Profile and the Security Formal Method

The goal of this section is to present the methodology we propose to deduce OrBAC security rules from the extended i* metamodel. It is divided into two activities:

- firstly, we need to reduce/simplify the i* modified meta-model by eliminating all entities and relations which have no correspondance in the OrBAC meta-model or are not necessary for our deduction process
- secondly, we analyse the reduced/simplified i* metamodel to identify configurations (or patterns) which imply a necessary OrBAC security rule.

The result of this process is the deduced OrBAC security policy (which is a set of OrBAC security rules) implied in the initial requirements model.

Therefore, this section is divided into two subsections: in the first subsection, we introduce such a reduction process in which we reduce the extended i* meta-model to focus specifically on OrBAC security. And, in the second subsection, we introduce a deduction process to deduce/extract OrBAC security policies.

Table 1. Correspondence between OrBAC concepts and concepts of our extended i*
metamodel

	OrBAC concepts	Concepts of our extended i*
OrBAC entities	Role	Role
	Activity	Task
	View	Resource
	Subject	Agent
	Action	Action [*]
	Object	Object [*]
	Context	Context [*]
OrBAC predicates	relevant_role, relevant_ activity, relevant_view	Not modelled because we consider only one organization
	Empower	Play
	Consider	Consider [*]
	Use	Use [*]
	Hold	Hold [*]
	Permission	Deducted rule [**]
	Prohibition	Deducted rule [**]
	Obligation	Deducted rule [**]
OrBAC additional mechanisms	Separation of entities and Priorization of rules	Not modelled because they need more information
	Delegation	Not modelled because it is close to an administrative notion
	Role hierarchies	IsA relation between roles
	Activity hierarchies	IsA relation between tasks [*]
	View hierarchies	IsA relation between resources [*]
	x	i* Entities: Boundary, Intentional Element, Goal, SoftGoal, Actor
	x	i* Relations: Dependency, And/Or Decomposition, Contribution, Means-End, Task Decomposition, Responsibility [*]

(*) New concept added to i* metamodel
(**) Deducted concept

5.1 I* Metamodel Reduction Process According to Necessary Conditions

Because i* is a requirements engineering method, its concepts and relations are
on the one hand, more general, and on the other hand, much less rigorous and
formal than those of OrBAC. Therefore, not all of its entities and relations can

correspond to those of OrBAC (see Table 1). To determine which i* metamodel entities and relations have no correspondance in the OrBAC metamodel or are not necessary for our deduction process, we define a reduction process.

This reduction process contains necessary criteria/conditions for the i* metamodel entities and relations to correspond well to the OrBAC ones. By applying these criteria/conditions, we obtain a reduced version of the i* metamodel which both corresponds better to the OrBAC metamodel and enables easier implementation of the model transformation between the two metamodels.

Criteria. In this section, we present a list of criteria to reduce the extended i* metamodel in order to preserve/obtain only the necessary entities and relations used to deduce/extract OrBAC security policies.

(1) *Optimisation criterion:*
    To reduce the number of i* metamodel entities to be processed, some entities should be eliminated in a such way so that the resulting metamodel is (functionally) similar with the initial metamodel. The similarity notion implies of course a loss of information, i.e. the entity that will be eliminated no longer exists in the model. However, for the resulted formal security model, this should have no impact because these entities do not have direct counterparts in the security metamodel.

(2) *Tree extraction criterion:*
    For implementation reasons, to be able to manipulate the i* models, which are graphs, we need to extract a tree structure.
    The main possibilities to extract a tree structure are around the concept of Task and its composition relation of TaskDecomposition, and around the concept of SoftGoal and its composition relation of AND/OR Decomposition. These options are analysed in the next section.

(3) *Insufficient information criterion:*
    Because i* is a requirements engineering method and therefore does not necessarily elicit rigorous and complete models, there may be information which is not sufficient to deduce formal security models/rules. We decide to discard such incomplete information.

Application of Criteria to I* Metamodel. In this section, we describe the application of the criteria identified in the previous section. For each criterion, we explain which entities of i* metamodel are implied.

(1) *Optimisation criterion:*
    According to the application of this criterion, we eliminate all the entities of extended i* we do not identify a direct counter-part in OrBAC. Then we eliminate the entities *SoftGoal, Goal, Intentional Element* and *Actor* using Table 1. Notice that *Boundary* is not eliminated. We deduce security rules using roles, so we need to know their limits to determine the security rules associated to these roles. These limits are represented by *Boundary* entities, so it is important to preserve them.

Some relations of the extended i* model are not related to OrBAC model: *Dependency, And/Or decomposition, Contribution, Means-end, Task decomposition* and *Responsibility*. According to the optimization criterion, we decide to eliminate these relations, except if:

- they can generate security rules
- they relate entities of the reduced model.

As we explained, *Responsibility* relations imply security properties. Hence, we should preserve them for the reduced model. *Task Decomposition* relations allow to decompose a task into a set of subtasks and resources, etc. If we eliminate these relations, we can lose security information (for example, rigths to use resources to perform tasks or subtasks). Therefore, we decide preserve *Task decomposition* relations. Finally, if we eliminate *Dependency* relations, we eliminate all relations between tasks of different roles. Hence, we can also lose security information. Thus, we decide to preserve them. In brief, we preserve the relations *Dependency, Task Decomposition* and *Responsibility*. And, we eliminate *And/Or decomposition, Contribution* and *Means-end*.

Notice that if a context is connected to a goal and this goal is eliminated. Then, this context will be connected to tasks related to this goal.

(2) *Tree extraction criterion:*

Remember that the main possibilities to extract a tree structure are around of *Task* and *SoftGoal* concepts. As the *SoftGoal* concept is eliminate applying the criterion 1, we choose the concept of *Task* for the tree extraction.

We call *"rootTasks"* the tasks associated directly to goals (not subgoals). The *rootTasks* establish the first level of tree structure. Our deduction process begins analyzing these tasks.

Every time an i* metamodel entity may be replaced/by-passed with a *Task* entity, this should be done.

Therefore, we modify *Dependency relations with Resource for Why* to point the Why ends towards Tasks that are decomposed into such Resources.

In the extended i* model, the relation *Dependency* is associated to *Intentional Element*. Once Intentional Element is eliminated, the concepts *Dependum, Why* and *How* associated to Intentional Element should be associated to *Resource* and *Task* in the reduced model.

(3) *Insufficient information criterion:*

For us, relations between *Tasks* and *Resources* can be used to deduce/generate possibly security rules. Hence, according to this criterion, we eliminate *Resources* have no relation to other Tasks. In a similar way, *Dependency relations with Resource for How* cannot be used to generate security rules.

Hence, we eliminate these Dependency relations.

The OrBAC rules are composed by contexts. *Dependency relations with Task for How* where the task is not associated to any context cannot generate security rules. Therefore, we eliminate these Dependency relations.

The Reduced I* Metamodel. After applying the previous criteria, we obtain a reduced i* metamodel, which is depicted in Fig. 3.

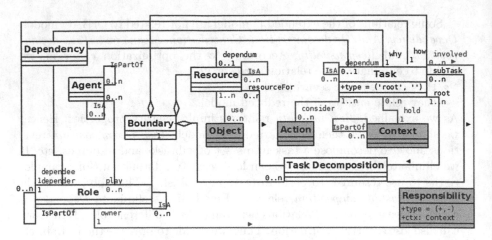

Fig. 3. The simplified/reduced i* metamodel used for the deduction of OrBAC security policies

Notice that, *Task decomposition* is associated to two entities: tasks (using *subtask* relation) and resources (using *resourceFor* relation). We add a constraint to restric the decomposition into only one entity (task decomposition can be subtask or resourceFor, not both). We use the same idea for Dependency relation which is associated to two dependum entities (tasks and resources).

In brief, the purpose of this section was to reduce the i* metamodel in order to preserve/obtain the necessary entities and relations for the deduction process of OrBAC security rules which will be explained in the next section.

5.2 Deduction Process to Extract/Deduce OrBAC Security Policies

In this section, we present the deduction process which extracts/deduces OrBAC security rules from system requirements. The main idea is to identify configurations (or patterns) in the simplified/reduced i* metamodel which imply a necessary OrBAC security rule. We base our approach on a previous work [23] in which the authors deduce RBAC security policies from SI* models.

As a result of the reduction process (Sect. 5.1), the model conforming to the reduced i* metamodel contains a graph of tasks, tasks (we call them *rootTasks*) which are composed of subtasks and ressources (see relations between entities task, task decomposition and ressource on Fig. 3). From this graph, for each rootTask, we extract a tree composed of its subtasks and ressources. Then, we search for relations (ressourceFor in Fig. 3) between a rootTask (and also subtask) and a ressource in these trees. These relations imply a security rule, either a permission, a prohibition or an obligation. We represent our tree extraction and analysis algorithm as a set of equations.

In i* models, the functionality of a system is established defining the functionality of roles implied in the system. Hence, the deduction of the security policy of the system is defined as the deduction of security policies for its roles:

$$secPol(system) = \sum_{x=1}^{m} secPol(role_x) \quad \text{where } \textbf{\textit{role}}_x \text{ is a role of "system"}$$

Moreover, roles have boundaries where a set of *task* and *resource* entities as well as a set of *task decomposition*, *dependency* and *responsibility* relations are defined (see Fig. 3). We analyse these relations to find *resourceFor* relations:

- *Task decompositions* and *Dependencies* use tasks as *root* entities and *why* entities respectively. Moreover, tasks are inside the boundaries of roles, therefore we infer that security policies for roles are defined as the deduction of security rules for theirs tasks. However, we distinguish two kinds of tasks: rootTasks and subTasks. The analyse of the tree structure starts from rootTasks (after that a recursive equation is called to analyse subTasks).
- *Responsibility* relations use roles as *owner* entities, so it implies that security policies for roles are defined as the deduction of security rules for their responsibilities.

Therefore, the security policy of the role is defined as follows:

$$secPol(role) = \underbrace{\sum_{x=1}^{n} secRule(rootTask_x)}_{(1)} \wedge \underbrace{\sum_{x=1}^{p} secRule(responsability_x)}_{(2)}$$

where **rootTask**$_x$ is the root task of **"role"** and **responsability**$_x$ is a responsibility of **"role"**

Security Rules for rootTasks. As we can see in Fig. 3, rootTasks can be decomposed in subTasks (*subTask* relation) and resources (*resourceFor* relation). In [23], *ResourceFor* relations are used to define RBAC permissions (a role is granted to perform a task on a resource). Remember that we add the *context* notion which is attached to tasks. Thus, we can infer/assume that roles perform tasks on resources in specific contexts. Therefore, this inference is used to define *OrBAC security rules*.

We analyse and extract OrBAC permissions from the tree structure (root-Tasks and subTasks) using the same idea presented in [23]. For this, we define *permAccess* as the recursive equation to analyse all subTasks in the tree structure. Thus, *secRule(rootTask)* is defined as *permAccess(rootTask, role, context)*, where *role* and *context* are associated to the *rootTask* (see Eq. 1).

In OrBAC, the dependency between security rules is defined using pre, post and ongoing obligations (see Sect. 3). In particular, a pre-obligation can be seen as a pre-requirement of a permission. In other words, the granted privilege of a permission depends on the priviledge of its pre-obligations. In i* models, this scenario of dependency can be depicted by subtask relations. A subtask relation stablishes that the fulfilment of the task depends on the fulfilment of its subtasks. Therefore, resourceFor relations related to *tasks* can imply *OrBAC permission* and resourceFor relations related to *subtasks* can imply *OrBAC pre-obligations*. Thus, we define the equation *permAccess* (see Eq. 1) as follows:

- If a task is associated to a resource then *permAccess* is defined as the generation of one OrBAC permission (OrBAC rule I) using this task and its resource. And a set of obligations related to task decompositions (or subtasks) and dependencies of the task are generated. For this, *obligAccess* (Eq. 1.1) and *obligAccessDep* (Eq. 1.2) are defined.

- Otherwise (the task has no resource), *permAccess* is defined as the addition of permissions for its *subtasks* using the same equation *permAccess* (as a loop to analyse all subtasks to define all permission rules).

obligAccess(superTask, subTask, role, context) is defined as follows (Eq. 1.1):

(1) $secRule(rootTask) = permAccess(rootTask, role, context)$

    where   **role** is the owner of **"rootTask"**

            context is the context attached to **"rootTask"**

    and   for $permAccess(task, role, context)$

      if **task** has a resource then

        $permAccess(task, role, context) = \{permission(role, task, resource, ctxTask)\}$ // OrBAC rule I

$$\wedge \sum_{x=1}^{y} obligAccess(task, task_x, role, ctxTask) \quad \wedge \sum_{x=1}^{u} obligAccessDep(task, dependum_x)$$
$$\text{(1.1)} \qquad\qquad\qquad\qquad\qquad \text{(1.2)}$$

        where   **resource** is the resource for **"task"**

          and   if **"task"** is not attached to any context then **ctxTask** is **context**

               else **ctxTask** is the context attached to **"task"**

               end if

          and   **task$_x$** is a subtask of **"task"**

          and   **dependum$_x$** is a dependum of **"task" (why element)**

      else // **task** has not any resource

$$permAccess(task, role, context) = \sum_{x=1}^{y} permAccess(task_x, role, context)$$

        where   if **"task"** is not attached to any context then **ctxTask** is **context**

              else **ctxTask** is the context attached to **"task"**

              end if

          and   **task$_x$** is a subtask of **"task"**

        // Notice that permissions for dependencies are not deduced to avoid duplicating security rules

      end if

(1.1) for $obligAccess(superTask, subTask, role, context)$

  if **subtask** has a resource then

    $obligAccess(superTask, subTask, role, context) = \{obligation(role, subtask, resource,$
$$access\_request(superTask), not\ ctxTask)\} \quad \text{// OrBAC rule II}$$

$$\wedge \sum_{x=1}^{r} obligAccess(subTask, task_x, role, ctxTask) \quad \wedge \sum_{x=1}^{s} obligAccessDep(subTask, dependum_x)$$

      where   **resource** is the resource for **"subTask"**

        and   if **"subtask"** is not attached to any context then **ctxTask** is **context**

              else **ctxTask** is the context attached to **"subtask"**

              end if

        and   **task$_x$** is a subtask of **"subTask"**

        and   **dependum$_x$** is a dependum of **"subTask" (why entity in the dependency)**

  else // **subtask$_x$** has not any resource

    $obligAccess(superTask, subTask, role, context) =$

$$\sum_{x=1}^{r} obligAccess(superTask, task_x, role, context) \quad \wedge \sum_{x=1}^{s} obligAccessDep(superTask, dependum_x)$$

      where   if **"subtask"** is not attached to any context then **ctxTask** is **context**

             else **ctxTask** is the context attached to **"subtask"**

             end if

        and   **task$_x$** is a subtask of **"subTask"**

       and   **dependum$_x$** is a dependum of **"subTask" (why element in the dependency)**

  end if

- If *subTask* is associated to a resource then *obligAccess* is defined as the generation of one OrBAC pre-obligation (OrBAC rule II) where the activation context is the access request realised by *superTask* and the violation context is not to obey the context of the equation (if the subtask is attached to a context we use it). And a set of obligations for the subtasks and dependencies of the subTask are generated using the same equations *obligAccess* (Eq. 1.1) and *obligAccessDep* (Eq. 1.2) (as a loop to analyse all subtasks and dependencies to define consecutive pre-obligations).
- Otherwise (subTask has no resource), *obligAccess* (Eq. 1.1) and *obligAccessDep* (Eq. 1.2) are called for the subtasks of *subTask* provoking a loop to find all pre-obligations. Notice that we use *superTask* (not *subTask*) in order to indicate that the task that called these subtasks is the one associated with the permission rule (generated above).

obligAccessDep(task, dependum (Eq. 1.2) is defined as the obligation for *taskHow* which is the task (*How entity*) linked to the *dependum* in the dependency relation. For this, we use *obligAccess* (Eq. 1.1) as follows:

(1.2) for *obligAccessDep(task, dependum)*
  $obligAccessDep(task, dependum) = obligAccess(task, taskHow, roleHow, contextHow)$
    where **taskHow** is related to **"dependum"**
    and **roleHow** is the owner of **"taskHow"**
    and **contextHow** is the context attached to **"taskHow"**

Security Rules for Responsibilities. As we mentioned, *responsibility relations* add or remove responsibilities for roles in order to perform tasks in unusual contexts. Unusual contexts can be seen as exceptional scenarios to the normal duty of the system. In spite of *responsibilities* being associated to exceptional scenarios, they are also part of the business of the system. We use them to deduce security rules using the equation *secRule(responsibility)* (Eq. 2). Notice that, in Eq. 2, *responsibility relations* are positive or negative:

- If the *type of responsibility* is positive then we define *secRule(responsibility)* as the generation of permissions of the task involved in the responsibility. We reuse *permAccess* to generate permissions and pre obligations (Eq. 1).

- Otherwise (the *type of responsibility* is negative), we define *secRule(responsibility)* as the generation of prohibitions of the task involved in the responsibility (using *prohibAccess*). *prohibAccess* (Eq. 2.1) is defined as follows:
  • If the involved task is associated to a resource then *prohibAccess* is defined as the generation of one OrBAC prohibition (OrBAC rule III) using this task and its resource. And a set of prohibitions for its subtasks is generated.
  • Otherwise (the involved task has not any resource), *prohibAccess* is defined as a set of prohibitions for its subtasks (as a loop to analyse all subtasks to find all prohibition rules).

Notice that, we only deduce permissions, prohibitions and pre obligations rules. We know that there are other kinds of obligations (post, ongoing and independent resource-control obligations). They are not considered in this paper because they are more complex to deduce, however they could be part of a future work.

(2) for $secRule(responsibility)$
  if **type of responsibility** $='+'$ **then** // positive resp.
    $secRule(responsability) =$
      $permAccess(task, role, context)$
        where   **task, role** and **context** are related to
                   "responsibility"
  else // negative responsibility
    $secRule(responsability) =$
      $prohibAccess(task, role, context)$ ... **(2.1)**
      where   **task, role** and **context** are related to
                    "responsibility"
end if

(2.1) for $prohibAccess(task, role, context)$
  if **task** has a resource **then**
    $prohibAccess(task, role, context) =$
      $\{prohibition(role, task, resource, ctxTask)\}$
                          // OrBAC rule III

$$\wedge \sum_{x=1}^{u} prohibAccess(task_x, role, ctxTask)$$

    where   **resource** is the resource for **"task"**
    and $task_x$ is a subtask of **"task"**
  else // **task** has not any resource
    $prohibAccess(task, role, context) =$

$$\sum_{x=1}^{u} prohibAccess(task_x, role, ctxTask)$$

    where $task_x$ is a subtask of **"task"**
    and   if **"task"** is not attached to any context then
              ctxTask is **context**
             else **ctxTask** is the context attached to **"task"**
              end if
end if

6 Case Study: A Medical System

In this section, we simulate the execution of the model transformation introduced previously. For this, we use a medical system where we have tried to cover all the possible cases to deduce security rules (see Fig. 4). Notice that the new entities and relations added to the i* metamodel are depicted in the second column of the legend (the grey and blue elements).

In this model, we can distinguish three roles: *physician, assistant* and *pharmacist*. A physician *cares patients which are him/her responsibility*, an assistant *manages appointments in working-hours* and a pharmacist *provides medicine in working-hours*.

To care patients, a physician needs to check the patient's appointment and to modify the medical record of the patient. After diagnosing the patient, the physician fulfills the record of duty control. On the other hand, the physician needs to consult the stock to prescribe the medicine for the patient. To manage appointments, an assistant needs to assists (helps) a patient to save an appointment for him/her. To provide medicine, a pharmacist modifies the stock of medicine.

Moreover, physicians are excluded to modify medical records during an audit context (see the negative responsibility of physician). In the same way, assistants are allowed to consult medicine in an urgency context (see the positive responsibility of assistant). On the other hand, the subject *John* plays the role *physician*, the action *updateFileMR* is considered as an implementation of the task *modify MR*, and the object *GeorgeMedRecord.doc* is used by the resource *medical record*.

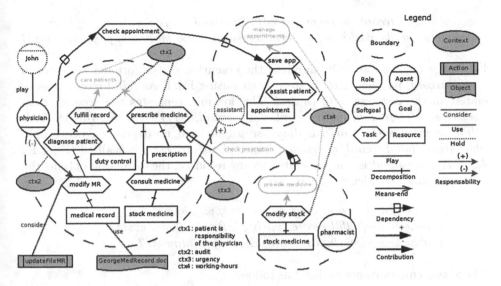

Fig. 4. The reduced Medical System model

6.1 Applying the Reduction Process

According to the reduction process explained in Sect. 5.1, in Fig. 4, we present the "reduced" model of the medical system.

Notice that goals such as *care patients*, *save appointments* and *provide medicine* are removed (criterion 1). The contexts which were attached to these goals point to their rootTasks.

Moreover, the task dependency *check prescription* is removed because is related to the *goal provide medicine* as Why entity (criterion 1).

6.2 Applying the Deduction Process

To deduce security policies we use the equations introduced in Sect. 5.2. Remember that the *security policy of a system* is the addition of security policies of it roles. And, the security policy of a role is a the addition of security rules of its rootTasks and responsibilities. Hence, we have:

$secPol(medSystem) = secPol(physician) + secPol(assistant) + secPol(pharmacist)$
where: $secPol(physician) = secRule(fulfill\ record) + secRule(prescribe\ medicine)$
$\qquad\qquad\qquad + secRule((-)responsibility),$
$\quad secPol(assistant) = secRule(save\ appointment) + secRule((+)responsibility),$
$\quad$ and $secPol(pharmacist) = secRule(modify\ stock).$

For *secRule(fulfill record)* and *secRule(prescribe medicine)*, we use Eq. 1 in the deduction process (we apply the same equation for all rootTasks of other roles: *save appointment* and *modify stock*):

$secRule(fulfill\ record) = permAccess(fulfill\ record, physician, ctx1)$

$secRule(prescribe\ medicine) = permAccess(prescribe\ medicine, physician, ctx1)$

For space reasons, we only simulate the execution of the formula *permAccess (fulfill record, physician, ctx1)* step by step using Eq. 1. As *fulfill record* is associated to the resource *duty control*. This formula generates the security rule: *"permission(physician, fulfill record, duty control, ctx1)"* (OrBAC rule I). And, *fulfill record* has the subTask *diagnose patient* and no depencies, hence only *obligAccess(fulfill record, diagnose patient, physician, ctx1)* is realised (Eq. 1.1). As *diagnose patient* is not associated to any resource, the deduction obtains the subtasks and dependencies for it. Thus:

$obligAccess(fulfill\ record, diagnose\ patient, physician, ctx1)$
$= obligAccess(fulfill\ record, modify\ MR, physician, ctx1)$
$+ obligAccessDep(fulfill\ record, check\ appointment).?$

These two equations are realised as follows:

- For *obligAccess(fulfill record, modify MR, physician, ctx1)* (Eq. 1.1), as *modify MR* is associated to the resource *medical record*, therefore *obligAccess* for *modify MR* generates the security rule: *"obligation(physician, modify MR, medical record, access_request(fulfill record), not ctx1)"* (OrBAC rule II) which means that a physician should modify a medical record when *fulfill record* requires it, and the patient is responsibility of the physician.
- For *obligAccessDep(fulfill record, check appointment)* (Eq. 1.2), as *check appointment* is related to the task *save appointment* of assistant, the equation *obligAccess(fulfill record, save app, assistant, ctx4)* is realised. Thus, this equation generates the security rule: *"obligation(assistant, save app, appointment, access_request(fulfill record), not ctx4)"* (OrBAC rule II) which means that an assistant should save the appointment when fulfill record required it during the working-hours. Notice that *save app* has the subtask *assist patient*, however this subtask has no resources or subtasks, so it does not influence the generation of security policies.

Therefore, *secRule(fulfill record)* is composed by three security rules: two pre obligations and one permission. We employ the same process for *secRule (prescribe medicine)* (see Eq. 1), and we obtain two security rules: *"permission(physician, prescribe medicine, prescription, ctx1)"*, *"obligation(physician, consult medicine, stock medicine, access_request(prescribe medicine), not ctx1)"*.

Notice that *physician* has a negative (-) responsibility associated to the task *modify MR* in the *audit* context (ctx2). To deduce security rules for this responsibility, we use *secRule(responsibility)* (Eq. 2). As the responsibility is negative, a set of prohibition (Eq. 2.1) are generated to the task *modify MR* and its subtasks (in this case there is no subtasks). Thus, the equation *prohibAccess(modify MR, physician, ctx2)* generates the security rule: *"prohibition(physician, modify MR, medical record, ctx2)"* (OrBAC rule III) which means that a physician is prohibited to modify medical records in the audit context.

Therefore, the security policy for the role physician is composed by six security rules: secRule(fulfill record) and secRule(prescribe medicine) (security policy for rootTasks) generate two permissions and three pre obligations. And secRule(negative responsibility) (security policy for responsibilities) generates one prohibition. We make the same analysis for the other roles. We finally obtain the following security policy: *perm1:permission (physician, fulfill record, duty control, ctx1)*, *oblig1:obligation (physician, modify MR, medical record, access_request(fulfill record), not ctx1)*, *oblig2: obligation (assistant, save app, appointment, access_request(fulfill record), not ctx4)*, *perm2: permission (physician, prescribe medicine, prescription,ctx1)* *oblig3: obligation (physician, consult medicine, stock medicine, access_request(prescribe medicine), not ctx1)*, *prohib1 prohibition (physician, modify MR, medical record, ctx2)*, *prem3: permission (assistant, save app, appointment, ctx4)*, *perm4: permission (assistant, consult medicine, stock medicine, ctx3)*, *perm5: permission (pharmacist, modify stock, stock medicine, ctx4)*.

We use MotOrBAC tool to evaluate our proposition. The idea was to create a prototype which generates files understood by MotOrBAC. These files contain the security policies deduced through our proposition. These security rules have been evaluated using MotOrBAC. In the evaluation of the set of previous security rules, a set of probable conflicts were detected. For example, there is a possible conflict between *oblig1* and *prohib1* when both contexts associated to them are activated at a same time. These conflicts can be solved using MotOrBAC or in a more specialized phase (e.g. design or implementation).

7 Conclusions

In this paper we presented an extension of the i* metamodel which allows to define system requirements more expressively in order to enable a model transformation to deduce access control policies based on OrBAC. In particular, we extend the i* metamodel with the notions of context, inheritance, responsibility and concrete entities.

Moreover, because the i* framework is intended to capture requirements in general, the i* models do not contain the same information than just that related to OrBAC model. Therefore, we proposed a reduction process in order to reduce the extended i* metamodel to focus specifically on OrBAC. After that, we analysed the simplified i* metamodel to identify configurations which imply OrBAC security rules. We proposed a deduction process (as a set of equations) using these configurations.

One of the main benefits of this work is allowing designers to define appropriate security policies by giving them the possibility to define contexts, inheritances, responsibilities and concrete entities such as objects and action not considered in the i* metamodel. Additionally, we give a matching between system requirements and OrBAC concepts.

Another benefit of this work, for us the most important benefit, is the result of the model transformation (reduction and deduction processes) is the deduced OrBAC security policy implied in the initial requirements model. It allows designers to evaluate early OrBAC security policies in order to detect and avoid potential problems which can be propagated to later phases of the systems development.

As perspectives, we must study security rules such as ongoing and post obligations, and obligations independent of resource-usage to include in our proposition. Finally, we are looking into implementing the model transformation into a language like QVT or ATL.

References

1. Anderson, R.: Security Engineering: A Guide to Building Dependable Distributed Systems. Wiley, New York (2001)
2. Kleppe, A., Warmer, J., Bast, W.: MDA Explained-the Model Driven Architecture: Practice and Promise. Addison-Wesley, Boston (2003)
3. Miége, A.: Definition of a formal framework for specifying security policies. The Or-BAC model and extensions, Ph.D. Thesis (2005)
4. Jürjens, J.: UMLsec: extending UML for secure systems development. In: Jézéquel, J.-M., Hussmann, H., Cook, S. (eds.) UML 2002. LNCS, vol. 2460, pp. 412–425. Springer, Heidelberg (2002)
5. Cuppens, F., Miège, A.: Modelling contexts in the Or-BAC model. In: 19th Annual Computer Security Applications Conference, December 2003
6. Cuppens, F., Cuppens-Boulahia, N., Miège, A.: Inheritance hierarchies in the Or-BAC model and application in a network environment. In: Second Foundations of Computer Security Workshop (FCS 2004) (2004)
7. Ben Ghorbel, M., Cuppens, F., Cuppens-Boulahia, N., Bouhoula, A.: Managing delegation in access control models. In: 15th International Conference on Advanced Computing and Communication (ADCOM 2007), Inde (2007)
8. Elrakaiby, Y., Cuppens, F., Cuppens-Boulahia, N.: Formal enforcement and management of obligation policies. Data Knowl. Eng. **71**(1), 127–147 (2012)
9. Autrel, F., Cuppens, F., Cuppens-Boulahia, N., Coma, C.: MotOrBAC 2: a security policy tool. In: Third Joint Conference on Security in Networks Architectures and Security of Information Systems (SARSSI) (2008)
10. Muñante, D., Gallon, L., Aniorté, P.: An approach based on Model-driven Engineering to define Security Policies using the access control model OrBAC. In: The Eight International Workshop on Frontiers in Availability, Reliability and Security (FARES) (2013)
11. Muñante, D., Gallon, L., Aniorté, P.: MoDELO: a MOdel-Driven sEcurity poLicy approach based on Orbac. In: 8ème Conférence sur la Sécurité des Architectures Réseaux et des Systèmes d'Information (SARSII) (2013)
12. Muñante, D., Chiprianov, V., Gallon, L., Aniorté, P.: A review of security requirements engineering methods with respect to risk analysis and model-driven engineering. In: Teufel, S., Min, T.A., You, I., Weippl, E. (eds.) CD-ARES 2014. LNCS, vol. 8708, pp. 79–93. Springer, Heidelberg (2014)
13. Lin, L., Nuseibeh, B., Ince, D., Jackson, M.: Using abuse frames to bound the scope of security problems. In: Proceedings of the 12th IEEE International Conference on Requirements Engineering (RE 2004), pp. 354–355. IEEE Computer Society (2004)
14. Sindre, G., Opdahl, A.L.: Eliciting security requirements with misuse cases. Requir. Eng. J. **10**(1), 34–44 (2005)

15. Sindre, G.: Mal-activity diagrams for capturing attacks on business processes. In: Heymans, P., Sawyer, P. (eds.) REFSQ 2007. LNCS, vol. 4542, pp. 355–366. Springer, Heidelberg (2007)
16. van Lamsweerde, A.: Elaborating security requirements by construction of intentional anti-models. In: Proceedings of the 26th International Conference on Software Engineering, pp. 148–157, 23–28 May 2004
17. Mouratidis, H., Giorgini, P.: Secure tropos: a security-oriented extension of the tropos methodology. Int. J. Softw. Eng. Knowl. Eng. 17(2), 285–309 (2007)
18. Elahi, G., Yu, E.: A goal oriented approach for modeling and analyzing security trade-offs. University of Toronto. Technical report (2007)
19. Anton, A.I., Earp, J.B.: Strategies for developing policies and requirements for secure electronic commerce systems. North Carolina State University. Technical report (2000)
20. Braber, F., Hogganvik, I., Lund, M.S., Stolen, K., Vraalsen, F.: Model-based security analysis in seven steps-a guided tour to the CORAS method. BT Technol. J. 25(1), 101–117 (2007)
21. Asnar, Y., Giorgini, Y.P., Massacci, F., Zannone, N.: From trust to dependability through risk analysis. In: Proceedings of the International Conference on Availability, Reliability and Security (AReS), pp. 19–26. IEEE Computer Society (2007)
22. Mayer, N., Rifaut, A., Dubois, E.: Towards a risk-based security requirements engineering framework. In: Proceedings of the 11th International Workshop on Requirements Engineering: Foundation for Software Quality (REFSQ 2005), in Conjunction with the 17th Conference on Advanced Information Systems Engineering (CAiSE 2005) (2005)
23. Massacci, F., Zannone, N.: A model-driven approach for the specification and analysis of access control policies. In: Moorsman, R., Tari, Z. (eds.) OTM 2008, Part II. LNCS, vol. 5332, pp. 1087–1103. Springer, Heidelberg (2008)
24. Ledru, Y., Richier, J., Idani, A., Labindh, M.: From KAOS to RBAC: a case study in designing access control rules from a requirements analysis. In: 6ème Conf. sur la Sécurité des Architectures Réseaux et des Systèmes d'Information (SARSSI 2011) (2011)
25. Mouratidis, H., Jürjens, J., Fox, J.: Towards a comprehensive framework for secure systems development. In: Martinez, F.H., Pohl, K. (eds.) CAiSE 2006. LNCS, vol. 4001, pp. 48–62. Springer, Heidelberg (2006)
26. Graa, M., Cuppens-Boulahia, N., Autrel, F., Azkia, H., Cuppens, F., Coatrieux, G., Cavalli, A., Mammar, A.: Using requirements engineering in an automatic security policy derivation process. In: Garcia-Alfaro, J., Navarro-Arribas, G., Cuppens-Boulahia, N., de Capitani di Vimercati, S. (eds.) DPM 2011 and SETOP 2011. LNCS, vol. 7122, pp. 155–172. Springer, Heidelberg (2012)
27. Hatebur, D., Heisel, M., Jürjens, J., Schmidt, H.: Systematic development of UMLsec design models based on security requirements. In: Giannakopoulou, D., Orejas, F. (eds.) FASE 2011. LNCS, vol. 6603, pp. 232–246. Springer, Heidelberg (2011)
28. Yu, E.: Modelling strategic relationships for process reengineering. Ph.D. thesis, University of Toronto (1995)
29. Elahi, G., Yu, E., Zannone, N.: A vulnerability-centric requirements engineering framework: analyzing security attacks, countermeasures, and requirements based on vulnerabilities. Requir. Eng. 15(1), 41–62 (2010)
30. Sandhu, J.R., Coyne, E.J., Feinstein, H.J., Youman, C.E.: Role-based access control models. IEEE Comput. 29, 38–47 (1996)

Optimal Proximity Proofs

Ioana Boureanu[1]([⊠]) and Serge Vaudenay[2]

[1] Akamai Technologies Limited, EMEA HQ, London, UK
icboureanu@gmail.com
http://people.itcarlson.com/ioana
[2] EPFL, Lausanne, Switzerland
http://lasec.epfl.ch

Abstract. Provably secure distance-bounding is a rising subject, yet an unsettled one; indeed, very few distance-bounding protocols, with formal security proofs, have been proposed. In fact, so far only two protocols, namely SKI (by Boureanu *et al.*) and FO (by Fischlin and Onete), offer all-encompassing security guaranties, i.e., resistance to distance-fraud, mafia-fraud, and terrorist-fraud. Matters like security, alongside with soundness, or added tolerance to noise do not always coexist in the (new) distance-bounding designs. Moreover, as we will show in this paper, *efficiency* and *simultaneous* protection against all frauds seem to be rather conflicting matters, leading to proposed solutions which were/are suboptimal. In fact, in this recent quest for provable security, efficiency has been left in the shadow. Notably, the tradeoffs between the security and efficiency have not been studied. In this paper, we will address these limitations, setting the "security vs. efficiency" record straight.

Concretely, by combining ideas from SKI and FO, we propose symmetric protocols that are efficient, noise-tolerant and—at the same time—provably secure against all known frauds. Indeed, our new distance-bounding solutions outperform the two aforementioned provably secure distance-bounding protocols. For instance, with a noise level of 5%, we obtain the same level of security as those of the pre-existent protocols, but we reduce the number of rounds needed from 181 to 54.

1 Introduction

As wireless technologies become more and more pervasive, being used daily in access control, remote unlocking credit-card payments and beyond, relay attacks also become a growing threat to the social acceptance of these techniques. It seems likely that nearly all wireless devices will eventually have to implement solutions to thwart these types of fraud. To defeat relay attacks, Brands and Chaum [12] introduced the notion of *distance-bounding protocols*. Distance bounding is a special problem of position-based cryptography [13]. Although there are many challenges to implement it, this can be achieved [11]. These protocols rely on information being local and incapable of travelling faster than the speed of light. So, in distance-bounding, an RFID reader can assess when

---

The full version of this paper is available on [10].

© Springer International Publishing Switzerland 2015
D. Lin et al. (Eds.): Inscrypt 2014, LNCS 8957, pp. 170–190, 2015.
DOI: 10.1007/978-3-319-16745-9_10

participants are close enough because the round-trip communication time must have been short enough. The whole idea of distance-bounding is that a *prover*, holding a key x, demonstrates that he is close to a *verifier* (who also knows this key x). The literature on distance-bounding considers several threat models.

- *Distance fraud* (DF): a far-away malicious prover tries to illicitly pass the protocol.
- *Mafia fraud* [16] (MF): a man-in-the-middle (MiM) adversary between a far-away honest prover and a verifier tries to exploit the prover's insights to make the verifier accept. (This generalizes relay attacks as not only does this adversary relay, but he may also modify the messages involved.)
- *Terrorist fraud* [16] (TF): a far-away malicious prover colludes with an adversary to make the verifier accept the adversary's rounds on behalf of this far-away prover, in such a way that the adversary gains no advantage to later pass the protocol on his own.
- *Impersonation fraud* [3]: An adversary tries to impersonate the prover to the verifier.
- *Distance hijacking* [15]: A far-away prover takes advantage of some honest, active provers (of which one is close) to make the verifier grant privileges for the far-away prover.

Avoine *et al.* [1] proposed one of the very first (semi-formal) model. Later, Dürholz *et al.* [17] proposed a formal model (herein called the *DFKO model*) based on exhaustive lists of impossible traces in protocols. Boureanu *et al.* [7,9,30] proposed a more complete model (herein called the *BMV model*) including the notion of time. Based on all these models, there were several variants and generalizations of these threats. The model in [7,9] factors all the previously enumerated common frauds into three possible threats:

- *Distance fraud.* This is the classical notion, but concurrent runs with many participants is additionally considered. I.e., it includes other possible provers (with other secrets) and verifiers. Consequently, this generalized distance fraud also includes distance hijacking.
- *Man-in-the-middle.* This formalization considers an attack working in two phases. During a *learning phase*, the adversary can interact with many honest provers and verifiers. Then, the *attack phase* contains a far away honest prover of given ID and possibly many other honest provers and other verifiers. The goal of the adversary is to make the verifier accept the proof with ID. Clearly, this generalizes mafia fraud (capturing relay attacks) and includes impersonation fraud.
- *Collusion fraud.* This formalization considers a far-away prover holding x who helps an adversary to make the verifier accept. This might be in the presence of many other honest participants. However, there should be no man-in-the-middle attack stemming from this malicious prover. I.e., one should not extract from this prover any advantage to (later) run a man-in-the-middle attack.

In Vaudenay [30], the last threat model is replaced by a notion coming from interactive proofs:

– *Soundness*. For all experiment with a verifier $\mathcal{V}$, there exists an extractor such that the following holds: if this extractor is given as input several views of all participants which were close to $\mathcal{V}$ in several executions and which made him accept therein, then this extractor reconstructs the secret x. This was further shown to generalize collusion-fraud resistance [30].

In Sect. 2, we refine these models in a more natural way, including at its basis a stronger, inner sense of interactive proofs. Indeed, distance-bounding (DB) should ideally behave like a traditional interactive proof system as it really is a *proof of proximity*. In this sense, it must satisfy: 1. **completeness** (i.e., an honest prover close to the verifier will certainly pass the protocol); 2. **soundness** (i.e., if the verifier accepts the protocol, then we could extract from close-by participants the information to define a successful prover); 3. **security** (i.e., no participant shall be able to extract some information from the honest prover to make the verifier accept). These properties are similar to what is required in *identification protocols*. They differ in that in DB we face the introduction of the notion of proximity.

More precisely, in the above approach, distance fraud (as in Definition 6) does not capture distance hijacking anymore, distance hijacking being now captured by soundness. This makes proofs simpler. To this end, we also formalize in Definition 8 security without a learning phase, and we extend in Definition 10 the definition of soundness in such a way that the extraction of the secret is no longer necessary.

There exist many distance-bounding protocols, but nearly all are broken in some way. The following protocols are all vulnerable to TF [19,24]: Hancke-Kuhn [21], Singelée-Preneel [28], Munilla-Peinado [25], Kim-Avoine [23], and Nikov-Vauclair [26]. Kim *et al.* [24] proved that the return channel of the verifier (i.e., whether the protocol succeeds or not) allows to do a MiM attack on the protocol in Tu-Piramuthu [29]. It is also applicable against the protocol in Reid *et al.* [27] as shown in Bay *et al.* [4]. Boureanu *et al.* [5] demonstrated that the security arguments of [2,21,24,27] were incorrect by constructing instances satisfying the assumptions by the authors and trivially insecure. Finally, Hancke [20] observed that noisy-resilience in nearly all protocols (including SwissKnife [24]) allowed to mount a TF. So, the problem of making provably secure distance bounding is of utmost importance. So far, only the SKI protocol [6–9] (built on the BMV model) and the Fischlin-Onete (FO) protocol [18] (built on the DFKO model) provide an all-encompassing proven security, i.e., they protect against all the above threats.

Organization. In Sect. 2, we advance revised security definitions for DB, rendering a more intuitive model, whilst maintaining backward compatibility; we also prove the latter preservation of results. In Sect. 3, we propose new, secure DB protocols DB1, DB2, and DB3. Section 3.5 considers the tradeoffs between security and efficiency, and presents the comparisons made in this sense. Results for SKI and FO are recalled and revisited in the full paper [10].

Contribution. The contribution of this paper is threefold:

- We build up on SKI [6–9] and FO [18,30] to propose DB1, DB2, and DB3, three new distance-bounding protocols which outperform both the SKI and the FO protocols.

  For instance, to offer a false acceptance rate of under 1 % and false rejection rate of under 1 %, at a noise level of 5 % during the rapid bit-exchange, DB1 (with parameter $q = 3$) requires 14/14/54 rounds for resistance to distance fraud / mafia fraud / terrorist fraud, respectively. For the same performance, SKI and FO require 84/48/181 and 84/84/? rounds[1], respectively. So, DB1 represents a *substantial improvement* in terms of efficiency, whilst maintaining provable security.

- When considering optimality amongst protocols requiring at least τ out of n correct rounds, no clock for the proer, and a challenge/response set of size q, we show security as follows:

	DF-resistance	MF-resistance	TF-resistance
DB1 $(q > 2)$	secure, optimal	secure, optimal	secure
DB2 $(q = 2)$	secure, suboptimal	secure, optimal	secure
DB3 $(q = 2)$	secure, optimal	secure, optimal	insecure

- For our security proofs, we build on the BMV model [7,9,30]. In doing so, we revisit the definition of mafia fraud / man-in-the-middle and the definition of terrorist fraud / collusion fraud. Thus, we provide a complete set of security definitions for distance bounding, capturing the previous notions, but being in line with the established theory behind interactive proofs.

Useful bounds for noisy communications. Following [7,9], to assert security in noisy communications, we will make use of the tail of the binomial distribution:

$$\mathsf{Tail}(n, \tau, \rho) = \sum_{i=\tau}^{n} \binom{n}{i} \rho^i (1 - \rho)^{n-i},$$

We recall that for any $\varepsilon, n, \tau, \rho$ such that $\frac{\tau}{n} < \rho - \varepsilon$, we have $\mathsf{Tail}(n, \tau, \rho) > 1 - e^{-2\varepsilon^2 n}$. For $\frac{\tau}{n} > \rho + \varepsilon$, we have $\mathsf{Tail}(n, \tau, \rho) < e^{-2\varepsilon^2 n}$. This comes from the Chernoff-Hoeffding bound [14,22].

2   Revised DB Security Model and Proofs

We now refine the security definitions and other tools from the BMV model [7,9,30]. In this section, we also discuss the links with the original notions.

In this paper, we concentrate on distance-bounding protocols based on symmetric cryptography (which is the overwhelmingly prevalent approach in DB).[2]

---

[1] As discussed herein, FO has an incomparable approach for TF-resistance in which the number of rounds is not relevant.

[2] Our model was recently extended to cover public-key distance-bounding [31,32].

Definition 1. *A (symmetric)* <u>distance-bounding protocol</u> *is a tuple* $(\mathcal{K}, P, V, B)$, *constructed of the following: a* <u>key domain</u> $\mathcal{K}$; *a two-party probabilistic polynomial-time (PPT) protocol* $(P(x), V(x))$, *where* P *is the* <u>proving algorithm</u>, V *is the* <u>verifying algorithm</u>, *and* x *is taken from* $\mathcal{K}$; *a distance bound* B. *At the end of the protocol, the verifier* $V(x)$ *sends a final message* Out_V. *This output denotes that the verifier* <u>accepts</u> *(*$\mathsf{Out}_V = 1$*) or* <u>rejects</u> *(*$\mathsf{Out}_V = 0$*).*

Informally, a distance-bounding protocol is complete if executing $P(x) \leftrightarrow V(x)$ on locations within a distance bounded by B makes $V(x)$ accept with overwhelming probability. The formalism is straightforward with the settings below.

We compare our protocols to *any* DB protocol that follows what we call the *common structure*.

Definition 2 (Common structure). *A DB protocol with the* <u>common structure</u> *based on parameters* $(n, \tau, \mathsf{num}_c, \mathsf{num}_r)$ *has some initialization and verification phases which do not depend on communication times.*[3] *These phases are separated by* n *rounds of timed challenge/response exchanges. This is called* <u>the distance bounding phase</u>. *A response is* <u>on time</u> *if the elapsed time between sending the challenge and receiving the response is at most* $2B$. *Provers don't measure time.*[4] *Challenges and responses are in sets of cardinality* num_c *and* num_r, *respectively.*

When the protocol follows the specified algorithms but messages during the distance bounding phase can be corrupted during transmission, we say that the protocol is <u>τ-complete</u> *if the verifier accepts if and only if at least* τ *rounds have a correct and on-time response.*

One can easily see that nearly *all* distance-bounding protocols in the literature fit this definition.

In practice, when the timed phase is subject to *noise*, we assume that there is a probability of p_{noise} that one round of challenge/response is corrupted. The probability that an honest prover, close to the verifier, passes the protocol is thus $\mathsf{Tail}(n, \tau, 1 - p_{\mathsf{noise}})$. So, with $\frac{\tau}{n} < 1 - p_{\mathsf{noise}}$ with a constant gap, the probability to fail is negligible, due to the Chernoff-Hoeffding bound [14,22].

Participants, Instances, Setup and Locations.

- In a DB protocol, participants can be a *prover*, a *verifier*, or *adversaries*. The prover and the verifier receive a key x which is randomly selected from the key space. We adopt a *static* adversarial model: i.e., at the beginning of the experiment, it is decided whether the prover is malicious or not. Participants have several *instances*. An instance has a *location*. It corresponds to the execution of a protocol during one session.
- A honest prover runs instances of the algorithm P denoted by $P(x)$. An instance of a malicious prover runs an arbitrary algorithm denoted by $P^*(x)$. **P** denotes the set of instances of the prover.

---

[3] The verification phase can be interactive or not.
[4] Provers have no clock. They are in a waiting state to receive the challenge and loose the notion of time while waiting.

– The verifier is honest without loss of generality.[5] He runs instances of the algorithm V denoted by $V(x)$. $\mathbf{V}$ denotes the set of instances of the verifier.
– Other participants are (without loss of generality) malicious and may run whatever algorithm, but with no initialized key. The set of such malicious participants is denoted $\mathbf{A}$. By contrast, a designated, one such instance is denoted $\mathcal{A}$.
– Locations are elements of a metric space.

Why a Single Identity? Our definition uses a single identity, without loss of generality. This is because provers or verifiers running the protocol with other identities (and keys independent of x) could be considered as elements of $\mathbf{A}$.

Definition 3 (DB Experiment). *An* experiment *exp for a distance-bounding protocol* $(\mathcal{K}, P, V, B)$ *is a setting* $(\mathbf{P}, \mathbf{V}, \mathbf{A})$ *with several instances of* participants, *at some* locations, *set up as above, and running an overall PPT sequence.*

In the above definition, the notion of experiment implies simultaneously several *different* entities: participants, physical locations, algorithms to be run by these participants and corruption states. As such, when used inside further definitions, the notion of experiment will implicitly or explicitly, upon the case, quantify over these entities.

We further assume that communicating from a location to another takes time equal to the distance. Indeed, no one can violate the fact that communication is limited by the speed of light. Adversaries can intercept some messages and replace them by others, but must adhere to the fact that computation is local.

Ideally, one should develop a formal model to define all these. This has actually been done in the BMV model [7,9]. In this paper, we keep the notions at the intuitive level, mainly due to space limitations, and since such a formal model would only be needed to prove the fundamental Lemma 4 below (which is proven in and adapted from [9, Lemma 1]). We rather take it axiomatically herein.

Lemma 4 (Fundamental Lemma). *Assume an experiment in which at some point a participant* $\mathcal{V}$ *broadcasts a message c, then waits for a response r. We let E be the event that the elapsed time between sending c and receiving r is at most $2B$. In the experiment,* Close *is the set of all participants (except $\mathcal{V}$) which are within a distance of up to B from $\mathcal{V}$, and* Far *is the set of all participants at a larger distance. For each user U, we consider his view $View_U$ just before the time when U can see the broadcast message c.*

We say that a message by U is independent[6] *from c if it was sent before U could see c, i.e., if it is the result of applying algorithm U on $View_U$, or on a prefix of it.*

There exists an algorithm Algo *with the following property. If E holds and r was sent from a participant in* Close, *we have $r = \mathsf{Algo}((View_U)_{U \in \mathsf{Close}}, c, w)$,*

---

[5] A "malicious verifier" running an algorithm $V^*(x)$ can be seen as a malicious prover running $V^*(x)$.
[6] we stress that this is a local definition of independence which is unrelated to statistical independence.

where w is the list of all messages independent from c which are not already in $(View_U)_{U \in \text{Close}}$ but seen[7] by any $U \in \text{Close}$. If E holds and r was sent from a participant in Far, then the message r is independent from c.

This lemma can be summarized as follows: a close-by participant cannot get online help from far away to answer correctly and in time to the challenge c.

Definition 5 (Distinguished Experiment). *We denote by* $\exp(\mathcal{V})$ *an experiment in which we fix a verifier instance* $\mathcal{V} = V(x)$ *from* **V**, *which we call* distinguished verifier. *Participants which are within a distance of at most* B *from a distinguished verifier* $\mathcal{V}$ *are called* close-by participants. *Others are called* far-away participants.

Participants can move during the experiment, but not faster than the transmission of information. For simplicity, we assume that far-away participants remain far away during the experiment.

Definition 6 (α-resistance to distance fraud). *We say that a distance-bounding protocol* α-resists to distance fraud *if for any distinguished experiment* $\exp(\mathcal{V})$ *where there is no participant close to* $\mathcal{V}$, *the probability that* $\mathcal{V}$ *accepts is bounded by* α.

Compared to [9], this definition is simplified and does not capture the notion of distance hijacking; therein, a far-away malicious $P^*(x)$ can make $\mathcal{V}$ accept by taking advantage of several honest provers which do not hold x but are close to $\mathcal{V}$. In [9], some close-by honest participants are allowed in the definition of distance fraud resistance. However, distance hijacking could generalize to the presence of any close-by honest participant who is running a protocol (for whatever honest reason) which could match (by some weird coincidence) the response function of the malicious prover. This is not captured by the definition of [9]. Nonetheless, in most of the cases, this bizarre situation can be ignored and we can concentrate on regular distance frauds. So, we simplified on purpose our Definition 6, excluding the more corner-case fraud of distance hijacking, as this simplifies the proofs quite a lot. Nonetheless, distance hijacking and other extensions of classical frauds will be captured by the notion of soundness, which we introduce below. Overall, we will treat all threats.

Theorem 7. *Given* $n, \tau, \text{num}_c, \text{num}_r$, *a DB protocol following the common structure cannot[8]* α-resist *to distance fraud for* α *lower than*

$$\text{Tail}\left(n, \tau, \max\left(\frac{1}{\text{num}_c}, \frac{1}{\text{num}_r}\right)\right).$$

---

[7] "Seen" means either received as being the destinator or by eavesdropping.

[8] In [33], a protocol with two bits of challenges and one bit of response achieving $\alpha = \text{Tail}(n, \tau, \frac{1}{3})$ is proposed. But it actually works with $\text{num}_r = 3$ as it allows response 0, response 1, and no response.

Proof. We construct a DF following *the early-reply strategy*: a malicious prover guesses with probability $\frac{1}{\mathsf{num}_c}$ the challenge c_i before it is emitted, and then he sends the response so that it arrives on time. The rest of the protocol is correctly simulated (with delay) after receiving the challenges. An incorrect guess would look like a round which was the victim of noise. So, the attack succeeds with probability $\mathsf{Tail}\left(n, \tau, \frac{1}{\mathsf{num}_c}\right)$. We can have a similar attack guessing the response r and succeeding with probability $\mathsf{Tail}\left(n, \tau, \frac{1}{\mathsf{num}_r}\right)$.   □

While the above definition protects verifiers against malicious provers, we need an extra notion to protect the honest prover against men-in-the-middle. This is as follows.

Definition 8 (β-secure distance-bounding protocol). *We say that a DB protocol is $\underline{\beta\text{-secure}}$ if for any distinguished experiment* $\exp(\mathcal{V})$ *where the prover is honest, and the prover instances are all far-away from* $\mathcal{V}$, *the probability that* $\mathcal{V}$ *accepts is bounded by β.*

Intuitively, this notion protects honest provers from identity theft. It implies that x cannot be extracted by a malicious participant; this is along the same lines as in zero-knowledge interactive protocols. This notion of security also captures resistance to relay attacks, mafia fraud, and man-in-the-middle attacks. The advantage of Definition 8 over the resistance to man-in-the-middle attacks, as it was defined in [7,9, Definition 4], is that we no longer need to formalize a *learning phase*, although we can easily show we capture these notions as well. Our definition is therefore simpler.

Theorem 9. *Given* $n, \tau, \mathsf{num}_c, \mathsf{num}_r$, *a DB protocol following the common structure cannot[9] be β-secure for β lower than*

$$\mathsf{Tail}\left(n, \tau, \max\left(\frac{1}{\mathsf{num}_c}, \frac{1}{\mathsf{num}_r}\right)\right).$$

Proof. We consider $\mathcal{V}$ and a far-away instance of the prover P, and a close-by MiM $\mathcal{A}$. In the initialization phase and the verification phase, $\mathcal{A}$ passively relays messages between $\mathcal{V}$ and P. During the challenge phase, and in the *pre-ask strategy*, $\mathcal{A}$ guesses the challenge before it is released and asks for the response to P on time so that he can later on answer to $\mathcal{V}$. Clearly, the attack succeeds with probability $\mathsf{Tail}\left(n, \tau, \frac{1}{\mathsf{num}_c}\right)$. We can have a similar attack with a *post-ask strategy* where $\mathcal{A}$ guesses the response at the same time he forwards the challenge to P. This succeeds with probability $\mathsf{Tail}\left(n, \tau, \frac{1}{\mathsf{num}_r}\right)$.[10]   □

The definition below is adapted from [30]. One difference is that γ' is no longer necessarily $1 - \mathsf{negl}$. It also considers extractors just passing the protocol,

---

[9] Same remark about [33] as in Theorem 7.

[10] Since provers loose the notion of time in the challenge phase, pre-ask and post-ask attacks cannot be detected.

instead of having to produce the secret; this is clearly more general. Our protocols herein will make the secret extractable though.

Definition 10 $((\gamma, \gamma', m)$-**soundness**). *We say that a DB protocol is* $\underline{(\gamma, \gamma', m)}$-$\underline{\text{sound}}$ *if for any distinguished experiment* $\exp(\mathcal{V})$ *in which* $\mathcal{V}$ *accepts with probability at least* γ, *there exists a PPT algorithm* $\mathcal{E}$ *called* extractor, *with the following property. By* $\mathcal{E}$ *running experiment* $\exp(\mathcal{V})$ *several times, in some executions denoted* $\exp_i(\mathcal{V})$, $i = 1, \dots, M$, *for* M *of expected value bounded by* m, *we have that*

$$\Pr\left[\mathsf{Out}_V = 1 : \mathcal{E}(\mathsf{View}_1, \dots, \mathsf{View}_M) \leftrightarrow \mathcal{V} \mid \mathsf{Succ}_1, \dots, \mathsf{Succ}_M\right] \geq \gamma',$$

where View_i *denotes the view of all close-by participants (except* $\mathcal{V}$) *and the transcript seen by* $\mathcal{V}$ *in the run* $\exp_i(\mathcal{V})$, *and* Succ_i *is the event that* $\mathcal{V}$ *accepts in the run* $\exp_i(\mathcal{V})$.

In other words, the extractor impersonates the prover to $\mathcal{V}$.[11] In more details, this means that having $\mathcal{V}$ accept in run $\exp_i(\mathcal{V})$ implies the following: a piece of x was given to the close-by participants and it is stored in View_i, and that m such independent pieces, on average, could allow $\mathcal{E}$ to impersonate $P(x)$ to $\mathcal{V}$. This notion is pretty strong as it could offer a guaranty against distance hijacking: a prover making such attack would implicitly leak his credentials.

3   New Highly Efficient, Symmetric Distance-Bounding Protocols

In the idea to outperform SKI and FO, we now advance a family of provably secure symmetric distance-bounding protocols, called DBopt. It includes DB1, DB2, and DB3. Indeed, we will see herein that DB1 is in fact optimal in terms of distance-fraud resistance and security with non-binary challenges. The DB2 and DB3 variants are motivated by the use of binary challenges, which is customary in distance-bounding designs. Whilst DB2 is suboptimal, it still performs well, almost always, i.e., better than SKI and FO. DB3 is optimal but not TF-resistant. The eager reader can directly inspect the performance/security graphs in Fig. 2, page 187, where we plot the (logs of) fraud-resistance thresholds, i.e., $-\log_2 \alpha$, $-\log_2 \beta$, and $-\log_2 \gamma$.

3.1   DBopt

We propose DBopt, a new family of symmetric distance-bounding protocols, as depicted in Fig. 1. It combines ideas taken from SKI [6–9] and the Swiss-Knife protocol [24] (as used by FO [18]). We use a security parameter s (the length of the secret x, i.e., $x \in \mathcal{K} = \mathbf{Z}_2^s$) and the following parameters based on s: the

---

[11] Note that cases where there is a close-by prover or a close-by verifier are trivial since they hold the secret x in their view.

number of rounds n, the length ℓ_{tag} of tag, a threshold τ, the nonce length ℓ_{nonce}, and a constant q which is a prime power, e.g., $q = 2$, $q = 3$, or $q = 4$. DBopt follows the *common structure* with parameters n, τ, and $num_c = num_r = q$.

As in SKI, we assume $L_\mu(x) = (\mu(x), \dots, \mu(x))$ for some function $x \mapsto \mu(x)$, but μ is not necessarily linear. Concretely, μ is a vector in $\mathbf{Z}_2^s$ and map a *fixed* injection from $\mathbf{Z}_2$ to $\mathsf{GF}(q)$. Hence, $\mu(x) = \mathsf{map}(\mu \cdot x)$ maps a bitstring x to a $\mathsf{GF}(q)$-representation of the bit obtained by the scalar product $\mu \cdot x$. We let $\mathcal{L}$ denote the set of all such possible L_μ mappings (map being fixed). The function f_x maps to different codomains, depending on its inputs: given two nonces N_P and N_V, $L_\mu \in \mathcal{L}$, and $b, c \in \mathsf{GF}(q)^n$, $f_x(N_P, N_V, L_\mu, b) \in \mathsf{GF}(q)^n$ and $f_x(N_P, N_V, L_\mu, b, c) \in \mathsf{GF}(q)^{\ell_{tag}}$.

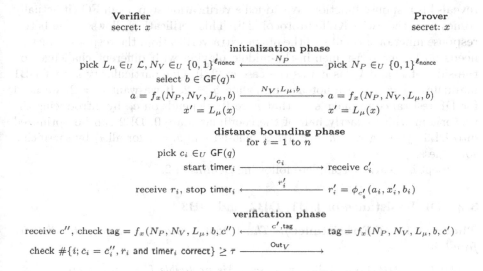

Fig. 1. The DBopt distance-bounding protocols

During the initialization, the participants exchange some nonces N_P, N_V, some $L_\mu \in \mathcal{L}$, and a vector b. The vector b could be fixed in the protocol, but is subject to some constraints as detailed below. V and P compute $a = f_x(N_P, N_V, L_\mu, b)$ and $x' = L_\mu(x)$. In the distance bounding phase, the response function is a linear function $r_i = \phi_{c_i}(a_i, x'_i, b_i)$ defined by the challenge c_i. The verification checks that the participants have seen the same challenges (based on the tag computed by $\mathsf{tag} = f_x(N_P, N_V, L_\mu, b, c)$), counts the number of rounds with a correct and timely response, and accepts if there are at least τ of them.

Clearly, the DBopt family is quite open to specific choices for q, map, b, and ϕ_c. We propose the instances DB1, DB2, and DB3. There are some specificities in each protocol which are summarized in the following table:

protocol	q	map	b	ϕ_{c_i}
DB1	$q > 2$	$\mathsf{map}(u) \neq 0$	no b used	$\phi_{c_i}(a_i, x'_i, b_i) = a_i + c_i x'_i$
DB2	$q = 2$	$\mathsf{map}(u) = u$	Hamming weight $\frac{n}{2}$	$\phi_{c_i}(a_i, x'_i, b_i) = a_i + c_i x'_i + c_i b_i$
DB3	$q \geq 2$	no map used	Hamming weight n	$\phi_{c_i}(a_i, x'_i, b_i) = a_i + c_i b_i$

Specifically, DB3 is the simplest protocol and is optimal, but it offers no soundness. DB2 works with binary challenges and responses, but it is not optimal. DB1 is optimal but needs $q \geq 3$ since it requires that map is injective from $\mathbf{Z}_2$ to $\mathsf{GF}(q)^*$.

Overall, DBopt is very similar to SKI. Like in SKI, the leak vector x' is fundamental for soundness: the vector x' encodes $\mu \cdot x$, which leaks if the prover reveals his response function. We added a verification step, as in FO (it actually comes from the Swiss-Knife protocol [24]). This verification allows to use better response functions: thanks to the above extra verification, the response function needs no longer resist men-in-the-middle playing with different challenges on the sides of P and V, as it was the case in [2,4]. One particularity is that DB1 mandates $x'_i \neq 0$ so cannot accommodate $q = 2$. If we want $q = 2$, we need for DF-resistance to make sure that r_i really depends on c_i, by introducing the vector b in which exactly half of the coordinates are 0. DB2 can be optimized into DB3 by using $r_i = a_i + c_i$ (so x' unused and $b_i = 1$ for all i) by sacrificing soundness.

DBopt is clearly τ-complete following Definition 2.

3.2   DF-Resistance of DB1, DB2, and DB3

Theorem 11 (DF-resistance). *The* DBopt *protocols* α-*resists to distance fraud for*

- *(DB1 and DB3)* $\alpha = \mathsf{Tail}(n, \tau, \frac{1}{q})$ *which is negligible for* $\frac{\tau}{n} > \frac{1}{q} + \mathsf{cte}$;
- *(DB2)* $\alpha = \mathsf{Tail}(\frac{n}{2}, \tau - \frac{n}{2}, \frac{1}{2})$ *which is negligible for* $\frac{\tau}{n} > \frac{3}{4} + \mathsf{cte}$.

Due to Theorem 7, DB1 and DB3 are optimal for DF-resistance. DB2 is clearly not optimal (as DB3 is better with the same $q = 2$). However, the bound is tight for DB2 as the DF guessing the response matches the α bound: the malicious prover always wins the rounds for which $x' = b_i$ (that is: exactly half of the rounds due to the Hamming weight of b) by sending the response in advance and passes with probability $\alpha = \mathsf{Tail}(\frac{n}{2}, \tau - \frac{n}{2}, \frac{1}{2})$.

Proof. We consider a distinguished experiment $\exp(\mathcal{V})$ with no close-by participant. Due to the distance, the answer r_i to $\mathcal{V}$ comes from far away. Thanks to Lemma 4, r_i is independent (in the sense of Lemma 4) from c_i. Since c_i is randomly selected when it is sent, r_i is statistically independent from c_i. For DB1, since $x'_i \neq 0$ by construction, r_i equals $a_i + c_i x'_i$ with probability $\frac{1}{q}$. The same goes for DB3. For DB2, thanks to the selection of b, this holds for exactly half of the rounds: those such that $x'_i + b_i \neq 0$. So, the probability to succeed in the experiment is bounded as stated.   □

3.3 Security of DB1, DB2, and DB3

As shown in [5], we cannot rely on the PRF assumption alone for DB1 or DB2, since the secret is used as a key of f_x and also outside f_x in x'. The circular-PRF assumption guarantees the PRF-ness of f, even when we encrypt a function $L_\mu(x)$ of the key. We new recall and extend the notion, to accommodate DB1 and DB2.

Definition 12 (Circular PRF). *We consider some parameters s, n_1, n_2, and q. Given $\tilde{x} \in \{0,1\}^s$, a function L from $\{0,1\}^s$ to $\mathsf{GF}(q)^{n_1}$, and a function F from $\{0,1\}^*$ to $\mathsf{GF}(q)^{n_2}$, we define an oracle $\mathcal{O}_{\tilde{x},F}$ by $\mathcal{O}_{\tilde{x},F}(y,L,A,B) = A \cdot L(\tilde{x}) + B \cdot F(y)$, using the dot product over $\mathsf{GF}(q)$. We assume that L is taken from a set of functions with polynomially bounded representation. Let $(f_x)_{x \in \{0,1\}^s}$ be a family of functions from $\{0,1\}^*$ to $\{0,1\}^{n_2}$. We say that the family f is a (ε, T)-circular-PRF if for any distinguisher limited to a complexity T, the advantage for distinguishing $\mathcal{O}_{x,f_x}$, $x \in_U \{0,1\}^s$, from $\mathcal{O}_{\tilde{x},F}$, $\tilde{x} \in_U \{0,1\}^s$, where F is uniformly distributed, is bounded by ε. We require two conditions on the list of queries:*

- *for any pair of queries (y,L,A,B) and (y',L',A',B'), if $y = y'$, then $L = L'$;*
- *for any $y \in \{0,1\}^*$, if (y,L,A_i,B_i), $i = 1,\ldots,\ell$ is the list of queries using this value y, then*

$$\forall \lambda_1,\ldots,\lambda_\ell \in \mathsf{GF}(q) \quad \sum_{i=1}^{\ell} \lambda_i B_i = 0 \implies \sum_{i=1}^{\ell} \lambda_i A_i = 0 \qquad (1)$$

over the $\mathsf{GF}(q)$-vector space $\mathsf{GF}(q)^{n_2}$ and $\mathsf{GF}(q)^{n_1}$.

This definition extends the one from [7,9] in the following sense: 1. the function L (the leak of x) is arbitrary instead of being linear; 2. this arbitrary function L now requires the first condition i.e., the same F-input implies the same leak function L. In [7,9], it was shown that the natural construction $f_x(y) = H(x,y)$ is circular-PRF in the random oracle model, with the definition from [7,9]. We can easily see that the same proof holds with Definition 12. It would be interesting to make other constructions without random oracles.

Theorem 13 (Security). *The DBopt protocols are β-secure for*

- *(DB1 and DB2) $\beta = \mathsf{Tail}(n,\tau,\frac{1}{q}) + \frac{r^2}{2}2^{-\ell_{\mathrm{nonce}}} + (r+1)\varepsilon + r2^{-\ell_{\mathrm{tag}}}$ when f is a (ε, T)-circular-PRF (as defined by Definition 12);*
- *(DB3) $\beta = \mathsf{Tail}(n,\tau,\frac{1}{q}) + \frac{r^2}{2}2^{-\ell_{\mathrm{nonce}}} + \varepsilon + 2^{-\ell_{\mathrm{tag}}}$ when f is a (ε, T)-PRF.*

There, r is the number of honest instances (of P or V) and T is a complexity bound on the experiment. β is negligible for $\frac{\tau}{n} > \frac{1}{q} + \mathsf{cte}$, r and T polynomially bounded, and ε negligible.

Based on that $\frac{r^2}{2}2^{-\ell_{\mathrm{nonce}}} + (r+1)\varepsilon + r2^{-\ell_{\mathrm{tag}}}$ (or the similar term for DB3) can be made negligible against β, DB1, DB2, and DB3 are *optimal* for security due to Theorem 9.

Proof. We consider a distinguished experiment $\exp(\mathcal{V})$ with no close-by $P(x)$, no $P^*(x)$, and where $\mathcal{V}$ accepts with probability p. We consider a game Γ_0 in which we simulate the execution of $\exp(\mathcal{V})$ and succeed if and only if Out_V by $\mathcal{V}$ is an acceptance message. Γ_0 succeeds with probability p.

First of all, we reduce to the same game Γ_1 whose success additionally requires that for every (N_P, N_V, L_μ) triplet, there is no more than one instance $P(x)$ and one instance $V(x)$ using this triplet. Since $P(x)$ is honest and selecting the ℓ_{nonce}-bit nonce N_P at random and the same for $V(x)$ selecting N_V, by looking at the up to $\frac{r^2}{2}$ pairs of $P(x)$'s or of $V(x)$'s and the probability that one selection of a nonce repeats, this new game succeeds with probability at least $p - \frac{r^2}{2}2^{-\ell_{\text{nonce}}}$.

Then, for DB1 and DB2, we outsource the computation of every $a_i + cx_i'$ to the oracle

$$\mathcal{O}_{x,f_x}(y, L_\mu, A, B) = (A \cdot L_\mu(x)) + (B \cdot f_x(y))$$

as in Definition 12, with $y = (N_P, N_V, L_\mu, b)$, $A \cdot L_\mu(x) = c(L_\mu(x))_i$, and $B \cdot f_x(y) = (f_x(y))_i$. I.e., $A_i = ce_i$ and $B_i = e_i$, where e_i is the vector having a 1 on its ith component and 0 elsewhere. This can be used with $c = c_i'$ by $P(x)$ (for computing r_i') or with $c = c_i$ by $V(x)$ (for verifying r_i). Similarly, the computation (by $P(x)$ or $V(x)$) of $\text{tag} = f_x(y)$ can be made by several calls of form $\mathcal{O}_{x,f_x}(y, L_\mu, 0, B)$. (We note that the y in this case has incompatible form with the y in the r_i computation.) So, every computation requiring x is outsourced. Note that queries to the same y must use the same L_μ since this is part of y. So, the first condition in Definition 12 to apply the circular-PRF assumption is satisfied. We consider the event E that there exists in the game some sequence (y, L_μ, A_j, B_j) of queries to $\mathcal{O}_{x,f_x}$ sharing the same (y, L_μ) and some λ_j's such that $\sum_j \lambda_j B_j = 0$ and $\sum_j \lambda_j A_j \neq 0$. We need to restrict to the event $\neg E$ to apply Definition 12. We consider the event E' that one instance in $\mathbf{V}$ receives a valid tag which was not computed by the prover $\mathbf{P}$ (i.e., it was forged).

Let c_i'' be the value received by $V(x)$ in the verification phase. We assume that V checks that tag is correct, timer_i is correct, and $c_i = c_i''$, then queries $\mathcal{O}_{x,f_x}(y, L_\mu, c_i e_i, e_i)$ only if these are correct. If E happens for some (y, L_μ), due to the property of Γ_1, each i has at most two queries. Since $B_j = e_{i_j}$, $\sum_j \lambda_j B_j = 0$ yields pairs of values j and j' such that $i_j = i_{j'} = i$, $A_j = c_i e_i$, $A_{j'} = c_i' e_i$, $B_j = B_{j'} = e_i$, and $\lambda_j + \lambda_{j'} = 0$. The event E implies that there exists one such pair such that $\lambda_j A_j + \lambda_{j'} A_{j'} \neq 0$. So, $c_i \neq c_i'$. But since V only queries if c_i'' and tag are correct, we have $c_i = c_i'' \neq c_i'$ and tag correct. So, $\mathcal{V}$ must have accepted some tag which was not computed by $P(x)$. So, E implies E'. We now show that $\Pr[E']$ is negligible.

We define Γ_2, the variant of Γ_1, which in turn requires that E' does not occur as an extra condition for success. We let E_j' be the event that tag_j, the jth value tag received by any $V(x)$ in $\mathbf{V}$ is forged. Let $\Gamma_{1,j}$ be the hybrid of Γ_1 stopping right after tag_j is received and succeeding if E_j' occurs but not $E_1', \dots, E_{j-1}'$.

Clearly, since $E_1' \cup \cdots \cup E_{j-1}'$ does not occur and we stop right after reception of tag_j, E cannot occur. (Remember that for E to occur for the first time upon a query to $\mathcal{O}_{x,f_x}$, there must be a prior tag which was forged.) So, the conditions to apply the circular-PRF security reduction in Definition 12 is satisfied in $\Gamma_{1,j}$. We apply the circular-PRF assumption and replace $\mathcal{O}_{x,f_x}$ by $\mathcal{O}_{\tilde{x},F}$, loosing some probability ε. We obtain a game $\Gamma_{2,j}$. Clearly, $\Gamma_{2,j}$ succeeds with probability bounded by $2^{-\ell_{\mathrm{tag}}}$ because F is random. So, $\Pr_{\Gamma_{1,j}}[\mathsf{success}] \leq \varepsilon + 2^{-\ell_{\mathrm{tag}}}$ in $\Gamma_{1,j}$.

So, $\Pr[E']$ is bounded by the sum of all $\Pr_{\Gamma_{1,j}}[\mathsf{success}]$, i.e. $\Pr_{\Gamma_1}[E'] \leq r\varepsilon + r2^{-\ell_{\mathrm{tag}}}$ since the number of hybrids is bounded by r. Hence, $\Pr_{\Gamma_2}[\mathsf{success}] \geq p - \frac{r^2}{2}2^{-\ell_{\mathrm{nonce}}} - r\varepsilon - r2^{-\ell_{\mathrm{tag}}}$.

Now, in the whole game Γ_2 where E' does not occur, we replace $\mathcal{O}_{x,f_x}$ by $\mathcal{O}_{\tilde{x},F}$ and obtain the simplified game Γ_3. We have $\Pr_{\Gamma_3}[\mathsf{success}] \geq p - \frac{r^2}{2}2^{-\ell_{\mathrm{nonce}}} - (r+1)\varepsilon - r2^{-\ell_{\mathrm{tag}}}$.

It is now easy to analyze the protocol Γ_3. Thanks to Lemma 4, the response is computed based on information from $P(x)$ (w in Lemma 4) which is independent (in the sense of Lemma 4) from the challenge. Either $P(x)$ was queried with a challenge before, but this could only match the correct one with probability $\frac{1}{q}$ and the adversary would fail with tag otherwise. Or, $P(x)$ leaked nothing about the response to this challenge, and the answer by the adversary can only be correct with probability $\frac{1}{q}$. In any case, his answer is correct with probability $\frac{1}{q}$. So, Γ_3 succeeds with probability up to $\mathsf{Tail}(n,\tau,\frac{1}{q})$.

To sum up, we have $p \leq \mathsf{Tail}(n,\tau,\frac{1}{q}) + \frac{r^2}{2}2^{-\ell_{\mathrm{nonce}}} + (r+1)\varepsilon + r2^{-\ell_{\mathrm{tag}}}$ for DB1 and DB2.

For DB3, we loose $\frac{r^2}{2}2^{-\ell_{\mathrm{nonce}}}$ from Γ_0 to Γ_1. In Γ_1, we apply the full PRF reduction and loose ε to obtain Γ_2 with a random function. We loose $2^{-\ell_{\mathrm{tag}}}$ more to assume that tag received by $\mathcal{V}$ was not forged in some Γ_3. Then, it is easy to see that either the prover was queried before c_i was known, but this will only succeed if c_i was correctly guessed, or it was queries after, but this will only succeed if the answer r_i was correctly guessed. So, Γ_3 succeeds with a probability bounded by $\mathsf{Tail}(n,\tau,\frac{1}{q})$. (Note that DB3 is insecure without the authenticating tag: the man-in-the-middle can just run the DB phase with the prover, deduce a, then answer all challenges from the verifier.) □

3.4 Soundness of DB1 and DB2

Theorem 14 (Soundness of DB1). *The DB1 scheme is $(\gamma, \gamma', s+2)$-sound for any $\gamma \geq \frac{q}{q-1}p_B$ and γ' such that $\gamma' = (1 - \gamma^{-1}p_B)^s$, where $p_B = \max_{a+b \leq n} p_B(a,b)$ and*

$$p_B(a,b) = \sum_{u+v \geq \tau - a} \binom{n-a-b}{u}\binom{b}{v}\left(1 - \frac{1}{q}\right)^{b+u-v}\left(\frac{1}{q}\right)^{n-a-b-u+v}$$

More precisely, any collusion fraud with a success probability $\gamma \geq \frac{p_B}{1 - \frac{1}{q} - \varepsilon}$ leaks one random $(\mu, \mu \cdot x)$ pair with probability at least $\frac{1}{q} + \varepsilon$. Assuming $p_B = p_B(0,0)$,[12] this compares γ to $\frac{q}{q-1}\mathsf{Tail}(n,\tau,\frac{q-1}{q})$.

---

[12] this is actually confirmed by experiment for the data we use.

For instance, for $\gamma = sp_B$ and $\frac{\tau}{n} > \frac{q}{q-1} + \mathsf{cte}$, γ is negligible and γ' is greater than a constant.

If we applied the same proof as for SKI from [30, Theorem 14], we would not get such a good result. We would rather obtain $\mathsf{Tail}(\frac{n}{2}, \tau - \frac{n}{2}, \frac{q-1}{q})$. So, our proof of Theorem 14 is substantially improved.

Proof. We consider a distinguished experiment $\exp(\mathcal{V})$ where $\mathcal{V}$ accepts with probability $p \geq \gamma$.

The verifier $\mathcal{V}$ has computed some a and x'. We apply Lemma 4. We let $\mathsf{Resp}_i(c)$ be the value of the response r_i arriving to $\mathcal{V}$ when c_i is replaced to c in the simulation. We show below that we can always compute $\mathsf{Resp}_i(c) - \mathsf{Resp}_i(c')$ for any (c, c') pair from a straightline simulation (i.e., without rewinding). Let View_i be the view of close-by participants $\mathcal{A}$ until the time before c_i arrives, and w_i be the extra information (independent from c_i, in the sense of Lemma 4) arriving from far-away. Due to Lemma 4, we have $\mathsf{Resp}_i(c) = \mathsf{Algo}(\mathsf{View}_i, c, w_i)$. So, we can easily compute $\mathsf{Resp}_i(c) - \mathsf{Resp}_i(c')$ without rewinding. The answer by a far-away participant is independent from c_i, so $\mathsf{Resp}_i(c) - \mathsf{Resp}_i(c') = 0$: we can compute $\mathsf{Resp}_i(c) - \mathsf{Resp}_i(c')$ as well.

We say that c is correct in the ith round if $\mathsf{Resp}_i(c) = a_i + cx'_i$. We let C_i be the set of correct c's for the ith round. We let S be the set of all i's such that $c_i \in C_i$. Finally, we let R (resp. R') be the set of all i's for which $\#C_i = q$ (resp. $\#C_i \leq 1$). I.e., all c's are correct in the ith round for $i \in R$ and at most one is correct for $i \in R'$.

By definition, the probability that $\#S \geq \tau$ is $p \geq \gamma$. We see that

$$\frac{\mathsf{Resp}_i(c) - \mathsf{Resp}_i(c')}{c - c'} = x'_i$$

if $i \in R$, for any $c \neq c'$. If the left-hand side leads to the same value ξ_i for each $c \neq c'$, we say that the round i votes for $x'_i = \xi_i$. If the (c, c') pairs do not lead to the same value in $\mathsf{GF}(q)$, we say that the round i does not vote. So, we can always compute the vote ξ_i from the views of close-by participants. The majority of the available $\mathsf{map}^{-1}(\xi_i)$ shall decode $\mu \cdot x$.

For DB1, we can prove that if the round i votes for some ξ_i such that $\xi_i \neq x'_i$, then we must have $i \in R'$. Indeed, if round i votes for some ξ_i and $\#C_i \geq 2$, it means that there exist two different challenges c and c' such that the responses $\mathsf{Resp}_i(c)$ and $\mathsf{Resp}_i(c')$ are correct. So, $\mathsf{Resp}_i(c) = a_i + cx'_i$ and $\mathsf{Resp}_i(c') = a_i + c'x'_i$. The vote ξ_i is $\frac{\mathsf{Resp}_i(c) - \mathsf{Resp}_i(c')}{c - c'}$ which is thus equal to x'_i. So, an incorrect vote cannot have two correct challenges: it must be for $i \in R'$. The majority of the votes does not give x'_i only when $\#R \leq \#R'$. So, we shall bound $\Pr[\#R \leq \#R']$.

Let $I, I' \subseteq \{1, \ldots, n\}$ such that $\#I \leq \#I'$ and $I \cap I'$ is empty. Let $p_B(a, b)$ be the probability that at least τ rounds succeed, when we know that a rounds succeed with probability 1, b rounds succeed with probability $\frac{1}{q}$, and the other succeed with probability $1 - \frac{1}{q}$. We have $\Pr[\#S \geq \tau, R = I, R' = I'] = \Pr[\#S \geq \tau | R = I, R' = I'] \Pr[R = I, R' = I']$ and $\Pr[\#S \geq \tau | R = I, R' = I'] \leq p_B(\#I, \#I') \leq p_B$ since we have $\#I$ correct rounds for sure and it remains to pick u correct challenges (out of at most $q-1$) among the $i \notin I \cup I'$ rounds, and v correct challenges (out of at most 1) among the $i \in I'$ rounds, for all u and v such that $u+v \geq \tau - \#I$. By summing over all choices for I and I', we obtain that $\Pr[\#S \geq \tau, \#R \leq \#R'] \leq p_B$ So, $\Pr[\#R > \#R' | \#S \geq \tau] \geq 1 - \gamma^{-1}p_B$. So, when the experiment succeeds, the extracting algorithm gets a random pair $(\mu, \mu \cdot x)$ with probability at least $1 - \gamma^{-1}p_B$. This is better than just guessing $\mu \cdot x$ when $\gamma > \frac{q}{q-1}p_B$.

We can do M many such accepting experiments, collect some $(\mu, \mu \cdot x)$ until we have s vector μ spanning $\mathsf{GF}(q)^s$, and reconstruct x with probability at least $\gamma' = \left(1 - \gamma^{-1}p_B\right)^s$. The probability that m samples in $\mathsf{GF}(q)^s$ do not generate this space is $p_m \leq q^{s-m}$ (the number of hyperplane, $q^s - 1$ times the probability that the m samples are all in this hyperplane, which is q^{-1} to the power m). So, the expected M until we generate the space is bounded by $s + \sum_{m \geq s} q^{s-m} \leq s+2$. Hence, after at most $s+2$ iterations on average, we can recover x by solving a linear system. This defines the extractor.

We can also push the extraction further when $1 - \gamma^{-1}p_B > \frac{1}{q}$ by solving an instance of the Learning Parity with Noise problem (LPN), which would still be feasible by the practical parameters s. Extraction can also work with a complexity overhead bounded by $\mathcal{O}(s^j)$ and a probability of at least $\gamma' = \mathsf{Tail}(s, s-j, 1 - \gamma^{-1}p_B)$, by finding at most j errors by exhaustive search or LPN solving algorithms.

The maximum $p_B = p_B(a,b)$ is always reached for $a = b$. Indeed, for all the values plotted in Fig. 2 with $n \geq 6$, we saw it was reached for $a = b = 0$. In this case, we have $p_B = p_B(0,0) = \overline{\mathsf{Tail}}(n, \tau, \frac{q-1}{q})$. ◻

Below, we prove that the result is tight for DB1 using $q = 3$. Whether this it tight for other q is an open question. Whether it is optimal for protocols following the common structure is also open.

DB1's tightness of the soundness proof. To show that the result is tight for DB1 with $q = 3$, we mount a terrorist fraud succeeding with probability $\gamma = \mathsf{Tail}(n, \tau, \frac{q-1}{q})$: let the malicious prover give to the adversary the tables for $c_i \mapsto r_i + e_i(c_i)$ for every round i. For each such i, randomly pick one entry for which $e_i(c_i)$ is a random *nonzero* value and let it be 0 for other, two entries. With such tables as a response function, the adversary passes the DB phase with probability γ. (Other phases are done by relaying messages.) Since the verifier accepts with negligible probability γ, the adversary learns as much as if Out_V was always set to 0.

For $q = 3$ and each i, based on random $a_i \in \mathsf{GF}(q)$, $x'_i \in \mathsf{GF}(q)^*$, and $c \mapsto e_i(c)$ as distributed above, we can easily see that the distribution of the transmitted table is independent from x'_i: for $x'_i = 1$, the table of $c \mapsto a_i + c x'_i$ defined by a random a_i is randomly picked from

$$\begin{pmatrix} 0 \mapsto 0 \\ 1 \mapsto 1 \\ 2 \mapsto 2 \end{pmatrix}, \begin{pmatrix} 0 \mapsto 1 \\ 1 \mapsto 2 \\ 2 \mapsto 0 \end{pmatrix}, \begin{pmatrix} 0 \mapsto 2 \\ 1 \mapsto 0 \\ 2 \mapsto 1 \end{pmatrix}.$$

When adding the random table $e_i(c)$, it becomes a uniformly distributed random table among those with an output set of cardinality 2. For $x'_i = 2$, the table of $a_i + cx'_i$ is randomly picked from

$$\begin{pmatrix} 0 \mapsto 0 \\ 1 \mapsto 2 \\ 2 \mapsto 1 \end{pmatrix}, \begin{pmatrix} 0 \mapsto 2 \\ 1 \mapsto 1 \\ 2 \mapsto 0 \end{pmatrix}, \begin{pmatrix} 0 \mapsto 1 \\ 1 \mapsto 0 \\ 2 \mapsto 2 \end{pmatrix},$$

but adding $e_i(c)$ leads to the same distribution as for $x'_i = 1$. So, the above attack does not leak and is a valid terrorist fraud. Theorem 14 essentially says that there is no valid terrorist fraud with a larger γ. So, the result is tight for DB1 with $q = 3$.

The same proof technique leads to the following result for DB2.

Theorem 15 (Soundness of DB2). *For $\frac{\tau}{n} > \frac{3}{4}$, the DB2 scheme is $(\gamma, \gamma', s + 2)$-sound for any $\gamma \geq 2\mathsf{Tail}(\frac{n}{2}, \tau - \frac{n}{2}, \frac{1}{2})$ and $\gamma' = (1 - \gamma^{-1}\mathsf{Tail}(\frac{n}{2}, \tau - \frac{n}{2}, \frac{1}{2}))^s$.*

Again, it is open whether this is optimal for a protocol with binary challenges. The bound is pretty tight for DB2: a malicious adversary could leak the $c_i \mapsto r_i$ tables for a random selection of half of the rounds, and leak the table with one bit flipped for the others. This will not leak x'_i and will pass with probability $\gamma = \mathsf{Tail}(\frac{n}{2}, \tau - \frac{n}{2}, \frac{1}{2})$.

3.5   Performance Comparisons

Figure 2 plots the resistance of DB1, DB2, and DB3 compared with SKI [6–9] and FO [18].[13] In these figures, we assume a noise level of $p_{\mathsf{noise}} = 5\%$ and we adjust τ in terms of the number of rounds n such that $\mathsf{Tail}(n, \tau, 1 - p_{\mathsf{noise}}) \approx 99\%$, for τ-completeness; i.e., we admit a false rejection rate if below 1%. We plot then $-\log_2 \alpha$, $-\log_2 \beta$, and $-\log_2 \gamma$ in terms of n, assuming that the residual terms (such as ε and $2^{-\ell_{\mathsf{tag}}}$ from the PRF and $2^{-\ell_{\mathsf{nonce}}}$ from the nonce) can be neglected. We used the following dominant security parameters:

protocol	α	β	γ
SKI	$\mathsf{Tail}(n, \tau, 3/4)$	$\mathsf{Tail}(n, \tau, 2/3)$	$\mathsf{Tail}(\frac{n}{2}, \tau - \frac{n}{2}, 2/3)$
FO	$\mathsf{Tail}(n, \tau, 3/4)$	$\mathsf{Tail}(n, \tau, 3/4)$	n/a
DB1	$\mathsf{Tail}(n, \tau, 1/q)$	$\mathsf{Tail}(n, \tau, 1/q)$	$\frac{q}{q-1}\mathsf{Tail}(n, \tau, 1 - 1/q)$
DB2	$\mathsf{Tail}(\frac{n}{2}, \tau - \frac{n}{2}, 1/2)$	$\mathsf{Tail}(n, \tau, 1/2)$	$\mathsf{Tail}(\frac{n}{2}, \tau - \frac{n}{2}, 1/2)$
DB3	$\mathsf{Tail}(n, \tau, 1/2)$	$\mathsf{Tail}(n, \tau, 1/2)$	n/a

As we can see, our protocols are better than SKI and FO on *all* curves.

---

[13] We take the FO protocol as described in [30] since the original one from [18] introduces two counters and has an incorrect parameter p_e. The one from [30] has been shown to provide an optimal expression for p_e.

DB3 is not plotted on the third graph since it is not sound. FO has an incomparable TF-resistance notion and is not plotted either. TF-resistance therein follows another philosophy: in order to pass a DB run, the FO protocol always leaks with a probability $\gamma' = \gamma$, no matter the number of rounds. Although this is an interesting idea, the price to pay is a much lower resistance to man-in-the-middle, as observed in [30].

Since we consider online attacks, security levels of 2^{-10} or 2^{-20} should suffice, i.e., better (online) security may be ambitious. We now report the minimal number of rounds to attain such security:

	security level 2^{-10}				security level 2^{-20}		
	DF	security	soundness		DF	security	soundness
SKI	84	48	181	SKI	151	91	315
FO	84	84	n/a	FO	151	151	n/a
DB1 $q = 3$	14	14	54	DB1 $q = 3$	24	24	92
DB1 $q = 4$	12	12	91	DB1 $q = 4$	20	20	152
DB2	69	24	79	DB2	123	43	131
DB3	24	24	n/a	DB3	43	43	n/a

Interpretation of results. As we can see in the table above, DB1 with $q = 4$ is the best choice for distance fraud and security. Unfortunately, its (non-tightly) proven soundness requires more rounds. DB1 with $q = 3$ seems to be the best compromise. But if we want to use binary challenges, we shall choose between DB2 (suboptimal for DF-resistance) and DB3 (not sound).

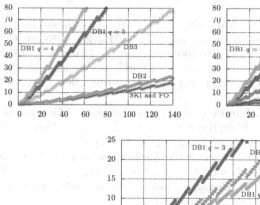

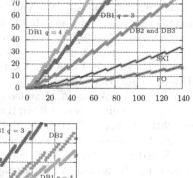

Fig. 2. Distance fraud resistance (top left) and security (top right), in equivalent bitlength, with respect to the number of rounds n. This assumes a τ-completeness level of 99 % and $p_{\text{noise}} = 5\,\%$. The bottom curve gives the soundness level. (Note that DB3 is not sound and that FO follows another TF-resistance philosophy.)

4   Conclusion

We provided the provably secure symmetric protocols DB1, DB2, and DB3 which require fewer rounds than the only two existing, provably secure protocols, SKI and FO. Prior to this, we have revised the formal model for distance-bounding protocols in a way which is closer to (the state of the art of) interactive proofs. We also studied optimality of all provably secure DB protocols, existing and advanced herein. Some open challenges remain: 1. identify an optimal and *sound* protocol for $\mathsf{num}_c = \mathsf{num}_r = 2$; 2. study the optimality of soundness; 3. implement these protocols.

References

1. Avoine, G., Bingöl, M., Kardas, S., Lauradoux, C., Martin, B.: A framework for analyzing RFID distance bounding protocols. J. Comput. Secur. **19**(2), 289–317 (2011)
2. Avoine, G., Lauradoux, C., Martin, B.: How secret-sharing can defeat terrorist fraud. In: ACM Conference on Wireless Network Security WISEC 2011, Hamburg, Germany, pp. 145–156. ACM (2011)
3. Avoine, G., Tchamkerten, A.: An efficient distance bounding RFID authentication protocol: balancing false-acceptance rate and memory requirement. In: Samarati, P., Yung, M., Martinelli, F., Ardagna, C.A. (eds.) ISC 2009. LNCS, vol. 5735, pp. 250–261. Springer, Heidelberg (2009)
4. Bay, A., Boureanu, I., Mitrokotsa, A., Spulber, I., Vaudenay, S.: The Bussard-Bagga and other distance-bounding protocols under attacks. In: Kutyłowski, M., Yung, M. (eds.) Inscrypt 2012. LNCS, vol. 7763, pp. 371–391. Springer, Heidelberg (2013)
5. Boureanu, I., Mitrokotsa, A., Vaudenay, S.: On the pseudorandom function assumption in (secure) distance-bounding protocols. In: Hevia, A., Neven, G. (eds.) LatinCrypt 2012. LNCS, vol. 7533, pp. 100–120. Springer, Heidelberg (2012)
6. Boureanu, I., Mitrokotsa, A., Vaudenay, S.: Secure and lightweight distance-bounding. In: Avoine, G., Kara, O. (eds.) LightSec 2013. LNCS, vol. 8162, pp. 97–113. Springer, Heidelberg (2013)
7. Boureanu, I., Mitrokotsa, A., Vaudenay, S.: Practical & provably secure distance-bounding. J. Comput. Secur. (JCS), IOS Press. Available as IACR Eprint 2013/465 report (2013, to appear). http://eprint.iacr.org/2013/465.pdf
8. Boureanu, I., Mitrokotsa, A., Vaudenay, S.: Towards secure distance bounding. In: Moriai, S. (ed.) FSE 2013. LNCS, vol. 8424, pp. 55–67. Springer, Heidelberg (2014)
9. Boureanu, I., Mitrokotsa, A., Vaudenay, S.: Practical & provably secure distance-bounding. In: Proceedings of ISC 2013 (to appear)
10. Boureanu, I., Vaudenay, S.: Optimal proximity proofs. IACR Eprint 2014/693 report (2014). http://eprint.iacr.org/2014/693.pdf
11. Boureanu, I., Vaudenay, S.: Challenges in distance-bounding. IEEE Secur. Priv. **13**(1), 41–48 (2015). doi:10.1109/MSP.2015.2
12. Brands, S., Chaum, D.: Distance-bounding protocols (extended abstract). In: Helleseth, T. (ed.) Advances in Cryptology — EUROCRYPT 1993. LNCS, vol. 765, pp. 344–359. Springer, Heidelberg (1994)

13. Chandran, N., Goyal, V., Moriarty, R., Ostrovsky, R.: Position based cryptography. In: Halevi, S. (ed.) CRYPTO 2009. LNCS, vol. 5677, pp. 391–407. Springer, Heidelberg (2009)

14. Chernoff, H.: A measure of asymptotic efficiency for tests of a hypothesis based on the sum of observations. Ann. Math. Stat. **23**(4), 493–507 (1952)

15. Cremers, C.J.F., Rasmussen, K.B., Schmidt, B., Čapkun, S.: Distance hijacking attacks on distance bounding protocols. In: IEEE Symposium on Security and Privacy S&P 2012, San Francisco, California, USA, pp. 113–127. IEEE Computer Society (2012)

16. Desmedt, Y.: Major security problems with the "unforgeable" (feige-)fiat-Shamir proofs of identity and how to overcome them. In: Congress on Computer and Communication Security and Protection Securicom 1988, Paris, France, pp. 147–159. SEDEP, Paris (1988)

17. Dürholz, U., Fischlin, M., Kasper, M., Onete, C.: A formal approach to distance-bounding RFID protocols. In: Lai, X., Zhou, J., Li, H. (eds.) ISC 2011. LNCS, vol. 7001, pp. 47–62. Springer, Heidelberg (2011)

18. Fischlin, M., Onete, C.: Terrorism in distance bounding: modeling terrorist-fraud resistance. In: Jacobson, M., Locasto, M., Mohassel, P., Safavi-Naini, R. (eds.) ACNS 2013. LNCS, vol. 7954, pp. 414–431. Springer, Heidelberg (2013)

19. Özhan Gürel, A., Arslan, A., Akgün, M.: Non-uniform stepping approach to RFID distance bounding problem. In: Garcia-Alfaro, J., Navarro-Arribas, G., Cavalli, A., Leneutre, J. (eds.) DPM 2010 and SETOP 2010. LNCS, vol. 6514, pp. 64–78. Springer, Heidelberg (2011)

20. Hancke, G.P.: Distance bounding for RFID: effectiveness of terrorist fraud. In: Conference on RFID-Technologies and Applications RFID-TA 2012, Nice, France, pp. 91 96. IEEE (2012)

21. Hancke, G.P., Kuhn, M.G.: An RFID distance bounding protocol. In: Conference on Security and Privacy for Emerging Areas in Communications Networks SecureComm 2005, Athens, Greece, pp. 67–73. IEEE (2005)

22. Hoeffding, W.: Probability inequalities for sums of bounded random variables. J. Am. Stat. Assoc. **58**, 13–30 (1963)

23. Kim, C.H., Avoine, G.: RFID distance bounding protocol with mixed challenges to prevent relay attacks. In: Garay, J.A., Miyaji, A., Otsuka, A. (eds.) CANS 2009. LNCS, vol. 5888, pp. 119–133. Springer, Heidelberg (2009)

24. Kim, C.H., Avoine, G., Koeune, F., Standaert, F.-X., Pereira, O.: The Swiss-knife RFID distance bounding protocol. In: Lee, P.J., Cheon, J.H. (eds.) Information Security and Cryptology ICISC 2008. LNCS, vol. 5461, pp. 98–115. Springer, Heidelberg (2009)

25. Munilla, J., Peinado, A.: Distance bounding protocols for RFID enhanced by using void-challenges and analysis in noisy channels. Wirel. Commun. Mob. Comput. **8**, 1227–1232 (2008)

26. Nikov, V., Vauclair, M.: Yet another secure distance-bounding protocol. In: Proceedings of SECRYPT 2008, Porto, Portugal, pp. 218–221. INSTICC Press (2008)

27. Reid, J., Nieto, J.M.G., Tang, T., Senadji, B.: Detecting relay attacks with timing-based protocols. In: ACM Symposium on Information, Computer and Communications Security ASIACCS 2007, Singapore, pp. 204–213. ACM (2007)

28. Singelée, D., Preneel, B.: Distance bounding in noisy environments. In: Stajano, F., Meadows, C., Capkun, S., Moore, T. (eds.) ESAS 2007. LNCS, vol. 4572, pp. 101–115. Springer, Heidelberg (2007)

29. Tu, Y.-J., Piramuthu, S.: RFID distance bounding protocols. In: Workshop on RFID Technology RFID 2007, Vienna, Austria, EURASIP (2007). http://www.eurasip.org/Proceedings/Ext/RFID2007
30. Vaudenay, S.: On modeling terrorist frauds. In: Susilo, W., Reyhanitabar, R. (eds.) ProvSec 2013. LNCS, vol. 8209, pp. 1–20. Springer, Heidelberg (2013)
31. Vaudenay, S.: Proof of proximity of knowledge. IACR Eprint 2014/695 report (2014). http://eprint.iacr.org/2014/695.pdf
32. Vaudenay, S.: Private and secure public-key distance bounding: application to NFC payment. In: Proceedings of Financial Cryptography 2015 (2015, to appear)
33. Youn, T.-Y., Hong, D.: Authenticated distance bounding protocol with improved FAR: beyond the minimal bound of FAR. IEICE Trans. Commun. **E97–B**(5), 930–935 (2014)

Lattice and Public Key Cryptography

Simpler CCA-Secure Public Key Encryption from Lossy Trapdoor Functions

Bei Liang[1,2,3(✉)], Rui Zhang[1,3], and Hongda Li[1,2,3]

[1] State Key Laboratory of Information Security, Institute of Information
Engineering of Chinese Academy of Sciences, Beijing, China
{liangbei,r-zhang,lihongda}@iie.ac.cn
[2] Data Assurance and Communication Security Research Center of Chinese
Academy of Sciences, Beijing, China
[3] University of Chinese Academy of Sciences, Beijing, China

Abstract. In STOC'08, Peikert and Waters presented a black-box construction of CCA-secure public key encryption (PKE) scheme from lossy trapdoor functions (LTDFs) [20], and they mentioned in the paper that their construction is a hybrid of Naor-Yung [18] and Canetti-Halevi-Katz [5], since the twin encryption technique and a strongly one-time signature are simultaneously used. It is well-known that a one-time signature brings either large ciphertext overhead if built from general assumptions like one-way functions, or additional computation cost during key generation/signing if built from number theoretic assumptions.

In this paper, we demonstrate that one can actually remove the one-time signature from the PW-scheme, and the resulting KEM can also be proved CCA-secure. However, the resulting KEM is not good enough, in particular, applying the known parameters choices of [20], one obtains a session key with length only sub-linear to the security parameter, thus not a suitable key for subsequent cryptographic tasks. We then to further into the analysis and manage to instantiate our KEM with standard assumptions to obtain a valid key.

Keywords: Lossy trapdoor functions · All-but-one lossy trapdoor functions · Chosen ciphertext security · KEM

1 Introduction

Chosen ciphertext security (CCA) [2,10,13,18,22,24] is widely accepted as the standard and proper security requirement for public key encryption (PKE) schemes. CCA-secure PKE is not only important for data security, it has also become a useful tools to obtain complex protocols. Historically, only a few constructions of CCA-security were known and they can be classified into two major

This research is supported by the National Natural Science Foundation of China (Grant No. 60970139) and the Strategic Priority Program of Chinese Academy of Sciences (Grant No. XDA06010702).

© Springer International Publishing Switzerland 2015
D. Lin et al. (Eds.): Inscrypt 2014, LNCS 8957, pp. 193–206, 2015.
DOI: 10.1007/978-3-319-16745-9_11

categories: Naor-Yung (including the universal hash proof system, designated verifier proof), namely, using proof of "well-formness" [6,18,26], and Canetii-Halevi-Katz (CHK), namely, transforms from identity/tag-based techniques [3–5].

In STOC'08, Peikert and Waters (PW) [20] introduced a new primitive called lossy trapdoor functions (LTDFs) and a variant called all-but-one lossy trapdoor functions (ABO-LTDFs). Combining these two functions, one can perform a hybrid of NY and CHK techniques to obtain CCA-security. The PW-scheme is interesting, not only because it admits the first CCA-secure encryption scheme from lattices, but also helps to understand how Naor-Yung and CHK techniques work in the scheme.

The PW-scheme mainly consists of two ingredients: lossy trapdoor functions (LTDFs) and All-But-One lossy trapdoor functions (ABO-LTDFs). An LTDF f is a public function with two modes of operations: One is the injective mode, with a suitable trapdoor for f, the entire input x can be efficiently recovered from $f(x)$. The other is the lossy mode, where f statistically loses a significant amount of information about its input, i.e., most outputs of f have many pre-images. An ABO-LTDF g is a further generalization of a LTDF, indexed by a branch b, which indicates whether g is injective or lossy.

In the "basic" PW-scheme, a ciphertext is $(b, c_1 = f(x), c_2 = g(b, x), c_3 = m \oplus h(x))$, where $x \in \{0, 1\}^n$ and b are sampled uniformly, and h is a randomness extractor. With the inverse f^{-1} of f as trapdoor, the decryption algorithm first recovers x, and checks whether $c_1 = f(x)$ and $c_2 = g(b, x)$. It aborts if either check fails, otherwise outputs $m = h(x) \oplus c_3$. Peikert and Waters addressed in [20] that the above construction is passively CCA-secure (or CCA1), since a ciphertext is clearly malleable. They further attach a one-time signature to the basic scheme, which upgrades its security to full CCA. In this sense, the PW-scheme also borrows the idea from the CHK-transform [5].

To remark, a one-time signature brings either long ciphertext overhead if constructed from one-way functions, or additional computation cost and possibly additional assumptions if constructed from number-theoretic assumptions, thus we conclude that if possible, the one-time signature should be removed. Actually, similar work was motivated for identity-based encryptions (IBEs), such as [3,4,29,30].

Related Work. Recently, Kiltz, Mohassel, and O'Neill [16] introduced the concept of adaptive trapdoor functions (ATDFs), and showed that ATDFs can be used to construct CCA-secure PKE schemes while it is weaker than LTDFs in the sense of black-box separation. In particular, they showed that one can achieve 1-bit CCA-secure encryption scheme without one-time signature from sufficiently lossy LTDFs and ABO-LTDFs. Compared with our method, their construction is rather theoretical and results in only a 1-bit KEM while assuming the underlying primitives should be more lossy. We note that is no general methods to have many-bits CCA-secure KEM without either narrowing the bandwidth or increasing computation costs. We conjecture the efficiency loss of their construction actually comes from ATDF as an additional step, which is crucial for their approach, thus it may not be easy to improve.

We also noticed that in another direction, Lai, Deng and Liu [17] (LDL) introduced chameleon ABO-LTDFs whose goal was to replace ABO-LTDFs and strongly unforgeable one-time signature hence improves the ciphertext overhead of the PW-scheme. In a closer view, the LDL-scheme is no more than an instantiation of a previously known technique for the separable TBE-to-PKE transform [29], with an observation that the basic PW-scheme is tag-based and satisfies separability. We note that a chameleon ABO-LTDFs always causes additional computation costs.

Therefore, the following question arises naturally: *Can we remove the one-time signature from the PW-scheme while avoid chameleon hashing?*

Our Motivation and Contribution. In this paper, we answer the above question in the affirmative. In particular, we construct a CCA-secure KEM from LTDFs and ABO-LTDFs solely. We note that Peikert and Waters have mentioned in their paper [21] that their basic encryption scheme shown above can be viewed as a KEM, while $h(x)$ is used as the session key. We remark that according to the parameters proposed in [20], the resulting session key is only sub-linear, and cannot be used by any symmetric encryption scheme to obtain enough security even if assuming a secure key derivation function (KDF).

Temporarily putting this problem aside, we observe that actually, the resulting session key $h(x)$ is not "easily malleable", since the decryption algorithm actually re-encrypts to check the validity of a ciphertext, which is also fixed, if the witness x and the branch b are fixed. However, this KEM is not CCA-secure yet, since it is still malleable.

To improve the above idea, we set $b = H(c_1)$, where H is a target collision-resistant (TCR) hash function. Now the ciphertext becomes $(f(x), g(H(c_1), x))$, and the session key is $h(x)$. In this way, f and g are then hedged together. Using a hybrid argument similar to that in [20], we can actually prove the CCA-security of our KEM. We postpone the detailed proof to later sections. In addition, by further careful analysis, we show $h(x)$ can be a good key if f and g are lossy enough. We make some further discussions about the system parameters and comparisons with previous ones, thus we are able to solve the problem of key length. It is well-known that together with CCA-secure data encapsulation mechanism (DEM), one can obtain CCA-secure hybrid PKE [25]. In this way, we are able to remove the one-time signature scheme from the PW-scheme while manage to avoid the chameleon hashing [17].

Finally, we remark that our technique is somehow close to BB1-KEM proposed by Boyen, Mei and Waters [4], where a selectively-secure IBE is transformed to a CCA-secure KEM. However, unlike BB1-KEM, our construction is generic for LTDFs and ABO-LTDFs, while their technique only apply to a few specific pairing-based IBE schemes.

2   Preliminaries

In this section, we review some useful notations and definitions.

Notations. Let $\mathbb{N}$ be the set of natural numbers. If M is a set, then $|M|$ denotes its size and $m \xleftarrow{R} M$ denotes the operation of picking an element m uniformly at random from M. We denote by λ a security parameter. For notational clarity we usually omit it as an explicit parameter. PPT denotes probabilistic polynomial time. Let $z \leftarrow A(x, y, \cdots)$ denote the operation of running an algorithm A with inputs $(x, y, \cdots)$ and output z. Let U_ℓ denote uniform distribution on ℓ-bit binary strings.

We say a function $\mathrm{negl}(\lambda)$ is *negligible* (in λ) if for $\lambda > k_0$ and $k_0 \in \mathbb{Z}$, $\mathrm{negl}(\lambda) < \lambda^{-c}$ for any constant $c > 0$. The statistical distance between two random variables X and Y over common domain D is $\triangle(X, Y) = \frac{1}{2} \sum_{z \in D} |\Pr[X = z] - \Pr[Y = z]|$. We say that X and Y are *statistical indistinguishable* if their statistical distance is negligible. It is obvious that statistical indistinguishability implies computational indistinguishability.

2.1   Lossy Trapdoor Functions

Define the following quantities as functions of the security parameter: $n(\lambda) = \mathrm{poly}(\lambda)$ represents the input length of a function and $k(\lambda) \leq n(\lambda)$ represents the *lossiness* of the function. For convenience, we also define the *residual leakage* $r(\lambda) = n(\lambda) - k(\lambda)$.

Definition 1 (Lossy Trapdoor Functions). *A collection of (n, k)-lossy trapdoor functions is given by a tuple of PPT algorithms $(S_{inj}, S_{lossy}, F, F^{-1})$ having the following properties.*

1. Easy to sample an injective function with trapdoor: $S_{inj}(1^\lambda)$ *outputs* (s, t) *where s is a function index and t is its trapdoor,* $F(s, \cdot)$ *computes an injective (deterministic) function* $f_s(\cdot)$ *over the domain* $\{0, 1\}^n$, *and* $F^{-1}(t, \cdot)$ *computes* $f_s^{-1}(\cdot)$.
2. Easy to sample a lossy function: $S_{lossy}(1^\lambda)$ *outputs s where s is a function index, and $F(s, \cdot)$ computes a (deterministic) function $f_s(\cdot)$ over the domain $\{0, 1\}^n$ whose image has size at most $2^r = 2^{n-k}$.*
3. Hard to distinguish injective from lossy: *The ensembles $\{s : (s, t) \leftarrow S_{inj}(1^\lambda)\}$ and $\{s : s \leftarrow S_{lossy}(1^\lambda)\}$ are computationally indistinguishable.*

2.2   All-But-One Lossy Trapdoor Functions

The notion of ABO-LTDFs is a richer abstraction of LTDFs. Informally, in an ABO collection, each function has an extra input called its branch. All of the branches are injective trapdoor functions (having the same trapdoor value), except for one branch which is lossy. Let $B = \{B_\lambda\}_{\lambda \in \mathbb{N}}$ be a collection of sets whose elements represent the branches.

Definition 2 (All-But-One Trapdoor Functions). *A collection of (n, k)-all-but-one lossy trapdoor functions with the branch collection B is given by a tuple of PPT algorithms $(S_{abo}, G_{abo}, G_{abo}^{-1})$ having the following properties:*

1. Sampling a trapdoor function with given lossy branch: *for any* $b^* \in B_\lambda$, $S_{abo}(b^*)$ *outputs* (s, t), *where s is a function index and t is its trapdoor.*
2. Evaluation of injective functions: *For any* $b \in B_\lambda$ *distinct from* b^*, $G_{abo}(s, b, \cdot)$ *computes an injective (deterministic) function* $g_{s,b}(\cdot)$ *over the domain* $\{0,1\}^n$, *and* $G_{abo}^{-1}(t, b, \cdot)$ *computes* $g_{s,b}^{-1}(\cdot)$.
3. Evaluation of lossy functions: $G_{abo}(s, b^*, \cdot)$ *computes a function* $g_{s,b^*}(\cdot)$ *over the domain* $\{0,1\}^n$ *whose image has size at most* $2^r = 2^{n-k}$.
4. Hidden lossy branch: *The ensembles* $\{s : (s,t) \leftarrow S_{abo}(b_0^*)\}_{\lambda \in \mathbb{N}, b_0^* \in B_\lambda}$ *and* $\{s : (s,t) \leftarrow S_{abo}(b_1^*)\}_{\lambda \in \mathbb{N}, b_1^* \in B_\lambda}$ *are computationally indistinguishable.*

2.3   Key Encapsulation Mechanism (KEM)

A *key encapsulation mechanism* $\mathcal{KEM}$ consists of the following PPT algorithms.

- A key generation algorithm KEM.Gen that on input 1^λ outputs a public/secret key pair (pk, sk). The public key pk defines a key space Keysp.
- An encryption algorithm KEM.Enc that on input 1^λ and a public key pk, outputs a pair (K, ψ), where K is a key and ψ is a ciphertext.
- A decryption algorithm KEM.Dec that on input 1^λ and a secret key sk, a string (in particular a ciphertext) ψ, outputs either a key K or the special symbol $\bot$.

The CCA-security for $\mathcal{KEM}$ is defined by the following game between a challenger and an adversary $\mathcal{A}$. The challenger runs the key generation algorithm KEM.Gen to obtain (pk, sk) and gives pk to $\mathcal{A}$. Then $\mathcal{A}$ can make decryption queries about some ciphertexts $\{\psi\}$ and the challenger responds with KEM.Dec (sk, ψ). After that $\mathcal{A}$ asks for the challenge query. Then the challenger runs

$$(K_0^*, \psi^*) \xleftarrow{R} \text{KEM.Enc}(1^\lambda, pk), \quad K_1^* \xleftarrow{R} \text{Keysp}, \quad \delta \xleftarrow{R} \{0,1\}.$$

Return the pair (K_δ^*, ψ^*) to $\mathcal{A}$. $\mathcal{A}$ still can make decryption queries about ciphertexts except for ψ^*. At the end of the game, $\mathcal{A}$ outputs a bit δ'. Let $\text{Adv}_{\mathcal{KEM},\mathcal{A}}^{\text{IND-CCA}}(\lambda) = |\Pr[\delta = \delta'] - 1/2|$ denote $\mathcal{A}$'s advantage in this game.

Definition 3. *We say that* $\mathcal{KEM}$ *is* IND-CCA *secure if* $\text{Adv}_{\mathcal{KEM},\mathcal{A}}^{\text{IND-CCA}}(\lambda)$ *is negligible for any PPT adversary* $\mathcal{A}$.

2.4   Hash Functions

A family of functions $\mathcal{H} = \{H : D \to R\}$ is called pairwise independent, if, for every distinct $x \neq x' \in D$ and every $y, y' \in R$,

$$\Pr_{h \leftarrow \mathcal{H}}[h(x) = y \wedge h(x') = y'] = 1/|R|^2.$$

A family of functions $\mathcal{H}' = \{H : D \to R\}$ is called target collision-resistant (TCR) family, if for any PPT $\mathcal{A}$, $\text{Adv}_{\mathcal{H}',\mathcal{A}}^{\text{TCR}}(\lambda)$ is negligible, where

$$\text{Adv}_{\mathcal{H}',\mathcal{A}}^{\text{TCR}}(\lambda) := \Pr[x \xleftarrow{R} D, H \xleftarrow{R} \mathcal{H}', y \leftarrow \mathcal{A}(x, H) : x \neq y \wedge H(x) = H(y)].$$

2.5   Extracting Randomness

The *min-entropy* of a random variable X is defined as $H_\infty(X) = -\log(\max_x \Pr[X = x])$. Dodis, Reyzin and Smith [9] defined *average min-entropy* of X given Y to be the logarithm of the average probability of the most likely value of X given Y:

$$\widetilde{H}_\infty(X|Y) = -\log(E_y[2^{-H_\infty(X|Y=y)}]).$$

They also proved that if Y has 2^ℓ possible values and Z is any random variable, then

$$\widetilde{H}_\infty(X|(Y,Z)) \geq \widetilde{H}_\infty(X|Z) - \ell.$$

In addition, we review the following useful lemma:

Lemma 4 ([9]). *Let X, Y be random variables such that $X \in \{0,1\}^n$, and $\widetilde{H}_\infty(X|Y) \geq \kappa$. Let $\mathcal{H}$ be a family of pairwise independent hash functions from $\{0,1\}^n \to \{0,1\}^\ell$. Then for $h \xleftarrow{R} \mathcal{H}$, we have*

$$\Delta\big((Y, h, h(X)), (Y, h, U_\ell)\big) \leq \epsilon,$$

as long as $\ell \leq \kappa - 2\log(1/\epsilon)$.

3   The Proposed Scheme

In this section, we present our KEM and analyze its security.

3.1   The Proposed CCA-Secure KEM

Let $(S_{inj}, S_{lossy}, F, F^{-1})$ be a collection of (n,k)-LTDFs, and let $(S_{abo}, G_{abo}, G_{abo}^{-1})$ be a collection of (n,k')-ABO-LTDFs having branch set $B = \{0,1\}^v$. We also require that $(n-k) + (n-k') \leq n - \kappa$, for some $\kappa = \kappa(n) = \omega(\log n)$. Let $\mathcal{H}$ be a family of pairwise independent hash functions from $\{0,1\}^n$ to $\{0,1\}^\ell$, where $\omega(\log \lambda) \leq \ell \leq \kappa - 2\log(1/\epsilon)$ for some negligible $\epsilon = \text{negl}(\lambda)$. Let $\mathcal{H}'$ be a family of TCR functions from $\{0,1\}^*$ to $\{0,1\}^v$.

Now, we describe our KEM $\mathcal{KEM} = (\text{KEM.Gen}, \text{KEM.Enc}, \text{KEM.Dec})$ as follows:

- KEM.Gen: takes as input 1^λ. Run $(s,t) \leftarrow S_{inj}(1^\lambda)$ and $(s',t') \leftarrow S_{abo}(1^\lambda, 0^v)$. Choose $h \xleftarrow{R} \mathcal{H}$ and $H \xleftarrow{R} \mathcal{H}'$. Finally, output the public key $pk = (s, s', h, H)$ and secret key $sk = (t, t', pk)$.
- KEM.Enc: takes as input 1^λ and pk. Choose $x \xleftarrow{R} \{0,1\}^n$ and compute

$$c_1 = F(s, x), \quad c_2 = G_{abo}(s', H(c_1), x), \quad \text{and } h(x).$$

Finally, output $(K, \psi) = (h(x), (c_1, c_2))$, where $K = h(x)$ is a key and $\psi = (c_1, c_2)$ is its ciphertext.

– KEM.Dec: takes as input 1^λ, sk and a string (in particular a ciphertext) ψ. Parse $\psi = (c_1, c_2)$. If fails, output $\bot$ and halt. Otherwise, compute $x = F^{-1}(t, c_1)$ and check whether

$$c_1 = F(s, x) \text{ and } c_2 = G_{abo}(s', H(c_1), x);$$

if not, output $\bot$ and halt. Finally, output $K = h(x)$.

The correctness of the KEM can be verified easily.

3.2 Security Analysis

Now we turn to the security proof. Formally, we have:

Theorem 5. *The KEM given in Sec. 3.1 is CCA-secure assuming that the LTDF, the ABO-LTDF, the pairwise independent hash functions and the TCR are secure.*

Proof. The proof uses a sequence of games ($Game_0$, $Game_1, \ldots, Game_5$) between a challenger and an adversary $\mathcal{A}$, where $Game_0$ is the standard CCA-game. We show that for $i = 0, \cdots, 4$, the differences of $\mathcal{A}$'s advantages in $Game_i$ and $Game_{i+1}$ are negligible. Finally, we show that $\mathcal{A}$ must have negligible advantage in $Game_5$. Therefore, we conclude that the KEM is IND-CCA secure.

Before describing the games, we define one global operation in the beginning of each game: the challenger samples $x^* \xleftarrow{R} \{0,1\}^n$ which will be used as the "randomness" to generate the challenge ciphertext ψ^*. We claim this change to the real IND-CCA game, namely $Game_0$, is harmless, since the view of $\mathcal{A}$ remains identical.

In the following, let X_i denote the event that the adversary wins in $Game_i$ ($i = 0, \cdots, 5$). Let q denote the total number of decryption queries made by $\mathcal{A}$.

$Game_0$: This game is the standard CCA-game. The challenger samples $x^* \xleftarrow{R} \{0,1\}^n$ and runs the key generation algorithm KEM.Gen to obtain (pk, sk) and gives pk to $\mathcal{A}$. Then $\mathcal{A}$ can make queries about ciphertexts $\{\psi\}$ and the challenger responds with KEM.Dec(sk, ψ). After that $\mathcal{A}$ asks for the challenge query. Then the challenger runs (using x^*)

$$(K_0^*, \psi^*) \leftarrow \text{KEM.Enc}(1^\lambda, pk), \quad K_1^* \xleftarrow{R} \text{Keysp}, \quad \delta \xleftarrow{R} \{0,1\}.$$

Return the pair (K_δ^*, ψ^*) to $\mathcal{A}$. $\mathcal{A}$ still can make decryption queries about ciphertexts except for ψ^*. At the end of the game, $\mathcal{A}$ outputs a bit δ'. If $\delta' = \delta$, we call $\mathcal{A}$ wins the game. Then we have

$$\text{Adv}_{\mathcal{KEM}, \mathcal{A}}^{\text{IND-CCA}}(\lambda) = |\Pr[X_0] - \frac{1}{2}|.$$

$Game_1$: This game is identical to $Game_0$ except that we make a small modification to the decryption oracle. When the adversary submits a ciphertext

$\psi = (c_1, c_2)$ satisfying $H(c_1) = H(c_1^*)$ for decryption, the challenger immediately aborts.

Let F denote the event that the adversary $\mathcal{A}$ makes decryption queries of the form $\psi = (c_1, c_1^*)$ with $H(c_1) = H(c_1^*)$. Then Game_0 and Game_1 proceed identically until event F occurs. We have

$$|\Pr[X_0] - \Pr[X_1]| \le \Pr[F],$$

by the difference lemma of [26]. If $c_1 = c_1^*$, then we have $c_2 = G_{abo}(s', H(c_1^*), x^*) = c_2^*$, since $c_1^* = F(s, x^*)$ is injective, which implies that $\psi = \psi^*$. Therefore, in event F, we only need to consider the case $c_1 \ne c_1^*$. In this case, we can easily use $\mathcal{A}$ to construct an adversary $\mathcal{B}$ who will find a collision of H. The success probability of $\mathcal{B}$ is at least same to that of $\mathcal{A}$'s querying one pair (c_1, c_1^*) satisfying $H(c_1) = H(c_1^*)$. Therefore, we have $\Pr[F] \le q \cdot \text{Adv}_{\mathcal{H}',\mathcal{B}}^{\text{TCR}}(\lambda)$. It follows that

$$|\Pr[X_0] - \Pr[X_1]| \le \Pr[F] \le q \cdot \text{Adv}_{\mathcal{H}',\mathcal{B}}^{\text{TCR}}(\lambda).$$

Since H is a TCR function, $|\Pr[X_0] - \Pr[X_1]|$ is negligible.

Game$_2$: This game is identical to Game_1 except that the ABO function in key generation algorithm KEM.Gen is chosen to have a lossy branch $b^* = H(c_1^*)$ rather than 0^v. Formally, in KEM.Gen(1^λ), the challenger firstly chooses $x^* \xleftarrow{R} \{0,1\}^n$, $H \xleftarrow{R} \mathcal{H}'$, runs $(s,t) \leftarrow S_{inj}(1^\lambda)$ and computes $c_1^* = F(s, x^*)$. Then it replaces $(s', t') \leftarrow S_{abo}(0^v)$ with $(s', t') \leftarrow S_{abo}(H(c_1^*))$.

A straightforward reduction to hidden lossy branch property of the ABO trapdoor functions yields

$$|\Pr[X_1] - \Pr[X_2]| = \text{negl}(\lambda).$$

Game$_3$: This game is identical to Game_2 except for another modification to the decryption oracle. When the adversary submits a ciphertext $\psi = (c_1, c_2)$ for decryption, the challenger computes $x = G_{abo}^{-1}(t', H(c_1), c_2)$ and checks that whether

$$c_1 = F(s, x) \text{ and } c_2 = G_{abo}(s', H(c_1), x).$$

If not, it responds $\bot$; else outputs $K = h(x)$.

In both games, when the adversary makes a valid decryption query of the form $\psi = (c_1, c_2)$, the challenger checks if $c_1 = F(s, x)$ and $c_2 = G_{abo}(s, H(c_1), x)$ for some x. It outputs $\bot$ if not. We remark that if $H(c_1) = H(c_1^*)$, the decryption oracle also outputs $\bot$. Therefore, $F(s, \cdot)$ and $G_{abo}(s', H(c_1), \cdot)$ are both injective, and there is a unique x such that $(c_1, c_2) = (F(s, x), G_{abo}(s', H(c_1), x))$. In Game_2, the challenger computes x by $F^{-1}(t, c_1)$, while in Game_3, one finds x by computing $G_{abo}^{-1}(t', H(c_1), c_2)$. Hence,

$$\Pr[X_2] = \Pr[X_3].$$

Game$_4$: This game is identical to Game_3 except that, in KEM.Gen, we replace the injective function with a lossy one, i.e., let $s \leftarrow S_{lossy}(1^\lambda)$. A straightforward reduction to the indistinguishability of injective functions from lossy functions of the lossy trapdoor functions collection yields

$$|\Pr[X_3] - \Pr[X_4]| = \text{negl}(\lambda).$$

Game$_5$: This game is identical to Game$_4$ except for the modification of challenge query. In particular, the challenger chooses $K^\dagger \xleftarrow{R} \text{Keysp}$ and gives $(\psi^*, K^\dagger)$ to $\mathcal{A}$ regardless of the value of δ. Obviously, we have

$$\Pr[X_5] = \frac{1}{2},$$

since now $K^\dagger$ is random and irrelevant to ψ^*.

In addition, we have the following lemma (we will give its proof shortly after the main proof).

Lemma 6. *The adversary's views in* Game$_4$ *and* Game$_5$ *are statistically indistinguishable. Hence,*

$$|\Pr[X_4] - \Pr[X_5]| = \text{negl}(\lambda).$$

Therefore, the advantage of $\mathcal{A}$ in the game is negligible. It follows that the KEM is IND-CCA secure, and this completes the proof.

Proof. We observe that, in Game$_4$ and Game$_5$, $F(s, \cdot)$ and $G_{abo}(s', H(c_1^*), \cdot)$ are both lossy functions with image sizes at most 2^{n-k} and $2^{n-k'}$. Therefore, $\psi^* = (c_1^*, c_2^*)$ can take at most $2^{n-k+n-k'} \leq 2^{n-\kappa}$ values. It follows that

$$\tilde{H}_\infty(x|c_1^*, c_2^*) \geq \tilde{H}_\infty(x) - (n - \kappa) = n - (n - \kappa) = \kappa.$$

Since $\ell \leq \kappa - 2\log(1/\epsilon)$, we have

$$\Delta((c_1^*, c_2^*, h, h(x)), (c_1^*, c_2^*, h, r')) \leq \epsilon = \text{negl}(\lambda),$$

where $r' \xleftarrow{R} \{0,1\}^\ell$. If $\delta = 1$, the views of $\mathcal{A}$ in two games are identical; else, the difference is statistically close. Thus, the adversary's views in two games are statistically indistinguishable.

4 Extensions and Discussions

4.1 Obtaining a Longer Session Key

In order to instantiate our scheme, it is necessary to construct a family of LTDFs and a family of ABO-LTDFs that have enough lossiness so that we can extract a session key with proper length. We carefully examine the LTDFs appeared in the literature [12,20,28] and find the following construction [12] which gives the required property.

Table 1. Efficiency comparison for our scheme with previously known CCA-secure PW-scheme and LDL-scheme. λ is the security parameter. ℓ is the keysize. n is the length of required randomness. We take into account two cases that the security parameters are 80 bits and 128 bits. The ciphertext overheads of our scheme are computed by $|F(s,x)| + |G_{abo}(s, H(c_1), x)| + |\text{MAC}| = 6\lambda + 6\lambda + |\text{MAC}|$, where $F(s,x)$ and $G_{abo}(s, H(c_1), x)$ are the LTDF and the ABO-LTDF given in [12], respectively. Instantiating one-time signature via number-theoretic assumption based schemes will either introduce additional assumptions and/or computation costs, so we don't take into account here. The ciphertext overheads of the LDL-scheme are computed by $|F(s,x)| + |F_{ch}(s', u, r, x)| + |r| = 6\lambda + 6\lambda + |\frac{1}{2}\lambda - 1|$, where $F(s,x)$ is the LTDF, $F_{ch}(s', u, r, x)$ is the chameleon ABO-LTDF given in [17] under the DCR assumption and $|r|$ is the size of branches set which is equal to $\log 2^{\lambda/2-1}$. The ciphertext overheads of above three scheme are computed under the DCR assumption. "$\checkmark$" in the "Online/Offline" column means that the scheme supports the ability of online/offline processing, whereas "$\times$" means that the scheme does not supports this ability.

| | λ (bit) | ℓ (bit) | n (bit) | Ciphertext | Security | Online/ |
			n (bit)	overhead (bit)	assumption	Offline
Our Scheme	80	114	434	1088	d-linear, SMA,	$\checkmark$
	128	186	704	1696	DCR, QR, LWE	
PW-Scheme	80	114	434	25600	d-linear, SMA,	$\checkmark$
	128	186	704	65536	DCR, QR, LWE	
LDL-Scheme	80	106	434	999	DCR	$\times$
	128	173	434	1599		

Theorem 7 ([12]). *For any polynomial $\tau = \tau(\lambda)$, there exist a collection of $(n, k) = ((\lambda - 1)\tau + \lambda/2 - 1, (\lambda - 1)\tau - \lambda/2 - 1)$-LTDFs and a collection of $(n, k') = ((\lambda - 1)\tau + \lambda/2 - 1, (\lambda - 1)\tau - \lambda/2 - 1)$-ABO-LTDFs with branches set $B = \{0, 1, \cdots, 2^{\lambda/2-1}\}$, provided that the decisional composite residuosity (DCR) assumption holds.*

Then we propose the following parameters for our scheme under the DCR-assumption. Let $\tau = 5$, $\epsilon = 2^{-\lambda}$, $\kappa = k + k' - n = (\lambda - 1)\tau - \frac{3}{2}\lambda - 1 = \frac{7}{2}\lambda - 6$. Then,

$$n = \frac{11}{2}\lambda - 6, \ \ell = \kappa - 2\log(1/\epsilon) = \frac{3}{2}\lambda - 6.$$

Efficiency comparison for our scheme with previously known CCA-secure PW-scheme and LDL-scheme is assembled in Table 1.

4.2   CCA-Secure PKE

It is well-known that a CCA-secure PKE scheme can be obtained from a CCA-secure KEM and a CCA-secure DEM [25]. In Table 1, we use a standard authenticated encryption (one-time pad plus MAC) as the DEM part for practical use. We note that by applying redundancy-free DEMs [19] (namely, block cipher operated in certain modes of operation), one can further reduce the ciphertext overhead as shown in Table 1 by at least λ-bits.

4.3   Comparisons with the PW-Scheme

For completeness, we present the original PW-scheme in Appendix A. In this scheme, Peikert and Waters used a strongly unforgeable one-time signature scheme for CCA-security which we manage to avoid. We remark that to obtain 80 bits security, an additional ciphertext overhead is enlarged by the size of a strongly one-time signature plus a verification key which sums up to approximately 25600 bits $(O(\lambda^2))$ [11] and the ciphertext overhead is approximate to 65536 bits for 128 bits security. Certainly, one can use number-theoretic one-time signature scheme such as DSA or Waters signature [27] etc., but it will either introduce additional assumptions and/or computation costs.

4.4   Comparisons with the LDL-Scheme

To explain the effectiveness of our methodology, we compare our scheme with that in [17] introduced by Lai, Deng and Liu (LDL-scheme, reviewed in Appendix B) since their scheme also removes the strongly unforgeable one-time signature of the PW-scheme and hence improves ciphertext overhead of the PW-scheme.

The LDL-scheme needs a variation of ABO function named chameleon ABO function to construct CCA-secure scheme. Chameleon ABO functions are relatively less-studied. Up to now, it seems that the chameleon ABO functions can only be constructed from homomorphic encryption scheme which needs some additional properties such as the message space $\mathcal{M}$ needs to be a finite field or commutative ring with multiplicative identity. In [17], Lai et al. only gave a concrete construction based on DCR assumption. On the other hand, our scheme only needs standard ABO-LTDFs which has been extensively studied and can be constructed based on kinds of assumptions such as DDH [20], LWE [20], d-linear [12], DCR [12], QR [12], Subgroup Membership Assumption (SMA) [28]. Finally, the LDL-scheme loses the ability of online/offline processing in the PW-scheme, while our scheme retains it.

5   Conclusion

In this paper, we demonstrate that one can actually remove the one-time signature from the PW-scheme, and the resulting KEM can also be proved CCA-secure. Moreover, we make an analysis about the system parameters and manage to instantiate our KEM with standard assumptions to obtain a valid key. We also compare our scheme with the PW-scheme introduced by Peikert and Waters which used a strongly unforgeable one-time signature scheme for CCA-security and the LDL-scheme introduced by Lai, Deng and Liu which is based on chameleon ABO functions for CCA-security.

Acknowledgements. We thank Jingyong Chang for discussion about the details of the works. We are also grateful to the anonymous reviewers for their helpful comments and suggestions.

A   The PW-Scheme

Let $(Gen, Sign, Ver)$ be a strongly unforgeable one-time signature scheme where the public verification keys are in $\{0,1\}^v$. Let $(S_{inj}, S_{lossy}, F, F^{-1})$ give a collection of (n,k)-lossy trapdoor functions, and let $(S_{abo}, G_{abo}, G_{abo}^{-1})$ give a collection of (n,k')-ABO lossy trapdoor functions having branches set $B = \{0,1\}^v$. We require that $(n-k) + (n-k') \le n - \kappa$, for some $\kappa = \kappa(n) = \omega(\log n)$. Let $\mathcal{H}$ be a universal family of hash functions from $\{0,1\}^n$ to $\{0,1\}^\ell$, where $0 < \ell \le \kappa - 2\log(1/\epsilon)$ for some negligible $\epsilon = \mathrm{negl}(\lambda)$. The message space is $\{0,1\}^\ell$. The CCA-secure scheme $(\mathcal{G}, \mathcal{E}, \mathcal{D})$ is as follows.

- $\mathcal{G}$: takes as input 1^λ. Run $(s,t) \leftarrow S_{inj}(1^\lambda)$ and $(s',t') \leftarrow S_{abo}(1^\lambda, 0^v)$. Choose $h \xleftarrow{R} \mathcal{H}$. Finally, output the public key $pk = (s, s', h)$ and secret key $sk = (t, t', pk)$.
- $\mathcal{E}$: takes as input 1^λ, pk and $m \in \{0,1\}^\ell$. It generates one-time signature key pair $(vk, sk_\sigma) \leftarrow Gen$, then choose $x \xleftarrow{R} \{0,1\}^n$ and compute

$$c_1 = F(s,x), \quad c_2 = G_{abo}(s', vk, x), \quad c_3 = m \oplus h(x).$$

  Finally, it signs the tuple (c_1, c_2, c_3) using sk_σ as $\sigma = Sign(sk_\sigma, (c_1, c_2, c_3))$. The ciphertext is

$$c = (vk, c_1, c_2, c_3, \sigma).$$

- $\mathcal{D}$: takes as input 1^λ, $sk = (t, t', pk)$ and a ciphertext $c = (vk, c_1, c_2, c_3, \sigma)$. It first checks that whether $Ver(vk, (c_1, c_2, c_3), \sigma) = 1$; if not, it outputs $\bot$. It then computes $x = F^{-1}(t, c_1)$, and checks that $c_1 = F(s,x)$ and $c_2 = G_{abo}(s', vk, x)$; if not, it outputs $\bot$. Finally, it outputs $m = c_3 \oplus h(x)$.

B   The LDL-Scheme

Let $(S_{inj}, S_{lossy}, F, F^{-1})$ be a collection of (n,k)-lossy trapdoor functions, and let $(S_{ch}, F_{ch}, F_{ch}^{-1}, CLB_{ch})$ be a collection of (n,k')-chameleon ABO lossy trapdoor functions having branches $\mathbb{A} \times \mathbb{B} := \{A_\lambda \times B_\lambda\}_{\lambda \in \mathbb{N}}$. Let $\mathcal{H}$ be a universal family of hash functions from $\{0,1\}^n$ to $\{0,1\}^\ell$. We also require that $(n-k) + (n-k') \le n - \kappa$, for some $\kappa = \kappa(n) = \omega(\log n)$, and $0 < \ell \le \kappa - 2\log(1/\epsilon)$ for some negligible $\epsilon = \mathrm{negl}(\lambda)$. The message space is $\{0,1\}^\ell$. The CCA-secure scheme $(\mathcal{G}', \mathcal{E}', \mathcal{D}')$ is as follows.

- $\mathcal{G}'$: takes as input 1^λ. Run $(s,t) \leftarrow S_{inj}(1^\lambda)$ and $(s',t') \leftarrow S_{ch}(1^\lambda)$. Choose $h \xleftarrow{R} \mathcal{H}$ and a collision-resistant hash function $H : \{0,1\}^* \to A_\lambda$. Finally, output the public key $pk = (s, s', h, H)$ and secret key $sk = (t, t', pk)$.
- $\mathcal{E}'$: takes as input 1^λ, pk and $m \in \{0,1\}^\ell$. It choose $x \xleftarrow{R} \{0,1\}^n$ and $r \xleftarrow{R} B_\lambda$, then compute

$$c_0 = h(x) \oplus m, c_1 = F(s,x), c_2 = F_{ch}(s', u, r, x),$$

  where $u = H(c_0, c_1)$. Finally, it outputs the ciphertext $c = (c_0, c_1, c_2, r)$.
- $\mathcal{D}'$: takes as input 1^λ, sk and a ciphertext $c = (c_0, c_1, c_2, r)$. It computes $x = F^{-1}(t, c_1)$ and $u = H(c_0, c_1)$. Then check whether $c_1 = F(s,x)$ and $c_2 = F_{ch}(s', u, r, x)$; if not, it outputs $\bot$. Finally, it outputs $m = c_0 \oplus h(x)$.

References

1. Abe, M., Cui, Y., Imai, H., Kiltz, E.: Efficient hybrid encryption from ID-based encryption. Des. Codes Crypt. **54**(3), 205–240 (2010)
2. Bellare, M., Desai, A., Pointcheval, D., Rogaway, P.: Relations among notions of security for public-key encryption schemes. In: Krawczyk, H. (ed.) CRYPTO 1998. LNCS, vol. 1462, pp. 26–45. Springer, Heidelberg (1998)
3. Boneh, D., Katz, J.: Improved efficiency for CCA-secure cryptosystems built using identity-based encryption. In: Menezes, A. (ed.) CT-RSA 2005. LNCS, vol. 3376, pp. 87–103. Springer, Heidelberg (2005)
4. Boyen, X., Mei, Q., Waters, B.: Direct chosen ciphertext security from identity-based techniques. In: ACM CCS 2005, pp. 320–329. ACM, New York (2005)
5. Canetti, R., Halevi, S., Katz, J.: Chosen-ciphertext security from identity-based encryption. In: Cachin, C., Camenisch, J.L. (eds.) EUROCRYPT 2004. LNCS, vol. 3027, pp. 207–222. Springer, Heidelberg (2004)
6. Cramer, R., Shoup, V.: A practical public key cryptosystem provably secure against adaptive chosen ciphertext attack. In: Krawczyk, H. (ed.) CRYPTO 1998. LNCS, vol. 1462, pp. 13–25. Springer, Heidelberg (1998)
7. Cramer, R., Shoup, V.: Design and analysis of practical public-key encryption schemes secure against adaptive chosen ciphertext attack. SIAM J. Comput. **33**(1), 167–226 (2003)
8. Cramer, R., Shoup, V.: Universal hash proofs and a paradigm for adaptive chosen ciphertext secure public-key encryption. In: Knudsen, L.R. (ed.) EUROCRYPT 2002. LNCS, vol. 2332, pp. 45–64. Springer, Heidelberg (2002)
9. Dodis, Y., Reyzin, L., Smith, A.: Fuzzy extractors: how to generate strong keys from biometrics and other noisy data. In: Cachin, C., Camenisch, J.L. (eds.) EUROCRYPT 2004. LNCS, vol. 3027, pp. 523–540. Springer, Heidelberg (2004)
10. Dolev, D., Dwork, C., Naor, M.: Nonmalleable cryptography. SIAM J. Comput. **30**(2), 391–437 (2000)
11. Even, S., Goldreich, O., Micali, S.: On-line/Off-line digital signatures. J. Cryptology **9**(1), 35–67 (1996)
12. Freeman, D.M., Goldreich, O., Kiltz, E., Rosen, A., Segev, G.: More constructions of lossy and correlation-secure trapdoor functions. J. Cryptology **26**(1), 39–74 (2013)
13. Goldwasser, S., Micali, S.: Probabilistic encryption. J. Comput. Syst. Sci. **28**(2), 270–299 (1984)
14. Hemenway, B., Ostrovsky, R.: On homomorphic encryption and chosen-ciphertext security. In: Fischlin, M., Buchmann, J., Manulis, M. (eds.) PKC 2012. LNCS, vol. 7293, pp. 52–65. Springer, Heidelberg (2012)
15. Hemenway, B., Ostrovsky, R.: Lossy trapdoor functions from smooth homomorphic hash proof systems. In: ECCC, vol. 16(127) (2009)
16. Kiltz, E., Mohassel, P., O'Neill, A.: Adaptive trapdoor functions and chosen-ciphertext security. In: Gilbert, H. (ed.) EUROCRYPT 2010. LNCS, vol. 6110, pp. 673–692. Springer, Heidelberg (2010)
17. Lai, J., Deng, R.H., Liu, S.: Chameleon all-but-one TDFs and their application to chosen-ciphertext security. In: Catalano, D., Fazio, N., Gennaro, R., Nicolosi, A. (eds.) PKC 2011. LNCS, vol. 6571, pp. 228–245. Springer, Heidelberg (2011)
18. Naor, M., Yung, M.: Public-key cryptosystems provably secure against chosen ciphertext attacks. In: STOC 1990, pp. 427–437. ACM, New York (1990)

19. Phan, D.H., Pointcheval, D.: About the security of ciphers (semantic security and pseudo-random permutations). In: Handschuh, H., Hasan, M.A. (eds.) SAC 2004. LNCS, vol. 3357, pp. 182–197. Springer, Heidelberg (2004)

20. Peikert, C., Waters, B.: Lossy trapdoor functions and their applications. In: STOC 2008, pp. 187–196. ACM, New York (2008)

21. Peikert, C., Waters, B.: Lossy trapdoor functions and their applications. Full version of [20]. http://www.cc.gatech.edu/~cpeikert/pubs/lossy_tdf.pdf

22. Rackoff, C., Simon, D.R.: Non-interactive zero-knowledge proof of knowledge and chosen ciphertext attack. In: Feigenbaum, J. (ed.) CRYPTO 1991. LNCS, vol. 576, pp. 433–444. Springer, Heidelberg (1992)

23. Rosen, A., Segev, G.: Chosen-ciphertext security via correlated products. In: Reingold, O. (ed.) TCC 2009. LNCS, vol. 5444, pp. 419–436. Springer, Heidelberg (2009)

24. Sahai, A.: Non-malleable non-interactive zero knowledge and adaptive chosen-ciphertext security. In: FOCS 1999, pp. 543–553. IEEE Computer Society Press, Los Alamitos (1999)

25. Shoup, V.: A proposal for an ISO standard for public key encryption (2001). http://eprint.iacr.org/2001/112

26. Shoup, V.: Sequences of games: a tool for taming complexity in security proofs. Cryptology ePrint Archive: Report 2004/332 (2004)

27. Waters, B.: Efficient identity-based encryption without random oracles. In: Cramer, R. (ed.) EUROCRYPT 2005. LNCS, vol. 3494, pp. 114–127. Springer, Heidelberg (2005)

28. Xue, H., Li, B., Lu, X., Jia, D., Liu, Y.: Efficient lossy trapdoor functions based on subgroup membership assumptions. In: Abdalla, M., Nita-Rotaru, C., Dahab, R. (eds.) CANS 2013. LNCS, vol. 8257, pp. 235–250. Springer, Heidelberg (2013)

29. Zhang, R.: Tweaking TBE/IBE to PKE transforms with chameleon hash functions. In: Katz, J., Yung, M. (eds.) ACNS 2007. LNCS, vol. 4521, pp. 323–339. Springer, Heidelberg (2007)

30. Zhang, J., Xie, X., Zhang, R., Zhang, Z.: A generic construction from selective-IBE to public-key encryption with non-interactive opening. In: Wu, C.-K., Yung, M., Lin, D. (eds.) Inscrypt 2011. LNCS, vol. 7537, pp. 195–209. Springer, Heidelberg (2012)

Attacking RSA with a Composed Decryption Exponent Using Unravelled Linearization

Zhangjie Huang[1,2,3], Lei Hu[1,2(✉)], and Jun Xu[1,2]

[1] State Key Laboratory of Information Security, Institute of Information
Engineering, Chinese Academy of Sciences, Beijing 100093, China
{zhjhuang,hu,jxu}@is.ac.cn
[2] Data Assurance and Communication Security Research Center, Chinese Academy
of Sciences, Beijing 100093, China
[3] University of Chinese Academy of Sciences, Beijing 100049, China

Abstract. Recently, Nitaj and Douh presented a new attack on RSA
with a composed decryption exponent. To be specific, they assumed that
the decryption exponent in RSA is of the form $d = Md_1 + d_0$ where M is
a known positive integer and d_0 and d_1 are two suitably small unknown
integers. They gave a lattice-based decryption exponent recovery attack
on this kind of RSA when the exponent d is under a larger bound than
the well-known one $N^{0.292}$ given by Boneh and Durfee. In this paper,
we reconsider the same problem and present a new attack by using the
unravelled linearization technique proposed by Herrmann and May at
Asiacrypt 2009. Our result is theoretically better than that of Nitaj and
Douh and more importantly, is more efficient in terms of the dimension
of lattice involved in the attack.

Keywords: RSA · Unravelled linearization · Coppersmith's method ·
Lattice basis reduction · LLL algorithm

1 Introduction

The RSA cryptosystem [15] is currently the most widely used public key cryptosystem. RSA involves a public exponent e and a private exponent d which are related
by the equation $ed \equiv 1 \pmod{\phi(N)}$ where modulus $N = pq$ is the product of
two large primes p and q and ϕ is Euler's totient function. Due to the complicated
exponentiation operations in RSA, one tends to use those e and d with some special
properties such as small in size or having low Hamming weight. This may introduce
some security risks and which should be taken into consideration seriously.

There has been a great deal of research on the security of RSA when the
private exponent d is small or some of its bits are exposed. It is a well-known
result that when $d < N^{\frac{1}{4}}$ (approximately), d can be recovered by a continued
fraction method from the public information N and e [17]. The result was then
improved to $d < N^{0.292}$ by Boneh and Durfee [2] using the celebrated method
of Coppersmith [4] for finding small roots of modular polynomial equations.

© Springer International Publishing Switzerland 2015
D. Lin et al. (Eds.): Inscrypt 2014, LNCS 8957, pp. 207–219, 2015.
DOI: 10.1007/978-3-319-16745-9_12

To study the security of RSA under partial key exposure, Boneh, Durfee and Frankel presented several attacks on RSA in 1998 [3]. They showed that if $d = Md_1 + d_0$ for known d_0 and $M \geq 2^{\frac{n}{4}}$, they can reconstruct all of d for an n-bit RSA modulus in polynomial time (in n and e). They also considered attacks knowing the most significant bits where Md_1 (as a whole) is exposed. Some extended results are given in [1,5].

Unlike the partial key exposure attacks where M and d_0 are known, Nitaj and Douh [14] proposed an attack on RSA under a new assumption that d is of the form $d = Md_1 + d_0$ where M is a known positive integer but d_0 and d_1 are two unknown and relatively small integers. To motivate this problem, consider the following situation. Now that letting d less than $N^{0.292}$ is insecure, one may think of using a larger private key d and keeping the cost of decryption/signing at a reasonable level. For example, one may use d of the form $d = 2^v d_1 + d_0$, where d_0 and d_1 are the least significant part and the most significant part of d respectively and v is some integer for $d_0 \ll 2^v$. Making d_0 and d_1 both be small but d large (larger than $N^{0.292}$) does not lose much efficiency but Boneh and Durfee's small private key attack is no longer applicable for this kind of d. Nitaj and Douh's result and our improved result show that one should take care to use even this kind of d.

Assuming that

$$e = N^\alpha, \ M = N^\beta, \ d_0 \leq N^\gamma \text{ and } d_1 \leq N^\delta,$$

Nitaj and Douh showed that if

$$\delta < \frac{5}{4} - \gamma - \frac{1}{4}\sqrt{12(\alpha + \beta - \gamma) + 3}, \tag{1}$$

then they can factor the modulus N in polynomial time. Their method is a direct application of the extended strategy of Jochemsz and May [9] for finding small roots of multivariate polynomials proposed in 2006, which itself is based on Coppersmith's method. Under this assumption on d, they concluded that they can attack RSA for some d whose values exceed $N^{0.292}$, the well-known bound of Boneh and Durfee. Some specific values of $(\alpha, \beta, \gamma, \delta)$ are given in Table 1. As we can see from the table, the size of d within the bound of the attack, which is roughly $N^{\beta+\delta}$, can be greater than the bound $N^{0.292}$.

Following the assumption of Nitaj and Douh on d, we present another lattice-based attack by using the unravelled linearization technique. The unravelled linearization technique was introduced by Herrmann and May at Asiacrypt 2009 [6] and they used this technique to obtain an elementary proof of Boneh and Durfee's bound $d < N^{0.292}$ and to achieve a better practical performance in cryptanalysis of RSA with small CRT exponents [7]. We use the same linearization as in [7] but in our problem we have to deal with a polynomial with one more variable. Finally, we are able to gain a similar improvement when applying the unravelled linearization technique to analyze RSA with a composed private exponent. We improve the bound (1) to

$$\delta < \frac{7}{6} - \gamma - \frac{1}{3}\sqrt{6(\alpha + \beta - \gamma) + 1}. \tag{2}$$

Table 1. Values of δ in terms of α, β and γ. The fourth column represents the values of δ computed from (1), which is the result given in [14], and the last column denotes the values of δ computed from (2) (see below), which is the result in this paper.

α $(\log_N(e))$	β $(\log_N(M))$	γ $(\log_N(d_0))$	δ (1) $(\log_N(d_1))$	δ (2) $(\log_N(d_1))$
1.0	0.5	0.1	0.037	0.044
1.0	0.4	0.1	0.071	0.077
1.0	0.3	0.2	0.043	0.047
1.0	0.3	0.1	0.107	0.112
1.0	0.25	0.25	0.031	0.034
0.75	0.5	0.3	0.001	0.003
0.75	0.4	0.2	0.101	0.103
0.75	0.3	0.2	0.141	0.143
0.75	0.25	0.25	0.133	0.134

The bound (2) is always better than that of Nitaj and Douh in theory and more importantly our method requires lattices of smaller dimensions compared to the lattices constructed in [14] for the same parameters $(\alpha, \beta, \gamma, \delta)$ in practice.

We summarize our result in the following theorem:

Theorem 1. *Let $N = pq$ be an RSA modulus with balanced prime factors p and q, i.e., p and q have the same bit length. Let e and d be the public exponent and the private exponent respectively. Assume d is of the form $d - Md_1 + d_0$ where M is a known positive integer and d_0, d_1 are two unknown integers. Suppose that $e = N^\alpha$, $M = N^\beta$, $d_0 \leq N^\gamma$ and $d_1 \leq N^\delta$. Then the modulus N can be factored in polynomial time if*

$$\delta < \frac{7}{6} - \gamma - \frac{1}{3}\sqrt{6(\alpha + \beta - \gamma) + 1},$$

for sufficiently large N.

The rest of this paper is organized as follows. In Sect. 2 we will give some basic results on lattices and briefly introduce Coppersmith's method. In Sect. 3, we will describe the unravelled linearization technique of Herrmann and May [7] for analyzing RSA with small private exponents. We will present our analysis on RSA with composed private exponents in Sect. 4. A comparison between Nitaj and Douh's method and ours will be given in Sect. 5.

2   Preliminaries

2.1   Lattices

An m-dimensional lattice L in $\mathbb{Z}^n$ $(m \leq n)$ is defined as the set of all integer linear combinations of m linearly independent (row) vectors $\{b_1, \ldots, b_m\}$ in

$\mathbb{Z}^n$. These vectors are called basis vectors, and are often denoted as a matrix: $B = (b_1^{\mathrm{T}}, \ldots, b_m^{\mathrm{T}})^{\mathrm{T}}$. Then the determinant of L can be computed as $\det(L) = \sqrt{\det(BB^{\mathrm{T}})}$.

Finding the shortest vector in a lattice is generally difficult, which is known as SVP in lattice theory. The goal of lattice reduction is to find a new lattice basis, whose basis vectors are short and almost orthogonal. The LLL algorithm [10] proposed by Lenstra, Lenstra and Lovász is the most famous lattice reduction algorithm, which returns a reduced basis in polynomial time. There are many variants of the LLL algorithm, such as the BKZ algorithm by Schnorr [16] and the L^2 algorithm by Nguyen and Stehlé [12]. The proof of the following fact can be found in [11].

Fact 1 (LLL). *Let L be a lattice spanned by the rows of $B = (b_1^{\mathrm{T}}, \ldots, b_m^{\mathrm{T}})^{\mathrm{T}}$. The LLL algorithm outputs a reduced basis $\{v_1, \ldots, v_m\}$ satisfying*

$$\|v_i\| \leq 2^{\frac{m(m-1)}{4(m-i+1)}} \det(L)^{\frac{1}{m-i+1}}, \ 1 \leq i \leq m$$

in polynomial time in m and in the bit length of the entries of the basis matrix B.

2.2   Coppersmith's Method

For the use of Coppersmith's method for finding the small roots of a modular polynomial equation, the reformulation given by Howgrave-Graham [8] is widely adopted. In general, Coppersmith's method consists of three basic steps:

1. Collect a set of polynomials which share the common desired roots modulo some integer.
2. Construct a lattice basis using these polynomials and then apply the LLL algorithm on the lattice.
3. Under certain conditions, some integer polynomials sharing the same small roots will be obtained and one can extract the roots from these integer polynomials using standard numerical methods for solving systems of polynomial equations.

The following lemma is due to Howgrave-Graham, which states that under which condition a modular equation holds over the integers. The norm of a polynomial $f(x_1, \ldots, x_n) = \sum a_{i_1,\ldots,i_n} x_1^{i_1} \ldots x_n^{i_n}$ is defined as $\|f(x_1, \ldots, x_n)\| = \sqrt{\sum |a_{i_1,\ldots,i_n}|^2}$.

Lemma 1 (Howgrave-Graham [8]). *Let $g(x_1, \ldots, x_n) \in \mathbb{Z}[x_1, \ldots, x_n]$ be a polynomial that consists of at most m monomials. Suppose that*

1. *$g(x_1^{(0)}, \ldots, x_n^{(0)}) \equiv 0 \pmod{b}$ for $|x_1^{(0)}| \leq X_1, \ldots, |x_n^{(0)}| \leq X_n$, and*
2. *$\|g(x_1 X_1, \ldots, x_n X_n)\| < \frac{b}{\sqrt{m}}$,*

then $g(x_1^{(0)}, \ldots, x_n^{(0)}) = 0$ holds over the integers.

Combining Howgrave-Graham's lemma with the LLL algorithm, we deduce that if

$$2^{\frac{m(m-1)}{4(m-i+1)}} \det(L)^{\frac{1}{m-i+1}} < \frac{b}{\sqrt{m}},$$

then the polynomials corresponding to the shortest i reduced basis vectors satisfy Howgrave-Graham's bound. The condition implies

$$\det(L) < 2^{-\frac{m(m-1)}{4}} \left(\frac{1}{\sqrt{m}}\right)^{m-i+1} b^{m-i+1}.$$

As in previous works, we ignore the terms that do not depend on b and simply check the condition $\det(L) < b^{m-i+1}$, which is called the enabling condition. In practice, this is convenient when b is large enough. After obtaining enough equations over the integers, we can solve the systems of equations by computing the Gröbner basis with respect to the lexicographic monomial ordering under the following heuristic assumption:

Assumption 1. *We can efficiently extract the desired roots by computing the Gröbner basis of the ideal generated by the polynomials corresponding to the first few LLL-reduced basis vectors.*

3   The Technique of Unravelled Linearization

Before we present our attack on RSA with a composed decryption exponent, we briefly introduce the usage of unravelled linearization technique in [7], where an elementary proof of Boneh and Durfee's bound $d < N^{0.292}$ was given.

In [7], the authors considered the problem of finding small roots of the two modular polynomials

$$f_0(x, y) = 1 + x(A + y) \mod e,$$
$$\tilde{f}_0(u, x) = u + Ax \mod e,$$

which are related to each other by the linearization relation $xy = u - 1$. They constructed a lattice basis using the following polynomials for a fixed integer m and $t \leq m$:

x-shifts: $\tilde{g}_{i,k}(u, x) = x^i \tilde{f}_0^k e^{m-k}$, for $k = 0, \ldots, m$ and $i = 0, \ldots, m - k$;   (3)

y-shifts: $\tilde{h}_{j,k}(u, x, y) = y^j \tilde{f}_0^k e^{m-k}$, for $j = 1, \ldots, t$ and $k = \left\lfloor \frac{m}{t} \right\rfloor j, \ldots, m$.   (4)

A lattice basis was constructed by using the coefficient vectors of $\tilde{g}_{i,k}(u\hat{U}, x\hat{X})$ and $\tilde{h}_{j,k}(u\hat{U}, x\hat{X}, y\hat{Y})$ as the basis vectors, where $\hat{U}$, $\hat{X}$ and $\hat{Y}$ are the upper bounds on the size of the roots. If they only use the x-shifts to construct a basis, the lattice basis is triangular. When adding the y-shifts, every occurrence of xy in the polynomials $\tilde{h}_{j,k}$ is substituted with $u - 1$. This process changes the monomials in the polynomials in the y-shifts. For some orderings on the polynomials and on the monomials these polynomials contain, they kept the property that every newly added polynomial introduces exactly one new monomial that did not appear in the basis before. Hence, the triangular structure of the basis matrix was retained, which makes it easy to compute the determinant of the lattice.

4   New Attack Using Unravelled Linearization

4.1   The Problem

First, we derive the polynomial which will be analyzed by using Coppersmith's method from the RSA key equation $ed \equiv 1 \pmod{\phi(N)}$, where $\phi(N) = (p-1)(q-1)$. Suppose $d = Md_1 + d_0$, we rewrite the key equation as

$$e(Md_1 + d_0) = 1 + k(p-1)(q-1),$$
$$k(N + 1 - (p+q)) - ed_0 + 1 = (eM)d_1.$$

for some k. Define the polynomial

$$f(x, y, z) = x(A + y) + ez + 1, \tag{5}$$

where $A = N+1$. The problem is to find the root $(x_0, y_0, z_0) = (k, -(p+q), -d_0)$ of the modular polynomial equation $f(x, y, z) = 0 \pmod{(eM)}$. Finding the root (x_0, y_0, z_0) is equivalent to factoring N, since p and q can be computed from $-(p+q)$ using $N = pq$ easily. For simplicity, we denote eM as M_e hereafter.

In order to use the unravelled linearization technique to construct a lattice basis, we use the same linearization for the polynomial $f(x, y, z)$ as in [7]. Define the linearization relation $xy = u - 1$. We get another polynomial

$$\bar{f}(u, x, z) = u + Ax + ez. \tag{6}$$

We will use the relation $xy = u - 1$ when we construct the lattice basis from some polynomials. Accordingly, we have $u_0 = x_0 y_0 + 1$ and also $\bar{f}(u_0, x_0, z_0) \equiv 0$ mod M_e.

Let $e = N^\alpha$, $M = N^\beta$, $d_0 \leq N^\gamma$ and $d_1 \leq N^\delta$, where $\gamma < \beta + \delta$. Then $d = Md_1 + d_0 < 2N^{\beta+\delta}$ and $k = (ed - 1)/\phi(N) < (2ed)/N < 4N^{\alpha+\beta+\delta-1}$. For $p < q < 2p$ in $N = pq$, it is not hard to see that $p + q < 5\sqrt{N}/2$. Thus, define

$$X = 4N^{\alpha+\beta+\delta-1}, \ Y = \frac{5\sqrt{N}}{2}, \ Z = N^\gamma \text{ and } U = 10N^{\alpha+\beta+\delta-\frac{1}{2}},$$

we have $|x_0| < X$, $|y_0| < Y$, $|z_0| < Z$ and $|u_0| < U$.

4.2   Description of the Attack

Now we present our construction. In our problem, we want to find the small root of $f(x, y, z) = x(A + y) + ez + 1$ (resp. $\bar{f}(u, x, z) = u + Ax + ez$), which contains one more variable than the polynomial $f_0(x, y) = 1 + x(A + y)$ (resp.

$\tilde{f}_0(u, x) = u + Ax)$ considered in [7]. Adapting the polynomials chosen in [7] for our problem, our polynomials contain two parts, the zx-shifts and the zy-shifts (see below), similarly. We have to carefully choose the two parts in our method. We will explain the ideas behind the choice for the polynomials in the remainder of this section.

Fixing a positive integer m and an integer $t \leq m$, whose value will be determined later, we choose the following polynomials:

zx-shifts: $\quad \bar{g}_{i,j,k}(u, x, z) = z^j x^i \bar{f}(u, x, z)^k M_e^{m-k}$,

for $j = m, \ldots, 0$, $k = 0, \ldots, m - j$ and $i = 0, \ldots, m - j - k$; (7)

zy-shifts: $\quad \bar{h}_{j,k,l}(u, x, y, z) = z^l y^j \bar{f}(u, x, z)^{k-l} M_e^{m-(k-l)}$,

for $j = 1, \ldots, t$, $k = \left\lfloor \frac{m}{t} \right\rfloor j, \ldots, m$ and $l = k, \ldots, 0$. (8)

Clearly, these polynomials share the same root (u_0, x_0, y_0, z_0) modulo M_e^m. When using these polynomials to form a lattice basis, we order these polynomials in such a way: the zx-shifts come first and then the zy-shifts. Among the zx-shifts, we order $\bar{g}_{i,j,k}$ by the indices (i, j, k): the outermost index is $j = m, \ldots, 0$, then $k = 0, \ldots, m - j$, and the innermost index is $i = 0, \ldots, m - j - k$. Similarly, among the zy-shifts, we order $\bar{h}_{j,k,l}$ by the indices (j, k, l): the outermost index is $j = 1, \ldots, t$, then $k = \left\lfloor \frac{m}{t} \right\rfloor j, \ldots, m$, and the innermost index is $l = k, \ldots, 0$.

In order to keep the final basis triangular, which will simplify the determinant calculations, we choose the polynomials along with some ordering, such that every polynomial introduces exactly one new monomial that does not present in the basis before. The zx-shifts and the zy-shifts we choose satisfy this requirement.

Let us look at the zx-shifts first. For any $j_0 \subset [0, m]$, the zx-shifts define a set of polynomials $\bar{g}_{i,j_0,k}$ for $k = 0, \ldots, m - j_0$ and $i = 0, \ldots, m - j_0 - k$. Note that, in our construction, we have $\bar{f}(u, x, z) = \tilde{f}_0(u, x) + ez$. Thus, we rewrite $\bar{g}_{i,j_0,k}$ as:

$$\bar{g}_{i,j_0,k} = z^{j_0} x^i \bar{f}^k M_e^{m-k} = z^{j_0} x^i (\tilde{f}_0 + ez)^k M_e^{m-k}$$

$$= z^{j_0} x^i \tilde{f}_0^k M_e^{m-k} + \underbrace{M_e^{m-k} x^i \sum_{i'=1}^{k} \binom{k}{i'} e^{i'} z^{j_0+i'} \tilde{f}_0^{k-i'}}_{}. \quad (9)$$

Note that the exponents of z in all the monomials in the summation part of (9) are in $[j_0 + 1, m]$. It is not hard to see that these monomials already appear in the polynomials $\bar{g}_{i,j,k}$ for $j \in [j_0 + 1, m]$. By our ordering on the zx-shifts, the polynomials $\bar{g}_{i,j,k}$ (for $j \in [j_0 + 1, m]$) come before the polynomials $\bar{g}_{i,j_0,k}$. Therefore, for arbitrary $k \in [0, m - j_0]$ and $i \in [0, m - j_0 - k]$, the polynomial $\bar{g}_{i,j_0,k}$ introduces new monomials in the term $z^{j_0} x^i \tilde{f}_0^k M_e^{m-k}$ in (9). Note that this term is very like the x-shifts in (3) except the powers of z and the constant factor. The polynomials in the x-shifts meet the requirement that every polynomial introduces only one new monomial. It follows that the terms

$z^{j_0} x^i \tilde{f}_0^k M_e^{m-k}$ also meet the requirement since they are the polynomials in the x-shifts each multiplied by z^{j_0} (ignore the constant factors). Consequently, the zx-shifts meet the requirement.

As for the zy-shifts, we will show how to order the polynomials such that they meet the requirement. We will focus on the monomials in polynomials, hence we will omit the constant factors in the zy-shifts for the ease of notation. We notice that, for arbitrary positive integers a, b and c, the polynomial $z^a y^b \bar{f}^c$ introduces exactly one new monomial $z^a y^b u^c$ if we already added the polynomials $z^{a+1} y^b \bar{f}^{c-1}$, $z^a y^{b-1} \bar{f}^{c-1}$ and $z^a y^{b-1} \bar{f}^c$. This can be seen from the following fact, where $\bar{f} = u + Ax + ez$ and $xy = u - 1$:

$$
\begin{aligned}
z^a y^b &\bar{f}^c \\
&= z^a y^{b-1} \bar{f}^{c-1} (y(u + Ax + ez)) \\
&= z^a y^{b-1} \bar{f}^{c-1} (yu + Au - A + eyz) \\
&= u z^a y^b \bar{f}^{c-1} + Au z^a y^{b-1} \bar{f}^{c-1} - A z^a y^{b-1} \bar{f}^{c-1} + e z^{a+1} y^b \bar{f}^{c-1}.
\end{aligned}
$$

It is clear that the monomials in the second term are all in $z^a y^{b-1} \bar{f}^c$, the monomials in the third term are all in $z^a y^{b-1} \bar{f}^{c-1}$, and the monomials in the fourth term are all in $z^{a+1} y^b \bar{f}^{c-1}$. Iterate the same expansion over the first term $u z^a y^b \bar{f}^{c-1}$ until the exponent of $\bar{f}$ is zero. At last, we will see that all monomials in $z^a y^b \bar{f}^c$ appear in $z^{a+1} y^b \bar{f}^{c-1}$, $z^a y^{b-1} \bar{f}^{c-1}$ or $z^a y^{b-1} \bar{f}^c$ except $z^a y^b u^c$. Along with the zx-shifts, it is not hard to show that the polynomials in the zy-shifts in (8) satisfy the above mentioned requirement. We use the same set of indices (j, k) as in [7] as shown in (8).

We construct a basis with the coefficient vectors of $\bar{g}_{i,j,k}(uU, xX, zZ)$ and $\bar{h}_{j,k,l}(uU, xX, yY, zZ)$ as its basis vectors and denote the lattice generated by the basis as L. When using the zy-shifts, every occurrence of xy is substituted with $u - 1$. Figure 1 shows an example of the basis for the parameters $m = 2$ and $t = 1$.

	z^2	z	zx	zu	1	x	x^2	u	ux	u^2	$z^2 y$	zyu	yu^2
$z^2 M_e^2$	$M_e^2 Z^2$												
$z M_e^2$	0	$M_e^2 Z$											
$zx M_e^2$	0	0	$M_e^2 ZX$										
$z \bar{f} M_e$	$e M_e Z^2$	0	$AM_e ZX$	$M_e ZU$									
M_e^2	0	0	0	0	M_e^2								
$x M_e^2$	0	0	0	0	0	$M_e^2 X$							
$x^2 M_e^2$	0	0	0	0	0	0	$M_e^2 X^2$						
$\bar{f} M_e$	0	$e M_e Z$	0	0	0	$AM_e X$	0	$M_e U$					
$x \bar{f} M_e$	0	0	$e M_e ZX$	0	0	0	$AM_e X^2$	0	$M_e UX$				
$\bar{f}^2$	$e^2 Z^2$	0	$2Ae ZX$	$2e ZU$	0	0	$A^2 X^2$	0	$2AUX$	U^2			
$z^2 y M_e^2$	0	0	0	0	0	0	0	0	0	0	$M_e^2 Z^2 Y$		
$zy \bar{f} M_e$	0	$-AM_e Z$	0	$AM_e ZU$	0	0	0	0	0	0	$e M_e Z^2 Y$	$M_e ZYU$	
$y \bar{f}^2$	0	$-2Ae Z$	0	$2Ae ZU$	0	$-A^2 X$	0	$-2AU$	$A^2 UX$	$2AU^2$	$e^2 Z^2 Y$	$2e ZYU$	YU^2

Fig. 1. Example: the lattice basis for $m = 2$ and $t = 1$

We then apply the LLL algorithm on the lattice L. In order to extract our desired root (u_0, x_0, y_0, z_0), we expect at least four polynomials which are algebraically independent. Since we have known that $x_0 y_0 + 1 = u_0$, we expect the polynomials corresponding to the first three LLL-reduced basis vectors satisfy Howgrave-Graham's condition. We will check the enabling condition $\det(L) < M_e^{m(\dim(L)-2)}$.

We leave the calculations of the dimension and the determinant of L in Appendix A. For a given integer m and $t = \tau m$, when m grows to infinity, we substitute the approximate values for X, Y, Z and U into the enabling condition and obtain

$$\frac{1}{24} \cdot (\alpha + \beta + \delta - 1) + \frac{1}{8}\tau^2 \cdot \frac{1}{2} + \frac{1}{24}(1 + 3\tau) \cdot \gamma + \frac{1}{24}(1 + 3\tau) \cdot (\alpha + \beta + \delta - \frac{1}{2})$$
$$+ \frac{1}{24}(3 + 5\tau) \cdot (\alpha + \beta) < \frac{1}{6}(1 + 2\tau) \cdot (\alpha + \beta),$$

which leads to

$$\delta < \frac{7}{6} - \gamma - \frac{1}{3}\sqrt{6(\alpha + \beta - \gamma) + 1},$$

in Theorem 1 and the optimized value of τ is $\frac{1}{2} - (\gamma + \delta)$.

5   Experiments

As the bounds (1) and (2) are asymptotic bounds when m tends toward infinity, in this section we compare the method of constructing lattice in [14] with ours from a practical point of view. We investigate the dimensions of lattice which are needed for the two methods for the same parameters $(\alpha, \beta, \gamma, \delta)$. Table 2 shows the running times for various parameter settings. In all experiments, we set e full-size, i.e., $e \approx N$. In practice, we found that the reduced lattice bases contain many more than three polynomials with (u_0, x_0, y_0, z_0) as their root. We included all these polynomials in the basis and we were able to compute the Gröbner basis in a few seconds. The desired root (u_0, x_0, y_0, z_0) was easily seen from the Gröbner basis and thus Assumption 1 is reasonable. Our experiments were run on a desktop PC with 1.8 GHz Intel Core i7-4500U CPU and 4 GB RAM.

To derive concrete lattice parameters m and t for a given parameter setting $(\alpha, \beta, \gamma, \delta)$, we use the exact formulae to calculate the dimension and determinant of lattices, like (11) and (12) in Appendix A. As in [13], we use

$$\|\boldsymbol{v}_i\| \approx \lambda^{\dim(L)} \det(L)^{\frac{1}{\dim(L)}} \tag{10}$$

to approximate the length of vectors in the LLL-reduced basis for L, where $\boldsymbol{v}_i$ is the i-th smallest vector in the LLL-reduced basis for L and the factor λ is an average factor in practice related to the output quality of the LLL algorithm. In our experiments, λ (for $i = 3$) never exceed 1.0 (see Table 2). So it is safe to ignore the term $\lambda^{\dim(L)}$.

For a specific parameter setting $(\alpha, \beta, \gamma, \delta)$ satisfying (2), we choose values for m and t in the following steps, such that our attack is achievable and the dimension of lattice is as small as possible:

Table 2. Parameters and the experimental results

N(bits)	α	β	γ	δ	Lattice paras & results					Lattice paras & results [14]				
					m	t	dim	LLL (s)	λ	m	t	dim	LLL (s)	λ
1024	1.0	0.30	0.10	0.06	4	1	40	4	0.936	6	1	112	627	0.863
1000	1.0	0.30	0.10	0.07	6	2	113	581	0.955	7	1	156	3082	0.949
1000	1.0	0.35	0.05	0.08	6	2	113	4373	0.881	7	1	156	9358	0.938
1600	1.0	0.35	0.10	0.06	8	2	209	48906	0.983	10	2	418	–	–
1000	1.0	0.40	0.08	0.05	7	2	165	22876	0.887	9	2	330	–	–
1600	1.0	0.40	0.10	0.04	8	2	209	188768	0.951	10	2	418	–	–
1000	1.0	0.45	0.05	0.05	7	2	165	13562	0.789	9	2	330	–	–
1000	1.0	0.50	0.08	0.01	7	2	165	7456	0.712	9	2	330	–	–
1000	1.0	0.55	0.05	0.01	7	2	165	9244	0.598	9	2	330	–	–

1. Choose a positive integer t. We start from one;
2. Solve the inequality $\det(L)^{\frac{1}{\dim(L)}} < M_e^m$ for m and choose the smallest positive integer from the solution if there is one;
3. If the above inequality gives no positive integer solutions, increase t by one and go to Step 2 again.

We also use this method to calculate the m and t which are needed for the attack of Nitaj and Douh. The exact expressions for the dimension and the determinant of lattice for their attack are given in Appendix B, which we extract from [14].

We reimplemented the attack of Nitaj and Douh for some parameters. As we can see from Table 2, the dimension of lattice that is needed for our attack is about half of the dimension of lattice in the attack of Nitaj and Douh. And the running times of the LLL reduction in our method are a few times less than that of Nitaj and Douh's method.

Acknowledgements. The authors would like to thank the anonymous reviewers for their helpful comments and suggestions. The work of this paper was supported by the National Key Basic Research Program of China (2013CB834203), the National Natural Science Foundation of China (Grants 61472417), the Strategic Priority Research Program of Chinese Academy of Sciences under Grant XDA06010702, and the State Key Laboratory of Information Security, Chinese Academy of Sciences.

A   Dimension and Determinant of the Lattice L

The dimension of the lattice L in Sect. 4.2 is

$$\dim(L) = \omega = \sum_{j=0}^{m}\sum_{k=0}^{m-j}\sum_{i=0}^{m-j-k} 1 + \sum_{j=1}^{t}\sum_{k=\lfloor\frac{m}{t}\rfloor j}^{m}\sum_{l=0}^{k} 1. \tag{11}$$

The determinant of L is

$$\det(L) = X^{s_x} Y^{s_y} Z^{s_z} U^{s_u} M_e^{s_e},\tag{12}$$

where the s_x, s_y, s_z, s_u and s_e are as follows:

$$s_x = \sum_{j=0}^{m} \sum_{k=0}^{m-j} \sum_{i=0}^{m-j-k} i,$$

$$s_y = \sum_{j=1}^{t} \sum_{k=\lfloor \frac{m}{t} \rfloor j}^{m} \sum_{l=0}^{k} j,$$

$$s_z = \sum_{j=0}^{m} \sum_{k=0}^{m-j} \sum_{i=0}^{m-j-k} j + \sum_{j=1}^{t} \sum_{k=\lfloor \frac{m}{t} \rfloor j}^{m} \sum_{l=0}^{k} l,$$

$$s_u = \sum_{j=0}^{m} \sum_{k=0}^{m-j} \sum_{i=0}^{m-j-k} k + \sum_{j=1}^{t} \sum_{k=\lfloor \frac{m}{t} \rfloor j}^{m} \sum_{l=0}^{k} (k-l),$$

$$s_e = \sum_{j=0}^{m} \sum_{k=0}^{m-j} \sum_{i=0}^{m-j-k} (m-k) + \sum_{j=1}^{t} \sum_{k=\lfloor \frac{m}{t} \rfloor j}^{m} \sum_{l=0}^{k} (m-(k-l)).$$

For sufficiently large m and $t = \tau m$, the above values can be rewritten as:

$$\omega = \frac{1}{6}(1+2\tau)m^3 + o(m^3), \qquad s_x = \frac{1}{24}m^4 + o(m^4),$$

$$s_y = \frac{1}{8}\tau^2 m^4 + o(m^4), \qquad s_z = \frac{1}{24}(1+3\tau)m^4 + o(m^4),$$

$$s_u = \frac{1}{24}(1+3\tau)m^4 + o(m^4), \qquad s_e = \frac{1}{24}(3+5\tau)m^4 + o(m^4).$$

B Dimension and Determinant of the Lattice in [14]

Denote the lattice in [14] as L'. For integers m and t, the dimension of L' is given as:

$$\dim(L') = \frac{1}{6}(m+1)(m+2)(m+3t+3),\tag{13}$$

and the determinant is given as:

$$\det(L') = \bar{X}^{n_x} \bar{Y}^{n_y} \bar{Z}^{n_z} M_e^{n_e},\tag{14}$$

where the n_x, n_y, n_z and n_e are as follows:

$$n_x = \frac{1}{24}m(m+1)(m+2)(m+4t+3),$$

$$n_y = \frac{1}{12}m(m+1)(m+2)(m+2t+3),$$

$$n_z = \frac{1}{24}(m+1)(m+2)(m^2+3m+4tm+6t^2+6t),$$

$$n_e = \frac{1}{24}m(m+1)(m+2)(3m+8t+9),$$

and $\bar{X}$, $\bar{Y}$ and $\bar{Z}$ are the bound of the roots.

References

1. Blömer, J., May, A.: New partial key exposure attacks on RSA. In: Boneh, D. (ed.) CRYPTO 2003. LNCS, vol. 2729, pp. 27–43. Springer, Heidelberg (2003)
2. Boneh, D., Durfee, G.: Cryptanalysis of RSA with private key d less than $N^{0.292}$. In: Stern, J. (ed.) EUROCRYPT 1999. LNCS, vol. 1592, pp. 1–11. Springer, Heidelberg (1999)
3. Boneh, D., Durfee, G., Frankel, Y.: An attack on RSA given a small fraction of the private key bits. In: Ohta, K., Pei, D. (eds.) ASIACRYPT 1998. LNCS, vol. 1514, pp. 25–34. Springer, Heidelberg (1998)
4. Coppersmith, D.: Small solutions to polynomial equations, and low exponent RSA vulnerabilities. J. Cryptology 10(4), 233–260 (1997)
5. Ernst, M., Jochemsz, E., May, A., de Weger, B.: Partial key exposure attacks on RSA up to full size exponents. In: Cramer, R. (ed.) EUROCRYPT 2005. LNCS, vol. 3494, pp. 371–386. Springer, Heidelberg (2005)
6. Herrmann, M., May, A.: Attacking power generators using unravelled linearization: when do we output too much? In: Matsui, M. (ed.) ASIACRYPT 2009. LNCS, vol. 5912, pp. 487–504. Springer, Heidelberg (2009)
7. Herrmann, M., May, A.: Maximizing small root bounds by linearization and applications to small secret exponent RSA. In: Nguyen, P.Q., Pointcheval, D. (eds.) PKC 2010. LNCS, vol. 6056, pp. 53–69. Springer, Heidelberg (2010)
8. Howgrave-Graham, N.: Finding small roots of univariate modular equations revisited. In: Darnell, M. (ed.) Cryptography and Coding. LNCS, vol. 1355, pp. 131–142. Springer, Heidelberg (1997)
9. Jochemsz, E., May, A.: A strategy for finding roots of multivariate polynomials with new applications in attacking RSA variants. In: Lai, X., Chen, K. (eds.) ASIACRYPT 2006. LNCS, vol. 4284, pp. 267–282. Springer, Heidelberg (2006)
10. Lenstra, A., Lenstra, H.W., Lovász, J.L.: Factoring polynomials with rational coefficients. Math. Ann. 261(4), 515–534 (1982)
11. May, A.: New RSA vulnerabilities using lattice reduction methods. Ph.D. thesis, University of Paderborn (2003)
12. Nguên, P.Q., Stehlé, D.: Floating-point LLL revisited. In: Cramer, R. (ed.) EUROCRYPT 2005. LNCS, vol. 3494, pp. 215–233. Springer, Heidelberg (2005)
13. Nguyên, P.Q., Stehlé, D.: LLL on theaverage. In: Hess, F., Pauli, S., Pohst, M. (eds.) ANTS 2006. LNCS, vol. 4076, pp. 238–256. Springer, Heidelberg (2006)

14. Nitaj, A., Douh, M.O.: A new attack on RSA with a composed decryption exponent. Int. J. Crypt. Inf. Secur. (IJCIS) **3**(4), 11–21 (2013)
15. Rivest, R.L., Shamir, A., Adleman, L.: A method for obtaining digital signatures and public-key cryptosystems. Commun. ACM **21**(2), 120–126 (1978)
16. Schnorr, C.: A hierarchy of polynomial time lattice basis reduction algorithms. Theor. Comput. Sci. **53**(23), 201–224 (1987)
17. Wiener, M.: Cryptanalysis of short RSA secret exponents. In: Quisquater, J.-J., Vandewalle, J. (eds.) EUROCRYPT 1989. LNCS, vol. 434, pp. 372–372. Springer, Heidelberg (1990)

Fully Homomorphic Encryption with Auxiliary Inputs

Fuqun Wang[1,2,3](✉) and Kunpeng Wang[1,2]

[1] State Key Laboratory of Information Security, Institute of Information
Engineering, Chinese Academy of Sciences, Beijing, China
{fqwang,kpwang}@is.ac.cn
[2] Data Assurance and Communication Security Research Center,
Chinese Academy of Sciences, Beijing, China
[3] University of Chinese Academy of Sciences, Beijing, China

Abstract. In this paper, we propose the first (leveled) fully homomorphic encryption (FHE) that remains secure even when the attacker
is equipped with *auxiliary inputs* – any computationally *hard-to-invert
function* of the secret key. It is more general than the tolerance of Berkoff
and Liu's leakage resilient fully homomorphic encryption, in which the
leakage is bounded by an *a priori* number of bits of the secret key. Specifically, we first compile the dual of Regev's public-key encryption scheme
proposed by Gentry, Peikert and Vaikuntanathan in 2008 into a fully
homomorphic encryption using Gentry, Sahai and Waters' *approximate
eigenvector* method. We then show that it is CPA (chosen-plaintext-
attack) secure in the presence of *hard-to-invert auxiliary inputs*, assuming
the hardness of learning with errors (LWE) problem.

Keywords: Fully homomorphic encryption · Leakage resilient cryptography · Learning with errors · Auxiliary inputs

1 Introduction

Fully Homomorphic Encryption. The notion of privacy homomorphism
(now called fully homomorphic encryption), which allows anyone to transform
a number of encrypted data $\mathsf{Enc}(b_1), \mathsf{Enc}(b_2), \cdots, \mathsf{Enc}(b_t)$ to a related encrypted
message $\mathsf{Enc}(f(b_1, b_2, \cdots, b_t))$ for any circuit f without decrypting them first
and revealing anything about $b_1, b_2, \cdots, b_t$ themselves, was proposed by Rivest,
Adleman and Dertouzos [33] in 1978. Until 2009, the first FHE scheme was
constructed by Gentry [19,20]. It soon becomes a hot point of research in cryptography for its prospect and potential in various applications including cloud
computing. A great many FHE schemes have been constructed, e.g., [4,6,7,10–
12,14,21,23,25,27,35] and references therein.

This work is supported in part by the National Nature Science Foundation of China
(Grant No. 61272040 and No. 61379137), and in part by the National Basic Research
Program of China (973 project) (Grant No. 2013CB338001).

© Springer International Publishing Switzerland 2015
D. Lin et al. (Eds.): Inscrypt 2014, LNCS 8957, pp. 220–238, 2015.
DOI: 10.1007/978-3-319-16745-9_13

Since ciphertexts contain errors that increase with homomorphic evaluation and result in failure of decryption to some extent, all FHEs (including leveled FHEs) need the encrypted secret key to *refresh* ciphertexts or *bootstrap* their (augmented) decryption circuits after several homomorphic operations. As *refreshing* or *bootstrapping* are considerably complex, these FHE schemes are not easy to understand.

In CRYPTO 2013, Gentry, Sahai and Waters [25] proposed the first leveled FHE (abbreviated GSW hereafter) without using the encrypted secret key, via *approximate eigenvector* trick. Their scheme applies Regev's public-key encryption (RPKE) [34] as a cornerstone. Based on GSW, several subsequent works have been studied swiftly. Brakerski and Vaikuntanathan [12] constructed the first FHE scheme as secure as LWE-based PKE, relied on the important observation that the asymmetric nature of matrix multiplication gives rise to a better trick for manipulating the noise growth. Alperin-Sheriff and Peikert [4] designed a faster bootstrapping process relied on a compact symmetric-key variant of GSW, which can bootstrap essentially any LWE-based FHE scheme.

In this work, following the fancy and novel techniques in [4, 12, 25], we will principally focus on dual of Regev's public key encryption proposed by Gentry, Peikert and Vaikuntanathan [24].

Leakage Resilient Cryptography. Leakage resilient cryptography, which aims to stand against secret-key leakage attacks, is more advanced than traditional modern cryptography which is built based on original assumption that secret key is generated with perfect random bits and can be stored and used perfectly. There are kinds of models of leakage resilient cryptography: Only Computation Leaks Information [18, 29], Bounded Leakage Model [3, 5, 30], Bounded Retrieval Model [1, 2], Continual Leakage Model [8, 16, 26] and Auxiliary Inputs Leakage [5, 15, 17].

Bounded leakage resilience allows an adversary to gain a bounded length of secret key, while auxiliary inputs security means that a scheme remains CPA secure in the presence of hard-to-invert auxiliary inputs. The latter can be seen as a generalization of the former in the sense that the auxiliary inputs may be a output-bounded function. In this paper, for the first time, we adapt the CPA security in the presence of hard-to-invert auxiliary inputs to fully homomorphic cryptosystems.

1.1 Motivation and Contribution

Motivation. A large number of traditional cryptosystems (e.g., public key encryption and digital signature) are proposed and proved to be secure under side-channel attacks [2, 3, 5, 8, 15–17, 26, 30]. As traditional cryptosystems, fully homomorphic encryption schemes also need to be designed more securely against kinds of side-channel attacks. Berkoff and Liu [7] showed that GSW with larger parameters is bounded adaptive leakage resilience. Their success prompts us to extend bounded leakage to other leakage models in fully homomorphic cryptography. Since a secure scheme with auxiliary inputs has excellent composition property as discussed in [15,17], we will mainly focus on the auxiliary inputs leakage model in this work. Constructing FHE schemes against other leakage models, e.g., continual leakage model, is left as an interesting future work.

More specially, conventional wisdom that shows the security of LWE-based cryptosystems via proving the public key indistinguishable from random, as explained by Berkoff and Liu [7], is no use to argue that they are yet secure against (adaptive) bounded memory leakage. The main reason is that the public key is a function of the secret key and an attacker, when it obtains both keys, can choose simply a relation function which reveals the relation between them. As the generalization of bounded memory leakage, auxiliary inputs leakage also faces this problem.

In 2009, Akavia, Goldwasser and Vaikuntanathan [3] argued that RPKE with appropriate parameters is bounded memory leakage, by showing directly the ciphertext indistinguishable from random. Following this, Berkoff and Liu [7] proved that GSW with larger parameters is also bounded memory leakage. Naturally, we would ask that *can we prove that* GSW *stands against auxiliary inputs leakage?*

We observe that Berkoff and Liu's techniques for proving the GSW against the bounded memory leakage are not useful here, because we even do not know that if RPKE is CPA secure against auxiliary inputs leakage or not to our knowledge. The crux of the problem is that the secret key is fully deterministic information-theoretically given the hard-to-invert auxiliary-input function of the secret key. So, we can not apply the notion of *min-entropy* of the secret key to argue auxiliary inputs leakage.

However, we observe that Dodis, Goldwasser, Kalai, Peikert and Vaikuntanathan [15] showed that dual of Regev's PKE with larger parameters resists against auxiliary inputs leakage, under the hardness of decisional LWE problem. Following this, we propose an FHE scheme compiling dual of Regev's PKE using approximate eigenvector method and argue that it is CPA secure against auxiliary inputs leakage.

Contribution. Our results are:

- **An FHE scheme based on DRPKE:** Although it is well known that an FHE scheme can be obtained from dual of Regev's public key encryption (DRPKE) as Brakerski and Gentry-Sahai-Waters mentioned in [10] and [25] respectively, it has still not appeared in literature to our knowledge. The main reason (drawback) is that the ciphertext size is larger than that in [10] or [25], which further reduces the efficiency that is the biggest bottleneck in fully homomorphic cryptosystems. However, in order to construct an FHE with auxiliary inputs, we have to present it with more details. Specially, we construct a leveled FHE scheme based on DRPKE, using *approximate eigenvector* method proposed by Gentry, Sahai and Waters [25]. Furthermore, the scheme (without key leakage) can make use of bootstrapping theorems of Brakerski and Vaikuntanathan [12] or Alperin-Sheriff and Peikert [4] to optimize the parameters or bootstrap and gain an unbounded FHE.

- **Leveled FHE against auxiliary inputs leakage:** The above scheme with
  "normal" parameters is not CPA secure with auxiliary inputs, but we can show
  that it is CPA secure against auxiliary inputs leakage via carefully setting the
  parameters. As far as we know, it is the first fully homomorphic encryption
  with auxiliary inputs as our main contribution in this work. Recall that Berkoff
  and Liu [7] argued the first FHE against bounded memory leakage. Our scheme
  against auxiliary inputs leakage is closely related to theirs, in the sense that
  auxiliary inputs leakage can be seen a more general leakage model as it implies
  bounded memory leakage.

  Moreover, we note that, although we consider that the GSW scheme (compiling the Regev's PKE) does not stand against auxiliary inputs leakage, we
  can design a symmetric-key FHE scheme against auxiliary inputs leakage from
  the symmetric variant of Regev's PKE. The important observation is that LWE
  problem itself is secure in the presence of auxiliary inputs, which is argued by
  Goldwasser, Kalai, Peikert and Vaikuntanathan [22]. We outline the symmetric-
  key FHE against auxiliary inputs leakage in Appendix A.

  We also note that our scheme against auxiliary inputs leakage is just a leveled
  FHE, meaning that we must set an *a priori* maximal circuit depth at the time
  of key generation. However, we can design an unbounded one, following Berkoff
  and Liu's novel idea, via compiling an unbounded multi-key FHE [27] and a
  leveled FHE against auxiliary inputs leakage on-the-fly. However, it is merely an
  unbounded FHE with inputs-bounded, since the number of secret keys of LTV
  is bounded by M set at the time of key generation. The construction is similar
  and thus is omitted. Constructing an unbounded FHE without input-bounded
  against auxiliary inputs leakage remains open.

1.2 Organization

We start with background and preliminary in Sect. 2. In Sect. 3, We present
formally a leveled FHE scheme from dual of Regev's PKE following approximate
eigenvector fashion and discuss concisely its basic property. We argue the new
(slightly modified) leveled FHE is CPA secure against auxiliary inputs leakage
in Sect. 4. In Sect. 5, we conclude this work.

2   Preliminaries

2.1   Notations

We will use the notations below in this work. Let boldface small letters (e.g.,
$\mathbf{x}, \mathbf{y}$) denote vectors and boldface capital letters (e.g., $\mathbf{A}, \mathbf{B}$) denote matrices.
The i-coordinate of $\mathbf{x}$ is written by x_i. We use $\langle \mathbf{x}, \mathbf{y} \rangle$ or $\mathbf{x} \cdot \mathbf{y}$ to denote the inner
product of two vectors. For any integer q, we define $\mathbb{Z}_q \triangleq (-q/2, q/2] \cap \mathbb{Z}$. For
arbitrary $y \in \mathbb{Z}$, we write $x = [y]_q$ to denote the unique value $x \in (-q/2, q/2]$
such that $x = y \pmod{q}$.

We use $s \overset{\$}{\leftarrow} \mathcal{S}$ to denote that s is drawn from a set $\mathcal{S}$ uniformly at random, and $d \leftarrow \mathcal{D}$ to denote that d is drawn from the distribution $\mathcal{D}$.

We let λ denote the *security parameter* in this paper. When we say a *negligible function* $negl(\lambda)$, it means a function that increases slower than λ^{-c} for any constant $c > 0$ and any sufficiently large value of λ. When we say an *event happens with overwhelming probability*, it means that it occurs with probability at least $1 - negl(\lambda)$ for some negligible function $negl(\lambda)$. We use $\mathcal{X} \approx_s \mathcal{Y}$ to denote *statistical indistinguishability* and $\mathcal{X} \approx_c \mathcal{Y}$ to denote *computational indistinguishability*. We write $y = \widetilde{O}_\lambda(x)$ if $y = O(x \cdot polylog(\lambda))$.

2.2 Homomorphism

We now present some definitions related to (fully) homomorphic encryption and Gentry's bootstrapping theorem adopted from [11,19]. We only consider bit-encryption scheme in this work.

Definition 1 (Homomorphic Encryption). *A (public-key) homomorphic encryption scheme* HE = (HE.KeyGen, HE.Enc, HE.Dec, HE.Eval) *is a quadruple of* PPT *algorithms as described below:*

- $(pk, evk, sk) \leftarrow$ HE.KeyGen(1^λ): *Output a public key pk, a public evaluation key evk and a secret key sk.*
- $c \leftarrow$ HE.Enc$_{pk}(b)$: *For a bit message* $b \in \{0,1\}$, *using the public key pk output a ciphertext c.*
- $b \leftarrow$ HE.Dec$_{sk}(c)$: *Using the secret key sk, decrypt a ciphertext c to a plaintext* $b \in \{0,1\}$.
- $c_f \leftarrow$ HE.Eval$_{evk}(f, c_1, c_2, \ldots, c_t)$: *Using the public evaluation key evk (and public parameters), apply an* NAND*-circuit* $f : \{0,1\}^t \rightarrow \{0,1\}$ *to* $c_1, c_2, \ldots, c_t$ *and output a ciphertext* c_f.

We remark that, in this work, the circuit f is represented by NAND-gates over $\mathbb{Z}_2$ in order to remain message small and is evaluated gate-by-gate. We also remark that any boolean circuit can be combined by NAND-gates, because NAND is a perfect set.

The notion of security we consider in Sect. 3 is CPA security, defined as follows.

Definition 2 (CPA Security). *An* HE *scheme is* CPA *secure if for any* PPT *attacker* $\mathcal{A}$ *it holds that*

$$\mathsf{ADV}_{HE,\mathcal{A}} \triangleq |\Pr[\mathcal{A}(pk, evk, \mathsf{HE.Enc}_{pk}(0)) = 1] - \Pr[\mathcal{A}(pk, evk, \mathsf{HE.Enc}_{pk}(1)) = 1]| = negl(\lambda)$$

where $(pk, evk, sk) \leftarrow$ HE.KeyGen(1^λ).

Now, we define the homomorphic properties. Remark that the definition of correctness of a scheme is implied by following homomorphic properties and hence is omitted.

Definition 3 ($\mathcal{C}$-homomorphic or Somewhat homomorphic). *Let $\mathcal{C} = \{\mathcal{C}_\lambda\}_{\lambda \in \mathbb{Z}}$ be a class of NAND-circuits. A scheme HE is $\mathcal{C}$-homomorphic (or somewhat homomorphic) if for any NAND-circuit $f \in \mathcal{C}$, and respective inputs $b_1, b_2, \ldots, b_t \in \{0, 1\}$, the following holds with overwhelming probability*

$$\Pr[\mathsf{HE.Dec}_{sk}(\mathsf{HE.Eval}_{evk}(f, c_1, c_2, \ldots, c_t)) \neq f(b_1, b_2, \ldots, b_t)] = \mathrm{negl}(\lambda),$$

where $(pk, evk, sk) \leftarrow \mathsf{HE.KeyGen}(1^\lambda)$ and $c_i \leftarrow \mathsf{HE.Enc}_{pk}(b_i)$.

Definition 4 (Compactness). *An HE scheme is compact if there exists a polynomial $g = g(\lambda)$ such that the output length of $\mathsf{HE.Eval}(\cdots)$ is bounded by $g(\lambda)$ bits (regardless of circuit f or the number of inputs).*

Definition 5 (Leveled FHE). *An HE scheme is a leveled fully homomorphic encryption if it gets an additional input 1^L in the $\mathsf{HE.KeyGen}$ algorithm, where $L = \mathrm{poly}(\lambda)$, and satisfies the definition for a compact, $\mathcal{C}$-homomorphic encryption scheme, where $\mathcal{C}$ is the set of all NAND-circuits of depth $\leq L$.*

Definition 6 (Unbounded FHE). *A homomorphic scheme HE is an unbounded (or pure) fully homomorphic encryption if it is both compact and homomorphic for the class of all circuits over $\mathbb{Z}_2$.*

Note that the main difference between leveled FHE and unbounded FHE is that the bit length of the evaluation key evk of the former is dependent on the depth L and that of the latter is not. However, the latter needs additional circular security assumption under the current state of the art.

Looking ahead, the scheme DRFHE (without key leakage) in Sect. 3 can be transformed into an unbounded FHE. We move on to define bootstrappable homomorphic encryption scheme and give bootstrapping theorem, since it is the only way to obtain an unbounded FHE as far as we know.

Definition 7 (Bootstrappable Homomorphic Encryption Scheme). *Let HE be $\mathcal{C}$-homomorphic, and let f_{NAND} be the augmented decryption function of the scheme defined as follows*

$$f_{\mathrm{NAND}}^{c_1, c_2}(sk) = \mathsf{HE.Dec}_{sk}(c_1) \ \mathrm{NAND} \ \mathsf{HE.Dec}_{sk}(c_2).$$

Then a $\mathcal{C}$-homomorphic scheme HE is bootstrappable if $f_{\mathrm{NAND}}^{c_1, c_2} \in \mathcal{C}$. Namely, the scheme enables to evaluate homomorphically its augmented decryption function $f_{\mathrm{NAND}}^{c_1, c_2}$.

Theorem 1 (Bootstrapping Theorem [19, 20]). *A bootstrappable homomorphic encryption scheme that can evaluate homomorphically its augmented decryption function and is also weakly circularly secure, can be converted into an unbounded (or pure) fully homomorphic encryption scheme.*

2.3 Vector Decomposition and Flatten

In homomorphic cryptography, we usually transform a vector into another one that makes its coefficients smaller and still maintains certain property. Our notation is mostly adapted from [6, 12, 25].

Vector Decomposition. Recall that q is an integer.

- BitDecomp$_q(\mathbf{u})$: This algorithm breaks a vector $\mathbf{u} \in \mathbb{Z}_q^n$ into its bit representation. Namely, let $u_i = \sum_{j=0}^{\lfloor \log q \rfloor} 2^j \cdot u_{i,j}$, where $u_{i,j} \in \{0,1\}$, and output a longer vector

$$(u_{1,\lfloor \log q \rfloor}, u_{1,\lfloor \log q \rfloor - 1}, \ldots, u_{1,0}, \ldots, u_{n,\lfloor \log q \rfloor}, u_{n,\lfloor \log q \rfloor - 1}, \ldots, u_{n,0}) \in \{0,1\}^{n \cdot (\lfloor \log q \rfloor + 1)}.$$

- Powersof2$_q(\mathbf{v})$: This algorithm converts a vector $\mathbf{v} \in \mathbb{Z}_q^n$ into a new one as follows

$$(2^{\lfloor \log q \rfloor} v_1, 2^{\lfloor \log q \rfloor - 1} v_1, \ldots, v_1, \ldots, 2^{\lfloor \log q \rfloor} v_n, 2^{\lfloor \log q \rfloor - 1} v_n, \ldots, v_n) \in \mathbb{Z}_q^{n \cdot (\lfloor \log q \rfloor + 1)}.$$

Lemma 1. *For all* $\mathbf{u}, \mathbf{v} \in \mathbb{Z}_q^n$, *we have*

$$\langle \mathsf{BitDecomp}_q(\mathbf{u}), \mathsf{Powersof2}_q(\mathbf{v}) \rangle = \langle \mathbf{u}, \mathbf{v} \rangle \bmod q.$$

We remark that this easily universalizes to decompositions with respect to bases other than the powers of two. For simplicity, we only describe bit decomposition case and often leave the subscript q out when there is no ambiguity in this paper.

In order to better manage noise increasing under homomorphic evaluation, Gentry, Sahai and Waters [25] proposed a flattening trick, which is very important to construct a leveled FHE without using evaluation key.

Flatten. Let $\mathbf{G} = \mathbf{g} \otimes \mathbf{I}_n \in \mathbb{Z}_q^{n \times (n \cdot (\lfloor \log q \rfloor + 1))}$, where $\mathbf{g} = (2^{\lfloor \log q \rfloor}, 2^{\lfloor \log q \rfloor - 1}, \ldots, 2, 1)$ and $\mathbf{I}_n$ denotes the n-dimensional identity matrix. We now define the algorithms BitDecomp^{-1} and Flatten.

- BitDecomp$_q^{-1}(\mathbf{u})$: For $\mathbf{u} \in \mathbb{Z}^{n \cdot (\lfloor \log q \rfloor + 1)}$, output $[\mathbf{G} \cdot \mathbf{u}]_q \in \mathbb{Z}_q^n$.
- Flatten$_q(\mathbf{u})$: For $\mathbf{u} \in \mathbb{Z}^{n \cdot (\lfloor \log q \rfloor + 1)}$, output BitDecomp$_q$(BitDecomp$_q^{-1}(\mathbf{u})$) $\in \{0,1\}^{n \cdot (\lfloor \log q \rfloor + 1)}$.

Lemma 2. *For all* $\mathbf{u} \in \mathbb{Z}_q^{n \cdot (\lfloor \log q \rfloor + 1)}, \mathbf{v} \in \mathbb{Z}_q^n$, *we have*

$$\langle \mathsf{Flatten}_q(\mathbf{u}), \mathsf{Powersof2}_q(\mathbf{v}) \rangle = \langle \mathbf{u}, \mathsf{Powersof2}_q(\mathbf{v}) \rangle \bmod q.$$

2.4 Gaussian Measures

It is well known that n-dimensional (continuous) Gaussian distribution can be expressed as the sum of n orthogonal 1-dimensional ones. So we only consider 1-dimensional Gaussian distribution as well as 1-dimensional discrete Gaussian distribution over the integers in this paper.

For any x and any $s > 0$, let

$$\rho_s(x) = e^{-\pi|x|^2/s^2}$$

be a Gaussian function scaled by s. Therefore, we can define continuous Gaussian distribution $\mathcal{D}_s$ is the distribution with probability density function proportional to ρ_s, and discrete Gaussian distribution $\mathcal{D}_{\mathbb{Z}+c,s}$ is the distribution supported on $\mathbb{Z} + c$ for some $c \in \mathbb{R}$, whose probability mass function is proportional to ρ_s.

The following lemma is proved by Brakerski et $al.$ [9]. It means that there exists an efficient PPT algorithm that samples exactly obeyed any (enough wide) discrete Gaussian distribution, in contrast to [24,32] where we only can sample within negligible statistical distance of a discrete Gaussian distribution.

Lemma 3. *For $c \in \mathbb{R}$, there exists a PPT algorithm $\lfloor\cdot\rceil_G$ that outputs a sample distributed according to $\mathcal{D}_{\mathbb{Z}+c,1}$.*

Now, we bound the absolute value of a sum of discrete Gaussian distribution in following lemmata, which can be found in [12,15].

Lemma 4. *If $u \leftarrow \mathcal{D}_s$, then with overwhelming probability, $|u| < s \cdot \omega(\sqrt{\log \lambda})$. Similarly, if $u \leftarrow \mathcal{D}_{\mathbb{Z}+c,s}$, then with overwhelming probability, $|u| < \max\{s, \omega(\sqrt{\log \lambda})\} \cdot \omega(\sqrt{\log \lambda}) = \tilde{O}_\lambda(s)$.*

Lemma 5. *Let $\mathbf{r} \in \{0,1\}^n$, $\mathbf{c} \in \mathbb{R}^n$ be arbitrary and $\mathbf{e} \leftarrow \mathcal{D}_{\mathbb{Z}+c,s}$. Then with overwhelming probability*

$$|\langle \mathbf{r}, \mathbf{e} \rangle| \leq \sqrt{n} \cdot \max\{s, \omega(\sqrt{\log \lambda})\} \cdot \omega(\sqrt{\log \lambda}) = \tilde{O}_\lambda(\sqrt{n} \cdot s).$$

Lemma 6. *Let $\mathbf{c} \in \mathbb{R}^n$ be arbitrary and $\mathbf{e} \leftarrow \mathcal{D}_{\mathbb{Z}|c,s}$, and let $\mathbf{r} \in \{0,1\}^n$ be dependent on $\mathbf{e}$. Then with overwhelming probability*

$$|\langle \mathbf{r}, \mathbf{c} \rangle| \leq n \cdot \max\{s, \omega(\sqrt{\log \lambda})\} \cdot \omega(\sqrt{\log \lambda}) = \tilde{O}_\lambda(n \cdot s).$$

We also need lemmata below to argue the scheme in Sect. 4 against auxiliary inputs leakage.

Lemma 7. *Let $\beta > 0$ and $q \in \mathbb{Z}$. Let $\mathbf{x} \in \mathbb{Z}^n$ and $\mathbf{y} \leftarrow \mathcal{D}^n_{\mathbb{Z},\beta q}$. Then $|\mathbf{x} \cdot \mathbf{y}| \leq \|\mathbf{x}\|_2 \cdot \beta q \cdot \omega(\sqrt{\log n})$ with overwhelming probability over the sample of $\mathbf{y}$.*

Lemma 8. *Let $\beta > 0$, $q \in \mathbb{Z}$ and $x \in \mathbb{Z}$. The statistical distance between the distribution $\mathcal{D}_{\mathbb{Z},\beta q}$ and $\mathcal{D}_{\mathbb{Z},\beta q} + x$ is at most $|x|/\beta q$.*

2.5   Learning with Errors (LWE)

The LWE problem was proposed by Regev [34] as a extension of "learning noisy parities". In this paper, we will mainly define the decisional learning with errors (DLWE) and give the relation to intractability of worst-case lattice problems.

Given positive integers $n = n(\lambda)$ and $q = q(\lambda) \geq 2$, a vector $\mathbf{s} \in \mathbb{Z}_q^n$, and a probability distribution χ over $\mathbb{Z}$, let $A_{\mathbf{s},\chi}$ be the distribution obtained by choosing uniformly a vector $\mathbf{a} \xleftarrow{\$} \mathbb{Z}_q^n$ and a noisy term $e \leftarrow \chi$, and outputting $(\mathbf{a}, [\langle \mathbf{a}, \mathbf{s} \rangle + e]_q) \in \mathbb{Z}_q^n \times \mathbb{Z}_q$. Now we define DLWE as follows.

Definition 8 (DLWE). *For n, q, χ defined above and an integer m, the DLWE$_{n,m,q,\chi}$ problem is, given m independent samples, to decide that, with non-negligible advantage, they are sampling from $A_{s,\chi}$ for a uniformly random and secret $s \in \mathbb{Z}_q^n$, or from the uniform distribution over $\mathbb{Z}_q^n \times \mathbb{Z}_q$. If there is not a priori bounded in the number of samples, we write DLWE$_{n,q,\chi}$ to denote the variant where the adversary can get an oracle access to $A_{s,\chi}$.*

For lattice dimension n and a number d, GapSVP$_\gamma$ is the promise problem of distinguishing whether a n-dimensional lattice has a vector shorter than d or no vector shorter than $\gamma \cdot d$. SIVP$_\gamma$ is the search problem of finding a set of "short" vectors with approximate factor γ.

The following two theorems state quantum and classical reductions from GapSVP or SIVP to DLWE$_{n,q,\chi}$, where $\chi = \mathcal{D}_{\mathbb{Z},\alpha q}$ for $\alpha \in (0,1)$ (Here we often write DLWE$_{n,q,\alpha}$ to denote DLWE$_{n,q,\chi}$).

Theorem 2 ([12,28,31,34]). *Let $q = q(n) \in \mathbb{N}$ be either a prime power or a product of co-prime powers of primes and $\alpha q \geq 2\sqrt{n}$. If there exists an efficient algorithm solving DLWE$_{n,q,\alpha}$ problem, we then have that,*

- *there is an efficient quantum algorithm that solves GapSVP$_{\widetilde{O}(n/\alpha)}$ and SIVP$_{\widetilde{O}(n/\alpha)}$ on any n-dimensional lattice.*
- *if $q \geq \widetilde{O}(2^{n/2})$, then there is an efficiently classical algorithm for GapSVP$_{\widetilde{O}(n/\alpha)}$ on any n-dimensional lattice.*

Theorem 3 ([9]). *Solving n-dimensional DLWE with poly(n) modulus implies an equally efficient algorithm to GapSVP or SIVP in dimension $\sqrt{n}$.*

2.6 Goldreich-Levin Theorem

Our leveled FHE with auxiliary inputs is heavily relied on the following extended variant of Goldreich-Levin theorem, which was given by Dodis *et al.* [15].

Theorem 4 ([15]). *Let S be a subset of $GF(q)$ for a prime q and $f : S^m \rightarrow \{0,1\}^*$ be an arbitrary function. For all $s \xleftarrow{\$} S^m, r \xleftarrow{\$} GF(q)^m, u \xleftarrow{\$} GF(q)$, if a distinguisher $\mathcal{D}$ runs in time T such that*

$$|\Pr[\mathcal{D}(r, \langle r, s \rangle, f(s)) = 1] - \Pr[\mathcal{D}(r, u, f(s)) = 1]| = \delta,$$

there then exists an algorithm $\mathcal{A}$ that runs in time $T' = T \cdot \text{poly}(m, |S|, 1/\delta)$ such that

$$\Pr[\mathcal{A}(f(s)) = s] \geq \frac{\delta^3}{512 \cdot m \cdot q^2}.$$

2.7 The Auxiliary Inputs Leakage Model for FHE

Definition 9. *A fully homomorphic encryption scheme FHE=(FHE.KeyGen, FHE.Enc, FHE.Dec, FHE.Eval) with plaintext space $\mathcal{M} = \{\mathcal{M}_n\}_{n \in \mathbb{N}}$ is CPA* *secure w.r.t. auxiliary inputs leakage from some family of efficiently computable functions $\mathcal{H}$ if for any PPT algorithm $\mathcal{A}$, any function $h \in \mathcal{H}$, and any sufficiently large $n \in \mathbb{N}$, it holds that*

$$\mathsf{ADV}_{\mathsf{FHE},\mathcal{A},h} \triangleq |\Pr[\mathsf{AIL}_0(\mathsf{FHE},\mathcal{A},n,h) = 1] - \Pr[\mathsf{AIL}_1(\mathsf{FHE},\mathcal{A},n,h) = 1]| = \mathrm{negl}(\lambda)$$

where $\mathsf{AIL}_b(\mathsf{FHE},\mathcal{A},n,h)$ is the output of the following game:

1. Setup. *The challenger generates $(sk, pk) \leftarrow \mathsf{FHE.keyGen}(1^\lambda)$ and sends pk to the adversary.*
2. AuxiliaryInputsLeakage. *The adversary $\mathcal{A}$ chooses a function $h \in \mathcal{H}$ and sends it to the challenger. The challenger replies with $h(pk, sk)$.*
3. Challenge. *The adversary sends m_0, m_1 to the challenger. The challenger selects $b \xleftarrow{\$} \{0,1\}$, computes $c \leftarrow \mathsf{FHE.Enc}_{pk}(m_b)$ and sends it to $\mathcal{A}$.*
4. Output. *The adversary outputs a guess $b' \in \{0,1\}$.*

In this paper, we will consider two classes of admissible functions $\mathcal{H}$ defined in [15]. Let $k = |sk|$ (bit length of the secret key). For $f(k) \geq 2^{-k}$, we can define $f(k)$-hard-to-invert auxiliary inputs function family as follows.

1. $\mathcal{H}_{un}(f(k)) = \{h : \{0,1\}^{|pk|+|sk|} \rightarrow \{0,1\}^* \mid \text{for all PPT } \mathcal{A}, \; \Pr[A(h(pk,sk)) = sk] \leq f(k)\}$
2. $\mathcal{H}_{pk-un}(f(k)) = \{h : \{0,1\}^{|pk|+|sk|} \rightarrow \{0,1\}^* \mid \text{for all PPT } \mathcal{A}, \; \Pr[A(pk,h(pk,sk)) = sk] \leq f(k)\}$

Remark that the former family $\mathcal{H}_{un}(f(k))$ is the class of polynomial-time computable functions h, such that given $h(pk, sk)$, no PPT adversary can find sk with probability better than $f(k)$, while the latter is the function family such that given $(pk, h(pk, sk))$, no PPT adversary can find sk with probability better than $f(k)$.

An FHE scheme is called $f(k)$-AI-CPA (auxiliary inputs CPA) secure, if it is CPA secure w.r.t. $\mathcal{H}_{un}(f(k))$. Similarly, An FHE scheme is called $f(k)$-wAI-CPA (weak auxiliary inputs CPA) secure, if it is CPA secure w.r.t. $\mathcal{H}_{pk-un}(f(k))$.

The following lemma shows that if public key is short, then we can first prove that an FHE is wAI-CPA secure and then obtain an FHE is AI-CPA secure with smaller function family.

Lemma 9 ([15]). *Let $t(k) = |pk|$ for a scheme FHE. If FHE is $f(k)$-wAI-CPA secure, then it is $(2^{-t(k)}f(k))$-AI-CPA secure.*

3   An FHE from DRPKE

First, in Sect. 3.1, we describe a leveled FHE (called DRFHE) derived from DRPKE, using Gentry-Sahai-Waters' *approximate eigenvector* method [25]. We then show concisely the correctness, security and homomorphism of DRFHE in Sect. 3.2. In next section, we will show that a variant of DRFHE is secure against sub-exponentially hard-to-invert auxiliary input functions, assuming the hardness of the LWE problem.

3.1 The Scheme: DRFHE

- DRFHE.Setup($1^\lambda, 1^L$): Select LWE parameters $q = q(\lambda, L)$, $n = n(\lambda, L)$, and $\chi = \mathcal{D}_{\mathbb{Z}, \alpha q}$ such that $\mathrm{LWE}_{n,q,\chi}$ attains at least 2^λ security, where λ is the security parameter, L is the maximum depth of circuits and χ is a discrete Gaussian distribution from which the noises are sampled. Select $m = O(n \log q)$ and let $prms = (n, m, q, \chi)$, $\ell = \lfloor \log q \rfloor + 1$ and $N = (m+1) \cdot \ell$.
- DRFHE.SKGen($prms$): Sample a vector $\mathbf{t} \xleftarrow{\$} \{0,1\}^m$. Let $sk = \mathbf{s} = (1, -\mathbf{t})$. Let $\mathbf{v} = \mathsf{Powersof2}(\mathbf{s})$.
- DRFHE.PKGen($prms, sk$): Choose a matrix $\mathbf{A} \xleftarrow{\$} \mathbb{Z}_q^{n \times m}$. Compute $\mathbf{u} = \mathbf{At}$. Set the public key $pk = \mathbf{P} = (\mathbf{u}, \mathbf{A})$. Note that $\mathbf{Ps} = \mathbf{0}$.
- DRFHE.Enc(pk, b): To encrypt a bit b, choose two matrices $\mathbf{R} \xleftarrow{\$} \mathbb{Z}_q^{N \times n}$ and $\mathbf{E} \leftarrow \chi^{N \times (m+1)}$ and output the ciphertext matrix

$$\mathbf{C} = \mathsf{Flatten}(b \cdot \mathbf{I}_N + \mathsf{BitDecomp}(\mathbf{R} \cdot \mathbf{P} + \mathbf{E})) \in \{0,1\}^{N \times N}.$$

- DRFHE.DecC($\mathbf{C}, sk$): Let $\mathbf{c}$ be the second row of $\mathbf{C}$. Output $b' = \lfloor [\langle \mathbf{c}, \mathbf{v} \rangle]_q \rfloor_2$, where the rounding function $\lfloor \cdot \rceil_2 : \mathbb{Z}_q \to \{0,1\}$ means that it outputs 0 if its argument is closer to 0 than to $2^{\ell-2}$ modulo q, otherwise outputs 1.
- DRFHE.NAND($\mathbf{C}_1, \mathbf{C}_2$): Given two ciphertext matrices $\mathbf{C}_1, \mathbf{C}_2$ for two plaintexts b_1, b_2, respectively, output $\mathsf{Flatten}\mathbf{I}_N - \mathbf{C}_1 \cdot \mathbf{C}_2)$.
- DRFHE.Eval($f, \mathbf{C}_1, \mathbf{C}_2, \ldots, \mathbf{C}_t$): apply an NAND-circuit $f : \{0,1\}^t \to \{0,1\}$ to t ciphertexts $\mathbf{C}_1, \mathbf{C}_2, \ldots, \mathbf{C}_t$, and output a ciphertext $\mathbf{C}_f$.
  Recall that the evaluator should execute any NAND-circuit gate-by-gate.

Remark 3.1. As the plaintexts involve in the growth of noises when evaluating homomorphically a circuit, we primarily consider the circuit combined by NANDs for controlling the plaintext small (the ciphertexts also involve in the growth of noises and Flatten insures that the entries of ciphertexts are small). Additionally, DRFHE also enable to evaluate circuits composed of additions and multiplications as Gentry-Sahai-Waters demonstrated in [25] (using an extended bit-by-bit decryption procedure for large plaintexts).

3.2 Analysis

Security. In order to show the semantic security of DRFHE under the average-case DLWE assumption (further under the worst-case GapSVP or SIVP assumption by theorem 2, 3), it suffices to show that $(\mathbf{P}, \mathbf{C}) \approx_c (\mathbf{U}, \mathbf{V})$, where $\mathbf{U}, \mathbf{V}$ are sampled uniformly at random. As Flatten and BitDecomp are deterministic algorithms, it is sufficient to prove that $(\mathbf{P}, \mathbf{RP} + \mathbf{E}) \approx_c (\mathbf{U}, \mathbf{V})$. It holds following the arguments in [24,25] and thus the details are omitted.

Correctness. Now, we estimate the level of noise in fresh ciphertext which is related with the decryption correctness of DRFHE.

For arbitrarily fresh ciphertext $\mathbf{C}$ encrypting b under public-key $\mathbf{P}$, by lemma 1, 2, it is very easy to see that

$$\mathbf{C} \cdot \mathbf{v} = b \cdot \mathbf{v} + \mathbf{RPs} + \mathbf{Es} = b \cdot \mathbf{v} + \mathbf{Es} \quad (\bmod q)$$

thus, $\mathbf{c}_2 \cdot \mathbf{v} = b \cdot v_2 + \mathbf{e}_2 \cdot \mathbf{s}$. If χ is $\tilde{O}_\lambda(\alpha q)$-bounded (it holds with overwhelming probability by lemma 4), we have $|\mathbf{e}_2 \cdot \mathbf{s}| \leq \tilde{O}_\lambda((\sqrt{m}+1) \cdot \tilde{O}_\lambda(\alpha q)) = (\sqrt{m}+1) \cdot \tilde{O}_\lambda(\alpha q)$ by lemma 5, 6. Since $v_2 = 2^{\ell-2} \geq q/4$, provided $(\sqrt{m}+1) \cdot \tilde{O}_\lambda(\alpha q) < q/8$, the correctness of decryption follows.

This results in the following definition about noise magnitude in any ciphertext $\mathbf{C}$.

Definition 10. *We define that* $\mathsf{Noise}_b = ||\mathbf{e}||_\infty$ *if* $\mathbf{e} \in \mathbb{Z}^N$ *is the noise vector in a ciphertext* $\mathbf{C}$ *encrypting* b *under public-key* $\mathbf{P}$ *such that* $\mathbf{C} \cdot \mathbf{v} = b \cdot \mathbf{v} + \mathbf{e}$ (mod q).

Homomorphism. Below, we argue that DRFHE can decrypt correctly after operating DRFHE.NAND and DRFHE.Eval for proper parameters.

Lemma 10. *For* $i = 1, 2$, *let* $b_i \in \{0, 1\}$, $\mathbf{C}_i \in \{0, 1\}^{N \times N}$ *and* $\mathbf{E}_i \in \mathbb{Z}^{N \times (m+1)}$ *such that* $\mathbf{C}_i \mathbf{v} = b_i \mathbf{v} + \mathbf{E}_i \mathbf{s}$ (mod q). *we then have with overwhelming probability*

$$\mathsf{Noise}_{b_1 \mathsf{NAND} b_2}(\mathsf{DRFHE.NAND}(\mathbf{C}_1, \mathbf{C}_2)) \leq \mathsf{Noise}_{b_1}(\mathbf{C}_1) + N \cdot \mathsf{Noise}_{b_2}(\mathbf{C}_2).$$

In other words, DRFHE.Dec *will be correct if* $(N+1)(\sqrt{m}+1) \cdot \tilde{O}_\lambda(\alpha q) < q/8$ *for fresh* $\mathbf{C}_i$.

Proof. Let $\mathbf{C}_{\mathsf{NAND}} = \mathsf{DRFHE.NAND}(\mathbf{C}_1, \mathbf{C}_2)$. Then we gain

$$\begin{aligned}
\mathbf{C}_{\mathsf{NAND}} \cdot \mathbf{v} &= \mathsf{Flatten}(\mathbf{I}_N - \mathbf{C}_1 \cdot \mathbf{C}_2) \cdot \mathbf{v} \\
&= (\mathbf{I}_N - \mathbf{C}_1 \cdot \mathbf{C}_2) \cdot \mathbf{v} \\
&= \mathbf{v} - \mathbf{C}_1 \cdot (\mathbf{C}_2 \cdot \mathbf{v}) \\
&= \mathbf{v} - \mathbf{C}_1 \cdot (b_2 \cdot \mathbf{v} + \mathbf{E}_2 \cdot \mathbf{s}) \\
&= \mathbf{v} - b_2 \cdot (b_1 \cdot \mathbf{v} + \mathbf{E}_1 \cdot \mathbf{s}) - \mathbf{C}_1 \cdot \mathbf{E}_2 \cdot \mathbf{s} \\
&= (1 - b_1 b_2) \cdot \mathbf{v} - (b_2 \cdot \mathbf{E}_1 \cdot \mathbf{s} + \mathbf{C}_1 \cdot \mathbf{E}_2 \cdot \mathbf{s})
\end{aligned}$$

Thus, by $b_2 \in \{0, 1\}$, $\mathbf{C}_1 \in \{0, 1\}^{N \times N}$ and the definition of Noise_b, we have

$$\mathsf{Noise}_{b_1 \mathsf{NAND} b_2}(\mathbf{C}_{\mathsf{NAND}}) \leq \mathsf{Noise}_{b_1}(\mathbf{C}_1) + N \cdot \mathsf{Noise}_{b_2}(\mathbf{C}_2)$$

which finishes the proof. □

By successively using lemma 10, it is much easier to get the below lemma, which states the scheme DRFHE can homomorphically evaluate a depth-L circuit of NANDs.

Theorem 5. *Let* n, q, χ *be the* LWE *parameters and* $b_i \in \{0, 1\}$, $\mathbf{C}_i \in \{0, 1\}^{N \times N}$ *and* $\mathbf{E}_i \in \mathbb{Z}^{N \times (m+1)}$ *such that* $\mathbf{C}_i \mathbf{v} = b_i \mathbf{v} + \mathbf{E}_i \mathbf{s}$ (mod q), $i \in [t] = \{1, 2, \ldots, t\}$. *For every depth-$L$ circuit of* NANDs f, *let* $\mathbf{C}_f \leftarrow \mathsf{DRFHE.Eval}(f, \mathbf{C}_1, \mathbf{C}_2, \ldots, \mathbf{C}_t)$. *If* $(1 + N)^L (\sqrt{m}+1) \cdot \tilde{O}_\lambda(\alpha q) < q/8$, *we then have*

$$\mathsf{DRFHE.Dec}(\mathbf{C}_f, sk) = f(b_1, b_2, \ldots, b_t)$$

with overwhelming probability over the randomness of all algorithms involved.

Remark 3.2. As we said in the introduction, the efficiency of DRFHE is moderately lower than GSW. The main reason is that the ciphertext size increases roughly by a factor $O(\log^2 q)$ (The ciphertext size of GSW is $(n+1)^2 \log^2 q$, while that of DRFHE is $(O(n \log q) + 1)^2 \log^2 q$). However, both DRFHE and GSW has quasi-additivity of behavior of noises under a sequence of asymmetric homomorphic multiplications. So, we can use bootstrapping theorems of Brakerski and Vaikuntanathan [12] or Alperin-Sheriff and Peikert [4] to bootstrap and gain an unbounded FHE with polynomial errors.

4   An FHE with Auxiliary Inputs

In this section, we present the leveled fully homomorphic encryption with auxiliary inputs (called DRFHEAI) by adjusting the parameters of DRFHE and show that it is AI-CPA secure under LWE assumption.

4.1   The Scheme: DRFHEAI

We only describe DRFHEAI.Setup and DRFHEAI.Enc because other algorithms are all as same as those in DRFHE.

- DRFHEAI.Setup($1^\lambda, 1^L$): Let the integer $n = n(\lambda, L)$, the prime $q \in (2^{n^\epsilon}, 2 \cdot 2^{n^\epsilon}]$ and the integer $m = ((n+3) \log q)^{1/\varepsilon}$ where $\epsilon, \varepsilon \in (0,1)$. Let $\ell = \lfloor \log q \rfloor + 1$ and $N = (m+1) \cdot \ell$. Let $\alpha = 2\sqrt{n}/q$ and $\beta = 1/(8\sqrt{nq})$.
- DRFHEAI.Enc(pk, b): To encrypt a bit b, choose two matrices $\mathbf{R} \xleftarrow{\$} \mathbb{Z}_q^{N \times n}$ and $\mathbf{E} = (\mathbf{e}, \mathbf{E}')$, where $\mathbf{e} \leftarrow \mathcal{D}_{\mathbb{Z}, \alpha q}^N$ and $\mathbf{E}' \leftarrow \mathcal{D}_{\mathbb{Z}, \beta q}^{N \times m}$. Output the ciphertext matrix

$$\mathbf{C} = \mathsf{Flatten}(b \cdot \mathbf{I}_N + \mathsf{BitDecomp}(\mathbf{R} \cdot \mathbf{P} + \mathbf{E})) \in \{0,1\}^{N \times N}.$$

Remark 4.1. The main differences between DRFHE and DRFHEAI are twofold: larger m and two encryption noises in the latter scheme. Both of them are set to show CPA security in the presence of auxiliary inputs leakage. Since the ciphertext sizes of both schemes are related with m, the ciphertext size of DRFHEAI increases roughly by a factor $O(n \log q)^{2/\varepsilon - 2}$ comparing to DRFHE.

Correctness and Homomorphic Property. In this section, we only show succinctly the correctness and the homomorphic property of DRFHEAI. The security will be proved in next section.

Since encryption noises $\mathbf{E} = (\mathbf{e}, \mathbf{E}')$ where $\mathbf{e} \leftarrow \mathcal{D}_{\mathbb{Z}, \alpha q}^N$ and $\mathbf{E}' \leftarrow \mathcal{D}_{\mathbb{Z}, \beta q}^{N \times m}$, for a fresh ciphertext $\mathbf{C}$ encrypting b, $\mathsf{Noise}_b(\mathbf{C}) \leq \widetilde{O}_\lambda(\alpha q) + \widetilde{O}_\lambda(\sqrt{m} \cdot \beta q) = \widetilde{O}_\lambda(\alpha + \sqrt{m}\beta) \cdot q$ by lemma 4, 5, 6. So, DRFHEAI.Dec can decrypt correctly for $\widetilde{O}_\lambda(\alpha + \sqrt{m}\beta) < 1/8$ with overwhelming probability.

Furthermore, after evaluating a depth-L circuit of NANDs, the noise magnitude will increase to $\widetilde{O}_\lambda(\alpha + \sqrt{m}\beta) \cdot q \cdot (N+1)^L$. Thus, DRFHEAI.Dec can decrypt correctly for $\widetilde{O}_\lambda(\alpha + \sqrt{m}\beta) \cdot (N+1)^L < 1/8$ with overwhelming probability.

4.2   DRFHEAI Against Auxiliary Inputs Leakage

In this section, we show that the scheme DRFHEAI is secure against sub-exponentially hard-to-invert auxiliary inputs, assuming the hardness of DLWE problem. In order to show it, we first prove the following lemma.

Lemma 11. *Let* $\mathbf{A} \overset{\$}{\leftarrow} \mathbb{Z}_q^{n \times m}$, $\mathbf{t} \overset{\$}{\leftarrow} \{0,1\}^m$, $\mathbf{r} \overset{\$}{\leftarrow} \mathbb{Z}_q^n$, $\mathbf{u} = \mathbf{At}$, $e \leftarrow \mathcal{D}_{\mathbb{Z},\alpha q}$, $e' \leftarrow \mathcal{D}_{\mathbb{Z},\beta q}^m$, $v \overset{\$}{\leftarrow} \mathbb{Z}_q$, $\mathbf{v}' \overset{\$}{\leftarrow} \mathbb{Z}_q^m$, *and* n, m, q, α, β *defined as in our* DRFHEAI *scheme. If the* $\mathrm{DLWE}_{n,m,q,\beta}$ *problem is hard, and for any fixed auxiliary-input function* h, *finding* $\mathbf{t}$ *is yet* $(q^n 2^{-m^\varepsilon})$-*hard given* $(\mathbf{A}, \mathbf{u}, h(\mathbf{A}, \mathbf{t}))$, *then,*

$$\mathcal{G}_{\mathrm{real}} \overset{\triangle}{=} (\mathbf{A}, \mathbf{u}, \mathbf{r} \cdot \mathbf{u} + e, \mathbf{r}\mathbf{A} + e', h(\mathbf{A}, \mathbf{t})) \approx_c \mathcal{G}_{\mathrm{ideal}} \overset{\triangle}{=} (\mathbf{A}, \mathbf{u}, v, \mathbf{v}', h(\mathbf{A}, \mathbf{t})).$$

Proof. We define a sequence of in-between hybrid games $\mathcal{G}_a, \mathcal{G}_b, \mathcal{G}_c$ as follows:

- $\mathcal{G}_a \overset{\triangle}{=} (\mathbf{A}, \mathbf{u}, (\mathbf{r}\mathbf{A} + e') \cdot \mathbf{t} + e, \mathbf{r}\mathbf{A} + e', h(\mathbf{A}, \mathbf{t}))$, where $\mathbf{t} \overset{\$}{\leftarrow} \{0,1\}^m$.
- $\mathcal{G}_b \overset{\triangle}{=} (\mathbf{A}, \mathbf{u}, \mathbf{v}' \cdot \mathbf{t} + e, \mathbf{v}', h(\mathbf{A}, \mathbf{t}))$, where $\mathbf{v}' \overset{\$}{\leftarrow} \mathbb{Z}_q^m$.
- $\mathcal{G}_c \overset{\triangle}{=} (\mathbf{A}, \mathbf{u}, v + e, \mathbf{v}', h(\mathbf{A}, \mathbf{t}))$, where $v \overset{\$}{\leftarrow} \mathbb{Z}_q$.

To show lemma 11, it suffices to show $\mathcal{G}_{\mathrm{real}} \approx_s \mathcal{G}_a \approx_c \mathcal{G}_b \approx_c \mathcal{G}_c \approx_c \mathcal{G}_{\mathrm{ideal}}$. It follows from several claims below.

Claim 1. $\mathcal{G}_{\mathrm{real}} \approx_s \mathcal{G}_a$.

Proof. The only difference between games $\mathcal{G}_{\mathrm{real}}$ and $\mathcal{G}_a$ is that $((\mathbf{r}\mathbf{A} + e') \cdot \mathbf{t} + e)$ takes the place of $(\mathbf{r} \cdot \mathbf{u} + e)$, where $\mathbf{t} \overset{\$}{\leftarrow} \{0,1\}^m$, $e \leftarrow \mathcal{D}_{\mathbb{Z},\alpha q}$ and $e' \leftarrow \mathcal{D}_{\mathbb{Z},\beta q}^m$. Note that $(\mathbf{r}\mathbf{A} + e') \cdot \mathbf{t} + e = \mathbf{r} \cdot \mathbf{u} + e' \cdot \mathbf{t} + e$. So we only need to show that the distribution of $e' \cdot \mathbf{t} + e$ is statistically indistinguishable from $\mathcal{D}_{\mathbb{Z},\alpha q}$. Since

$$e' \cdot \mathbf{t}/(\alpha q) \le \|e'\|_2 \cdot \|\mathbf{t}\|_2/(\alpha q) \le \sqrt{m} \cdot \beta q \cdot \omega(\sqrt{\log n})/(\alpha q) \le 16 \cdot \sqrt{m} \cdot n \cdot \omega(\sqrt{\log n})/\sqrt{q},$$

by lemma 7, is negligible, we gain by lemma 8 that above two distributions are indistinguishable. The claim follows.   □

Claim 2. $\mathcal{G}_a \approx_c \mathcal{G}_b$.

Proof. The only difference between games $\mathcal{G}_a$ and $\mathcal{G}_b$ is that $(\mathbf{r}\mathbf{A} + e')$ is replaced by $\mathbf{v}'$, where $\mathbf{v}' \overset{\$}{\leftarrow} \mathbb{Z}_q^m$. By assumption that the $\mathrm{DLWE}_{n,m,q,\beta}$ problem is hard, the advantage of distinguishing above two distributions is a negligible function of n.   □

Claim 3. $\mathcal{G}_b \approx_c \mathcal{G}_c$.

Proof. The only difference between games $\mathcal{G}_b$ and $\mathcal{G}_c$ is that $(v + e)$ substitutes $(\mathbf{v}' \cdot \mathbf{t} + e)$, where $v \overset{\$}{\leftarrow} \mathbb{Z}_q$. Since e only appears in this place, it is sufficient to show that the distributions below are indistinguishable

$$\mathcal{G}_b' \overset{\triangle}{=} (\mathbf{A}, \mathbf{u}, \mathbf{v}' \cdot \mathbf{t}, \mathbf{v}', h(\mathbf{A}, \mathbf{t})) \approx_c \mathcal{G}_c' \overset{\triangle}{=} (\mathbf{A}, \mathbf{u}, v, \mathbf{v}', h(\mathbf{A}, \mathbf{t})).$$

This implies that we only need to reduce the work of inverting h to the work of obtaining a non-negligible distinguishing advantage between $\mathcal{G}'_b$ and $\mathcal{G}'_c$. For the sake of contradiction, assume that the distinguishing advantage between $\mathcal{G}'_b$ and $\mathcal{G}'_c$ is $\delta(n)$, which is non-negligible for infinitely many n's.

Since $q^{n+3} = 2^{-m^\varepsilon}$ and $512 \cdot m/(\delta^3 \cdot q) < 1$ for large enough n, by theorem 4 (Goldreich-Levin theorem), there exists an algorithm that, given $pk = (\mathbf{A}, \mathbf{u})$, inverts $h(\mathbf{A}, \mathbf{t})$ with probability more than

$$\frac{\delta^3}{512 \cdot m \cdot q^2} = q^n \cdot \frac{\delta^3 \cdot q}{512 \cdot m \cdot q^{n+3}} > q^n \cdot 2^{-m^\varepsilon}.$$

The contradiction follows. $\square$

Claim 4. $\mathcal{G}_c \approx_c \mathcal{G}_{\text{ideal}}$.

This claim is clear from $v \xleftarrow{\$} \mathbb{Z}_q$ and $e \leftarrow \mathcal{D}_{\mathbb{Z},\alpha q}$.

It thus holds that $\mathcal{G}_{\text{real}} \approx_s \mathcal{G}_a \approx_c \mathcal{G}_b \approx_c \mathcal{G}_c \approx_c \mathcal{G}_{\text{ideal}}$, which finishes the proof. $\square$

Now, we show our main theorem.

Theorem 6. *Let the parameters n, m, q, α, β be the same as in the leveled fully homomorphic encryption scheme DRFHEAI presented above. If the $\text{DLWE}_{n,m,q,\beta}$ problem is hard, then DRFHEAI is 2^{-m^ε}-AI-CPA secure (when $\mathbf{A}$ is a common system parameter).*

Proof. Note that the length of "user-specific" public-key $\mathbf{u}$ is $n \log q$ bits, by lemma 9, it is sufficient to show that DRFHEAI is $(q^n 2^{-m^\varepsilon})$-wAI-CPA secure. For any fixed auxiliary-input function h, finding $\mathbf{t}$ is yet $(q^n 2^{-m^\varepsilon})$-hard given $(\mathbf{A}, \mathbf{u}, h(\mathbf{A}, \mathbf{t}))$. We consider a PPT attacker $\mathcal{A}$ with advantage $\delta = \delta(n) = \text{ADV}_{\mathcal{A},h}(n)$ at playing the AIL game.

Recall that the attacker's view is $(\mathbf{P}, \mathbf{C}_b, h(\mathbf{P}, \mathbf{s}))$, where $\mathbf{C}_b$ is an encryption of message bit $b \in \{0, 1\}$. Let $\mathbf{C}'_b = \text{BitDecomp}^{-1}(\mathbf{C}_b) = \text{BitDecomp}^{-1}(b \cdot \mathbf{I}_N) + \mathbf{RP} + \mathbf{E}$. It is sufficient to consider a PPT attacker who plays the AIL game with $\mathbf{C}'_b$, because the algorithm BitDecomp^{-1} is deterministic. Then, a PPT attacker's view is actually $(\mathbf{P}, \text{BitDecomp}^{-1}(b \cdot \mathbf{I}_N) + \mathbf{RP} + \mathbf{E}, h(\mathbf{P}, \mathbf{s}))$. Therefore, it suffices to show $(\mathbf{P}, \mathbf{RP} + \mathbf{E}, h(\mathbf{P}, \mathbf{s})) \approx_c (\mathbf{P}, \mathbf{V} \xleftarrow{\$} \mathbb{Z}_q^{N \times (m+1)}, h(\mathbf{P}, \mathbf{s}))$.

Recall that $\mathbf{P} = (\mathbf{u}, \mathbf{A})$ where $\mathbf{A} \xleftarrow{\$} \mathbb{Z}_q^{n \times m}$, $\mathbf{t} \xleftarrow{\$} \{0, 1\}^m$, $\mathbf{u} = \mathbf{At}$, $\mathbf{E} = (\mathbf{e}, \mathbf{E}')$ where $\mathbf{e} \leftarrow \mathcal{D}_{\mathbb{Z},\alpha q}^N$ and $\mathbf{E}' \leftarrow \mathcal{D}_{\mathbb{Z},\beta q}^{N \times m}$, and $\mathbf{s} = (1, -\mathbf{t})$. So, for $\mathbf{v} \xleftarrow{\$} \mathbb{Z}_q^N$ and $\mathbf{V} \xleftarrow{\$} \mathbb{Z}_q^{N \times (m+1)}$, we define

$$\mathcal{G}_{\text{AIL}} = (\mathbf{A}, \mathbf{u}, \mathbf{Ru} + \mathbf{e}, \mathbf{RA} + \mathbf{E}', h(\mathbf{A}, \mathbf{t})), \mathcal{G}_{\text{IDEAL}} = (\mathbf{A}, \mathbf{u}, \mathbf{v}, \mathbf{V}, h(\mathbf{A}, \mathbf{t})).$$

We will go to show that $\mathcal{G}_{\text{AIL}} \approx_c \mathcal{G}_{\text{IDEAL}}$. Recall that random matrix $\mathbf{R} \xleftarrow{\$} \mathbb{Z}_q^{N \times n}$, we can view it as an assemblage of N independent uniformly random n-vectors $\mathbf{r}_i \xleftarrow{\$} \mathbb{Z}_q^n$. Hence,

$$\mathcal{G}_{\text{AIL}} = (\mathbf{A}, \mathbf{u}, \{\mathbf{r}_i \cdot \mathbf{u} + e_i\}_{i \in [N]}, \{\mathbf{r}_i \cdot \mathbf{A} + \mathbf{e}'_i\}_{i \in [N]}, h(\mathbf{A}, \mathbf{t})).$$

We consider a sequence of hybrid games $\mathcal{G}_i, 0 \leq i \leq N$, as defined below:

$$\mathcal{G}_i = (\mathbf{A}, \mathbf{u}, \{v_j\}_{j \leq i}, \{\mathbf{r}_j \cdot \mathbf{u} + e_j\}_{j \geq i+1}, \{\mathbf{v}_j\}_{j \leq i}, \{\mathbf{r}_j \cdot \mathbf{A} + \mathbf{e}'_j\}_{j \geq i+1}, h(\mathbf{A}, \mathbf{t})),$$

where $v_j \xleftarrow{\$} \mathbb{Z}_q, \mathbf{v}_j \xleftarrow{\$} \mathbb{Z}_q^m, 1 \leq j \leq i$.

It is prone to check that $\mathcal{G}_0 = \mathcal{G}_{\text{AIL}}$ and $\mathcal{G}_N = \mathcal{G}_{\text{IDEAL}}$. To show theorem 6, it is sufficient to prove that $\mathcal{G}_i \approx_c \mathcal{G}_{i+1}$, for all $i, 0 \leq i \leq N - 1$.

By lemma 11 that presents that for a singleton vector $\mathbf{r} \xleftarrow{\$} \mathbb{Z}_q^n$,

$$\mathcal{G}_{\text{real}} = (\mathbf{A}, \mathbf{u}, \mathbf{r} \cdot \mathbf{u} + e, \mathbf{r}\mathbf{A} + \mathbf{e}', h(\mathbf{A}, \mathbf{t})) \approx_c \mathcal{G}_{\text{ideal}} = (\mathbf{A}, \mathbf{u}, v, \mathbf{v}', h(\mathbf{A}, \mathbf{t})),$$

it suffices to reduce the task of distinguishing between $\mathcal{G}_{\text{real}}$ and $\mathcal{G}_{\text{ideal}}$ to the task of distinguishing $\mathcal{G}_i$ from $\mathcal{G}_{i+1}$.

The reduction algorithm $\mathcal{B}$ that achieves this simulates the view of the attacker $\mathcal{A}$ as follows. Given an input $\mathcal{G} = (\mathbf{A}, \mathbf{u}, w, \mathbf{w}', h(\mathbf{A}, \mathbf{t})$, for $1 \leq j \leq i$, samples $v_j \xleftarrow{\$} \mathbb{Z}_q, \mathbf{v}_j \xleftarrow{\$} \mathbb{Z}_q^m$, and for $i+2 \leq j \leq N$, samples $\mathbf{r}_j \xleftarrow{\$} \mathbb{Z}_q^n, e_j \leftarrow \mathcal{D}_{\mathbb{Z}, \alpha q}$, $\mathbf{e}'_j \leftarrow \mathcal{D}_{\mathbb{Z}, \beta q}^m$, and draws up the following distribution

$$\mathcal{G}' = (\mathbf{A}, \mathbf{u}, \{v_j\}_{j \leq i}, w, \{\mathbf{r}_j \cdot \mathbf{u} + e_j\}_{j \geq i+2}, \{\mathbf{v}_j\}_{j \leq i}, \mathbf{w}', \{\mathbf{r}_j \cdot \mathbf{A} + \mathbf{e}'_j\}_{j \geq i+2]}, h(\mathbf{A}, \mathbf{t})).$$

Thus, the distribution $\mathcal{G}'$ equals to $\mathcal{G}_i$ if $\mathcal{G} = \mathcal{G}_{\text{real}}$, while the distribution $\mathcal{G}'$ equals to $\mathcal{G}_{i+1}$ if $\mathcal{G} = \mathcal{G}_{\text{ideal}}$. It follows that the advantage of $\mathcal{B}$ is equal to that of $\mathcal{A}$. This finishes the proof. □

5   Conclusions

In this work, we first compiled an FHE scheme from dual of Regev's PKE, using *approximata eigenvector* method. we then argued that it is CPA secure against auxiliary inputs leakage by carefully selecting the parameters. As far as we know, it is the first FHE scheme against auxiliary inputs leakage. Since the ciphertext in this scheme is larger than that in GSW, we consider that it is an important problem to show that GSW or other efficient FHEs stand against auxiliary inputs leakage.

Acknowledgement. The authors would like to thank anonymous reviewers for their helpful comments and suggestions.

A   A Symmetric-Key FHE with Auxiliary Inputs

We first recall an important result from [22] which claims that the standard LWE assumption implies that binary-LWE is secure even in the presence of hard-to-invert auxiliary inputs.

Lemma 12. ([22] Theorem 5). *Set $k \geq \log q$. Let $\mathcal{H}$ be the class of all functions $h : \{0, 1\}^n \to \{0, 1\}^*$ that are 2^{-k}-hard-to-invert, i.e., given $h(\mathbf{s})$, no PPT algorithm can find $\mathbf{s}$ with probability better than 2^{-k}.*

For any super-polynomial $q = q(\lambda)$, any $m = \text{poly}(n)$, and any $\alpha, \beta \in (0,1)$ such that $\beta/\alpha = \text{negl}(\lambda)$, it holds that

$$(\mathbf{A}, \mathbf{A}^T\mathbf{s} + \mathbf{e}, h(\mathbf{s})) \approx_c (\mathbf{A}, \mathbf{u}, h(\mathbf{s})),$$

where $\mathbf{A} \xleftarrow{\$} \mathbb{Z}_q^{n \times m}$, $\mathbf{s} \xleftarrow{\$} \{0,1\}^m$, $\mathbf{u} \xleftarrow{\$} \mathbb{Z}_q^m$ and $\mathbf{e} \leftarrow \mathcal{D}_{\mathbb{Z},\alpha q}^m$, assuming the $\text{LWE}_{n',m,q,\beta}$ assumption, where $n' \triangleq \frac{k - \omega(\log \lambda)}{\log q}$.

The symmetric-key FHE with auxiliary inputs SKFHE is described below.

– SKFHE.Setup($1^\lambda, 1^L$): Select LWE parameters $n = n(\lambda, L)$, super-polynomial $q = q(\lambda, L)$, and $\chi = \mathcal{D}_{\mathbb{Z},\alpha q}$, where λ is the security parameter, L is the maximum multiplicative depth of circuits and some $\alpha \in (0,1)$. Set $\ell = \lfloor \log q \rfloor + 1$ and $m = (n+1) \cdot \ell$, and let $prms = (n, m, q, \chi)$.
– SKFHE.SKGen($prms$): Sample a vector $\mathbf{t} \xleftarrow{\$} \{0,1\}^n$. Let $sk = \mathbf{s} = (1, -\mathbf{t})$. Let $\mathbf{v} = \text{Powersof2}(\mathbf{s})$.
– SKFHE.Enc(pk, b): To encrypt symmetrically a bit $b \in \{0,1\}$, sample an error vector $\mathbf{e} \leftarrow \mathcal{D}_{\mathbb{Z},\alpha q}^m$ and a matrix $\mathbf{A} \xleftarrow{\$} \mathbb{Z}_q^{n \times m}$, and output the ciphertext matrix

$$\mathbf{C} = \text{Flatten}(b \cdot \mathbf{I}_N + \text{BitDecomp}([\mathbf{At} + \mathbf{e}||\mathbf{A}])) \in \{0,1\}^{N \times N}.$$

– SKFHE.Dec($\mathbf{C}, sk$): Let $\mathbf{c}$ be the second row of $\mathbf{C}$. Output $b' = \lfloor [\langle \mathbf{c}, \mathbf{v} \rangle]_q \rceil_2$, where the rounding function $\lfloor \cdot \rceil_2 : \mathbb{Z}_q \rightarrow \{0,1\}$ means that it outputs 0 if its argument is closer to 0 than to $2^{\ell-2}$ modulo q, otherwise outputs 1.
– SKFHE.NAND($\mathbf{C}_1, \mathbf{C}_2$): Given two ciphertext matrices $\mathbf{C}_1, \mathbf{C}_2$ for two plaintexts b_1, b_2, respectively, output $\text{Flatten}(\mathbf{I}_N - \mathbf{C}_1 \cdot \mathbf{C}_2)$.
– SKFHE.Eval($f, \mathbf{C}_1, \mathbf{C}_2, \ldots, \mathbf{C}_t$): apply a NAND-circuit $f : \{0,1\}^t \rightarrow \{0,1\}$ to t ciphertexts $\mathbf{C}_1, \mathbf{C}_2, \ldots, \mathbf{C}_t$, and output a ciphertext $\mathbf{C}_f$.

Analysis. The scheme SKFHE is essentially as same as the symmetric-key FHE in [12] except a larger modulus q. So, the correctness and homomorphic property of SKFHE follow the analysis in [12]. Here, we only show that it is CPA secure even in the presence of hard-to-invert auxiliary inputs assuming standard LWE assumption.

Theorem 7. *Let $prms = (n, m, q, \chi)$ be the parameters of SKFHE described above. For any $k = k(\lambda)$ and any function $h : \{0,1\}^n \rightarrow \{0,1\}^*$ that is 2^{-k}-hard-to-invert, if the standard $\text{LWE}_{n',m,q,\beta}$ assumption is hard where $n' \triangleq \frac{k - \omega(\log \lambda)}{\log q}$ and where β satisfies $\beta/\alpha = \text{negl}(\lambda)$, the scheme SKFHE then is CPA secure with auxiliary inputs $h(\mathbf{s})$.*

In order to show theorem 7, by lemma 12, it is sufficient to prove the following lemma, which says that the scheme SKFHE is CPA secure against auxiliary inputs $h(\mathbf{s})$ if $(\mathbf{A}, \mathbf{A}^T\mathbf{s} + \mathbf{e}, h(\mathbf{s})) \approx_c (\mathbf{A}, \mathbf{u}, h(\mathbf{s}))$ for appropriate parameters.

Lemma 13. *Let $prms = (n, m, q, \chi)$ be the parameters of SKFHE described above. For any $k = k(\lambda)$ and any function $h : \{0,1\}^n \rightarrow \{0,1\}^*$ that is 2^{-k}-hard-to-invert, the scheme SKFHE is CPA secure with auxiliary inputs $h(\mathbf{s})$, if $(\mathbf{A}, \mathbf{A}^T\mathbf{s} + \mathbf{e}, h(\mathbf{s})) \approx_c (\mathbf{A}, \mathbf{u}, h(\mathbf{s}))$, where $\mathbf{A} \xleftarrow{\$} \mathbb{Z}_q^{n \times m}$, $\mathbf{s} \xleftarrow{\$} \{0,1\}^m$, $\mathbf{u} \xleftarrow{\$} \mathbb{Z}_q^m$ and $\mathbf{e} \leftarrow \mathcal{D}_{\mathbb{Z},\alpha q}^m$.*

Proof. The proof of lemma 13 is similar as the proof of lemma 4 in [22] and hence is omitted.   □

References

1. Alwen, J., Dodis, Y., Naor, M., Segev, G., Walfish, S., Wichs, D.: Public-key encryption in the bounded-retrieval model. In: Gilbert, H. (ed.) EUROCRYPT 2010. LNCS, vol. 6110, pp. 113–134. Springer, Heidelberg (2010)
2. Alwen, J., Dodis, Y., Wichs, D.: Leakage-resilient public-key cryptography in the bounded-retrieval model. In: Halevi, S. (ed.) CRYPTO 2009. LNCS, vol. 5677, pp. 36–54. Springer, Heidelberg (2009)
3. Akavia, A., Goldwasser, S., Vaikuntanathan, V.: Simultaneous hardcore bits and cryptography against memory attacks. In: Reingold, O. (ed.) TCC 2009. LNCS, vol. 5444, pp. 474–495. Springer, Heidelberg (2009)
4. Alperin-Sheriff, J., Peikert, C.: Faster bootstrapping with polynomial error. In: Garay, J.A., Gennaro, R. (eds.) CRYPTO 2014, Part I. LNCS, vol. 8616, pp. 297–314. Springer, Heidelberg (2014)
5. Brakerski, Z., Goldwasser, S.: Circular and leakage resilient public-key encryption under subgroup indistinguishability. In: Rabin, T. (ed.) CRYPTO 2010. LNCS, vol. 6223, pp. 1–20. Springer, Heidelberg (2010)
6. Brakerski, Z., Gentry, C., Vaikuntanathan, V.: Fully homomorphic encryption without bootstrapping, pp. 309–325. In: ITCS (2012)
7. Berkoff, A., Liu, F.-H.: Leakage resilient fully homomorphic encryption. In: Lindell, Y. (ed.) TCC 2014. LNCS, vol. 8349, pp. 515–539. Springer, Heidelberg (2014)
8. Brakerski, Z., Kalai, Y.T., Katz, J., Vaikuntanathan, V.: Overcoming the hole in the bucket: public-key cryptography resilient to continual memory leakage, pp. 501–510. In: FOCS (2010)
9. Brakerski, Z., Langlois, A., Peikert, C., Regev, O., Stehlé, D.: Classal hardness of learning with errors, pp. 575–584. In: STOC (2013)
10. Brakerski, Z.: Fully homomorphic encryption without modulus switching from classical gapSVP. In: Safavi-Naini, R., Canetti, R. (eds.) CRYPTO 2012. LNCS, vol. 7417, pp. 868–886. Springer, Heidelberg (2012)
11. Brakerski, Z., Vaikuntanathan, V.: Efficient fully homomorphic encryption from (standard) LWE, pp. 97–106. In: FOCS (2011)
12. Brakerski, Z., Vaikuntanathan, V.: Lattice-based FHE as secure as PKE, pp. 1–12. In: ITCS (2014)
13. Coron, J.-S., Lepoint, T., Tibouchi, M.: Scale-invariant fully homomorphic encryption over the integers. In: Krawczyk, H. (ed.) PKC 2014. LNCS, vol. 8383, pp. 311–328. Springer, Heidelberg (2014)
14. Coron, J.-S., Mandal, A., Naccache, D., Tibouchi, M.: Fully homomorphic encryption over the integers with shorter public keys. In: Rogaway, P. (ed.) CRYPTO 2011. LNCS, vol. 6841, pp. 487–504. Springer, Heidelberg (2011)
15. Dodis, Y., Goldwasser, S., Tauman Kalai, Y., Peikert, C., Vaikuntanathan, V.: Public-key encryption schemes with auxiliary inputs. In: Micciancio, D. (ed.) TCC 2010. LNCS, vol. 5978, pp. 361–381. Springer, Heidelberg (2010)
16. Dodis Y., Haralambiev K., Lopez-Alt A., Wichs D.: Cryptography against continuous memory attacks, pp. 511–520. In: FOCS (2010)
17. Dodis, Y., Kalai, Y.T., Lovett, S.: On cryptography with auxiliary inputs, pp. 621–630. In: STOC (2009)

18. Faust, S., Kiltz, E., Pietrzak, K., Rothblum, G.N.: Leakage-resilient signatures. In: Micciancio, D. (ed.) TCC 2010. LNCS, vol. 5978, pp. 343–360. Springer, Heidelberg (2010)
19. Gentry, C.: A Fully homomorphic encryption scheme. Ph.D. thesis, Stanford University (2009). http://crypto.stanford.edu/craig
20. Gentry, C.: Fully homomorphic encryption using ideal lattices, pp. 169–178. In: STOC (2009)
21. Gentry, C., Halevi, S.: Implementing gentry's fully-homomorphic encryption scheme. In: Paterson, K.G. (ed.) EUROCRYPT 2011. LNCS, vol. 6632, pp. 129–148. Springer, Heidelberg (2011)
22. Goldwasser, S., Kalai, Y.T., Peikert, C., Vaikuntanathan, V.: Robustness of the learning with errors assumption, pp. 230–240. In: ICS (2010)
23. Goldwasser, S., Kalai, Y.T., Popa, R.A., Vaikuntanathan, V., Zeldovich, N.: How to run turing machines on encrypted data. In: Canetti, R., Garay, J.A. (eds.) CRYPTO 2013, Part II. LNCS, vol. 8043, pp. 536–553. Springer, Heidelberg (2013)
24. Gentry, C., Peikert, C., Vaikuntanathan, V.: Trapdoors for hard lattices and new cryptographic constructions, pp. 197–206. In: STOC (2008)
25. Gentry, C., Sahai, A., Waters, B.: Homomorphic encryption from learning with errors: conceptually-simpler, asymptotically-faster, attribute-based. In: Canetti, R., Garay, J.A. (eds.) CRYPTO 2013, Part I. LNCS, vol. 8042, pp. 75–92. Springer, Heidelberg (2013)
26. Juma, A., Vahlis, Y.: Protecting cryptographic keys against continual leakage. In: Rabin, T. (ed.) CRYPTO 2010. LNCS, vol. 6223, pp. 41–58. Springer, Heidelberg (2010)
27. López-Alt, A., Tromer, E., Vaikuntanathan, V.: On-the-fly multiparty computation on the cloud via multikey fully homomorphic encryption, pp. 1219–1234. In: STOC (2012)
28. Micciancio, D., Peikert, C.: Trapdoors for lattices: simpler, tighter, faster, smaller. In: Pointcheval, D., Johansson, T. (eds.) EUROCRYPT 2012. LNCS, vol. 7237, pp. 700–718. Springer, Heidelberg (2012)
29. Micali, S., Reyzin, L.: Physically observable cryptography. In: Naor, M. (ed.) TCC 2004. LNCS, vol. 2951, pp. 278–296. Springer, Heidelberg (2004)
30. Naor, M., Segev, G.: Public-key cryptosystems resilient to key leakage. In: Halevi, S. (ed.) CRYPTO 2009. LNCS, vol. 5677, pp. 18–35. Springer, Heidelberg (2009)
31. Peikert, C.: Public key cryptosystems from the worst-case shortest vector problem, pp. 333–32. In: STOC (2009)
32. Peikert, C.: An efficient and parallel gaussian sampler for lattices. In: Rabin, T. (ed.) CRYPTO 2010. LNCS, vol. 6223, pp. 80–97. Springer, Heidelberg (2010)
33. Rivest, R., Adleman, L., Dertouzos, M.: On sata banks and privacy homomorphisms. In: Foundations of Secure Computation, pp. 169–179 (1978)
34. Regev O.: On Lattices, Learning with Errors, Random Linear Codes, and Cryptography. In: STOC, pp. 84–93 (2005)
35. van Dijk, M., Gentry, C., Halevi, S., Vaikuntanathan, V.: Fully homomorphic encryption over the integers. In: Gilbert, H. (ed.) EUROCRYPT 2010. LNCS, vol. 6110, pp. 24–43. Springer, Heidelberg (2010)

Trapdoors for Ideal Lattices with Applications

Russell W.F. Lai[1]([✉]), Henry K.F. Cheung[2], and Sherman S.M. Chow[1]

[1] Department of Information Engineering, The Chinese University of Hong Kong,
Sha Tin, New Territories, Hong Kong
{wflai,sherman}@ie.cuhk.edu.hk
[2] Department of Systems Engineering and Engineering Management,
The Chinese University of Hong Kong, Sha Tin, New Territories, Hong Kong
kfcheung@se.cuhk.edu.hk

Abstract. There is a lack of more complicated ideal-lattice-based cryptosystems which require the use of lattice trapdoors, for the reason that currently known trapdoors are either only applicable to general lattices or not well-studied in the ring setting. To facilitate the development of such cryptosystems, we extend the notion of lattice trapdoors of Micciancio and Peikert (Eurocrypt '12) into the ring setting with careful justification. As a demonstration, we use the new trapdoor to construct a new hierarchical identity-based encryption scheme, which allows us to construct public-key encryption with chosen-ciphertext security, signatures, and public-key searchable encryption.

Keywords: Ideal lattices · Trapdoors · Identity-based encryption

1 Introduction

Lattice-based cryptography is a promising alternative to create cryptosystems that are secure even against quantum adversaries. Many powerful primitives including fully-homomorphic encryption [1–4], homomorphic signatures [5,6], multilinear map [7], (hierarchical) identity-based encryption [8,9] (which is also useful for achieving other cryptographic goals like public-key encryption with chosen-ciphertext security, signatures, and public-key searchable encryption), and much more can be realized by lattices. Security reductions of some of these constructions are directly based on the now well-studied (ring-)LWE (learning with errors) or (ring-)SIS (short integer solutions) problems, which are both as hard as the corresponding worst-case (ideal) lattice problems.

Hard Lattice Problems. An instance of the LWE problem is defined by a random n by m integer matrix A and a vector b, where $b = A^T s + e \bmod q$ for some

This work is supported by grants 439713, 14201914 from Research Grants Council (RGC), and grants 4055018, 4930034 from The Chinese University of Hong Kong. Sherman Chow is supported by the Early Career Award from RGC. Part of the work was done while the second author is with Department of Information Engineering.

© Springer International Publishing Switzerland 2015
D. Lin et al. (Eds.): Inscrypt 2014, LNCS 8957, pp. 239–256, 2015.
DOI: 10.1007/978-3-319-16745-9_14

secret vector s and small noise vector e. The problem is to find the vector s. As a "dual" problem to LWE, an instance of the SIS problem is defined by the same random matrix A, where one is asked to find a short vector x so that $Ax = 0 \bmod q$.

The "ring" versions of LWE and SIS, named ring-LWE and ring-SIS respectively, are specific instances of LWE and SIS respectively defined for some structured matrix A to be explained below.

Ideal Lattices. In ideal lattices, or the so called "ring setting", the matrix A above is required to have some additional algebraic structures. One commonly used example is to interpret each column of A as coefficients of a degree-$(n-1)$ polynomial $p(x)$, and require that $xp(x) \bmod (x^n + 1)$ is also contained in some column of A. In such case, the matrix multiplications by A are equivalent to polynomial multiplications. We can therefore view each vector v as an element $\mathbf{v}$ in the ring $R_q = \mathbb{Z}_q[x]/\langle x^n + 1 \rangle$, and each n-by-n sub-matrix A_i in A a ring element $\mathbf{a}_i$ in R_q. As the (ring-)LWE and (ring-)SIS problems have such simple forms, the operations performed in the corresponding cryptosystems are rather efficient.

Due to the algebraic structure of ideal lattices, cryptosystems based on ideal lattices (with security based on the ring-LWE or ring-SIS assumptions) are more efficient than their counterparts in general lattices: (1) The size of some parameters, which are originally matrices, is reduced by a factor of n, as each n-by-n sub-matrix is now represented as a ring element; (2) The multiplications of ring elements in R_q can be implemented on hardware by a variant of Fourier transform [3].

Lattice Trapdoors. For more complicated primitives, a "trapdoor" is generated together with a random lattice so that, while it is still hard for the adversary to solve the (ring-)LWE or (ring-)SIS problems, the problems become easily solvable with the help of the trapdoor.

Initiated by the work of Gentry *et al.* [10], a (old-type) trapdoor [10,11] of (the lattice defined by) a matrix A is a short basis of the lattice $\Lambda_q^\perp(A)$, which contains all the vectors x such that $Ax = \mathbf{0} \bmod q$. Using the trapdoor, one can sample short vectors x so that $Ax = u \bmod q$ for any target vector u. Moreover, the owner of the trapdoor of A can "delegate" the trapdoor of an extended matrix (A, B) for any matrix B.

Micciancio and Peikert [12] developed a new type of trapdoors for general lattices which is simpler and more efficient to use when compared to the old trapdoors. A new-type trapdoor of A is a matrix T with small norm so that $A \begin{bmatrix} T \\ I \end{bmatrix} = HG$ for some invertible matrix H, which is referred as the tag of the trapdoor, and a nicely structured matrix G called the primitive matrix, where the inversions of SIS (also known as "Gaussian sampling") and LWE involving G are easy and efficient.[1] At the high-level sense, T can be considered as a secret

---

[1] We switch the notation from the original R in [12] to T to avoid clashing of notations in the later sections.

transformation from A to G which reduces the originally difficult inversions of SIS and LWE involving A to the much easier inversions involving G. Notice that the new-type trapdoors have the additional ability to invert LWE, which is not the case for the old-type trapdoors. In addition, the size of the new-type trapdoors is much (at least 4 times) smaller than that of the old-type.

However, despite the increase of efficiency, there is a lack of cryptosystems in ideal lattices that require the use of trapdoors. One possible reason for this is that, the trapdoors introduced by Gentry et al. [10] and improved by Alwen and Peikert [11] are based on general lattices. Stehlé [13] attempted to extend the trapdoor algorithms to ideal lattices, but the result is based on a non-standard ideal-LWE assumption which, unlike the ring-LWE assumption, does not have search-to-decision reduction. Later, Micciancio and Peikert [12] introduced a new notion of lattice trapdoors which have even greater functionality, namely, to invert not only SIS but also LWE. More importantly, the new trapdoors can be translated to the ring setting, as mentioned in [12] but unfortunately without much details.

Our Contributions. In this work, we extend the trapdoors from Micciancio and Peikert [12] to the ring setting. As a result, the sizes of the "primitive vectors", the public vectors and trapdoors are reduced by a factor of n. As in other recent cryptosystems [2,3,14] that are based on ring-LWE, we work with the "preferred" choice of ring $R := \mathbb{Z}[x]/\langle x^n + 1 \rangle$ where n is a power of 2. For such choice of ring R, the general strategy of transforming the trapdoors to the ring setting is to interpret each n by n submatrix in the construction of [12] as a ring element in R. By breaking down elements in R in terms of the "power basis" $1, x, x^2, \ldots, x^{n-1}$, we show that some of the algorithms in [12] can be reused. We also justify the correctness of such transformation carefully by replacing certain theorems and lemmas by those proven in the ring setting.

Finally, we demonstrate the power of the new trapdoors by constructing a new identity-based encryption (IBE) scheme which improves the IBE scheme constructed by Agrawal et al. [8] in three aspects, namely, being ideal-lattice-based, having reduced trapdoor size, and being secure against chosen-ciphertext attack.

2   Preliminary

Notations. Let λ be the security parameter. Let $f(x) = f_\lambda(x) \in \mathbb{Z}[x]$ be a polynomial of degree $n = n(\lambda)$. Let $q = q(\lambda) \in \mathbb{Z}$ be a prime integer, $p = p(\lambda) \in \mathbb{Z}_q^*$ be relatively prime to q. Let $R := \mathbb{Z}[x]/\langle f(x) \rangle$ and $R_q := R/qR$. Let χ be a distribution over the ring R. "$|$" denotes row concatenation of vectors or matrices. If S is a set, then $x \leftarrow S$ denotes the sampling of a uniformly random element x from S. If X is a distribution, then $x \leftarrow X$ denotes the sampling of a random element x according to the distribution X. If $\mathcal{A}$ is an algorithm, then $x \leftarrow \mathcal{A}$ means that x is the output of the algorithm $\mathcal{A}$. To distinguish between elements, vectors and matrices of $\mathbb{Z}$ and R, we follow the notations listed in

Table 1. Notations of elements, vectors and matrices of $\mathbb{Z}$ and R

	Element	Vector	Matrix
Integers $\mathbb{Z}$	a	$\boldsymbol{a}$	A
Ring R	$\mathbf{a}$	$\mathbf{a}$	$\mathbf{A}$

Table 1. We denote the k-by-k identity matrix over R by $\mathbf{I}_k$ and the k-by-l zero matrix over R by $\mathbf{0}_{k \times l}$. Without further specifications, $\|\boldsymbol{x}\|$ denotes the L_2 norm of the vector $\boldsymbol{x}$ and is extended naturally to $\|\mathbf{x}\|$ via the coefficient embedding.

2.1 Lattice Background

Statistical Distance. Let X and Y be two random variables taking values in some finite set Ω. The statistical distance $\Delta(X; Y)$ is defined as

$$\Delta(X; Y) := \frac{1}{2} \sum_{s \in \Omega} |\Pr[X = s] - \Pr[Y = s]|.$$

We say that the ensembles of random variables $X(\lambda)$ and $Y(\lambda)$ are statistically close if $\Delta(X; Y)$ is a negligible function in λ.

Integer Lattices. We consider three types of integer lattices. For an integer modulus q, $A \in \mathbb{Z}_q^{n \times m}$ and $u \in \mathbb{Z}_q^n$, define:

$$\Lambda_q(A^T) := \{\boldsymbol{x} \in \mathbb{Z}^m : \exists \, \boldsymbol{s} \in \mathbb{Z}_q^n \text{ s.t. } A^T \boldsymbol{s} = \boldsymbol{x} \bmod q\}$$
$$\Lambda_q^{\perp}(A) := \{\boldsymbol{x} \in \mathbb{Z}^m : A\boldsymbol{x} = \mathbf{0} \bmod q\}$$
$$\Lambda_q^{\boldsymbol{u}}(A) := \{\boldsymbol{x} \in \mathbb{Z}^m : A\boldsymbol{x} = \boldsymbol{u} \bmod q\}$$

Note that for any $\boldsymbol{t} \in \Lambda_q^{\boldsymbol{u}}(A)$, $\Lambda_q^{\boldsymbol{u}}(A) = \Lambda_q^{\perp}(A) + \boldsymbol{t}$ is a shift of $\Lambda_q^{\perp}(A)$.

Ideal Lattices. Correspondingly, we consider three types of ideal lattices. For an integer modulus q, $\mathbf{a} \in R_q^k$ and $\mathbf{u} \in R_q$, define:

$$\Lambda_q(\mathbf{a}) := \{\mathbf{x} \in R^k : \exists \, \mathbf{s} \in R_q \text{ s.t. } \mathbf{as} = \mathbf{x} \bmod q\}$$
$$\Lambda_q^{\perp}(\mathbf{a}^T) := \{\mathbf{x} \in R^k : \mathbf{a}^T \mathbf{x} = \mathbf{0} \bmod q\}$$
$$\Lambda_q^{\mathbf{u}}(\mathbf{a}^T) := \{\mathbf{x} \in R^k : \mathbf{a}^T \mathbf{x} = \mathbf{u} \bmod q\}$$

Note that for any $\mathbf{t} \in \Lambda_q^{\mathbf{u}}(\mathbf{a}^T)$, $\Lambda_q^{\mathbf{u}}(\mathbf{a}^T) = \Lambda_q^{\perp}(\mathbf{a}^T) + \mathbf{t}$ is a shift of $\Lambda_q^{\perp}(\mathbf{a}^T)$.

Discrete Gaussian. Let $L \subset \mathbb{Z}^n$, $c \in \mathbb{R}^n$, $\sigma \in \mathbb{R}^+$. Define:

$$\rho_{\sigma,c}(x) = \exp(-\pi \frac{\|x - c\|^2}{\sigma^2}) \text{ and } \rho_{\sigma,c}(L) = \sum_{x \in L} \rho_{\sigma,c}(x).$$

The discrete Gaussian distribution over L with center c and parameter σ is defined as

$$\forall x \in L, \mathcal{D}_{L,\sigma,c}(x) = \frac{\rho_{\sigma,c}(x)}{\rho_{\sigma,c}(L)}.$$

For $c = 0$, denote $\rho_{\sigma,0}$ as ρ_σ and $\mathcal{D}_{L,\sigma,0}$ as $\mathcal{D}_{L,\sigma}$.

Gram-Schmidt Norm. Let $T = \{t_1, \ldots, t_k\} \subset \mathbb{R}^m$ be a set of real vectors, and $\|T\|$ denotes the L_2-norm of the longest vector in T, *i.e.*, $\|T\| := \max_{j=1}^k \|t_j\|$, $\widetilde{T} := \{\widetilde{t}_1, \ldots, \widetilde{t}_k\}$ denotes the Gram-Schmidt orthogonalization of the vectors $\widetilde{t}_1, \ldots, \widetilde{t}_k$ taken in that order. $\|\widetilde{T}\|$ is called the Gram-Schmidt norm of T.

2.2 Assumptions

The learning with errors (LWE) problem defined by Regev [15] is now a well-studied hard problem that is as hard as some worst-case lattice hard problems such as the shortest vector problem (SVP), via either quantum or classical reductions [15,16]. An LWE instance is defined by a matrix $A \in \mathbb{Z}_q^{n \times m}$ and a vector $b \in \mathbb{Z}_q^m$. The search version of LWE is to find a secret vector $s \in \mathbb{Z}^n$ so that $A^T s + e = b$ for some short error vector $e \in \mathbb{Z}^m$. The decision version is to decide whether such a pair of (A, b) comes from the uniform distribution or the LWE distribution, *i.e.* $A^T s + e = b$ for some short error vector $e \in \mathbb{Z}^m$.

To define LWE in the ring setting, namely ring-LWE, A is restricted to have a certain algebraic structure. We interpret the entries in a column of A as the coefficients of a degree-$(n-1)$ polynomial $p(x)$, and require that the vector given by the coefficients of $xp(x) \bmod f(x)$, for some degree-n polynomial $f(x)$, is also contained in some column of A. Assuming that $m = nk$ for some integer k, we can then interpret the i-th n-by-n sub-matrix of A as a ring element $\mathbf{a}_i$ in R_q, the vector s as $\mathbf{s}$ in R, e as $(\mathbf{e}_1, \ldots, \mathbf{e}_k)^T$ in R^k and b as $(\mathbf{b}_1, \ldots, \mathbf{b}_k)^T$ in R_q^k. Multiplications between a sub-matrix of A and the vector s correspond to the multiplications of the ring elements $\mathbf{a}_i$ and $\mathbf{s}$. Apparently, the search version of ring-LWE is then to find $\mathbf{s}$ given $\{\mathbf{a}_i, \mathbf{b}_i = \mathbf{a}_i \mathbf{s} + \mathbf{e}_i\}_{i=1}^k$. The decision version of ring-LWE is to distinguish the distribution of the given samples from the uniform distribution. For the detailed definitions and reductions related to ring-LWE, we refer to the comprehensive work of Lyubashevsky *et al.* [3,14].

In this work, we further restrict the polynomial $f(x)$ such that $f(x) = x^n + 1$, which is the preferred choice in recent ring-LWE-based cryptosystems [2,3,14] due to its simplicity. In this case, the ring-LWE assumption has a much simpler form. This special case is named as polynomial LWE (PLWE) by Brakerski and Vaikuntanathan [2].

Definition 1 *(PLWE assumption [2]). For all $\lambda \in \mathbb{N}$, $l = \mathsf{poly}(\lambda)$, $\mathbf{a}_i, \mathbf{u} \leftarrow R_q$, $\mathbf{e}_i, \mathbf{s} \leftarrow \chi$, the $\mathsf{PLWE}_{f,q,\chi}^{(l)}$ assumption states that the distribution of $\{(\mathbf{a}_i, \mathbf{a}_i \mathbf{s} + p\mathbf{e}_i)\}_{i=1}^l$ is computationally indistinguishable from the distribution of $\{(\mathbf{a}_i, \mathbf{u}_i)\}_{i=1}^l$.*

Theorem 1 *[2, Theorem 1]. Let λ be the security parameter. Let $k \in \mathbb{N}$ and let $m = 2^{\lfloor \log \lambda \rceil}$ be a power of two. Let $\Phi_m(x) = x^n + 1$ be the m-th cyclotomic polynomial of degree $n = \varphi(m) = m/2$. Let $\sigma \geq \omega(\sqrt{\log n})$ be a real number, and let $q \equiv 1 \pmod{m}$ be a prime integer. Let $R = \mathbb{Z}[x] = \langle \Phi_m(x) \rangle$. Then there is a randomized reduction from $(n^2 q/r) \cdot (n(l+1)/\log(n(l+1)))^{1/4}$-approximate R-SVP to $\mathsf{PLWE}_{\Phi_m,q,\chi}^{(l)}$ where $\chi = \mathcal{D}_{\mathbb{Z}^n,\sigma}$ is the discrete Gaussian distribution. The reduction runs in time $\mathsf{poly}(n,q,l)$.*

3   Primitive Vectors in Ideal Lattices

Although Micciancio and Peikert [12] mentioned that their trapdoors can be extended to ideal lattices, they did not explain it in details. In this section, we first extend their notion of primitive matrices for general lattices to the notion of primitive vectors (of ring elements) for ideal lattices. We will then show in Sect. 4 how to use the primitive vectors to generate trapdoors for ideal lattices. The general strategy used in these sections is to interpret each n-by-n submatrices in the notion of trapdoors for general lattices as ring elements in R_q.

3.1   Construction of Primitive Vectors

Recall from [12] that a matrix $G \in \mathbb{Z}_q^{n \times m}$ is primitive if its columns generate all of $\mathbb{Z}_q^n$, *i.e.*, $G \cdot \mathbb{Z}^m = \mathbb{Z}_q^n$. For some nicely structured primitive matrices, LWE inversion and Gaussian sampling can be done efficiently. Given such a primitive matrix, the crux of the trapdoor generation algorithm is to perform a random transform on the primitive matrix.

As mentioned in [12, Sect. 4.3], the primitive vector $\mathbf{g} = (1, 2, \ldots, 2^{k-1})^T$ in R_q^k can be used in the ring setting to replace the previous primitive matrix G by interpreting the values in the ring R_q instead of $\mathbb{Z}_q$. Furthermore, the inversion and Gaussian sampling algorithms can be obtained in the ring setting as well.

Intuitively, to obtain a primitive vector in the ring setting, we need to find a primitive matrix (in the general lattices setting) in which each n-by-n submatrix is rotational, *i.e.*, a column is obtained by shifting the previous column by one entry and adding a negative sign to the first entry. One way is to permute the columns of the previous primitive matrix G to obtain such a structure. An example of G is as follows [12]:

$$G := \begin{bmatrix} \cdots g^T \cdots & & & \\ & \cdots g^T \cdots & & \\ & & \ddots & \\ & & & \cdots g^T \cdots \end{bmatrix}, g^T = \begin{bmatrix} 1 \ 2 \ 4 \ldots 2^{k-1} \end{bmatrix} \in \mathbb{Z}_q^{1 \times k}$$

We permute columns of G so that identical terms forms n-by-n diagonal block matrices. As a result, we obtain:

$$G' := [I_n | 2I_n | \ldots | 2^{k-1} I_n].$$

Since G' is obtained by permutation of columns of G, G' is still primitive. By the same permutation on the basis S of $\Lambda_q^\perp(G)$, we obtain the basis S' of $\Lambda_q^\perp(G')$, where

$$S' := \begin{bmatrix} 2I_n & & & & q_0 I_n \\ -I_n & 2I_n & & & q_1 I_n \\ & -I_n & & & q_2 I_n \\ & & \ddots & & \vdots \\ & & & 2I_n & q_{k-2}I_n \\ & & & -I_n & q_{k-1}I_n \end{bmatrix} \in \mathbb{Z}^{nk \times nk}.$$

The matrices G' and S' correspond to the collections of vectors of ring elements $\mathbf{g} = (1, 2, \ldots, 2^{k-1})^T \in R_q^k$ and $(\mathbf{s}_1, \ldots, \mathbf{s}_k) \in R_q^{k \times k}$ respectively, where $\mathbf{s}_i = (0, \ldots, 0, 2, -1, 0, \ldots, 0)^T$ for $i < k$, and $\mathbf{s}_k = (q_0, q_1, \ldots, q_{k-1})^T$, where $q = \sum_{i=0}^{k-1} 2^i q_i$ and $q_i \in \{0, 1\}$.

Theorem 2 summarizes the result of primitive vectors in the ring setting, with explanation deferred to the later subsections.

Theorem 2. *For any $q = \sum_{i=0}^{k-1} 2^i q_i < 2^k$ where $q_i \in \{0,1\}$ and $k \geq 1$, there exists $\mathbf{g} = (1, 2, \ldots, 2^{k-1})^T \in R_q^k$ and $\mathbf{S} = (\mathbf{s}_1, \ldots, \mathbf{s}_k) \in R_q^{k \times k}$ (thus $\mathbf{g}^T \mathbf{S} = \mathbf{0}_{1 \times k} \in R_q^k$), such that:*

- *We have $\|\tilde{\mathbf{s}}_i\| < \sqrt{5}$ in the coefficient embedding.*
- *The storage requirement of $\mathbf{g}$ and $\mathbf{S}$ are further reduced by a factor of n compared to their counterparts in general lattices.*
- *Inverting $\alpha_{\mathbf{g}}(\mathbf{z}, \mathbf{e}) := \mathbf{g}\mathbf{z} + \mathbf{e} \bmod q$ can be performed in quasilinear $O(n \cdot \log^c n)$ time for any $\mathbf{z} \in R$ and any $\mathbf{e} \in q \cdot \mathbf{B}^{-T} \cdot [-\frac{1}{2}, \frac{1}{2})^{nk}$, where $\mathbf{B}$ can denote either $\mathbf{S}$ or $\tilde{\mathbf{S}}$. Moreover, the algorithm is perfectly parallelizable, running in polylogarithmic $O(\log^c n)$ time using n processors.*
- *Preimage sampling for $\beta_{\mathbf{g}}(\mathbf{x}) = \mathbf{g}^T \mathbf{x} \bmod q$ with Gaussian parameter $\sigma \geq \|\tilde{\mathbf{S}}\| \cdot w(\sqrt{\log n})$ can be performed in quasilinear $O(n \cdot \log^c n)$ time, or parallel polylogarithmic $O(\log^c n)$ time using n processors.*

3.2 Inversion for Primitive Vectors

Given a PLWE instance $\mathbf{b} = \mathbf{g}\mathbf{z} + \mathbf{e}$, which is equivalent to $\mathbf{b}_i = 2^i \mathbf{z} + \mathbf{e}_i$ for $i = 0, \ldots, k-1$, we can expand $\mathbf{b}_i$, $\mathbf{z}$ and $\mathbf{e}_i$ in terms of the power basis $1, x, x^2, \ldots, x^{n-1}$ so that the problem is equivalent to solving $b_{ij} = 2^i z_j + e_{ij}$ independently for $j = 0, \ldots, n-1$, where $\mathbf{b}_i = \sum_{j=0}^{n-1} b_{ij} x^j$, $\mathbf{z} = \sum_{j=0}^{n-1} z_j x^j$ and $\mathbf{e}_i = \sum_{j=0}^{n-1} e_{ij} x^j$. Recombining the terms according to i, the problem becomes solving $\boldsymbol{b}_j = \boldsymbol{g} z + \boldsymbol{e}_j$ where $\boldsymbol{g} = (1, 2, \ldots, 2^{k-1})^T \in \mathbb{Z}_q^k$. Let $S = (\boldsymbol{s}_1, \ldots, \boldsymbol{s}_k)$ be a basis of $\Lambda_q^\perp(\boldsymbol{g}^T)$. Then $V = qS^{-T} = (\boldsymbol{v}_1, \ldots, \boldsymbol{v}_k)$ is a basis of $\Lambda_q(\boldsymbol{g})$. We can then use Babai's nearest plane algorithm to recover $z \in \mathbb{Z}_q$ from $\boldsymbol{b} = \boldsymbol{g} z + \boldsymbol{e}$:

Algorithm 3. *[17]* **Babai's Nearest Plane Algorithm**
   Input: $\{v_1, \ldots, v_k\}$ *a basis of* $\Lambda_q(g)$, b.
   Output: z *and* e.

1. *Compute Gram-Schmidt basis* $v_1^*, \ldots, v_k^*$.
2. *For* $j = k \to 1$:
   (a) *Compute* $l_j = <b_j, v_j^*> / <v_j^*, v_j^*>$.
   (b) *Set* $b_{j-1} = b_j - (l_j - \lfloor l_j \rceil) v_j^* - \lfloor l_j \rceil v_j$.
3. *Return* $z = \sum_{j=1}^{k} \lfloor l_j \rceil c_j \bmod q$, $e = b - gz$, *where* $v_j = c_j g \bmod q$.

3.3 Gaussian Sampling for Primitive Vectors

We first recall that the goal of Gaussian sampling in [12] is to sample a vector from $\Lambda_q^u(G)$. This can be done by repeating n times of the sampling from $\Lambda_q^{u_j}(g^T)$ for a desired syndrome $u_j \in \mathbb{Z}_q$, where $j = 0, \ldots, n-1$.

For the later task, there are two extreme approaches and one hybrid approach. In the one extreme, we first pre-compute a large set of samples from $\mathcal{D}_{\mathbb{Z}^k, \sigma}$ and bucket them according to the different values of u. The sampling algorithm simply draws one sample from the appropriate bucket. This approach requires large storage so that each bucket can be filled with sufficient number of samples. The other extreme exploits the fact that if q is a power of 2, then we have the orthogonalized basis $\tilde{S}_k = 2I_k$. In this case, there is a simple and efficient way to perform Babai's nearest plane algorithm [17]. In this algorithm, we first pre-compute two large sets of samples from $\mathcal{D}_{2\mathbb{Z}, \sigma}$ and $\mathcal{D}_{2\mathbb{Z}+1, \sigma}$. The sampling algorithm draws each coefficient of x one by one from the appropriate set. This approach requires less storage space but takes k steps to complete. Naturally, there is a hybrid approach that pre-computes samples from $\mathcal{D}_{\mathbb{Z}^l, \sigma}$ for some $l < k$ and fills in the coefficients of x in blocks of l.

To perform Gaussian sampling in the ring setting, we can of course use the sampling algorithm for general lattices and perform the permutation mentioned above to the preimage. More formally, recall that our task is to sample a vector of ring elements $\mathbf{x}$ from $\Lambda_q^{\mathbf{u}}(\mathbf{g}^T) = \{\mathbf{x} \in R^k : \mathbf{g}^T \mathbf{x} = \mathbf{u} \bmod q\}$ where $\mathbf{u} \in R_q$. That is, $\sum_{i=0}^{k-1} 2^i \mathbf{x}_i = \mathbf{u}$. By expanding $\mathbf{x}_i$ in the power basis $1, x, x^2, \ldots, x^{n-1}$, this is equivalent to $\sum_{i=0}^{k-1} 2^i x_{ij} = u_j$ for $j = 0, 1, \ldots, n-1$, where $\mathbf{x}_i = \sum_{j=0}^{n-1} x_{ij} x^j$ and $u = \sum_{j=0}^{n-1} u_j x^j$. Thus, we can use the same sampling algorithms for each equation $\sum_{i=0}^{k-1} 2^i x_{ij} = u_j$ in the ring setting, since x_{ij} and u_j are integers modulo q. However, notice that the reduction of ring-LWE in [2,3,14] requires that $q = 1 \bmod 2n$, which means that q cannot be a power of 2. Therefore, practically we can only use the first approach for Gaussian sampling in the ring setting.

4 Trapdoors in Ideal Lattices

Analogous to the trapdoors for general lattices defined in [12], we extend the notion to the ring setting. This includes the derivation of old-type trapdoors from

Gentry *et al.* [10] (Sect. 4.1), the generation of (new-type) trapdoors (Sect. 4.2), ring-LWE inversion (Sect. 4.3), Gaussian sampling (Sect. 4.4) and trapdoors delegation (Sect. 4.5).

Definition 2. *Let* $\mathbf{a} \in R_q^{l+k}$ *and* $\mathbf{g} \in R_q^k$. *A* $\mathbf{g}$-*trapdoor for* $\mathbf{a}$ *is a collection of linearly independent vectors of ring elements* $\mathbf{R} = (\mathbf{r}_1, \dots, \mathbf{r}_k) \in R_q^{l \times k}$ *such that* $\mathbf{a}^T \begin{bmatrix} \mathbf{R} \\ \mathbf{I}_k \end{bmatrix} = \mathbf{h}\mathbf{g}^T$, *for some non-zero ring element* $\mathbf{h} \in R_q$. $\mathbf{h}$ *is referred as the tag or label of the trapdoor. The qulity of the trapdoor is measured by its largest singular value* $s_1(\mathbf{R})$, *which is computed as the largest singular value of the matrix obtained by interpreting* $\mathbf{R}$ *as a matrix in* $\mathbb{Z}_q^{ln \times kn}$.

4.1   Derivation of Old Trapdoors

Lemma 1. *Let* $\mathbf{g} \in R_q^k$ *and* $\mathbf{S} = (\mathbf{s}_1, \dots, \mathbf{s}_k) \in R^{k \times k}$ *be linearly independent with* $\mathbf{g}^T \mathbf{s}_i = \mathbf{0} \in R_q$ *for* $i = 1, \dots, k$. *Let* $\mathbf{a} \in R_q^{l+k}$ *have trapdoor* $\mathbf{R} = (\mathbf{r}_1, \dots, \mathbf{r}_k) \in R^{k \times k}$ *with tag* $\mathbf{h} \in R_q$. *Then the lattice* $\Lambda_q^{\perp}(\mathbf{a}^T)$ *is generated by*

$$\mathbf{S_a} = \begin{bmatrix} \mathbf{I}_l & \mathbf{R} \\ \mathbf{0}_{k \times l} & \mathbf{I}_k \end{bmatrix} \begin{bmatrix} \mathbf{I}_l & \mathbf{0}_{l \times k} \\ \mathbf{W} & \mathbf{S} \end{bmatrix},$$

where $\mathbf{W} \in R^{k \times l}$ *is an arbitrary solution to* $\mathbf{g}^T \mathbf{W} = -\mathbf{h}^{-1}\mathbf{a}^T [\mathbf{I}_l | \mathbf{0}_{l \times k}]^T \mod q$. *Moreover, the basis* $\mathbf{S_a}$ *satisfies* $\|\widetilde{\mathbf{S_a}}\| \le s_1 \left(\begin{bmatrix} \mathbf{I}_l & \mathbf{R} \\ \mathbf{0}_{k \times l} & \mathbf{I}_k \end{bmatrix} \right) \cdot \|\widetilde{\mathbf{S}}\| \le (s_1(\mathbf{R}) + 1) \cdot$

$\|\widetilde{\mathbf{S}}\|$, *when* $\mathbf{S_a}$ *is orthogonalized in suitable order and interpreted as a matrix in* $\mathbb{Z}^{[(l+k)n] \times [(l+k)n]}$ *by the coefficient embedding.*

Proof. Compared to the derivation in general lattices, the non-trivial part is to construct a matrix $\mathbf{W}$ of ring elements, or equivalently, a matrix W consisting of $k \times l$ blocks of $n \times n$ rotational matrices. Otherwise, the rest of the proof follows the proof of [12, Lemma 5.3]. To construct such a matrix $\mathbf{W}$, let $\mathbf{a} = (\mathbf{a}_1, \mathbf{a}_2, \dots, \mathbf{a}_{l+k})^T \in R_q^{l+k}$ and let

$$\mathbf{W} = \begin{bmatrix} \mathbf{w}_{1,1} & \cdots & \mathbf{w}_{1,l} \\ \vdots & \ddots & \vdots \\ \mathbf{w}_{k,1} & \cdots & \mathbf{w}_{k,l} \end{bmatrix}$$

where $\mathbf{w}_{i,j} \in R_q$.

Now, $\mathbf{g}^T \mathbf{W} = -\mathbf{h}^{-1}\mathbf{a}^T [\mathbf{I}_l | \mathbf{0}_{l \times k}]^T \mod q$ implies

$$[1|2|\dots|2^{k-1}]\mathbf{W} = [1|2|\dots|2^{k-1}] \begin{bmatrix} \mathbf{w}_{1,1} & \cdots & \mathbf{w}_{1,l} \\ \vdots & \ddots & \vdots \\ \mathbf{w}_{k,1} & \cdots & \mathbf{w}_{k,l} \end{bmatrix} = -\mathbf{h}^{-1} [\mathbf{a}_1 \, \mathbf{a}_2 \, \dots \, \mathbf{a}_l].$$

This equation implies that for each $j = 1, \ldots, l$, we need to independently solve

$$[\mathbf{1}|\mathbf{2}|\ldots|\mathbf{2}^{k-1}] \begin{bmatrix} \mathbf{w}_{1,j} \\ \vdots \\ \mathbf{w}_{k,j} \end{bmatrix} = -\mathbf{h}^{-1}\mathbf{a}_i \in R_q.$$

By expanding $\mathbf{w}_{i,j}$ and $\mathbf{a}_j$ with respect to the power basis $1, x, x^2, \ldots, x^{n-1}$, the problem is equivalent to solving the system for each coefficient independently. $\square$

Although the derivation of the old-type trapdoors for ideal lattices is merely theoretical, it solves an open problem in [10] which asked how trapdoors can be generated together with random looking ideal lattices.

4.2   Generation of New Trapdoor

As in [12], the derivation of old trapdoors from the new trapdoors is just a proof of concept and will not be used in the rest of this work. In this subsection, we extend their trapdoors for general lattices to our ring version in Algorithm 4.

Algorithm 4. ringGenTrap$^{\mathcal{D}}(\mathbf{a}_0, \mathbf{h})$
Input:

- *a vector of ring elements* $\mathbf{a}_0 = (\mathbf{a}_1, \ldots, \mathbf{a}_l)^T \in R_q^l$;
- *a non-zero ring element* $\mathbf{h} \in R_q$;
- *a distribution* $\chi^{l \times k}$ *over* $R^{l \times k}$. *(If no particular* $\mathbf{a}_0$, $\mathbf{h}$ *are given as input, then the algorithm may choose them itself, e.g. picking* $\mathbf{a}_0 \leftarrow R_q^l$ *uniformly, and setting* $\mathbf{h} = 1$.)

Output:

- *a vector of ring elements* $\mathbf{a} = (\mathbf{a}_0^T, \mathbf{a}_1^T)^T \in R_q^{l+k}$;
- *a trapdoor* $\mathbf{R} = (\mathbf{r}_1, \ldots, \mathbf{r}_k) \in R^{l \times k}$ *with tag* $\mathbf{h} \in R_q$.

1. *Choose a collection of linearly independent vectors of ring elements* $\mathbf{R} = (\mathbf{r}_1, \ldots, \mathbf{r}_k) \in R^{l \times k}$ *from distribution* $\chi^{l \times k}$,
2. *Output* $\mathbf{a} = (\mathbf{a}_0^T, \mathbf{h}\mathbf{g}^T - \mathbf{a}_0^T\mathbf{R})^T \in R_q^{l+k}$ *and trapdoor* $\mathbf{R} \in R^{l \times k}$.

Moreover, the distribution of $\mathbf{a}$ *is close to uniform (either statistically or computationally) as long as the distribution of* $(\mathbf{a}_0^T, -\mathbf{a}_0^T\mathbf{R})$ *is.*

The correctness of Algorithm 4 is immediate. To show that the distribution of $(\mathbf{a}_0^T, -\mathbf{a}_0^T\mathbf{R})$ is close to uniform, we need to show that the distribution of $\mathbf{a}_0^T\mathbf{R}$ is close to uniform and hence is independent to that of $\mathbf{a}_0$, or equivalently the distribution of $\mathbf{a}_0^T\mathbf{r}_i$ is close to uniform and independent to that of $\mathbf{a}_0$ for all i. As for the trapdoors for general lattices, the uniformity of $\mathbf{a}$ can be instantiated to be either statistical by using a regularity lemma or computational by the ring-LWE assumption.

Lemma 2. *[3, Sect. 7] (Regularity Lemma) Let $\mathbf{a}_i \leftarrow R_q$ and $\mathbf{r}_i \leftarrow \chi$ for $i = 1, \ldots, l$. Then $\mathbf{b} = \sum_{i=1}^{l} \mathbf{a}_i \mathbf{r}_i$ is within $2^{-\Omega(n)}$ statistical distance to the uniform distribution over R_q. Moreover, the case where $l = 2$ corresponds to the normal form of ring-LWE.*

4.3 ring-LWE Inversion from New Trapdoors

Given a trapdoor $\mathbf{R}$ for $\mathbf{a} \in R_q^{l+k}$ and a $\mathsf{PLWE}_{f,q,\chi}^{(l)}$ instance $\mathbf{b} = \mathbf{a}s + \mathbf{e} \bmod q$, the ring-LWE inversion algorithm given in Algorithm 5 is to find the solution s to the instance.

Algorithm 5. $\mathsf{ringInvert}^{\mathcal{O}}(\mathbf{R}, \mathbf{a}, \mathbf{b})$
Input:

- *an oracle $\mathcal{O}$ for inverting the function $\alpha_{\mathbf{g}}(s', \mathbf{e}')$ when $\mathbf{e}' \in R^k$ is suitably small;*
- *a vector of ring element $\mathbf{a} \in R_q^{l+k}$;*
- *g-trapdoor $\mathbf{R} \in R^{l \times k}$ for $\mathbf{a}$ with tag $\mathbf{h}$;*
- *vector $\mathbf{b} = \mathbf{a}s + \mathbf{e}$ for any $s \in R_q$ and suitably small $\mathbf{e} \in R^{l+k}$.*

Output: s *and* $\mathbf{e}$.

1. *Get $(s', \mathbf{e}') \leftarrow \mathcal{O}([\mathbf{R}^T | \mathbf{I}_k]\mathbf{b})$.*
2. *return $s = \mathbf{h}^{-1}s'$ and $\mathbf{e} = \mathbf{b} - \mathbf{a}s$ (interpreted as a vector in R^{l+k} with where each entry has coefficients in $[-\frac{q}{2}, \frac{q}{2})$).*

The correctness of Algorithm 5 is indicated by Theorem 6 stated below.

Theorem 6. *Suppose that $\mathcal{O}$ in Algorithm 5 correctly inverts $\alpha_{\mathbf{g}}(s', \mathbf{e}')$ for any small error vector $\mathbf{e}' \in \mathcal{D}_{\mathbb{Z}^n, \sigma\sqrt{l\sigma^2 \cdot \omega(\log n) + k}}^k$. Then for any $s \in R_q$ and $\mathbf{e} \leftarrow \chi^{l+k}$, Algorithm 5 correctly inverts $\alpha_{\mathbf{a}}(s, \mathbf{e})$ with overwhelming probability over the choice of $\mathbf{e}$.*

Proof. We first show that $\mathbf{b}^T \begin{bmatrix} \mathbf{R} \\ \mathbf{I}_k \end{bmatrix}$ gives a correct input to the oracle $\mathcal{O}$.

$$
\begin{aligned}
\mathbf{b}^T \begin{bmatrix} \mathbf{R} \\ \mathbf{I}_k \end{bmatrix} &= (\mathbf{a}^T s + \mathbf{e}^T) \begin{bmatrix} \mathbf{R} \\ \mathbf{I}_k \end{bmatrix} \\
&= \mathbf{a}^T s \begin{bmatrix} \mathbf{R} \\ \mathbf{I}_k \end{bmatrix} + \mathbf{e}^T \begin{bmatrix} \mathbf{R} \\ \mathbf{I}_k \end{bmatrix} \\
&= s[\mathbf{a}_0^T | \mathbf{h}\mathbf{g}^T - \mathbf{a}_0^T \mathbf{R}] \begin{bmatrix} \mathbf{R} \\ \mathbf{I}_k \end{bmatrix} + \mathbf{e}^T \begin{bmatrix} \mathbf{R} \\ \mathbf{I}_k \end{bmatrix} \\
&= s(\mathbf{a}_0^T \mathbf{R} + \mathbf{h}\mathbf{g}^T - \mathbf{a}_0^T \mathbf{R}) + \mathbf{e}^T \begin{bmatrix} \mathbf{R} \\ \mathbf{I}_k \end{bmatrix} \\
&= \mathbf{g}^T \mathbf{h} s + \mathbf{e}^T \begin{bmatrix} \mathbf{R} \\ \mathbf{I}_k \end{bmatrix} \\
&= \mathbf{g}^T s' + \mathbf{e}^T \begin{bmatrix} \mathbf{R} \\ \mathbf{I}_k \end{bmatrix}
\end{aligned}
$$

Now we need to show that $\mathbf{e'} = \mathbf{e}^T \begin{bmatrix} \mathbf{R} \\ \mathbf{I}_k \end{bmatrix}$ has the appropriate distribution.
Consider

$$\mathbf{e}'_j = \sum_{i=1}^{l} \mathbf{e}_i \mathbf{r}_{ij} + \sum_{i=l+1}^{l+k} \mathbf{e}_i \quad \forall j = 1, \ldots, k$$

where $\mathbf{e}'_j = j$-th component of $\mathbf{e'}$, $\mathbf{e}_i = i$-th component of $\mathbf{e}$, $\mathbf{r}_{ij} = ij$-th component of $\mathbf{R}$. Since each entry of $\mathbf{e}$ and $\mathbf{R}$ are sampled from $\chi = \mathcal{D}_{\mathbb{Z}^n,\sigma}$, then the distribution of $\mathbf{e}_i \mathbf{r}_{ij}$ is statistically close to $\mathcal{D}_{\mathbb{Z}^n,\sigma^2 \cdot \omega(\sqrt{\log n})}$ [3, Lemma 8.7]. Hence, the distribution of $\mathbf{e}'_j$ is statistically close to $\mathcal{D}_{\mathbb{Z}^n,\sigma\sqrt{l\sigma^2 \cdot \omega(\log n)+k}}$ [3, Lemma 8.6]. Therefore, the distribution of $\mathbf{e'}$ has the correct distribution. $\qquad \square$

4.4   Gaussian Sampling from New Trapdoors

Given a trapdoor $\mathbf{R}$ for $\mathbf{a} \in R_q^{l+k}$ and $\mathbf{u} = \beta_{\mathbf{a}}(\mathbf{x}) = \mathbf{a}^T \mathbf{x} \bmod q$, the Gaussian Sampling algorithm given in Algorithm 7 is to find the solution $\mathbf{x}$ to the instance.

Algorithm 7. ringSampleD$^{\mathcal{O}}(\mathbf{R}, \mathbf{a}_0, \mathbf{h}, \mathbf{u}, \sigma)$
Input:
Offline phase:

- *an oracle $\mathcal{O}(\mathbf{v})$ for Gaussian sampling over a desired coset $\Lambda_q^{\mathbf{v}}(\mathbf{g}^T)$ with parameter σ, where $\mathbf{v} \in R_q$;*
- *a vector of ring elements $\mathbf{a}_0 \in R_q^l$;*
- *a trapdoor $\mathbf{R} \in R^{l \times k}$;*
- *a Gaussian parameter σ.*

Online phase:

- *a non-zero tag $\mathbf{h} \in R_q$ defining $\mathbf{a} = (\mathbf{a}_0^T, \mathbf{hg}^T - \mathbf{a}_0^T \mathbf{R})^T \in R_q^{l+k}$ ($\mathbf{h}$ may instead be provided in the offline phase, if it is known);*
- *syndrome $\mathbf{u} \in R_q$.*

Output: A vector $\mathbf{x}$ drawn from a distribution statistically close to $\mathcal{D}_{\Lambda_q^{\mathbf{v}}(\mathbf{a}_0^T),\sigma'}$ for some Gaussian parameter σ'.
Offline phase:

1. *Choose fresh perturbations $\mathbf{p}_1 \leftarrow \chi_1^l$ and $\mathbf{p}_2 \leftarrow \chi_2^k$ for some distributions χ_1 and χ_2 over R.*
2. *Compute $\mathbf{w}_0 = \mathbf{a}_0^T(\mathbf{p}_1 - \mathbf{R}\mathbf{p}_2) \in R_q$ and $\mathbf{w}_1 = \mathbf{g}^T \mathbf{p}_2 \in R_q$.*

Online phase:

1. *Let $\mathbf{v} \leftarrow \mathbf{h}^{-1}(\mathbf{u} - \mathbf{w}_0) - \mathbf{w}_1 = \mathbf{h}^{-1}(\mathbf{u} - \mathbf{a}^T \mathbf{p}) \in R_q$, and choose $\mathbf{z} \leftarrow \mathcal{D}_{\Lambda_q^{\mathbf{v}}(\mathbf{g}^T),\sigma}$ by calling $\mathcal{O}(\mathbf{v})$.*
2. *Return $\mathbf{x} \leftarrow \begin{bmatrix} \mathbf{p}_1 \\ \mathbf{p}_2 \end{bmatrix} + \begin{bmatrix} \mathbf{R} \\ \mathbf{I}_k \end{bmatrix} \mathbf{z}$.*

Theorem 8. *Algorithm 7 is correct.*

Proof. Let $\mathbf{x} \leftarrow \mathsf{ringSampleD}^{\mathcal{O}}(\mathbf{R}, \mathbf{a}_0, \mathbf{h}, \mathbf{u}, \sigma)$. Then

$$\mathbf{a}^T\mathbf{x} = [\mathbf{a}_0^T | \mathbf{h}\mathbf{g}^T - \mathbf{a}_0^T\mathbf{R}] \left(\begin{bmatrix} \mathbf{p}_1 \\ \mathbf{p}_2 \end{bmatrix} + \begin{bmatrix} \mathbf{R} \\ \mathbf{I}_k \end{bmatrix} \mathbf{z} \right)$$

$$= \mathbf{a}_0^T\mathbf{p}_1 + \mathbf{a}_0^T\mathbf{R}\mathbf{z} + \mathbf{h}\mathbf{g}^T\mathbf{p}_2 - \mathbf{a}_0^T\mathbf{R}\mathbf{p}_2 + \mathbf{h}\mathbf{g}^T\mathbf{z} - \mathbf{a}_0^T\mathbf{R}\mathbf{z}$$

$$= \mathbf{a}_0^T(\mathbf{p}_1 - \mathbf{R}\mathbf{p}_2) + \mathbf{h}\mathbf{g}^T\mathbf{p}_2 + \mathbf{h}\mathbf{v}$$

$$= \mathbf{w}_0 + \mathbf{h}\mathbf{w}_1 + \mathbf{u} - \mathbf{w}_0 - \mathbf{h}\mathbf{w}_1$$

$$= \mathbf{u}$$

Now, consider $\mathbf{x} = \begin{bmatrix} \mathbf{p}_1 \\ \mathbf{p}_2 \end{bmatrix} + \begin{bmatrix} \mathbf{R} \\ \mathbf{I}_k \end{bmatrix} \mathbf{z}$. Since each entry of $\mathbf{R}$ and $\mathbf{z}$ are sampled from $\chi = \mathcal{D}_{\mathbb{Z}^n,\sigma}$, and each entry of $\mathbf{p}_1$ and $\mathbf{p}_2$ are sampled from $\chi_1 = \mathcal{D}_{\mathbb{Z}^n,\sigma^2 \cdot \omega\sqrt{\log n}}$ and $\chi_2 = \mathcal{D}_{\mathbb{Z}^n,\sigma\sqrt{\sigma^2(k+1)\cdot\omega(\log n)-1}}$, respectively, then the distributions of all entries of $\mathbf{x}$ are statistically close to $\chi' = \mathcal{D}_{\mathbb{Z}^n,\sigma'}$, where $\sigma' = \sigma^2\sqrt{k+1} \cdot \omega(\sqrt{\log n})$. [3, Lemma 8.6 & 8.7]. $\qquad\square$

4.5 Trapdoors Delegation

Using the trapdoor of $\mathbf{a}$, there is an efficient trapdoor delegation algorithm given in Algorithm 9 that generates a trapdoor for the vector $(\mathbf{a}^T, \mathbf{a}_1^T)^T$.

Algorithm 9. $\mathsf{ringDelTrap}^{\mathcal{O}}(\mathbf{a}' = (\mathbf{a}^T, \mathbf{a}_1^T)^T, \mathbf{h}', \sigma)$
Input:

– *an oracle $\mathcal{O}$ for discrete Gaussian sampling over cosets of $\Lambda_q^\perp(\mathbf{a}^1)$ with parameter σ';*
– *a vector of ring elements $\mathbf{a}' = (\mathbf{a}^T, \mathbf{a}_1^T)^T \in R_q^{m+k}$;*
– *a non-zero ring element $\mathbf{h}' \in R_q$.*

Output: a trapdoor $\mathbf{R} \in R^{(m+k)\times k}$ for $\mathbf{a}'$ with tag $\mathbf{h}'$.

– *Using $\mathcal{O}$, sample each column of $\mathbf{R}$ independently from a discrete Gaussian with parameter σ' over the appropriate cosets of $\Lambda_q^\perp(\mathbf{a}^T)$, so that $\mathbf{a}^T\mathbf{R} = \mathbf{h}'\mathbf{g}^T - \mathbf{a}_1^T$.*

5 INDr-ID-CCA-Secure (H)IBE in Ideal Lattices

5.1 Identity-Based Encryption

Identity-based encryption (IBE) is a generalization of public-key encryption [18]. In an IBE, to encrypt a message to an identity *id*, the encrypter does not need to lookup the public key for the intended identity *id*. Instead, the encryption algorithm simply takes the public parameters, the identity *id* and the message as input and outputs a ciphertext encrypting the message to *id*. The identity *id* obtains its secret key derived from the master secret key through the key generation algorithm for decrypting all ciphertexts encrypted to *id*. Formally, the syntax of IBE is defined as follows.

Definition. An identity-based encryption (IBE) scheme consists of four PPT algorithms (Setup, Extract, Encrypt, Decrypt). The Setup algorithm outputs a public parameter PP and a master key MK. Using MK, the Extract algorithm extracts a secret key SK_{id} for an identity id. Unlike public-key encryption, the Encrypt algorithm in IBE can encrypt messages directly to an identity id. The user with identity id uses her secret key SK_{id} to Decrypt a ciphertext.

Hierarchical identity-based encryption (HIBE) is an extension of IBE such that an identity $ID = [id_1|\ldots|id_d]$ is a hierarchy of identities with depth d. There is an additional algorithm Derive that inputs a secret key $SK_{[id_1|\ldots|id_{j-1}]}$ in the $(j-1)$-th level and an identity $ID = [id_1|\ldots|id_j]$ in the j-th level and outputs the secret key SK_{ID} for identity ID.

Security. In addition to the (adaptive) chosen-plaintext-attack (CPA(2)) security or (adaptive) chosen-ciphertext-attack (CCA(2)) security as in public-key encryption, an (H)IBE scheme should also be secure against chosen-identity-attack (ID). A weaker security model called selective-identity-attack (sID) is also considered, where the adversary must choose the identity she is going to perform CPA(2) or CCA(2) before receiving the public parameter. The indistinguishability (IND) of ciphertexts under the combinations of attacks in the two sets of variants give us eight security model, namely, IND-ID-CCA(2), IND-sID-CCA(2), IND-ID-CPA(2) and IND-sID-CPA(2). In [8], a stronger security guarantee in which the ciphertexts are indistinguishable from random (INDr) elements in the ciphertext space is considered. This implies both semantic security (of the plaintext) and recipient anonymity.

From now on, we will focus on the INDr-ID-CCA security, which is modeled as a security game between a PPT simulator and a PPT adversary. In this game, the simulator first generates the public parameters PP and passes them to the adversary. The adversary is then granted the rights to query the secret keys SK_{id} for polynomially many identities id of its choice, and the rights to query the decryption of any ciphertext of its choice. After that, the adversary issues a challenge message to be encrypted to the identity id^*, whose secret key has never been queried before. The simulator replies by either encrypting the challenge message to id^* or generating a uniformly random ciphertext. The adversary wins the game if it can guess which among two ways the ciphertext is generated.

An INDr-ID-CCA2-secure HIBE of depth d can be obtained by combining an INDr-ID-CPA-secure HIBE of depth $d+1$ and a strong one-time signature scheme [19]. The rough idea is to encrypt the message to the "identity" $[id_1|id_2|\ldots|id_d|vk]$ where vk is the verification key of the one-time signature, and sign the ciphertext using the one-time signature. In particular, from IBE, we obtain a CCA2-secure public-key encryption scheme. We will omit the details here.

Applications. Most earlier (H)IBE schemes are realized by pairings. Readers can refer to [20, 21] for reviews of those. Agrawal *et al.* proposed a lattice-based (H)IBE scheme in the standard model [8]. The ciphertext of their scheme can be proven to be a random element in the ciphertext space, which implies receipt anonymity against user attacks. An anonymous IBE can be used to obtain a public-key searchable encryption scheme [22]. By applying Naor's transformation [18], we can also obtain a signature scheme from an IBE scheme.

5.2   Construction

Using the trapdoors for ideal lattices developed above, the CCA-secure public-key encryption provided in [12] and the INDr-ID-CPA-secure (H)IBE scheme in [8], we construct an INDr-ID-CCA-secure (H)IBE in ideal lattices. We only present the basic IBE scheme below, because the HIBE scheme can be obtained trivially by defining the Derive algorithm of HIBE to be the same as the Extract algorithm of the basic IBE in our case.

The basic IBE scheme is constructed as follows:

- Setup(1^λ)
    - Sample $\mathbf{a}_{-1} \leftarrow R_q^l$, $\mathbf{a}_0, \mathbf{a}_1, \ldots, \mathbf{a}_t \leftarrow R_q^k$ and $\mathbf{h} \leftarrow R_q \setminus \{0\}$.
    - Sample $(\mathbf{a}, \mathbf{R}_{MK}) \leftarrow \mathsf{ringGenTrap}^{\mathcal{D}}(\mathbf{a}_{-1}, \mathbf{h})$.
    - Output $PP = (\mathbf{a}, \mathbf{a}_0, \mathbf{a}_1, \ldots, \mathbf{a}_t)$ and $MK = (\mathbf{h}, \mathbf{R}_{MK})$.
- Extract(PP, MK, id)
    - Sample $\mathbf{h}_{id} \leftarrow R_q$.
    - Set $\mathbf{a}_{id} = \mathbf{a}_0 + \sum_{i=1}^t id_i \mathbf{a}_i$ and $\mathbf{f}_{id, \mathbf{h}_{id}} = (\mathbf{a}^T, \mathbf{a}_{id}^T)^T$.
    - Sample $\mathbf{R}_{id} \leftarrow \mathsf{ringDelTrap}^{\mathcal{O}}(\mathbf{f}_{id, \mathbf{h}_{id}}, \mathbf{h}_{id}, \sigma)$.
    - Output $SK_{id} = (\mathbf{h}_{id}, \mathbf{R}_{id})$.
- Encrypt($PP, id, \mathbf{m} \in R_p^k$)
    - Sample $\mathbf{h}' \leftarrow R_q$.
    - Set $\mathbf{a}_{id, \mathbf{h}'} = \mathbf{a}_{id} + \mathbf{h}'\mathbf{g}$.
    - Sample $\mathbf{s} \leftarrow \chi$, $\mathbf{e}_0 \leftarrow \chi^{l+k}$ and $\mathbf{R}_i \leftarrow \chi^{l \times k}$ for $i = 0, 1, \ldots, t$.
    - Set $\mathbf{R} = \mathbf{R}_0 + \sum_{i=1}^t id_i \mathbf{R}_i$.
    - Set $\mathbf{e}_1^T = -\mathbf{e}_0^T \mathbf{R}$.
    - Compute $\mathbf{u} = \mathbf{a}_0 \mathbf{s} + p\mathbf{e}_0$ and $\mathbf{v} = \mathbf{a}_{id, \mathbf{h}'}\mathbf{s} + p\mathbf{e}_1 + \mathbf{m}$.
    - Output $CT = (\mathbf{h}', \mathbf{u}, \mathbf{v})$.
- Decrypt(PP, SK_{id}, CT)
    - Output $\perp$ if $\mathbf{h}' = -\mathbf{h}_{id}$.
    - Set $\mathbf{f}_{id, (\mathbf{h}_{id}+\mathbf{h}')} = (\mathbf{a}^T, \mathbf{a}_{id}^T + \mathbf{h}'\mathbf{g}^T)^T = (\mathbf{a}^T, (\mathbf{h}_{id} + \mathbf{h}')\mathbf{g}^T - \mathbf{a}^T \mathbf{R}_{id})^T$.
    - Compute $(\mathbf{s}, \mathbf{e}) \leftarrow \mathsf{ringInvert}^{\mathcal{O}}(\mathbf{R}_{id}, \mathbf{f}_{id, (\mathbf{h}_{id}+\mathbf{h}')}, (\mathbf{u}^T, \mathbf{v}^T)^T)$.
    - Compute $(\mathbf{0}_{1 \times k}, \mathbf{m}^T)^T = \mathbf{e} \bmod p$.
    - Output $\mathbf{m}$.

Theorem 10. *By the* $\mathsf{PLWE}_{f,q,\chi}^{(l)}$ *assumption, the (H)IBE scheme stated above is INDr-ID-CCA-secure.*

Proof. The simulation strategy is a result of combining those from [8,12]. The simulator is given a PLWE instance $(\mathbf{a}, \mathbf{b})$. It chooses the rest of the public parameters $\mathbf{a}_0, \ldots, \mathbf{a}_t$ as follows:

- It samples $\mathbf{h}^* \leftarrow R_q$.
- It samples $\mathbf{R}_i \leftarrow \chi^{l \times k}$ and let $\mathbf{a}_i^T = \mathbf{h}_i \mathbf{g}^T - \mathbf{a}^T \mathbf{R}_i$, where $\mathbf{h}_0 = 1 - \mathbf{h}^*$ and $\mathbf{h}_i \leftarrow R_q$, for $i = 0, \dots, t$.

Since the distribution of $\mathbf{a}$ is uniform and each entry of $\mathbf{R}_i$ is sampled from the distribution χ, by the regularity lemma (Lemma 2), the distribution of $\mathbf{a}_i$ is also uniform for all i.

To answer the queries for secret key of id, it simply returns $(\mathbf{h}_{id}, \mathbf{R}_{id})$ where $\mathbf{h}_{id} = 1 + \sum_{i=1}^{t} id_i \mathbf{h}_i - \mathbf{h}^*$ and $\mathbf{R}_{id} = \mathbf{R}_0 + \sum_{i=1}^{t} id_i \mathbf{R}_i$. Note that $\mathbf{a}_{id}^T = \mathbf{a}_0^T + \sum_{i=1}^{t} id_i \mathbf{a}_i^T = \mathbf{h}_{id} \mathbf{g}^T - \mathbf{a}^T \mathbf{R}_{id}$.

It answers the decryption queries $(id, CT = (\mathbf{h}, \mathbf{u}, \mathbf{v}))$ using the tag $\mathbf{h}_{id}$, the trapdoor $\mathbf{R}_{id}$ and the Decrypt algorithm. As long as $\mathbf{h} \neq \mathbf{h}^*$ or $-\mathbf{h}_{id} \neq \mathbf{h}^*$, the simulator can still simulate faithfully. Since $\mathbf{a}_i$ is uniformly random in the view of the adversary, $\mathbf{h}^*$ and $\mathbf{h}_i$ are all hidden from the adversary for all $i = 0, 1, \dots, t$. Therefore the event $-\mathbf{h}_{id} = \mathbf{h} = \mathbf{h}^*$ only happens with negligible probability.

Finally, the challenge ciphertext for $(id^*, \mathbf{m}^*)$ is generated as $(\mathbf{h}^*, \mathbf{u}^*, \mathbf{v}^*)$ where $\mathbf{u}^* = \mathbf{b}$ and $\mathbf{v}^* = (-\mathbf{b}^T \mathbf{R}_{id^*} + \mathbf{m}^{*T})^T$.

Suppose that the $(\mathbf{a}, \mathbf{b})$ given in the PLWE instance is uniform, then the distribution of the challenge ciphertext $(\mathbf{h}^*, \mathbf{u}^*, \mathbf{v}^*)$ is also uniform. Otherwise, suppose $\mathbf{b} = \mathbf{a}s + p\mathbf{e}$ for some s and $\mathbf{e}$ sampled from the appropriate distributions, then

$$
\begin{aligned}
\mathbf{v}^{*T} &= -\mathbf{b}^T \mathbf{R}_{id^*} + \mathbf{m}^{*T} \\
&= -s\mathbf{a}^T \mathbf{R}_{id^*} - p\mathbf{e}^T \mathbf{R}_{id^*} + \mathbf{m}^{*T} \\
&= s(\mathbf{a}_{id^*}^T - \mathbf{h}_{id^*} \mathbf{g}^T) - p\mathbf{e}^T \mathbf{R}_{id^*} + \mathbf{m}^{*T}
\end{aligned}
$$

By [8, Lemma 24], we have $\mathbf{h}_{id^*} = -\mathbf{h}^*$ with non-negligible probability. In such case, we have

$$
\mathbf{v}^{*T} = s(\mathbf{a}_{id^*}^T + \mathbf{h}^* \mathbf{g}^T) - p\mathbf{e}^T \mathbf{R}_{id^*} + \mathbf{m}^{*T}
$$

which is distributed identically as valid ciphertexts do. □

6   Concluding Remarks

We detailed how to generate trapdoors for ideal lattices. We then use it to construct a new (H)IBE scheme. Our scheme has several improvement over that constructed by Agrawal *et al.* [8]:

- Our scheme is based on ideal lattices, therefore the size of the public parameters, master key and the identity secret key are reduced by a factor of n.
- Using the new trapdoor delegation algorithm, the size of the identity secret key grows linearly, rather than quadratically, in the depth of the hierarchy.
- Our scheme is secure against chosen-chiphertext-attack.

References

1. Gentry, C.: Fully homomorphic encryption using ideal lattices. In: STOC, pp. 169–178 (2009)
2. Brakerski, Z., Vaikuntanathan, V.: Fully homomorphic encryption from ring-LWE and security for key dependent messages. In: Rogaway, P. (ed.) CRYPTO 2011. LNCS, vol. 6841, pp. 505–524. Springer, Heidelberg (2011)
3. Lyubashevsky, V., Peikert, C., Regev, O.: A toolkit for ring-LWE cryptography. In: Johansson, T., Nguyen, P.Q. (eds.) EUROCRYPT 2013. LNCS, vol. 7881, pp. 35–54. Springer, Heidelberg (2013)
4. Gentry, C., Sahai, A., Waters, B.: Homomorphic encryption from learning with errors: conceptually-simpler, asymptotically-faster, attribute-based. In: Canetti, R., Garay, J.A. (eds.) CRYPTO 2013, Part I. LNCS, vol. 8042, pp. 75–92. Springer, Heidelberg (2013)
5. Boneh, D., Freeman, D.M.: Homomorphic signatures for polynomial functions. In: Paterson, K.G. (ed.) EUROCRYPT 2011. LNCS, vol. 6632, pp. 149–168. Springer, Heidelberg (2011)
6. Boneh, D., Freeman, D.M.: Linearly homomorphic signatures over binary fields and new tools for lattice-based signatures. In: Catalano, D., Fazio, N., Gennaro, R., Nicolosi, A. (eds.) PKC 2011. LNCS, vol. 6571, pp. 1–16. Springer, Heidelberg (2011)
7. Garg, S., Gentry, C., Halevi, S.: Candidate multilinear maps from ideal lattices. In: Johansson, T., Nguyen, P.Q. (eds.) EUROCRYPT 2013. LNCS, vol. 7881, pp. 1–17. Springer, Heidelberg (2013)
8. Agrawal, S., Boneh, D., Boyen, X.: Efficient lattice (H)IBE in the standard model. In: Gilbert, H. (ed.) EUROCRYPT 2010. LNCS, vol. 6110, pp. 553–572. Springer, Heidelberg (2010)
9. Agrawal, S., Boneh, D., Boyen, X.: Lattice basis delegation in fixed dimension and shorter-ciphertext hierarchical IBE. In: Rabin, T. (ed.) CRYPTO 2010. LNCS, vol. 6223, pp. 98–115. Springer, Heidelberg (2010)
10. Gentry, C., Peikert, C., Vaikuntanathan, V.: Trapdoors for hard lattices and new cryptographic constructions. In: STOC, pp. 197–206 (2008)
11. Alwen, J., Peikert, C.: Generating shorter bases for hard random lattices. Theory Comput. Syst. 48(3), 535–553 (2011)
12. Micciancio, D., Peikert, C.: Trapdoors for lattices: simpler, tighter, faster, smaller. In: Pointcheval, D., Johansson, T. (eds.) EUROCRYPT 2012. LNCS, vol. 7237, pp. 700–718. Springer, Heidelberg (2012)
13. Stehlé, D., Steinfeld, R., Tanaka, K., Xagawa, K.: Efficient public key encryption based on ideal lattices. In: Matsui, M. (ed.) ASIACRYPT 2009. LNCS, vol. 5912, pp. 617–635. Springer, Heidelberg (2009)
14. Lyubashevsky, V., Peikert, C., Regev, O.: On ideal lattices and learning with errors over rings. In: Gilbert, H. (ed.) EUROCRYPT 2010. LNCS, vol. 6110, pp. 1–23. Springer, Heidelberg (2010)
15. Regev, O.: On lattices, learning with errors, random linear codes, and cryptography. J. ACM 56(6), 34:1–34:40 (2009)
16. Brakerski, Z., Langlois, A., Peikert, C., Regev, O., Stehlé, D.: Classical hardness of learning with errors. In: STOC, pp. 575–584 (2013)
17. Babai, L.: On Lovász' Lattice Reduction and the Nearest Lattice Point Problem (Shortened Version). In: Mehlhorn, K. (ed.) STACS 1985. LNCS, vol. 182, pp. 13–20. Springer, Heidelberg (1985)

18. Boneh, D., Franklin, M.K.: Identity-based encryption from the Weil pairing. SIAM J. Comput. **32**(3), 586–615 (2003)
19. Boneh, D., Canetti, R., Halevi, S., Katz, J.: Chosen-ciphertext security from identity-based encryption. SIAM J. Comput. **36**(5), 1301–1328 (2007)
20. Chow, S.S.M.: Removing escrow from identity-based encryption. In: Jarecki, S., Tsudik, G. (eds.) PKC 2009. LNCS, vol. 5443, pp. 256–276. Springer, Heidelberg (2009)
21. Chow, S.S.M.: New privacy-preserving architectures for identity-/attribute-based encryption. Ph.D. thesis, New York University (2010)
22. Abdalla, M., Bellare, M., Catalano, D., Kiltz, E., Kohno, T., Lange, T., Malone-Lee, J., Neven, G., Paillier, P., Shi, H.: Searchable encryption revisited: consistency properties, relation to anonymous IBE, and extensions. J. Cryptol. **21**(3), 350–391 (2008)

Block Cipher and Hash Function

Speeding Up the Search Algorithm for the Best Differential and Best Linear Trails

Zhenzhen Bao[✉], Wentao Zhang, and Dongdai Lin

State Key Laboratory of Information Security, Institute of Information Engineering,
Chinese Academy of Sciences, Beijing, China
{baozhenzhen,zhangwentao,ddlin}@iie.ac.cn

Abstract. For judging the resistance of a block cipher to differential cryptanalysis or linear cryptanalysis it is necessary to establish an upper bound on the probability of the best differential or the bias of the best linear approximation. However, getting a tight upper bound is not a trivial problem. We attempt it by searching for the best differential and the best linear trails, which is a challenging task in itself. Based on some previous works, new strategies are proposed to speed up the search algorithm, which are called *starting from the narrowest point, concretizing and grouping search patterns*, and *trialling in minimal changes order* strategies. The efficiency of the resulting improved algorithms allows us to state that the probability (bias) of the *best* 4-round differential (linear) trail in NOEKEON is 2^{-51} (2^{-25}) and the probability (bias) of the best 10-round (11-round) differential (linear) trail is at most 2^{-131} (2^{-71}). For SPONGENT, the *best* differential trails for certain number of rounds in the permutation functions with width $b \in \{88, 136, 176, 240\}$ are found. That allows us to update some results presented by its designers.

Keywords: Differential cryptanalysis · Linear cryptanalysis · Differential trail · Linear trail · Search algorithm · Optimization · NOEKEON · SPONGENT

1 Introduction

Differential cryptanalysis (DC) [1] and linear cryptanalysis (LC) [2] are two of the most powerful attacks against modern block ciphers in which an adversary exploits good differentials or good linear approximations. The first step in a differential or a linear attack consists in finding differentials or linear approximations of the cipher with probabilities or bias as high as possible. In most cases, differential trails with highest probability and linear approximation trails with largest bias can be used to estimate the power of the corresponding attack. Differential (linear) trails consist of a sequence of differences (approximations) through the rounds of the primitive and those with the highest probability (the largest bias) are called the best. However, the problem of searching best trails is not trivial, because of the great cardinality of the set of candidates [3,4].

© Springer International Publishing Switzerland 2015
D. Lin et al. (Eds.): Inscrypt 2014, LNCS 8957, pp. 259–285, 2015.
DOI: 10.1007/978-3-319-16745-9_15

For many block ciphers, such as AES, NOEKEON and PRESENT, researchers prefer counting the minimal number of active S-Boxes to get the upper bound of the best probability (bias) of differential (linear) trails [5–8]. In this method, concepts such as branch number and structures of the linear layer are used. Furthermore, tools using MILP are developed [9]. However, those approaches could only provide a kind of differential trails without the instantiated actual differences or without the knowledge of exact probabilities of those trails. Remarkably, authors of [10] use a variant of Dijkstra's algorithm which is essentially a breadth-first search to efficiently find all best truncated differential trails with minimal number of active S-Boxes and instantiate them with actual differences. This method is very powerful, however, on one hand, it may fail to find the best differential trail which does not have the minimal number of active S-Boxes, and on the other hand it is powerless in the case of bit-oriented ciphers. Specifically, although this breadth-first approach is in polynomial time in the number of rounds, it is exponential in the state. Thus, for ciphers using large number of small S-Boxes, which is typically 4 bits wide, and have weak alignment, an intermediate state tends to large, and a PC cannot store all the intermediate state, that is, we cannot choose a breadth-first strategy. Thus, we seek for depth-first method to find the best trails.

In 1994, Matsui proposed a branch-and-bound depth-first search algorithm making it possible to effectively find the best differential trails and linear approximation trails of DES [11]. Unfortunately, his method is not fast enough for some other cryptosystems like FEAL. Consequently, improvements on Matsui's algorithm were studied by Moriai et al. [3] and Aoki et al. [12]. The work in [3] was based on analyzing the dominant factor of search complexity and it introduced the concept of search patterns in order to reduce unnecessary search candidates. The authors successfully obtained new results on best linear approximations for FEAL by applying the proposed search algorithm. In [12], Aoki et al. further optimized the search algorithm. They presented good results of the search for the best differential trails of FEAL using a pre-search strategy.

Recently, automatic tools for searching for differential trails in ARX ciphers are relatively mature [13,14]. One of them [14] is also extended from Matsui's algorithm. However, due to the fact that it uses a partial, rather than the full DDT, their algorithm is not guaranteed to find the best differential trail. To the best of our knowledge, there is no application of those tools which are designed for ARX ciphers to Sbox-based ciphers.

For modern Sbox-based ciphers, expanding block size and good diffusion cause the probabilities of the best trails of very short rounds to be tiny. Generally, the smaller the probability of a best trail, the longer the time of a search will be. The heuristic search algorithm described in [4] might be helpful, but could hardly satisfy the cryptographers to fully estimate the vulnerability of a modern cipher to DC and LC. For designers who need to repeatedly apply the search algorithm to their draft ciphers to choose the best possible components and to decide a proper number of rounds, and for attackers who want to obtain large sets of trails with probabilities as high as possible and with rounds as many as possible, it is profitable to further optimize the search algorithm.

In this paper, we focus on this problem and aim to speed up the depth-first search algorithm for the best actual differential and linear trails.

The target objects are of Sbox-based iterated block cipher [5] in which all intermediate rounds use the same round transformation. We only consider those iterated ciphers with round keys being added to the state by means of XOR operation which is very common in modern block ciphers.

1.1 Our Contributions

We present three new optimization strategies to speed up the search algorithm for the best trails, which are called *starting from the narrowest point*, *concretizing and grouping search patterns*, and *trialling in minimal changes order* strategies.

- *Starting from the narrowest point* is very helpful to reduce complexity to a great extent by raising the threshold to the candidates at the earliest phases of the search procedure and maximizing shareable work at those phases.
- *Concretizing and grouping search patterns* further maximizes the scope of shared works and collects more information on search patterns to filter out invalid ones, while keeping the memory requirement appropriate.
- *Trialling in minimal changes order* utilizes the locality of the nonlinear layer and the linearity of the linear layer, to tame the brute force search to behave in a systematical and efficient manner.

Experimental results show that, the first two strategies bring a speed up by a factor of around 740–2800, which can be seen in Table 2. Considering the profit brought by the third strategy, the resulting improved algorithm has around 1500–5000 speedup ratio for the experimental subject.

Our final improved algorithm has been applied to search for the best differential and best linear trails in a block cipher named NOEKEON which was designed by Joan Daemen et al. [6]. The efficiency of the improved algorithm allows us to find out the *best* trails, thus to state that probability (bias) of the best 4-round differential (linear) trail in NOEKEON is 2^{-51} (2^{-25}). Additionally, probability (bias) of the best 5-round and 6-round trails are $\leq 2^{-65}$ ($\leq 2^{-32}$) and $\leq 2^{-80}$ ($= 2^{-40}$) respectively. That allows us to claim that the probability (bias) of the best 10-round (11-round) differential (linear) trail in NOEKEON is at most 2^{-131} (2^{-71}). The results are summarized in Table 3. Besides, we found out the longest linear trail holding with bias larger then 2^{-65}, which is a 9-round trail with bias 2^{-62}. These improved positive results contribute to the estimation of the security of NOEKEON against differential (linear) cryptanalysis.

We have also used this final improved algorithm to search for the best differential trails in the permutation functions of SPONGENT, a hash function. We found out the *best* differential trails for variants with width $b \in \{88, 136, 176, 240\}$ for certain number of rounds, and update some results presented by its designers in [16]. Some of the results are summarized in Table 4.

These strategies are also useful for us to search for the best differential and best linear trails for other primitives and helpful to search for best multiple differential (multi-dimensional linear) distinguishers.

By the way, all of the experiments and results in this paper are timed and obtained on a PC with Intel(R) Core(TM) i5-4570S 2.90 GHz CPU, and 4 GB RAM, using single-thread program in C.

1.2 Organization

The paper is organized as follows. Some preliminaries and symbolic conventions are presented in Sect. 2. In Sect. 3, we introduce and briefly discuss three previous works including Matsui's algorithm, Moriai et al.'s algorithm and Aoki et al.'s algorithm. Section 4 establishes our overall strategies and basic principles. The three optimization strategies *starting from the narrowest point, concretizing and grouping search patterns* and *trialling in minimal changes order* are covered in Sects. 5–7, which provide the justification and experimental results on the efficiency. Finally, new results on best trails in a block cipher NOEKEON and in the permutation functions of a hash function SPONGENT are shown in Sect. 8. In Sect. 9, we conclude our algorithm and prospect for further improvement.

2   Notations and Preliminaries

For convenience, we will explain the optimization strategies with SPN ciphers with non-linear layer being a parallel execution of 4×4 - S-Boxes in our mind. While, proposed strategies are also applicable to ciphers of Feistel structure and with larger S-Boxes.

Because of the duality between the search for the best differential trails and the search for the best linear trails [11], we will explain the optimization strategies from the perspective of differential.

A more natural way will be used to characterize the power of trails - the weight of differential trail which is the sum of the weight of round differentials, where the latter is the negative of its binary logarithm of its probability [15, Chap. 5].

id_r: the input difference of the r-th round differential[1]
od_r: the output difference of the r-th round differential
p_r: the probability of the r-th round differential
w_r: the weight of the r-th round differential, $w_r = - \log_2 p_r$
$w_{(id_r, od_r)}$: the weight of the r-th round differential (id_r, od_r)
w^r: the weight of a r-round differential trail, $w^r = \sum_{i=1}^{r} w_i$, where w_i is the
      weight of the i-th round differential composing that r-round differential trail
Bw^r: the weight of the best r-round differential trail
Bwc^r: the candidate of Bw^r
ASN: the abbreviation of *Number of Active S-Boxes*
asn_r: ASN at the S-Layer of the r-th round[2]

---

[1] We index the rounds begin with 1, i.e. $1 \leq r \leq n$, where n is the number of rounds of a block cipher.

[2] When using the *starting from the narrowest point strategy*, we index the rounds relatively to the narrowest point.

3 Previous Works

3.1 Matsui's Algorithm

Matsui's algorithm [11] works by induction on the number of rounds n and derives the best n-round weight Bw^n from the knowledge of all best r-round weight Bw^r ($1 \leq r \leq n - 1$). The original search algorithm targets DES. Here, we summarize Matsui's algorithm for SPN ciphers. The framework consists of the recursive procedures described in Algorithm 1. In Algorithm 1, Bwc^n holds the temporary approximation of the value of Bw^n. It is an upper bound of Bw^n and improved in a decreasing manner during the search bounded by conditions $\sum_{i=1}^{r} w_i + Bw^{n-r} \leq Bwc^n$ ($1 \leq r \leq n - 1$). When all of the possible paths have been traversed, Bwc^n turns to be the exact value of Bw^n.

Algorithm 1. Matsui's Algorithm

```
 1: Bwcⁿ ← an upper bound of Bwⁿ          14:     for all candidate of od_r do
 2: procedure ROUND-1                      15:         w_r ← w_(id_r,od_r)
 3:     for all candidate of od₁ do        16:         if ∑ʳ_{i=1} w_i + Bwⁿ⁻ʳ ≤ Bwcⁿ then
 4:         w₁ ← min_{id₁}(w_(id₁,od₁))    17:             ROUND-i(r + 1)
 5:         if w₁ + Bwⁿ⁻¹ ≤ Bwcⁿ then      18:         end if
 6:             ROUND-i(2)                  19:     end for
 7:         end if                          20:     else
 8:     end for                             21:         w_n ← min_{od_n}(w_(id_n,od_n))
 9:     Exit the program                    22:         if ∑ⁿ_{i=1} w_i < Bwcⁿ then
10: end procedure                           23:             Bwcⁿ ← ∑ⁿ_{i=1} w_i
11: procedure ROUND-i(r) (2 ≤ r ≤ n)        24:         end if
12:     id_r ← od_{r-1}                     25:     end if
13:     if r < n then                       26: end procedure
```

3.2 Moriai et al.'s Algorithm

Moriai et al.'s program [3] is based on Matsui's algorithm. The concept of *search patterns* was introduced to detecting the unnecessary and impossible search candidates.

Definition 1 (Search Pattern [3]). *An n-round search pattern used in the search for the best differential trail is a vector of n values of weights, which is denoted as $\mathbb{W}^n = (w_1, w_2, \dots, w_n)$, where w_i is the weight of the i-th round differential ($1 \leq i \leq n$). Let $|\mathbb{W}^n| \equiv \sum_{i=1}^{n} w_i$.*

Given n and Bwc^n which is a lower bound of Bw^n, their algorithm first generates all possible patterns[3] using Bw^r ($1 \leq r \leq n - 1$). It then examines whether there is a differential trail fitting one of the patterns. If none of the patterns has a real trail, another candidate for Bw^n is similarly trialled. Their algorithm is summarized as Algorithm 2. There are two improvements in Algorithm 2 compared with Algorithm 1. As for the first improvement, knowledge of all weights

---

[3] For simplicity, we sometimes address *search patterns* with the term *patterns*.

of best r-round trails is more sufficiently used by observing that $\forall r, i$ $(1 \le r \le n - 1, 1 \le i \le n - r + 1)$, $\sum_{j=i}^{i+r-1} w_j \ge Bw^r$. Thus it can delete more non-existent candidates. As for the second improvement which targets involutory ciphers, concept of search patterns is used and patterns are classified into two equivalent classes, the class which has more candidates is deleted and thus it can delete duplicate candidates and reduce the computation complexity.

Algorithm 2. Moriai et al.'s Algorithm

1: $Bwc^n \leftarrow$ a lower bound of Bw^n
2: **while** *true* **do**
3:     Generate all search patterns $(w_1, w_2, \ldots, w_n)$ which make the following hold:
     1. $\sum_{i=1}^{n-r} w_i + Bw^r \le Bwc^n$, for $1 \le r \le n - 1$
     2. $\forall r, i$ $(1 \le r \le n - 1, \ 1 \le i \le n - r + 1)$, $\sum_{j=i}^{i+r-1} w_j \ge Bw^r$.
4:     Discard either search pattern $(w_1, w_2, \ldots, w_n)$ or $(w_n, w_{n-1}, \ldots, w_1)$ whichever has more search candidates.
5:     Search the differential trails corresponding to the search patterns as Algorithm 1 with all the inequalities replaced by equalities.
6:     **if** find out a trail **then** $Bw^n \leftarrow Bwc^n$ and Exit
7:     **end if**
8:     $Bwc^n \leftarrow Bwc^n + 1$
9: **end while**

3.3 Aoki et al.'s Algorithm

In Algorithm 2 restrictions are merely based on weights of the best, while patterns of the best and patterns of r-round trails which are not the best cannot be used. Aoki et al.'s [12] considered the patterns themselves. They checked about the existence of the combined patterns using information about search patterns of differential trails with various weights by a pre-search strategy. Their algorithm is summarized as Algorithm 3.

Algorithm 3. Aoki et al.'s Algorithm

1: **procedure** PRE-SEARCH
2:     Search r-round $(r < n)$ differential trails with various weights, and compile information to the extent possible whether or not the search pattern exist for each round and weight.
3: **end procedure**
4: **procedure** SEARCH
5:     Do as Algorithm 2, but discard the search patterns which do not exist using the information from the pre-search phase.
6: **end procedure**

---

4   Overall Strategy and Basic Principle in Our Work

Based on the three previous works, our program works by inducting on the number of rounds n, using concept of the search patterns and doing pre-search. For n-round cipher, w^n is initialized with a lower bound of Bw^n and it is increased by an unit until exceeding the range of weight we considered. During this procedure, Bw^n is determined when the first time an n-round differential trail being found out. For each temporary value of w^n, we generate the corresponding search pattern set using Bw^r ($1 \leq r \leq n-1$) as in Algorithm 2 and using the information about existence of r-round ($1 \leq r \leq n-1$) search patterns collected during the preceding search phases. Within each search pattern set, the search traverses in a depth first manner. In Appendix A, framework of our search approach is shown in Algorithm 4. Algorithms 5–8 formalize the search procedure deploying the *starting from the narrowest point, concretizing and grouping search patterns*, and *trialling in minimal changes order* strategies which will be explained in the following Sects. 5–7.

5   Starting from the Narrowest Point Strategy

It has been shown in [3] that for Feistel ciphers the complexity of search for the best n-round trails is dominated by the number of candidates in procedures of the first two rounds. Similarly, for SPN ciphers, we found that the complexity of search is dominated by the number of candidates of the first two rounds (i.e., the first two layers in the depth-first search procedure). Generally, the number of candidates of the first two rounds greatly depends on the weight of the first round, the smaller the weight, the less the number of candidates. Thus, reducing the number of candidates at the first two layers will be much helpful. In this section, we propose our first strategy, i.e., *starting from the narrowest point* strategy. By using this strategy, the number of candidates at the first two layers can be reduced greatly.

5.1   Proposal and Justification of the Starting from the Narrowest Point Strategy

We organize the search patterns using a kind of balance trees, with roots starting from the *narrowest point* instead of the first point of the search patterns, see Definition 2 and Example 1 for clarify.

Definition 2 (Narrowest Point and Relative-Index Form). *Given an n-round search pattern* $\mathbb{W}^n = (w_1, w_2, \ldots, w_n)$, *suppose there are k minimal components* $w_{x_1}, w_{x_2}, \ldots, w_{x_k}$, *i.e.* $w_{x_i} = w_{min} \equiv \min(w_1, w_2, \ldots, w_n)$ *for* $1 \leq i \leq k$. *Let* $\mathrm{nxt}(x) \equiv \begin{cases} x+1 & 1 \leq x \leq n-1 \\ n-1 & x = n \end{cases}$, *and* $v_{min} \equiv \min(w_{\mathrm{nxt}(x_1)}, w_{\mathrm{nxt}(x_2)}, \ldots, w_{\mathrm{nxt}(x_k)})$. *Suppose w_m is the first component of $\mathbb{W}^n$ which satisfies $w_m = w_{min}$ and $w_{\mathrm{nxt}(m)} = v_{min}$, we call the index m the **narrowest point**, and say w_m lies* at the **narrowest point** of $\mathbb{W}^n$. If we index each component w_x of $\mathbb{W}^n$ with the relative distance between x and m (i.e., $x - m$), $\mathbb{W}^n$ can be rewritten as $\breve{\mathbb{W}}^n = (\breve{w}_{-m+1}, \ldots, \breve{w}_{-1}, \breve{w}_0, \breve{w}_1, \ldots, \breve{w}_{n-m})$, where $\breve{w}_{x-m} = w_x$. We call $\breve{\mathbb{W}}^n$ the **relative-index form** of $\mathbb{W}^n$ and define the relative index of w_x as $rix(w_x) = rix(\breve{w}_{x-m}) = x - m$, $1 \leq x \leq n$.

A search pattern $\mathbb{W}^n$ is placed at the search tree in its relative-index form $\breve{\mathbb{W}}^n$ as depicted in Fig. 1.

Example 1. Considering a set of search patterns with $|\mathbb{W}^3| = 30$ of 3-round NOEKEON: $\mathbb{S} =\{$ (2, 14, 14), (14, 2, 14), (4, 13, 13), (13, 4, 13), (6, 11, 13), (13, 6, 11), (6, 12, 12), (12, 6, 12) $\}$, Fig. 2 depicts two ways to organize the search patterns in $\mathbb{S}$. In Fig. 2a, search patterns in $\mathbb{S}$ are organized from the first point, and they are organized from the narrowest point in Fig. 2b.

There are three main reasons why *starting from the narrowest point strategy* can help to greatly reduce the search complexity:

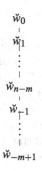

Fig. 1. Placing a search pattern at the search tree in its relative-index form

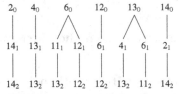
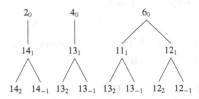

(a) Organizing from the first points(index components relative to the round-index of the first component)

(b) Organizing from the narrowest points (index components relative to the round-index of the narrowest point)

Fig. 2. Organizing the search patterns in Example 1 in two ways. Nodes in trees are elements representing one-round weights in search patterns. The subscript in each node represent the index of the point relative to the starting point.

1. Firstly, the range of the minimal value in a search pattern set is narrower than the range of an arbitrary value. In general, the range of the allowed minimal value which composes a sum is narrower than the range of the allowed arbitrary value. Here is a simple example. Assume we need to partition 11 to 4 positive numbers x_1, x_2, x_3, x_4, then the smallest number must equal to 1 or 2, while x_1 can be any number between 1 and 8. In our situation, take Example 1 again, for the search pattern set with $|W^3| = 30$ of 3-round NOEKEON, set of values at the narrowest point is $\{2, 4, 6\}$, while set of values at the first point is $\{2, 4, 6, 12, 13, 14\}$. By organizing search patterns from the narrowest points, more nodes and longer prefix-paths can be shared. From Fig. 2 we can deduce that sharing is maximized among search patterns by organizing from the narrowest point:
   - There are only 7 nodes at the first two layers of the latter structure Fig. 2b, while 14 nodes at that of the former Fig. 2a.
   - More patterns can share search prefix-pathes in the latter structure Fig. 2b than in the former Fig. 2a.
2. The second reason which makes the *starting from the narrowest point strategy* work better is that the smaller the weight, the less the number of candidate round differentials. For most block ciphers, the number of candidate round differentials with larger weight is much more than that with smaller weight. That is because, for a single S-Box, take the 4×4-bit S-Box of NOEKEON for example, the number of differential pairs with weight 3 is 72 while the number of differential pairs with weight 2 is only 24. Moreover, round differentials with smaller weight usually have less active S-Boxes. In the starting round, the less the number of active S-Boxes, the less the number of candidate round differentials. Table 1 shows the explosive increase of number of one-round candidates with the increase of one-round weight.
3. The third reason is that by starting from the narrowest point, restriction are more stringent, backtracking in invalid path could arise early. This dues to the diffusion property within small number of rounds of the cipher. Take NOEKEON for example and limit the number of active S-Boxes of one round being less than 18, if the preceding round is with one active S-Box, there are 12 trails[4] propagating through the succeeding round, and if the preceding round is with two active S-Boxes, there are 981 trails. The number of trails that could propagate through the succeeding round increase explosively with the increase of the number of active S-Boxes in the preceding round, which can be seen in [6, Appendix A]. Combining with the fact that number of differential pairs with weight 3 is much more than that with weight 2 for active S-Boxes, smaller weight of the preceding round leads a narrower range of possible value of weight of the succeeding round, and less number of differential trails which could propagate through the succeeding rounds. Accordingly, for a given search pattern, if the search procedure starts from the narrowest point, there is much less candidate differential trails within the starting

---

[4] The number is up to rotation equivalence for NOEKEON.

two rounds and to search in the search patterns which do not exist in reality, the workload taken by the search procedure before it being blocked at some nodes in search trees will be much smaller than that of starting from an arbitrary point.

Table 1. Numbers of candidates for one-round differential under various weight

w^1	2	3	4	5
CN	24	72	$\binom{32}{2} \times 24^2$	$\binom{32}{2} \times 2 \times 24 \times 72$
	$\approx 2^{4.58}$	$\approx 2^{6.17}$	$\approx 2^{18.12}$	$\approx 2^{20.71}$

w^1	6	7	8	9
CN	$\binom{32}{2} \times 72^2$ $+ \binom{32}{3} \times 24^3$	$\binom{32}{3} \times 3 \times 24^2 \times 72$	$\binom{32}{3} \times 3 \times 24 \times 72^2$ $+ \binom{32}{4} \times 24^4$	$\binom{32}{3} \times 72^3$ $+ \binom{32}{4} \times 4 \times 72$ $\times 24^3$
	$\approx 2^{26.08}$	$\approx 2^{29.20}$	$\approx 2^{33.68}$	$\approx 2^{37.08}$

CN: Numbers of one-round candidates.

Suppose there are 32 identical 4×4-bit S-Boxes in one round of the cipher and we ignore the rotation equivalence.

Suppose number of differential pairs with weight 3 is 72 and that with weight 2 is 24 in the DDT of an S-Box.

5.2 Experimental Results of Starting from the Narrowest Point Strategy

Table 2 includes the experimental results comparison between search with the *starting from the narrowest point strategy* and search without the strategy. It can be seen that hundreds of speed up ratio are reached by adopting the *starting from the narrowest point strategy* from column "Time-First", column "Time-Narrowest" and column "Ratio-(F/N)".

At the end of this section, we would like to point out the following.

When the starting point is the first round, it is sufficient to test only one of the input difference compatible with each output difference under the first round weight. Similarly, it is sufficient to test whether there exist one output difference compatible with the input difference under the weight of the last round. We call this the *free ends equivalent effect*.

When the narrowest point is internal round, *free ends equivalent effect* should also be considered. There are both forward and backward propagations in branches of search trees. For the last round of the forward propagation and last round of the backward propagation, which are the real last round and the first round of the current n-round cipher respectively, it is sufficient to test only one of the output differences compatible with the input difference. Besides, similar with the above discussed *free ends equivalent effect*, there is a *starting point equivalence*. Specifically, when output difference of the starting round is fixed, there are many input differences of this round compatible with it. However, forward branches need to be traversed under only one of the compatible input differences of the starting round instead of all.

6   Concretizing and Grouping Search Patterns Strategy

It has been proved experimentally that the pre-search strategy is very powerful to filter out non-existent search patterns. However, as containers of results of pre-search, search patterns turn to be too abstract to store enough information on the underlying trails. To collect more information from preceding search phases, search patterns should be concretized to a proper extent while keeping the storage requirement appropriate. Thus, we propose the *concretizing and grouping search patterns strategy*. Besides, *starting point equivalence* can exist among various weights of narrowest point. That equivalence should be further considered to maximize shareable works at early phases of the procedure which dominate the search complexity.

6.1   Proposal and Justification of the Concretizing and Grouping Search Patterns Strategy

Concretizing: First, we append information of possible number of active S-Boxes at the narrowest point to each search pattern. Thus, a search pattern turns to be several concretized search patterns. For a search pattern $\mathbb{W}^n = (w_1, \ldots, w_m, \ldots, w_n)$ and its relative-index form $\breve{\mathbb{W}}^n = (\breve{w}_{-m+1}, \ldots, \breve{w}_0, \ldots, \breve{w}_{n-m})$, its concretized search patterns are $[\breve{\mathbb{W}}^n] = \{(\breve{w}_{-m+1}, \ldots, \binom{[asn]}{\breve{w}_0}, \ldots, \breve{w}_{n-m}) | asn \in [asn\_min, asn\_max]\}$ where $[asn\_min, asn\_max]$ is the range of possible ASN of round-differential at the narrowest round with round-weight $\breve{w}_0$.

Grouping: We then group the concretized search patterns according to the number of active S-Boxes at the narrowest point. For two search patterns having same possible ASN at the narrowest point:

1. $\mathbb{W}_1^n = (w_{1,1}, \ldots, w_{1,m_1}, \ldots, w_{1,n})$ and its relative-index form $\breve{\mathbb{W}}_1^n = (\breve{w}_{1,-m_1+1}, \ldots, \breve{w}_{1,0}, \ldots, \breve{w}_{1,n-m_1})$, and one of its concretized pattern
$$\breve{\mathbb{W}}_1^n = (\breve{w}_{1,-m_1+1}, \ldots, \binom{[asn]}{\breve{w}_{1,0}}, \ldots, \breve{w}_{1,n-m_1})$$

2. $\mathbb{W}_2^n = (w_{2,1}, \ldots, w_{2,m_2}, \ldots, w_{2,n})$ and its relative-index form $\breve{\mathbb{W}}_2^n = (\breve{w}_{2,-m_2+1}, \ldots, \breve{w}_{2,0}, \ldots, \breve{w}_{2,n-m_2})$, and one of its concretized pattern
$$\breve{\mathbb{W}}_2^n = (\breve{w}_{2,-m_2+1}, \ldots, \binom{[asn]}{\breve{w}_{2,0}}, \ldots, \breve{w}_{2,n-m_2}),$$

suppose

$$\left(\begin{array}{c}[asn]\\ \{\breve{w}_{1,0},\breve{w}_{2,0}\}_0\end{array}\right)$$

$$v_i^{\{\breve{w}_{1,0},\breve{w}_{2,0}\}}$$

– $w_{1,m_1} \neq w_{2,m_2}$ and

– $w_{1,\mathrm{nxt}(m_1)} = w_{2,\mathrm{nxt}(m_2)} = v$,

– $\mathrm{rix}(w_{1,\mathrm{nxt}(m_1)}) = \mathrm{rix}(w_{2,\mathrm{nxt}(m_2)}) = i$ where $i \in \{1,-1\}$ and

– $w_{1,\mathrm{nxt}(\mathrm{nxt}(m_1))} \neq w_{2,\mathrm{nxt}(\mathrm{nxt}(m_2))}$,

we group them as $\breve{w}_{1,-m_1+1}^{\{\breve{w}_{1,0}\}}$ $\breve{w}_{2,-m_2+1}^{\{\breve{w}_{2,0}\}}$.

Figure 3a depicts the way how we concretize the search patterns by an example set of search patterns with $|\mathbb{W}^3| = 35$ of 3-round NOEKEON. Figure 3b depicts the way how we group the search patterns by the same example set of search patterns taken in Fig. 3a.

Then the search is processed group by group, instead of value by value of weight starting from the narrowest point. Grouping search patterns in such a way can bring two advantages.

1. Firstly, more specified knowledge of the search patterns will be learned during the pre-search phase. Information on allowed number of active S-Boxes at the narrowest point of a search pattern are stored in memory and can be used to filter search patterns in later process. For example, in Fig. 3b, search pattern $(16,6,13)$ is included in the tree with $[3]^5$ as the root node. Since a round differential with weight 6 is possible to have 2 active S-Boxes, $(16,6,13)$ should have been included in the tree with $[2]$ as the root node as well. However, during the former search of 2-round cipher, we learn that search pattern $(6,13)$ does not exist when round differential with weight 6 has 2 active S-Boxes. Thus, $(16,6,13)$ can be deleted in the tree with $[2]$ as the root. Besides, since the allowed number of active S-Boxes at the narrowest point are usually small, memory requirement stays appropriate.
2. Secondly, once the search patterns are grouped according to the allowed number of active S-Boxes and starting from the output difference in the narrowest point, searches can share the forward propagation prefixes among different search patterns with various narrowest point weight. For example, in Fig. 3b search pattern $(13,9,13)$, $(14,8,13)$, $(15,7,13)$ and $(16,6,13)$ share the same prefix $([3],13_1)$ when we search starting from an output difference with 3 active S-Boxes at the narrowest point.

---

[5] For simplicity, we use the number in square bracket to represent the root node (eg. $[3]$ is the shortening of $\left(\begin{array}{c}[3]\\ \{6,7,8,9\}_0\end{array}\right)$).

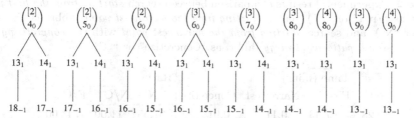

(a) Concretizing search patterns by appending information of possible number of active S-Boxes at the narrowest point

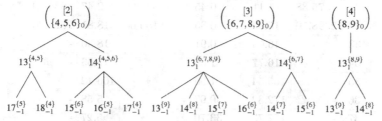

(b) Grouping search patterns by number of active S-Boxes at the narrowest point. Superscript numbers in brace represent the narrowest point weight of the patterns in which the node belongs to. For example, search pattern $(14, 8, 13)$ is included in the branch with $\left(\begin{smallmatrix} [3] \\ \{6,7,8,9\}_0 \end{smallmatrix} \right)$ as the root node, $13_1^{\{6,7,8,9\}}$ as the first layer node, and $14_{-1}^{\{8\}}$ as the leaf. It is also included in the branch with $\left(\begin{smallmatrix} [4] \\ \{8,9\}_0 \end{smallmatrix} \right)$ as the root node, $13_1^{\{8,9\}}$ as the first layer node, and $14_{-1}^{\{8\}}$ as the leaf.

Fig. 3. Concretizing and grouping search patterns. Number in square bracket appended at root node of each tree represents the possible number of active S-Boxes at the narrowest point. Subscripts represent indices of the points relative to the starting points.

6.2 Experimental Results of Concretizing and Grouping Search Patterns

Table 2 summarizes the experimental results comparison between search with the *concretizing and grouping search patterns strategy* and searches without the strategy. Column "Ratio-(N/C)" shows a tens of times speedup ratio. A hundreds of speedup ratio are achieved combined with the efficiency brought by the first strategy shown as in column "Ratio-(F/C)".

7 Trialling in Minimal Changes Order Strategy

Once the set of search patterns is created, to obtain a differential trail, we only need to simply generate and concatenate round differentials under fixed round weights, if there exist any. By looking up the differential distribution table (DDT) of S-Box in the S-Layers and by executing the P-Layers between continuous rounds on the differences, round differentials can be constructed and connected to a differential trial. However, for NOEKEON and SPONGENT, execution of the

Table 2. Experimental results comparison between search *starting from the first point* (abbr. as "First" or "F"), search *starting from the narrowest point* (abbr. as "Narrowest" or "N"), and search *starting from the narrowest point* with the *concretizing and grouping search patterns strategy* (abbr. as "Concretize" or "C")

w^3	Time (mins)			Ratio		
	First	Narrowest	Concretize	F/N	N/C	F/C
28	7.43	0.11	0.01	67.55	11.00	743.00
29	8.01	0.98	0.01	8.17	98.00	801.00
30	375.60	1.10	0.44	341.45	2.50	853.64
31	375.88	1.11	0.45	338.63	2.47	835.29
32	2398.50	15.99	0.85	150.00	18.81	2821.76
33	-	16.65	0.91	-	18.30	-
34	-	16.77	1.08	-	15.53	-
35	-	165.54	1.56	-	106.12	-
36	-	172.82	30.97	-	5.58	-
37	-	177.73	33.70	-	5.27	-

Ratio: Ratio between two kinds of time.
Rows are separated by weight of 3-round trails in NOEKEON.
All of the experiments are done with the *trialling in minimal changes order strategy.*
Weight range of pre-search information of 2-round cipher is [8, 31].

P-Layer will become the most costly part of the search process. That is because executing the P-Layer by looking up big tables which is a more suitable way of implementation in the case of searching for differential trails, the cost of P-Layer can be up to 10 times of the cost of generating a new candidate differential at the S-Layer. What is more, each replacement of candidate differential in a single S-Box at the S-Layer, calls for the replacement of differential at the P-Layer in full scope.

We avoid the full execution of the P-Layer considering that there is locality of individual S-Box within S-Layer and linearity of P-Layer, which makes local calculation feasible when generate round differentials. Further more, we propose the *trialling in minimal changes order strategy* to minimize the number of local calculation, thus to minimize the cost of generating round differentials. The following are explicit explanations.

If we can pre-calculate all 128-bit outputs corresponding to local nonzero inputs (let us take 4-bit for example hereafter) of P-Layer, we can get the output differences corresponding to the input differences which is locally active by simple XOR operations instead of the costly execution of P-Layer. Linear operation (XOR) on 128-bit (256-bit) data can resorting to the SSE and AVX instructions.

P-Layer operations can be further removed completely by planting the above 4×128-bit differences tables of P-Layer to the DDT of S-Box. We call each 128-bit SP-Layer output difference caused by a 4-bit (located at a single S-Box) S-Layer input difference the "128-bit contribution difference" to the 128-bit round output difference. We generate table of 4-bit input differences and their 128-bit contribution differences which is called *contribution differential distribution table* (CDDT).

To minimize the cost of generating new candidate differences, we generate the new from the old with minimal local changes by removing and adding 128-bit contribution differences which can be done by looking up the CDDT and by simple XOR instructions. The following shows how we achieve the least number of looking up table and XOR instructions.

Candidate round differentials are characterized by the *weight patterns of active S-Boxes* and *indices* of their 4-bit candidate differences within each active S-Box. By *weight patterns of active S-Boxes*, we mean the possible compositions of partition one-round weight into weights of active S-Boxes. Take one-round differential weight as 10 and number of active S-Boxes as 4 for example, weight patterns of active S-Boxes are (3322),(2323),(2332),(3223),(3232) and (2233), which are restricted 4-compositions of 10. We then run through all the candidate output differences by enumerate the weight patterns of active S-Boxes with a light algorithm extended from [17] and run through all the indices within each weight pattern with an algorithm named "Loopless reflected mixed-radix Gray generation" in [18] to achieve minimal changes and least XOR operations. Generally, generating a new candidate only cost two XOR-operations. An example can be seen in Example 2.

Example 2. Following is an example of *trialling in minimal changes order strategy*. Assume there are 4 active S-Boxes in the round input difference, and input differences of the four active S-Boxes are 0x2, 0x1, 0x4 and 0x8. We need to run through all the compatible round output difference with round weight equals to 10. To minimize the cost, we try the weight patterns of 4 active S-Boxes (3322),(2323),(2332),(3223),(3232) and (2233) in an order as show in Fig. 4. Within a weight pattern, take (2233) for example, assume num(0x2, 2) = 3, num(0x1, 2) = 2, num(0x4, 3) = 3 and num(0x8, 3) = 2, where num(id, w) denotes the number of candidate output differences given input difference id and weight w of a single S-Box. We run through candidates by the indices in an order as shown in Fig. 5.

That bring us at least double times speedup. An intuition understanding for the *trialling in minimal changes order strategy* is that, we utilize the small change effect and large scale effect. Small change effect means that generating the new from the old might be much cheaper than generate from nothing if the changes are subtle. Large scale effect means that doing things in large scale can be more economical and efficient. The cost to systemically finish the whole is much less than the sum of cost to separately finish each individual.

$$(3322) \Rightarrow (3\underset{\smile}{2}3\underset{\smile}{2}) \Rightarrow (\underset{\smile}{2}3\underset{\smile}{3}2) \Rightarrow (23\underset{\smile}{2}\underset{\smile}{3}) \Rightarrow (3\underset{\smile}{2}2\underset{\smile}{3}) \Rightarrow (\underset{\smile}{2}2\underset{\smile}{3}3)$$

Fig. 4. An example of trialling weight patterns of active S-Boxes in minimal changes order. Bold numbers with underbreves are items exchanged from the former weight pattern. There are only 2 changes at each step.

$$1111 \Rightarrow 111\underbrace{2} \Rightarrow 11\underbrace{2}2 \Rightarrow 112\underbrace{1} \Rightarrow 11\underbrace{3}1 \Rightarrow 113\underbrace{2} \Rightarrow 1\underbrace{2}32 \Rightarrow 123\underbrace{1}$$
$$\Downarrow$$
$$222\underbrace{1} \Leftarrow 22\underbrace{2}2 \Leftarrow 221\underbrace{2} \Leftarrow \underbrace{2}211 \Leftarrow 121\underbrace{1} \Leftarrow 12\underbrace{1}2 \Leftarrow 1\underbrace{2}22 \Leftarrow 122\underbrace{1}$$
$$\Downarrow$$
$$223\underbrace{1} \Rightarrow 2232 \Rightarrow 2\underbrace{1}32 \Rightarrow 213\underbrace{1} \Rightarrow 21\underbrace{2}1 \Rightarrow 2122 \Rightarrow 211\underbrace{2} \Rightarrow 2111$$
$$\Downarrow$$
$$323\underbrace{1} \Leftarrow 3\underbrace{2}32 \Leftarrow 313\underbrace{2} \Leftarrow 31\underbrace{3}1 \Leftarrow 3121 \Leftarrow 31\underbrace{2}2 \Leftarrow 311\underbrace{2} \Leftarrow \underbrace{3}111$$
$$\Downarrow$$
$$32\underbrace{2}1 \Rightarrow 322\underbrace{2} \Rightarrow 32\underbrace{1}2 \Rightarrow 321\underbrace{1}$$

Fig. 5. An example of trialling candidates within a weight pattern in the mixed-radix Gray code order. Bold numbers with a underbreve are the unique item changed from the former.

8   Results on Best Trails of NOEKEON and SPONGENT

8.1   Object Cipher - A Block Cipher NOEKEON

NOEKEON is a self-inverse block cipher with a block and key length of 128-bit. It is a 16 rounds iterated cipher with a round transformation composed of transformations Theta, Pi1, Gamma, Pi2 and XORing a Working Key. The round transformation can be split into two parts - nonlinear part Gamma and linear part Lambda = (Pi1 ∘ Theta ∘ Pi2). Gamma can be specified as the S-layer which is a parallel execution of 32 4 × 4-bit identical S-Boxes. Lambda can be seen as the P-layer. A full description of NOEKEON can be found in [6]. By searching the complete space of 4-round trails (both linear and differential) with less than 24 active S-Boxes, the designers can guarantee that there are no 4-round differential (linear) trails with a predicted probability (bias) above 2^{-48} (2^{-25}).

In this work, by adopting the proposed three optimization strategies and making use of the symmetry properties, these statements are confirmed and further refined. They turn to be as follow: of all 4-round differential (linear) trails, the *best* has a probability (bias) equals to 2^{-51} (2^{-25}). Figure 6 in Appendix B shows one of the best 4-round differential trails. It takes 21 (1.2) hours to systematically investigate whether 4-round differential (linear) trails of weight up to 51 (25) exist on the formerly mentioned PC. Table 3 summarizes more results about differential (linear) trails of NOEKEON we have achieved.

Besides, *best* 6-round and 9-round linear trails with bias 2^{-40} and 2^{-62} are found out and Fig. 7 in Appendix B shows one of them. The 9-round linear trail is the longest one holding with bias larger than 2^{-65}. By observing that the internal part in the best 6-round linear trail is iterative on a 2-round trail which is also a sub-trail in the best 9-round, a 10-round linear trail with bias 2^{-68} can be constructed.

According to the results in Table 3, the probability (bias) of best 10-round (11-round) differential (linear) trails in NOEKEON is at most 2^{-131} (2^{-71}). As mentioned in [6], for a DC attack to exist, there must be a predictable difference propagation over all but a few rounds with a probability significantly larger than 2^{-127}, and LC attacks are possible if there are predictable input-output correlation values (2 times of bias) over all but a few rounds significantly larger than 2^{-64}, thus we can benefit from these results that exclude classical DC (LC) attacks on NOEKEON.

8.2   Object Cipher - A Hash Function SPONGENT

SPONGENT is a lightweight hermetic sponge hash function with a PRESENT-type permutation [16]. There are 13 variants with 11 kinds of permutation width. In this work, variants with permutation width $b \in \{88, 136, 176, 240\}$ are considered.

By applying the three proposed optimization strategies, *best* differential trails corresponding to the number of rounds in [16, Table 3] are searched. The results suggest that finding out the unconditional best trails could help to establish more tight upper bound on the probability of the best differential than that provided by finding the trails with minimal number of active S-Boxes. New results are

Table 3. Comparison between results from specification of NOEKEON and results from this work. Entries with * are the updates due to this work. Note that the trails we found are confirmed to be the best.

| #Rounds | NOEKEON-differential | | | | NOEKEON-linear | | | |
| | Spec. | | This | | Spec. | | This | |
	ASN	Prob.	ASN	Prob.	ASN	Bias	ASN	Bias
1	1	2^{-2}	1	2^{-2}	1	2^{-2}	1	2^{-2}
2	4	2^{-8}	4	2^{-8}	4	2^{-5}	4	2^{-5}
3	-	-	*13	*2^{-28}	-	-	*13	*2^{-14}
4	-	$\leq 2^{-48}$	*22	*2^{-51}	-	$\leq 2^{-25}$	*21	*2^{-25}
5	-	-	-	*$\leq 2^{-65}$	-	-	-	*$\leq 2^{-32}$
6	-	-	-	*$\leq 2^{-80}$	-	-	*33	*2^{-40}

Table 4. Comparison between results from specification of SPONGENT and results from this work. Entries with * are the updates due to this work.

| | $b = 88$ | | | | $b = 136$ | | | |
| #Rounds | Spec. | | This | | Spec. | | This | |
	ASN	Prob	ASN	Prob	ASN	Prob	ASN	Prob
5	10	2^{-21}	10	*2^{-20}	10	2^{-22}	10	*2^{-20}
10	20	2^{-47}	20	2^{-47}	24	2^{-60}	*22	*2^{-55}
15	30	2^{-74}	30	2^{-74}	40	2^{-101}	*43	*2^{-96}

| | $b = 176$ | | | | $b = 240$ | | | |
| #Rounds | Spec. | | This | | Spec. | | This | |
	ASN	Prob	ASN	Prob	ASN	Prob	ASN	Prob
5	10	2^{-21}	10	*2^{-20}	10	2^{-21}	10	*2^{-20}
10	20	2^{-50}	20	*2^{-46}	20	2^{-43}	20	2^{-43}
15	30	2^{-79}	30	2^{-79}	30	2^{-66}	30	2^{-66}

listed in Table 4. Figures 8 and 9 in Appendix B depicts two of the updated best trails. Besides, longer differential trails are found, which could correct and update the results listed in [16, Table 4]:

- For variants with $b = 88$, probability of *best* 17-round (18-round) differential trail is 2^{-86} (2^{-94}), which was found out within 1 min, and shown in Fig. 10 in Appendix B.
- For variants with $b = 240$, probability of *best* 44-round differential trail is 2^{-196}, which was found within 1 min. By observing the results up to 44-round, we can conclude the following: $Bw^6 = 30$ and for $r \geq 7$, $Bw^r = \begin{cases} Bw^{r-1} + 4 & \text{if } r \text{ is even} \\ Bw^{r-1} + 5 & \text{if } r \text{ is odd} \end{cases}$. An observation is that there is a 2-round iterative trail with weight pattern $(4, 5)$ composing the best trails, as shown in Fig. 11 in Appendix B.
- For variants with $b = 176$, *best* 17-round (18-round) differential trail with weight 91 (99) was found within 50 min, and shown in Fig. 12 in Appendix B.

9 Conclusion and Future Work

We improved the search algorithm for the best differential and best linear trails by introducing three optimization strategies. Those strategies reduce the complexity to a great extent by organizing candidate search patterns properly, collecting more information during preceding procedures and trialling in good order, which allowed us to find out best trails more efficiently. At the end, we briefly overview future work.

- We trial a search pattern in an order of $(\breve{w}_0, \breve{w}_1, \ldots, \breve{w}_{n-m}, \breve{w}_{-1}, \ldots, \breve{w}_{-m+1})$. As an anonymous reviewers suggested, it might also be interesting to consider the order $(\breve{w}_0, \breve{w}_1, \breve{w}_{-1}, \breve{w}_2, \breve{w}_{-2}, \ldots)$.
- Experimental result on NOEKEON shows that search patterns that cannot be filtered out by the information collected in preceding procedures are usually with fat paunches, while the existing search patterns are usually with narrow waists. That can be understood considering the diffusion property of the target cipher. How to use this empirical knowledge to add heuristics to the search algorithm remains unclear.
- By avoiding detailed properties of the target ciphers, our algorithm is general to some extent, while remaining space for further improvement by utilizing more special properties of the object ciphers.
- Strategies proposed in this paper are also helpful to generate all the trails up to a given weight. Thus, they can be adopted when search for the best multiple differential (multi-dimensional linear) distinguishers. While, we haven't adopted them to the case of related-key differential.

Acknowledgement. Many thanks go to the anonymous reviewers for many useful comments and suggestions. The research presented in this paper is supported by the National Natural Science Foundation of China (No.61379138), the "Strategic Priority Research Program" of the Chinese Academy of Sciences (No.XDA06010701).

A   Our Search Algorithm

Algorithm 4. Our Search Approach Part 1 - Framework of Our Search Approach

---

1: **for** $n \leftarrow 1$, $RoundN$ **do** $\triangleright$ n is the current number of round and $RoundN$ is the total of that considered for the cipher. All of the following variables are global which can be directly accessed in each procedure. Superscript n on the shoulders of those variables is changed with the value of n, thus for different values of n, they are different variables.

2:    $w\_l^n \leftarrow$ a lower bound for Bw^n

3:    $n\_w^n \leftarrow$ the number of extra values of weights larger than Bw^n $\triangleright$ Determine the amount of pre-search information of n-round cipher, there is no pre-search information of n-round cipher if $n\_w^n = -1$

4:    $w\_u^n \leftarrow 0$ $\triangleright$ w_u^n will be update as the upper bound for values of weights at the end of procedure SearchForRounds, and thus $[w\_l^n, w\_u^n]$ is the range of values of w^n under which we will completely examine the existence of the search patterns.

5:    $Bw^n \leftarrow \infty$

6:    $w^n \leftarrow w\_l^n$ $\triangleright$ w^n is a global temp variable holding current weight of n-round

7:    $ExistentPatterns^n \leftarrow \varnothing$ $\triangleright$ Trees of existent search patterns under $w^n \in [Bw^n, w\_u^n]$

8:    $ExistentPatternsCG^n \leftarrow \varnothing$ $\triangleright$ Trees of existent concretized patterns under $w^n \in [Bw^n, w\_u^n]$

9:    $SearchPatterns^n \leftarrow \varnothing$ $\triangleright$ Temporary trees of search patterns

10:    $SearchPatternsCG^n \leftarrow \varnothing$ $\triangleright$ Temporary trees of concretized search patterns

11:    SEARCHFORROUNDS(n)

12:    Next-n: $\triangleright$ A tag for long jump

13: **end for**

14: **procedure** SEARCHFORROUNDS(n)

15:    $i \leftarrow -1$

16:    **while** $i \leq n\_w^n$ **do**

17:        GENERATESEARCHPATTERNS $\triangleright$ Generate, filter and formalize search patterns

18:        ORGANIZESEARCHPATTERNS $\triangleright$ Concretize and group search patterns

19:        SEARCHFROMTHENARROWEST $\triangleright$ Search in trees of organized search patterns

20:        $SearchPatterns^n \leftarrow \varnothing$ $\triangleright$ Clear trees of search patterns

21:        $SearchPatternsCG^n \leftarrow \varnothing$ $\triangleright$ Clear trees of concretized search patterns

22:        $w^n \leftarrow w^n + 1$

23:        **if** $Bw^n \neq \infty$ **then** $i \leftarrow i + 1$

24:        **end if**

25:    **end while**

26:    $ExistentPatterns^n \leftarrow_{gather} ExistentPatternsCG^n$ $\triangleright$ Gather information on existence of search patterns $\breve{\mathbb{W}}^n$ according to the information on existence of corresponding concretized search patterns $\breve{W}^n$.

27:    $w\_u^n \leftarrow Bw^n + n\_w^n$

28: **end procedure**

Algorithm 5. Our Search Approach Part 2 - Generate and Filter, Concretize and Group Search Patterns

---

29: **procedure** GENERATESEARCHPATTERNS
30:     **while** partition of w^n could generate a new n-composition **do**
31:         Partition w^n into n components to form a new possible search pattern $\mathbb{W}^n = (w_1, w_2, \ldots, w_n)$, which make the following hold:
32:             1. $|\mathbb{W}^n| = \sum_{i=1}^{n} w_i = w^n$,
33:             2. $\sum_{i=1}^{n-r} w_i + Bw^r \leq w^n$ for $\forall r \ (1 \leq r \leq n-1)$, and
34:             3. $w^r \geq Bw^r$ and $\mathbb{W}^r = (w_i, w_{i+1}, \ldots, w_{i+r-1}) \in ExistentPatterns^r$ if $w^r \in [Bw^r, w\_u^r]$, for $\forall r, i \ (1 \leq r \leq n-1, \ 1 \leq i \leq n-r+1)$, where $w^r = \sum_{j=i}^{i+r-1} w_j$.
35:         Find the narrowest point of $\mathbb{W}^n$, let it be m.   ▷ If the cipher is involution, let $\bar{\mathbb{W}}^n = (w_n, w_{n-1}, \ldots, w_1)$ and $\bar{w}_{\bar{m}}$ be the narrowest point of $\bar{\mathbb{W}}^n$. If $w_{\mathrm{nxt}(m)} = \bar{w}_{\mathrm{nxt}(\bar{m})}$, let $\breve{\bar{\mathbb{W}}}^n$ be the relative-index form of $\bar{\mathbb{W}}^n$, if $\breve{\bar{\mathbb{W}}}^n \in SearchPatterns^n$, discard $\mathbb{W}^n$ and continue. Else, if $w_{\mathrm{nxt}(m)} > \bar{w}_{\mathrm{nxt}(\bar{m})}$, discard $\mathbb{W}^n$ and $\mathbb{W}^n \leftarrow \bar{\mathbb{W}}^n$.
36:         Turn $\mathbb{W}^n = (w_1, w_2, \ldots, w_n)$ into its relative-index form $\breve{\mathbb{W}}^n = (\breve{w}_{-m+1}, \ldots, \breve{w}_0, \ldots, \breve{w}_{n-m})$.
37:         $SearchPatterns^n \leftarrow_{insert} \breve{\mathbb{W}}^n$
38:     **end while**
39: **end procedure**

40: **procedure** ORGANIZESEARCHPATTERNS
41:     **for all** $\breve{\mathbb{W}}^n = (\breve{w}_{-m+1}, \ldots, \breve{w}_0, \ldots, \breve{w}_{n-m}) \in SearchPatterns^n$ **do**
42:         $asn\_min \leftarrow$ the minimal possible ASN determined by $\breve{w}_0$
43:         $asn\_max \leftarrow$ the maximal possible ASN determined by $\breve{w}_0$
44:         **for** $asn \leftarrow asn\_min, asn\_max$ **do**
45:             **for all** $r, \hat{r}, (2 \leq r < n, 1 \leq \hat{r} \leq r)$ **do**
46:                 **if** $\breve{w}_0$ is at the narrowest point of reduced search pattern $(\breve{w}_{-\hat{r}+1}, \ldots, \breve{w}_0, \ldots, \breve{w}_{r-\hat{r}})$ **then**
47:                     Let $\breve{W}^r \leftarrow (\breve{w}_{-\hat{r}+1}, \ldots, \binom{[asn]}{\breve{w}_0}, \ldots, \breve{w}_{r-\hat{r}})$ and $\breve{w}^r \leftarrow \sum_{i=-\hat{r}+1}^{r-\hat{r}} \breve{w}_i$, then
48:                         **if** $\breve{w}^r \in [Bw^r, w\_u^r]$ and $\breve{W}^r \notin ExistentPatternsCG^r$ **then** goto next asn
49:                         **end if**
50:                 **end if**
51:             **end for**
52:             $SearchPatternsCG^n \leftarrow_{groupingInto} \breve{W}^n = (\breve{w}_{-m+1}, \ldots, \binom{[asn]}{\breve{w}_0}, \ldots, \breve{w}_{n-m})$
53:         **end for**
54:     **end for**
55: **end procedure**

Algorithm 6. Our Search Approach Part 3 - Search in Trees of Organized Search Patterns

56: **procedure** SEARCHFROMTHENARROWEST

57:     **for all** $rootnode \leftarrow \left(\dfrac{[asn]}{\{\breve{w}_{1,0}, \breve{w}_{2,0}, \ldots, \breve{w}_{k,0}\}_0} \right) \in SearchPatternsCG^n$ **do**

58:         $asn_0 \leftarrow asn$

59:         $asn\_min_1 \leftarrow$ the allowed minimal ASN determined by the minimal weight of the first forward rounds succeeding the current $rootnode$

60:         $asn\_max_1 \leftarrow$ the allowed maximal ASN determined by the maximal weight of the first forward rounds succeeding the current $rootnode$

61:         $asn\_min_{-1} \leftarrow$ the allowed minimal ASN determined by the minimal weight of the first backward rounds succeeding the forward branches of the current $rootnode$

62:         $asn\_max_{-1} \leftarrow$ the allowed maximal ASN determined by the maximal weight of the first backward rounds succeeding the forward branches of the current $rootnode$

63:         **while** $\exists$ new candidate round-output-differential od_0 **do**

64:             Generate a new od_0 with the *trailling in minimal changes order strategy*, which makes the following hold:

65:             1. there are asn_0 active S-Boxes at the narrowest round, and

66:             2. $asn_1 \in [asn\_min_1, asn\_max_1]$, where asn_1 is ASN at the first succeeding forward round, which is computed from od_0.

67:             **while** $\exists$ new candidate round-input-differential id_0 **do**

68:                 Generate a new id_0 with the *trailling in minimal changes order strategy*, which makes the following hold:

69:                 1. id_0 is compatible with od_0 with $w_{(id_0,od_0)} \in \{\breve{w}_{1,0}, \breve{w}_{2,0}, \ldots, \breve{w}_{k,0}\}$,

70:                 2. $asn_{-1} \in [asn\_min_{-1}, asn\_max_{-1}]$, where asn_{-1} is ASN at the first backward round, which is computed from id_0.

71:                 $idListAtNarrowestPoint^{w(id_0,od_0)} \leftarrow_{insert} id_0$

72:             **end while**

73:             **if** $\exists$ succeeding forward branches under $node_r$ **then**

74:                 SEARCHFORWARD(od_0, 1)

75:             **end if**

76:             **for all** $\breve{w}_{i,0} \in \{\breve{w}_{1,0}, \breve{w}_{2,0}, \ldots, \breve{w}_{k,0}\}$ **do**

77:                 **if** $\exists$ succeeding backward branches with $\breve{w}_{i,0}$ as the narrowest point under $node_r$ **then**   ▷ For simplicity henceforth, we say *branches with $\breve{w}_{i,0}$ as the narrowest point* if there are nodes carry superscript including $\breve{w}_{i,0}$ on their shoulders.

78:                 **for all** $id_0 \in idListAtNarrowestPoint^{\breve{w}_{i,0}}$ **do**

79:                     SEARCHBACKWARD($\breve{w}_{i,0}$, id_0, -1)

80:                 **end for**

81:                 **end if**

82:             **end for**

83:         **end while**

84:     **end for**

85: **end procedure**

Algorithm 7. Our Search Approach Part 4 - Search in Forward Branches

---

86: **procedure** SEARCHFORWARD(id_r, r)

87:     **for all** $node_r \leftarrow \breve{w}_r^{\{\breve{w}_{1,0},\ldots,\breve{w}_{j,0}\}} \in$ forward branches succeeding the preceding node in $SearchPatternsCG^n$ **do**

88:         **if** $\exists$ succeeding forward branches under $node_r$ **then**

89:             **while** $\exists$ new candidate od_r **do**

90:                 Generate a new od_r with the *trailling in minimal changes order strategy*, which makes the following hold:

91:                     1. od_r is compatible with id_r with $w_{(id_r,od_r)} = \breve{w}_r$,

92:                     2. $asn_{r+1} \in [asn\_min_{r+1}, asn\_max_{r+1}]$, where asn_{r+1} is ASN at the succeeding forward round, which is computed from od_r, and asn_min_{r+1} (or asn_max_{r+1}) is determined by the minimal (or maximal) weight of nodes on the succeeding forward branches of $node_r$.

93:                 SEARCHFORWARD(od_r, $r + 1$)

94:             **end while**

95:         **end if**

96:         **if** $\exists$ succeeding backward branches under $node_r$ or $\nexists$ succeeding branches under $node_r$ **then**

97:             **if** $\exists$ an od_r compatible with id_r with $w_{(id_r,od_r)} = \breve{w}_r$ **then**

98:                 **if** $\exists$ succeeding backward branches under $node_r$ **then**

99:                     **for all** $\breve{w}_{i,0} \in \{\breve{w}_{1,0},\ldots,\breve{w}_{j,0}\}$, $1 \le i \le j$ **do**

100:                         **if** $\exists$ succeeding backward branches with $\breve{w}_{i,0}$ as the narrowest point under $node_r$ **then**

101:                             **for all** $id_0 \in idListAtNarrowestPoint^{\breve{w}_{i,0}}$ **do**

102:                                 SEARCHBACKWARD($\breve{w}_{i,0}$, id_0, -1)

103:                             **end for**

104:                         **end if**

105:                     **end for**

106:                 **else**

107:                     **if** this is the first time get an n-round differential trail **then** $Bw^n \leftarrow w^n$

108:                     **end if**

109:                     **if** $n\_w^n = -1$ **then goto** Next-n

110:                     **end if**

111:                     $\breve{W} \leftarrow_{delete} SearchPatternsCG^n$ ▷ Delete current concretized search pattern from trees of concretized search patterns under current w^n

112:                     $ExistentPatternsCG^n \leftarrow_{insert} \breve{W}$ ▷ Save current concretized search pattern to trees of actually existent concretized search patterns under $w^n \in [Bw^n, w\_u^n]$

113:                 **end if**

114:             **end if**

115:         **end if**

116:     **end for**

117: **end procedure**

Algorithm 8. Our Search Approach Part 5 - Search in Backward Branches

---

118: **procedure** SEARCHBACKWARD($\breve{w}_{i,0}$, od_r, r)

119:    **for all** $node_r \leftarrow \breve{w}_r^{\{\breve{w}_{1,0},\ldots,\breve{w}_{j,0}\}} \in$ backward branches succeeding the preceding node in $SearchPatternsCG^n$ **do**

120:        **if** $\breve{w}_{i,0} \in \{\breve{w}_{1,0}, \ldots, \breve{w}_{j,0}\}$ **then**

121:            **if** $\exists$ succeeding backward branches under $node_r$ **then**

122:                **while** $\exists$ new candidate id_r **do**

123:                    Generate a new id_r with the *trailling in minimal changes order strategy* making the following hold:

124:                        1. id_r is compatible with od_r with $w_{(id_r, od_r)} = \breve{w}_r$,

125:                        2. $asn_{r-1} \in [asn\_min_{r-1}, asn\_max_{r-1}]$, where asn_{r-1} is ASN at the succeeding backward round, which is computed from id_r, and asn_min_{r-1} (or asn_max_{r-1}) is determined by the minimal (or maximal) weight of nodes on the succeeding backward branches of $node_r$.

126:                    SEARCHBACKWARD($\breve{w}_{i,0}$, id_r, $r-1$)

127:                **end while**

128:            **else**

129:                **if** $\exists id_r$ compatible with od_r with $w_{(id_r, od_r)} = \breve{w}_r$ **then**

130:                    **if** this is the first time get an n-round differential trail **then** $Bw^n \leftarrow w^n$

131:                    **end if**

132:                    **if** $n\_w^n = -1$ **then** goto Next-n

133:                    **end if**

134:                    $\breve{W} \leftarrow_{delete} SearchPatternsCG^n$ ▷ Delete current concretized search pattern from trees of concretized search patterns under current w^n

135:                    $ExistentPatternsCG^n \leftarrow_{insert} \breve{W}$ ▷ Save current concretized search pattern to trees of actually existent concretized search patterns under $w^n \in [Bw^n, w\_u^n]$

136:                **end if**

137:            **end if**

138:        **end if**

139:    **end for**

140: **end procedure**

B   Examples of Best Trails

R	id of Gamma	od of Gamma	w_r
1	0x0000000000000001000020210c280001	0x00000000000000c0000408e0182000c	17
2	0x00000000000000c0000008e0000000c	0x0000000000000001000000210000000 1	8
3	0x000000000040000000400800004000 00	0x00000000020000000200400003000 00	12
4	0x00000000020004000210400003000 00	0x000000000800020008c020000800000	14

Fig. 6. A best 4-round differential trail with weight 51 in NOEKEON

R	*im* of Gamma	*om* of Gamma	Bias
1	0x000000000000000800004c0100000001	0x000000000000000500009201000000001	2^{-6}
2	0x000000000000000500009a2100000001	0x000000000000000500009a2100000001	2^{-9}
3	0x000000000000000500009201000000001	0x000000000000000500009201000000001	2^{-7}
4	0x000000000000000500009a2100000001	0x000000000000000500009a2100000001	2^{-9}
5	0x000000000000000500009201000000001	0x000000000000000500009201000000001	2^{-7}
6	0x000000000000000500009a2100000001	0x000000000000000800004cc100000001	2^{-7}

Fig. 7. A best 6-round linear trail with bias 2^{-40} in NOEKEON

R	S-Layer *id*	S-Layer *od*	w_r
1	0x0000000000000000000003030000030003	0x0000000000000000000002020000020002	8
2	0x0000000000000000000000005044400000000	0x0000000000000000000000040cc00000000	6
3	0x000000c0000000b0000000000000000000	0x0000000800000008000000000000000000	6
4	0x0202000000000000000000000000000000	0x0a0a000000000000000000000000000000	4
5	0x5000000000000000500000000000000000	0x4000000000000000400000000000000000	4
6	0x0000000020001000000000000000000000	0x0000000050005000000000000000000000	5
7	0x0000000002200000000000000002200000	0x0000000000aa000000000000000005500000	10
8	0x0030000000000006000300000000000060	0x0020000000000002000200000000000040	10
9	0x0000000000000002200110000000000000	0x0000000000000000550055000000000000	10
10	0x0000000000003300000000000000033000	0x0000000000002200000000000000022000	8
11	0x00000000000000000000c0006000000000	0x0000000000000000000010001000000000	5
12	0x0000000000000000000000000000002200	0x000000000000000000000000000000aa00	4
13	0x0000000030000000000000003000000000	0x0000000020000000000000002000000000	4
14	0x0000000000000000001000080000000000	0x0000000000000000003000090000000000	4
15	0x0000010000000000000002000000008400	0x000003000000000000000a000000009c00	8

Fig. 8. A best 15-round differential trial with weight 96 in SPONGENT with $b = 136$

R	S-Layer *id*	S-Layer *od*	w_r
1	0x000000000005000500000000000000000000000000	0x000000000040004000000000000000000000000000	4
2	0x000000000000022000000000000000000000000000	0x000000000000066000000000000000000000000000	6
3	0x000000000000060000000006000000000000000000	0x000000000000010000000001000000000000000000	4
4	0x000000000000000000000000000000000020040000	0x000000000000000000000000000000000060060000	6
5	0x000000000000000000090000000009000000000000	0x000000000000000000080000000008000000000000	4
6	0x000000801000000000000000000000000000000000	0x000000030300000000000000000000000000000000	5
7	0x000000000000000000005000000000500000000000	0x000000000000000000000040000000040000000000	4
8	0x000000000000001002000000000000000000000000	0x000000000000000050050000000000000000000000	5
9	0x000000000000009000000000000009000000000000	0x000000000000008000000000000008000000000000	4
10	0x000100000400000000000000000000000000000000	0x000300000c0000000000000000000000000000000000	4

Fig. 9. A best 10-round differential trial with weight 46 in SPONGENT with $b = 176$

R	S-Layer id	S-Layer od	w_r
1	0x9009000000000000900900	0x8008000000000000800800	8
2	0x9009000000000000000000	0x8008000000000000000000	4
3	0x8800000000000000000000	0x9900000000000000000000	4
4	0xc000000000000000300000	0x1000000000000000200000	5
5	0x0000000000000008200000	0x0000000000000005500000	6
6	0x0000000006000000000060	0x0000000001000000000010	4
7	0x0000000000000000001002	0x0000000000000000005005	5
8	0x0000000000900000000009	0x0000000000200000000004	6
9	0x0000000000100200000000	0x0000000000500500000000	5
10	0x0000000090000000000900	0x0000000080000000000800	4
11	0x0080100000000000000000	0x0050500000000000000000	5
12	0x000000a0000000000a0000	0x0000001000000000010000	6
13	0x0000000000000000008010	0x0000000000000000005050	5
14	0x000000000a0000000000a	0x0000000000100000000001	6
15	0x0000000000000000000801	0x0000000000000000000505	5
16	0x0000000000500000000005	0x0000000000400000000004	4
17	0x0000000080100000000000	0x0000000090300000000000	4

Fig. 10. A best 17-round differential trial with weight 86 in SPONGENT with $b = 88$

R		Trail	w_r
1	S-Layer id	0x00000000000000000000000000000000000009000000000000009000000000000	4
	S-Layer od	0x00000000000000000000000000000000000008000000000000008000000000000	
2	S-Layer id	0x00000000100100000000000000000000000000000000000000000000000000000	4
	S-Layer od	0x00000000300300000000000000000000000000000000000000000000000000000	
3	S-Layer id	0x00000000000000000000000000000000000009000000000000009000000000000	4
	S-Layer od	0x00000000000000000000000000000000000008000000000000008000000000000	
4	S-Layer id	0x00000000800100000000000000000000000000000000000000000000000000000	5
	S-Layer od	0x00000000300300000000000000000000000000000000000000000000000000000	
⋮	⋮	2-round iterative trails as (R3, R4) where R3 and R4 are the differential of the third and the fourth round	⋮
43	S-Layer id	0x00000000000000000000000000000000000009000000000000009000000000000	4
	S-Layer od	0x00000000000000000000000000000000000008000000000000008000000000000	
44	S-Layer id	0x00000000800100000000000000000000000000000000000000000000000000000	4
	S-Layer od	0x00000000900300000000000000000000000000000000000000000000000000000	

Fig. 11. A best 44-round differential trial with weight 196 in SPONGENT with $b = 240$

R	S-Layer id	S-Layer od	w_r
1	0x0000000000000000000000000000330000000000000033	0x0000000000000000000000000000220000000000000022	8
2	0x0000000000000000000000000003000300000000000000	0x0000000000000000000000000002000200000000000000	4
3	0x0000000000000000000000000000088000000000000000	0x0000000000000000000000000000099000000000000000	4
4	0x0000000600000000000000000000000000000006000	0x0000000100000000000000000000000000000002000	5
5	0x0000000000000000000000000000000801000000000	0x0000000000000000000000000000000303000000000	5
6	0x0000000000000000000000000000000a0000000000a00	0x0000000000000000000000000000000010000000000100	6
7	0x0000000000000000000000000000000000000002004	0x0000000000000000000000000000000000000006006	6
8	0x0000000000000000000090000000009000000000000	0x0000000000000000000040000000004000000000000	6
9	0x0000000000000004008000000000000000000000000	0x0000000000000003003000000000000000000000000	6
10	0x0000000000000000000000090000000009000000000	0x0000000000000000000000020000000002000000000	6
11	0x0000000000000000000000002004000000000000000	0x0000000000000000000000006006000000000000000	6
12	0x0000000000000000090000000009000000000000000	0x0000000000000000080000000008000000000000000	4
13	0x0000020040000000000000000000000000000000000	0x0000060060000000000000000000000000000000000	6
14	0x0000000000009000000000090000000000000000000	0x0000000000002000000000020000000000000000000	6
15	0x0000000000000000000080100000000000000000	0x0000000000000000000050500000000000000000	5
16	0x0000000000000005000000000000000000050000	0x0000000000000004000000000000000000040000	4
17	0x0000000000000040000100000000000000000000	0x00000000000000c0000300000000000000000000	4
18	0x0001000000000010000000000080000000000800000	0x0003000000000030000000000090000000000900000	8

Fig. 12. A best 18-round differential trial with weight 99 in SPONGENT with $b = 176$

References

1. Biham, E., Shamir, A.: Differential cryptanalysis of DES-like cryptosystems. In: Menezes, A., Vanstone, S.A. (eds.) CRYPTO 1990. LNCS, vol. 537, pp. 2–21. Springer, Heidelberg (1991)
2. Matsui, M.: Linear cryptanalysis method for DES cipher. In: Helleseth, T. (ed.) EUROCRYPT 1993. LNCS, vol. 765, pp. 386–397. Springer, Heidelberg (1994)
3. Ohta, K., Moriai, S., Aoki, K.: Improving the search algorithm for the best linear expression. In: Coppersmith, D. (ed.) CRYPTO 1995. LNCS, vol. 963, pp. 157–170. Springer, Heidelberg (1995)
4. Collard, B., Standaert, F.-X., Quisquater, J.-J.: Improved and multiple linear cryptanalysis of reduced round serpent. In: Pei, D., Yung, M., Lin, D., Wu, C. (eds.) Information Security and Cryptology. LNCS, vol. 4990, pp. 51–65. Springer, Heidelberg (2008)
5. Daemen, J., Rijmen, V.: The Design of Rijndael - AES - The Advanced Encryption Standard. Springer, Heidelberg (2002)
6. Daemen, J., Peeters, M., Van Assche, G., Rijmen, V.: Nessie Proposal: The Block Cipher NOEKEON. Nessie submission (2000)
7. Bogdanov, A., Knudsen, L.R., Leander, G., Paar, C., Poschmann, A., Robshaw, M.J.B., Seurin, Y., Vikkelsoe, C.: PRESENT: an ultra-lightweight block cipher. In: Paillier, P., Verbauwhede, I. (eds.) CHES 2007. LNCS, vol. 4727, pp. 450–466. Springer, Heidelberg (2007)

8. Biryukov, A., Nikolić, I.: Automatic search for related-key differential character-istics in byte-oriented block ciphers: application to AES, Camellia, Khazad and others. In: Gilbert, H. (ed.) EUROCRYPT 2010. LNCS, vol. 6110, pp. 322–344. Springer, Heidelberg (2010)

9. Mouha, N., Wang, Q., Gu, D., Preneel, B.: Differential and linear cryptanalysis using mixed-integer linear programming. In: Wu, C.-K., Yung, M., Lin, D. (eds.) Inscrypt 2011. LNCS, vol. 7537, pp. 57–76. Springer, Heidelberg (2012)

10. Fouque, P.-A., Jean, J., Peyrin, T.: Structural evaluation of AES and chosen-key distinguisher of 9-round AES-128. In: Canetti, R., Garay, J.A. (eds.) CRYPTO 2013, Part I. LNCS, vol. 8042, pp. 183–203. Springer, Heidelberg (2013)

11. Matsui, M.: On correlation between the order of S-boxes and the strength of DES. In: De Santis, A. (ed.) EUROCRYPT 1994. LNCS, vol. 950, pp. 366–375. Springer, Heidelberg (1995)

12. Aoki, K., Kobayashi, K., Moriai, S.: Best differential characteristic search of FEAL. In: Biham, E. (ed.) FSE 1997. LNCS, vol. 1267, pp. 41–53. Springer, Heidelberg (1997)

13. Leurent, G.: Construction of differential characteristics in ARX designs application to skein. In: Canetti, R., Garay, J.A. (eds.) CRYPTO 2013, Part I. LNCS, vol. 8042, pp. 241–258. Springer, Heidelberg (2013)

14. Biryukov, A., Velichkov, V.: Automatic search for differential trails in ARX ciphers. In: Benaloh, J. (ed.) CT-RSA 2014. LNCS, vol. 8366, pp. 227–250. Springer, Heidelberg (2014)

15. Daemen, J.: Cipher and hash function design strategies based on linear and differ-ential cryptanalysis. Doctoral Dissertation, March 1995, K.U.Leuven (1995)

16. Bogdanov, A., Knezevic, M., Leander, G., Toz, D., Varici, K., Verbauwhede, I.: SPONGENT: the design space of lightweight cryptographic hashing. IEEE Trans. Comput. 62(10), 2041–2053 (2013)

17. Ehrlich, G.: Loopless Algorithms for Generating Permutations, Combinations, and Other Combinatorial Configurations. Journal of the ACM 20(3), 500–513 (1973)

18. Knuth, D.E.: The Art of Computer Programming. Introduction to Combinatorial Algorithms and Boolean Functions, vol. 4. Addison Wesley, Upper Saddle River (2008)

The Boomerang Attacks
on BLAKE and BLAKE2

Yonglin Hao$^{(\boxtimes)}$

Department of Computer Science and Technology,
Tsinghua Universtiy, Beijing 100084, China
haoyl12@mails.tsinghua.edu.cn

Abstract. In this paper, we study the security margins of hash functions BLAKE and BLAKE2 against the boomerang attack. We launch boomerang attacks on all four members of BLAKE and BLAKE2, and compare their complexities. We propose 8.5-round boomerang attacks on both BLAKE-512 and BLAKE2b with complexities 2^{464} and 2^{474} respectively. We also propose 8-round attacks on BLAKE-256 with complexity 2^{198} and 7.5-round attacks on BLAKE2s with complexity 2^{184}. We verify the correctness of our analysis by giving practical 6.5-round Type I boomerang quartets for each member of BLAKE and BLAKE2. According to our analysis, some tweaks introduced by BLAKE2 have increased its resistance against boomerang attacks to a certain extent. But on the whole, BLAKE still has higher a secure margin than BLAKE2.

1 Introduction

Cryptographic hash functions (simply referred as hash functions) are playing a significant role in the modern cryptology. They are indispensable in achieving secure systems such as digital signatures, message authentication codes and so on. In the cryptanalysis of hash functions, one of the greatest breakthrough was made by Wang et al. in 2005 when they successfully launched collision attacks on widely used hash functions MD5 [1] and SHA-1 [2]. After that, the analytic methods against hash functions have been greatly improved which threatens the security of existing hash functions. To cope with this situation, NIST proposed the transition from SHA-1 to SHA-2. Furthermore, NIST also launched the SHA-3 competition to develop a new hash standard. After years' analysis, five proposals entered the final round of SHA-3 and the one named Keccak became the new SHA-3 standard in 2012 [3].

The BLAKE hash function [4] was one of the five finalists of the SHA-3 competition [5]. Although it was not selected as the SHA-3 standard, along with the other finalists, BLAKE is assumed to be a very strong hash function with high security margin and very good performance in software.

BLAKE2 [6] is a new family of hash functions based on BLAKE. According to [6], the main objective of BLAKE2 is to provide a number of parameters for use in applications without the need of additional constructions and modes, and

© Springer International Publishing Switzerland 2015
D. Lin et al. (Eds.): Inscrypt 2014, LNCS 8957, pp. 286–310, 2015.
DOI: 10.1007/978-3-319-16745-9_16

also to speed-up even further the hash function to a level of compression rate close to MD5.

Ever since its proposal, BLAKE has attracted a considerable amount of cryptanalysis, such as impossible differential attack [7], differential attack [8], collision, preimage [9] etc. There is also a boomerang distinguisher on BLAKE-32 given by Biryukov et al. in [10] but some incompatible problems were pointed out by Leurent in [11]. Despite of the incompatibilities, [10] indicates that the boomerang method may have good efficiency in analyzing the BLAKE family. Recently, Bai et al. have given the first valid 7-round and 8-round boomerangs for BLAKE-256 [12].

As to BLAKE2, Guo et al. [13] have given a thorough security analysis of it. In their paper, they applied almost all the existing attacks on BLAKE to BLAKE2. According to their results, the tweaks introduced by BLAKE2, if analyzed separately, reduce the security of the version in some theoretical attacks. Some cryptanalysis methods manage to reach more rounds for BLAKE2 than BLAKE. BLAKE seems to have better resistance than BLAKE2 against various cryptanalysis methods. However, [13] did not evaluate the security margin of the two hash function families under the boomerang method and this is what we are going to do in this paper.

The original boomerang attack was introduced by Wagner in 1999 [14] as a tool for the cryptanalysis of block ciphers. It is an adaptive chosen plaintext and ciphertext attack utilizing differential cryptanalysis. Later, Kelsey et al. [15] developed the original version into a chosen plaintext attack called the amplified boomerang attack. Developments were also made by Biham et al. in [16,17].

During the past few years, the idea of the boomerang attack has been applied to many hash functions. Biryukov et al. [10] and Lamberger et al. [18] independently applied the boomerang attack to BLAKE-32 and SHA-256. The SHA-256 result was later improved by Biryukov et al. in [19]. Ever after, we saw the boomerang results on many hash functions such as SIME-512 [20], HAVAL [21], RIPEMD-128/160 [22], HAS-160 [21], Skein-256/512 [23,24], SM3 [25,26] and BLAKE-256 [12]. The boomerang attack has become a common tool for analyzing various hash functions.

Our contribution. We reevaluate the boomerang attack on BLAKE-256 in [12] and apply the method to the keyed permutations of all BLAKE and BLAKE2 members namely BLAKE-256, BLAKE-512, BLAKE2s and BLAKE2b. We construct boomerang distinguishers for 8.5-round keyed permutation of BLAKE-512 and BLAKE2b (both from round 2.5 to 11). The complexity for attacking BLAKE-512 is 2^{464} and that for BLAKE2b is 2^{474}. We also present 7.5-round attack on BLAKE2s (round 2.5 to 10) with complexity 2^{184}. Besides, we lower the complexity of the 8-round BLAKE-256 result in [12] from 2^{200} to 2^{198} with slight modification of the differential characteristic. We present our boomerang results along with previous ones in Table 1. As can be seen, some tweaks introduced by BLAKE2 have surprisingly increased its resistance against boomerang attacks to a certain extent. But, since BLAKE has more rounds, the secure margin of BLAKE is still higher than that of BLAKE2.

Table 1. All existing boomerang results on BLAKE and BLAKE2.

Hash function	Target	Rounds	Time	Source
BLAKE-256	CF	6	2^{102}	[10]
	CF	6.5[a]	2^{184}	
	CF	7[a]	2^{232}	
	KP	6	$2^{11.75}$	
	KP	7[a]	2^{122}	
	KP	8[a]	2^{242}	
	KP	7	2^{37}[b]	[12]
	KP	8	2^{200}	
	KP	**8**	2^{198}	**This paper**
BLAKE2s	**KP**	**7.5**	2^{184}	**This paper**
BLAKE-512	**KP**	**8.5**	2^{464}	**This paper**
BLAKE2b	**KP**	**8.5**	2^{474}	**This paper**

KP: Keyed Permutation
CF: Compression Function
[a]: there are some incompatible problems in their attacks
[b]: this is the complexity for the Type III boomerang while others are of Type I.

Organization of the Paper. In Sect. 2, we briefly introduce the round functions of BLAKE and BLAKE2, and provide the overview of the boomerang attack. Section 3 describes the way that we deduce the differential characteristics and the process of building the boomerang distinguishers. Finally, we conclude our paper in Sect. 4.

2   Preliminary

In the first part of this section, we make a brief introduction of the two families of hash functions, BLAKE and BLAKE2. Since our boomerang analysis mainly focus on the keyed permutation of BLAKE and BLAKE2, which excludes the Initialization and Finalization procedures, we only introduce the round functions in this section. We refer the readers to [4,6] for information about initialization and finalization phases. We also give some notations that are used through this paper.

In the second part of this section, we review the procedure of the boomerang attack on hash functions and give some definitions that we use in the description of our attacks.

2.1   The Round Functions of BLAKE and BLAKE2

BLAKE and BLAKE2 share many similarities. As the successor of BLAKE, BLAKE2 has a 32-bit version (BLAKE2s) and a 64-bit version (BLAKE2b), corresponding to BLAKE-256 and BLAKE-512 of BLAKE respectively. Both BLAKE and BLAKE2 process 16-word message blocks. However, differences can be witnessed at every level including internal permutation, compression function, and hash function construction. Some notations have to be introduced first:

$\leftarrow$ variable assignment;
$+$ modular 2^{32} or 2^{64} addition (according to the word length);
$-$ modular 2^{32} or 2^{64} subtraction (according to the word length);
$\oplus$ bitwise exclusive or;
$\lll n$ cyclic shift n bits towards the most significant bit;
$\ggg n$ cyclic shift n bits towards the least significant bit;
$\wedge$ bitwise AND operation for words.

The Round functions of both BLAKE and BLAKE2 process a state of 16 64-bit or 32-bit words represented by a 4×4 matrix as follows:

$$V = \begin{pmatrix} v_0 & v_1 & v_2 & v_3 \\ v_4 & v_5 & v_6 & v_7 \\ v_8 & v_9 & v_{10} & v_{11} \\ v_{12} & v_{13} & v_{14} & v_{15} \end{pmatrix}.$$

In the remainder of this paper, we denote the 16-word intermediate state by the capital letters such as V, TV and M for message block. Single 64-bit or 32-bit words are denoted by small letters such as v, tv and m for message words. We also refer to the i-th bit of a word v ($i = 0, \cdots 31$ or 63 from the least significant to the most significant) as $v[i]$.

Once the state V is initialized, V is processed by several rounds (10, 12, 14, 16 for BLAKE2s, BLAKE2b, BLAKE-256, BLAKE512 respectively) of G functions, which means computing

$$G_0(v_0, v_4, v_8, v_{12}), G_1(v_1, v_5, v_9, v_{13}), G_2(v_2, v_6, v_{10}, v_{14}), G_3(v_3, v_7, v_{11}, v_{15})$$
$$G_4(v_0, v_5, v_{10}, v_{15}), G_5(v_1, v_6, v_{11}, v_{12}), G_6(v_2, v_7, v_8, v_{13}), G_7(v_3, v_4, v_9, v_{14})$$

where $G_i(a, b, c, d), i = 0, \cdots, 7$ differ among BLAKE2s, BLAKE2b, BLAKE-256, BLAKE512 and are all listed in Table 2. The σ_r in Step 1 and 5 of the G_i function in Table 2 belongs to the set of permutations as defined in Table 3. At round $r > 9$, the permutation used is $\sigma_{r \bmod 10}$ (for example, if $r = 11$, the permutation $\sigma_{11 \bmod 10} = \sigma_1$ is used).

Since we need detailed analysis of the intermediate states, we further breakdown the round functions. We denote the state after r rounds of iterations by V^r ($r = 0, 1, \cdots$). Then, TV^r is acquired after the first 4 steps of $G_{0,\cdots,3}$ and $V^{r+0.5}$ is computed after $G_{0,\cdots,3}$ are completed. Similarly, we can compute $TV^{r+0.5}$

Table 2. The G_i Functions of BLAKE-256, BLAKE2s, BLAKE-512, BLAKE2b

Step	BLAKE-256	BLAKE2s
1	$a = a + b + (m_{\sigma_r(2i)} \oplus c_{\sigma_r(2i+1)})$	$a = a + b + m_{\sigma_r(2i)}$
2	$d = (d \oplus a) \ggg 16$	$d = (d \oplus a) \ggg 16$
3	$c = c + d$	$c = c + d$
4	$b = (b \oplus c) \ggg 12$	$b = (b \oplus c) \ggg 12$
5	$a = a + b + (m_{\sigma_r(2i+1)} \oplus c_{\sigma_r(2i)})$	$a = a + b + m_{\sigma_r(2i+1)}$
6	$d = (d \oplus a) \ggg 8$	$d = (d \oplus a) \ggg 8$
7	$c = c + d$	$c = c + d$
8	$b = (b \oplus c) \ggg 7$	$b = (b \oplus c) \ggg 7$
Step	BLAKE-512	BLAKE2b
1	$a = a + b + (m_{\sigma_r(2i)} \oplus c_{\sigma_r(2i+1)})$	$a = a + b + m_{\sigma_r(2i)}$
2	$d = (d \oplus a) \ggg 32$	$d = (d \oplus a) \ggg 32$
3	$c = c + d$	$c = c + d$
4	$b = (b \oplus c) \ggg 25$	$b = (b \oplus c) \ggg 24$
5	$a = a + b + (m_{\sigma_r(2i+1)} \oplus c_{\sigma_r(2i)})$	$a = a + b + m_{\sigma_r(2i+1)}$
6	$d = (d \oplus a) \ggg 16$	$d = (d \oplus a) \ggg 16$
7	$c = c + d$	$c = c + d$
8	$b = (b \oplus c) \ggg 11$	$b = (b \oplus c) \ggg 63$

from $V^{r+0.5}$ by applying steps 1,2,3,4 of $G_{4,\cdots,7}$ and further compute V^{r+1} by finishing $G_{4,\cdots,7}$. This representation is illustrated as (1) and (2).

$$G_{0,\cdots,3} : V^r \xrightarrow{\text{Steps } 1,\cdots,4} TV^r \xrightarrow{\text{Steps } 5,\cdots,8} V^{r+0.5} \tag{1}$$

$$G_{4,\cdots,7} : V^{r+0.5} \xrightarrow{\text{Steps } 1,\cdots,4} TV^{r+0.5} \xrightarrow{\text{Steps } 5,\cdots,8} V^{r+1} \tag{2}$$

In this way, we can refer to any intermediate state word of any round easily.

2.2 The Boomerang Attack

About the boomerang attack on hash functions, we mainly review the known-related-key boomerang method given in [19]. We consider the compression function, denoted by CF, as $CF(M, K) = E(M, K) + M$ and that it can be decomposed into two sub-functions as $CF = CF_1 \circ CF_0$. In this way, we can start from the middle steps since M and the key K can be chosen randomly [19,23]. Then we have a backward (top) differential characteristic $(\beta, \beta_k) \to \alpha$ with probability p for CF_0^{-1}, and a forward (bottom) differential characteristic $(\gamma, \gamma_k) \to \delta$ with probability q for CF_1. Finally, we can launch the known-related-key boomerang attack with these two differential characteristics as follows:

Table 3. The definition of σ_r where $r = 0, \cdots, 9$.

σ_0	0	1	2	3	4	5	6	7	8	9	10	11	12	13	14	15
σ_1	14	10	4	8	9	15	13	6	1	12	0	2	11	7	5	3
σ_2	11	8	12	0	5	2	15	13	10	14	3	6	7	1	9	4
σ_3	7	9	3	1	13	12	11	14	2	6	5	10	4	0	15	8
σ_4	9	0	5	7	2	4	10	15	14	1	11	12	6	8	3	13
σ_5	2	12	6	10	0	11	8	3	4	13	7	5	15	14	1	9
σ_6	12	5	1	15	14	13	4	10	0	7	6	3	9	2	8	11
σ_7	13	11	7	14	12	1	3	9	5	0	15	4	8	6	2	10
σ_8	6	15	14	9	11	3	0	8	12	2	13	7	1	4	10	5
σ_9	10	2	8	4	7	6	1	5	15	11	9	14	3	12	13	0

1. Choose randomly a intermediate state (X_1, K_1) and compute $(X_i, K_i), i = 2, 3, 4$ by $X_3 = X_1 \oplus \beta$, $X_2 = X_1 \oplus \gamma$, $X_4 = X_3 \oplus \gamma$, and $K_3 = K_1 \oplus \beta_k$, $K_2 = K_1 \oplus \gamma_k$, $K_4 = K_3 \oplus \gamma_k$.
2. Compute backward from (X_i, K_i) and obtain P_i by $P_i = CF_0^{-1}(X_i, K_i)$ $(i = 1, 2, 3, 4)$.
3. Compute forward from (X_i, K_i) and obtain C_i by $C_i = CF_1(X_i, K_i)$ $(i = 1, 2, 3, 4)$.
4. Check whether $P_1 \oplus P_3 = P_2 \oplus P_4 = \alpha$ and $C_1 \oplus C_2 = C_3 \oplus C_4 = \delta$.

It can be deduced that $P_1 \oplus P_3 = P_2 \oplus P_4 = \alpha$ and $C_1 \oplus C_2 = C_3 \oplus C_4 = \delta$ hold with probability at least p^2 in the backward direction and q^2 in the forward direction. Therefore, the attack succeeds with probability $p^2 q^2$ when assuming that the differential characteristics are independent.

According H. Yu et al. in [24], for a n-bit random permutation, there are three types of boomerang distinguishers:

- Type I: A quartet satifies $P_1 \oplus P_3 = P_2 \oplus P_4 = \alpha$ and $C_1 \oplus C_2 = C_3 \oplus C_4 = \delta$ for fixed differences α and δ. In this case, the generic complexity is 2^n.
- Type II: Only $C_1 \oplus C_2 = C_3 \oplus C_4$ is satisfied (This property is also called zero-sum or second-order differential collision). In this case, the complexity for obtaining such a quartet is $2^{n/3}$ [27].
- Type III: A quartet satisfied $P_1 \oplus P_3 = P_2 \oplus P_4$ and $C_1 \oplus C_2 = C_3 \oplus C_4$. In this case, the best known still takes time $2^{n/2}$.

We only study the Type I boomerang distinguisher in this paper. Besides, the complexity 2^{37} of the 7-round boomerang in [12] is actually the complexity for a Type III boomerang attack. The Type I complexity for the 7-round attack should be $2^{2 \times (1+4+16+1)} = 2^{44}$ according to their methods.

3   The Boomerang Attacks on BLAKE and BLAKE2

In this section, we describe our boomerang attacks on BLAKE and BLAKE2. We only illustrate our strategies by comparing BLAKE-512 and BLAKE2b while those of BLAKE-256 and BLAKE2s can be deduced accordingly. Details are presented in Appendix A.

3.1   Construction of Differential Characteristics

The very first step for the boomerang attack is constructing two differential characteristics with high probability. Since BLAKE and BLAKE2 are ARX hash functions (only use three simple operations namely **Modular Add** "+", **Rotation** "$\ggg$" and **XOR** "$\oplus$"), we can use the XOR difference and deduce the difference linearly by considering the only nonlinear operation "+" as similar linear operation "$\oplus$".

The XOR difference in this paper is represented in two forms as follow:

- **Hex form:** such as $\Delta v = 0x8003$ indicates that bits $v[0, 1, 15]$ of the word v are active (having non-zero XOR difference).
- **Numeric form:** such as $\Delta v = (15, 1, 0)$ is equivalent to $\Delta v = 0x8003$ in hex form. Besides, if $\Delta v = 0x0$ in hex form, we denoted by $\Delta v = \phi$ in numeric form.

The numeric form is mainly used to describe the differential characteristics since it has better outlook and can save some space. But in practice, we use the hex form to linearly deduce differential characteristics. For example, in the G function of BLAKE-512, we have

$$ta = a + b + (m_i \oplus c_j)$$

where c_j is constant. Suppose we have acquired the differences Δa, Δb and Δm_i, we can deduce Δta as

$$\Delta ta = \Delta a \oplus \Delta b \oplus \Delta m_i.$$

Once we have determined the difference of the message block ΔM and that of a intermediate state ΔV^r ($r = 0, 0.5, 1, \cdots$), we can linearly extend the difference backward and forward.

We construct the two differential characteristics for the boomerang attack, where the top differential characteristic is from round 2.5 to 6.5 and bottom differential difference is from 6.5 to 11. We denote the difference of the top by $\Delta^t V^r$ ($r \in [2.5, 6.5]$) and that of the bottom by $\Delta^b V^r$ ($r \in [6.5, 11]$). Similarly, the difference for the message block is denoted as $\Delta^t M$ in the top characteristic and $\Delta^b M$ in the bottom characteristic. The main procedures for our characteristic construction can be summarized as follows:

Import Difference: We first import simple difference to message block $\Delta^b M$ ($\Delta^t M$) and the intermediate state $\Delta^b V^8$ ($\Delta^t V^4$).

Linear Extension: After we have determined $\Delta^b M$ ($\Delta^t M$) and $\Delta^b V^8$ ($\Delta^t V^4$), we extend the difference backward to round 6.5 (2.5) and forward to round 11 (6.5) to acquire the whole bottom (top) differential characteristic.

Construct the Bottom Differential Characteristic. In order to lower the complexity, we only import 1-bit differences to both $\Delta^b M$ and $\Delta^b V^8$. The selection of active bits is based on **Observation 1** in [10].

We found that m_{11} of the 16 message words, namely $m_0, \cdots, m_{15}$, appears at Step 1 in G_2 at round 8 and also appears at Step 5 in G_4 at round 9. So, the first step of our construction is importing 1-bit difference to m_{11} and v_2^8 as

$$\Delta^b m_{11} = \Delta^b v_2^8 = (63). \tag{3}$$

In this way, according to **Observation 1** in [10], we can pass round 8 and 9 with probability 2^{-1}. Then, we set $\Delta^b m_i = \phi$ ($i \in \{0, 1, \cdots, 15\} \setminus \{11\}$) and $\Delta^b v_j^8 = \phi$ ($j \in \{0, 1, \cdots, 15\} \setminus \{2\}$). Now that $\Delta^b M$ and $\Delta^b V^8$ are settled, we can linearly extend the difference backward to $\Delta^b V^{6.5}$ and forward to $\Delta^b V^{11}$. This method can be applied to both BLAKE-512 and BLAKE2b. We present the bottom characteristics of BLAKE-512 and BLAKE2b as Tables 4 and 5 in Appendix A respectively.

For BLAKE-256 and BLAKE2s, we can also import difference to $\Delta^b M$ and $\Delta^b V^8$ as

$$\Delta^b m_{11} = \Delta^b v_2^8 = (31). \tag{4}$$

and linearly deduce the whole bottom differential characteristics. The differential characteristic for BLAKE-256 mounts to round 10.5 and BLAKE2s reaches round 10 since it only has 10 rounds in total according to [6]. Refer to Tables 6 and 7 in Appendix A for detailed descriptions.

Construct the Top Differential Characteristic. The top differential characteristic starts from $\Delta^t V^{2.5}$ and ends at $\Delta^t V^{6.5}$. The strategy of constructing the top differential characteristic is similar to that of its bottom counterpart. We found that m_5 appears at Step 1 in G_1 at round 4 and also appears at Step 5 in G_5 at round 5, so we decide to import the 1-bit difference at m_5 and v_1^4. We assign that

$$\Delta^t m_5 = \Delta^t v_1^4 = (y), \text{ where } y \in \{0, \cdots, 63\}. \tag{5}$$

and that $\Delta^b m_i = \phi$ ($i \in \{0, \cdots, 15\} \setminus \{5\}$) and $\Delta^b v_j^4 = \phi$ ($j \in \{0, \cdots, 15\} \setminus \{1\}$). Then, we can linearly extend the difference backward and forward. The position of the active bit y in (5) requires careful selection. In order to avoid incompatible problems and enhance the efficiency of the attack, y must meet the following conditions:

1. When linearly extend the difference from $\Delta^t V^4(y)$ to $\Delta^t V^{6.5}(y)$, make sure that

$$\Delta^b v_i^{6.5} \wedge \Delta^t v_i^{6.5}(y) = \text{0x0, for all } i \in \{0, \cdots, 15\}. \qquad (6)$$

This restriction avoid the contradictions in the intersection part of the two differential characteristics.

2. **(Only for BLAKE-512)**. Make sure that the constants c_{10} and c_7 satisfies:

$$c_{10}[y] = \neg c_7[y]. \qquad (7)$$

According to the linear extension, we have $\Delta^t v_1^{3.5} = \phi$. It requires $(m_5 \oplus c_{10})[y] = \neg(m_5 \oplus c_7)[y]$, so (7) must be satisfied.

3. **(Only for BLAKE2b)**. When linearly extend to $\Delta^t V^{3.5}$, $\Delta^t v_1^{3.5}$ should be set to

$$\Delta^t v_1^{3.5} = \Delta^t m_5 + \Delta^t m_5$$

instead of 0x0. Because BLAKE2b omit the use of constant, the difference can not be eliminated at $v_1^{3.5}$.

The available ys satisfying conditions 1 and 2 compose a set $\mathbb{X}_{512}$, and those satisfying conditions 1 and 3 compose a set $\mathbb{X}_{2b}$. According to our analysis, $\mathbb{X}_{512}$ has 13 elements and $\mathbb{X}_{2b}$ has 40 elements. We present $\mathbb{X}_{512}$ and $\mathbb{X}_{2b}$ along with the corresponding top differential characteristics in Tables 8 and 9 in Appendix A.

Using the same method, we can also acquire the available ys for BLAKE-256 ($\mathbb{X}_{256}$) and BLAKE2s ($\mathbb{X}_{2s}$). We illustrate $\mathbb{X}_{256}$ and $\mathbb{X}_{2s}$ along with their characteristics in Tables 10 and 11 in Appendix A.

3.2 Finding the Boomerang Quartet Using Message Modification Technique

The goal of our boomerang attack is to find a quartet, denoted by $(_aV^{2.5}, {}_bV^{2.5}, {}_cV^{2.5}, {}_dV^{2.5})$, and the message blocks $(_aM, {}_bM, {}_cM, {}_dM)$ that satisfies

$$_aV^{2.5} \oplus {}_cV^{2.5} = {}_bV^{2.5} \oplus {}_dV^{2.5} = \Delta^t V^{2.5} \qquad (8)$$

$$_aM \oplus {}_cM = {}_bM \oplus {}_dM = \Delta^t M \qquad (9)$$

$$_aM \oplus {}_bM = {}_cM \oplus {}_dM = \Delta^b M \qquad (10)$$

and, after 8.5 rounds, the corresponding quartet $(_aV^{11}, {}_bV^{11}, {}_cV^{11}, {}_dV^{11})$ satisfies

$$_aV^{11} \oplus {}_bV^{11} = {}_cV^{11} \oplus {}_dV^{11} = \Delta^b V^{11}.$$

We start by searching for appropriate $_aV^{6.5}$ and $_aM$. Once $_aV^{6.5}$ is determined, $_bV^{6.5}$, $_cV^{6.5}$ and $_dV^{6.5}$ can be settled directly since

$$_aV^{6.5} \oplus {}_cV^{6.5} = {}_bV^{6.5} \oplus {}_dV^{6.5} = \Delta^t V^{6.5} \qquad (11)$$

$$_aV^{6.5} \oplus {}_bV^{6.5} = {}_cV^{6.5} \oplus {}_dV^{6.5} = \Delta^b V^{6.5} \qquad (12)$$

Once $_aM$ is determined, $_bM$, $_cM$ and $_dM$ can also be determined according to (9) and (10). The step of finding the quartet is as follows:

1. Construct an intermediate state, denoted by $V^{6.5}$, and a message block, denoted by M, by setting the values of their 16 words randomly.
2. Compute backward to TV^6 and V^6, and forward to $TV^{6.5}, V^7$. During the process, if one of bit conditions, which are deduced from the top and bottom characteristics, is violated, we can fix it by modifying the words of $V^{6.5}$ or M. This process is called the "message modification".
3. After all conditions between round 6 and 7 are satisfied, we assign that $_aV^{6.5} \leftarrow V^{6.5}$ and $_aM \leftarrow M$. We also assign corresponding values to $_bV^{6.5}$, $_cV^{6.5}$, $_dV^{6.5}$ according to (11, 12) and to $_bM$, $_cM$, $_dM$ according to (9, 10).
4. Having acquired $(_aV^{6.5}, _bV^{6.5}, _cV^{6.5}_dV^{6.5})$ and $(_aM, _bM, _cM, _dM)$, we compute backward to round 2.5. During the process, we check whether the differences of the intermediate states conform to the top differential characteristic. Once a contradiction is detected, go back to 1.
5. Compute forward from round 6.5 to round 11. During the computation, we check whether differences of the intermediate states conform to the bottom differential characteristic. Once a contradiction is detected, go back to 1. Otherwise, output the quartet $(_aV^{11}, _bV^{11}, _cV^{11}, _dV^{11})$.

Complexity analysis. For all 4 members of BLAKE and BLAKE2, there are 30 conditions in $\Delta^b V^6 \rightarrow \Delta^t V^{6.5}$. 29 of them can be fixed using the message modification technique. All two conditions in $\Delta^t V^6 \rightarrow \Delta^t V^{5.5}$ can be fixed as well. Similarly, all 40 conditions in $\Delta^b V^{6.5} \rightarrow \Delta^b V^7$ and 2 out of 6 conditions in $\Delta^b V^7 \rightarrow \Delta^b V^{7.5}$ can be fixed. Then, we analyze the four members separately as follows:

BLAKE-512: In the bottom characteristic, there are 4 unfixed conditions in $\Delta^b V^7 \rightarrow \Delta^b V^{7.5}$, 1 in $\Delta^b V^{9.5} \rightarrow \Delta^b V^{10}$, 24 in $\Delta^b V^{10} \rightarrow \Delta^b V^{10.5}$ and 138 in $\Delta^b V^{10.5} \rightarrow \Delta^b V^{11}$, which is 167 in total. In the top characteristics, the situation is as follows: 1 unfixed condition in $\Delta^t V^{4.5} \rightarrow \Delta^t V^4$, 2 in $\Delta^t V^4 \rightarrow \Delta^t V^{3.5}$, 11 in $\Delta^t V^{3.5} \rightarrow \Delta^t V^3$ and 51 in $\Delta^t V^3 \rightarrow \Delta^t V^{2.5}$, which is 65 in total. So, the complexity of the boomerang attack on BLAKE-512 is $2^{(167+65) \times 2} = 2^{464}$.

BLAKE2b: In the bottom characteristic, there are 4 unfixed conditions in $\Delta^b V^7 \rightarrow \Delta^b V^{7.5}$, 1 in $\Delta^b V^{9.5} \rightarrow \Delta^b V^{10}$, 24 in $\Delta^b V^{10} \rightarrow \Delta^b V^{10.5}$ and 124 in $\Delta^b V^{10.5} \rightarrow \Delta^b V^{11}$, which is 153 in total. The top differential characteristic is slightly different from BLAKE-512 after finishing the procedure $\Delta^t V^{6.5} \rightarrow \Delta^t V^4$. There are 3 unfixed conditions in $\Delta^t V^4 \rightarrow \Delta^t V^{3.5}$, 13 in $\Delta^t V^{3.5} \rightarrow \Delta^t V^3$ and 67 in $\Delta^t V^3 \rightarrow \Delta^t V^{2.5}$. So the number of unfixed conditions in the top characteristic enhances to $1 + 3 + 13 + 67 = 84$. The complexity of the boomerang attack on BLAKE2b is $2^{(153+84) \times 2} = 2^{474}$.

BLAKE-256: Similar to BLAKE-512, the bottom characteristic of BLAKE-256, terminated at round 10.5, has $4 + 1 + 24 = 29$ unfixed conditions ($\Delta^b V^{6.5} \rightarrow \Delta^t V^{10.5}$). For the top characteristic of BLAKE-256, if we choose the active bit position $y = 20 \in \mathbb{X}_{256}$, which is also the case of [12], there should be 71 unfixed conditions and the complexity of this 8-round boomerang attack is $2^{(29+71)\times 2} = 2^{200}$. However, if we choose $y = 28 \in \mathbb{X}_{256}$, 1 condition in $\Delta^t V^3 \to \Delta^t V^{2.5}$ can be eliminated and the complexity of the attack can lower to $2^{(29+70)\times 2} = 2^{198}$.

BLAKE2s: Similar to BLAKE2b, the bottom characteristic for BLAKE2s, terminated at round 10, has $4+1 = 5$ unfixed conditions. The top characteristic has 88 unfixed conditions. So the complexity of this 7.5-round boomerang attack for BLAKE2s is $2^{(5+88)\times 2} = 2^{186}$. Like BLAKE-256, if we choose $y = 28 \in \mathbb{X}_{2s}$, we can eliminate 1 condition in $\Delta^t V^3 \to \Delta^t V^{2.5}$ and lower the complexity by 2^2 to 2^{184}.

Practical Verifications. For each member of BLAKE and BLAKE2, we present a 6.5 round (from round 3.5 to round 10) Type I boomerang quartet based on our characteristics and present it in Appendix B. In order to show the structural difference between BLAKE and BLAKE2, we use the examples with the same message difference, which means: for BLAKE-256 and BLAKE2s, $\Delta^t m_5 = (28)$ ($y = 28 \in \mathbb{X}_{256} \cap \mathbb{X}_{2s}$) and $\Delta^b m_{11} = (31)$; for BLAKE-512 and BLAKE2b, $\Delta^t m_5 = (9)$ ($y = 9 \in \mathbb{X}_{512} \cap \mathbb{X}_{2b}$) and $\Delta^b m_{11} = (63)$.

4 Conclusion

In this paper, we compare the security margin of BLAKE and BLAKE2 under the boomerang attack model. We deduce valid differential characteristics and present boomerang attacks on keyed permutations of BLAKE-512, BLAKE2b, BLAKE-256 and BLAKE2s. According to our analysis, the boomerang method can mount to similar rounds for BLAKE and BLAKE2. For the same number of rounds, the complexities for attacking BLAKE2 are slightly higher than those for BLAKE, which indicates that some tweaks introduced by BLAKE2, aiming at enhancing efficiency and flexibility, have accidentally reinforced the resistance against the boomerang attack. However, since BLAKE has more rounds than BLAKE2, the security margin of BLAKE is still higher than that of BLAKE2. This result is in accordance with the assumptions of the designers.

Acknowledgement. This work has been supported by the National Natural Science Foundation of China (Grant No. 61133013) and by 973 Program (Grant No. 2013CB834205).

A The Bottom and Top Differential Characteristics for BLAKE and BLAKE2

Table 4. The bottom characteristic for BLAKE-512. $\Delta^b m_{11} = (63)$.

r	Difference (Numeric Form)	Cond	r	Difference (Numeric Form)	Cond
6.5	$\Delta^b v_0^{6.5} = (63, 42, 35, 10, 3)$	-	10.5	$\Delta^b v_0^{10.5} = (63, 6)$	24
	$\Delta^b v_1^{6.5} = (63, 31)$			$\Delta^b v_1^{10.5} = (43, 36, 11)$	
	$\Delta^b v_2^{6.5} = (63, 47, 24, 15)$			$\Delta^b v_2^{10.5} = (22)$	
	$\Delta^b v_3^{6.5} = (60, 56, 40, 31,$			$\Delta^b v_3^{10.5} = (54)$	
	$\quad 24, 10, 8)$				
	$\Delta^b v_4^{6.5} = (60, 56, 47, 40, 35,$			$\Delta^b v_4^{10.5} = (59, 43, 36, 20, 4)$	
	$\quad 24, 15, 8)$				
	$\Delta^b v_5^{6.5} = (63, 35, 24, 3)$			$\Delta^b v_5^{10.5} = (57, 48, 41, 32, 16, 9, 0)$	
	$\Delta^b v_7^{6.5} = (63, 47, 24, 15)$			$\Delta^b v_6^{10.5} = (59, 36, 11)$	
	$\Delta^b v_8^{6.5} = (63, 47, 31, 15)$			$\Delta^b v_7^{10.5} = (52, 43, 27, 4)$	
	$\Delta^b v_{10}^{6.5} = (24)$			$\Delta^b v_8^{10.5} = (54, 47, 31, 15)$	
	$\Delta^b v_{11}^{6.5} = (63, 31)$			$\Delta^b v_9^{10.5} = (59, 52, 27, 20, 4)$	
	$\Delta^b v_{13}^{6.5} = (63)$			$\Delta^b v_{10}^{10.5} = (47, 6)$	
	$\Delta^b v_{14}^{6.5} = (35, 31, 10)$			$\Delta^b v_{11}^{10.5} = (63, 38, 15)$	
	$\Delta^b v_{15}^{6.5} = (56, 42, 31, 24, 10)$			$\Delta^b v_{12}^{10.5} = (54, 47, 15)$	
7	$\Delta^b v_0^7 = (63, 24)$	40 (40 fixed)		$\Delta^b v_{13}^{10.5} = (59, 52, 27, 20)$	
	$\Delta^b v_1^7 = (63, 31)$			$\Delta^b v_{14}^{10.5} = (6)$	
	$\Delta^b v_2^7 = (63)$			$\Delta^b v_{15}^{10.5} = (63, 38)$	
	$\Delta^b v_4^7 = (63, 24)$		11	$\Delta^b v_0^{11} = (63, 57, 48, 41, 22, 16, 13, 9, 6)$	138
	$\Delta^b v_5^7 = (63, 31)$			$\Delta^b v_1^{11} = (63, 61, 59, 43, 38, 34,$	
	$\Delta^b v_5^7 = (63, 31)$			$\quad 22, 13, 11, 2)$	
	$\Delta^b v_8^7 = (63)$			$\Delta^b v_2^{11} = (54, 52, 50, 27, 18, 11, 6, 4)$	
	$\Delta^b v_9^7 = (63, 47, 31, 15)$			$\Delta^b v_3^{11} = (61, 59, 54, 50, 36, 20, 18, 13, 4)$	
	$\Delta^b v_{10}^7 = (63)$			$\Delta^b v_4^{11} = (59, 57, 55, 39, 34, 23, 9, 7, 2)$	
	$\Delta^b v_{13}^7 = (47, 15)$			$\Delta^b v_5^{11} = (59, 50, 27, 14, 2)$	
	$\Delta^b v_{14}^7 = (63, 31)$			$\Delta^b v_6^{11} = (55, 39, 34, 32, 23, 20, 11, 7, 2, 0)$	
7.5	$\Delta^b v_2^{7.5} = (63)$	6 (2 fixed)		$\Delta^b v_7^{11} = (59, 55, 39, 36, 32, 25, 23,$	
	$\Delta^b v_8^{7.5} = (63)$			$\quad 20, 16, 11, 9, 7, 4)$	
	$\Delta^b v_{13}^{7.5} = (63, 31)$			$\Delta^b v_8^{11} = (54, 47, 36, 34, 31, 27, 20, 15,$	
				$\quad 11, 2)$	
8	$\Delta^b v_2^8 = (63)$	0		$\Delta^b v_9^{11} = (61, 45, 43, 34, 20, 6, 4, 2)$	
···	ϕ	0		$\Delta^b v_{10}^{11} = (61, 38, 32, 25, 22, 6, 0)$	
				$\Delta^b v_{11}^{11} = (61, 50, 45, 43, 38, 31, 18)$	
				$\Delta^b v_{12}^{11} = (63, 61, 50, 47, 45, 43, 31, 27,$	
				$\quad 22, 18, 11)$	
10	$\Delta^b v_0^{10} = (63)$	1		$\Delta^b v_{13}^{11} = (54, 52, 34, 20, 2)$	
	$\Delta^b v_5^{10} = (36)$			$\Delta^b v_{14}^{11} = (61, 59, 45, 43, 38, 36, 34,$	
	$\Delta^b v_{10}^{10} = (47)$			$\quad 22, 11, 6, 4, 2)$	
	$\Delta^b v_{15}^{10} = (47)$			$\Delta^b v_{15}^{11} = (61, 48, 47, 41, 22, 16, 9, 6)$	

Table 5. The bottom characteristic for BLAKE2b. $\Delta^b m_{11} = (63)$.

r	Difference (Numeric Form)	Cond	r	Difference (Numeric Form)	Cond
6.5	$\Delta^b v_0^{6.5} = (63, 62, 54, 30, 22)$	-	10.5	$\Delta^b v_0^{10.5} = (63, 7)$	24
	$\Delta^b v_1^{6.5} = (63, 31)$			$\Delta^b v_1^{10.5} = (56, 48, 24)$	
	$\Delta^b v_2^{6.5} = (63, 47, 23, 15)$			$\Delta^b v_2^{10.5} = (23)$	
	$\Delta^b v_3^{6.5} = (62, 55, 46, 39, 31, 23, 7)$			$\Delta^b v_3^{10.5} = (55)$	
	$\Delta^b v_4^{6.5} = (55, 47, 46, 39, 23,$			$\Delta^b v_4^{10.5} = (56, 48, 32, 16, 8)$	
	$\quad 22, 15, 7)$				
	$\Delta^b v_5^{6.5} = (63, 54, 23, 22)$			$\Delta^b v_5^{10.5} = (57, 41, 33, 25, 17, 9, 1)$	
	$\Delta^b v_7^{6.5} = (63, 47, 23, 15)$			$\Delta^b v_6^{10.5} = (48, 24, 8)$	
	$\Delta^b v_8^{6.5} = (63, 47, 31, 15)$			$\Delta^b v_7^{10.5} = (56, 40, 16, 0)$	
	$\Delta^b v_{10}^{6.5} = (23)$			$\Delta^b v_8^{10.5} = (55, 47, 31, 15)$	
	$\Delta^b v_{11}^{6.5} = (63, 31)$			$\Delta^b v_9^{10.5} = (40, 32, 16, 8, 0)$	
	$\Delta^b v_{13}^{6.5} = (63)$			$\Delta^b v_{10}^{10.5} = (47, 7)$	
	$\Delta^b v_{14}^{6.5} = (62, 31, 22)$			$\Delta^b v_{11}^{10.5} = (63, 39, 15)$	
	$\Delta^b v_{15}^{6.5} = (62, 55, 31, 30, 23)$			$\Delta^b v_{12}^{10.5} = (55, 47, 15)$	
7	$\Delta^b v_0^7 = (63, 23)$	40 (40 fixed)		$\Delta^b v_{13}^{10.5} = (40, 32, 8, 0)$	
	$\Delta^b v_1^7 = (63, 31)$			$\Delta^b v_{14}^{10.5} = (7)$	
	$\Delta^b v_2^7 = (63)$			$\Delta^b v_{15}^{10.5} = (63, 39)$	
	$\Delta^b v_4^7 = (63, 23)$		11	$\Delta^b v_0^{11} = (63, 41, 33, 23, 17, 15,$	124
				$\quad 9, 7, 1)$	
	$\Delta^b v_5^7 = (63, 31)$			$\Delta^b v_1^{11} = (56, 48, 39, 24, 23, 16, 15, 8)$	
	$\Delta^b v_8^7 = (63)$			$\Delta^b v_2^{11} = (55, 40, 32, 24, 16, 7)$	
	$\Delta^b v_9^7 = (63, 47, 31, 15)$			$\Delta^b v_3^{11} = (63, 55, 48, 16, 15, 8, 0)$	
	$\Delta^b v_{10}^7 = (63)$			$\Delta^b v_4^{11} = (49, 48, 16, 8, 1)$	
	$\Delta^b v_{13}^7 = (47, 15)$			$\Delta^b v_5^{11} = (50, 40, 16, 8, 0)$	
	$\Delta^b v_{14}^7 = (63, 31)$			$\Delta^b v_6^{11} = (57, 49, 48, 33, 32, 25,$	
7.5	$\Delta^b v_2^{7.5} = (63)$	6 (2 fixed)		$\quad 24, 17, 16, 1)$	
	$\Delta^b v_8^{7.5} = (63)$			$\Delta^b v_7^{11} = (57, 48, 41, 32, 24, 17,$	
				$\quad 16, 8, 1)$	
	$\Delta^b v_{13}^{7.5} = (63, 31)$			$\Delta^b v_8^{11} = (55, 47, 40, 32, 31,$	
				$\quad 24, 16, 15)$	
8	$\Delta^b v_2^8 = (63)$	0		$\Delta^b v_9^{11} = (63, 56, 48, 47, 32, 7)$	
$\cdots$	ϕ	0		$\Delta^b v_{10}^{11} = (63, 57, 49, 39, 25, 23, 7)$	
				$\Delta^b v_{11}^{11} = (63, 56, 47, 39, 32, 31, 0)$	
10	$\Delta^b v_0^{10} = (63)$	1		$\Delta^b v_{12}^{11} = (56, 40, 32, 31, 24, 23, 0)$	
	$\Delta^b v_5^{10} = (48)$			$\Delta^b v_{13}^{11} = (55, 48, 32, 16, 0)$	
	$\Delta^b v_{10}^{10} = (47)$			$\Delta^b v_{14}^{11} = (63, 56, 47, 39, 24, 23, 8, 7)$	
	$\Delta^b v_{15}^{10} = (47)$			$\Delta^b v_{15}^{11} = (63, 47, 41, 33, 23, 9, 7, 1)$	

Table 6. The bottom characteristic for BLAKE-256. $\Delta^b m_{11} = (31)$.

r	Difference (Numeric Form)	Cond	r	Difference (Numeric Form)	Cond
6.5	$\Delta^b v_0^{6.5} = (31, 22, 18, 6, 2)$	-	8	$\Delta^b v_2^8 = (31)$	0
	$\Delta^b v_1^{6.5} = (31, 15)$		$\cdots$	ϕ	0
	$\Delta^b v_2^{6.5} = (31, 23, 11, 7)$		10	$\Delta^b v_0^{10} = (31)$	1
	$\Delta^b v_3^{6.5} = (30, 27, 19, 15, 11, 6, 3)$			$\Delta^b v_5^{10} = (16)$	
	$\Delta^b v_4^{6.5} = (30, 27, 23, 19, 18, 11, 7, 3)$			$\Delta^b v_{10}^{10} = (23)$	
	$\Delta^b v_5^{6.5} = (31, 18, 11, 2)$			$\Delta^b v_{15}^{10} = (23)$	
	$\Delta^b v_7^{6.5} = (31, 23, 11, 7)$		10.5		24
	$\Delta^b v_8^{6.5} = (31, 23, 15, 7)$				
	$\Delta^b v_{10}^{6.5} = (11)$			$\Delta^b v_0^{10.5} = (31, 3)$	
	$\Delta^b v_{11}^{6.5} = (31, 15)$			$\Delta^b v_1^{10.5} = (20, 16, 4)$	
	$\Delta^b v_{13}^{6.5} = (31)$			$\Delta^b v_2^{10.5} = (11)$	
	$\Delta^b v_{14}^{6.5} = (18, 15, 6)$			$\Delta^b v_3^{10.5} = (27)$	
	$\Delta^b v_{15}^{6.5} = (27, 22, 15, 11, 6)$			$\Delta^b v_4^{10.5} = (28, 20, 16, 8, 0)$	
7	$\Delta^b v_0^7 = (31, 11)$	40 (40 fixed)		$\Delta^b v_5^{10.5} = (29, 25, 21, 17, 13, 5, 1)$	
	$\Delta^b v_1^7 = (31, 15)$			$\Delta^b v_6^{10.5} = (28, 16, 4)$	
	$\Delta^b v_2^7 = (31)$			$\Delta^b v_7^{10.5} = (24, 20, 12, 0)$	
	$\Delta^b v_4^7 = (31, 11)$			$\Delta^b v_8^{10.5} = (27, 23, 15, 7)$	
	$\Delta^b v_5^7 = (31, 15)$			$\Delta^b v_9^{10.5} = (28, 24, 12, 8, 0)$	
	$\Delta^b v_8^7 = (31)$			$\Delta^b v_{10}^{10.5} = (23, 3)$	
	$\Delta^b v_9^7 = (31, 23, 15, 7)$			$\Delta^b v_{11}^{10.5} = (31, 19, 7)$	
	$\Delta^b v_{10}^7 = (31)$			$\Delta^b v_{12}^{10.5} = (27, 23, 7)$	
	$\Delta^b v_{13}^7 = (23, 7)$			$\Delta^b v_{13}^{10.5} = (28, 24, 12, 8)$	
	$\Delta^b v_{14}^7 = (31, 15)$			$\Delta^b v_{14}^{10.5} = (3)$	
7.5	$\Delta^b v_2^{7.5} = (31)$	6 (2 fixed)		$\Delta^b v_{15}^{10.5} = (31, 19)$	
	$\Delta^b v_8^{7.5} = (31)$				
	$\Delta^b v_{13}^{7.5} = (31, 15)$				

Table 7. The bottom characteristic for BLAKE2s. $\Delta^b m_{11} = (31)$.

r	Difference (Numeric Form)	Cond	r	Difference (Numeric Form)	Cond
6.5		-	7	$\Delta^b v_0^7 = (31, 11)$	40 (40 fixed)
				$\Delta^b v_1^7 = (31, 15)$	
				$\Delta^b v_2^7 = (31)$	
	$\Delta^b v_0^{6.5} = (31, 22, 18, 6, 2)$			$\Delta^b v_4^7 = (31, 11)$	
	$\Delta^b v_1^{6.5} = (31, 15)$			$\Delta^b v_5^7 = (31, 15)$	
	$\Delta^b v_2^{6.5} = (31, 23, 11, 7)$			$\Delta^b v_8^7 = (31)$	
	$\Delta^b v_3^{6.5} = (30, 27, 19, 15, 11, 6, 3)$			$\Delta^b v_9^7 = (31, 23, 15, 7)$	
	$\Delta^b v_4^{6.5} = (30, 27, 23, 19, 18, 11, 7, 3)$			$\Delta^b v_{10}^7 = (31)$	
	$\Delta^b v_5^{6.5} = (31, 18, 11, 2)$			$\Delta^b v_{13}^7 = (23, 7)$	
	$\Delta^b v_7^{6.5} = (31, 23, 11, 7)$			$\Delta^b v_{14}^7 = (31, 15)$	
	$\Delta^b v_8^{6.5} = (31, 23, 15, 7)$		7.5	$\Delta^b v_2^{7.5} = (31)$	6 (2 fixed)
	$\Delta^b v_{10}^{6.5} = (11)$			$\Delta^b v_8^{7.5} = (31)$	
	$\Delta^b v_{11}^{6.5} = (31, 15)$			$\Delta^b v_{13}^{7.5} = (31, 15)$	
	$\Delta^b v_{13}^{6.5} = (31)$		8	$\Delta^b v_2^8 = (31)$	0
	$\Delta^b v_{14}^{6.5} = (18, 15, 6)$		$\cdots$	ϕ	0
	$\Delta^b v_{15}^{6.5} = (27, 22, 15, 11, 6)$		10	$\Delta^b v_0^{10} = (31)$	1
				$\Delta^b v_5^{10} = (16)$	
				$\Delta^b v_{10}^{10} = (23)$	
				$\Delta^b v_{15}^{10} = (23)$	

Table 8. The top characteristic for BLAKE-512. Message difference is $\Delta^t m_5 = (y)$ where $y \in \mathbb{X}_{512}$

$\mathbb{X}_{512} = \{5, 9, 18, 20, 22, 29, 34, 38, 41, 45, 48, 52, 54\}$	
r — Difference (Numeric Form)	Cond
2.5 $\Delta^t v_0^{2.5} = (y + 32)$	51
$\Delta^t v_1^{2.5} = (y + 48, y + 25, y + 16)$	
$\Delta^t v_2^{2.5} = (y + 41, y + 25, y + 11, y + 9, y + 61, y + 57)$	
$\Delta^t v_3^{2.5} = (y + 43, y + 36, y + 25, y + 11, y + 4)$	
$\Delta^t v_4^{2.5} = (y + 36, y + 4)$	
$\Delta^t v_5^{2.5} = (y)$	
$\Delta^t v_6^{2.5} = (y + 48, y + 25, y + 16, y)$	
$\Delta^t v_7^{2.5} = (y + 48, y + 41, y + 36, y + 32, y + 25, y + 16, y + 9, y, y + 61, y + 57)$	
$\Delta^t v_8^{2.5} = (y + 48, y + 32, y + 16, y)$	
$\Delta^t v_9^{2.5} = (y + 25)$	
$\Delta^t v_{10}^{2.5} = (y + 32, y + 16)$	
$\Delta^t v_{11}^{2.5} = (y + 48, y + 32, y + 16)$	
$\Delta^t v_{12}^{2.5} = (y + 32)$	
$\Delta^t v_{13}^{2.5} = (y + 48, y + 36, y + 32, y + 16, y + 11, y)$	
$\Delta^t v_{14}^{2.5} = (y + 43, y + 25, y + 11, y + 57)$	
$\Delta^t v_{15}^{2.5} = (y + 48)$	
3 $\Delta^t v_0^3 = (y + 32, y)$	11
$\Delta^t v_3^3 = (y + 25)$	
$\Delta^t v_4^3 = (y + 32, y)$	
$\Delta^t v_7^3 = (y + 25, y)$	
$\Delta^t v_8^3 = (y + 48, y + 32, y + 16, y)$	
$\Delta^t v_{11}^3 = (y)$	
$\Delta^t v_{12}^3 = (y + 48, y + 16)$	
$\Delta^t v_{15}^3 = (y)$	
3.5 $\Delta^t v_{11}^{3.5} = (y)$	2
$\Delta^t v_{12}^{3.5} = (y + 32, y)$	
4 $\Delta^t v_1^4 = (y)$	1
$\cdots$ ϕ	2 (2 fixed)
6 $\Delta^t v_1^6 = (y)$	30 (29 fixed)
$\Delta^t v_6^6 = (y + 37)$	
$\Delta^t v_{11}^6 = (y + 48)$	
$\Delta^t v_{12}^6 = (y + 48)$	
6.5 $\Delta^t v_0^{6.5} = (y, y + 55)$	-
$\Delta^t v_1^{6.5} = (y + 7, y)$	
$\Delta^t v_2^{6.5} = (y + 44, y + 37, y + 12)$	
$\Delta^t v_3^{6.5} = (y + 23)$	
$\Delta^t v_4^{6.5} = (y + 53, y + 44, y + 37, y + 28, y + 5)$	
$\Delta^t v_5^{6.5} = (y + 44, y + 37, y + 21, y + 5, y + 60)$	
$\Delta^t v_6^{6.5} = (y + 49, y + 42, y + 33, y + 17, y + 10, y + 1, y + 58)$	
$\Delta^t v_7^{6.5} = (y + 37, y + 12, y + 60)$	
$\Delta^t v_8^{6.5} = (y + 48, y + 39, y + 16, y)$	
$\Delta^t v_9^{6.5} = (y + 48, y + 32, y + 16, y + 55)$	
$\Delta^t v_{10}^{6.5} = (y + 53, y + 28, y + 21, y + 5, y + 60)$	
$\Delta^t v_{11}^{6.5} = (y + 48, y + 7)$	
$\Delta^t v_{12}^{6.5} = (y + 48, y + 39, y)$	
$\Delta^t v_{13}^{6.5} = (y + 48, y + 16, y + 55)$	
$\Delta^t v_{14}^{6.5} = (y + 53, y + 28, y + 21, y + 60)$	
$\Delta^t v_{15}^{6.5} = (y + 7)$	

Table 9. The top characteristic for BLAKE2b. Message difference is $\Delta^t m_5 = (y)$ where $y \in \mathbb{X}_{2b}$

$\mathbb{X}_{2b} = \{0, 1, 2, 3, 4, 8, 9, 10, 11, 12, 16, 17, 18, 19, 20, 24, 25, 26, 27,$
$28, 32, 33, 34, 35, 36, 40, 41, 42, 43, 44, 48, 49, 50, 51, 52, 56, 57, 58, 59, 60\}$

r	Difference (Numeric Form)	Cond
2.5	$\Delta^t v_0^{2.5} = (y + 32)$	67
	$\Delta^t v_1^{2.5} = (y + 48, y + 24, y + 16, y + 1)$	
	$\Delta^t v_2^{2.5} = (y + 47, y + 40, y + 33, y + 24, y + 8, y + 1, y + 63, y + 56)$	
	$\Delta^t v_3^{2.5} = (y + 31, y + 25, y + 24, y + 23, y + 63, y + 55)$	
	$\Delta^t v_4^{2.5} = (y + 25, y + 23, y + 1, y + 55)$	
	$\Delta^t v_5^{2.5} = (y)$	
	$\Delta^t v_6^{2.5} = (y + 48, y + 24, y + 16, y)$	
	$\Delta^t v_7^{2.5} = (y + 48, y + 47, y + 40, y + 33, y + 32, y + 24, y + 23,$	
	$\qquad\qquad y + 16, y + 8, y + 1, y, y + 56)$	
	$\Delta^t v_8^{2.5} = (y + 49, y + 48, y + 33, y + 32, y + 17, y + 16, y + 1, y)$	
	$\Delta^t v_9^{2.5} = (y + 24, y + 1)$	
	$\Delta^t v_{10}^{2.5} = (y + 32, y + 16)$	
	$\Delta^t v_{11}^{2.5} = (y + 48, y + 32, y + 16, y + 1)$	
	$\Delta^t v_{12}^{2.5} = (y + 33, y + 32, y + 1)$	
	$\Delta^t v_{13}^{2.5} = (y + 49, y + 48, y + 32, y + 23, y + 17, y + 16, y, y + 63)$	
	$\Delta^t v_{14}^{2.5} = (y + 31, y + 24, y + 1, y + 63, y + 56)$	
	$\Delta^t v_{15}^{2.5} = (y + 48)$	
3	$\Delta^t v_0^3 = (y + 32, y)$	13
	$\Delta^t v_1^3 = (y + 1)$	
	$\Delta^t v_3^3 = (y + 24)$	
	$\Delta^t v_4^3 = (y + 32, y)$	
	$\Delta^t v_7^3 = (y + 24, y)$	
	$\Delta^t v_8^3 = (y + 48, y + 32, y + 16, y)$	
	$\Delta^t v_9^3 = (y + 1)$	
	$\Delta^t v_{11}^3 = (y)$	
	$\Delta^t v_{12}^3 = (y + 48, y + 16)$	
	$\Delta^t v_{13}^3 = (y + 33, y + 1)$	
	$\Delta^t v_{15}^3 = (y)$	
3.5	$\Delta^t v_1^{3.5} = (y + 1)$	3
	$\Delta^t v_{11}^{3.5} = (y)$	
	$\Delta^t v_{12}^{3.5} = (y + 32, y)$	
4	$\Delta^t v_1^4 = (y)$	1
$\cdots$	ϕ	2 (2 fixed)
6	$\Delta^t v_1^6 = (y)$	30 (29 fixed)
	$\Delta^t v_6^6 = (y + 49)$	
	$\Delta^t v_{11}^6 = (y + 48)$	
	$\Delta^t v_{12}^6 = (y + 48)$	

(Continued)

Table 9. (Continued)

$\mathbb{X}_{2b} = \{0,1,2,3,4,8,9,10,11,12,16,17,18,19,20,24,25,26,27,$
$28,32,33,34,35,36,40,41,42,43,44,48,49,50,51,52,56,57,58,59,60\}$

r	Difference (Numeric Form)	Cond
6.5	$\Delta^t v_0^{6.5} = (y, y+56)$	-
	$\Delta^t v_1^{6.5} = (y+8, y)$	
	$\Delta^t v_2^{6.5} = (y+49, y+25, y+57)$	
	$\Delta^t v_3^{6.5} = (y+24)$	
	$\Delta^t v_4^{6.5} = (y+49, y+41, y+17, y+1, y+57)$	
	$\Delta^t v_5^{6.5} = (y+49, y+33, y+17, y+9, y+57)$	
	$\Delta^t v_6^{6.5} = (y+42, y+34, y+26, y+18, y+10, y+2, y+58)$	
	$\Delta^t v_7^{6.5} = (y+49, y+25, y+9)$	
	$\Delta^t v_8^{6.5} = (y+48, y+40, y+16, y)$	
	$\Delta^t v_9^{6.5} = (y+48, y+32, y+16, y+56)$	
	$\Delta^t v_{10}^{6.5} = (y+41, y+33, y+17, y+9, y+1)$	
	$\Delta^t v_{11}^{6.5} = (y+48, y+8)$	
	$\Delta^t v_{12}^{6.5} = (y+48, y+40, y)$	
	$\Delta^t v_{13}^{6.5} = (y+48, y+16, y+56)$	
	$\Delta^t v_{14}^{6.5} = (y+41, y+33, y+9, y+1)$	
	$\Delta^t v_{15}^{6.5} = (y+8)$	

Table 10. The top characteristic for BLAKE-256. Message difference is $\Delta^t m_5 = (y)$ where $y \in \mathbb{X}_{256}$.

$\mathbb{X}_{256} = \{20, 28\}$

r	Difference (Numeric Form)	Cond
2.5	$\Delta^t v_0^{2.5} = (y+16)$	54/53[a]
	$\Delta^t v_1^{2.5} = (y+8, y+24, y+12)$	
	$\Delta^t v_2^{2.5} = (y+7, y+4, y+31, y+28, y+20, y+12)$	
	$\Delta^t v_3^{2.5} = (y+7, \mathbf{y+3}, y+23, y+19, y+12)$	
	$\Delta^t v_4^{2.5} = (\mathbf{y+3}, y+19)$	
	$\Delta^t v_5^{2.5} = (y)$	
	$\Delta^t v_6^{2.5} = (y+8, y, y+24, y+12)$	
	$\Delta^t v_7^{2.5} = (y+8, y+4, y, y+31, y+28, y+24, y+$ $20, y+19, y+16, y+12)$	
	$\Delta^t v_8^{2.5} = (y+8, y, y+24, y+16)$	
	$\Delta^t v_9^{2.5} = (y+12)$	
	$\Delta^t v_{10}^{2.5} = (y+8, y+16)$	
	$\Delta^t v_{11}^{2.5} = (y+8, y+24, y+16)$	
	$\Delta^t v_{12}^{2.5} = (y+16)$	
	$\Delta^t v_{13}^{2.5} = (y+8, y+7, y, y+24, y+19, y+16)$	
	$\Delta^t v_{14}^{2.5} = (y+7, y+28, y+23, y+12)$	
	$\Delta^t v_{15}^{2.5} = (y+24)$	

(Continued)

Table 10. (*Continued*)

$\mathbb{X}_{256} = \{20, 28\}$		
r	Difference (Numeric Form)	Cond
3	$\Delta^t v_0^3 = (y, y+16)$	14
	$\Delta^t v_3^3 = (y+12)$	
	$\Delta^t v_4^3 = (y, y+16)$	
	$\Delta^t v_7^3 = (y, y+12)$	
	$\Delta^t v_8^3 = (y+8, y, y+24, y+16)$	
	$\Delta^t v_{11}^3 = (y)$	
	$\Delta^t v_{12}^3 = (y+8, y+24)$	
	$\Delta^t v_{15}^3 = (y)$	
3.5	$\Delta^t v_{11}^{3.5} = (y)$	2
	$\Delta^t v_{12}^{3.5} = (y, y+16)$	
4	$\Delta^t v_1^4 = (y)$	1
$\cdots$	ϕ	2 (2 fixed)
6	$\Delta^t v_1^6 = (y)$	30 (29 fixed)
	$\Delta^t v_6^6 = (y+17)$	
	$\Delta^t v_{11}^6 = (y+24)$	
	$\Delta^t v_{12}^6 = (y+24)$	
6.5	$\Delta^t v_0^{6.5} = (y, y+28)$	-
	$\Delta^t v_1^{6.5} = (y+4, y)$	
	$\Delta^t v_2^{6.5} = (y+5, y+21, y+17)$	
	$\Delta^t v_3^{6.5} = (y+12)$	
	$\Delta^t v_4^{6.5} = (y+1, y+25, y+21, y+17, y+13)$	
	$\Delta^t v_5^{6.5} = (y+9, y+1, y+29, y+21, y+17)$	
	$\Delta^t v_6^{6.5} = (y+6, y+2, y+30, y+26, y+22, y+18, y+14)$	
	$\Delta^t v_7^{6.5} = (y+5, y+29, y+17)$	
	$\Delta^t v_8^{6.5} = (y+8, y, y+24, y+20)$	
	$\Delta^t v_9^{6.5} = (y+8, y+28, y+24, y+16)$	
	$\Delta^t v_{10}^{6.5} = (y+9, y+1, y+29, y+25, y+13)$	
	$\Delta^t v_{11}^{6.5} = (y+4, y+24)$	
	$\Delta^t v_{12}^{6.5} = (y, y+24, y+20)$	
	$\Delta^t v_{13}^{6.5} = (y+8, y+28, y+24)$	
	$\Delta^t v_{14}^{6.5} = (y+9, y+29, y+25, y+13)$	
	$\Delta^t v_{15}^{6.5} = (y+4)$	

[a]: If $y = 28$, the condition $v_3^{2.5}[y+3] = \neg v_4^{2.5}[y+3]$ in $\Delta^t V^3 \to \Delta^t V^{2.5}$ can be eliminated.

Table 11. The top characteristic for BLAKE2s. Message difference is $\Delta^t m_5 = (y)$ where $y \in \mathbb{X}_{2s}$.

$\mathbb{X}_{2s} = \{0, 4, 8, 12, 16, 20, 24, 28\}$		
r	Difference (Numeric Form)	Cond
2.5	$\Delta^t v_0^{2.5} = (y + 16)$	68
	$\Delta^t v_1^{2.5} = (y + 8, y + 1, y + 24, y + 12)$	
	$\Delta^t v_2^{2.5} = (y + 7, y + 4, y + 1, y + 31, y + 28, y + 20, y + 17, y + 12)$	
	$\Delta^t v_3^{2.5} = (y + 7, y + 3, y + 23, y + 19, y + 13, y + 12)$	
	$\Delta^t v_4^{2.5} = (y + 3, y + 1, y + 19, y + 13)$	
	$\Delta^t v_5^{2.5} = (y)$	
	$\Delta^t v_6^{2.5} = (y + 8, y, y + 24, y + 12)$	
	$\Delta^t v_7^{2.5} = (y + 8, y + 4, y + 1, y, y + 31, y + 28, y + 24, y + 20,$	
	$\qquad\qquad y + 19, y + 17, y + 16, y + 12)$	
	$\Delta^t v_8^{2.5} = (y + 9, y + 8, y + 1, y, y + 25, y + 24, y + 17, y + 16)$	
	$\Delta^t v_9^{2.5} = (y + 1, y + 12)$	
	$\Delta^t v_{10}^{2.5} = (y + 8, y + 16)$	
	$\Delta^t v_{11}^{2.5} = (y + 8, y + 1, y + 24, y + 16)$	
	$\Delta^t v_{12}^{2.5} = (y + 1, y + 17, y + 16)$	
	$\Delta^t v_{13}^{2.5} = (y + 9, y + 8, y + 7, y, y + 25, y + 24, y + 19, y + 16)$	
	$\Delta^t v_{14}^{2.5} = (y + 7, y + 1, y + 28, y + 23, y + 12)$	
	$\Delta^t v_{15}^{2.5} = (y + 24)$	
3	$\Delta^t v_0^3 = (y, y + 16)$	16
	$\Delta^t v_1^3 = (y + 1)$	
	$\Delta^t v_3^3 = (y + 12)$	
	$\Delta^t v_4^3 = (y, y + 16)$	
	$\Delta^t v_7^3 = (y, y + 12)$	
	$\Delta^t v_8^3 = (y + 8, y, y + 24, y + 16)$	
	$\Delta^t v_9^3 = (y + 1)$	
	$\Delta^t v_{11}^3 = (y)$	
	$\Delta^t v_{12}^3 = (y + 8, y + 24)$	
	$\Delta^t v_{13}^3 = (y + 1, y + 17)$	
	$\Delta^t v_{15}^3 = (y)$	
3.5	$\Delta^t v_1^{3.5} = (y + 1)$	3
	$\Delta^t v_{11}^{3.5} = (y)$	
	$\Delta^t v_{12}^{3.5} = (y, y + 16)$	
4	$\Delta^t v_1^4 = (y)$	1
$\cdots$	ϕ	2 (2 fixed)

(Continued)

Table 11. (*Continued*)

$\mathbb{X}_{2s} = \{0, 4, 8, 12, 16, 20, 24, 28\}$	
r	Difference (Numeric Form)	Cond
6	$\Delta^t v_1^6 = (y)$	30 (29 fixed)
	$\Delta^t v_6^6 = (y + 17)$	
	$\Delta^t v_{11}^6 = (y + 24)$	
	$\Delta^t v_{12}^6 = (y + 24)$	
6.5	$\Delta^t v_0^{6.5} = (y, y + 28)$	-
	$\Delta^t v_1^{6.5} = (y + 4, y)$	
	$\Delta^t v_2^{6.5} = (y + 5, y + 21, y + 17)$	
	$\Delta^t v_3^{6.5} = (y + 12)$	
	$\Delta^t v_4^{6.5} = (y + 1, y + 25, y + 21, y + 17, y + 13)$	
	$\Delta^t v_5^{6.5} = (y + 9, y + 1, y + 29, y + 21, y + 17)$	
	$\Delta^t v_6^{6.5} = (y + 6, y + 2, y + 30, y + 26, y + 22, y + 18, y + 14)$	
	$\Delta^t v_7^{6.5} = (y + 5, y + 29, y + 17)$	
	$\Delta^t v_8^{6.5} = (y + 8, y, y + 24, y + 20)$	
	$\Delta^t v_9^{6.5} = (y + 8, y + 28, y + 24, y + 16)$	
	$\Delta^t v_{10}^{6.5} = (y + 9, y + 1, y + 29, y + 25, y + 13)$	
	$\Delta^t v_{11}^{6.5} = (y + 4, y + 24)$	
	$\Delta^t v_{12}^{6.5} = (y, y + 24, y + 20)$	
	$\Delta^t v_{13}^{6.5} = (y + 8, y + 28, y + 24)$	
	$\Delta^t v_{14}^{6.5} = (y + 9, y + 29, y + 25, y + 13)$	
	$\Delta^t v_{15}^{6.5} = (y + 4)$	

[a]: If $y = 28$, the condition $v_3^{2.5}[y + 3] = \neg v_4^{2.5}[y + 3]$ in $\Delta^t V^3 \rightarrow \Delta^t V^{2.5}$ can be eliminated.

B 6.5-Round Examples for BLAKE and BLAKE2

The main difference between BLAKE-256 and BLAKE2s (BLAKE-512 and BLAKE2b) is at $\Delta^t v_1^{3.5}$, where $\Delta^t v_1^{3.5} = (29)$ for BLAKE-2s ($\Delta^t v_1^{3.5} = (10)$ for BLAKE-2b) and ϕ for BLAKE-256 (BLAKE-512). We specifically emphasize this part with bold dark format.

Table 12. Example for 6.5-round BLAKE-256 with $y = 28 \in \mathbb{X}_{256} \bigcap \mathbb{X}_{2s}$.

ΔM	$\Delta^b m_{11} = (31),\ \Delta^t m_5 = (28)$							
$_aM$	0x932a5d7f	0xa2625330	0x46a9466f	0xae3052a3	0xbf9a6338	0xd4167790	0x7bf0ef5e	0x4ef572ba
	0x308dc96d	0x23b415c3	0x6fb64798	0xa75b42e8	0x3cb6d30e	0xb56003b4	0x7a4db777	0x715b79a
$_bM$	0x932a5d7f	0xa2625330	0x46a9466f	0xae3052a3	0xbf9a6338	0xd4167790	0x7bf0ef5e	0x4ef572ba
	0x308dc96d	0x23b415c3	0x6fb64798	0x275b42e8	0x3cb6d30e	0xb56003b4	0x7a4db777	0x715b79a
$_cM$	0x932a5d7f	0xa2625330	0x46a9466f	0xae3052a3	0xbf9a6338	0xc4167790	0x7bf0ef5e	0x4ef572ba
	0x308dc96d	0x23b415c3	0x6fb64798	0xa75b42e8	0x3cb6d30e	0xb56003b4	0x7a4db777	0x715b79a
$_dM$	0x932a5d7f	0xa2625330	0x46a9466f	0xae3052a3	0xbf9a6338	0xc4167790	0x7bf0ef5e	0x4ef572ba
	0x308dc96d	0x23b415c3	0x6fb64798	0x275b42e8	0x3cb6d30e	0xb56003b4	0x7a4db777	0x715b79a
$\Delta^t V^{3.5}$	$\Delta^t v_1^{3.5} = \phi,\ \Delta^t v_{11}^{3.5} = (28),\ \Delta^t v_{12}^{3.5} = (28,2)$							
$_aV^{3.5}$	0x7ce3001a	0x5f257eb	0x7cb1b540	0xf5f76e6	0x62eba0a0	0x8723a3b3	0x3a617d3b	0x616c91a2
	0xf2e28cd6	0x2dd8b157	0x888f9a21	0x6074df04	0x370f729f	0xeecddee4	0x7f42197f	0x36ace0f3
$_bV^{3.5}$	0xce7042ae	0xc394a0c1	0xbdedbda1	0xbf9d773f	0x7fdd9e46	0xdefe6c9e	0xf9985a99	0x2e67c857
	0x8903f293	0xfc2ed055	0xbcb66021	0x5ac97fd7	0xa42a029b	0x60de7589	0x637162de	0xfd1bd434
$_cV^{3.5}$	0x7ce3001a	0x5f257eb	0x7cb1b540	0xf5f76e6	0x62eba0a0	0x8723a3b3	0x3a617d3b	0x616c91a2
	0xf2e28cd6	0x2dd8b157	0x888f9a21	0x7074df04	0x270f629f	0xeecddee4	0x7f42197f	0x36ace0f3
$_dV^{3.5}$	0xce7042ae	0xc394a0c1	0xbdedbda1	0xbf9d773f	0x7fdd9e46	0xdefe6c9e	0xf9985a99	0x2e67c857
	0x8903f293	0xfc2ed055	0xbcb66021	0x4ac97fd7	0xa42a029b	0x60de7589	0x637162de	0xfd1bd434
$\Delta^b V^{10}$	$\Delta^b v_0^{10} = (31),\ \Delta^b v_5^{10} = (23),\ \Delta^b v_{18}^{10} = (23)$							
$_aV^{10}$	0x9920f4d5	0x7d8a6621	0xc7139615	0x205a3fce	0x4ded77e1	0x1ed1c43f	0x6e8efedc	0xf6f4fe72
	0x6e17623b	0x4cd8bea2	0xfe2149af	0xd2f8e09c	0x53b6139c	0x3972162e	0xd4f82167	0x4d1b2a46
$_bV^{10}$	0x1920f4d5	0x7d8a6621	0xc7139615	0x205a3fce	0x4ded77e1	0x1ed0c43f	0x6e8efedc	0xf6f4fe72
	0x6e17623b	0x4cd8bea2	0xfea149af	0xd2f8e09c	0x53b6139c	0x3972162e	0xd4f82167	0x4d1b2a46
$_cV^{10}$	0x5a870d65	0x8d12db5	0x537127c9	0xabdb13a9	0xcaf27105	0x17ef5f49	0x66721638	0x8f333fbf
	0xccdc1196	0x3d9aaba6	0x84ee030c	0xda86539	0x976348e3	0xfde7c240	0x1df99dc8	0x568a818c
$_dV^{10}$	0xda870d65	0x8d12db5	0x537127c9	0xabdb13a9	0xcaf27105	0x17ee5f49	0x66721638	0x8f333fbf
	0xccdc1196	0x3d9aaba6	0x846e030c	0xda86539	0x976348e3	0xfde7c240	0x1df99dc8	0x560a818c

Table 13. Example for 6.5-round BLAKE2s with $y = 28 \in \mathbb{X}_{256} \bigcap \mathbb{X}_{2s}$.

ΔM	$\Delta^b m_{11} = (31),\ \Delta^t m_5 = (28)$							
$_aM$	0xce9f1cc6	0x7f3a9b64	0x9e9ddc55	0x4553fa8c	0x2ef4ad99	0x33a0533a	0x8b1d785c	0xc7f56492
	0xe5b2b205	0xd44f69a1	0x2d83e500	0x18b03f68	0x13d0c628	0x15fce9f2	0x9108f878	0xc477ca04
$_bM$	0xce9f1cc6	0x7f3a9b64	0x9e9ddc55	0x4553fa8c	0x2ef4ad99	0x33a0533a	0x8b1d785c	0xc7f56492
	0xe5b2b205	0xd44f69a1	0x2d83e500	0x98b03f68	0x13d0c628	0x15fce9f2	0x9108f878	0xc477ca04
$_cM$	0xce9f1cc6	0x7f3a9b64	0x9e9ddc55	0x4553fa8c	0x2ef4ad99	0x23a0533a	0x8b1d785c	0xc7f56492
	0xe5b2b205	0xd44f69a1	0x2d83e500	0x18b03f68	0x13d0c628	0x15fce9f2	0x9108f878	0xc477ca04
$_dM$	0xce9f1cc6	0x7f3a9b64	0x9e9ddc55	0x4553fa8c	0x2ef4ad99	0x23a0533a	0x8b1d785c	0xc7f56492
	0xe5b2b205	0xd44f69a1	0x2d83e500	0x98b03f68	0x13d0c628	0x15fce9f2	0x9108f878	0xc477ca04
$\Delta^t V^{3.5}$	$\Delta^t v_1^{3.5} = (29),\ \Delta^t v_{11}^{3.5} = (28),\ \Delta^t v_{12}^{3.5} = (28,2)$							
$_aV^{3.5}$	0x71177c4a	0x456e63aa	0x63bc0484	0xe348f6a9	0xfa5c62fe	0x1229c0a3	0x12ea25d0	0xd7a6a55f
	0x3ca79134	0x6ccc6e48	0x2bd29e5	0xc386b1	0x86f12557	0x414c79f1	0x3fb6c33	0x4baef1a0
$_bV^{3.5}$	0xaae6286d	0x1af8dcfe	0x70a74337	0xa293966a	0xe35d9b23	0xe74273b3	0xfb967985	0xc16500a7
	0x57a589c8	0x5edbf5ae	0x66de7b25	0x15c8f5ff	0xd730836	0x357d6100	0x3ae77969	0x54a834da
$_cV^{3.5}$	0x71177c4a	0x656e63aa	0x63bc0484	0xe348f6a9	0xfa5c62fe	0x1229c0a3	0x12ea25d0	0xd7a6a55f
	0x3ca79134	0x6ccc6e48	0x2bd29e5	0x10c386b1	0x96f13557	0x414c79f1	0x3fb6c33	0x4baef1a0
$_dV^{3.5}$	0xaae6286d	0x1af8dcfe	0x70a74337	0xa293966a	0xe35d9b23	0xe74273b3	0xfb967985	0xc16500a7
	0x57a589c8	0x5edbf5ae	0x66de7b25	0x5c8f5ff	0x1d731836	0x357d6100	0x3ae77969	0x54a834da
$\Delta^b V^{10}$	$\Delta^b v_0^{10} = (31),\ \Delta^b v_5^{10} = (16),\ \Delta^b v_{10}^{10} = (23),\ \Delta^b v_{15}^{10} = (23)$							
$_aV^{10}$	0x945cf52e	0x422107ab	0x3a682330	0x2f8bd4f1	0xeead389	0x21e907ec	0x17138a07	0xae021462
	0x229a3e13	0x3c623c2c	0x64327d4a	0xf1d0e09a	0x5df5abad	0x1be8464a	0x7890983a	0x85288868
$_bV^{10}$	0x145cf52e	0x422107ab	0x3a682330	0x2f8bd4f1	0xeead389	0x21e807ec	0x17138a07	0xae021462
	0x229a3e13	0x3c623c2c	0x64b27d4a	0xf1d0e09a	0x5df5abad	0x1be8464a	0x7890983a	0x85a88868
$_cV^{10}$	0xc136da56	0xe91ba476	0xfa9ad265	0x6b4d2f9e	0x68ef06c8	0x9ab4757a	0xe63456e0	0x8818e9d4
	0x5da1784c	0x57ecd14b	0xcb0788b8	0xf3148edf	0xa19d7f24	0xf17b5303	0x9ec70b70	0x2f763872
$_dV^{10}$	0x4136da56	0xe91ba476	0xfa9ad265	0x6b4d2f9e	0x68ef06c8	0x9ab5757a	0xe63456e0	0x8818e9d4
	0x5da1784c	0x57ecd14b	0xcb8788b8	0xf3148edf	0xa19d7f24	0xf17b5303	0x9ec70b70	0x2ff63872

Table 14. Example for 6.5-round BLAKE-512 with $y = 9 \in \mathbb{X}_{512} \cap \mathbb{X}_{2b}$.

ΔM	$\Delta^b m_{11} = (63),\ \Delta^t m_5 = (9)$			
$_aM$	0x9c1860c444a6a9f4	0xc95a712fd5a29b72	0x6e5c6811448b300f	0x5c0af45531e396d3
	0x679dee5280c15ad0	0x329f5347ccb9bf64	0x297828d3ec89e9d0	0xa55ffc029ea78609
	0xef01f63ec485f87d	0x86560936e36d9dff	0xfd9674bb724d62e0	0x9c03f6f64a96659f
	0xe3666bd816053d27	0xe4669665a4a0a440	0x1cbf0c93a121eb09	0x65a6a90ac809c019
$_bM$	0x9c1860c444a6a9f4	0xc95a712fd5a29b72	0x6e5c6811448b300f	0x5c0af45531e396d3
	0x679dee5280c15ad0	0x329f5347ccb9bf64	0x297828d3ec89e9d0	0xa55ffc029ea78609
	0xef01f63ec485f87d	0x86560936e36d9dff	0xfd9674bb724d62e0	0x1c03f6f64a96659f
	0xe3666bd816053d27	0xe4669665a4a0a440	0x1cbf0c93a121eb09	0x65a6a90ac809c019
$_cM$	0x9c1860c444a6a9f4	0xc95a712fd5a29b72	0x6e5c6811448b300f	0x5c0af45531e396d3
	0x679dee5280c15ad0	0x329f5347ccb9bd64	0x297828d3ec89e9d0	0xa55ffc029ea78609
	0xef01f63ec485f87d	0x86560936e36d9dff	0xfd9674bb724d62e0	0x9c03f6f64a96659f
	0xe3666bd816053d27	0xe4669665a4a0a440	0x1cbf0c93a121eb09	0x65a6a90ac809c019
$_dM$	0x9c1860c444a6a9f4	0xc95a712fd5a29b72	0x6e5c6811448b300f	0x5c0af45531e396d3
	0x679dee5280c15ad0	0x329f5347ccb9bd64	0x297828d3ec89e9d0	0xa55ffc029ea78609
	0xef01f63ec485f87d	0x86560936e36d9dff	0xfd9674bb724d62e0	0x1c03f6f64a96659f
	0xe3666bd816053d27	0xe4669665a4a0a440	0x1cbf0c93a121eb09	0x65a6a90ac809c019
$\Delta^t V^{3.5}$	$\Delta^t \mathbf{v}_1^{3.5} = \phi,\ \Delta^t v_{11}^{3.5} = (9),\ \Delta^t v_{12}^{3.5} = (41, 9)$			
$_aV^{3.5}$	0xc87af7255a6ec986	0xc59be5b07a4418d7	0x5295eb179fee042c	0x4f87d569d171c685
	0xc1c24f85f094b263	0xbc711b20878eb4ea	0x1cda016fcf08ee93	0x878f439bd1398fec
	0x982d7a384b8549bb	0x29cd6958f1a234c3	0xb81579cd9c3cff45	0xbfba600ee495e360
	0x4d5e10f24eba6506	0x4f8a20a0c7164ef8	0x4156d917e0e33e7b	0x8f204cb6dc806747
$_bV^{3.5}$	0x57926274c228f656	0x5ac46fa843cda867	0x936f1f621381dad4	0xbd0f73ec836d47bc
	0xbac8918094537e74	0x1edec058ea817875	0xc5bf41aeadf39382	0x4149082191041e60
	0x9fd575b7fe10ace3	0x8fed3642acc17d51	0x1ded33ae6ee468ba	0x5365299759c0a42
	0x89f06ef09e1612ee	0xe597ede91683a2d8	0x380825eb30587c4f	0xff48c413164455c3
$_cV^{3.5}$	0xc87af7255a6ec986	0xc59be5b07a4418d7	0x5295eb179fee042c	0x4f87d569d171c685
	0xc1c24f85f094b263	0xbc711b20878eb4ea	0x1cda016fcf08ee93	0x878f439bd1398fec
	0x982d7a384b8549bb	0x29cd6958f1a234c3	0xb81579cd9c3cff45	0xbfba600ee495e160
	0x4d5e10f24eba6506	0x4f8a20a0c7164ef8	0x4156d017e0e33e7b	0x8f204cb6dc806717
$_dV^{3.5}$	0x57926274c228f656	0x5ac46fa843cda867	0x936f1f621381dad4	0xbd0f73ec836d47bc
	0xbac8918094537e74	0x1edec058ea817875	0xc5bf41aeadf39382	0x4149082191041e60
	0x9fd575b7fe10ace3	0x8fed3642acc17d51	0x1ded33ae6ee468ba	0x5365299759c0842
	0x89f06cf09e1610ee	0xe597ede91683a2d8	0x389825cb39587e4f	0xff48c413164455c3
$\Delta^b V^{10}$	$\Delta^b v_0^{10} = (63),\ \Delta^b v_5^{10} = (36),\ \Delta^b v_{10}^{10} = (47),\ \Delta^b v_{15}^{10} = (47)$			
$_aV^{10}$	0x1b404ab31fbe9343	0xc01ae4355f49855f	0xf52deb99e6d25dee	0xba1e74d813d9e09c
	0x1d4142ceee078181	0x8c7261a65899559	0x780312586191c134	0x86c7c29f8161a9ac
	0x77f4ec97a373e3dd	0x7068ac849086f0c3	0xfc3c0163cdc3f7b9	0x52d68b2940599cfa
	0x59ad1c82831be8f7	0x74d99e11568eb396	0x3552275c6ddcf7a3	0x8dfe0979b5e83dbd
$_bV^{10}$	0x9b404ab31fbe9343	0xc01ae4355f49855f	0xf52deb99e6d25dee	0xba1e74d813d9e09c
	0x1d4142ceee078181	0x8c7260a65899559	0x780312586191c134	0x86c7c29f8161a9ac
	0x77f4ec97a373e3dd	0x7068ac849086f0c3	0xfc3c8163cdc3f7b9	0x52d68b2940599cfa
	0x59ad1c82831be8f7	0x74d99e11568eb396	0x3552275c6ddcf7a3	0x8dfe8979b5e83dbd
$_cV^{10}$	0xcc0a78ca6c133737	0xa6a12a75a2ab0a78	0xaaff3e032bf0964f	0x6a833f52c06326f8
	0x1571fbe8468d6869	0x224b394014f172d8	0x72a0866c8eb1dfcc	0x4af2b98060eea9bb
	0xe7f5b1b201006785	0xa57c9190f805d201	0xdea0ecffe0219e24	0xbbec25c771762bfb
	0xd312a8ab8e4df740	0xd9a366032739ede2	0xb8d5bfa962e8d684	0xb122b4542c543d9d
$_dV^{10}$	0x4c0a78ca6c133737	0xa6a12a75a2ab0a78	0xaaff3e032bf0964f	0x6a833f52c06326f8
	0x1571fbe8468d6869	0x224b395014f172d8	0x72a0866c8eb1dfcc	0x4af2b98060eea9bb
	0xe7f5b1b201006785	0xa57c9190f805d201	0xdea06cffe0219e24	0xbbec25c771762bfb
	0xd312a8ab8e4df740	0xd9a366032739ede2	0xb8d5bfa962e8d684	0xb12234542c543d9d

Table 15. Example for 6.5-round BLAKE2b with $y = 9 \in \mathbb{X}_{512} \bigcap \mathbb{X}_{2b}$.

ΔM	$\Delta^b m_{11} = (63),\ \Delta^t m_5 = (9)$			
${}_aM$	0x3cec6965bf357a5	0x3efa6687e114e70d	0x6fe9d72277e832e4	0x60574e830fad0b27
	0x1bad3b4b1257079e	0x43b8e8ebf1bc4557	0xc553a639b52984b0	0x95bd9c03c94695e5
	0xc4e9f58d840c74c9	0x2186128d765d51b0	0x10bc4fee175e6c82	0x18ddcb4d4ac938ee
	0x5cf2d8b6cf1ea3ce	0x3ec5aa659dacedf5	0xadf91c482e6b4506	0xa34876d149007c7b
${}_bM$	0x3cec6965bf357a5	0x3efa6687e114e70d	0x6fe9d72277e832e4	0x60574e830fad0b27
	0x1bad3b4b1257079e	0x43b8e8ebf1bc4557	0xc553a639b52984b0	0x95bd9c03c94695e5
	0xc4e9f58d840c74c9	0x2186128d765d51b0	0x10bc4fee175e6c82	0x98ddcb4d4ac938ee
	0x5cf2d8b6cf1ea3ce	0x3ec5aa659dacedf5	0xadf91c482e6b4506	0xa34876d149007c7b
${}_cM$	0x3cec6965bf357a5	0x3efa6687e114e70d	0x6fe9d72277e832e4	0x60574e830fad0b27
	0x1bad3b4b1257079e	0x43b8e8ebf1bc4757	0xc553a639b52984b0	0x95bd9c03c94695e5
	0xc4e9f58d840c74c9	0x2186128d765d51b0	0x10bc4fee175e6c82	0x18ddcb4d4ac938ee
	0x5cf2d8b6cf1ea3ce	0x3ec5aa659dacedf5	0xadf91c482e6b4506	0xa34876d149007c7b
${}_dM$	0x3cec6965bf357a5	0x3efa6687e114e70d	0x6fe9d72277e832e4	0x60574e830fad0b27
	0x1bad3b4b1257079e	0x43b8e8ebf1bc4757	0xc553a639b52984b0	0x95bd9c03c94695e5
	0xc4e9f58d840c74c9	0x2186128d765d51b0	0x10bc4fee175e6c82	0x98ddcb4d4ac938ee
	0x5cf2d8b6cf1ea3ce	0x3ec5aa659dacedf5	0xadf91c482e6b4506	0xa34876d149007c7b
$\Delta^t V^{3.5}$	$\Delta^t \mathbf{v}_1^{3.5} = (10),\ \Delta^t v_{11}^{3.5} = (9),\ \Delta^t v_{12}^{3.5} = (41, 9)$			
${}_aV^{3.5}$	0xa8f431bca7166664	0x2bce47208c2b479d	0x2554f082eb89d530	0x12b06bc7f71ebe12
	0x5d733d5fa41457fc	0xae2b3d68d8adfe2f	0xe03c7fa88285b93d	0xe134f22af656a9d9
	0x8bcd47a74a5e35a2	0x21098bd0acfbc078	0x9d0ddd6c2403d2ab	0xf0dbb0c6a9a392c5
	0xd72aa227f3c2a651	0x406e07f8eec1929f	0x863da54a0653fe1f	0xefb750af7de2c392
${}_bV^{3.5}$	0x63b1930b9a252aff	0xd754470ae2a5de96	0x1b39d8f987ec3762	0x201afad51a642cb1
	0x1d5c8e5fb50c1c68	0x709103f9ba538f43	0xb847dad7a1bf8a56	0xa59f9b63902edb4
	0x40d96db5d9d3b546	0x332aed26d86aceaa	0x424eaab611c9c6f	0x802b683db9ac54b9
	0x110cd82fdac384dd	0xa93fe8a10201b57b	0x49eed3d94b17685a	0xcdf2a00fd5300651
${}_cV^{3.5}$	0xa8f431bca7166664	0x2bce47208c2b439d	0x2554f082eb89d530	0x12b06bc7f71ebe12
	0x5d733d5fa41457fc	0xae2b3d68d8adfe2f	0xe03c7fa88285b93d	0xe134f22af656a9d9
	0x8bcd47a74a5e35a2	0x21098bd0acfbc078	0x9d0ddd6c2403d2ab	0xf0dbb0c6a9a390c5
	0xd72aa027f3c2a451	0x406e07f8eec1929f	0x863da54a0653fe1f	0xefb750af7de2c392
${}_dV^{3.5}$	0x63b1930b9a252aff	0xd754470ae2a5da96	0x1b39d8f987ec3762	0x201afad51a642cb1
	0x1d5c8e5fb50c1c68	0x709103f9ba538f43	0xb847dad7a1bf8a56	0xa59f9b63902edb4
	0x40d96db5d9d3b546	0x332aed26d86aceaa	0x424eaab611c9c6f	0x802b683db9ac56b9
	0x110cda2fdac386dd	0xa93fe8a10201b57b	0x49eed3d94b17685a	0xcdf2a00fd5300651
$\Delta^b V^{10}$	$\Delta^b v_0^{10} = (63),\ \Delta^b v_5^{10} = (48),\ \Delta^b v_{10}^{10} = (47),\ \Delta^b v_{15}^{10} = (47)$			
${}_aV^{10}$	0x96ace3d164600933	0x6785c14493444a3d	0xadc3b5f6dbc8c992	0xada06d115f42653a
	0xcb06b797a6152dbe	0xf701f3e0f76be4cb	0xf4baf3238d75bdb6	0xb71965677688de57
	0xaa494db2c0d12db8	0x10ab8d9652485fcf	0xb97f5a3ef869239f	0x560aff2ec6a0d95f
	0x1597013f79b484d1	0x182beacffdc6ec05	0x6802644a544f6271	0x59ac761a17acecca
${}_bV^{10}$	0x16ace3d164600933	0x6785c14493444a3d	0xadc3b5f6dbc8c992	0xada06d115f42653a
	0xcb06b797a6152dbe	0xf700f3e0f76be4cb	0xf4baf3238d75bdb6	0xb71965677688de57
	0xaa494db2c0d12db8	0x10ab8d9652485fcf	0xb97fda3ef869239f	0x560aff2ec6a0d95f
	0x1597013f79b484d1	0x182beacffdc6ec05	0x6802644a544f6271	0x59acf61a17acecca
${}_cV^{10}$	0xe8d4f6a3aa68e9d6	0x1ba5272a94ed608d	0x51b3a429d5ee6873	0x50af4c1bb7b31dd2
	0x738835de6bff309d	0xc5fc88e668afef14	0x1671fea856c55b2d	0xd04b446c31b59a1b
	0x8f120d94bae51fa1	0x5be58c40a2d2c0a9	0xe9c1de5ac5992a67	0xa307fd45e31b7817
	0xbd4864acd0f2e4bc	0x4a8a43605d94a9b4	0x16e63ec7c12bc056	0x30e48769ae169de0
${}_dV^{10}$	0x68d4f6a3aa68e9d6	0x1ba5272a94ed608d	0x51b3a429d5ee6873	0x50af4c1bb7b31dd2
	0x738835de6bff309d	0xc5fd88e668afef14	0x1671fea856c55b2d	0xd04b446c31b59a1b
	0x8f120d94bae51fa1	0x5be58c40a2d2c0a9	0xe9c15e5ac5992a67	0xa307fd45e31b7817
	0xbd4864acd0f2e4bc	0x4a8a43605d94a9b4	0x16e63ec7c12bc056	0x30e40769ae169de0

References

1. Wang, X., Yu, H.: How to break MD5 and other hash functions. In: Cramer, R. (ed.) EUROCRYPT 2005. LNCS, vol. 3494, pp. 19–35. Springer, Heidelberg (2005)
2. Wang, X., Yin, Y.L., Yu, H.: Finding collisions in the full SHA-1. In: Shoup, V. (ed.) CRYPTO 2005. LNCS, vol. 3621, pp. 17–36. Springer, Heidelberg (2005)
3. Bertoni, G., Daemen, J., Peeters, M., Assche, G.: The keccak reference. Submission to NIST (Round 3) 13 (2011)
4. Aumasson, J.P., Henzen, L., Meier, W., Phan, R.C.W.: SHA-3 proposal blake. Submission to NIST (2008)
5. Chang, S.j., Perlner, R., Burr, W.E., Turan, M.S., Kelsey, J.M., Paul, S., Bassham, L.E.: Third-round report of the SHA-3 cryptographic hash algorithm competition. Citeseer (2012)
6. Aumasson, J.-P., Neves, S., Wilcox-O'Hearn, Z., Winnerlein, C.: BLAKE2: Simpler, smaller, fast as MD5. In: Jacobson, M., Locasto, M., Mohassel, P., Safavi-Naini, R. (eds.) ACNS 2013. LNCS, vol. 7954, pp. 119–135. Springer, Heidelberg (2013)
7. Aumasson, J.-P., Guo, J., Knellwolf, S., Matusiewicz, K., Meier, W.: Differential and invertibility properties of BLAKE. In: Hong, S., Iwata, T. (eds.) FSE 2010. LNCS, vol. 6147, pp. 318–332. Springer, Heidelberg (2010)
8. Dunkelman, O., Khovratovich, D.: Iterative differentials, symmetries, and message modification in blake-256. In: ECRYPT2 Hash Workshop, vol. 2011 (2011)
9. Ji, L., Liangyu, X.: Attacks on round-reduced blake. Technical Report, Citeseer (2009)
10. Biryukov, A., Nikolić, I., Roy, A.: Boomerang attacks on BLAKE-32. In: Joux, A. (ed.) FSE 2011. LNCS, vol. 6733, pp. 218–237. Springer, Heidelberg (2011)
11. Leurent, G.: Arxtools: A toolkit for arx analysis. In: The Third SHA-3 Candidate Conference (2012)
12. Bai, D., Yu, H., Wang, G., Wang, X.: Improved boomerang attacks on round-reduced sm3 and blake-256 (2013). http://eprint.iacr.org/
13. Guo, J., Karpman, P., Nikolić, I., Wang, L., Wu, S.: Analysis of BLAKE2. In: Benaloh, J. (ed.) CT-RSA 2014. LNCS, vol. 8366, pp. 402–423. Springer, Heidelberg (2014)
14. Wagner, D.: The boomerang attack. In: Knudsen, L.R. (ed.) FSE 1999. LNCS, vol. 1636, pp. 156–170. Springer, Heidelberg (1999)
15. Kelsey, J., Kohno, T., Schneier, B.: Amplified boomerang attacks against reduced-round MARS and serpent. In: Schneier, B. (ed.) FSE 2000. LNCS, vol. 1978, pp. 75–93. Springer, Heidelberg (2001)
16. Biham, E., Dunkelman, O., Keller, N.: The rectangle attack - rectangling the serpent. In: Pfitzmann, B. (ed.) EUROCRYPT 2001. LNCS, vol. 2045, pp. 340–357. Springer, Heidelberg (2001)
17. Biham, E., Dunkelman, O., Keller, N.: Related-key boomerang and rectangle attacks. In: Cramer, R. (ed.) EUROCRYPT 2005. LNCS, vol. 3494, pp. 507–525. Springer, Heidelberg (2005)
18. Lamberger, M., Mendel, F.: Higher-order differential attack on reduced SHA-256. IACR Cryptology ePrint Archive 2011, 37 (2011)
19. Biryukov, A., Lamberger, M., Mendel, F., Nikolić, I.: Second-order differential collisions for reduced SHA-256. In: Lee, D.H., Wang, X. (eds.) ASIACRYPT 2011. LNCS, vol. 7073, pp. 270–287. Springer, Heidelberg (2011)

20. Mendel, F., Nad, T.: Boomerang distinguisher for the SIMD-512 compression function. In: Bernstein, D.J., Chatterjee, S. (eds.) INDOCRYPT 2011. LNCS, vol. 7107, pp. 255–269. Springer, Heidelberg (2011)

21. Sasaki, Y., Wang, L., Takasaki, Y., Sakiyama, K., Ohta, K.: Boomerang distinguishers for full HAS-160 compression function. In: Hanaoka, G., Yamauchi, T. (eds.) IWSEC 2012. LNCS, vol. 7631, pp. 156–169. Springer, Heidelberg (2012)

22. Sasaki, Y., Wang, L.: Distinguishers beyond three rounds of the RIPEMD-128/-160 compression functions. In: Bao, F., Samarati, P., Zhou, J. (eds.) ACNS 2012. LNCS, vol. 7341, pp. 275–292. Springer, Heidelberg (2012)

23. Leurent, G., Roy, A.: Boomerang attacks on hash function using auxiliary differentials. In: Dunkelman, O. (ed.) CT-RSA 2012. LNCS, vol. 7178, pp. 215–230. Springer, Heidelberg (2012)

24. Yu, H., Chen, J., Wang, X.: The boomerang attacks on the round-reduced skein-512. In: Knudsen, L.R., Wu, H. (eds.) SAC 2012. LNCS, vol. 7707, pp. 287–303. Springer, Heidelberg (2013)

25. Kircanski, A., Shen, Y., Wang, G., Youssef, A.M.: Boomerang and slide-rotational analysis of the SM3 hash function. In: Knudsen, L.R., Wu, H. (eds.) SAC 2012. LNCS, vol. 7707, pp. 304–320. Springer, Heidelberg (2013)

26. Bai, D., Yu, H., Wang, G., Wang, X.: Improved boomerang attacks on SM3. In: Boyd, C., Simpson, L. (eds.) ACISP. LNCS, vol. 7959, pp. 251–266. Springer, Heidelberg (2013)

27. Wagner, D.: A generalized birthday problem. In: Yung, M. (ed.) CRYPTO 2002. LNCS, vol. 2442, pp. 288–304. Springer, Heidelberg (2002)

Second Preimage Analysis of Whirlwind

Riham AlTawy and Amr M. Youssef[✉]

Concordia Institute for Information Systems Engineering, Concordia University,
Montréal, Québec, Canada
youssef@ciise.concordia.ca

Abstract. Whirlwind is a keyless AES-like hash function that adopts the Sponge model. According to its designers, the function is designed to resist most of the recent cryptanalytic attacks. In this paper, we evaluate the second preimage resistance of the Whirlwind hash function. More precisely, we apply a meet in the middle preimage attack on the compression function which allows us to obtain a 5-round pseudo preimage for a given compression function output with time complexity of 2^{385} and memory complexity of 2^{128}. We also employ a guess and determine approach to extend the attack to 6 rounds with time and memory complexities of 2^{496} and 2^{112}, respectively. Finally, by adopting another meet in the middle attack, we are able to generate n-block message second preimages of the 5 and 6-round reduced hash function with time complexity of 2^{449} and 2^{505} and memory complexity of 2^{128} and 2^{112}, respectively.

Keywords: Cryptanalysis · Hash functions · Meet in the middle · Second preimage attack · Whirlwind

1 Introduction

Building a cryptographic primitive based on an existing component model has a very important advantage other than the possibility of sharing optimized components in a resource constrained environment. Namely, the advantage of adopting a model that has took its fair share of cryptanalysis and is still going strong. Consequently, the new primitive is expected to inherit most of the good qualities and the underlying features. The Advanced Encryption Standard (AES) wide trail strategy [8] has proven solid resistance to standard differential and linear attacks over more than a decade. This fact has made AES-like primitives an attractive alternative to dedicated constructions. Besides the ISO standard Whirlpool [21], we have seen a strong inclination towards proposing AES-like hash functions during the SHA-3 competition [20] (e.g., the SHA-3 finalists Grøstl [9] and JH [28], and LANE [12]). Additionally, Stribog [16] the new Russian hash standard, officially known as GOST R 34.11-2012 [1], is also among the recently proposed AES-like hash functions. This shift in the hash functions design concepts is due to the fact that Wang et al. attacks [26,27] are most effective on Add-Rotate-Xor (ARX) based hash functions where one can find differential patterns that propagate with acceptable probabilities. Moreover, these attacks take advantage

© Springer International Publishing Switzerland 2015
D. Lin et al. (Eds.): Inscrypt 2014, LNCS 8957, pp. 311–328, 2015.
DOI: 10.1007/978-3-319-16745-9_17

of the weak message schedules of most of ARX-designs. Hence, using message modification techniques [27], significant reduction in the attack complexity can be achieved.

Whirlwind is a keyless AES-like hash function that adopts the Sponge model [7]. It is proposed by Baretto et al. in 2010 as a response to the recent cryptanalytic attacks that have improved significantly during the SHA-3 competition. Sharing the designers of Whirlpool, Whirlwind design is inspired by Whirlpool and takes into account the recent development in hash function cryptanalysis, particularly the rebound attack [17]. In fact, the designers add more security features as a precaution against possible improvements. The most important features are adopting an extended Sponge model where the compression function operates on $2n$-bit state and outputs an n-bit chaining value, and employing 16×16-bit Sboxes. Using large Sboxes aims to decrease the probability of a given differential trail. Unlike Whirlpool, the Whirlwind compression function has no independent mixing of the chaining value. In the latter, the chaining value is processed independently and mixed with the message state at an XOR transformation. The presence of the key schedule has been exploited as an additional degree of freedom by cryptanalysts. Consequently, it has contributed to many improvements of the inbound phase of the rebound attack [14,18]. These improvements have enabled the attack to cover more rounds. Accordingly, for the designers of Whirlwind, eliminating both the key schedule and the interaction between the message and the compression function output via the feedforward, and employing large Sboxes, limit both the effect and scope of the rebound attack to a great extent. However, from our perspective, some of these features made one of our meet in the middle (MitM) pseudo preimage attacks on the compression function easier and with lower complexity than that on Whirlpool [29]. More precisely, with the absence of the key schedule, using large Sboxes and the output truncation has enabled us to find an execution separation such that the matching probability can be balanced with the available forward and backward starting values as will be discussed in Sects. 4 and 5.

Aoki and Sasaki proposed the meet in the middle preimage attack [5] following the work of Laurent on MD4 [15]. Afterwards, the first MitM preimage attack on the AES block cipher in hashing modes was proposed by Sasaki in FSE 2011 [22]. He applied the attack on Whirlpool and a 5-round pseudo preimage attack on the compression function was presented and used for a second preimage attack on the whole hash function in the same work. In the sequel, Wu et al. [29] formalized the approach and employed a time-memory trade off to improve the time complexity of the 5-round attack on the Whirlpool compression function. Moreover, they applied the MitM pseudo preimage attack on Grøstl and adapted the attack to produce pseudo preimages of the reduced hash function. Afterwards, a pseudo preimage attack on the 6-round Whirlpool compression function and a memoryless preimage attack on the reduced hash function were proposed in [24]. Finally, AlTawy and Youssef, combined MitM pseudo preimages of the compression function of Stribog with a multicollision attack to generate preimages of the reduced hash function [2].

In this work, we investigate the security of Whirlwind and its compression function, assessing their resistance to the MitM preimage attacks. Employing the partial matching and initial structure concepts [22], we present a pseudo preimage attack on the compression function reduced to 5 out of 12 rounds. More precisely, we present an execution separation for the compression function that balances the forward and backward starting values with the corresponding matching probability [29]. Furthermore, we employ a guess and determine approach [24] to guess parts of the state. This approach helps in maintaining partial state knowledge for one more round. Consequently, we are able to extend the attack by one more round. In spite of the compression function truncated output, the proposed 6-round execution separation maximizes the overall probability of the attack by balancing the chosen number of starting values and the guess size. Finally, we show how to generate n-block messages second preimages of the Whirlwind hash function using the presented pseudo preimage attacks on the compression function.

The rest of the paper is organized as follows. In the next section, the description of the Whirlwind hash function along with the notation used throughout the paper are provided. A brief overview of the MitM preimage attack and the used approaches are given in Sect. 3. Afterwards, in Sects. 4 and 5, we provide detailed description of the attacks and their corresponding complexity. In Sect. 6, we show how second preimages of the hash function are generated using the attacks presented in Sects. 4 and 5. Finally, the paper is concluded in Sect. 7.

2   Whirlwind Description

Whirlwind [6] is a keyless AES-like hash function that adopts a Sponge-like model. The function employs a 12-round compression function which operates on 1024-bit state. The internal state is represented by an 8×8 matrix S of 16-bit (word) elements where each element is indexed by its position in row i and column j.

$$
\begin{bmatrix}
S_{0,0} & S_{0,1} & S_{0,2} & S_{0,3} & S_{0,4} & S_{0,5} & S_{0,6} & S_{0,7} \\
S_{1,0} & S_{1,1} & S_{1,2} & S_{1,3} & S_{1,4} & S_{1,5} & S_{1,6} & S_{1,7} \\
S_{2,0} & S_{2,1} & S_{2,2} & S_{2,3} & S_{2,4} & S_{2,5} & S_{2,6} & S_{2,7} \\
S_{3,0} & S_{3,1} & S_{3,2} & S_{3,3} & S_{3,4} & S_{3,5} & S_{3,6} & S_{3,7} \\
S_{4,0} & S_{4,1} & S_{4,2} & S_{4,3} & S_{4,4} & S_{4,5} & S_{4,6} & S_{4,7} \\
S_{5,0} & S_{5,1} & S_{5,2} & S_{5,3} & S_{5,4} & S_{5,5} & S_{5,6} & S_{5,7} \\
S_{6,0} & S_{6,1} & S_{6,2} & S_{6,3} & S_{6,4} & S_{6,5} & S_{6,6} & S_{6,7} \\
S_{7,0} & S_{7,1} & S_{7,2} & S_{7,3} & S_{7,4} & S_{7,5} & S_{7,6} & S_{7,7}
\end{bmatrix}
$$

An element $S_{i,j}$ can be seen as 4×1 matrix of 4-bit nibbles.

$$
S_{i,j} = \begin{bmatrix} S_{i,j,0,0} \\ S_{i,j,0,1} \\ S_{i,j,1,0} \\ S_{i,j,1,1} \end{bmatrix}.
$$

Accordingly, each row of the state matrix S is in fact 4×8 4-bit nibble matrix. The reason for the switch between the 16-bit elements and the 4-bit nibbles is due to the fact that the adopted round transformations operate on different fields $(GF(2^{16})$ and $GF(2^4))$. More precisely, the round function updates the state by applying the following four transformations:

- γ: A nonlinear bijective mapping over $GF(2^{16})$. This substitution layer works on the 16-bit elements where it replaces each 16-bit element by its multiplicative inverse over $GF(2^{16})$ and zero is replaced by itself.
- θ: A linear transformation that mixes rows. It works by applying the linear transformations λ_0 and λ_1 on the 4-bit nibble elements. Hence, if each state row is 4×8 4-bit nibble matrix, the updated row is:

$$
\theta(S_i) = \begin{cases} \lambda_0(S_{i,*,0,0}) = S_{i,*,0,0} \cdot M_0 \\ \lambda_1(S_{i,*,0,1}) = S_{i,*,0,1} \cdot M_1 \\ \lambda_1(S_{i,*,1,0}) = S_{i,*,1,0} \cdot M_1 \\ \lambda_0(S_{i,*,1,1}) = S_{i,*,1,1} \cdot M_0, \end{cases}
$$

where $*$ denotes the column index, $M_0 = \text{dyadic}(0x5, 0x4, 0xA, 0x6, 0x2, 0xD, 0x8, 0x3)$, $M_1 = \text{dyadic}(0x5, 0xE, 0x4, 0x7, 0x1, 0x3, 0xF, 0x8)$, and $\text{dyadic}(m)$ denotes the MDS dyadic matrix M corresponding to the sequence m over $GF(2^4)$, i.e., $M_{i,j} = m_{i \oplus j}$. Nevertheless, θ inherits the optimal diffusion properties of its underlying transformations. However, these transformations cannot be directly applied on elements of $GF(2^{16})$ through simple matrix multiplication as this requires the use of a linearized polynomial.
- τ: A transposition layer where the 16-bit 8×8 matrix is transposed.
- σ^r: A linear transformation where that 16-bit state is XORed with a round dependant constant state C^r.

As depicted in Fig. 1, the compression function $\phi(h, m)$ operates on 512-bit message block m and 512-bit chaining value h, both represented by 8×4 matrices of 16-bit elements. The internal state S is initialized such that its first four columns are set to h and the last four columns are set to m. The state is then updated by applying the four transformations for twelve rounds. Finally, the last four columns of the last state are truncated and the input chaining value h is XORed with the first four columns of the last state to generate the compression function output.

Whirlwind employs a finalization step. More precisely, after processing all the message blocks, an extra compression function call with a null message block is adopted. If the desired output size is $\log_2(N)$ bits, the output of the finalization step is then reduced modulo N. Whirlwind also uses an adaptable initialization value where the IV used to process the first message block depends on the desired reduction value N. To compute the initial value h_0, the reduction value is converted to an 8×4 matrix, thus $h_0 = \phi(0, N)$. To compute the hash of a given message M, it is first padded by 1 followed by zeros to obtain a bit string whose length is an odd multiple of 256, and finally with the 256-bit right justified

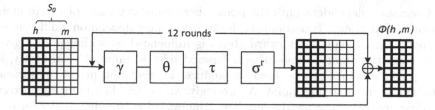

Fig. 1. The compression function ϕ.

binary representation of $|M|$. The padded message is then divided into t 512-bit blocks: $m_0, m_1, ... m_{t-1}$. Finally, the message blocks are processed as follows:

$$h_0 = \phi(0, N),$$
$$h_i = \phi(h_{i-1}, m_{i-1}), \text{ for } i = 1, 2, ..., t,$$
$$h_{t+1} = \phi(h_t, 0).$$

The output $H(M)$ is equal to $h_{t+1} \bmod N$. For further details, the reader is referred to [6,25].

2.1 Notation

Let S be (8×8) 16-bit state denoting the internal state of the function. The following notation is used in our attacks:

- S_i: The message state at the beginning of round i.
- S_i^U: The message state after the U transformation at round i, where $U \in \{\gamma, \theta, \tau, \sigma^r\}$.
- $S_i[r, c]$: A word at row r and column c of state S_i.
- $S_i[\text{row } r]$: Eight words located at row r of state S_i.
- $S_i[\text{col } c]$: Eight words located at column c of state S_i.

3 MitM Preimage Attacks

Given a compression function CF that processes a chaining value h and a message block m, a preimage attack on CF is defined as follows: given h and x, where x is the compression function output, find m such that $CF(h, m) = x$. However, in a pseudo preimage attack, only x is given and one must find h and m such that $CF(h, m) = x$. The effect of a pseudo preimage attack on the compression function by itself is not important. However, these attacks can be used to build a preimage or second preimage attacks on the whole hash function [19]. As demonstrated in Sect. 6, pseudo preimages of the Whirlwind compression function can be utilized to compose an n-block second preimages of the hash function.

The main concept of the proposed MitM attacks is to divide the attacked execution rounds at the starting point into two independent executions that proceed in opposite directions (forward and backward chunks). The two executions must remain independent until the point where matching takes place. To maintain the independence constraint, each execution must depend on a different set of inputs, e.g., if only the forward chunk is influenced by a change in a given input, then this input is known as a forward neutral input. Consequently, all of its possible values can be used to produce different outputs of the forward execution at the matching point. Accordingly, all neutral inputs for each execution direction attribute for the number of independent starting values for each execution. Hence, the output of the forward and the backward executions can be independently calculated and stored at the matching point. Similar to all MitM attacks, the matching point is where the outputs of the two separated chunks meet to find a solution from both the forward and backward directions that satisfies both executions. While for block ciphers, having a matching point is achieved by employing both the encryption and decryption oracles, for hash function, this is accomplished by adopting the cut and splice technique [5] which utilizes the employed mode of operation. In other words, given the compression function output, this technique chains the input and output states through the feedforward as we can consider the first and last states as consecutive rounds. Subsequently, the overall attacked rounds behave in a cyclic manner and one can find a common matching point between the forward and backward executions and consequently can also select any starting point.

The MitM preimage attack has been applied to MD4 [5,10], MD5 [5], HAS-160 [11], and all functions of the SHA family [3,4,10]. The attack exploits the fact that all the previously mentioned functions are ARX-based and operate in the Davis-Mayer (DM) mode, where the state is initialized by the chaining value and some of the expanded message blocks are used independently in each round. Thus, one can determine which message blocks affect each execution for the MitM attack. However, several AES-like hash functions operate in the Miyaguchi-Preneel mode, where the input message is fed to the initial state which undergoes a chain of successive transformations. Consequently, the process of separating independent executions becomes relatively more complicated. Cryptanalysts are forced to adopt a pseudo preimage attack when the compression function operates in Davis-Mayer mode. This is due to the fact that the main execution takes place on a state initialized by the chaining value. Subsequently, using the cut and splice technique enforces changes in the first state through the feedforward. Additionally, even if function operates in the Miyaguchi-Preneel mode, attempting a MitM preimage attack usually generates pseudo preimages when the complexity of finding a preimage is higher than the available degrees of freedom in the message. Consequently, the chaining value is utilized as a source of randomization to satisfy the number of multiple restarts required by the attack. As a result, we end up with a pseudo preimage rather than a preimage of the compression function output.

This class of attacks has witnessed significant improvements since its inception. Most of these attacks aim to make the starting and matching points span over more than one round transformation and hence increase the number of the overall attacked rounds. More precisely, the initial structure approach [22,23]

provides the means for the starting point to cover a few successive transformations where words in the states belong to both the forward and backward chunks.
Although neutral words of both chunks are shared within the initial structure,
independence of both executions is achieved in the rounds at the edges of the initial structure. Additionally, the partial matching technique [5] allows only parts
of the state to be matched at the matching point. This method is used to extend
the matching point further and makes use of the fact that round transformations
may update only parts of the state. Thus the remaining unchanged parts can
be used for matching. This approach is highly successful in ARX-based hash
functions which are characterized by the slow diffusion of their round update
functions and so some state variables remain independent in one direction while
execution is in the opposite direction. The unaffected parts of the states at each
chunk are used for partial matching at the matching point. However, in AES-
like hash functions, full diffusion is achieved after two rounds and this approach
can be used to extend the matching point of two states for a limited number of
transformations. Once a partial match is found, the inputs of both chunks that
resulted in the matched values are selected and used to evaluate the remaining undetermined parts of the state at the matching point to check for a full
state match. Figure 2 illustrates the MitM preimage attack approaches for the
Whirlwind compression function. The red and blue arrows denote the forward
and backward executions on the message state, respectively. S_0 is the first state
initialized by h and m and S_t is the last attacked state.

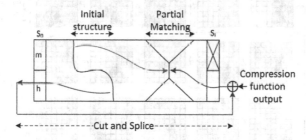

Fig. 2. MitM preimage attack techniques customized for Whirlwind operation (Color
figure online).

In the next section, we apply the techniques discussed in this section to
generate a 5-round pseudo preimage of the Whirlwind compression function.

4   A Pseudo Preimage of the 5-Round Compression Function

To proceed with the attack, we first need to separate the two execution chunks
around the initial structure. More precisely, we divide the five attacked rounds of
execution into a 2-round forward chunk and a 2-round backward chunk around
a starting point (initial structure). The proposed chunk separation is shown

in Fig. 3. Our choices of forward and backward starting values in the initial structure determine the complexity of the attack. Specifically, we try to balance the number starting values in each direction and the number of known words at the matching point at the end of each chunk. The total number of starting values in both directions should produce candidate pairs at the matching point to satisfy the matching probability. For further clarification, we first explain how the initial structure is constructed. The main idea is to have maximum state knowledge at the start of each execution chunk. This can be achieved by choosing several words as neutral so that the number of corresponding output words of the θ and θ^{-1} transformations at the start of both chunk that are constant or relatively constant is maximized. A relatively constant word is a word at the state directly after the initial structure whose value depends on the value of the neutral words in one execution direction but remains constant from the opposite execution perspective. The initial structure for the 5-round MitM preimage attack on the compression function of Whirlwind is shown in Fig. 4.

Following Fig. 4, our aim is to have five constants in the three lowermost rows in state d and determine the available values of the corresponding blue rows that make them hold. The values of the three lowermost blue rows are the available

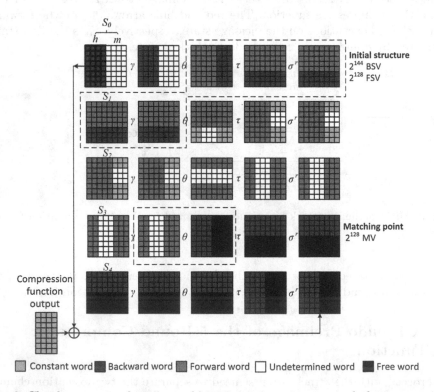

Fig. 3. Chunk separation for a 5-round MitM pseudo preimage attack the compression function. BSV: Backward starting value, FSV: Forward starting value, MV: Matching value (Color figure online).

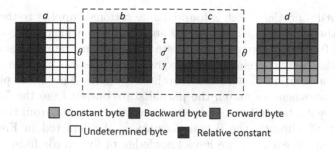

Fig. 4. Initial structure used in the attack on the 5-round compression function (Color figure online).

backward starting values. For each row, we randomly choose the five constant words in $d[\text{row } 7]$ and then determine the values of blue words in $c[\text{row } 7]$ so that after applying θ on $c[\text{row } 7]$, the chosen values of the five constants hold. Since the linear mapping is applied on the 4-bit nibbles, we need twenty constant nibbles in $d[\text{row } 7]$. This can be achieved by maintaining twenty variable nibbles in $c[\text{row } 7]$ to solve a system of twenty equations when the other twelve nibbles are fixed. Accordingly, for any of the last three rows in state c, we can randomly choose any three blue words and compute the remaining five so that the output of θ maintains the previously chosen five constant words at $d[\text{row } 7]$. To this end, we have nine free (blue) words, three for each row in state c. Thus the number of backward starting values is 2^{144} which means that we can start the backward execution by 2^{144} different starting values and hence 2^{144} different output values at the matching point S_3^θ. Similarly, we choose 32 constant words in state a and for each row in state b we randomly choose four red nibbles and compute the other sixteen red nibbles such that after the θ^{-1} transformation we get the predetermined constants at each row in a. However, the value of the four shaded blue words in each row of state a depends also on the three blue words in the rows of state b. We call these bytes relative constants because their final values cannot be determined until the backward execution starts and these values are different for each backward execution iteration. Specifically, their final values are the predetermined constants acting as an offset XORed with the corresponding blue nibbles multiplied by M_0^{-1} or M_1^{-1} coefficients. In the sequel, we have eight free words (one for each row in b) which means 2^{128} forward starting value and hence 2^{128} different input values to the matching point S_3^γ.

As depicted in Fig. 3, the forward chunk starts at S_1^θ and ends at S_3^γ which is the input state to the matching point. The backward chunk starts at S_0^γ and ends after the feedforward at S_3^θ which is the output state of the matching point. The red words are the neutral ones for the forward chunk and after choosing them in the initial structure, all the other red words can be independently calculated. White words in the forward chunk are the ones whose values depend on the neutral words of the backward chunk which are the blue words in the initial structure. Accordingly, their values are undetermined, i.e., these words cannot be evaluated until a partial match is found. Same rationale applies to the backward chunk and the blue words. Grey bytes are constants which can be either the compression function output or the chosen constants in the initial structure.

To find the pseudo preimage of the given compression function output, we have to find a solution that satisfies both executions. This takes place at the matching point where we match the partial state output from the forward execution at S_3^γ with the full state (due to truncation) output from the backward execution at S_3^θ through the θ transformation. As depicted in Fig. 3, at the matching point, in each row we have knowledge of five words from the forward execution and four words from the backward execution. Since the linear mapping is performed on 4-bit nibbles, we can form sixteen 4-bit linear equations using twelve 4-bit unknowns and match the resulting forward and backward values through the remaining four 4-bit equations. More precisely, we use the following equation to compute the first 4-bit nibble row in the first state row $b_{0,j,0,0}$ through the linear transformation λ_0 given the 4-bit nibble input row $a_{0,j,0,0}$. For ease of notation, we denote the first 4-bit nibble in a word located in the first row and column j as a_j (i.e., $a_{0,j,0,0} = a_j$). We use a similar notation for b.

$$
\begin{bmatrix} a_0 & a_1 & \overline{a_2} & \overline{a_3} & \overline{a_4} & a_5 & a_6 & a_7 \end{bmatrix}
\begin{bmatrix}
0x5 & 0x4 & 0xA & 0x6 \\
0x4 & 0x5 & 0x6 & 0xA \\
0xA & 0x6 & 0x5 & 0x4 \\
0x6 & 0xA & 0x4 & 0x5 \\
0x2 & 0xD & 0x8 & 0x3 \\
0xD & 0x2 & 0x3 & 0x8 \\
0x8 & 0x3 & 0x2 & 0xD \\
0x3 & 0x8 & 0xD & 0x2
\end{bmatrix}
= \begin{bmatrix} b_0 & b_1 & b_2 & b_3 \end{bmatrix}
$$

Because we are evaluating the first 4-bit nibble in the output words and the four rightmost column of the output state are truncated, we only use half of the dyadic matrix M_0. In the above equation, we use the overline to denote the unknown first 4-bit nibbles at the first row words. More precisely, there are three unknown nibbles a_2, a_3, and a_4 in the input and all the nibbles in the output are known. Accordingly, given the λ_0 transformation linear matrix M_0, we can form four linear equations to compute b_0, b_1, b_2, and b_3. Then we evaluate the values of the three unknown nibbles a_2, a_3, and a_4 from three out of the four equations and substitute their values in the remaining one. With probability 2^{-4} the right hand side of the remaining equation is equal to the corresponding known backward nibble. Hence, the matching size per 4-bit nibble row is 2^4 and since we have four 4-bit nibble rows per word row, the matching size is 2^{16} for state row, Thus, the matching probability for the whole state is 2^{-128}. The choice of the number of forward and backward starting values directly affects the matching probability as their number determines the number of red and blue words at a given state row at the matching point. If the number of blue and red words are not properly chosen at the initial structure, we can reach no matching value. More precisely, we cannot have a matching value if the total number of red and blue words in a given row at the matching point is less than or equal to eight. In what follows we summarize the attack steps:

1. We randomly choose the constants in states S_1^θ and S_0^γ at forward and backward output of the initial structure.
2. For each forward starting value fw_i in the 2^{128} forward starting values at S_0^γ, we evaluate the forward matching value fm_i at S_3^γ and store (fw_i, fm_i) in a lookup table T.
3. For each backward starting value bw_j in the 2^{144} backward starting values in S_1^γ, we evaluate the backward matching value bm_j at S_3^θ and check if there exists an $fm_i = bm_j$ in T. If found, then these solutions partially match through the linear transformations and the full match should be checked using the matched starting points fw_i and bw_i. If a full match exists, then output the chaining value and the message M_i, else go to step 1.

To minimize the attack complexity, the number of the starting values of both execution and the matching value must be kept as close as possible to each other. In the chunk separation shown in Fig. 3, the number of forward and backward starting values, and the matching values are 2^{128}, 2^{144}, and 2^{128}, respectively. To further explain the complexity of the attack, we consider the attack procedure. After step 2, we have 2^{128} forward matching values at S_3^γ and we need 2^{128} memory to store them. At the end of step 3, we have 2^{144} backward matching values at S_3^θ. Accordingly, we get $2^{128+144} = 2^{272}$ candidate pairs for partial matching. Since the probability of a partial match is 2^{-128}, we expect 2^{144} pairs to partially match. The probability that a partially matching pair results in a full match is the probability that the matching forward and backward starting values generates the three unknown columns in S_3^γ equal to the ones that resulted from the partial match. This probability is equal to $2^{24 \times 16} = 2^{-384}$. As we have 2^{144} partially matching pairs, we expect $2^{144-384} = 2^{-240}$ pairs to fully match. Thus we need to repeat the attack 2^{240} times to get one fully matching pair. The time complexity for one repetition of the attack is 2^{128} for the forward computation, 2^{144} for the backward computation, and 2^{144} to test if the partially matching pairs fully match. Consequently, the overall complexity of the attack is $2^{240}(2^{128} + 2^{144} + 2^{144}) \approx 2^{385}$ time and 2^{128} memory.

5   Extending the Attack by One More Round

The wide trail strategy adopted by Whirlwind implies that one unknown word leads to a full unknown state after two rounds. Consequently, in the previous 5-round attack, the matching point is chosen exactly two rounds away from the initial structure in each direction. Attempting to go one more round in either directions always fails because at the end of each chunk execution the state has undetermined bytes at each row. Consequently, applying the linear transformation θ to such state results in a full loss of state knowledge and matching cannot be achieved. To maintain partial state knowledge, we adopt a guess and determine approach [24]. Hence, we can probabilistically guess some of the undetermined row words in the state before the linear transformation in either direction. Thus, we maintain knowledge of some state rows after the linear transformation θ which are used for matching. Due to truncation and the large size of Sboxes, we

have to carefully choose both starting values in the initial structure to minimize
the number of guessed words as much as possible and to result in an acceptable
number of correctly guessed matching pairs. The proposed chunk separation for
the 6-round MitM pseudo preimage attack is shown in Fig. 5. In order to be able
extend the attack by one extra round in the forward direction, we guess the six
unknown words (yellow words) in state S_4^γ. As a result, we can reach state S_5^γ
with three determined columns where the matching takes place.

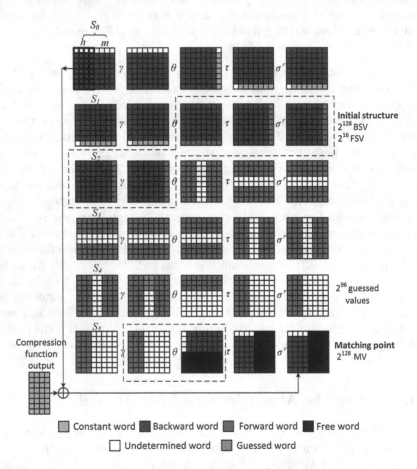

Fig. 5. Chunk separation for a 6-round MitM pseudo preimage attack on the compression function. BSV: Backward starting value, FSV: Forward starting value, MV:
Matching value (Color figure online).

The chosen separation and guessed values maximize the attack probability
by carefully selecting the forward, backward, and guessed bit values. We aim to
increase the number of starting forward values and keep the number of backward
and matching values as close as possible and larger than the number of guessed

values. For our attack, the chosen number for the forward and backward starting values, and the guessed values are 2^{16}, 2^{128}, and 2^{96}, respectively. Setting these parameters fixes the number of matching values to 2^{128}. In what follows, we give an overview of the attack procedure and complexity based on the above chosen parameters:

1. We first start by randomly choosing the constants in S_1^γ and S_2^θ at the edges of the initial structure.
2. For each forward starting value fw_i and guessed value g_i in the 2^{16} forward starting values and the 2^{96} guessed values, we compute the forward matching value fm_i at S_5^γ and store (fw_i, g_i, fm_i) in a lookup table T.
3. For each backward starting value bw_j in the 2^{128} backward starting values, we compute the backward matching value bm_j at S_5^θ and check if there exists an $fm_i = bm_j$ in T. If found, then a partial match exists and the full match should be checked using the matched forward, guessed, and backwards starting values fw_i, g_i, and bw_i. If a full match exists, then we output the chaining value h_i and the message m_i, else go to step 1.

The complexity of the attack is evaluated as follows: after step 2, we have $2^{16+96} = 2^{112}$ forward matching values which need 2^{112} memory for the look up table. At the end of step 3, we have 2^{128} backward matching values. Accordingly, we get $2^{112+128} = 2^{240}$ partial matching candidate pairs. Since the probability of a partial match is 2^{-128} and the probability of a correct guess is 2^{-96}, we expect $2^{240-128-96} = 2^{16}$ correctly guessed partially matching pairs. Due to truncation, we are interested only in the uppermost four rows at the matching point. More precisely, we want the partially matching starting value to result in the correct values on the twenty four unknown words in both S_4^γ and S_4^θ that make the blue and red words hold. The probability that the latter condition takes place is $2^{24\times-16} = 2^{-384}$. Consequently, the expected number of fully matching pairs is 2^{-368} and hence we need to repeat the attack 2^{368} times to get a full match. The time complexity for one repetition is 2^{112} for the forward computation, 2^{128} for the backward computation, and 2^{16} to check that partially matching pairs fully match. The overall complexity of the attack is $2^{368}(2^{112}+2^{128}+2^{16}) \approx 2^{496}$ time and 2^{112} memory.

6 Second Preimage of the Hash Function

In this section, we show how the previously presented pseudo preimage attacks on the Whirlwind compression function can be utilized to generate second preimages for the whole hash function. The last two compression function calls in the Whirlwind hash function differ than the previous ones, hence they are considered a final step in the execution of the function. In this step, the first compression function call operates on the padded message, and the state of the second compression function call is initialized by the chaining value and an 8×4 all zero message. Accordingly, attempting to use our pseudo preimage attacks to invert the final compression function call does not result in the expected all zero message and if extended can rarely satisfy the correct padding. Consequently, using these attacks to generate preimages does not work. However, if we can get the correct chaining values for the last two compression function calls such that when both the correct padding and null message are used we get the target compression function output, then we can use our pseudo preimage attack to get the right messages. This requirements can be fulfilled if we consider a second preimage attack. When one attempts a second preimage attack, one is given a hash function H that operates with an initial value IV and a message block m. Then, one must find m' such that $H_{IV}(m) = H_{IV}(m')$. When we consider a second preimage attack, using the give message m, we can know exactly the input chaining values for the last two compression function calls such that we get the desired hash function output. We only need to find another equal length message m' that is, given the IV, generates the chaining value required by the padding compression function call. Our attack is an n-block second preimage attack $(n \geq 2)$ where given an n-block message m, we generate another n-block message m' such that both messages hash to the same value. More precisely, to build m', we copy the finalization step of m and use our pseudo preimage attacks along with another meet in the middle attack to search for m'. For illustration, we are using 2-block messages to describe our attack. As depicted in Fig. 6, the attack is divided into three stages:

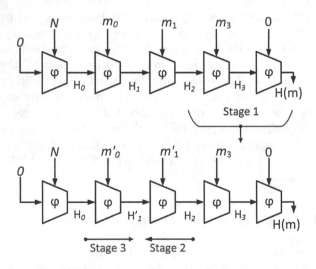

Fig. 6. Second preimage attack on the hash function.

1. Given a 2-block message $m = m_0 \| m_1$ and the truncation value N, we compute the adaptable initialization vector $H_0 = \phi(0, N)$, compose the padding message $m_2 = 1 \| 0^{500} \| 1 \| 0^{10}$, and hash m and get the desired $H(m)$. This process is shown

in the upper hash function execution in Fig. 6. To begin building m', we copy the last compression function calls with there chaining values. Specifically, we consider H_2 to be the output of the compression function call operating on the second block m'_1 of the massage we are searching for.

2. In this stage, given H_2, we produce 2^p pseudo preimages for the second message block compression function call. The output of this step is 2^p pairs of a candidate chaining value H'_1 and a candidate second message block m'_1. We store these resulting candidate pairs (H'_1, m'_1) in a table T.

3. To this end, we try to search for the first message block m'_0 such that using the initial vector H_0, $\phi(H_0, m'_0)$ produces one of the chaining values H'_1 in the table T. In the sequel, we randomly choose m'_0, compute H'_1 and check if it exists in T. As T contains 2^p entries, it is expected to find a match after 2^{512-p} evaluations of the following compression function call with random m'_0 each time:

$$H'_1 = \phi(H_0, m'_0)$$

Once a matching H'_1 value is found in T, the chosen m'_0 is the first message block and the corresponding m'_1 is the second message block such that $m' = m'_0 \| m'_1$ and $H(m) = H(m')$.

The time complexity of the attack is evaluated as follows: we need $2^p \times$ (complexity of pseudo preimage attack) in stage 2, and 2^{512-p} evaluations of one compression function call for the MitM attack at stage 3. The memory complexity for the attack is as follows: 2^p states to store the pseudo preimages for the MitM in stage 2, in addition to the memory complexity of the pseudo preimage attack on the compression function which is 2^{128} or 2^{112} for the 5-round or 6-round compression function. Since the time complexity is highly influenced by p, we have chosen $p = 64$ for the 5-round attack and $p = 8$ for the 6-round attack to obtain the maximum gain. Accordingly, 2-block second preimages for 5-round Whirlwind hash function are produced with a time complexity of $2^{64+385} + 2^{512-64} \approx 2^{449}$ and memory complexity of $2^{128} + 2^{64} \approx 2^{128}$. The time complexity for the 6-round attack is $2^{8+496} + 2^{512-8} \approx 2^{505}$ and the memory complexity is $2^{112} + 2^8 \approx 2^{112}$.

7   Conclusion

In this paper, we have analyzed Whirlwind and its compression function with respect to preimage attacks. We have shown that with a carefully balanced chunk separation, pseudo preimages for the 5-round reduced compression function are generated. Additionally, we have adopted a guess and determine technique and we were able to extend the 5-round attack by one more round. Finally, using another MitM attack, we utilized the compression function pseudo preimage attacks to produce 5 and 6-round hash function n-block second preimages.

Whirlwind is proposed to improve the Whirlpool design. While, the new improvements limit the extent of rebound attacks significantly, they do not consider MitM preimage attacks. It should be noted that the elimination of the compression function key schedule and using large Sboxes in the same time made our attacks possible. Indeed, with the large state, the chosen constants in the initial structure are enough to satisfy the number of restarts required by the attack complexity. On the other hand, for Whirlpool, the available freedom in the internal state only cannot by itself fulfill the attack complexity. Also, while the adopted truncation and feedforward prohibit interaction between the input message block and the output state thus limiting the ability of difference cancellation, it enhanced the full matching probability, particularly, if we can have full state knowledge at one side of the matching point like our 5-round attack. It is interesting to note that if the adopted model follows the exact Sponge construction where the message is XORed to the internal state and truncation is performed in the finalization step, thus the compression function always maintains a state larger than the hash function output size, our compression function attacks would not work. It should also be noted that the switch between $GF(2^{16})$ and $GF(2^4)$ in different round transformations does not only alleviate potential concerns regarding algebraic attacks but also enhances the resistance of the function to integral attacks [13]. More precisely, the integral properties that are preserved by the substitution layer are shared independently among nibbles by the following linear transformation for the span of one round only. Finally, we know that the presented results do not directly impact the practical security of the Whirlwind hash function. However, they are first steps in the public cryptanalysis of its proposed design concepts with respect to second preimage resistance.

Acknowledgment. The authors would like to thank the anonymous reviewers for their valuable comments and suggestions that helped improve the quality of the paper. This work is supported by the Natural Sciences and Engineering Research Council of Canada (NSERC) and Le Fonds de Recherche du Québec - Nature et Technologies (FRQNT).

References

1. The National Hash Standard of the Russian Federation GOST R 34.11-2012. Russian Federal Agency on Technical Regulation and Metrology report (2012). https://www.tc26.ru/en/GOSTR34112012/GOST_R_34_112012_eng.pdf
2. AlTawy, R., Youssef, A.M.: Preimage attacks on reduced-round stribog. In: Pointcheval, D., Vergnaud, D. (eds.) AFRICACRYPT. LNCS, vol. 8469, pp. 109–125. Springer, Heidelberg (2014)
3. Aoki, K., Guo, J., Matusiewicz, K., Sasaki, Y., Wang, L.: Preimages for step-reduced SHA-2. In: Matsui, M. (ed.) ASIACRYPT 2009. LNCS, vol. 5912, pp. 578–597. Springer, Heidelberg (2009)
4. Aoki, K., Sasaki, Y.: Meet-in-the-middle preimage attacks against reduced SHA-0 and SHA-1. In: Halevi, S. (ed.) CRYPTO 2009. LNCS, vol. 5677, pp. 70–89. Springer, Heidelberg (2009)
5. Aoki, K., Sasaki, Y.: Preimage attacks on one-block MD4, 63-step MD5 and more. In: Avanzi, R.M., Keliher, L., Sica, F. (eds.) SAC 2008. LNCS, vol. 5381, pp. 103–119. Springer, Heidelberg (2009)

6. Barreto, P., Nikov, V., Nikova, S., Rijmen, V., Tischhauser, E.: Whirlwind: a new cryptographic hash function. Des. Codes Crypt. **56**(2–3), 141–162 (2010)
7. Bertoni, G., Daemen, J., Peeters, M., Van Assche, G.: On the indifferentiability of the sponge construction. In: Smart, N.P. (ed.) EUROCRYPT 2008. LNCS, vol. 4965, pp. 181–197. Springer, Heidelberg (2008)
8. Daemen, J., Rijmen, V.: The Design of Rijndael: AES- The Advanced Encryption Standard. Springer, Berlin (2002)
9. Gauravaram, P., Knudsen, L.R., Matusiewicz, K., Mendel, F., Rechberger, C., Schläffer, M., Thomsen, S.S.: Grøstl a SHA-3 candidate. NIST submission (2008)
10. Guo, J., Ling, S., Rechberger, C., Wang, H.: Advanced meet-in-the-middle preimage attacks: first results on full tiger, and improved results on MD4 and SHA-2. In: Abe, M. (ed.) ASIACRYPT 2010. LNCS, vol. 6477, pp. 56–75. Springer, Heidelberg (2010)
11. Hong, D., Koo, B., Sasaki, Y.: Improved preimage attack for 68-step HAS-160. In: Lee, D., Hong, S. (eds.) ICISC 2009. LNCS, vol. 5984, pp. 332–348. Springer, Heidelberg (2010)
12. Indesteege, S.: The Lane hash function. Submission to NIST (2008). http://www.cosic.esat.kuleuven.be/publications/article-1181.pdf
13. Knudsen, L.R., Wagner, D.: Integral cryptanalysis. In: Daemen, J., Rijmen, V. (eds.) FSE 2002. LNCS, vol. 2365, pp. 112–127. Springer, Heidelberg (2002)
14. Lamberger, M., Mendel, F., Rechberger, C., Rijmen, V., Schläffer, M.: Rebound distinguishers: results on the full whirlpool compression function. In: Matsui, M. (ed.) ASIACRYPT 2009. LNCS, vol. 5912, pp. 126–143. Springer, Heidelberg (2009)
15. Leurent, G.: MD4 is not one-way. In: Nyberg, K. (ed.) FSE 2008. LNCS, vol. 5086, pp. 412–428. Springer, Heidelberg (2008)
16. Matyukhin, D., Rudskoy, V., Shishkin, V.: A perspective hashing algorithm. In: RusCrypto (2010). (in Russian)
17. Mendel, F., Rechberger, C., Schläffer, M., Thomsen, S.S.: The rebound attack: cryptanalysis of reduced Whirlpool and Grøstl. In: Dunkelman, O. (ed.) FSE 2009. LNCS, vol. 5665, pp. 260–276. Springer, Heidelberg (2009)
18. Mendel, F., Rechberger, C., Schläffer, M., Thomsen, S.S.: Rebound attacks on the reduced Grøstl hash function. In: Pieprzyk, J. (ed.) CT-RSA 2010. LNCS, vol. 5985, pp. 350–365. Springer, Heidelberg (2010)
19. Menezes, A.J., Van Oorschot, P.C., Vanstone, S.A.: Handbook of Applied Cryptography. CRC Press, Boca Raton (2010)
20. NIST. Announcing request for candidate algorithm nominations for a new cryptographic hash algorithm (SHA-3) family. Federal Register, vol. 72(212) November 2007. http://csrc.nist.gov/groups/ST/hash/documents/FR_Notice_Nov07.pdf
21. Rijmen, V., Barreto, P.S.L.M.: The Whirlpool hashing function. NISSIE submission (2000)
22. Sasaki, Y.: Meet-in-the-middle preimage attacks on AES hashing modes and an application to whirlpool. In: Joux, A. (ed.) FSE 2011. LNCS, vol. 6733, pp. 378–396. Springer, Heidelberg (2011)
23. Sasaki, Y., Aoki, K.: Finding preimages in full MD5 faster than exhaustive search. In: Joux, A. (ed.) EUROCRYPT 2009. LNCS, vol. 5479, pp. 134–152. Springer, Heidelberg (2009)
24. Sasaki, Y., Wang, L., Wu, S., Wu, W.: Investigating fundamental security requirements on whirlpool: improved preimage and collision attacks. In: Wang, X., Sako, K. (eds.) ASIACRYPT 2012. LNCS, vol. 7658, pp. 562–579. Springer, Heidelberg (2012)

25. Tischhauser, E.W.: Mathematical aspects of symmetric-key cryptography. Ph.D. thesis, Katholieke Universiteit Leuven, May 2012. http://www.cosic.esat.kuleuven.be/publications/thesis-201.pdf
26. Wang, X., Yin, Y.L., Yu, H.: Finding collisions in the full SHA-1. In: Shoup, V. (ed.) CRYPTO 2005. LNCS, vol. 3621, pp. 17–36. Springer, Heidelberg (2005)
27. Wang, X., Yu, H.: How to break MD5 and other hash functions. In: Cramer, R. (ed.) EUROCRYPT 2005. LNCS, vol. 3494, pp. 19–35. Springer, Heidelberg (2005)
28. Wu, H.: The hash function JH (2011). http://www3.ntu.edu.sg/home/wuhj/research/jh/jh-round3.pdf
29. Wu, S., Feng, D., Wu, W., Guo, J., Dong, L., Zou, J.: (Pseudo) Preimage attack on round-reduced Grøstl hash function and others. In: Canteaut, A. (ed.) FSE 2012. LNCS, pp. 127–145. Springer, Heidelberg (2012)

Boomerang Attack on Step-Reduced SHA-512

Hongbo Yu$^{(\boxtimes)}$ and Dongxia Bai

Department of Computer Science and Technology, Tsinghua University,
Beijing 100084, China
yuhongbo@mail.tsinghua.edu.cn, baidx10@mails.tsinghua.edu.cn

Abstract. SHA-2 (SHA-224, SHA-256, SHA-384 and SHA-512) is hash function family issued by the National Institute of Standards and Technology (NIST) in 2002 and is widely used all over the world. In this work, we analyze the security of SHA-512 with respect to boomerang attack. Boomerang distinguisher on SHA-512 compression function reduced to 48 steps is proposed, with a practical complexity of 2^{51}. A practical example of the distinguisher for 48-step SHA-512 is also given. As far as we know, it is the best practical attack on step-reduced SHA-512.

Keywords: SHA-512 · Hash functions · Boomerang attack

1 Introduction

Cryptographic hash functions play an important role in modern cryptology. In 2005, many notable hash functions, including MD5 and SHA-1, were broken by Wang *et al.* [32,33]. Since these breakthrough results, many cryptographers have been convinced that these widely used hash functions can no longer be considered secure. Hash functions have been the target in lots of cryptanalytic attacks and cryptanalysis against hash functions has been improved significantly. People not only evaluate the three classical security requirements (preimage resistance, 2nd preimage resistance and collision resistance), but also consider all properties different from the expectation of a random oracle, such as (semi-) free-start collisions, near-collisions, boomerang distinguishers, etc. This is an important progress of the cryptanalysis for hash functions, since the security margin can be measured.

In recent years, the SHA-3 competition [23] organized by NIST has attracted more attention from the cryptographic community. However, as commonly used algorithms in many applications, SHA-2 still deserves much detailed analysis to get a good view on its security. In this paper, we present boomerang attack on the reduced-step SHA-512.

Hongbo Yu is Supported by 973 program (No. 2013CB834205), the National Natural Science Foundation of China (Nos. 61133013 and 61373142), the Tsinghua University Initiative Scientific Research Program (No. 20111080970).

D. Lin et al. (Eds.): Inscrypt 2014, LNCS 8957, pp. 329–342, 2015.
DOI: 10.1007/978-3-319-16745-9_18

Related Work. In the last few years, the security of SHA-256/512 against several attacks has been discussed in many papers. In [12] Isobe and Shibutani presented preimage attacks on SHA-256 and SHA-512 reduced to 24 steps. It was improved by Aoki *et al.* to 43-step SHA-256 and 46-step SHA-512 in [1]. Then Guo *et al.* gave advanced meet-in-the-middle preimage attacks on 42-step SHA-256/512 [10]. Later Khovratovich *et al.* applied biclique to preimages and extended attacks to 45 steps on SHA-256 and 50 steps on SHA-512 [15]. Note that all these attacks only slightly faster than generic attack complexity 2^{256}.

With respect to collision resistance, Mendel *et al.* presented the first collision attack on SHA-256 reduced to 18 steps in [21]. Then in [25] Nikolić and Biryukov improved the collision techniques and constructed a practical collision for 21 steps and a semi-free-start collision for 23 steps of SHA-256. This was later extended to 24 steps on SHA-256 and SHA-512 by Sanadhya and Sarkar [26], and Indesteege *et al.* [11]. Then Mendel *et al.* improved the semi-free-start collisions on SHA-256 from 24 to 32 steps and gave a collision attack for 27 steps, which are all practical [19]. The best known collision attacks on SHA-256 so far are semi-free-start collisions for 38 and collisions for 31 out of 64 steps by Mendel *et al.* in [20]. Recently, Eichlseder *et al.* presented semi-free-start collisions for SHA-512 on up to 38 steps in [8]. Compared with the preimage attacks, all these attacks have practical complexities.

At the rump session of Eurocrypt 2008, Yu and Wang presented non-randomness of SHA-256 reduced to 39 steps [34], and gave a practical example of 33 steps. In [11], Indesteege *et al.* show nonrandom behavior of the SHA-256 compression function in the form of free-start near-collisions for up to 31 steps. In [17], Lamberger and Mendel gave a second-order differential collision on 46 steps of SHA-256 compression function. Later, Biryukov *et al.* extended the result in [17] by one round and presented a practical attack on 47 steps of SHA-256 in [7] in which the application of the attack strategy to SHA-512 was discussed, but no detailed differentials and example were given.

Our Contribution. In this work, the boomerang attack is used to show non-random properties for 48 (out of 80) steps of SHA-512 and an example of a confirming quartet is given. To the best of our knowledge, this is the best practical attack on reduced SHA-512. The summary of previous results and ours on SHA-512 are given in Table 1.

Outline. The structure of this paper is as follows. We give a short description of SHA-512 in Sect. 2. Section 3 summaries boomerang attack on hash functions. Then we present our boomerang attack on 48-step SHA-512 in Sect. 4. Finally, a conclusion of the paper is given in Sect. 5.

2   Description of SHA-2

The SHA-2 (SHA-224, SHA-256, SHA-384 and SHA-512) hash function family is standardized by the National Institute of Standards and Technology (NIST), and

Table 1. Summary of the attacks on SHA-512

Attack type	Target	Steps	Time	Source
Preimage attack	HF	24	2^{480}	[12]
	HF	42	2^{501}	[1]
	HF	46	$2^{511.5}$	
	HF	42	$2^{494.6}$	[10]
	HF	50	$2^{511.5}$	[15]
Pseudo-preimage attack	HF	24	2^{480}	[12]
	HF	46	2^{509}	[1]
	HF	57	2^{511}	[15]
Collision	HF	24	2^{53}	[11]
	HF	24	$2^{22.5}$	[26]
Semi-free-start collision	HF	38	$2^{40.5}$	[8]
Boomerang attack	CF	48	2^{51}	Sect. 4

adopts the Merkle-Damgård structure [22]. This section gives a short description of SHA-512. A complete specification can be found in [24].

SHA-512 is an iterated hash function that processes 1024-bit input message blocks and produces a 512-bit hash value. The compression function of SHA-512 consists of a message expansion function and a state update function.

The message expansion function splits the 1024-bit message block into 16 words $m_i, i = 0, \ldots, 15$, and expands them into 80 64-bit message words w_i as follows:

$$w_i = \begin{cases} m_i, & 0 \leq i \leq 15, \\ \sigma_1(w_{i-2}) + w_{i-7} + \sigma_0(w_{i-15}) + w_{i-16}, & 16 \leq i \leq 79, \end{cases}$$

where the functions $\sigma_0(X)$ and $\sigma_1(X)$ are given by

$$\sigma_0(X) = (X \ggg 1) \oplus (X \ggg 8) \oplus (X \gg 7),$$
$$\sigma_1(X) = (X \ggg 19) \oplus (X \ggg 61) \oplus (X \gg 6).$$

The state update function updates 8 64-bit chaining values $v_i = (a_i, b_i, \ldots, h_i)$ in 80 steps using the 64-bit word w_i as follows:

$$t_1 = h_i + \Sigma_1(e_i) + F_1(e_i, f_i, g_i) + k_i + w_i,$$
$$t_2 = \Sigma_0(a_i) + F_0(a_i, b_i, c_i),$$
$$a_{i+1} = t_1 + t_2, \ b_{i+1} = a_i, \ c_{i+1} = b_i, \ d_{i+1} = c_i,$$
$$e_{i+1} = d_i + t_1, \ f_{i+1} = e_i, \ g_{i+1} = f_i, \ h_{i+1} = g_i,$$

where k_i is a step constant and the function $F_0, F_1, \Sigma_0, \Sigma_1$ are defined as follows:

$$F_0(X, Y, Z) = (X \wedge Y) \oplus (Y \wedge Z) \oplus (X \wedge Z),$$
$$F_1(X, Y, Z) = (X \wedge Y) \oplus (\neg X \wedge Z),$$
$$\Sigma_0(X) = (X \ggg 28) \oplus (X \ggg 34) \oplus (X \ggg 39),$$
$$\Sigma_1(X) = (X \ggg 14) \oplus (X \ggg 18) \oplus (X \ggg 41).$$

After 80 steps, the final hash value is computed by adding the output values to the initial state variables.

3   Boomerang Distinguishers of Hash Functions

The boomerang attack was introduced by Wagner in 1999 [31] against block ciphers. It treats the cipher as a cascade of two sub-ciphers, and uses short differentials in each sub-cipher. These differentials are combined in an adaptive chosen plaintext and ciphertext attack to exploit properties of the cipher that has high probability. Then Kelsey et al. [14] further developed it into a chosen plaintext attack called the amplified boomerang attack, and later it was developed by Biham et al. [4] into the rectangle attack. In [5], Biham et al. combined the boomerang (and the rectangle) attack with related-key differentials and proposed the related-key boomerang and rectangle attacks, which use the related-key differentials instead of the single-key differentials.

In recent years, the idea has been applied to hash functions as part of the new and useful hash function results. The first application presented in [13] used the idea of boomerang attack for the message modification technique in the collision attack for SHA-1. However, we note that this work does not build a boomerang property for a hash function to distinguish the hash functions from a random oracle, but only use the boomerang attack as a neutral bits tool for message modifications. The standard applications of boomerang attack to hash function were independently proposed by Biryukov et al. on BLAKE [6] and Lamberger and Mendel on the SHA-2 family [17]. In their works, the boomerang attacks are standalone distinguishers and work in the same way as for block ciphers - by producing the quartet of plaintexts and ciphertexts (input chaining values and output chaining values). Boomerang distinguishers have also been applied to SIMD-512 [18], HAVAL [27], RIPEMD [28], HAS-160 [29], Skein [9,35] and SM3 [3,16].

Let H be the compression function of a hash function, H_0 and H_1 be two sub-ciphers: $H = H_1 \circ H_0$. The boomerang attack for a compression function can be summarized as follows.

- Choose a random chaining value $v^{(1)}$ and a message $w^{(1)}$, compute $v^{(2)} = v^{(1)} + \beta, v^{(3)} = v^{(1)} + \gamma, v^{(4)} = v^{(3)} + \beta$ and $w^{(2)} = w^{(1)} + \beta_w, w^{(3)} = w^{(1)} + \gamma_w, w^{(4)} = w^{(3)} + \beta_w$. We get a quartet $S = \{(v^{(i)}, w^{(i)}) | i = 1, 2, 3, 4\}$.
- Compute backward from the quartet S using H_0^{-1} to obtain the initial values IV_1, IV_2, IV_3 and IV_4.
- Compute forward from the quartet S using H_1 to obtain the output values h_1, h_2, h_3 and h_4.
- Check whether $IV_2 - IV_1 = IV_4 - IV_3 = \alpha$ and $h_3 - h_1 = h_4 - h_2 = \delta$ are fulfilled.

The complexity of the boomerang attack is summarized in [9,35] as follows.

- Type I: A quartet satisfies $IV_2 - IV_1 = IV_4 - IV_3 = \alpha$ and $h_3 - h_1 = h_4 - h_2 = \delta$ for fixed α and δ. In this case, the generic complexity is 2^n where n is the size of hash value.

– Type II: Only $h_3 - h_1 = h_4 - h_2$ are required. This property is also called a second-order differential collision in [7]. In this case, the complexity for obtaining such a quartet is $2^{n/3}$ using Wagner's generalized birthday attack [30].

– Type III: A quartet satisfies $IV_2 - IV_1 = IV_4 - IV_3$ and $h_3 - h_1 = h_4 - h_2$. This property is also called a zero-sum distinguisher in [2]. In this case, the best known attack still takes time $2^{n/2}$.

4   The Boomerang Attack on Reduced SHA-512

In this section, we apply the boomerang attack to the SHA-512 compression function reduced to 48 steps. The basic idea of our attack is to connect two short differential paths in a quartet. The first step of our attack is to find two short differentials with high probabilities and the middle connection part in the middle does not contain any contradictions. Secondly, we derive the sufficient conditions for the messages and chaining variables for the steps in the middle. Thirdly, we satisfy the conditions in the middle steps by modifying the chaining variables and the message words. Finally, after the message modification, we search the right quartets that pass the verification of the distinguisher.

4.1   Step-Reduced Differential Paths

As shown in Tables 2 and 3, we present the two differential paths used to construct the boomerang distinguisher on 48-step SHA-512, where the top differential path is from step 23 to 1, and the bottom one is from step 24 to 48. To describe the differential paths, we utilize the XOR difference $\Delta a = a \oplus a'$, and $\Delta a : i(1 \leq i \leq 64)$ is used to denote that the i-th bit of a is different from the i-th bit of a', and the rest of the bits of a and a' are the same.

We start from the middle states of the distinguisher quartet S, and the differences of the message words w_i and the chaining variables $v_{23} = (a_{23}, b_{23}, \ldots, h_{23})$ of the top differential path are selected as follows:

– $\Delta w_7 : 64, \Delta w_{22} : 56, 57, 63, \Delta w_i = 0 (0 \leq i \leq 21, i \neq 7)$, if the top path has the differences in message words with this form, 18 steps (step 22 to 5) can be passed with probability 1 so that the path of this type has higher probability than any other ones not following this strategy.

– $\Delta a_{23} : 56, 63, \Delta e_{23} : 56, 63$, these differences are decided by the choice of differences of the message words above. In order to cancel the differences of message words, we derive the differences of all these chaining variables.

Now for the bottom differential path, we choose the differences as follows:

– $\Delta w_{23} : 41, \Delta w_{32} : 41, \Delta w_{47} : 33, 40, \Delta w_i = 0 (24 \leq i \leq 46, i \neq 32)$, thus we can pass 14 steps (step 33 to 46) for free similarly.

– $\Delta b_{23} : 2, 7, 13, 23, 27, 64, \Delta c_{23} : 41, \Delta e_{23} : 5, 13, \Delta f_{23} : 2, 7, \Delta g_{23} : 41, \Delta h_{23} = \Sigma_1(\Delta e_{23})$, according to the differences of message words above and also considering the compatibility with the top differential path in the middle steps, the differences of chaining variables in the bottom path can be derived with some sufficient conditions given in part of Tables 6 and 7.

Table 2. The top differential path used for boomerang attack on SHA-512.

Step	Δw_i	Δa	Δb	Δc	Δd	Δe	Δf	Δg	Δh
		64			23, 25, 30, 36, 46, 50	0	0	28, 36	25, 30
1			64			23, 46, 50			28, 36
2				64			23, 46, 50		
3					64			23, 46, 50	
4						64			23, 46, 50
5							64		
6								64	
7									64
8		64							
9-22									
23	56, 57, 63	56, 63				56, 63			

Table 3. The bottom differential path used for boomerang attack on SHA-512.

Step	Δw_i	Δa	Δb	Δc	Δd	Δe	Δf	Δg	Δh
23	33, 40		2, 7, 13, 23, 27, 64	41		5, 13	2, 7	41	$\Sigma_1(\Delta e_{23})$
24	41			2, 7, 13, 23, 27, 64	41		5, 13	2, 7	41
25		41			2, 7, 13, 23, 27, 64			5, 13	2, 7
26			41			23, 27, 64			5, 13
27				41			23, 27, 64		
28					41			23, 27, 64	
29						41			23, 27, 64
30							41		
31								41	
32									41
33	41								
34-47									
48	33, 40	33, 40	0	0	0	33, 40	0	0	0

4.2 Message Differences

Let $w_i^{(1)}$ and $w_i^{(2)}$ ($0 \leq i \leq 15$) be two 1024-bit messages whose differences are shown in Table 2. In order to carry out the message modification in the middle steps (steps 23–32), we also need to determine the specific differences $w_i^{(1)} \oplus w_i^{(2)}$ ($23 \leq i \leq 32$).

For convenience, let $\Delta w_i^{(1,2)}$ denote the XOR difference of $w_i^{(1)}$ and $w_i^{(2)}$. According to the message expansion, we can compute the message differences $\Delta w_i^{(1,2)}$, $(22 \leq i \leq 47)$ as follows.

$$\Delta w_{22}^{(1,2)} = (\sigma_1(w_{20}^{(1)}) + w_{15}^{(1)} + \sigma_0(w_7^{(1)}) + w_8^{(1)}) \oplus (\sigma_1(w_{20}^{(2)}) + w_{15}^{(2)} + \sigma_0(w_7^{(2)}) + w_8^{(2)})$$
$$\Delta w_{23}^{(1,2)} = (\sigma_1(w_{21}^{(1)}) + w_{16}^{(1)} + \sigma_0(w_8^{(1)}) + w_9^{(1)}) \oplus (\sigma_1(w_{21}^{(2)}) + w_{16}^{(2)} + \sigma_0(w_8^{(2)}) + w_9^{(2)})$$
$$\dots = \dots$$
$$\Delta w_{47}^{(1,2)} = (\sigma_1(w_{45}^{(1)}) + w_{40}^{(1)} + \sigma_0(w_{32}^{(1)}) + w_{31}^{(1)}) \oplus (\sigma_1(w_{45}^{(2)}) + w_{40}^{(2)} + \sigma_0(w_{32}^{(2)}) + w_{31}^{(2)})$$

Since $\sigma_0(w_7^{(1)}) \oplus \sigma_0(w_7^{(2)}) = \sigma_0(\Delta w_7^{(1,2)}) = 0x4180000000000000$ and $\Delta w_{22}^{(1,2)} = 0x4180000000000000$. The first equation holds if

$$w_{22,56}^{(1)} = w_{7,57}^{(1)} \oplus w_{7,63}^{(1)} \oplus w_{7,64}^{(1)}, \tag{1}$$

$$w_{22,57}^{(1)} = w_{7,58}^{(1)} \oplus w_{7,64}^{(1)} \oplus w_{7,1}^{(1)}, \tag{2}$$

$$w_{22,63}^{(1)} = w_{7,64}^{(1)} \oplus w_{7,9}^{(1)}. \tag{3}$$

In the same way, we set the differences $\Delta w_i^{(1,2)}$ $(23 \leq i \leq 32)$ in the Table 4 and deduce the sufficient conditions on $w^{(1)}$ in Table 6 to meet the message expansion. Because we don't need to fulfill the message modifications in steps 34 to 48, the message differences $\Delta w_i^{(1,2)}$ in these steps can keep free.

For the bottom path, the message differences $\Delta w_i^{(1,3)}$, $\Delta w_i^{(2,4)}$ $(22 < i < 47)$ are set in Table 3. In order to get $w_{47}^{(3)} - w_{47}^{(1)} = w_{47}^{(4)} - w_{47}^{(2)}$, according to the message expansion

$$w_{47} = w_{31} + \sigma_0(w_{32}) + w_{40} + \sigma_1(w_{45}),$$

Table 4. Message differences in steps 23 to 33.

i	$\Delta w^{(1,2)}$	$\Delta w^{(1,3)}$
22	4180000000000000	0000008100000000
23	8000000000000000	0000010000000000
24	0502081000000002	0
25	0200100000000004	0
26	2804080001020010	0
27	1008000002000020	0
28	00825520891408a1	0
29	c184220110080140	0
30	0504804080200408	0
31	0001008000400800	0
32	2ab100a291089050	0000010000000000

the following three equations must be satisfied.

$$w_{32,41}^{(1)} \oplus w_{32,48}^{(1)} \oplus w_{32,47}^{(1)} = w_{32,41}^{(2)} \oplus w_{32,48}^{(2)} \oplus w_{32,47}^{(2)} \tag{4}$$

$$w_{32,34}^{(1)} \oplus w_{32,41}^{(1)} \oplus w_{32,40}^{(1)} = w_{32,34}^{(2)} \oplus w_{32,41}^{(2)} \oplus w_{32,40}^{(2)} \tag{5}$$

$$w_{32,35}^{(1)} \oplus w_{32,42}^{(1)} \oplus w_{32,41}^{(1)} = w_{32,35}^{(2)} \oplus w_{32,42}^{(2)} \oplus w_{32,41}^{(2)} \tag{6}$$

The message difference $\Delta w_{32}^{(1,2)}$ we selected in Table 4 happens to meet the Eqs. (4)–(6). Otherwise, we can adjust it.

Extend the messages $w_i^{(3)}$ and $w_i^{(4)}$ ($22 \leq i \leq 47$) in the backward direction. If we want to get $\Delta w_i^{(1,3)} = \Delta w_i^{(2,4)}$ ($0 \leq i \leq 22$), the following three equations must be satisfied.

$$w_{22,56}^{(3)} = w_{7,57}^{(3)} \oplus w_{7,63}^{(3)} \oplus w_{7,64}^{(3)} \tag{7}$$

$$w_{22,57}^{(3)} = w_{7,58}^{(3)} \oplus w_{7,64}^{(3)} \oplus w_{7,1}^{(3)} \tag{8}$$

$$w_{22,63}^{(3)} = w_{7,64}^{(3)} \oplus w_{7,9}^{(3)} \tag{9}$$

4.3   Message Modification

Here message modification technique [33] can be used to modify the message words and chaining variables to satisfy the conditions of the middle steps to significantly improve the complexity of our attack.

For the middle steps (23 to 33) of the boomerang distinguisher, by modifying some certain message words and chaining variables, we can fulfill all the conditions of one side and part of conditions of the other side of the bottom path. After the message modification, the conditions of step 23 in the top differential path can hold with probability 1, and the conditions of steps 24 to 33 in the bottom can hold with probability at least 2^{-40}.

4.4   Sketch of the Attack

We divide our attack into two phases: the first phase is to find the right message words $w_{22}^{(1)}, \ldots, w_{32}^{(1)}$ and chaining variables $v_{23}^{(1)}$ so that the bottom paths of both sides in steps 23 to 33 hold; the second phase is to search $w_{17}^{(1)}, \ldots, w_{21}^{(1)}$ so that we can find a distinguisher quart. The sketch of attack is as follows.

1. Randomly select eleven 64-bit message words $w_i^{(1)}$ ($22 \leq i \leq 32$), and a 512-bit chaining variables $v_{23}^{(1)} = (a_{23}^{(1)}, b_{23}^{(1)}, \ldots, h_{23}^{(1)})$. Modify the messages $w_i^{(1)}$ ($22 \leq i \leq 32$) to meet the conditions in Table 6. Compute $v_i^{(1)}$ ($23 \leq i \leq 33$). Modify $v_{23}^{(1)}$ and $w_i^{(1)}$ ($22 \leq i \leq 32$) so that $v_i^{(1)}$ ($24 \leq i \leq 33$) satisfy all the conditions in Table 7.
2. Let $w_i^{(2)} = w_i^{(1)} \oplus \Delta w_i^{(1,2)}$, $w_i^{(3)} = w_i^{(1)} \oplus \Delta w_i^{(1,3)}$, $w_i^{(4)} = w_i^{(2)} \oplus \Delta w_i^{(1,3)}$ ($22 \leq i \leq 32$). The message differences $\Delta w_i^{(1,2)}$ and $\Delta w_i^{(1,3)}$ are defined in Table 4. Compute $v_i^{(j)}$ ($j = 2, 3, 4; 23 \leq i \leq 33$). Check whether $v_{33}^{(1)} \oplus v_{33}^{(3)} = v_{33}^{(2)} \oplus v_{33}^{(4)} = 0$. If yes, goto the next step. Otherwise, go back to step 1.

Table 5. Example of a quart satisfying $H(IV^{(3)}, M^{(3)}) - H(IV^{(1)}, M^{(1)}) - H(IV^{(4)}, M^{(4)}) + H(IV^{(2)}, M^{(2)}) = 0$ for 48 steps of the SHA-512 compression function.

$IV^{(1)}$	d51d68d22cd614bb	ad109f079123bc43	3e30194750de9356	b934d669f648b886
	2788083c8af206a4	f53a6844e79ca3ff	83333924f0fb45ee	aeca4ed80990f3c1
$IV^{(2)}$	551d68d22cd614bb	ad109f079123bc43	be30194750de9356	3936f6621588b886
	a788083c8af206a4	753a4844e79ca3ff	0333591cf87b45ee	2ec84edfead0f3c1
$IV^{(3)}$	3ca41aa7cc2ed702	d28a0787d13ece62	aaa0ccee378c5884	45960268826fa783
	126c152e3ed3c3d8	90227712dcb66469	c96f7308aa86be3c	5adecf0ca7c8cff9
$IV^{(4)}$	bca41aa7cc2ed702	d28a0787d13ece62	2aa0ccee378c5884	c5982260a1afa783
	926c152e3ed3c3d8	10225712dcb66469	496f9300b206be3c	dadccf148908cff9
$M^{(1)}$	7897cf7f1c02fa18	c0e30c69c197577d	f6016b4df4a5101b	44cf12bc7c5f7f89
	d28a43112a41160f	a481e26554edd575	8a4f5ecd8ee90f42	0c10896df299f0a3
	8bd715591505422b	82f9e09643a6f94e	8ae783224a988778	d858b794e8b95a4a
	d98d2e211f08b5e3	3185a2321c2013d0	493b7695ecb8bc63	40dde2bb03f050f7
$M^{(2)}$	7897cf7f1c02fa18	c0e30c69c197577d	f6016b4df4a5101b	44cf12bc7c5f7f89
	d28a43112a41160f	a481e26554edd575	8a4f5ecd8ee90f42	8c10896df299f0a3
	8bd715591505422b	82f9e09643a6f94e	8ae783224a988778	d858b794e8b95a4a
	d98d2e211f08b5e3	3185a2321c2013d0	493b7695ecb8bc63	40dde2bb03f050f7
$M^{(3)}$	2ec928b5e9b2bae2	da67703373f8f947	c4c2b463d9c34453	a4d359b70a54809d
	829416361d1acc84	49208682435343aa	8a4c7b5efe34b2e8	d6bcd7d0a70c5663
	ef5a6123cadba871	a134d5cebfae6e21	e32944037719f06e	81033c0b86b9f18e
	9d4d5849a78a6aa9	4634d6dd6a193ca7	783f014e5106c88e	bcd2f996a68b63f7
$M^{(4)}$	2ec928b5e9b2bae2	da67703373f8f947	c4c2b463d9c34453	a4d359b70a54809d
	829416361d1acc84	49208682435343aa	8a4c7b5efe34b2e8	56bcd7d0a70c5663
	ef5a6123cadba871	a134d5cebfae6e21	e32944037719f06e	81033c0b86b9f18e
	9d4d5849a78a6aa9	4634d6dd6a193ca7	783f014e5106c88e	bcd2f996a68b63f7

3. Select five 64-bit message words $w_i^{(1)}$ $(17 \leq i \leq 21)$ randomly. Let $w_i^{(2)} = w_i^{(1)} (17 \leq i \leq 21)$. Compute $w_i^{(1)}$ and $w_i^{(2)}$ $(33 \leq i \leq 47, 0 \leq i \leq 16)$ in forward and backward directions separately. Let $w_i^{(3)} = w_i^{(1)}$ and $w_i^{(4)} = w_i^{(2)}$ when $33 \leq i \leq 37$. Compute $w_i^{(3)}$ and $w_i^{(4)}$ when $38 \leq i \leq 47$ and $0 \leq i \leq 21$ by the message expansion.

4. Compute $v_{22}^{(j)}, v_{21}^{(j)}, \ldots, v_0^{(j)}$ $(j = 1, 2, 3, 4)$ in backward direction and $v_{34}^{(j)}$, $v_{35}^{(j)}, \ldots, v_{48}^{(j)}$ $(j = 1, 2, 3, 4)$ in forward direction. Check whether $v_0^{(2)} - v_0^{(1)} = v_0^{(4)} - v_0^{(3)}$ and $v_{48}^{(2)} - v_{48}^{(1)} = v_{48}^{(4)} - v_{48}^{(3)}$. If yes, output $w_i^{(j)}$ $(j = 1, 2, 3, 4; 0 \leq i \leq 15)$ and $v_1^{(j)}$ $(j = 1, 2, 3, 4)$. Otherwise, go to step 3.

4.5 Complexity of the Attack

Based on the two differential paths and the message modification technique, we construct a 48-step boomerang distinguisher for SHA-512 compression function.

The middle steps (23 to 33) of the boomerang distinguisher hold with probability 2^{-40}. Besides, the probability of steps 22 to 1 of the top differential path is about 2^{-45} and for steps 34 to 48 of the bottom path is 1. The probability of the message expansion is 2^{-6}. Hence, the complexity of the 48-step attack is $2^{40} + 2^{45} \times 2^6 \approx 2^{51}$ if we only get a zero-sum distinguisher, while the generic one is 2^{256}.

The practical complexity of our attack leads to a practical boomerang distinguisher on up to 48-step compression function of SHA-512, and we are able to find its corresponding boomerang quartets. An example of 48-step boomerang distinguisher for SHA-512 compression function is given in Table 5.

5 Conclusion

In this work, we propose two step-reduced differential paths with high probabilities of SHA-512 and build a boomerang distinguisher for the compression function of SHA-512 up to 48 steps out of 80 steps with practical complexity 2^{51}, and the example of boomerang quartet is aslo presented. Our attack is the best practical result on SHA-512 to date.

Appendix

Table 6. The message conditions in $w_{22}^{(1)} - w_{32}^{(1)}$.

Message	Conditions
$w_{22}^{(1)}$	$w_{22,15}^{(1)} = w_{22,14}^{(1)}$, $w_{22,44}^{(1)} = w_{22,43}^{(1)} \oplus w_{22,35}^{(1)} \oplus w_{22,34}^{(1)} \oplus w_{22,15}^{(1)} \oplus w_{22,14}^{(1)}$, $w_{22,48}^{(1)} = w_{22,47}^{(1)} \oplus w_{22,14}^{(1)} \oplus w_{22,15}^{(1)} \oplus w_{22,6}^{(1)} \oplus w_{22,5}^{(1)}$, $w_{22,57}^{(1)} = w_{22,56}^{(1)} \oplus 1$
$w_{23}^{(1)}$	$w_{23,41}^{(1)} = w_{22,33}^{(1)} \oplus w_{23,34}^{(1)} \oplus w_{23,40}^{(1)}$, $w_{23,42}^{(1)} = w_{23,40}^{(1)} \oplus w_{23,35}^{(1)} \oplus w_{23,34}^{(1)} \oplus 1$, $w_{23,48}^{(1)} = w_{23,40}^{(1)} \oplus w_{23,41}^{(1)} \oplus w_{23,47}^{(1)} \oplus 1$
$w_{24}^{(1)}$	$w_{24,2}^{(1)} = w_{22,21}^{(1)} \oplus w_{22,63}^{(1)} \oplus w_{22,8}^{(1)}$, $w_{24,37}^{(1)} = w_{22,57}^{(1)} \oplus w_{22,35}^{(1)} \oplus w_{22,44}^{(1)}$, $w_{24,37}^{(1)} = w_{22,57}^{(1)} \oplus w_{22,35}^{(1)} \oplus w_{22,44}^{(1)}$, $w_{24,44}^{(1)} = w_{22,63}^{(1)} \oplus w_{22,41}^{(1)} \oplus w_{22,50}^{(1)}$, $w_{24,50}^{(1)} = w_{22,6}^{(1)} \oplus w_{22,48}^{(1)} \oplus w_{22,57}^{(1)}$, $w_{24,57}^{(1)} = w_{22,12}^{(1)} \oplus w_{22,54}^{(1)} \oplus w_{22,63}^{(1)}$, $w_{24,59}^{(1)} = w_{22,15}^{(1)} \oplus w_{22,57}^{(1)}$
$w_{25}^{(1)}$	$w_{25,3}^{(1)} = w_{23,22}^{(1)} \oplus w_{23,64}^{(1)} \oplus w_{23,9}^{(1)}$, $w_{25,45}^{(1)} = w_{23,64}^{(1)} \oplus w_{23,42}^{(1)} \oplus w_{23,51}^{(1)}$, $w_{25,58}^{(1)} = w_{23,13}^{(1)} \oplus w_{23,55}^{(1)} \oplus w_{23,64}^{(1)}$
$w_{26}^{(1)}$	$w_{26,5}^{(1)} = w_{24,24}^{(1)} \oplus w_{24,2}^{(1)} \oplus w_{24,11}^{(1)}$, $w_{26,18}^{(1)} = w_{24,37}^{(1)} \oplus w_{24,15}^{(1)} \oplus w_{24,24}^{(1)}$, $w_{26,25}^{(1)} = w_{24,44}^{(1)} \oplus w_{24,22}^{(1)} \oplus w_{24,31}^{(1)}$, $w_{26,44}^{(1)} = w_{24,63}^{(1)} \oplus w_{24,41}^{(1)} \oplus w_{24,50}^{(1)}$, $w_{26,51}^{(1)} = w_{24,6}^{(1)} \oplus w_{24,48}^{(1)} \oplus w_{24,57}^{(1)}$, $w_{26,60}^{(1)} = w_{24,15}^{(1)} \oplus w_{24,57}^{(1)}$, $w_{26,62}^{(1)} = w_{24,17}^{(1)} \oplus w_{24,59}^{(1)}$
$w_{27}^{(1)}$	$w_{27,6}^{(1)} = w_{25,25}^{(1)} \oplus w_{25,3}^{(1)} \oplus w_{25,12}^{(1)}$, $w_{27,26}^{(1)} = w_{25,45}^{(1)} \oplus w_{25,23}^{(1)} \oplus w_{25,32}^{(1)}$, $w_{27,52}^{(1)} = w_{25,7}^{(1)} \oplus w_{25,49}^{(1)} \oplus w_{25,58}^{(1)}$, $w_{27,57}^{(1)} = w_{27,19}^{(1)} \oplus w_{27,39}^{(1)} \oplus w_{27,6}^{(1)} \oplus w_{24,44}^{(1)} \oplus 1$, $w_{27,61}^{(1)} = w_{25,16}^{(1)} \oplus w_{25,58}^{(1)}$

(*Continued*)

Table 6. (*Continued*)

$w_{28}^{(1)}$	$w_{28,1}^{(1)} = w_{26,20}^{(1)} \oplus w_{26,62}^{(1)} \oplus w_{26,7}^{(1)}$, $w_{28,6}^{(1)} = w_{26,25}^{(1)} \oplus w_{26,3}^{(1)} \oplus w_{26,12}^{(1)}$, $w_{28,8}^{(1)} = w_{26,27}^{(1)} \oplus w_{26,5}^{(1)} \oplus w_{26,14}^{(1)}$, $w_{28,12}^{(1)} = w_{26,31}^{(1)} \oplus w_{26,9}^{(1)} \oplus w_{26,18}^{(1)}$, $w_{28,19}^{(1)} = w_{26,38}^{(1)} \oplus w_{26,16}^{(1)} \oplus w_{26,25}^{(1)}$, $w_{28,21}^{(1)} = w_{26,40}^{(1)} \oplus w_{26,18}^{(1)} \oplus w_{26,27}^{(1)}$, $w_{28,25}^{(1)} = w_{26,44}^{(1)} \oplus w_{26,22}^{(1)} \oplus w_{26,31}^{(1)}$, $w_{28,28}^{(1)} = w_{26,47}^{(1)} \oplus w_{26,25}^{(1)} \oplus w_{26,34}^{(1)}$, $w_{28,32}^{(1)} = w_{26,51}^{(1)} \oplus w_{26,29}^{(1)} \oplus w_{26,38}^{(1)}$, $w_{28,38}^{(1)} = w_{26,57}^{(1)} \oplus w_{26,35}^{(1)} \oplus w_{26,44}^{(1)}$, $w_{28,41}^{(1)} = w_{26,60}^{(1)} \oplus w_{26,38}^{(1)} \oplus w_{26,47}^{(1)}$, $w_{28,43}^{(1)} = w_{26,62}^{(1)} \oplus w_{26,40}^{(1)} \oplus w_{26,49}^{(1)}$, $w_{28,45}^{(1)} = w_{26,64}^{(1)} \oplus w_{26,42}^{(1)} \oplus w_{26,51}^{(1)}$, $w_{28,47}^{(1)} = w_{26,2}^{(1)} \oplus w_{26,44}^{(1)} \oplus w_{26,53}^{(1)}$, $w_{28,50}^{(1)} = w_{26,5}^{(1)} \oplus w_{26,47}^{(1)} \oplus w_{26,56}^{(1)}$, $w_{28,56}^{(1)} = w_{26,11}^{(1)} \oplus w_{26,53}^{(1)} \oplus w_{26,62}^{(1)}$
$w_{29}^{(1)}$	$w_{29,7}^{(1)} = w_{27,26}^{(1)} \oplus w_{27,4}^{(1)} \oplus w_{27,13}^{(1)}$, $w_{29,9}^{(1)} = w_{27,28}^{(1)} \oplus w_{27,6}^{(1)} \oplus w_{27,15}^{(1)}$, $w_{29,14}^{(1)} = w_{22,56}^{(1)} \oplus w_{24,59}^{(1)}$, $w_{29,15}^{(1)} = w_{22,57}^{(1)} \oplus w_{27,59}^{(1)} \oplus 1$, $w_{29,20}^{(1)} = w_{27,39}^{(1)} \oplus w_{27,17}^{(1)} \oplus w_{27,26}^{(1)}$, $w_{29,21}^{(1)} = w_{22,63}^{(1)} \oplus w_{24,2}^{(1)} \oplus w_{29,8}^{(1)} \oplus 1$, $w_{29,29}^{(1)} = w_{27,48}^{(1)} \oplus w_{27,26}^{(1)} \oplus w_{27,35}^{(1)}$, $w_{29,33}^{(1)} = w_{27,52}^{(1)} \oplus w_{27,30}^{(1)} \oplus w_{27,39}^{(1)}$, $w_{29,42}^{(1)} = w_{27,61}^{(1)} \oplus w_{27,39}^{(1)} \oplus w_{27,48}^{(1)}$, $w_{29,43}^{(1)} = w_{22,56}^{(1)} \oplus w_{24,37}^{(1)} \oplus w_{29,34}^{(1)}$, $w_{29,44}^{(1)} = w_{22,56}^{(1)} \oplus w_{29,34}^{(1)} \oplus w_{29,43}^{(1)} \oplus w_{22,57}^{(1)} \oplus w_{29,35}^{(1)} \oplus 1$, $w_{29,46}^{(1)} = w_{27,1}^{(1)} \oplus w_{27,43}^{(1)} \oplus w_{27,52}^{(1)}$, $w_{29,47}^{(1)} = w_{22,56}^{(1)} \oplus w_{24,50}^{(1)} \oplus w_{29,5}^{(1)}$, $w_{29,48}^{(1)} = w_{22,56}^{(1)} \oplus w_{29,6}^{(1)} \oplus w_{22,57}^{(1)} \oplus w_{29,5}^{(1)} \oplus w_{29,47}^{(1)}$, $w_{29,50}^{(1)} = w_{24,44}^{(1)} \oplus w_{29,41}^{(1)} \oplus w_{22,63}^{(1)}$, $w_{29,51}^{(1)} = w_{27,6}^{(1)} \oplus w_{27,48}^{(1)} \oplus w_{27,57}^{(1)}$, $w_{29,54}^{(1)} = w_{24,57}^{(1)} \oplus w_{29,12}^{(1)} \oplus w_{22,63}^{(1)}$, $w_{29,55}^{(1)} = w_{24,57}^{(1)} \oplus w_{29,13}^{(1)} \oplus w_{27,19}^{(1)} \oplus w_{27,61}^{(1)} \oplus 1$, $w_{29,56}^{(1)} = w_{22,56}^{(1)}$, $w_{29,57}^{(1)} = w_{22,57}^{(1)}$, $w_{29,58}^{(1)} = w_{29,6}^{(1)} \oplus w_{29,48}^{(1)} \oplus w_{29,57}^{(1)} \oplus w_{29,7}^{(1)} \oplus w_{29,49}^{(1)} \oplus 1$, $w_{29,63}^{(1)} = w_{22,63}^{(1)}$, $w_{29,64}^{(1)} = w_{27,19}^{(1)} \oplus w_{27,61}^{(1)}$
$w_{30}^{(1)}$	$w_{30,4}^{(1)} = w_{28,23}^{(1)} \oplus w_{28,1}^{(1)} \oplus w_{28,10}^{(1)}$, $w_{30,11}^{(1)} = w_{28,30}^{(1)} \oplus w_{28,8}^{(1)} \oplus w_{28,17}^{(1)}$, $w_{30,22}^{(1)} = w_{28,41}^{(1)} \oplus w_{28,19}^{(1)} \oplus w_{28,28}^{(1)}$, $w_{30,32}^{(1)} = w_{28,51}^{(1)} \oplus w_{28,29}^{(1)} \oplus w_{28,38}^{(1)}$, $w_{30,33}^{(1)} = w_{30,11}^{(1)} \oplus w_{30,20}^{(1)} \oplus w_{30,32}^{(1)} \oplus w_{30,10}^{(1)} \oplus w_{30,19}^{(1)} \oplus 1$, $w_{30,39}^{(1)} = w_{28,58}^{(1)} \oplus w_{28,36}^{(1)} \oplus w_{28,45}^{(1)}$, $w_{30,42}^{(1)} = w_{25,45}^{(1)} \oplus w_{25,3}^{(1)} \oplus w_{30,22}^{(1)} \oplus w_{30,9}^{(1)} \oplus w_{28,6}^{(1)} \oplus w_{28,48}^{(1)} \oplus w_{28,57}^{(1)}$, $w_{30,45}^{(1)} = w_{30,23}^{(1)} \oplus w_{30,32}^{(1)} \oplus w_{30,44}^{(1)} \oplus w_{30,22}^{(1)} \oplus w_{30,31}^{(1)} \oplus 1$, $w_{30,48}^{(1)} = w_{28,3}^{(1)} \oplus w_{28,45}^{(1)} \oplus w_{28,54}^{(1)}$, $w_{30,51}^{(1)} = w_{28,6}^{(1)} \oplus w_{28,48}^{(1)} \oplus w_{28,57}^{(1)}$, $w_{30,52}^{(1)} = w_{30,30}^{(1)} \oplus w_{30,39}^{(1)} \oplus w_{30,51}^{(1)} \oplus w_{30,29}^{(1)} \oplus w_{30,38}^{(1)}$, $w_{30,54}^{(1)} = w_{30,32}^{(1)} \oplus w_{30,41}^{(1)} \oplus w_{30,51}^{(1)} \oplus w_{30,29}^{(1)} \oplus w_{30,38}^{(1)} \oplus 1$, $w_{30,57}^{(1)} = w_{28,12}^{(1)} \oplus w_{28,54}^{(1)} \oplus w_{28,63}^{(1)}$, $w_{30,59}^{(1)} = w_{28,14}^{(1)} \oplus w_{28,56}^{(1)}$, $w_{30,64}^{(1)} = w_{25,3}^{(1)} \oplus w_{30,22}^{(1)} \oplus w_{30,9}^{(1)} \oplus 1$
$w_{31}^{(1)}$	$w_{31,12}^{(1)} = w_{29,31}^{(1)} \oplus w_{29,9}^{(1)} \oplus w_{29,18}^{(1)}$, $w_{31,23}^{(1)} = w_{29,42}^{(1)} \oplus w_{29,20}^{(1)} \oplus w_{29,29}^{(1)}$, $w_{31,40}^{(1)} = w_{29,59}^{(1)} \oplus w_{29,37}^{(1)} \oplus w_{29,46}^{(1)}$, $w_{31,49}^{(1)} = w_{29,4}^{(1)} \oplus w_{29,46}^{(1)} \oplus w_{29,55}^{(1)}$
$w_{32}^{(1)}$	$w_{32,5}^{(1)} = w_{30,24}^{(1)} \oplus w_{30,2}^{(1)} \oplus w_{30,11}^{(1)}$, $w_{32,7}^{(1)} = w_{30,26}^{(1)} \oplus w_{30,4}^{(1)} \oplus w_{30,13}^{(1)}$, $w_{32,13}^{(1)} = w_{30,33}^{(1)} \oplus w_{30,11}^{(1)} \oplus w_{30,20}^{(1)}$, $w_{32,16}^{(1)} = w_{30,35}^{(1)} \oplus w_{30,13}^{(1)} \oplus w_{30,22}^{(1)}$, $w_{32,20}^{(1)} = w_{30,39}^{(1)} \oplus w_{30,17}^{(1)} \oplus w_{30,26}^{(1)}$, $w_{32,25}^{(1)} = w_{30,45}^{(1)} \oplus w_{30,23}^{(1)} \oplus w_{30,32}^{(1)}$, $w_{32,29}^{(1)} = w_{30,48}^{(1)} \oplus w_{30,26}^{(1)} \oplus w_{30,35}^{(1)}$, $w_{32,32}^{(1)} = w_{30,51}^{(1)} \oplus w_{30,29}^{(1)} \oplus w_{30,38}^{(1)} \oplus 1$, $w_{32,34}^{(1)} = w_{30,51}^{(1)} \oplus w_{30,29}^{(1)} \oplus w_{30,38}^{(1)} \oplus 1$, $w_{32,38}^{(1)} = w_{30,57}^{(1)} \oplus w_{30,35}^{(1)} \oplus w_{30,44}^{(1)}$, $w_{32,40}^{(1)} = w_{30,59}^{(1)} \oplus w_{30,37}^{(1)} \oplus w_{30,46}^{(1)}$, $w_{32,47}^{(1)} = w_{23,47}^{(1)} \oplus 1$, $w_{32,49}^{(1)} = w_{30,4}^{(1)} \oplus w_{30,46}^{(1)} \oplus w_{30,55}^{(1)}$, $w_{32,53}^{(1)} = w_{30,8}^{(1)} \oplus w_{30,50}^{(1)} \oplus w_{30,59}^{(1)}$, $w_{32,54}^{(1)} = w_{30,9}^{(1)} \oplus w_{30,51}^{(1)} \oplus w_{30,60}^{(1)}$, $w_{32,56}^{(1)} = w_{30,11}^{(1)} \oplus w_{30,53}^{(1)} \oplus w_{30,62}^{(1)}$, $w_{32,58}^{(1)} = w_{27,58}^{(1)}$, $w_{32,60}^{(1)} = w_{30,15}^{(1)} \oplus w_{30,57}^{(1)}$, $w_{32,62}^{(1)} = w_{30,17}^{(1)} \oplus w_{30,59}^{(1)}$

Table 7. The conditions of chaining variables in the middle steps.

Steps	Conditions
23	$a_{23,56}^{(1)} = w_{22,56}^{(1)} \oplus 1$, $a_{23,62}^{(1)} = w_{22,63}^{(1)} \oplus a_{23,3}^{(1)} \oplus w_{22,56}^{(1)} \oplus a_{23,4}^{(1)} \oplus a_{23,57}^{(1)}$, $a_{23,63}^{(1)} = w_{22,63}^{(1)}$
	$b_{23,41}^{(1)} = a_{23,41}^{(1)}$
	$c_{23,2}^{(1)} = a_{23,2}^{(1)}$, $c_{23,7}^{(1)} = a_{23,7}^{(1)}$, $c_{23,13}^{(1)} = a_{23,13}^{(1)}$, $c_{23,23}^{(1)} = a_{23,23}^{(1)}$, $c_{23,27}^{(1)} = a_{23,27}^{(1)}$, $c_{23,63}^{(1)} = b_{23,63}^{(1)} \oplus 1$, $c_{23,64}^{(1)} = a_{23,64}^{(1)}$
	$e_{23,13}^{(1)} = b_{23,13}^{(1)} \oplus 1$, $e_{23,56}^{(1)} = w_{22,56}^{(1)} \oplus 1$, $e_{23,63}^{(1)} = w_{22,63}^{(1)}$
	$f_{23,2}^{(1)} = b_{23,2}^{(1)} \oplus 1$, $f_{23,7}^{(1)} = b_{23,7}^{(1)} \oplus 1$
	$g_{23,41}^{(1)} = c_{23,41}^{(1)}$, $g_{23,63}^{(1)} = f_{23,63}^{(1)}$
24	$a_{24,2}^{(1)} = b_{24,2}^{(1)}$, $a_{24,7}^{(1)} = b_{24,7}^{(1)}$, $a_{24,13}^{(1)} = b_{24,13}^{(1)}$, $a_{24,23}^{(1)} = b_{24,23}^{(1)}$, $a_{24,27}^{(1)} = b_{24,27}^{(1)}$, $a_{24,41}^{(1)} = b_{24,41}^{(1)}$, $a_{24,63}^{(1)} = a_{23,63}^{(1)} \oplus 1$, $a_{24,64}^{(1)} = b_{24,64}^{(1)}$
	$e_{24,2}^{(1)} = 1$, $e_{24,5}^{(1)} = 0$, $e_{24,7}^{(1)} = 1$, $e_{24,13}^{(1)} = 0$
25	$a_{25,41}^{(1)} = h_{24,41}^{(1)}$, $a_{25,36}^{(1)} = a_{25,30}^{(1)} \oplus h_{24,41}^{(1)} \oplus h_{25,2}^{(1)} \oplus 1$, $a_{25,46}^{(1)} = a_{25,35}^{(1)} \oplus h_{24,41}^{(1)} \oplus h_{25,7}^{(1)} \oplus 1$, $a_{25,52}^{(1)} = a_{25,47}^{(1)} \oplus h_{24,41}^{(1)} \oplus d_{25,13}^{(1)}$
	$e_{25,5}^{(1)} = 1$, $e_{25,13}^{(1)} = 0$, $e_{25,23}^{(1)} = f_{25,23}^{(1)} \oplus 1$, $e_{25,27}^{(1)} = f_{25,27}^{(1)}$, $e_{25,64}^{(1)} = f_{25,64}^{(1)}$
26	$e_{26,23}^{(1)} = d_{25,23}^{(1)}$, $e_{26,27}^{(1)} = d_{25,27}^{(1)}$, $e_{26,41}^{(1)} = c_{26,41}^{(1)}$, $e_{26,46}^{(1)} = e_{26,23}^{(1)} \oplus e_{26,19}^{(1)} \oplus e_{23,5}^{(1)} \oplus 1$, $e_{26,54}^{(1)} = e_{26,31}^{(1)} \oplus e_{26,27}^{(1)} \oplus e_{23,13}^{(1)} \oplus 1$, $e_{26,64}^{(1)} = e_{26,37}^{(1)} \oplus e_{26,41}^{(1)} \oplus e_{26,23}^{(1)} \oplus g_{26,23}^{(1)} \oplus 1$
27	$e_{27,23}^{(1)} = 0$, $e_{27,27}^{(1)} = 0$, $e_{27,64}^{(1)} = 0$
	$a_{27,41}^{(1)} = b_{27,41}^{(1)}$
28	$e_{28,23}^{(1)} = 1$, $e_{28,27}^{(1)} = 1$, $e_{28,41}^{(1)} = f_{28,41}^{(1)}$, $e_{28,64}^{(1)} = 1$
29	$e_{29,41}^{(1)} = d_{28,41}^{(1)}$, $e_{29,45}^{(1)} = e_{29,41}^{(1)} \oplus e_{29,4}^{(1)} \oplus h_{29,27}^{(1)} \oplus 1$, $e_{29,18}^{(1)} = e_{29,14}^{(1)} \oplus d_{28,41}^{(1)} \oplus h_{29,64}^{(1)} \oplus 1$, $e_{29,64}^{(1)} = e_{29,37}^{(1)} \oplus d_{28,41}^{(1)} \oplus h_{29,23}^{(1)} \oplus 1$
30	$e_{30,41}^{(1)} = 0$
31	$e_{31,41}^{(1)} = 1$

References

1. Aoki, K., Guo, J., Matusiewicz, K., Sasaki, Y., Wang, L.: Preimages for step-reduced SHA-2. In: Matsui, M. (ed.) ASIACRYPT 2009. LNCS, vol. 5912, pp. 578–597. Springer, Heidelberg (2009)
2. Aumasson, J.-P., Meier, W.: Zero-sum Distinguishers for Reduced Keccak-f and for the Core Functions of Luffa and Hamsi (2009). http://131002.net/data/papers/AM09.pdf
3. Bai, D., Yu, H., Wang, G., Wang, X.: Improved boomerang attacks on SM3. In: Boyd, C., Simpson, L. (eds.) ACISP 2013. LNCS, vol. 7959, pp. 251–266. Springer, Heidelberg (2013)
4. Biham, E., Dunkelman, O., Keller, N.: The rectangle attack - rectangling the serpent. In: Pfitzmann, B. (ed.) EUROCRYPT 2001. LNCS, vol. 2045, pp. 340–357. Springer, Heidelberg (2001)

5. Biham, E., Dunkelman, O., Keller, N.: Related-key boomerang and rectangle attacks. In: Cramer, R. (ed.) EUROCRYPT 2005. LNCS, vol. 3494, pp. 507–525. Springer, Heidelberg (2005)
6. Biryukov, A., Nikolić, I., Roy, A.: Boomerang attacks on BLAKE-32. In: Joux, A. (ed.) FSE 2011. LNCS, vol. 6733, pp. 218–237. Springer, Heidelberg (2011)
7. Biryukov, A., Lamberger, M., Mendel, F., Nikolić, I.: Second-order differential collisions for reduced SHA-256. In: Lee, D.H., Wang, X. (eds.) ASIACRYPT 2011. LNCS, vol. 7073, pp. 270–287. Springer, Heidelberg (2011)
8. Eichlseder, A., Mendel F., Schläffer, M.: Branching heuristics in differential collision search with applications to SHA-512. In: FSE 2014 (accepted paper)
9. Leurent, G., Roy, A.: Boomerang attacks on hash function using auxiliary differentials. In: Dunkelman, O. (ed.) CT-RSA 2012. LNCS, vol. 7178, pp. 215–230. Springer, Heidelberg (2012)
10. Guo, J., Ling, S., Rechberger, C., Wang, H.: Advanced meet-in-the-middle preimage attacks: first results on full tiger, and improved results on MD4 and SHA-2. In: Abe, M. (ed.) ASIACRYPT 2010. LNCS, vol. 6477, pp. 56–75. Springer, Heidelberg (2010)
11. Indesteege, S., Mendel, F., Preneel, B., Rechberger, C.: Collisions and other nonrandom properties for step-reduced SHA-256. In: Avanzi, R.M., Keliher, L., Sica, F. (eds.) SAC 2008. LNCS, vol. 5381, pp. 276–293. Springer, Heidelberg (2009)
12. Isobe, T., Shibutani, K.: Preimage attacks on reduced tiger and SHA-2. In: Dunkelman, O. (ed.) FSE 2009. LNCS, vol. 5665, pp. 139–155. Springer, Heidelberg (2009)
13. Joux, A., Peyrin, T.: Hash functions and the (amplified) boomerang attack. In: Menezes, A. (ed.) CRYPTO 2007. LNCS, vol. 4622, pp. 244–263. Springer, Heidelberg (2007)
14. Kelsey, J., Kohno, T., Schneier, B.: Amplified boomerang attacks against reducedround MARS and serpent. In: Schneier, B. (ed.) FSE 2000. LNCS, vol. 1978, pp. 75–93. Springer, Heidelberg (2001)
15. Khovratovich, D., Rechberger, C., Savelieva, A.: Bicliques for preimages: attacks on Skein-512 and the SHA-2 family. In: Canteaut, A. (ed.) FSE 2012. LNCS, vol. 7549, pp. 244–263. Springer, Heidelberg (2012)
16. Kircanski, A., Shen, Y., Wang, G., Youssef, A.M.: Boomerang and slide-rotational analysis of the SM3 hash function. In: Knudsen, L.R., Wu, H. (eds.) SAC 2012. LNCS, vol. 7707, pp. 304–320. Springer, Heidelberg (2013)
17. Lamberger, M., Mendel, F., Higher-Order Differential Attack on Reduced SHA-256. Cryptology ePrint Archive: Report 2011/037 (2011)
18. Mendel, F., Nad, T.: Boomerang distinguisher for the SIMD-512 compression function. In: Bernstein, D.J., Chatterjee, S. (eds.) INDOCRYPT 2011. LNCS, vol. 7107, pp. 255–269. Springer, Heidelberg (2011)
19. Mendel, F., Nad, T., Schläffer, M.: Finding SHA-2 characteristics: searching through a minefield of contradictions. In: Lee, D.H., Wang, X. (eds.) ASIACRYPT 2011. LNCS, vol. 7073, pp. 288–307. Springer, Heidelberg (2011)
20. Mendel, F., Nad, T., Schläffer, M.: Improving local collisions: new attacks on reduced SHA-256. In: Johansson, T., Nguyen, P.Q. (eds.) EUROCRYPT 2013. LNCS, vol. 7881, pp. 262–278. Springer, Heidelberg (2013)
21. Mendel, F., Pramstaller, N., Rechberger, C., Rijmen, V.: Analysis of step-reduced SHA-256. In: Robshaw, M. (ed.) FSE 2006. LNCS, vol. 4047, pp. 126–143. Springer, Heidelberg (2006)
22. Menezes, A.J., van Oorschot, P.C., Vanstone, S.A.: Handbook of Applied Cryptography. CRC Press, Boca Raton (1997)

23. National Institute of Standards and Technology: Announcing Request for Candidate Algorithm Nominations for a New Cryptographic Hash Algorithm (SHA-3) Family. Federal Register **27**(212), 62212–62220 (2007)
24. National Institute of Standards and Technology: FIPS PUB 180–3: Secure Hash Standard. Federal Information Processing Standards Publication 180–3, U.S. Department of Commerce, October 2008
25. Nikolić, I., Biryukov, A.: Collisions for step-reduced SHA-256. In: Nyberg, K. (ed.) FSE 2008. LNCS, vol. 5086, pp. 1–15. Springer, Heidelberg (2008)
26. Sanadhya, S.K., Sarkar, P.: New collision attacks against up to 24-step SHA-2. In: Chowdhury, D.R., Rijmen, V., Das, A. (eds.) INDOCRYPT 2008. LNCS, vol. 5365, pp. 91–103. Springer, Heidelberg (2008)
27. Sasaki, Y.: Boomerang distinguishers on MD4-family: first practical results on full 5-Pass HAVAL. In: Miri, A., Vaudenay, S. (eds.) SAC 2011. LNCS, vol. 7118, pp. 1–18. Springer, Heidelberg (2012)
28. Sasaki, Y., Wang, L.: 2-Dimension Sums: Distinguishers Beyond Three Rounds of RIPEMD-128 and RIPEMD-160, February 2012. http://eprint.iacr.org/2012/049.pdf
29. Sasaki, Y., Wang, L., Takasaki, Y., Sakiyama, K., Ohta, K.: Boomerang distinguishers for full HAS-160 compression function. In: Hanaoka, G., Yamauchi, T. (eds.) IWSEC 2012. LNCS, vol. 7631, pp. 156–169. Springer, Heidelberg (2012)
30. Wagner, D.: A generalized birthday problem. In: Yung, M. (ed.) CRYPTO 2002. LNCS, vol. 2442, pp. 288–303. Springer, Heidelberg (2002)
31. Wagner, D.: The boomerang attack. In: Knudsen, L.R. (ed.) FSE 1999. LNCS, vol. 1636, pp. 156C–169. Springer, Heidelberg (1999)
32. Wang, X., Yin, Y.L., Yu, H.: Finding collisions in the full SHA-1. In: Shoup, V. (ed.) CRYPTO 2005. LNCS, vol. 3621, pp. 17–36. Springer, Heidelberg (2005)
33. Wang, X., Yu, H.: How to break MD5 and other hash functions. In: Cramer, R. (ed.) EUROCRYPT 2005. LNCS, vol. 3494, pp. 19–35. Springer, Heidelberg (2005)
34. Wang, X., Yu, H.: Non-randomness of 39-step SHA-256. Presented at rump session of EUROCRYPT (2008)
35. Yu, H., Chen, J., Wang, X.: The boomerang attacks on the round-reduced skein-512. In: Knudsen, L.R., Wu, H. (eds.) SAC 2012. LNCS, vol. 7707, pp. 287–303. Springer, Heidelberg (2013)

Collision Attack on 4-Branch, Type-2 GFN Based Hash Functions Using Sliced Biclique Cryptanalysis Technique

Megha Agrawal, Donghoon Chang, Mohona Ghosh$^{(\boxtimes)}$,
and Somitra Kumar Sanadhya

Indraprastha Institute of Information Technology, Delhi (IIIT-D), Delhi, India
{meghaa,donghoon,mohonag,somitra}@iiitd.ac.in

Abstract. In this work, we apply the sliced biclique cryptanalysis technique to show 8-round collision attack on a hash function H based on 4-branch, Type-2 Generalized Feistel Network (Type-2 GFN). This attack is generic and works on 4-branch, Type-2 GFN with any parameters including the block size, type of round function, the number of S-boxes in each round and the number of SP layers inside the round function. We first construct a 8-round distinguisher on 4-branch, Type-2 GFN and then use this distinguisher to launch 8-round collision attack on *compression functions* based on Matyas-Meyer-Oseas (MMO) and Miyaguchi-Preneel (MP) modes. The complexity of the attack on 128-bit compression function is 2^{56}. The attack can be directly translated to collision attack on MP and MMO based *hash functions* and pseudo-collision attack on Davies-Meyer (DM) based *hash functions*. When the round function F is instantiated with double SP layer, we show the first 8 round collision attack on 4-branch, Type-2 GFN with double SP layer based compression function. The previous best attack on this structure was a 6-round near collision attack shown by Sasaki at Indocrypt'12. His attack cannot be used to generate full collisions on 6-rounds and hence our result can be regarded the best so far in literature on this structure.

Keywords: Sliced biclique cryptanalysis · Hash functions · Collision attack · Generalized Feistel Network · Double SP layer

1 Introduction

Feistel structure is one of the basic building blocks of block ciphers and block ciphers based constructions. A Feistel network divides the input message into two sub-blocks (or two branches). Generalized Feistel Networks (GFN) are variants of Feistel networks with more than two branches, i.e., a k-branch GFN partitions the input message into k sub-blocks. They are sometimes favored over traditional Feistel scheme due to their high parallelism, simple design and suitability for low cost implementations. Many types of generalized Feistel schemes have been proposed and studied by researchers, e.g., unbalanced Feistel network [26],

© Springer International Publishing Switzerland 2015
D. Lin et al. (Eds.): Inscrypt 2014, LNCS 8957, pp. 343–360, 2015.
DOI: 10.1007/978-3-319-16745-9_19

alternating Feistel Network [2], type-1, type-2 and type-3 Feistel network [34] etc. Type-2, GFN in particular has seen wide adoption in well known block ciphers such as RC6 [23], SHAvite3 [3], CLEFIA [28], HIGHT [13] etc. Security analysis of generalized Feistel network [4,11,27,30,32] has been an active area of research for past many years. In fact, a comprehensive study done by Bogdanov et al. in [6] suggests that Type-2 GFN and its variants are more robust and secure against differential and linear cryptanalysis as compared to Type-1 GFN. Hence, we choose Type-2 GFN (shown in Fig. 1) as the basis for our study.

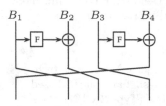
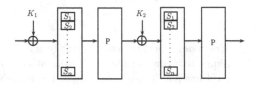

Fig. 1. 4 branch, Type-2 Generalized Feistel Structure with right cyclic shift.

Fig. 2. Double SP Function.

Biclique cryptanalysis technique has garnered considerable interest amongst cryptographic community in the past couple of years. This approach, which is a variant of meet-in-the-middle attack, was first introduced by Khovratovich et al. in [16] for preimage attack on hash functions Skein and SHA-2. The concept was taken over by Bogdanov et al. [5] to successfully cryptanalyze full round AES and has been subsequently adopted to break many other block ciphers such as ARIA [33], SQUARE [19], TWINE [7], HIGHT [12], PRESENT [1] etc. All these biclique related attacks are carried out under the "unknown key settings" where the key used is unknown to the attacker and the main motive is to recover the secret key. However, this may not always be the case. Particularly, in the case of block cipher based hash modes such as Matyas-Meyer-Oseas (MMO) and Miyuguchi-Preneel (MP), initial vector IV (which acts as the key to the underlying block cipher) is a fixed public constant assumed to be known apriori to the attacker. Such scenarios are called "known key settings" in the attack model. Under such conditions, the aim of the attacker is to find a property which distinguishes known key instantiations of target block cipher from random permutations [17,20]. These settings are considered much stronger from the attacker's point of view since he unwillingly loses some degree of freedom reducing chances of carrying out actual generic attacks such as finding full collisions. Until recently, most of the collision attacks on hash functions under MMO and MP modes were restricted to variants of generic attack such as pseudo-collisions [18] and near collisions [29]. In [15], Khovratovich used biclique technique to mount actual collision and preimage attacks on Grøstl and Skein under known key settings. He proposed a variant of classical biclique technique used in [5] to carry out his attack. He termed this variant as *sliced biclique* technique (details of which are discussed in Sect. 3.3). Though the results of this work are quite interesting, yet they have not been studied further. Although the security of GFN have been studied earlier under known key settings [8,9,14,24,25], all these previous studies have utilized rebound attack technique [21] for their cryptanalysis. These factors motivated us to investigate the use of sliced biclique framework to study Type-2, GFN based constructions under known key settings.

It is generally desired that round function F inside a generalized Feistel network should provide good diffusion and confusion properties. This is often realized by implementing F as a *substitution-permutation network* (nonlinear S-box transformation followed by linear permutation) as part of the round function design. There is a general belief that increasing the number of active S-boxes provides more security against certain attacks. In [6], Bogdanov and Shibutani stressed on the importance of double SP (substitution-permutation) layers in the round function of Feistel networks as opposed to the single SP layer in the traditional design. They analyzed several designs such as single SP, double SP, SPS (substitution-permutation-substitution) and multiple SP layers and showed that double SP (shown in Fig. 2) layer achieves maximum security with respect to proportion of active S-boxes in all S-boxes involved against differential and linear cryptanalysis. They especially compared double SP structure with single SP and showed that for Type-1 and Type-2 GFNs, proportion of linearly and differentially active S-boxes in double SP instantiations is 50 % and 33 % higher respectively as compared to the single SP instantiation. Their research advocated a possibility of designing more efficient and secure block cipher based constructions using double SP layer. In [24], Sasaki presented a 7-round distinguisher attack on 4-branch, type-2 GFN with double SP layer and a 6-round near collision attack on the compression function based on the same structure. Kumar et al. [8] further improved the distinguishing attack on 4-branch, type-2 GFN with double SP layer by showing a 8-round distinguisher for the same. However, the form of truncated differential trails followed in [8,24] cannot be used to launch collision attack when the above GFN structure is instantiated in compression function modes under known key settings.

Our Contributions. The main contributions of this work are as follows:

1. We apply sliced biclique technique to construct a 8-round distinguisher on 4-branch, Type-2 Generalized Feistel Network.
2. We use the distinguisher so constructed to demonstrate a 8-round collision attack on 4-branch, Type-2 GFN based compression functions (in MMO and MP mode) under known key settings with a complexity of 2^{56} (on 128-bit hash output). The attack can be directly translated to collision attacks on MMO and MP mode based hash functions and pseudo-collision attacks on Davies-Meyer (DM) mode based hash functions.
3. When the round function F is instantiated with double SP layer, we demonstrate the first 8-round collision attack on 4-branch, Type-2 GFN with double SP layer.

4. We investigate CLEFIA which is a real world-implementation of 4-branch, Type-2 GFN and demonstrate an 8-round collision attack on CLEFIA based hash function with a complexity of 2^{56}.

The paper is organized as follows. In Sect. 2 we give the notations used in our paper followed by Sect. 3 which explains the important preliminaries. In Sect. 4, we present our distinguishing attack on 8 rounds of 4 branch, Type-2 GFN under fixed key settings. We use this distinguishing attack to show collision attack on 4-branch, Type-2 GFN based compression function in Sect. 5 followed by extension of this attack to hash functions in Sect. 6. Finally in Sect. 7, we summarize and conclude our work. The collision attack on CLEFIA based hash function is discussed in Appendix A.

2 Notation

We consider 4-branch, type-2, generalized Feistel network for our attack. Following notation is followed in the rest of the paper.

N : Input message size (in bits)
n : Message word size (in bits) which is input to each branch, i.e., $n = N/4$
$R : Round R
$R_p : p^{th} word in round R. Each round has 4 words corresponding to 4 partitions of 4-branch GFN, i.e., $1 \leq p \leq 4$
$R_p^l : l^{th} block of word p in round R

3 Preliminaries

In this section, we give a brief overview of the key concepts used in our cryptanalysis technique to facilitate better understanding.

3.1 Type-2 Generalized Feistel Network (GFN) Instantiated with Double SP Layer

One round of Type-2 GFN is shown in Fig. 3. A GFN with 4 branches divides the input B into four equal parts $[B_1, B_2, B_3, B_4]$. A round of Type-2 GFN with left cyclic shift outputs $[F(B_1) \oplus B_2, B_3, F(B_3) \oplus B_4, B_1]$ for some keyed nonlinear function F [6]. On the other hand, a round of Type-2 GFN with right cyclic shift outputs $[F(B_3) \oplus B_4, B_1, F(B_1) \oplus B_2, B_3]$ (shown in Fig. 1) for round function F.

The round transformation function F when defined by non-linear S-box layer followed by a permutation layer P exhibits substitution permutation structure. The permutation P is generally implemented using standard MDS matrix [22,31].

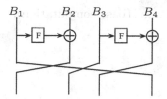

Fig. 3. 4-branch, Type-2, Generalized Feistel Network with left cyclic shift.

If this SP structure is applied twice one after another then it is called double SP, as shown in Fig. 2. Few reasons favoring double SP over single SP function are as follows [6]:

- The second S-box in double SP provides larger number of active S boxes when differential and linear attacks are applied.
- The second permutation layer in double SP structure limits the differential effect, i.e., number of differential trails resulting in same differential is smaller as compared to round function having single permutation layer.

3.2 t-bit Partial Target Preimage Attack

Let the output of a hash function H with initial chaining value IV and message M be denoted by h, i.e., $h = \mathrm{H}(IV, M)$. In this attack, when the attacker is given t-bits of h, his aim is to find a message M' such that the hash output $h' = \mathrm{H}(IV, M')$ matches these t-bits of h and at the same positions. The other bits of hash output $\mathrm{H}(IV, M')$ are generated randomly.

3.3 Sliced Biclique Cryptanalysis

In this section, we describe sliced biclique cryptanalysis technique to show preimage attack on hash function. Later, we use this preimage attack to launch collision attack. We consider MMO mode for our explanation. In the MMO mode $H = E_{IV}(M) \oplus M$, where IV is the initial chaining value acting as the key for the block cipher E, M is the message and H is the hash value produced. Since we assume IV to be public and hence known to the attacker, the cipher E becomes a simple permutation, i.e., $H = E(M) \oplus M$. Sliced biclique technique can then be applied for preimage search as follows.

The attacker first selects an internal intermediate state Q and partitions the full state space into sets of size 2^{2d} represented as $Q_{i,j}$ for some suitable range of i and j. Each set is defined by its base state $Q_{0,0}$ which is randomly selected by the attacker. Let f be a sub-permutation within E which maps $Q_{i,j}$ to another set of intermediate states $P_{i,j}$, i.e., $Q_{i,j} \underset{f}{\to} P_{i,j}$. These $Q_{i,j}$ and $P_{i,j}$ are obtained using 2^d Δ_i and ∇_j differentials as follows:

1. $Q_{0,0} \xrightarrow{f} P_{0,0}$ .... (base computation),
2. $Q_{i,0} = Q_{0,0} \oplus \Delta_i,$
3. $Q_{i,0} \xrightarrow{f} P_{i,0},$
4. $Q_{0,j} = Q_{0,0} \oplus \nabla_j,$
5. $Q_{0,j} \xrightarrow{f} P_{0,j},$
6. $Q_{i,j} = Q_{i,0} \oplus \nabla_j,$
7. $P_{i,j} = P_{0,j} \oplus \Delta_i,$ where $0 \leq i, j \leq 2^d - 1.$

It has been shown in [15] that $Q_{i,j} \xrightarrow{f} P_{i,j}$ forms a biclique, if Δ_i and ∇_j trails are non-interleaving, i.e., they do not share any active non-linear component between them.[1] The parameter d is called the dimension of the biclique. Each $Q_{0,0}$ defines one biclique structure consisting of 2^{2d} intermediate states.

To find a valid preimage M, the attacker then applies meet-in-the-middle (MITM) technique in the rest of the rounds. In the MITM stage, the attacker chooses an internal state $v \in \{E \setminus f\}$ and computes its value both in the forward direction as a function of P (denoted as $\overrightarrow{v_{i,j}}$) and in the backward direction as a function of Q (denoted as $\overleftarrow{v_{i,j}}$) respectively for every (i,j) pair. This process is shown in Fig. 4.

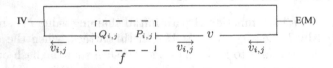

Fig. 4. Biclique Attack.

To compute $\overleftarrow{v}$ in the backward direction, the value of $E(M)$ is required (as shown in Fig. 4) which can be easily calculated by $E(M) = H \oplus M$. To reduce the complexity of the attack, the attacker tries to choose the state v such that in the forward direction it only depends on j and in the backward direction it only depends on i, i.e., states $Q_{i,j}$ and $P_{i,j}$ form a *sliced biclique* if the following conditions hold [15][2]:

$$\forall i, j : \qquad \overrightarrow{v_{i,j}} = \overrightarrow{v_{0,j}},$$
$$\forall i, j : \qquad \overleftarrow{v_{i,j}} = \overleftarrow{v_{i,0}}.$$

Let $\overrightarrow{v_{0,j}} = \overrightarrow{v_j}$ and $\overleftarrow{v_{i,0}} = \overleftarrow{v_i}$. Finally, the attacker checks if:

$$\exists i, j : \qquad \overrightarrow{v_j} = \overleftarrow{v_i}.$$

---

[1] It is not necessary for independent biclique/sliced biclique attack to have Δ and ∇ differentials start from distinct ends of the subcipher. The only requirement that is essential is that both trails should be non-interleaving.

[2] In the traditional biclique key recovery attack in [5], this special restriction on v is not required.

If such an (i, j) pair exists, the corresponding $Q_{i,j}$ becomes the preimage candidate. If not, then the attacker picks up another set of states with different base value $Q_{0,0}$ and repeats the whole procedure.

Complexity of the attack. The sliced biclique attack comprises of 2 phases - biclique construction phase and MITM phase. Let the block cipher E consist of y rounds and the number of rounds covered in the biclique phase be x. This implies the number of rounds covered in the MITM phase is $y - x = z$. For each set of messages, in the biclique phase, since all $\Delta_i \neq \nabla_j$ and Δ_i trails are independent of ∇_j trails, the construction of biclique is simply reduced to computation of Δ_i and ∇_j trails independently which requires no more than 2.2^d computations of f, i.e.,

$$\text{Complexity of biclique phase} = 2^d \times \frac{x}{y} + 2^d \times \frac{x}{y} = 2^{d+1} \times \frac{x}{y}.$$

Similarly, in the MITM phase, the attacker needs to call each of $\overrightarrow{v_j}$ and $\overleftarrow{v_i}$ for 2^d times, i.e., a total of 2^{d+1} times. Let the number of rounds covered in the forward and backward direction be a and b respectively. Hence,

$$\text{Complexity of MITM phase} = 2^d \times \frac{a}{y} + 2^d \times \frac{b}{y} = 2^d \times \frac{a + b(= z)}{y} = 2^d \times \frac{y - x}{y}.$$

It is now easy to check that the overall complexity of sliced biclique preimage attack for one set of messages does not require more than 2^d full computations of E, i.e.,

$$\text{Total Complexity} = 2^{d+1} \times \frac{x}{y} + 2^d \times \frac{y - x}{y} = 2^d \times (1 + \frac{x}{y}) \approx 2^d \text{ since, } x \ll y.$$

If m bicliques are constructed, then the total cost is $m \times 2^d$. For further reading on sliced biclique and classical bicliques one can refer to [15] and [5,16] respectively.

4 Distinguishing Attack on 4-Branch, Type-2 GFN Based Permutation Using Sliced Biclique Cryptanalysis Technique

In this section, we present a 8-round distinguisher on permutation E_k (where k is the key) which is a 8-round, 4-branch, Type-2 Generalized Feistel Network. We assume that the S-box layer has good differential property and the P-layer implements standard MDS matrix.[3] We also assume that the key k (that is IV in the overlying hash function construction) is a fixed constant. The distinguishing property used by the distinguisher is as follows:

---

[3] In this line of work, implementation of P-layer as a standard MDS matrix having optimal branch number is believed to be a good design choice [6,14,24,25].

Distinguishing Property. Let E_k be a block cipher with message size $N = 128$-bits. The aim of the adversary is to collect 2^{16} (plaintext, ciphertext) pairs such that the XOR of the lower 16 bits of the third word in the plaintext and the lower 16 bits of the third word in the ciphertext (where each word is of size 32-bits) is always a 16-bit constant value chosen by the attacker, i.e.,

$$(\text{plaintext})_3^2 \oplus (\text{ciphertext})_3^2 = \text{constant} \tag{1}$$

where, —constant— = 16-bits.[4]

In case of random permutation. When E_k is a random permutation, the probability that any (plaintext, ciphertext) pair satisfies the desired property (as mentioned in Eq. 1) is approximately 2^{-16}. This means that the expected time complexity to generate one such (plaintext, ciphertext) pair is 2^{16}. Hence, expected time complexity to generate 2^{16} such (plaintext, ciphertext) pairs is 2^{32}.

In case of E instantiated with 4-branch, Type-2, GFN. For the illustration of our attack, we consider $N = 128$-bit and $n = 32$-bit each. The attacker first chooses a random base value $Q_{0,0}$ (as discussed in Sect. 3.3). Let $\Delta_i = (\bar{0}\bar{0} \mid i\bar{0} \mid \bar{0}\bar{0} \mid \bar{0}\bar{0})$ and $\nabla_j = (\bar{0}\bar{0} \mid \bar{0}j \mid \bar{0}\bar{0} \mid \bar{0}\bar{0})$ where $(0 \leq i, j \leq 2^{16} - 1)$ be the Δ and ∇ differences injected in Round 4. Here each $\bar{0}$ represents 0^{16}. The propagation of Δ_i trail (marked as '—' in green) and ∇_j trail (marked as '-' in red) is shown in Figs. 5 and 6 respectively. In these figures, the four words shown in each round are the corresponding inputs to four branches at each round. In ∇_j trail, the attacker first injects the given j difference in \$$4_2^2$ word only. As the ∇_j trail propagates as shown in Fig. 6, \$$4_1$ and \$$4_4$ words are subsequently affected. The dimension of this biclique is $d=16$.

It is easy to check that Δ_i and ∇_j trails are independent and do not share any non-linear components (shown in Fig. 7) between them in rounds 4 and 5.

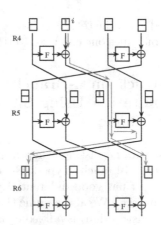

Fig. 5. Δ_i difference injection in Round 4 and its propagation (Color figure online).

---

[4] Here $(plaintext)_3^2$ denotes second block of third word of plaintext as described in Sect. 2. The term $(ciphertext)_3^2$ can be understood similarly.

Thus, a 2-round biclique (consisting of $2^{2d} = 2^{32}$ messages) is formed where the biclique covers rounds 4 and 5. Now the aim of the attacker is to find a matching variable v which only depends on Δ_i trail in one direction and ∇_j trail in the other direction (as discussed in Sect. 3.3). Hence, from round 6 only ∇_j trail is propagated in the forward direction and from round 3 only Δ_i trail is propagated in the backward direction (as shown in Fig. 8). At the end of 8th round it can be seen that $\$1_3^2$ (marked in yellow in Fig. 8) in the backward direction is not affected by Δ_i trail (i.e., will be affected by ∇_j trail only) and $\$8_3^2$ (marked in yellow in Fig. 8) in the forward direction remains unaffected by ∇_j trail (i.e., will be affected by Δ_i trail only). Through feed forward operation, 16 bits of $\$1_3^2$ can then be matched with 16 bits of $\$8_3^2$. Hence, in this attack we choose $\$8_3^2$ to be our matching variable v and $—v— = 16$ which is denoted by t.

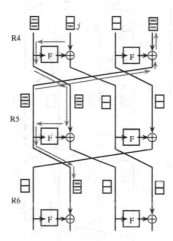

Fig. 6. ∇_j difference injection in Round 4 and its propagation (Color figure online).

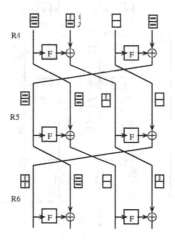

Fig. 7. 2-round biclique placed in Round 4 - 5.

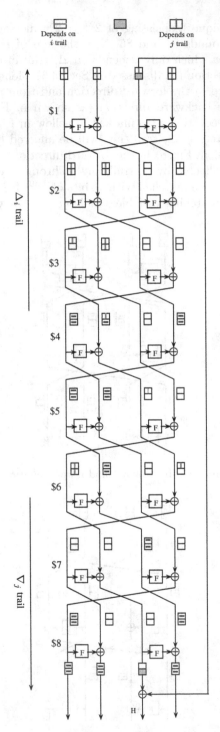

Fig. 8. Matching in 8 rounds of 4-branch Type-2 GFN with right cyclic shift (Color figure online).

Once the matching variable v is obtained, as mentioned above, through our biclique attack, $2^{2d} = 2^{32}$ (plaintext, ciphertext) pairs are generated in a set. Out of these 2^{2d} (plaintext, ciphertext) pairs, there exists $2^{2d-t} = 2^{16}$ (plaintext, ciphertext) pairs which match on matching variable v. In other words, if we XOR the lower 16 bits of the third word in the plaintext and the lower 16 bits of the third word in the ciphertext (i.e., at positions \$1$_3^2$ and \$8$_3^2$ respectively).

Equation 1 will always be satisfied. These 2^{16} (plaintext, ciphertext) pairs will be generated with a computational complexity of $2^d = 2^{16}$ (as discussed in § 3.3) which is lower than the computational complexity of 2^{32} in case of random permutation. Hence, a valid distinguisher for E when instantiated with 4-branch, Type-2, GFN is constructed.

Similarly, our attack can be applied to messages of other sizes as well. In Table 1, we report the complexity values for our distinguisher attack on message inputs of different size.

Table 1. Complexity values for our distinguishing attack on message inputs of different size. Here N represents the input message size in bits and #(P-C) pairs represent the number of plaintext-ciphertext pairs needed for our attack. The number of plaintext-ciphertext pairs depends on the size of matching variable v.

N	n	#(P-C) pairs	Complexity of our attack	Complexity of random permutation
64	16	2^8	2^8	2^{16}
256	64	2^{32}	2^{32}	2^{64}
512	128	2^{64}	2^{64}	2^{128}

5   Collision Attack on 4-Branch, Type-2, GFN Based Compression Function

The distinguisher constructed in the previous section can be used to launch collision attack on 4-branch, Type-2, GFN based compression function as described below. Here the compression function is assumed to be in MMO mode and the output is assumed to be of $N = 128$-bits.

- The attacker first chooses a t-bit constant of his choice.
- In the above attack, the attacker then finds a matching variable v, where —v— $\leq t$. In our attack, —v— $= t = 16$ bits.
- There are $2^{2d} = 2^{32}$ messages in a biclique set. Out of these 2^{2d} messages, only 2^{2d-t} messages will match on v. This means that out of 2^{32} messages only 2^{16} messages will survive the MITM phase.
- In other words, it can be said that the attacker has generated 2^{16} t-bit partial target preimages with these t-bits equal to an arbitrarily chosen constant selected in first step.

- These 2^{16} t-bit partial target preimages collide on $t = 16$ bits. Hence, if the attacker generates $2^{(N-t)/2}$ such preimages which collide on t-bits, there exists a colliding pair with high probability which collide on the remaining $N - t$ bits as well. Thus, the attacker will generate $2^{(128-16)/2} = 2^{56}$ such t-bit partial target preimages to obtain a collision on complete hash output H with high probability.
- Now, one sliced biclique generates 2^{16} t-bit partial target preimages. Hence, to generate 2^{56} such preimages, the attacker needs to construct $2^{56-16} = 2^{40}$ sliced bicliques (or, $2^{(N-t)/2-(2d-t)}$ bicliques where, $2^{(N-t)/2} = 2^{56}$ and $2^{(2d-t)} = 2^{16}$).

Complexity of the collision attack. Since the computational complexity of performing sliced biclique attack once is $2^d = 2^{16}$ (as discussed in Sect. 3.3), hence computational complexity of running sliced biclique attack 2^{40} times is $2^{40} \times 2^{16} = 2^{56}$. Therefore, given IV, the complexity to find a pair of messages (M, M') such that $CF(IV, M) = CF(IV, M')$, when CF (i.e., compression function) is instantiated with 8-rounds of 4-branch type-2 GFN is 2^{56} ($< 2^{64}$ brute-force attack). Here, compression function output is of 128-bits size. In general, the complexity of the attack is given by the following formula:

$$\text{Complexity} = 2^{\frac{(N-t)}{2} - (2d-t)} \times 2^d.$$

For the purpose of illustration, we show the cost of our attack for various message sizes in Table 2.

Table 2. Complexity values for our collision attack on message inputs of different size. Here N represents the input message size in bits, n represents the branch word size in bits and t represents the size of matching variable v in bits. In our attack $d = t$ always.

N	n	t	Rounds	Complexity of our attack	Brute force complexity
64	16	8	8	2^{28}	2^{32}
128	32	16	8	2^{56}	2^{64}
256	64	32	8	2^{112}	2^{128}
512	128	64	8	2^{224}	2^{256}

Since we need to store all partial preimages to find the colliding pair, memory required is of the order of 2^{56} (for 128-bit output). However, it is mentioned in [15] that memoryless equivalents of these attacks do exist. In Appendix A, we show the collision attack on CLEFIA which is a real world implementation of 4-branch, Type-2, GFN.

Collision Attack on 4-branch Type-2 GFN with Double SP layer. The above attack technique is generic and independent of the internal F-function

Table 3. Comparison of our results with previous cryptanalytic results on 4-branch, Type-2, GFN with double SP layer.

Rounds	Attack type	Reference
6	Near collisions	[24]
7	Distinguishing	[24]
8	Distinguishing	[8]
8	Distinguishing	This work, § 4
8	Full collisions	This work, § 5

structure. Hence, if we instantiate the round function F with double SP-layer, the above attack can be directly translated to 8-round collision attack on 4-branch, Type-2 GFN with double SP layer based compression function with a complexity of 2^{56}. This betters the 6-round near collision attack on the same structure shown by Sasaki in [24]. In Table 3 we compare our result with the previous cryptanalysis results on 4-branch, Type-2 GFN with double SP layer.

As discussed above, since the attack technique is generic, presence of multiple SP layers in the round function F does not provide any extra resistance against sliced biclique attack as compared to double SP layer. In fact, in our collision attack neither the attack complexity nor the the number of rounds attacked change if double SP layer is replaced by multiple SP layers. This is in contrast to attacks such as rebound attacks [21], where the number of SP layers inside the round function F influence the number of rounds attacked [8,9,14,24,25] in Generalized Feistel Networks.

6 Collision Attack on Hash Functions

In this attack, given the IV, the aim of the attacker is to find a pair of messages (M, M') such that $H(M) = H(M')$. To do so, the attacker first finds two

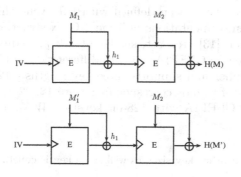

Fig. 9. Collision Attack.

messages M_1 and M_1' which collide to same hash value h_1 using collision attack technique described in Sect. 5 with a complexity of 2^{56}. Now he concatenates any message M_2 with M_1 and M_1' (as shown in Fig. 9) such that $H(M_1\|M_2) = H(M_1'\|M_2)$. Message M_2 can also be chosen such that it satisfies padding restrictions (where length of input message is appended at the end) if required. In this way, collision attack can be carried out on 4-branch, Type-2, GFN with double SP layer based hash function with a complexity of 2^{56}. Since we assume known key settings (i.e., key part to the underlying block cipher is known to the attacker), hence this attack can be used to generate collisions in MP and MMO based hash functions but pseudo collisions in DM based hash functions.

7   Conclusions

In this work, we apply the sliced biclique technique to show collision attack on 8-rounds of 4-branch, type-2 GFN. When it is instantiated with double SP layer, we present the first 8-round collision on 4-branch, type-2 GFN with double SP layer. It would be interesting to apply sliced biclique technique to attack other potential targets. One possible extension can be to apply this attack technique on 2-branch, Type-2 GFN such as Shavite-3 etc.

Acknowledgements. The authors would like to thank the anonymous reviewers for their valuable comments as it helped in improving the quality of the paper.

A   8-Round Collision Attack on CLEFIA Based Compression Function

In this section, we investigate CLEFIA which is a real world-implementation of 4-branch, Type-2 GFN. In the attacks discussed in Sects. 4 and 5, we considered 4-branch, Type-2 GFN with double SP layer where right cyclic shift is applied on the message sub-blocks at the end of each round. This was done to facilitate direct comparison with previous results [8,24] on the same structure. However in [34], Type-2 GFN's have been defined with left cyclic shift and is followed in all the practical implementations of Type-2 GFN structure - e.g., RC6 [23], CLEFIA [28], HIGHT [13] etc. Yet, similar attack procedure (as discussed in Sect. 5) can be applied on CLEFIA but with different Δ_i and ∇_j trails. CLEFIA is a 128-bit block cipher and supports three key lengths - 128-bit, 192-bit and 256-bit. The number of rounds correspondingly are 18, 22 and 26. Here, in this section, we examine CLEFIA with 128-bit keysize.[5] WK_0 and WK_1 represent

---

[5] The attack works on other key sizes as well since key is constant under known key settings.

the whitening keys at the start of the cipher. Each round has two 32-bit round keys RK_{2i-2} and RK_{2i-1} (where, $1 \leq i \leq 18$).

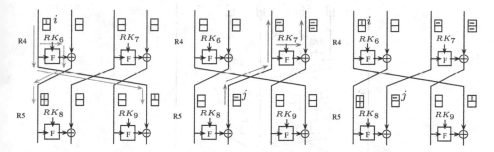

Fig. 10. Δ_i difference injection in Round 4 and its propagation (Color figure online)

Fig. 11. ∇_j difference injection in Round 5 and its propagation (Color figure online)

Fig. 12. 1-round biclique placed in Round 4

In this attack, let $\Delta_i = (i\bar{0} \mid \bar{0}\bar{0} \mid \bar{0}\bar{0} \mid \bar{0}\bar{0})$ be the Δ difference injected in Round 4 and $\nabla_j = (\bar{0}\bar{0} \mid j\bar{0} \mid \bar{0}\bar{0} \mid \bar{0}\bar{0})$ be the ∇ difference injected in Round 5 where $(0 < i, j < 2^{16} - 1)$. Here each $\bar{0}$ represents 0^{16}. The attacker first chooses a random base value $Q_{0,0}$ and then injects the Δ_i and ∇_j differences accordingly. The propagation of Δ_i trail (marked as '—' in green) and ∇_j trail (marked as '-' in red) is shown in Figs. 10 and 11 respectively. The dimension of this biclique is $d=16$. It is easy to check that Δ_i and ∇_j trails are independent and do not share any non-linear components (shown in Fig. 12) between them in round 4. Thus a 1-round biclique (consisting of $2^{2d} = 2^{32}$ messages) is formed in 4 round.

From round 5 only ∇_j trail is propagated in the forward direction and from round 3 only Δ_i trail is propagated in the backward direction (as shown in Fig. 13). At the end of 8th round it can be seen that 1_3^2 (marked in yellow in Fig. 13) in the backward direction is not affected by Δ_i trail and 8_3^2 (marked in yellow in Fig. 13) in the forward direction remains unaffected by ∇_j trail. Through feed forward operation, 16 bits of 1_3^2 can then be matched with 16 bits of 8_3^2. Hence, in this attack we choose 8_3^2 to be our matching variable v. The steps of collision attack for CLEFIA are exactly the same as discussed in Sects. 5 and 6. Therefore, we can generate collisions in 8-rounds of CLEFIA based hash function with a complexity of 2^{56}.

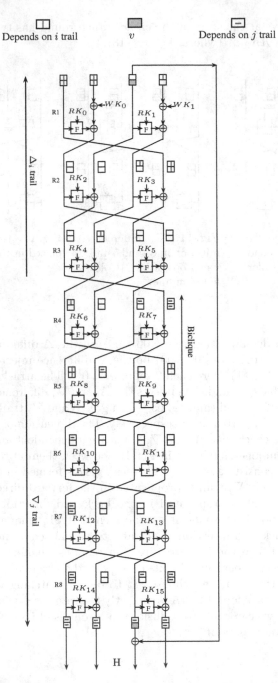

Fig. 13. Matching in 8 rounds of CLEFIA (Color figure online)

References

1. Abed, F., Forler, C., List, E., Lucks, S., Wenzel, J.: Biclique cryptanalysis of PRESENT, LED, And KLEIN. Cryptology ePrint Archive, Report 2012/591 (2012). http://eprint.iacr.org/2012/591
2. Anderson, R.J., Biham, E.: Two practical and provably secure block ciphers: BEARS and LION. In: Gollmann [10], pp. 113–120
3. Biham, E., Dunkeman, O.: The SHAvite-3 Hash Function. Submission to NIST SHA-3 competition. www.cs.technion.ac.il/orrd/SHAvite-3/
4. Bogdanov, A.: On the differential and linear efficiency of balanced Feistel networks. Inf. Process. Lett. **110**(20), 861–866 (2010)
5. Bogdanov, A., Khovratovich, D., Rechberger, C.: Biclique cryptanalysis of the full AES. In: Lee, D.H., Wang, X. (eds.) ASIACRYPT 2011. LNCS, vol. 7073, pp. 344–371. Springer, Heidelberg (2011)
6. Bogdanov, A., Shibutani, K.: Generalized Feistel networks revisited. Des. Codes Cryptogr. **66**(1–3), 75–97 (2013)
7. Çoban, M., Karakoç, F., Boztaş, Ö.: Biclique cryptanalysis of TWINE. In: Pieprzyk, J., Sadeghi, A.-R., Manulis, M. (eds.) CANS 2012. LNCS, vol. 7712, pp. 43–55. Springer, Heidelberg (2012)
8. Chang, D., Kumar, A., Sanadhya, S.: Security analysis of GFN: 8-round distinguisher for 4-branch type-2 GFN. In: Paul, G., Vaudenay, S. (eds.) INDOCRYPT 2013. LNCS, vol. 8250, pp. 136–148. Springer, Heidelberg (2013)
9. Dong, L., Wenling, W., Shuang, W., Zou, J.: Known-key distinguishers on type-1 Feistel scheme and near-collision attacks on its hashing modes. Front. Comput. Sci. **8**(3), 513–525 (2014)
10. Gollmann, D. (ed.): FSE 1996. LNCS, vol. 1039. Springer, Heidelberg (1996)
11. Hoang, V.T., Rogaway, P.: On generalized Feistel networks. In: Rabin, T. (ed.) CRYPTO 2010. LNCS, vol. 6223, pp. 613–630. Springer, Heidelberg (2010)
12. Hong, D., Koo, B., Kwon, D.: Biclique attack on the full HIGHT. In: Kim, H. (ed.) ICISC 2011. LNCS, vol. 7259, pp. 365–374. Springer, Heidelberg (2012)
13. Hong, D., et al.: HIGHT: a new block cipher suitable for low-resource device. In: Goubin, L., Matsui, M. (eds.) CHES 2006. LNCS, vol. 4249, pp. 46–59. Springer, Heidelberg (2006)
14. Kang, H., Hong, D., Moon, D., Kwon, D., Sung, J., Hong, S.: Known-key attacks on generalized Feistel schemes with SP round function. IEICE Trans. **95–A**(9), 1550–1560 (2012)
15. Khovratovich, D.: Bicliques for permutations: collision and preimage attacks in stronger settings. In: Wang, X., Sako, K. (eds.) ASIACRYPT 2012. LNCS, vol. 7658, pp. 544–561. Springer, Heidelberg (2012)
16. Khovratovich, D., Rechberger, C., Savelieva, A.: Bicliques for preimages: attacks on Skein-512 and the SHA-2 family. In: Canteaut, A. (ed.) FSE 2012. LNCS, vol. 7549, pp. 244–263. Springer, Heidelberg (2012)
17. Knudsen, L.R., Rijmen, V.: Known-key distinguishers for some block ciphers. In: Kurosawa, K. (ed.) ASIACRYPT 2007. LNCS, vol. 4833, pp. 315–324. Springer, Heidelberg (2007)
18. Li, J., Isobe, T., Shibutani, K.: Converting meet-in-the-middle preimage attack into pseudo collision attack: application to SHA-2. In: Canteaut, A. (ed.) FSE 2012. LNCS, vol. 7549, pp. 264–286. Springer, Heidelberg (2012)
19. Mala, H.: Biclique cryptanalysis of the block cipher SQUARE. Cryptology ePrint Archive, Report 2011/500 (2011). http://eprint.iacr.org/2011/500

20. Mendel, F., Peyrin, T., Rechberger, C., Schläffer, M.: Improved cryptanalysis of the reduced Grøstl compression function, ECHO permutation and AES block cipher. In: Jacobson Jr., M.J., Rijmen, V., Safavi-Naini, R. (eds.) SAC 2009. LNCS, vol. 5867, pp. 16–35. Springer, Heidelberg (2009)
21. Mendel, F., Rechberger, C., Schläffer, M., Thomsen, S.S.: The rebound attack: cryptanalysis of reduced whirlpool and Grøstl. In: Dunkelman, O. (ed.) FSE 2009. LNCS, vol. 5665, pp. 260–276. Springer, Heidelberg (2009)
22. Rijmen, V., Daemen, J., Preneel, B., Bosselaers, A., De Win, E.: The cipher SHARK. In: Gollmann [10], pp. 99–111
23. Rivest, R.L., Robshaw, M.J. B., Yin, Y.L.: RC6 as the AES. In: AES Candidate Conference, pp. 337–342 (2000)
24. Sasaki, Y.: Double-SP is weaker than Single-SP: rebound attacks on Feistel ciphers with several rounds. In: Galbraith, S., Nandi, M. (eds.) INDOCRYPT 2012. LNCS, vol. 7668, pp. 265–282. Springer, Heidelberg (2012)
25. Sasaki, Y., Yasuda, K.: Known-key distinguishers on 11-round Feistel and collision attacks on its hashing modes. In: Joux, A. (ed.) FSE 2011. LNCS, vol. 6733, pp. 397–415. Springer, Heidelberg (2011)
26. Schneier, B., Kelsey, J.: Unbalanced Feistel networks and block cipher design. In: Gollmann, D. (ed.) FSE 1996. LNCS, vol. 1039, pp. 121–144. Springer, Heidelberg (1996)
27. Shirai, T., Shibutani, K.: Improving immunity of feistel ciphers against differential cryptanalysis by using multiple MDS matrices. In: Roy, B., Meier, W. (eds.) FSE 2004. LNCS, vol. 3017, pp. 260–278. Springer, Heidelberg (2004)
28. Shirai, T., Shibutani, K., Akishita, T., Moriai, S., Iwata, T.: The 128-Bit blockcipher CLEFIA (extended abstract). In: Biryukov, A. (ed.) FSE 2007. LNCS, vol. 4593, pp. 181–195. Springer, Heidelberg (2007)
29. Su, B., Wu, W., Wu, S., Dong, L.: Near-collisions on the reduced-round compression functions of skein and BLAKE. In: Heng, S.-H., Wright, R.N., Goi, B.-M. (eds.) CANS 2010. LNCS, vol. 6467, pp. 124–139. Springer, Heidelberg (2010)
30. Suzaki, T., Minematsu, K.: Improving the generalized Feistel. In: Hong, S., Iwata, T. (eds.) FSE 2010. LNCS, vol. 6147, pp. 19–39. Springer, Heidelberg (2010)
31. Vaudenay, S.: On the need for multipermutations: cryptanalysis of MD4 and SAFER. In: Preneel, B. (ed.) FSE 1994. LNCS, vol. 1008, pp. 286–297. Springer, Heidelberg (1995)
32. Wenling, W., Zhang, W., Lin, D.: Security on generalized Feistel scheme with SP round function. Int. J. Netw. Secur. 3(3), 215–224 (2006)
33. Chen, S.Z., Xu, T.M.: Biclique attack of the full ARIA-256. Cryptology ePrint Archive, Report 2012/011 (2012). http://eprint.iacr.org/2012/011
34. Zheng, Y., Matsumoto, T., Imai, H.: On the construction of block ciphers provably secure and not relying on any unproved hypotheses. In: Brassard, G. (ed.) CRYPTO 1989. LNCS, vol. 435, pp. 461–480. Springer, Heidelberg (1990)

Rig: A Simple, Secure and Flexible Design for Password Hashing

Donghoon Chang, Arpan Jati, Sweta Mishra$^{(\boxtimes)}$, and Somitra Kumar Sanadhya

Indraprastha Institute of Information Technology, Delhi (IIIT-D), Delhi, India
{donghoon,arpanj,swetam,somitra}@iiitd.ac.in

Abstract. Password Hashing, a technique commonly implemented by a server to protect passwords of clients, by performing a one-way transformation on the password, turning it into another string called the hashed password. In this paper, we introduce a secure password hashing framework *Rig* which is based on secure cryptographic hash functions. It provides the flexibility to choose different functions for different phases of the construction. The design of the scheme is very simple to implement in software and is flexible as the memory parameter is independent of time parameter (no actual time and memory trade-off) and is strictly sequential (difficult to parallelize) with comparatively huge memory consumption that provides strong resistance against attackers using multiple processing units. It supports client-independent updates, i.e., the server can increase the security parameters by updating the existing password hashes without knowing the password. *Rig* can also support the server relief protocol where the client bears the maximum effort to compute the password hash, while there is minimal effort at the server side. We analyze *Rig* and show that our proposal provides an exponential time complexity against the low-memory attack.

Keywords: Password · Password hashing · GPU attack · Cache-timing attack · Client-independent update · Server-relief technique

1 Introduction

A password is a secret word or string of characters which is used by a principal to prove her identity as an authentic user to gain access to a resource. Being secret, passwords cannot be revealed to other users of the same system. In order to ensure the confidentiality of the passwords even when the authentication data is somehow leaked from the server, passwords are never stored in clear, but transformed into an illegible form and then stored. Specifically, 'Password Hashing' is the technique which performs a one-way transformation on a password and turns it into another string, called the 'hashed' password. Strong password protection, i.e., a technique of password hashing that makes brute force attack on password guessing infeasible, either in software or by using GPUs (Graphics Processing Unit), is essential to protect the user security and identity. Thus any working password hashing scheme should be resistant to brute force attack.

© Springer International Publishing Switzerland 2015
D. Lin et al. (Eds.): Inscrypt 2014, LNCS 8957, pp. 361–381, 2015.
DOI: 10.1007/978-3-319-16745-9_20

Password hashing is an active topic of interest in cryptography community and a competition on password hashing is going on [1]. Currently, the significant constructions for password hashing are PBKDF2 [10], Bcrypt [12] and Scrypt [11]. All of these do not satisfy most of the necessary requirements mentioned at the competition page [1]. PBKDF2 (NIST standard) consumes very less memory as it was mainly designed to derive keys from a seed (password). Bcrypt uses fixed memory (4 KB) for its implementation. Scrypt is not simple (different internal modules) and not flexible (time and memory parameters are dependent) and susceptible to cache timing attack (discussed in Sect. 5).

Specifically, the rate at which an attacker can guess passwords is a key factor in determining the strength of the password hashing scheme. Current requirements [1] for a secure password hashing scheme are the following:

- The construction should be slow to resist password guessing attack but should have a fast response time to prove the authenticity of the user.
- It should have a simple design and should be easy to implement (coding, testing, debugging, integration), i.e., the algorithm should be simple in the sense of clarity and concise with less number of internal components and primitives.
- It should be flexible and scalable, i.e., if memory and time are not dependent then one would be able to scale any of the parameters to get required performance.
- Cryptographic security [1]: The construction should behave as a random function (random-looking output, one-way, collision resistant, immune to length extension, etc.).
- Resistant to GPU attack: A typical GPU has lots of processing cores but has limited amount of memory for each single core. It is quite efficient for an attacker to utilize all the available processing cores with limited memory to run brute-force attack over the password choices. Use of comparatively huge memory per password hash by the password hashing construction can restrict the use of GPU. Therefore, the design should have large memory consumption to force comparatively slow and costly hardware implementation that can resist the GPU attack.
- Leakage Resilience: The construction should protect against information extraction from physical implementation, i.e., the scheme should not leak information about the password due to cache timing or memory leakage, while supporting any length of password.
- The construction should have the ability to transform an existing hash to a different cost setting (client independent update, explained in Sect. 5) without knowledge of the password.
- It is good if the construction provides server relief technique where the client performs most of the computations for password hashing and the server puts minimal effort with minimal use of resources, to reduce the load of the server. This property needs a secure protocol to maintain the security of the hash computation (discussed in Sect. 5).

The most challenging threat faced by any password hashing scheme is the existence of cheap, massively parallel hardware such as Graphics Processing Units (GPUs), Application-Specific Integrated Circuits (ASICs) and Field-Programmable Gate Arrays (FPGAs). Using such efficient hardware, an adversary with multiple computing units can easily try multiple different passwords in parallel. To prevent such attempts we need to slow down password hash computation and ensure that there is little parallelism in the design. One way to achieve this is to use a 'Sequential memory-hard' algorithm, a term first introduced with the design of 'Scrypt' [11], a password hashing scheme. The main design principle of Scrypt is that it asymptotically uses almost as many memory locations as it uses operations to slow down the password-hash computation. Memory is *relatively* expensive, so, a typical GPU or other cheap massively-parallel hardware with lots of cores can only have a limited amount of memory for each single core. Hence an attacker with access to such hardware will still not be able to utilize all the available processing cores due to the lack of sufficient memory and will be forced to have an (almost) sequential implementation of the password hashing scheme.

In this document we propose *Rig*, a password hashing scheme which aims to address the above mentioned requirements. *Rig* is based on cryptographic (secure) hash functions and is very simple to implement in software. It is flexible as the memory parameter is independent of time parameter (no actual time and memory trade-off) and is strictly sequential (difficult to parallelize) with comparatively huge memory consumption that provides strong resistance against attackers using multiple processing units. It supports client-independent password hash up-gradation without the need of the actual password. This feature helps the server to increase the security parameters to calculate the password hash to reduce the constant threats of technological improvements, specifically in the field of hardware. *Rig* provides protection against the extraction of information from cache-timing attack and prevents denial-of-service attack if implemented to provide server-relief technique. We analyze *Rig* and show that our proposal provides an exponential time complexity against memory-free attack. It gives the flexibility to choose different functions for different phases of the construction and we denote the general construction of *Rig* as Rig $[H_1, H_2, H_3]$. In this work we provide two variants of Rig $[H_1, H_2, H_3]$. A strictly sequential variant, Rig [Blake2b, BlakeCompress, Blake2b] and the other variant, Rig [BlakeExpand, BlakePerm, Blake2b] which improves the performance by performing memory operations in larger chunks.

The rest of the document is organised as follows. In Sect. 2 we present the important preliminaries necessary for understanding the specification. This is followed by the introduction of significant hardwares used as attack platform in Sect. 3. The specification and design rationale of the scheme are presented in Sects. 4 and 5 respectively. Subsequently, the implementation aspects and performance analysis are presented in Sects. 6 and 7. Finally, in Sects. 8 and 9, we provide the security analysis of the scheme and the conclusions of the paper respectively.

2   Preliminaries

The techniques used in our construction are discussed below.

- **Binary 64-bit mapping:** It is a 64-bit binary representation of the decimal value. The binary number

$$a_{n-1}2^{n-1} + a_{n-2}2^{n-2} + \cdots + a_0$$

  is denoted as $a_{n-1}a_{n-2}\cdots a_0$ where $a_i \in \{0,1\}$ and n is the number of digits to the left of the binary (radix) point. In our construction we use $n = 64$ and we denote $\text{binary}_{64}(x)$ for 64-bit binary representation of the value x.
- **Bit reversal permutation** [7,9] **(br):** It is implemented to permute the indices of an array of $n = 2^k$ elements where $k \in \mathbb{N}$. We explain the steps of the permutation through Algorithm 1 below.

  The example of a bit reversal permutation applied on an array of $m = 2^3$ elements where $k = 3$ and indices are $0, 1, \cdots, 7$ is given below.
  $br[000, 001, 010, 011, 100, 101, 110, 111] = [000, 100, 010, 110, 001, 101, 011, 111]$
  $= br[0], br[1], br[2], br[3], \cdots, br[7]$.

Algorithm 1. Bit reversal permutation (br)

Input: Indices of an array A of $n = 2^k$ elements where $k \in \mathbb{N}$ and
  indices are: $0, 1, 2, \cdots, n-1$
Output: Permuted indices of array A as: $br[0], br[1], br[2], \cdots, br[n-1]$
  1 for $i = 0$ to $n-1$
  2 | $(i)_{bin_k} = i_{k-1}i_{k-2}\cdots i_1 i_0 = \sum_{j=0}^{k-1} 2^j i_j$
  3 | $\triangleright$ $(i)_{bin_k} = k$-bit binary representation of value i
  4 | $br[i] = \sum_{j=0}^{k-1} 2^j i_{k-1-j}$
  5 return $br[0], br[1], br[2], \cdots, br[n-1]$

---

3   Attack Platforms: Significant Hardwares

According to Moore's Law [13], the number of transistors on integrated circuits doubles approximately every two years. This has indeed been the case over the history of computing hardware. Following this law, hardware is becoming more and more powerful with time. This happens to be the most prominent threat for existing password hashing schemes. Consequently, there is a need to raise the cost of brute force attack by controlling the performance of the massively parallel hardware available.

An important electronic circuit, **Graphics Processing Unit** (GPU), and their highly parallel structure makes them more effective than general-purpose CPUs for algorithms where processing of large blocks of data is done in parallel. An **Application-Specific Integrated Circuit** (ASIC), is an integrated circuit (IC) which can be customized with memory chips to implement a dedicated design. An ASIC can not be altered after final design hence the designers need to be certain of their design when it is implemented in ASIC. On the other hand,

Field Programmable Gate Arrays (FPGAs) are programmable integrated circuits and consist of an array of logic elements together with an interconnected network and memory chips, providing high-performance. A designer can test her design on an FPGA before implementing it on an ASIC.

Both ASIC and FPGAs can be configured to perform password hashing with highly optimized performance. The cost of implementation on FPGA is cheaper than ASICs if the number of units of the hardware required is small. Therefore, one can easily use parallel FPGAs to increase the rate of password guessing. RIVYERA FPGA cluster is an example of a very powerful and cost optimized hardware. It can hash 3,56,352 passwords per second by using PBKDF2 (NIST standard, Password Based Key Derivation Function 2) with SHA-512 and 512-bit derived key length [6]. This high performance is possible on the FPGA because PBKDF2 does not consume high memory for password hashing. Comparing FPGAs with GPUs (Graphics processing units), the authors of [6] provide results of the same implementation on 4 Tesla C2070 GPUs as 1,05,351 passwords per second. ASIC is better than FPGA purely on performance in terms of number of hashes per second. However, FPGA is preferable when cost is considered with the speed of hashing. Following Moore's law, the speed of hardware is likely to increase by almost a factor of two in less than two years. However, as processor speeds continue to outpace memory speeds [8], the gap between processor and memory performance increases by about 50 % per year [4]. Thus, there is a need to minimize the effects of such high performance hardware. Hence, we need a password hashing algorithm which consumes comparatively large memory to prevent parallel implementation.

4   Specification

Our construction is described in Fig. 1. Following is the step-by-step description of Algorithm 2 which explains our construction *Rig*.

1. First we need to fix the following parameters:
   - pwd = The user password of any length.
   - s = The salt value of any length.
   - n = The number of iterations required to perform iterative transformation phase.
   - m_c = The memory count from which the memory-cost is defined as: $m = 2^{m_c}$, i.e., m denotes the number of items to be stored in the memory. The value of m is updated as: $m_{i+1} = 2 \times m_i$ at each round.
   - r = The number of rounds for the setup phase followed by iterative transformation phase and output generation phase.
   - l = The output length of the password hash.
   - t = The number of bits retained from hash output after truncation. Used with a function $\text{trunc}_t(x) = x \gg (|x| - t)$, where x is the hash output.

2. **Initialization Phase:** We map the parameters, namely the values: password length pwd_l, salt length s_l, n and the output length, l to a 64-bit binary value

using binary$_{64}$ mapping. We create the value x as the concatenation ($\|$) of the above mentioned parameters as

$$x = \text{pwd} \parallel \text{binary}_{64}(pwd_l) \parallel s \parallel \text{binary}_{64}(s_l) \parallel \text{binary}_{64}(n) \parallel \text{binary}_{64}(l)$$

and compute $H_1(x) = \alpha$ where H_1 is the underlying hash function. We use α for further calculations in the setup phase.

3. **Setup Phase:** We initialize h_0 with the value of π after the decimal point. We take as many digits of π as desired to ensure that $|h_0| = |\alpha|$. The values h_0 and α are used to initialize two arrays k and a and further $m - 1$ values of the arrays are iteratively calculated as shown in the Fig. 1. First t-bits of each hash output are stored in the array k.

   The large number of calls to the underlying hash function are guaranteed to have different inputs by the use of different counter values. H_2 denotes the underlying hash function.

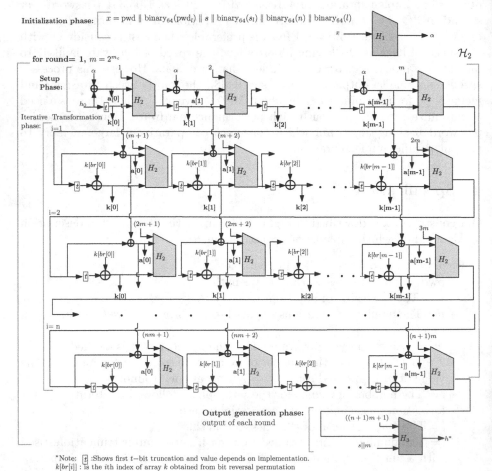

*Note: $\boxed{t}$:Shows first t-bit truncation and value depends on implementation.
$k[br[i]]$: is the ith index of array k obtained from bit reversal permutation

Fig. 1. Graphical representation of the proposed construction.

4. **Iterative Transformation Phase:** This phase is designed to make constant use of the stored array values and to update them. Here we modify each element of the arrays a and k, n-times where n is the number of iterations.

   Array a is accessed sequentially where values of array k are accessed using bit reversal permutation explained in Algorithm 1. We denote the index of array k obtained applying bit-reversal permutation as: $br[j]$, $0 \leq j \leq m - 1$.

5. **Output Generation Phase:** After execution of the setup phase and iterative transformation phase sequentially, we calculate one more hash, denoted by H_3 to get the output of each round. If round $= 1$, this output is considered as the password hash.

Note: The output is an l-bit value. The algorithm stores the output as the hashed password. Our construction allows for storing a truncated portion of the hash output as well. If this is desired we can take one of the following two approaches.

(a) The user may run the complete algorithm as described above and truncate the final output after r rounds to the desired length. This approach does not support client-independent update.

(b) To support client-independent updates the user can choose a length for truncation which is sufficient to claim brute-force security. Then append some constant value, we suggest the hexadecimal value of π after first 64-bytes of decimal point. Take as many digits as desired to make the output length of each round equal to the length of α of the setup phase. So this way one can reduce the storage requirement for password hashes at the server.

Rig. A password hashing scheme

Algorithm 2. Rig [H_1, H_2, H_3] Construction

Input: Password (pwd), Password length (pwd_l), Salt (s), Salt length (s_l), No. of iterations (n),
   Memory count (m_c), No. of bits to be retained from hash output of the setup phase (t),
   Output length (l), No. of rounds (r)
Output: l-bit hash value $h_r{}^*$ obtained after r rounds
   1   ▷ Initialization phase: generates α from password
   2   Initialize: a random salt (s) of atleast 16-bytes, number of iterations (n),
           value of memory count m_c where $m = 2^{m_c}$, value 1
   3   $x = \text{pwd} \parallel \text{binary}_{64}(pwd_l) \parallel s \parallel \text{binary}_{64}(s_l) \parallel \text{binary}_{64}(n) \parallel \text{binary}_{64}(l)$ ▷ concatenation: $\parallel$
   4   $\alpha = H_1(x)$ ▷ H_1 : underlying hash function
   5   **for** round 1 to r
   6   ▷ Initialization of Setup phase: Creates two arrays k and a
         where $|k| = |a| = m$ where $m = 2^{(round-1)} \times 2^{m_c}$
   7     $h_0 = $ initialized with the value of π after decimal, and $|h_0| = |\alpha|$
   8     $a[0] = \alpha \oplus h_0$, $k[0] = trunc_t(h_0)$
   9   **for** $i = 1$ to m
 10. $h_i = H_2(i \parallel a[i-1] \parallel k[i-1])$ ▷ H_2 : underlying hash function
 11   **if** $i \neq m$
 12        $a[i] = \alpha \oplus h_i$
 13        $k[i] = trunc_t(h_i)$ ▷ retains the first t-bits of the hash output
 14   ▷ Initialization of Iterative Transformation phase
 15   **for** $i = 1$ to n
 16   **for** $j = 1$ to m
 17        $a[j-1] = a[j-1] \oplus h_{\{im+j-1\}}$
 18        $br[j-1] = $ index value of array k obtained using
                bit reversal permutation
 19   ▷ initialize a temporary array $|k_{temp}| = |k|$
 20        $k_{temp}[j-1] = k[br[j-1]] \oplus trunc_t(h_{im+j-1})$
 21        $h_{im+j} = H_2((im+j) \parallel a[j-1] \parallel k_{temp}[j-1])$
 22       $k = k_{temp}$
 23   ▷ Output generation phase
 24     $h_{round}{}^* = (H_3((n+1)m + 1) \parallel h_{(n+1)m} \parallel s \parallel \text{binary}_{64}(m))$
 25   **if** round $< r$
 26       $\alpha = h_{round}{}^*$

5   Design Rationale

Existing password hashing schemes are not simple and do not fulfill the necessary requirements as discussed in Sect. 1. We have tried to design a solution which overcomes the known disadvantages of existing schemes (PBKDF2 [10], Bcrypt [12] and Scrypt [11]). The primary concerns against existing proposals are their complex design and their inefficiency to prevent hardware threats. We have tried to strengthen our design by considering the necessary requirements as mentioned in Sect. 1.

1. **Initialization Phase:** We have used concatenation of password, salt, 64-bit value of pwd_l, s_l, n and l as input to increase the size of input. This resists brute force dictionary attack.

2. **Setup Phase:** In this phase we initialize h_0 with π, as we want to have a random sequence and π is not known to have any pattern in the sequence of digits after the decimal point. We generate the values that are required to be stored and repeatedly accessed throughout the remaining phases. This ensures that a large memory requirement criteria for a password hashing scheme is satisfied which neutralizes the threat of using recent technological trends, such as GPUs, ASICs etc. We use different counter values for each hash computation to make all hash inputs different. This reduces collisions and hence makes the output different.

   For array k there is a flexibility to vary the bit storage by taking first t-bits of the hash output where t is taken to be close to the hash-length but not equal to the hash-length. This fulfills the demand of huge memory while at the same time ensures sequential hash calculation and forces an attacker to compute the hash at run-time thus slowing him down. Further, it also allows to extend the scope of implementation in that a low memory device may keep very few bits of the hash values stored but may increase the number of iterations. This ensures that Rig can be implemented in resource constrained devices.

3. **Iterative Transformation Phase:** To make the storage requirement compulsary, this phase progresses sequentially, accessing and updating all stored values at each iteration. Here again, we use different counters for hash input for the same reason as mentioned above. In this phase the memory access pattern is made password independent to reduce the chance of cache timing attack which we have explained later in this section.

4. **Output Generation Phase:** This is the last phase of each round. We reuse the salt value as input to make the collision attack difficult.

   Apart from that the output of each round can be truncated to a desired length. This is optionally mentioned to handle the situations when it is required to reduce the server storage per password.

The other important criteria taken into account in the design of the scheme are the following:

5. **Simplicity and Flexibility:** Symmetry (as setup phase and iterative transformation phase follows similar structure) in the design of Rig enhances the

overall clarity of the scheme. An earlier password hashing scheme Scrypt [11] uses PBKDF2 (internally calls HMAC-SHA256) and ROMix (internally calls BlockMix and uses Salsa20/8). Unlike Scrypt, *Rig* uses only a single primitive (a cryptographically secure hash function). This makes our scheme easier to understand and easy to implement (coding, testing, debugging and integration).

In our scheme the memory parameter is independent of the time parameter. This flexibility in design choice allows a user to scale any of these parameters to get the required performance. On the other hand we have the flexibility for the choice of the functions H_1, H_2 and H_3 (see Fig. 1), but proper selection of the primitives are required to maintain the overall design properties and security. Therefore there can be multiple variants of *Rig* aimed at different implementations or scenarios.

6. **Random Output:** Our scheme calls a hash function repeatedly. We use different counters for each of these hash calls to ensure that no input to the hash function is repeated. The security of *Rig* relies on the prevention of preimage and collision attacks against the underlying hash function. Use of any state-of-the-art hash function (e.g. any finalist of SHA-3 competition) ensures the security of our scheme. We use Blake2b [3] in demonstrating the performance of our scheme later in this paper, although any other hash function could easily be used instead.

With the property of different input, different output and same input, same output, our scheme mimics the Random Oracle Model. This provides theoretical justification of the security of *Rig*.

7. **Client-independent Update** [7]: Our design supports client independent update, i.e., server can increase the security parameter without knowing the password. This is possible if we fix the value of n (number of iterations) and increase the number of rounds r. Each round of the algorithm doubles the memory consumption m from the previous round and hence increases the security parameter. This is possible because the output of each round can be treated as the value α at the next round and then can easily follow the Algorithm 2 to produce the output of the next round. The idea of client independent update of the security parameter m is fulfilled by the following way. The value of m is updated at each next round $i + 1$ (say) from its previous round i as: $m_{i+1} = 2 \times m_i$.

The overall procedure is: output of each round is the input to the next. Each round gives full hash maintaining all requirements of a good password hashing technique. By increasing the number of rounds, the scheme increases the required memory and time hence increases the security parameter without the interference of the client.

8. **Resistance Against Cache-timing Attack:** In our construction, to access the memory which is stored in an array k, we use bit-reversal permutation, which is independent of the password used. If a password dependent permutation is used and if the array can be stored in the cache while accessing the values, an attacker can trace the access pattern observing the time difference in each access of the array index. This helps the adversary to filter

the passwords that follows similar memory access pattern and to make a list of feasible passwords. Therefore, a password hashing scheme should have password-independent memory access patterns and to follow this requirement we use bit reversal permutation as in [7].

9. **Server-relief Hashing:** Current requirement of a password hashing technique is that it should be slow and should demand comparatively large memory to implement. But this requirement may put extra load on server. Therefore we need a protocol to divide the load between the client and the server. The idea is provided in [7] and our construction supports this property following the protocol as mentioned below:

First the authentication server provides the salt to the client. The client performs the initialization phase, setup phase and iterative transformation phase (see Algorithm 2), and sends the end result to the server. The server computes the output generation phase and produces the final hash. This way we can easily reduce the load of the server.

Note: In this case, an attacker acting as a client, can repeatedly send some random data without following the computations of the proposed algorithm to the server. Here, the attacker can easily get the access with a correct guess. But, the complexity of the random guess will be equivalent to the brute-force complexity i.e. 2^n, where n is the output length of the underlying hash function. Therefore this can not be a feasible attack.

6   Implementation Aspects

This proposed construction for password hashing can be implemented efficiently on a wide range of processors. However, the same implementation will require huge number of computations if dedicated hardware such as ASIC or FPGA is used with limited memory.

Our design allows the flexibility to utilize less storage with increased number of calculations if we retain few bits of the intermediate hash computation after truncation and increase the number of iterations n. This way, *Rig* can be efficient on low memory devices.

We designed Rig to have a highly flexible structure. By changing the functions H_1, H_2 and H_3 (see Fig. 1) we can completely change the overall design properties. From side channel resistance to GPU or ASIC/FPGA resistance, any property can be achieved by the proper selection of the above primitives. Therefore there can be multiple variants aimed at different implementations or scenarios. As mentioned before, we describe the general construction of *Rig* as Rig $[H_1,H_2,H_3]$, where we can design/choose the functions H_1, H_2, H_3 for implementing different variants of *Rig*.

We have designed and implemented two versions of *Rig* as follows:

1. **Rig [Blake2b, BlakeCompress, Blake2b].** This variant is strictly sequential. Full Blake2b is used for H_1 and H_3 while the first round of the compression function of Blake2b is used for H_2 (and we call it *BlakeCompress*).

We have removed the constants in the 'G' function of Blake2b as suggested by the Blake authors in [3] to improve the overall performance. This version does a large number of random reads and writes and as a result it is strictly bounded by memory latency.

2. **Rig [BlakeExpand, BlakePerm, Blake2b].** This variant is designed to improve the performance by performing memory operations in larger chunks. The functions H_1 and H_2 are parallelized internally and the idea of handling larger chunk size improves the performance significantly without changing the overall sequential nature and memory-hardness of *Rig*. It also makes this variant of *Rig* much more difficult to execute efficiently in GPUs and FPGA/ASIC (explained in Sects. 6.4 and 6.5). We implemented the functions H_1 as 'BlakeExpand' and H_2 as 'BlakePerm'. These functions are explained later. The function H_3 uses full Blake2b.

6.1 Design of Rig [Blake2b, BlakeCompress, Blake2b]

This strictly sequential variant follows the general construction of *Rig* as explained in Sect. 4. The functions H_1 and H_3 implements Blake2b (full hash). The function H_2 is implemented using first round of Blake2b compression function.

6.2 Design of Rig [BlakeExpand, BlakePerm, Blake2b]

The optimized variant of Rig uses an expansion function BlakeExpand to expand the state and a compression function BlakePerm to compress the state. Full Blake2b is used to hash the output state after the iterative-transformation phase to obtain the final hash. The design aspects are described below.

6.2.1 Design of the BlakeExpand Expansion Function.
The BlakeExpand function is a very simple function which expands the input x to a fixed size of 8KiB. The function BlakeExpand is an instantiation of H_1. The input x passes through 128 individual instances of Blake2b (full hash) each appended by a counter as $x_i = x \parallel i$, for $0 \le i \le 127$ and produces the output $\alpha = \alpha_0 \parallel \alpha_1 \parallel \cdots \parallel \alpha_{127}$ where each α_i is of length 512 bits, i.e., 64 bytes. This construction ensures that the output of the function is random and the randomness depends solely on the cryptographic strength of Blake2b. Since this function needs to be executed only once, it has negligible impact on the performance of the overall *Rig* construction.

6.2.2 Design of BlakePerm Function.
We provide the design considerations for the function BlakePerm before the description of the design.

Design Considerations for BlakePerm Function. The DRAM memory latency is the limiting factor for the entire design of *Rig*. Initialization and copying data takes over 70 percent of the total run-time. In order to improve the overall performance one trivial optimization would be to increase the size of chunks in which the read and write operations are performed. The latest high performance processor offerings from Intel and AMD influenced many of the design decisions as they would be the most common target platform. There are several design considerations like:

- **L1 Cache Size.** This is one of the major factors because the L1 cache has the lowest latency of around 1–1.5 ns (as few as 3 clocks). Therefore it is important that the work piece (chunk) fits within this size for high performance in case of a compute intensive task.
- **L1/L2/L3 Cache Line Size.** A typical modern processor has cache line size of 64 bytes. Therefore working in multiple of 64 bytes with preferably aligned memory access is the best strategy. The problem with working with other non-multiple sizes is that there would be a lot of extra accesses and split loads and stores, which will dramatically reduce performance in computation intensive tasks. The Blake2b compression function nicely fits this requirement as it compresses 128 bytes to 64 bytes. The only other requirement is an implementation detail for setting the proper aligned memory access while memory allocation. As a result, in our implementation we have zero split load/stores.
- **DRAM Latency.** The memory latency is the primary limiting factor in algorithms having random reads and writes. DRAM generally has latency values of 250+ clock cycles. One strategy to get around this problem would be to perform reads and writes in larger chunks. If the chunk size is large enough, the performance hit due to latency can become significantly small. We tested against various sizes from 2 KiB to 64 KiB and observed that 16 KiB is a good size; and it also fits the L1 cache. We have, as a result chosen chunk size of 16 KiB for the H_2 function.

Design of the BlakePerm Function. The BlakePerm function is a compression function which compresses 16 KiB of data to 8 KiB. It is a two step function as described below.

1. **Compression.** It compresses the data using a single round Blake2b compression function. A single such compression function compresses 128 bytes to 64 bytes, as a result we need 128 such functions to compress 16 KiB to 8 KiB.
2. **Permutation.** A permutation layer is needed to mix the compressed data so that the bit-relations are spread evenly among the output bits over several rounds. Even though, any random permutation can be chosen in such a scenario, a permutation of the form: output $O_i = (I_i \times A + B) \mod C$ was chosen, where $0 \leq I_i \leq 1023$. The values $A = 109$, $B = 512$ and $C = 1024$ were chosen carefully after a series of experiments and diffusion tests. The permutation works on words of 8 bytes at a time, as a result the total addresses would be $8192/8 = 1024$. The value B = 512 is the number by which the overall permutation is cyclically rotated, it is half of the total size. The value of A is the most critical, even though the only requirement for a permutation is that A should be co-prime with C. The value of A strongly affects the overall characteristics and it needs to be carefully selected. It takes around 5 rounds for all

the output bits to be fully affected by a single change in the input data. Since H_2 function is sequentially applied hundreds of times, the function BlakePerm produces complete avalanche and is cryptographically strong.

6.3   Parallelization

The design of *Rig* is sequential and therefore it is impossible to parallelize the overall implementation. As a result we chose to parallelize the H_1 and H_2 functions. The most critical function which affects the performance of the overall design is the H_2 function. The optimized variant Rig [BlakeExpand, BlakePerm, Blake2b] takes an input of 16 KiB and compresses it using 128 Blake2b compression functions. These 128 operations can be done in parallel to improve the performance without affecting the overall sequential nature of *Rig*. H_1 can similarly be parallelized but it has negligible effect on the overall performance.

6.4   GPU Resistance

We designed *Rig* to have side-channel resistance, in pursuit of which we had to choose password-independent memory access patterns. Such memory-access patterns are harder to protect against GPU attacks. Modern GPUs have very strict requirements for memory accesses and very small cache sizes per core, as a result small random reads and writes dramatically reduce performance.

While designing H_2 of Rig [BlakeExpand, BlakePerm, Blake2b], we chose a permutation which causes reads and writes at significantly varying distances. Combined with the bit-reversal permutation used in *Rig* at the iterative transformation phase, the overall design is hard to parallelize efficiently.

As the H_2 function is pluggable, a new function can be easily added which performs small password-dependent memory accesses and make the design significantly GPU resistant. But, any such function would break the strict side-channel resistance.

6.5   ASIC/FPGA Resistance

The *Rig* construction is strictly sequential and is therefore non-parallelizable. The compression function H_2 (BlakePerm) as explained above, can be parallelized. But, the size of the inputs and outputs (16 KiB to 8 KiB) which needs 128 parallel instances of Blake2b compression function is too large for implementations with a large number of simultaneous *Rig* instances.

Even though there can be a lot of possibilities of implementations with varying numbers of compression functions, the overall space requirement still remains high.

The biggest problem in case of ASIC resistance would however come from the memory latency and bandwidth of the DRAM needed for storage of the extremely large state (several hundred megabytes to a few gigabytes). Even though the compression functions consume less power because of their simplicity, the latency and very high memory bandwidth requirements would make parallel implementations on ASIC prohibitively expensive. For example, for a single instance of Rig [BlakeExpand, BlakePerm, Blake2b] having $n = 4$ (5 memory passes), and 1 GiB of state, the bandwidth on a standard PC exceeds 7.37 GiB/s as shown in Table 2.

7 Performance Analysis

The reference implementation of *Rig* has been done in C language on an Intel Core i7-4770 CPU with 16 GB RAM at 2400 MHz. For the implementation of the first round of the compression function of Blake2b[1] for the function H_2, we use AVX2 instructions. Specifically these AVX2 instructions are used to parallelize the implementation of first round G-function of Blake2b. The following Tables (1 and 2) show the performance figure in terms of 'memory hashing speed' and 'DRAM bandwidth' for different values of parameter n (number of iterations).

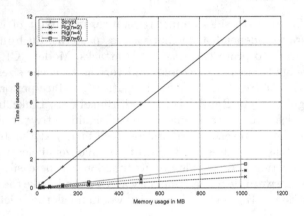

Fig. 2. Performance of Rig (at different value of n) and Scrypt.

7.1 Suggested Parameters

The performance figures provided in Tables 1 and 2 show that, as expected, the memory hashing speed for Rig [BlakeExpand, BlakePerm, Blake2b] is significantly higher than that of the strictly sequential variant.

Due to the wide spectrum of possible uses it is very difficult to suggest optimal values for parameters which suits every possible implementation scenario. However, we can suggest values for common applications. For the parameter 'n' (number of iterations) we suggest values higher than 3. This means that one should have at least four passes over memory (including setup phase). For some scenarios this may be increased to make low memory attacks prohibitively expensive.

---

[1] The idea of using reduced-round Blake2b is inspired from [2,5].

Table 1. Performance of RIG [Blake2b, BlakeCompress, Blake2b]

```
-----------------------------------------------------------------
|1)RIG[Blake2b, BlakeCompress, Blake2b]:Memory Hashing Speed(MiB/s)|
-----------------------------------------------------------------
| M =>    |  15 M |  30 M |  60 M | 120 M | 240 M | 480 M | 960 M   |
-----------------------------------------------------------------
| n =  2 |   681 |   684 |   663 |   675 |   681 |   674 |   659   |
| n =  4 |   383 |   383 |   381 |   388 |   388 |   391 |   392   |
| n =  6 |   268 |   270 |   267 |   275 |   272 |   258 |   270   |
| n =  8 |   211 |   210 |   210 |   212 |   210 |   201 |   209   |
| n = 10 |   174 |   173 |   170 |   173 |   170 |   174 |   170   |
-----------------------------------------------------------------
|  Memory Bandwidth (GiB/s)                                       |
-----------------------------------------------------------------
| n =  2 | 3.552 | 3.566 | 3.458 | 3.517 | 3.551 | 3.515 | 3.436   |
| n =  4 | 3.596 | 3.595 | 3.575 | 3.642 | 3.640 | 3.666 | 3.677   |
| n =  6 | 3.638 | 3.665 | 3.624 | 3.737 | 3.693 | 3.500 | 3.666   |
| n =  8 | 3.740 | 3.731 | 3.736 | 3.758 | 3.728 | 3.562 | 3.707   |
| n = 10 | 3.826 | 3.789 | 3.721 | 3.796 | 3.737 | 3.811 | 3.736   |
-----------------------------------------------------------------
```

Table 2. Performance of RIG [BlakeExpand, BlakePerm, Blake2b]

```
-----------------------------------------------------------------
|2)RIG[BlakeExpand, BlakePerm, Blake2b]:Memory Hashing Speed(MiB/s)|
-----------------------------------------------------------------
| M =>    |  64 M | 128 M | 256 M | 512 M | 1 GiB | 2 GiB | 4 GiB   |
-----------------------------------------------------------------
| n =  2 |  1377 |  1345 |  1307 |  1326 |  1315 |  1312 |  1318   |
| n =  4 |   858 |   846 |   829 |   845 |   835 |   838 |   833   |
| n =  6 |   621 |   621 |   617 |   618 |   606 |   610 |   619   |
| n =  8 |   500 |   485 |   481 |   490 |   485 |   489 |   489   |
| n = 10 |   403 |   398 |   399 |   402 |   404 |   404 |   402   |
-----------------------------------------------------------------
|  DRAM Memory Bandwidth (GiB/s)                                  |
-----------------------------------------------------------------
| n =  2 | 6.728 | 6.575 | 6.389 | 6.478 | 6.429 | 6.412 | 6.439   |
| n =  4 | 7.547 | 7.441 | 7.295 | 7.431 | 7.347 | 7.374 | 7.329   |
| n =  6 | 7.888 | 7.898 | 7.847 | 7.858 | 7.709 | 7.750 | 7.873   |
| n =  8 | 8.309 | 8.070 | 8.003 | 8.152 | 8.072 | 8.123 | 8.126   |
| n = 10 | 8.282 | 8.179 | 8.205 | 8.251 | 8.291 | 8.293 | 8.265   |
-----------------------------------------------------------------
```
n = no. of iterations, performance figures averaged over 20-iterations.

The memory count value (m_c) would depend strongly on the requirement and the actual use-case. For a server client architecture where the clients are expected to have enough free RAM, the value can be set to use few tens of megabytes to a few hundred megabytes. In a mobile environment, this can be further reduced to allow for clients with smaller memories. In the case the algorithm is to be used as a proof-of-work test, large memory requirements of a few gigabytes combined with a large 'n' value can be set. It is important to keep 'n' high (as high as 6–10) in case the overall memory cost is very small.

The performance of Scrypt (with suggested parameters [11]) and the results from Table 2 are depicted in Fig. 2. The graph shows the memory processing rate when consumable memory to compute the password hash is fixed. The comparison shows that the memory consumption of Scrypt is comparatively small with time. Scrypt takes approximately 6 seconds for 512 MB memory while Rig at $n = 2$ and $n = 4$ takes approximately 0.389 seconds and 0.613 seconds respectively.

8   Security Analysis

Rig satistifies the basic requirement of a non-invertible design for password hashing because of the following reasons: (i) the iterative use of underlying primitive, the (secure) cryptographic hash function and (ii) the initial hashing of password with random salt and other parameters and the final use of salt with chaining data. This makes recovering password from the hashed output quite challenging.

Another important point is the simple, sequential and symmetric design of the scheme. The simplicity makes it easy to understand and sequential design makes the parallel implementation hard and prevents significant speed up by the use of multiple processing units.

Flexibility of the design and the independence of the selection of memory parameter from time parameter makes it unique from existing constructions.

8.1   Resistance Against Low Memory Attack

Attacker's Approach: An attacker running multiple instances of Rig may try to do the calculations using smaller part of the memory (low memory) or almost no-memory (memory-free) to reduce the memory cost per password guess. This approach may allow parallel implementations of independent password guesses, utilizing almost all the available processing cores. This may not give advantage over single password hash computation but may increase the overall throughput of password guessing as compared to the legitimate implementation of the algorithm. Next we explain how feasible the low-memory or memory-free attack approach is, from the attacker's point of view.

Attack Scenario Varying the Required Storage Values: We emphasize that the goal of analyzing the complexity of low memory attack is to show the approximate impact on the overall processing cost to implement the algorithm Rig. Our construction needs to store two arrays a and k as shown in Fig. 1.

Therefore we try to compute the time complexity when most of these array values are not stored. The cost of calculation for the values of array a are dominated by the cost of array k. Therefore, for the simplicity of the evaluation we consider the calculation cost for array k.

To vary the required storage at each iteration, we assume that **we store t consecutive values,** $0 \leq t \leq m - 1$, of both the arrays at iterative transformation phase. This assumption is without loss of generality as we can easily calculate the index value of array k from the bit reversal permutation explained in Sect. 2. We also store the hash chaining values after each iteration.

Effect of Bit-Reversal Permutation on Low Memory Scenario: We use the bit-reversal permutation to shuffle the access of the array k. The effect of this yields exponential complexity for the low memory scenario. This is because at every step we update the values of array k and each updated value depends on all previous values. Let at iteration i, $1 \leq i \leq n$, $k[j]$, $0 \leq j \leq m - 1$, is the required value that is not stored. Then we need to compute the value $k[j]$ at all previous $i - 1$ iterations and as the access was not sequential, it is difficult to calculate the exact complexity. Hence, we compute the expected time complexity of a password hashing for memory constrained scenario.

Low Memory Attack Complexity: The algorithm *Rig* can be computed with time complexity $O((n + 1)mr)$ and space complexity $O(m)$ where $2m$ is the required number of stored values, n is the number of iterations used and r is the number of rounds. An attacker using reduced memory storage (i.e., 0 to $m - 2$ stored values) will require a time complexity of $O(r \times m^{n+1})$ for a single password computation.

The detailed analysis of the low memory attack complexity can be found in the full version of the paper uploaded on Cryptology eprint archive.

8.2   Resistance Against Collision Attack

In the design of *Rig* (see Fig. 3) we define three different functions, H_1, H_2 and H_3. The input of H_1 is x where x is the concatenation of password, 64-bit value of password length, salt, 64-bit value of salt length, 64-bit value of n (number of iterations) and 64-bit value of the output length of password hash. The output of H_1 is α. The function $\mathcal{H}_2$ signifies the repetitive computation of function H_2 at setup phase and iterative transformation phase (see Fig. 1) and generates the output c which is the output of iterative transformation phase. Therefore the inputs of $\mathcal{H}_2$ are α, m_c and n, where m_c is the number of memory count and n, the number of iterations used. Finally H_3 takes the concatenation of a 64-bit value which is the function of m_c and n, output of $\mathcal{H}_2$, the value salt and 64-bit value of 2^{m_c} and produces the output of password hash. Here we are considering round $r = 1$ (w.l.o.g). This is because, different values of round, say r and r' implies collision of H_3.

Theorem 1. *Collision for* Rig *means*

i. collision for H_1, or

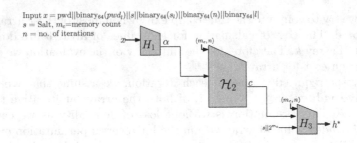

Input $x = $ pwd$||$binary$_{64}(pwd_t)||s||$binary$_{64}(s_t)||$binary$_{64}(n)||$binary$_{64}|l|$
$s = $ Salt, m_c=memory count
$n = $ no. of iterations

Fig. 3. Rig$[H_1, H_2, H_3](x, m_c, n, s)$, where $\mathcal{H}_2$ signifies repetitive use of H_2.

  ii. *collision for $\mathcal{H}_2$ when $\alpha \neq \alpha'$ and $(m_c, n) = (m'_c, n')$ for two different password hash computations, where $m_c = m'_c = $ the memory count and $n = n' = $ number of iterations, or*
iii. *collision for H_3.*

Proof. We analyse the collision of *Rig* with five possible cases as shown in Fig. 4. Specifically, we include all possible conditions that results in collision of H_1 or collision of $\mathcal{H}_2$ or collision of H_3 which implies the overall collision of *Rig*.

  CaseI.   Collision Rig $[H_1, H_2, H_3]$ if $(s, m_c, n) = (s', m'_c, n') \implies$ Collision H_1: The construction of *Rig* is such that if we get collision at H_1 for two different inputs say, x and x' and if $(s, m_c, n) = (s', m'_c, n')$ then it implies collision of $\mathcal{H}_2$ which implies collision at H_3, i.e., collision of Rig $[H_1, H_2, H_3]$.

  CaseII. Collision Rig $[H_1, H_2, H_3]$ if $(s, m_c, n) = (s', m'_c, n')$ and $\alpha \neq \alpha' \implies$ Collision $\mathcal{H}_2$: For two different inputs x, x' if $\alpha \neq \alpha'$ then collision of $\mathcal{H}_2 \implies$ collision of H_3 when $(s, m_c, n) = (s', m'_c, n')$ for respective inputs.

  CaseIII. Collision Rig $[H_1, H_2, H_3]$ if $(s, m_c, n) = (s', m'_c, n')$, $\alpha \neq \alpha'$ and $c \neq c'$ $\implies$ Collision H_3: For two different inputs x, x' if $\alpha \neq \alpha'$ and $c \neq c'$ and if $(s, m_c, n) = (s', m'_c, n')$ then collision Rig $[H_1, H_2, H_3] \implies$ collision H_3, i.e., collision is due to H_3.

  CaseIV. Collision Rig $[H_1, H_2, H_3]$ if $(s, m_c, n) \neq (s', m'_c, n') \implies$ Collision H_3: For two different inputs x, x' if $\alpha \neq \alpha'$ and $c \neq c'$ and if $(s, m_c, n) \neq (s', m'_c, n')$ then collision Rig $[H_1, H_2, H_3] \implies$ collision H_3.

  CaseV.   Collision Rig $[H_1, H_2, H_3]$ if $x = x'$ and $m_c \neq m'_c \implies$ Collision H_3: For two different password hash calculations if $m_c \neq m'_c$ for the same input x then collision of Rig $[H_1, H_2, H_3]$ is for collision at H_3.

Therefore these five cases describe how collisions of *Rig* implies collision of H_1 or $\mathcal{H}_2$ or H_3. Hence *Rig* is collision resistant if H_1 or $\mathcal{H}_2$ or H_3 are collision resistant.   □

The implementations of the provided variants are expected to be secure against collision attack. The detailed analysis is included in the full version of the paper uploaded on Cryptology eprint archive at https://eprint.iacr.org/2015/009.pdf.

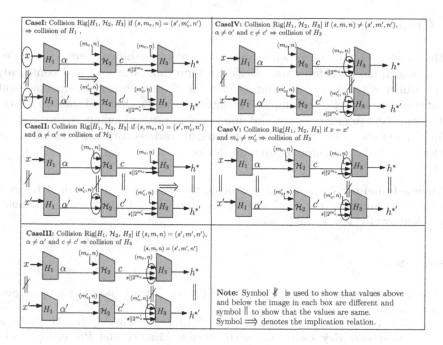

Fig. 4. Five cases showing collisions for Rig $[H_1, H_2, H_3]$.

8.3 Resistance Against Cache-Timing Attack

As discussed in Sect. 5, our construction accesses array k (see Sect. 4) using a bit reversal permutation which is password independent. Hence cache timing attack is not possible on our costruction. Further, since the only primitive used in our scheme is a secure hash function, the security of our scheme can be formulated in terms of the security of the underlying hash function. With the current state-of-the-art we have the possibility of using SHA-3 implementation, or even any of the other finalists of SHA-3 competition, which are resilient to side-channel attacks. Thus our scheme resists cache timing attacks.

8.4 Resistance Against Denial-of-Service Attack

In computing, a denial-of-service (DoS) attack is an attempt to make a machine or network resource unavailable to its intended users. This is possible by making the server busy injecting lots of request for some resource consuming calculation. It is quite easy if the server uses some slow password hashing technique for authentication. To handle such situations, the server-relief technique can provide some relief to the server from heavy calculations as the client will do the heavy part of the algorithm. This way we can reduce the chances of DoS attacks with slow password hashing schemes.

9 Conclusions

In this work, we have proposed a secure password hashing scheme *Rig* based on cryptographic hash functions. Besides supporting necessary and commonly used features of a password hashing scheme, *Rig* also supports client-independent updates and server relief technique. The flexibility in the choice of the two important parameters (memory count and number of iterations) enhances the scope of its implementation. *Rig* can be implemented in various software and hardware platforms including resource constraint devices.

Acknowledgments. We would like to thank the anonymous reviewers of Inscrypt and the contributors to the PHC mailing list (specially Bill Cox) whose comments helped improve the paper significantly.

References

1. Password Hashing Competition (PHC) (2014). https://password-hashing.net/index.html
2. Almeida, L.C., Andrade, E.R., Barreto, P.S.L.M., Simplício, Jr., M.A.: Lyra: Password-Based Key Derivation with Tunable Memory and Processing Costs. IACR Cryptology ePrint Archive 2014:30 (2014)
3. Aumasson, J.-P., Neves, S., Wilcox-O'Hearn, Z., Winnerlein, C.: BLAKE2: Simpler, smaller, fast as MD5. In: Jacobson, M., Locasto, M., Mohassel, P., Safavi-Naini, R. (eds.) ACNS 2013. LNCS, vol. 7954, pp. 119–135. Springer, Heidelberg (2013)
4. Carvalho, C.: The gap between processor and memory speeds. In: Proceedings of IEEE International Conference on Control and Automation (2002)
5. Daemen, J., Rijmen, V.: A new MAC construction ALRED and a specific instance ALPHA-MAC. In: Gilbert, H., Handschuh, H. (eds.) FSE 2005. LNCS, vol. 3557, pp. 1–17. Springer, Heidelberg (2005)
6. Dürmuth, M., Güneysu, T., Kasper, M., Paar, C., Yalcin, T., Zimmermann, R.: Evaluation of standardized password-based key derivation against parallel processing platforms. In: Foresti, S., Yung, M., Martinelli, F. (eds.) ESORICS 2012. LNCS, vol. 7459, pp. 716–733. Springer, Heidelberg (2012)
7. Forler, C., Lucks, S., Wenzel, J.: The Catena Password Scrambler, Submission to Password Hashing Competition (PHC) (2014)
8. Gray, J., Shenoy, P.: Rules of Thumb in Data Engineering. Technical Report, MS-TR-99-100, Microsoft Research, Advanced Technology Division. December 1999, Revised March 2000
9. Lengauer, T., Tarjan, R.E.: Upper and lower bounds on time-space tradeoffs. In: Proceedings of the 11h Annual ACM Symposium on Theory of Computing, April 30 – May 2, 1979, Atlanta, Georgia, USA, pp. 262–277 (1979)
10. Burr, W., Turan, M.S., Barker, E., Chen, L.: NIST: Special Publication 800–132, Recommendation for Password-Based Key Derivation. Computer Security Division Information Technology Laboratory. http://csrc.nist.gov/publications/nistpubs/800-132/nist-sp800-132.pdf

11. Percival, C.: Stronger key derivation via sequential memory-hard functions. In: BSDCon (2009). http://www.bsdcan.org/2009/schedule/attachments/87_scrypt.pdf
12. Provos, N., Mazières, D.: A future-adaptable password scheme. In: USENIX Annual Technical Conference, FREENIX Track, pp. 81–91. USENIX (1999)
13. Schaller, R.R.: Moore's law: past, present, and future. IEEE Spectrum, June 1997

Authentication and Encryption

Efficient Hardware Accelerator for AEGIS-128 Authenticated Encryption

Debjyoti Bhattacharjee[1]($\boxtimes$) and Anupam Chattopadhyay[2]

[1] Indian Statistical Institute, Kolkata, India
debjyoti08911@gmail.com
[2] School of Computer Engineering, NTU, Singapore, Singapore
anupam@ntu.edu.sg

Abstract. Security of transaction is of paramount importance in modern world of ubiquitous computing and data movement. To provide a framework of standard authenticated encryption techniques, CAESAR contest has been announced recently. Multiple entries in this contest are based on AES, which has been also, a popular choice as a primitive for authenticated encryption in the past. In this paper, we perform in-depth study of efficient hardware implementation for AES-based AEGIS-128 authenticated encryption, a prominent entry in the CAESAR contest. Through a complete study of possible throughput-area improvement techniques, we report multiple design points ranging from a high throughput of 121.07 Gbps design to a low-area implementation of 18.72 KGE, using commercial synthesis flows and 65 nm ASIC technology. We believe our results will serve as important design metric for the CAESAR contest as well as for efficient AEGIS-128 deployment.

1 Introduction

AEGIS is an authenticated encryption algorithm in the currently ongoing CAESAR [4] competition. AEGIS-128 processes 80-byte state vector in a single state update operation as shown in the Fig. 1 of the algorithm. Each state update uses 5 AES rounds in parallel to update the 80-byte state.

Among nearly 50 entries reported to CAESAR, nearly 1/5th are based on AES [1]. Besides this current contest, high-performance and area-efficient AES-based authenticated encryption schemes have been reported in the literature [2,3]. This makes AEGIS an important candidate in the current portfolio of CAESAR. Furthermore, no attacks on AEGIS have been reported so far, though this is not true for all AES-based authenticated encryption schemes, e.g., ALE [3,20]. In this paper, we study AEGIS for efficient implementations from the perspective of area and throughput. The key computational block, i.e., the state update is studied closely via theoretical analyses to reveal potential improvement options. Whenever possible, we resort to the rich literature of efficient AES implementation [5,10] to gain maximum improvement.

The rest of the paper is organized as following. In the Sect. 2, the potential throughput improvement and area minimization techniques are discussed from a

© Springer International Publishing Switzerland 2015
D. Lin et al. (Eds.): Inscrypt 2014, LNCS 8957, pp. 385–402, 2015.
DOI: 10.1007/978-3-319-16745-9_21

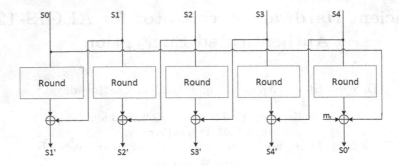

Fig. 1. AEGIS StateUpdate128 operation block diagram

theoretical perspective. Section 3 discusses the hardware implementation based on these analyses. The experimental results are reported and benchmarked with other authenticated encryption implementations in the Sect. 4. Section 5 concludes the paper with outline of future works. To make the paper self-contained, the AEGIS algorithm is briefly reviewed in the Appendix A.

2 Theoretical Analysis of AEGIS for Improving Cycle-per-Byte

We analyze the performance of AEGIS-128 from a theoretical perspective with respect to cycles-per-byte (cpb). Let $msglen$ and $adlen$ be the length of plaintext and associated data respectively. We assume 128-bit tag generation by AEGIS-128. Let n be the number of clock cycles required for completing one StateUpdate128 (kindly refer to Appendix A for details) operation. Thus, we need

- 10 StateUpdate128 operations for initialization
- $\lceil\frac{adlen}{128}\rceil$ number of StateUpdate128 operations for processing associated data
- $\lceil\frac{msglen}{128}\rceil$ number of StateUpdate128 operations for encryption
- 7 StateUpdate128 operations for finalization.

Therefore the cpb of AEGIS-128 for encryption of $msglen$-bits of plaintext and processing $adlen$-bits of associated data with tag generation is

$$cpb = n(10 + \lceil\frac{adlen}{128}\rceil + \lceil\frac{msglen}{128}\rceil + 7)/(\frac{msglen}{8}) \qquad (1)$$

2.1 Pipelining

We analyze pipelining as a method to improve on the throughput of the AEGIS-128 implementation. Based on the optimized AES implementations, the AES round operation is split into four pipeline stages, namely SBOX, SR, MC and ARK. SBOX stage performs AES-SubBytes operation using Rijndael S-box, SR stage performs AES-ShiftRows step, MC stage performs AES-MixColumns step and ARK stage performs AES-AddRoundKey step. The choice of number of pipeline stages to be four is governed by two factors. First, this enables splitting the effective critical path of the design into approximately four equal subparts, thereby, theoretically we can effectively increment the maximum operating frequency 4× in comparison to the non pipelined implementation of the algorithm. Second, the aforementioned four-stage distribution does not cause any data dependency and therefore, stalls are avoided. We refrain from using a 5-stage pipeline due to the extra burden of pipeline registers, which will contribute significantly to the area of the design. The scheduling of pipeline is discussed in detail in the Sect. 3. For processing the first 128-bit part of the current state, the architecture requires for 4 clock cycles for initial loading of the pipeline. For processing the consecutive 128-bit part of the state, a single clock cycle is required. Therefore, cycle per byte can be calculated as follows.

1. Initialization = 4+4+9×5 = 53 clock cycles (Initial 8 clock cycles are required for the first StateUpdate128, and each of the consecutive StateUpdate128 operations require 5 cycles.).
2. Processing of authenticated data = $5 \times \lceil \frac{adlen}{128} \rceil$ clock cycles.
3. Encryption = $5 \times \lceil \frac{msglen}{128} \rceil$ clock cycles.
4. Finalization = $5 \times 7 = 35$ clock cycles.

Therefore cpb with pipelined architecture is

$$cpb_{pipe} = (88 + 5(\lceil \frac{adlen}{128} \rceil + \lceil \frac{msglen}{128} \rceil)) / \frac{msglen}{8} \qquad (2)$$

The corresponding area required for implementation of the architecture can be approximated as following.

$$area_{pipe} = (A_{round} + A_{pipeline\_register} + A_{state\_register} + A_{next\_state\_register}), \quad (3)$$

where A_{round} is the area required for one AES round, $A_{pipeline\_register}$ is the area needed for the pipeline registers. By efficient implementation of AES, area of A_{round} can be reduced. For low-area AEGIS design, we focused on the reduction of the state registers, $A_{state\_register}$ and $A_{next\_state\_register}$ which are individually of size 80-bytes. The optimizations are discussed in the following Sect. 3.

2.2 Parallelism

AEGIS-128 algorithm offers the possibility of processing the entire 80-byte state in parallel by using 5 AES-Round cores in parallel. The parallel architecture thereby offers a very high throughput. It requires a single clock cycle to complete the StateUpdate128 operation.

The throughput of parallel architecture can be derived as follows.

1. Initialization = 10 clock cycles
2. Processing of authenticated data = $\lceil \frac{adlen}{128} \rceil$ clock cycles

3. Encryption = $\lceil \frac{msglen}{128} \rceil$ clock cycles
4. Finalization = 7 clock cycles

The cpb of parallel architecture is thus

$$cpb_{parallel} = (17 + \lceil \frac{adlen}{128} \rceil + \lceil \frac{msglen}{128} \rceil) / \frac{msglen}{8} \qquad (4)$$

The area projection of the parallel architecture is

$$area_{parallel} = (5 \times A_{round} + A_{state\_register}) \qquad (5)$$

Though, the parallelization increases the combinational logic by requiring 5 AES round operations, the state registers can be reduced significantly. The updated state can be directly written to the current state register as shown in Fig. 9.

2.3 Unrolling

We can unroll the design for obtaining a even lower cpb in comparison to the parallel architecture mentioned in Sect. 2.2. Let n be the unroll factor. Then we can compute the cpb as follows. In one clock cycle, n StateUpdate128 operations can be completed. Total number of StateUpdate128 operations required is $s_{num} = 10 + \lceil \frac{adlen}{128} \rceil + \lceil \frac{msglen}{128} \rceil + 7$. Therefore total number of clock cycles required till finalization is given as $\lceil \frac{s_{num}}{n} \rceil$. Hence, we have the following.

$$cpb_{unrolled} = \lceil \frac{s_{num}}{n} \rceil / \frac{msglen}{8} \qquad (6)$$

The area projection of the unrolled architecture is

$$area_{unrolled} = n \times (5 \times A_{round} + A_{state\_register}) \qquad (7)$$

Thus, the area-efficiency (η) can be determined as following.

$$\eta_{unrolled}^{area} = \frac{throughput_{unrolled}}{area_{unrolled}} = \frac{n \times \frac{msglen}{8}}{n \times (5 \times A_{round} + A_{state\_register})} = \eta_{parallel}^{area} \qquad (8)$$

The area efficiency $\eta_{unrolled}^{area}$ of this design is same as that of the parallel architecture $\eta_{parallel}^{area}$. Since we do not gain on area-efficiency in comparison to the parallel architecture mentioned in Sect. 2.2, we did not implement the unrolled architecture.

3 Accelerator Implementation of AEGIS-128

Based on the theoretical analysis, the accelerator is implemented for meeting different design constraints. The theoretical guidelines for improving cpb are used, while possibilities for reducing the storage are explored. We could achieve

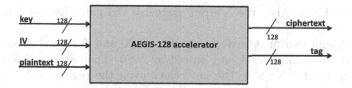

Fig. 2. Top-level Structure of AEGIS-128 core

the projected cpb for all the optimization possibilities. For all the design variants, the same top-level structure, as shown in the Fig. 2 is maintained.

For all the designs, the current 80-byte state is stored in 128-bit registers, denoted as Si, where $0 \leq i \leq 4$. The next state is represented by Si', where $0 \leq i \leq 4$. The next state is represented by NSi, where $0 \leq i \leq 4$. It should be noted that S_i represents the 80-byte state of AEGIS-128 and should not be confused with the 128-bit state registers Si.

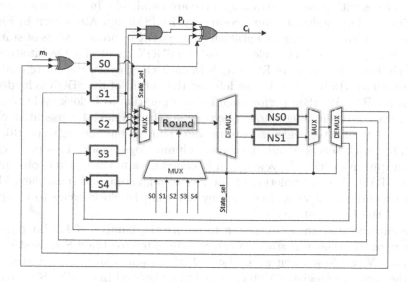

Fig. 3. Base implementation of AEGIS-128 core

3.1 Base Implementation

In this implementation, a single AES Round is executed in a clock cycle processing 128-bit of current state. It takes input from the state registers $S0 \cdots S4$ depending on the last 128-bit state processed by the AES Round. It takes 5 clock cycles to complete a single StateUpdate128 operation. The architecture is shown in Fig. 3. We have not shown the control logic and signals in the architecture diagrams for clarity. Also, the 5-input exclusive OR gates for tag generation is skipped.

In this design two next state registers - $NS0$ and $NS1$ are used. Initially, 128-bit state $S4$ is processed and stored in $NS0$. Then state $S0$ is processed and saved in register $NS1$. In the next clock cycle,the register $NS0$ is saved to register $S0$ and state $S1$ is processed and the result is saved to register $NS0$. In the following clock cycle, the register $NS1$ is saved to register $S1$ and state $S2$ is processed and the result is saved to register $NS1$. Then, $S3$ is processed and the result is directly written to state register $S4$, and the next state registers $NS0$ and $NS1$ are written to state registers $S2$ and $S3$ respectively. By this simple optimization, 3 out of 5 next state registers can be optimized away.

3.2   Design Points for Area Optimization

Design $A0_1$

It is often necessary to have cryptographic hardware accelerators to run on low power devices with strict constraints on the area available. In order to serve such scenarios, we implemented a very compact AEGIS design $A0_1$ shown in Fig. 4. We used the Canright's implementation [5] of SBOX to process 8 bits of state at a time. It requires 16 clock cycles to load the SBOX result of a 128-bit state Si to the sixteen 8-bit registers R0$\cdots$R15. In each clock cycle, the entire chain of registers R0 to R15 is shifted to the left and the output of the SBOX is loaded to the register R15. We then perform shift rows operation in 1 clock cycle, since it just involves data shuffling among the registers. In the next 4 consecutive clock cycles, we left shift the registers column-wise, shifting out the registers [R0, R4, R8, R12], which serves as input to mix columns logic block. This produces a 32-bit output in each clock cycle which is loaded back to the register column[R3, R7, R11, R15]. After completion of the mix column step, add round key(ARK) step is executed 8-bit at a time thereby requiring 16 clock cycles to complete update of the 128-bit state Si.

We implemented an optimization to reduce the number of 128-bit registers necessary to store the next state. When $S0$ is updated to obtain $S'1$, we store $S'1$ in register NSO. Now when we update $S1$, the result is stored to register $NS0$ while the content of register $NS0$ is loaded to register $S1$ in parallel. Similarly, in the next two clock cycle, $S2$ and $S3$ are updated respectively. During the clock cycle, in which $S4$ is updated, the result of update i.e. $S0'$ is stored directly to register $S0$ and the value of register $NS0$ is saved to the register $S4$. The number of clock cycles required by state update operation is reduced, by performing ARK on the contents of 8-bit register R0, left shift by one the entire chain of registers R1 to R15, and loading the 8-bit SBOX result from processing the next 128-bit state to the register R15 in a single clock cycle. Hence, the SBOX operation and ARK operation can be performed in parallel. Thereby, only for the first 128-bit state update, we require $16 + 1 + 4 + 16 = 37$ clock cycles and for all the following 128-bit state update, $1 + 4 + 16 = 21$ clock cycles are required.

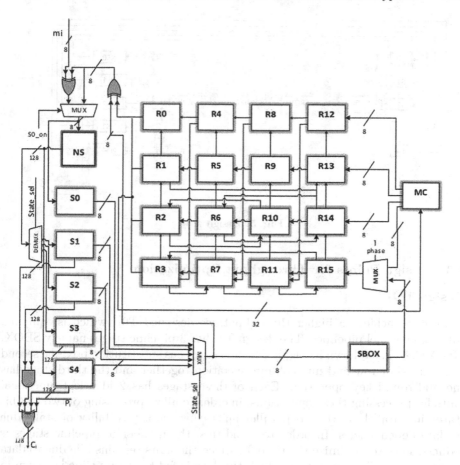

Fig. 4. Design A0$_1$

Design AO$_2$

Design A0$_2$ implements the AEGIS algorithm using a 4-stage pipeline as shown in Fig. 5. The pipelined implementation improves on the maximum operating frequency of the AEGIS core over the base implementation. The design has the operations substitution-bytes, shift-rows, mix-columns and add-round-key in the stages SBOX, SR, MC and ARK respectively. A0$_1$ has a 128-bit datapath. Each stage requires a single clock cycle to process its input. The order in which the 128-bit state Si is processed is shown in Fig. 6. If the first state processed in a StateUpdate128 operation is Si, then in the next StateUpdate128 operation, the first state to be processed is $S(i+1)\%5$. Such an ordering of processing the states eliminates the possibility of any pipeline stalls due to data dependencies. The first StateUpdate128 operation requires 8 (4 clock cycles to fill the pipeline and process $S0$ and then next 4 clock cycles for states $S1, S2, S3, S4$) clock cycles. Subsequent StateUpdate128 operations requires 5 clock cycles to complete.

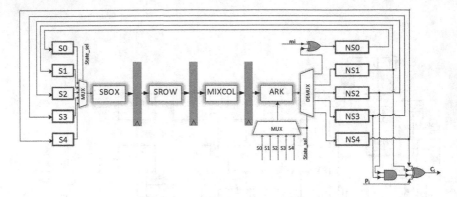

Fig. 5. Design $A0_2$

3.3   Design Points for Throughput Optimization

Design TO_1

In order to achieve a higher throughput, we designed TO_1, which is equipped with two parallel pipelines. The design TO_1 has 3 pipeline stages namely SBOX, SR+MC,ARK as shown in Fig. 7. The first stage has substitution box, the second stage has shift row and mix column operation together and the third stage has the add round key operation. Each of these stages has 2 identical functional units for processing these operations, in order to allow processing of two 128-bit states in parallel. With two parallel units, there is a possibility of stalls due to data dependencies. In order to avoid this, the number of pipeline stages is reduced so as to minimize the time for which the units remains idle due to data dependencies. The schedule in which the data is fed to the pipeline is shown in Fig. 8. During the update of the first four 128-bit part of the state, we achieve 100 % hardware utilization by processing two 128-bit in parallel. For the 5th 128-bit of the state, due to data dependency, we can start processing only one 128-bit state and therefore one of the pipelined Round functional unit remains idle.

For the first StateUpdate128 operation, $S0, S1$ is processed in the first 3 clock cycles, $S2, S3$ get processed by the next clock cycle, and $S5$ is completely processed in the next clock cycle. Thereafter, it requires 3 clock cycles to complete one StateUpdate128 operation. For obtaining the maximum utilization of the hardware, we suggest to use the following schedule for processing the states Si. During the first stage of StateUpdate128 operation, if Si and $Si + 1$ are processed, then during the first stage of the next StateUpdate operation, states $Si + 1$ and $Si + 2$ are to be processed. The schedule is depicted in Fig. 8. This implementation provides a throughput improvement of 1.44× in comparison to the implementation $A0_2$ while having a corresponding area increase of 1.4×.

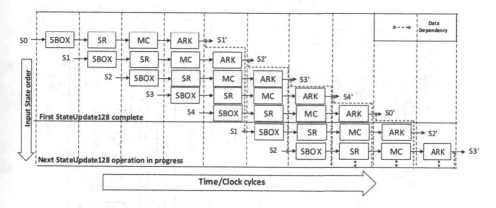

Fig. 6. Pipeline Schedule of design AO_2

Design TO_2

In order to push the throughput even further, we designed the high performance AEGIS-128 hardware accelerator TO_2 shown in Fig. 9. Five Round operations process the entire 80-byte state S in parallel in a single clock cycle. This implementation directly writes back the updated state value into the state registers, without the need for any intermediate register and has greatly reduced control logic when compared to the other implementations. It requires a single clock cycle for the StateUpdate128 operation to complete in this implementation.

4   Results and Benchmarkings

At first, we report the performances of the different design points explained in the previous sections. This is coupled with a comparison with the software implementation performance, reported by the authors [19]. Next, we report CAESAR entries, where performance data is available. It is expected that all the CAESAR entries will go through a detailed, fair benchmarking process in terms of their hardware performance[1]. However, not all the entries have reported hardware/software performance figures yet. Moreover, performance reported at diverse technology nodes and PVT corners make the process of comparison extremely daunting. Finally, AEGIS performance is compared to the latest hardware implementation results of authenticated encryption modes of AES. As detailed in a recent analysis for CAESAR entries' competitiveness [9], we perform the comparison with AES-CCM and AES-GCM modes' implementations.

The hardware accelerators have been implemented with Language for Instruction Set Architecture (LISA) [6], which allows a cycle-accurate description to be modeled using high-level languages. Part of the design uses block cipher constructions, for which the tool reported in [8] is used. Synopsys Processor Designer [18]

---

[1] We are happy to share the RTL implementation to inquisitive researchers for this purpose.

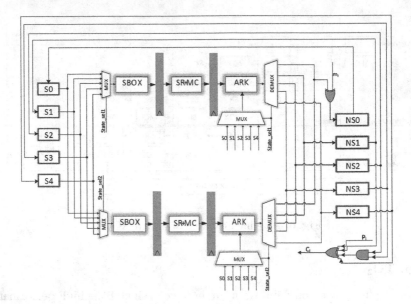

Fig. 7. Design TO_1

tool was used to generate register-transfer level (RTL) implementation for the accelerators. We generated timing-optimized and area-optimized RTL automatically from the high-level description. High-level design entry allowed multiple design iterations within a short span of time. On the other hand, possibility to include RTL code directly in the high-level description kept the scope for low-level optimizations where needed. The circuits have been synthesized and evaluated with Synopsys Design Compiler, version H-2013.03-SP1 using the 65-nm CMOS Faraday technology library. For power estimation, switching activity files are back-annotated from gate-level simulation. Detailed performance results are presented in Table 1. Throughput, Area (indicated by KGE), area-efficiency, total power (in mW) and energy-efficiency (joule/bit) are the performance metrics used to evaluate the performance of the accelerators. Unless mentioned otherwise, a message length of 1024 bytes for calculating throughput.

The performance of AEGIS-128 hardware accelerators in cycles per byte is shown in Table 2. The data is also plotted graphically in Fig. 10. The cycles per byte for design AO_1 are relatively high when compared to other designs,hence are not shown in the plot.The prime reason is due to the fact that AO_1 is a highly area optimized design. Considering high throughput, we have been able to achieve very low cpb values of 0.07 for message length of 4096 bytes for design TO_2.

It can be noted that, in comparison to the software performance of AEGIS-128, reported for processors with special instruction sets for AES [19], each one of our hardware accelerators are faster, except the area-optimized design AO_1. For 4096-byte messages, the performance of our designs - base implementation,

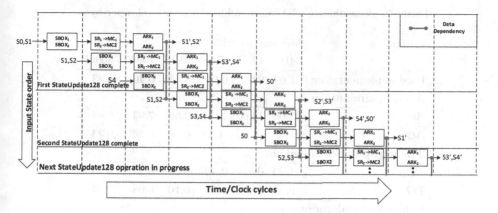

Fig. 8. Pipeline Schedule for design TO_1

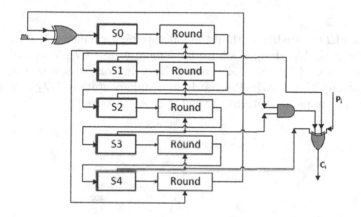

Fig. 9. Architecture of Design TO_2

Table 1. Performance of AEGIS-128 architectures

Design	Optimization	Clock frequency (MHz)	Area (KGE)	Throughput (Gbps)	Efficiency (Gbps/ kGE)	Total power (mW)	Energy efficiency (pJ/bit)
Base	Speed	915	59.34	17.04	0.29	74.41	4.07
implementation	Area	915	57.01	17.04	0.30	64.65	1.06
AO_1	Speed	1435	20.55	1.35	0.07	21.96	15.17
	Area	1410	18.72	1.32	0.07	21.58	15.17
AO_2	Speed	2010	60.88	37.44	0.61	33.08	0.82
	Area	1962	56.65	36.55	0.65	34.21	0.87
TO_1	Speed	1725	88.91	53.55	0.60	28.33	0.49
	Area	1700	84.71	52.77	0.62	28.33	0.50
TO_2	Speed	1300	172.72	121.07	0.70	232.75	1.79
	Area	1300	172.72	121.07	0.70	232.75	1.79

Table 2. Performance of AEGIS architectures in cycles per byte

Design	Message length					
	64B	128B	256B	512B	1024B	4096B
Base implementation	1.64	0.98	0.64	0.48	0.40	0.33
Area optimization						
AO_1	33.06	19.66	12.95	9.60	7.93	6.67
AO_2	1.69	1.00	0.66	0.48	0.40	0.33
Throughput optimization						
TO_1	1.03	0.61	0.40	0.29	0.24	0.20
TO_2	0.33	0.20	0.13	0.10	0.08	0.07
Software implementation						
AEGIS(EA)[19]	3.37	1.99	1.30	0.96	0.8	0.66

AO_2, TO_1 and TO_2 are better than the software performance [19] of the AEGIS-128 implemented. The designs - base implementation, AO_2, TO_1 and TO_2 are faster by 2×, 2×, 3.3× and 9.4× respectively in comparison to the software implementation. The AEGIS-128 hardware implementation TO_2 is faster than all the software implementations of the CAESAR entries [4].

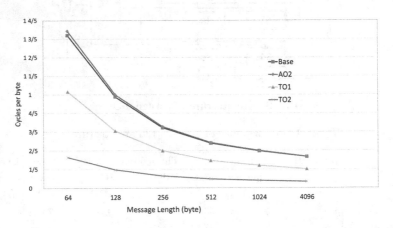

Fig. 10. Cycles per byte vs message length for AEGIS-128 architectures

We highlight the key points of the performance results. At first, we would like to state that design AO_1 has the lowest area of about 18 kGE, among all the designs of AEGIS-128. By sharing of resources, we are able to reduce the area like using a single SBOX unit to process 8-bits of state in a clock cycle. Similarly, we used a single unit for mix column for processing 32-bits in a clock cycle.

The design AO_2 has a very similar area in comparison to the BASE implementation. Due to pipelining, the maximum operating frequency of $A0_2$ is at least 1.88× more than the base implementation, resulting in a improved throughput which 2.14× of that of BASE implementation.

The design TO_1 is the most energy efficient design for AEGIS-128 with an energy efficiency of 0.50 pJ/bit. It has a throughput of 1.4× of the design AO_2. The reason for this value to be lower than 2 is due to the fact, the data dependencies prohibits the full utilization of both the pipelined Round cores at all the times.

The design TO_2 has the highest performance in comparison of to rest of the designs. It achieves a maximum throughput of 121.07 Gbps at 1300 MHz. This design also achieves the highest efficiency with respect to area with a value of 0.7 Gbps/kGE. The prime reason being the fully parallel cores required the least amount of control circuitry leading to savings in area and with all the 5 Round cores running in parallel without any data dependencies, it is able to achieve highest throughput.

4.1   Reported Performance Results of CAESAR Entries

We summarize the known implementations and their results without drawing any hard comparisons.

- ICEPOLE v1 [11]: ICEPOLE implementation reported a throughput of about 42 Gbps and 38.78 Gbps for implementation using Xilinx Virtex 6 and Altera Stratix IV FPGA.
- Minalpher v1 [15]:
  1. Minealpher High Speed Core: The area and timing optimized cores have a throughput of 6.1 Gbps and 9.9 Gbps respectively with an area efficiency of 426.37 Mbps/kGE and 596.16 Mbps/kGE respectively.
  2. Minalpher Low-area coprocessor: The area and timing optimized cores have an area of 2.81 KGE and 3.36 KGE respectively at 45 nm technology node.
- SCREAM &iSCREAM [7]: SCREAM-10(2R/cycle) offers the maximum throughput of 5.19 Gbps with an area of $17292\,\mu m^2$ KGE while SCREAM-10(1R/cycle) offers throughput of 4.74 Gbps with an area of $12951\,\mu m^2$. iSCREAM-12(2R/cycle) offers the maximum throughput of 4.41 Gbps with an area of $17024\,\mu m^2$ while iSCREAM-12(1R/cycle) offers throughput of 3.79 Gbps with an area of $13375\,\mu m^2$. Both the implementations are reported for 65 nm technology node. In the same technology node, for the sake of a very approximate reference, the smallest area for our AEGIS-128 implementation takes $23961\,\mu m^2$ with a throughput of 1.32 Gbps.

4.2   Comparison with AES-based Authenticated Encryption Schemes

Two recently reported efficient hardware implementations for AES-GCM and AES-CCM authenticated encryptions have following performance figures.

- AES-GCM [12]: AES-GCM required 20.5 KGE area and has a throughput of 2.62 Gbps using 90 nm technology node. In comparison to AES-GCM,.
- AES-CCM [13]: AES-CCM implementation with the highest throughput offers maximum throughput of 12.35 Gbps and has 875 KGE area using 65 nm technology node.

In contrast, AEGIS-128 lowest area implementation has 18.72 KGE with a throughput of 1.32 Gbps and the highest throughput of AEGIS-128 architecture is 121.07 Gbps with an area of 172.72 KGE. Clearly, AEGIS-128 offers significantly higher throughput than that could be achieved with AES variants. It can be noted that, the speed achievable with AEGIS is considerably higher than that reported in accelerator implementations for prominent block/stream ciphers in recent times [14, 16, 17]. At the same time, for a low-area implementation, AEGIS offers a very competitive solution.

5   Conclusion

In this paper, we performed a detailed implementation study of AEGIS authenticated encryption algorithm. The optimizations for throughput and area are derived from a theoretical analysis. Subsequently, diverse design points with various performance metrics, e.g., area, throughput, area-efficiency, energy-efficiency and power are proposed. These design points are implemented using a high-level design platform and their pre-layout implementation reports using an ASIC technology library are reported. The implementation results are shown to be competitive in terms of area and significantly superior in terms of throughput, when compared with state-of-the-art authenticated encryption protocols.

A   AEGIS-128

We briefly describe the AEGIS-128 algorithm [19]. AEGIS-128 uses a 128-bit key and 128-bit initialization vector to encrypt and authenticate a message. The associated data length and plain text length are less than 2^{64} bits. The authentication tag length is less than or equal to 128-bits. The use of 128-bit tag length is strongly recommended.

The state update function StateUpdate128 of AEGIS-128 updates a 80-byte state S_i using a 16-byte message block m_i. StateUpdate128 is stated in Algorithm 1. The block diagram of StateUpdate128 function is shown in Fig. 1. The function Round is an AES encryption round as shown in Fig. 11.

A.1   Encryption and Generation of Tag of AEGIS-128

AEGIS-128 has 4 phases for encryption. We describe each phase concisely. We use the same notations that are defined in the original paper with the description of AEGIS-128 [19].

Algorithm 1. StateUpdate128 function of AEGIS-128

1: **function** STATEUPDATE128(S_i , m_i)
2:      $S_{i+1,0}$ = Round($S_{i,4}$, $S_{i,0} \oplus m_i$)
3:      $S_{i+1,1}$ = Round($S_{i,0}$, $S_{i,1}$)
4:      $S_{i+1,2}$ = Round($S_{i,1}$, $S_{i,2}$)
5:      $S_{i+1,3}$ = Round($S_{i,2}$, $S_{i,3}$)
6:      $S_{i+1,4}$ = Round($S_{i,3}$, $S_{i,4}$)
7:      return S_{i+1}
8: **end function**

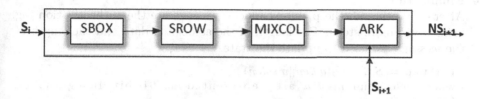

Fig. 11. Round function used by StateUpdate128

1 Initialization of AEGIS-128

Initialization of AEGIS-128 consists of loading the key and IV into the state, and running the cipher for 10 steps with the key and IV being used as message.

$$S_{-10,0} = K_{128} \oplus IV_{128}$$
$$S_{-10,1} = const_1$$
$$S_{-10,2} = const_0$$
$$S_{-10,3} = K_{128} \oplus const_0$$
$$S_{-10,4} = K_{128} \oplus const_1$$

for $i = -10$ to -1
{
    if ($i\%2==0$)
        $m_i = K_{128}$
    else
        $m_i = K_{128} \oplus IV_{128}$

    S_{i+1} = StateUpdate128 (S_i , m_i)
}

2 Processing the Authenticated Data

The associated data AD is used to update the state. If the last associated data block is not a full block, use 0 bits to pad it to 128 bits, and the padded full block is used to update the state. Note that if adlen = 0, the state will not be updated.

    S_{i+1} = StateUpdate128 (S_i , AD_i)
}

3 Encryption of Plaintext Data

If the last plaintext block is not a full block, use 0 bits to pad it to 128 bits, and the padded full block is used to update the state. But only the partial block is encrypted. Note that if msglen = 0, the state will not get updated, and there is no encryption.

```
= 0 to v-1 {
    C_i = P_i ⊕ S_{u+i,1} ⊕ S_{u+i,4} ⊕ (S_{u+i,2} & S_{u+i,3})
    S_{u+i+1} = StateUpdate128 (S_{u+i}, AD_i)
}
```

4 Finalization

After encrypting all the plaintext blocks, we generate the authentication tag using seven more steps. The length of the associated data and the length of the message are used to update the state.

Let $\text{tmp} = S_{u+v,3} \oplus (adlen \| msglen)$,
where $adlen$ and $msglen$ are represented as 64−bit integers.
for $i = u + v$ to $u + v + 6$
    $S_{i+1} = \text{StateUpdate128}(S_i, tmp)$
$T' = \bigoplus_{i=0}^{4} S_{u+v+7,i}$

The authentication tag T consists of the first t bits of T'.

A.2   Decryption and Verification of AEGIS-128

The exact values of key size, IV, size, and tag-size should be known to the decryption and verification processes. The decryption starts with the initialization and the processing of authenticated data. Then the ciphertext is decrypted as follows:

− If the last ciphertext block is not a full block, decrypt only the partial ciphertext block. The partial plaintext block is padded with 0 bits, and the padded full plaintext block is used to update the state.
− For $i = 0$ to $v − 1$, Perform decryption and update the state.

The finalization in the decryption process is the same as that in the encryption process. It is emphasized that if the verification fails, the ciphertext and the newly generated authentication tag should not be given as output; otherwise, the state of AEGIS-128 is vulnerable to known-plaintext or chosen-ciphertext attacks (using a fixed IV).

References

1. Announcing the ADVANCED ENCRYPTION STANDARD (AES). Federal Information Processing Standards Publication 197. United States National Institute of Standards and Technology (NIST). 26 November 2001
2. Bogdanov, A., Lauridsen, M.M., Tischhauser, E.: AES-Based Authenticated Encryption Modes in Parallel High-Performance Software. https://eprint.iacr.org/2014/186.pdf. Accessed on 27 July 2014
3. Bogdanov, A., Mendel, F., Regazzoni, F., Rijmen, V., Tischhauser, E.: ALE: AES-based lightweight authenticated encryption. In: Moriai, S. (ed.) FSE 2013. LNCS, vol. 8424, pp. 447–466. Springer, Heidelberg (2014)
4. CAESAR Submissions. http://competitions.cr.yp.to/caesar.html
5. Canright, D., Batina, L.: A very compact "perfectly masked" S-Box for AES. In: Bellovin, S.M., Gennaro, R., Keromytis, A.D., Yung, M. (eds.) ACNS 2008. LNCS, vol. 5037, pp. 446–459. Springer, Heidelberg (2008)
6. Chattopadhyay, A., Meyr, H., Leupers, R.: LISA: a uniform ADL for embedded processor modelling, implementation and software toolsuite generation. In: Mishra, P., Dutt, N. (eds.) Processor Description Languages, pp. 95–130. Morgan Kaufmann, Boston (2008)
7. Grosso, V., Leurent, G., Standaert, F., Varici, K., Durvaux, F., Gaspar, L., Kerckhof, S.: SCREAM & iSCREAM Side-Channel Resistant Authenticated Encryption with Masking. http://competitions.cr.yp.to/round1/screamv1.pdf. Accessed on 12 July 2014
8. Khalid, A., Hassan, M., Chattopadhyay, A., Paul, G.: RAPID-FeinSPN: a rapid prototyping framework for feistel and SPN-based block ciphers. In: Bagchi, A., Ray, I. (eds.) ICISS 2013. LNCS, vol. 8303, pp. 169–190. Springer, Heidelberg (2013). doi:10.1007/978-3-642-45204-813
9. Kim, H., Kim, K.: Who can survive in CAESAR competition at round-zero? In: The 31th Symposium on Cryptography and Information Security Kagoshima, Japan, 21–24 January 2014
10. Moradi, A., Poschmann, A., Ling, S., Paar, C., Wang, H.: Pushing the limits: a very compact and a threshold implementation of AES. In: Paterson, K.G. (ed.) EUROCRYPT 2011. LNCS, vol. 6632, pp. 69–88. Springer, Heidelberg (2011)
11. Morawiecki, P., Gaj, K., Homsirikamol, E., Matusiewicz, K., Pieprzyk, J., Rogawski, M., Srebrny, M., Wojcik, M.: ICEPOLE v1. http://competitions.cr.yp.to/round1/icepolev1.pdf. Accessed on 12 July 2014
12. Mozaffari-Kermani, M., Reyhani-Masoleh, A.: Efficient and high-performance parallel hardware architectures for the AES-GCM. IEEE Trans. Comput. 61(8), 1165–1178 (2012)
13. Nguyen, D.K., Lanante, L., Ochi, H.: High throughput resource saving hardware implementation of AES-CCM for robust security network. J. Autom. Control Eng. 1(3), 250–254 (2013)
14. Paul, G., Chattopadhyay, A.: Designing stream ciphers with scalable data-widths: a case study with HC-128. Springer J. Crypt. Eng. 4(2), 135–143 (2014). doi:10.1007/s13389-014-0071-0
15. Sasaki, Y., Todo, Y., Aoki, K., Naito, Y., Sugawara, T., Murakami, Y., Matsui, M., Hirose, S.: Minalpher v1. http://competitions.cr.yp.to/round1/minalpherv1.pdf. Accessed on 12 July 2014
16. Sen Gupta, S., Chattopadhyay, A., Khalid, A.: Designing integrated accelerator for stream ciphers with structural similarities. Crypt. Commun. Discrete Struct. Boolean Funct. Sequences 5(1), 19–47 (2013). doi:10.1007/s12095-012-0074-6

17. Sen Gupta, S., Chattopadhyay, A., Sinha, K., Maitra, S., Sinha, B.P.: High performance hardware implementation for RC4 stream cipher. IEEE Trans. Comput. **62**(4), 730–743 (2012)
18. Synopsys Processor Designer. http://www.synopsys.com/systems/blockdesign/processordev/pages/default.aspx
19. Wu, H., Preneel, B.: AEGIS: a fast authenticated encryption algorithm. In: Lange, T., Lauter, K., Lisoněk, P. (eds.) SAC 2013. LNCS, vol. 8282, pp. 185–202. Springer, Heidelberg (2014)
20. Wu, S., Wu, H., Huang, T., Wang, M., Wu, W.: Leaked-state-forgery attack against the authenticated encryption algorithm ALE. In: Sako, K., Sarkar, P. (eds.) ASIACRYPT 2013, Part I. LNCS, vol. 8269, pp. 377–404. Springer, Heidelberg (2013)

Fully Collusion-Resistant Traceable Key-Policy Attribute-Based Encryption with Sub-linear Size Ciphertexts

Zhen Liu[1]($\boxtimes$), Zhenfu Cao[2], and Duncan S. Wong[1]

[1] City University of Hong Kong, Hong Kong SAR, China
zhenliu7-c@my.cityu.edu.hk, duncan@cityu.edu.hk
[2] East China Normal University, Shanghai, China
zfcao@sei.ecnu.edu.cn

Abstract. Recently a series of expressive, secure and efficient Attribute-Based Encryption (ABE) schemes, both in key-policy flavor and ciphertext-policy flavor, have been proposed. However, before being applied into practice, these systems have to attain traceability of malicious users. As the decryption privilege of a decryption key in Key-Policy ABE (resp. Ciphertext-Policy ABE) may be shared by multiple users who own the same access policy (resp. attribute set), malicious users might tempt to leak their decryption privileges to third parties, for financial gain as an example, if there is no tracing mechanism for tracking them down. In this work we study the traceability notion in the setting of Key-Policy ABE, and formalize Key-Policy ABE supporting fully collusion-resistant blackbox traceability. An adversary is allowed to access an arbitrary number of keys of its own choice when building a decryption-device, and given such a decryption-device while the underlying decryption algorithm or key may not be given, a blackbox tracing algorithm can find out at least one of the malicious users whose keys have been used for building the decryption-device. We propose a construction, which supports both fully collusion-resistant blackbox traceability and high expressivity (i.e. supporting any monotonic access structures). The construction is fully secure in the standard model (i.e. it achieves the best security level that the conventional non-traceable ABE systems do to date), and is efficient that the fully collusion-resistant blackbox traceability is attained at the price of making ciphertexts grow only sub-linearly in the number of users in the system, which is the most efficient level to date.

Keywords: Attribute-based encryption · Key-policy · Blackbox traceability · Efficiency

1 Introduction

Attribute-based encryption (ABE), as a promising tool for fine-grained access control on encrypted data, has attracted much attention since its introduction by Sahai and Waters [25] in 2005, and work has been done to achieve better

© Springer International Publishing Switzerland 2015
D. Lin et al. (Eds.): Inscrypt 2014, LNCS 8957, pp. 403–423, 2015.
DOI: 10.1007/978-3-319-16745-9_22

expressivity, security and efficiency, in both key-policy flavor [1,7,10,14,22–24,28] and ciphertext-policy flavor [6,9,11,15,16,22,24,27]. Due to their high expressivity of access policy and efficient one-to-many encryption, both Key-Policy ABE (KP-ABE) and Ciphertext-Policy ABE (CP-ABE) have extensive applications. For example, in a KP-ABE system (CP-ABE proceeds the other way around), each private key is associated with an access policy over descriptive attributes issued by an authority, each ciphertext is associated with an attribute set specified by the encryptor, and if and only if the attribute set of a ciphertext satisfies the access policy, the private key can decrypt the ciphertext. In a pay-TV system the television broadcaster can encrypt the broadcast using the descriptive attributes, such as the name of the program ("The Big Bang Theory"), the genre ("drama"), the season, the episode number, the year, the month, etc. And a subscriber may determine and pay for his subscribing policy, such as "(The Big Bang Theory OR Criminal Minds) AND 2014".

Recently, the expressivity, security and efficiency of ABE have been relatively well developed. The KP-ABE systems in [14,22] and the CP-ABE systems in [15,16,22][1] are highly expressive (i.e. supporting any monotonic access structures), fully secure in the standard model, and satisfactorily efficient. However, to apply these systems into practice, the traceability of malicious users is needed. In an ABE system in general, as a decryption privilege could be possessed by multiple users who own the same access policy (in KP-ABE) or attribute set (in CP-ABE), malicious users might tempt to leak their decryption privileges to third parties, for financial gain as an example, if there is no tracing mechanism for finding these malicious users out. For example, both user Alex with access policy "(The Big Bang Theory OR Criminal Minds) AND 2014" and user Bob with access policy "(The Big Bang Theory OR CSI) AND 2014" might be the malicious user who builds and sells a decryption blackbox that can decrypt the ciphertexts generated under attributes {The Big Bang Theory, 2014}.

While all the aforementioned ABE systems suffer from this problem, some recent attempts [13,17–20,26,29] have been made to achieve traceability. Specifically, there are two levels of traceability: (1) given a well-formed decryption key, a *Whitebox* tracing algorithm can find out the original key owner; and (2) given a decryption-device while the underlying decryption algorithm or key may not be given, a *Blackbox* tracing algorithm, which treats the decryption-device as an oracle, can find out at least one of the malicious users whose keys have been used for building the decryption-device. Furthermore, a system is said to support *fully collusion-resistant blackbox traceability* if an adversary can access an arbitrary number of keys (in other words, when an arbitrary number of malicious users collude) when building the decryption-device, and is said to support *t-collusion-resistant blackbox traceability*, if an adversary is restricted from getting more than t decryption keys when building the decryption-device. While the Blackbox Traceable CP-ABE scheme in [19] is highly expressive, fully secure in the standard model, and efficient in achieving fully collusion-resistant

---

[1] Reference [14] is the full version of [15], where [15] proposed expressive, fully secure and efficient CP-ABE schemes, [14] further proposed an expressive, fully secure and efficient KP-ABE scheme additionally.

Table 1. Comparison with existing traceable KP-ABE schemes

	Adaptively secure	Highly expressive	Fully collusion-resistant traceable	Overhead for traceability
[29]	×	✓	×	$O(\log \mathcal{K})$
[26]	×	×	×	$O(t^2 \log \mathcal{K} + \log(1/\epsilon))^a$
[13]	×	✓	✓	$O(\mathcal{K})$
This paper	✓	✓	✓	$O(\sqrt{\mathcal{K}})$

[a] Reference [26] is only t-collusion-resistant traceable and the large overhead $O(t^2 \log \mathcal{K} + \log(1/\epsilon))$ makes the scheme impractical. Furthermore, to achieve fully collusion-resistant traceability (i.e., $t = \mathcal{K}$, the number of users in the system), the overhead of the scheme will be $O(\mathcal{K}^2 \log \mathcal{K} + \log(1/\epsilon))$. ϵ is the probability of error that a colluder is not traced.

blackbox traceability at the expense of overhead sub-linear in the number of users, there is no traceable KP-ABE scheme achieving comparable expressivity, security and efficiency. In particular, (1) the blackbox traceability of [29] is 1-collusion-resistant (i.e. cannot resist collusion attack); (2) Reference [26] only supports single threshold policy and t-collusion-resistant blackbox traceability; (3) the fully collusion-resistant blackbox traceable predicate encryption scheme in [13] implies an expressive KP-ABE scheme (at the cost of converting monotonic access structure into DNF, which will result in larger ciphertext size), but the overhead for traceability is linear in the number of users; and (4) all the schemes in [13, 26, 29] are only selectively secure.

1.1 Our Results

In this paper, we first formalize the fully collusion-resistant blackbox traceability notions for expressive KP-ABE, then we formalize a simpler primitive called Augmented KP-ABE and show that a secure Augmented KP-ABE can be directly transformed into a Traceable KP-ABE. With such a transformation, to obtain a fully collusion-resistant blackbox traceable KP-ABE scheme, we propose an Augmented KP-ABE construction, implying a traceable KP-ABE construction that is fully secure in the standard model, highly expressive in supporting any monotonic access structures, and efficient in achieving fully collusion-resistant blackbox traceability at the expense of having the ciphertext size be sub-linear in the number of users in the system. In Table 1 we compare our traceable KP-ABE scheme with existing traceable KP-ABE schemes in literature.

Paper Organization. In Sect. 2 we formalize the fully collusion-resistant blackbox traceability notions for expressive KP-ABE. Then in Sect. 3, we propose a primitive called Augmented KP-ABE and define its security notions using message-hiding and index-hiding games, and show that an Augmented KP-ABE with message-hiding and index-hiding properties implies a secure KP-ABE with traceability. Finally in Sect. 4, we propose a concrete construction of Augmented KP-ABE and prove its message-hiding and index-hiding properties in the standard model.

2   KP-ABE with Traceability

We first review the definition of KP-ABE which is based on conventional (non-traceable) KP-ABE (e.g. [10,14]) with the exception that in our 'functional' definition, we explicitly assign and identify users using unique indices, and let $\mathcal{K}$ be the number of users in a KP-ABE system. Then we introduce the fully collusion-resistant traceability definition against attributes-specific decryption blackbox, which reflects most practical applications.

2.1   KP-ABE

Before defining KP-ABE system, we first provide some background about access policy in the context of KP-ABE.

Definition 1 (Access Structure). *[2] Let $\mathcal{U} = \{a_1, a_2, \ldots, a_n\}$ be a set of attributes. A collection $\mathbb{A} \subseteq 2^{\mathcal{U}}$ is monotone if $\forall B, C : B \in \mathbb{A}$ and $B \subseteq C$ imply $C \in \mathbb{A}$. An access structure (resp., monotone access structure) is a collection (resp., monotone collection) $\mathbb{A}$ of non-empty subsets of $\mathcal{U}$, i.e., $\mathbb{A} \subseteq 2^{\mathcal{U}} \setminus \{\emptyset\}$. The sets in $\mathbb{A}$ are called authorized sets, and the sets not in $\mathbb{A}$ are called unauthorized sets. Also, for an attribute set $S \subseteq \mathcal{U}$, if $S \in \mathbb{A}$ then we say S satisfies the access structure $\mathbb{A}$, otherwise we say S does not satisfy $\mathbb{A}$.*

Unless stated otherwise, by an access structure we mean a monotone access structure for the rest of this paper.

For simplicity, for a positive integer, for example n, we use the notation $[n]$ to denote the set $\{1, 2, \ldots, n\}$. A Key-Policy ABE (KP-ABE) scheme consists of the following four algorithms:

Setup$(\lambda, \mathcal{U}, \mathcal{K}) \to$ (PP, MSK). The algorithm takes as input a security parameter $\lambda \in \mathbb{N}$, the attribute universe (i.e., the set of attributes) $\mathcal{U}$, and the number of users $\mathcal{K}$ in the system, it outputs a public parameter PP and a master secret key MSK.

KeyGen(PP, MSK, $\mathbb{A}$) $\to$ SK$_{k,\mathbb{A}}$. The algorithm takes as input PP, MSK, and an access structure $\mathbb{A}$, and outputs a private key SK$_{k,\mathbb{A}}$, which is assigned and identified by a unique index $k \in [\mathcal{K}]$.

Encrypt(PP, M, S) $\to CT_S$. The algorithm takes as input PP, a message M, and an attribute set $S \subseteq \mathcal{U}$, and outputs a ciphertext CT_S. S is implicitly included in CT_S.

Decrypt(PP, CT_S, SK$_{k,\mathbb{A}}$) $\to M$ or $\bot$. The algorithm takes as input PP, a ciphertext CT_S associated with an attribute set S, and a private key SK$_{k,\mathbb{A}}$. If S satisfies $\mathbb{A}$, the algorithm outputs message M, otherwise it outputs $\bot$ indicating the failure of decryption.

Correctness. For all access structures $\mathbb{A} \subseteq 2^{\mathcal{U}} \setminus \{\emptyset\}$, $k \in [\mathcal{K}]$, $S \subseteq \mathcal{U}$, and messages M: if (PP, MSK) $\leftarrow$ Setup$(\lambda, \mathcal{U}, \mathcal{K})$, SK$_{k,\mathbb{A}} \leftarrow$ KeyGen(PP, MSK, $\mathbb{A}$), $CT_S \leftarrow$ Encrypt(PP, M, S), and S *satisfies* $\mathbb{A}$, we have Decrypt(PP, CT_S, SK$_{k,\mathbb{A}}$) $= M$.

Security. The security of the above KP-ABE scheme is defined using the following message-hiding game, which is a typical semantic security game and is based on that for conventional KP-ABE [10,14] security against adaptive adversaries, except that each key is explicitly identified by a unique index.

$\mathsf{Game}_{\mathsf{MH}}$. The message-hiding game is defined between a challenger and an adversary $\mathcal{A}$ as follows:

Setup. The challenger runs $\mathsf{Setup}(\lambda, \mathcal{U}, \mathcal{K})$ and gives the public parameter PP to $\mathcal{A}$.

Phase 1. For $i = 1$ to Q_1, $\mathcal{A}$ adaptively submits (index, access structure) pair $(k_i, \mathbb{A}_{k_i})$ to the challenger, and the challenger responds with $\mathsf{SK}_{k_i, \mathbb{A}_{k_i}}$.

Challenge. $\mathcal{A}$ submits two equal-length messages M_0, M_1 and an attribute set S^*. The challenger flips a random coin $b \in \{0, 1\}$, and sends $CT_{S^*} \leftarrow \mathsf{Encrypt}(\mathsf{PP}, M_b, S^*)$ to $\mathcal{A}$.

Phase 2. For $i = Q_1 + 1$ to Q, $\mathcal{A}$ adaptively submits (index, access structure) pair $(k_i, \mathbb{A}_{k_i})$ to the challenger, and the challenger responds with $\mathsf{SK}_{k_i, \mathbb{A}_{k_i}}$.

Guess. $\mathcal{A}$ outputs a guess $b' \in \{0, 1\}$ for b.

$\mathcal{A}$ wins the game if $b' = b$ under the **restriction** that S^* does not satisfy any of the queried access structures $\mathbb{A}_{k_1}, \ldots, \mathbb{A}_{k_Q}$. The advantage of $\mathcal{A}$ is defined as $\mathsf{MHAdv}_{\mathcal{A}} = |\Pr[b' = b] - \frac{1}{2}|$.

Definition 2. *A $\mathcal{K}$-user KP-ABE scheme is secure if for all probabilistic polynomial time (PPT) adversaries $\mathcal{A}$ the advantage $\mathsf{MHAdv}_{\mathcal{A}}$ is a negligible function of λ.*

It is worth noticing that: (1) although the index of each user private key is assigned by the KeyGen algorithm, to capture the security that an attacker can adaptively choose keys to corrupt, the above security model allows the adversary to specify the index when he makes a key query, i.e., for $i = 1$ to Q, the adversary submits (index, access structure) pair $(k_i, \mathbb{A}_{k_i})$ to query a private key for access structure $\mathbb{A}_{k_i}$, and the challenger will assign k_i to be the index of the private key, where $Q \leq \mathcal{K}$, $k_i \in [\mathcal{K}]$, and $k_i \neq k_j \ \forall 1 \leq i \neq j \leq Q$ (this is to guarantee that each user/key can be *uniquely* identified by an index); and (2) for $k_i \neq k_j$ we do not require $\mathbb{A}_{k_i} \neq \mathbb{A}_{k_j}$, i.e., different users/keys may have the same access structure. We remark that these two points apply to the rest of the paper.

Remark: Compared with a conventional (non-traceable) KP-ABE [10,14], the above definition has the same Encrypt and Decrypt functionality, and almost the same Setup and KeyGen with only slight differences: predefining the number of users $\mathcal{K}$ in Setup and assigning each user a unique index $k \in [\mathcal{K}]$. Presetting the number of users is indeed a tradeoff but is also a necessary cost for achieving blackbox traceability. We stress that in practice, this should not incur much concern, and all the existing blackbox traceable systems (e.g. [4,5,8,13,19,26]) have the same setting. Also being consistent with the conventional definition of KP-ABE, the user indices are not used in normal encryption (i.e. the encryptors do not need to know the indices of any users in order to encrypt) and different users (with different indices) may have the same access policy. In summary, a secure KP-ABE system defined as above has all the appealing properties that a conventional KP-ABE system [10,14] has, that is, fully collusion-resistant security, fine-grained access control on encrypted data, and efficient one-to-many encryption. The unique index of each user/private key is to uniquely identify the users and allow the traceability.

2.2   KP-ABE Traceability

An attributes-specific decryption blackbox $\mathcal{D}$ in the setting of KP-ABE is viewed as a probabilistic circuit that can decrypt ciphertexts generated under some specific attribute set. *In particular, an attributes-specific decryption blackbox $\mathcal{D}$ is described with an attribute set $S_\mathcal{D}$ and a non-negligible probability value ϵ (i.e. $\epsilon = 1/f(\lambda)$ for some polynomial f), and this blackbox $\mathcal{D}$ can decrypt the ciphertexts generated under $S_\mathcal{D}$ with probability at least ϵ.* Such an attributes-specific decryption blackbox reflects most practical scenarios. In particular, once a decryption blackbox is found being able to decrypt some ciphertext with non-negligible probability (regardless of how this is found, for example, an explicit description of the blackbox's decryption ability is given, or the law enforcement agency finds some clue), we can regard it as an attributes-specific decryption blackbox with the corresponding attribute set (which is associated to the ciphertext).[2] And for a decryption blackbox, if multiple attribute sets are found that corresponding ciphertexts can be decrypted by this blackbox with non-negligible probability, we can regard the blackbox as multiple attributes-specific decryption blackboxes, each with a different attribute set.

We now define a tracing algorithm against an attributes-specific decryption blackbox as follows.

$\mathsf{Trace}^\mathcal{D}(\mathsf{PP}, S_\mathcal{D}, \epsilon) \to \mathbb{K}_T \subseteq [\mathcal{K}]$. *This is an oracle algorithm that interacts with an attributes-specific decryption blackbox $\mathcal{D}$. By given the public parameter PP, an attribute set $S_\mathcal{D}$, and a probability value ϵ, the algorithm runs in time polynomial in λ and $1/\epsilon$, and outputs an index set $\mathbb{K}_T \subseteq [\mathcal{K}]$ which identifies the set of malicious users. Note that ϵ has to be polynomially related to λ, i.e. $\epsilon = 1/f(\lambda)$ for some polynomial f.*

The following tracing game captures the notion of **fully collusion-resistant traceability** against attributes-specific decryption blackbox. In the game, the adversary targets to build a decryption blackbox $\mathcal{D}$ that can decrypt ciphertexts generated under some attribute set $S_\mathcal{D}$ with non-negligible probability. The tracing algorithm, on the other side, is designed to extract the index of at least one of the malicious users whose decryption keys have been used for constructing $\mathcal{D}$.

---

[2] Note that in the setting of predicate encryption [12], which can informally be regarded as a KP-ABE system with attribute-hiding property, the decryption blackbox [13] is also modeled similarly, i.e., the tracing algorithm takes as input an attribute I and a decryption blackbox $\mathcal{D}$ that decrypts ciphertexts associated with the attribute I.

Game$_{TR}$. The **tracing game** is defined between a challenger and an adversary $\mathcal{A}$ as follows:

Setup. The challenger runs Setup$(\lambda, \mathcal{U}, \mathcal{K})$ and gives the public parameter PP to $\mathcal{A}$.

Key Query. For $i = 1$ to Q, $\mathcal{A}$ adaptively submits (index, access structure) pair $(k_i, \mathbb{A}_{k_i})$ to the challenger, and the challenger responds with SK$_{k_i, \mathbb{A}_{k_i}}$.

Decryption Blackbox Generation. $\mathcal{A}$ outputs a decryption blackbox $\mathcal{D}$ associated with an attribute set $S_{\mathcal{D}}$ and a non-negligible probability value ϵ.

Tracing. The challenger runs Trace$^{\mathcal{D}}$(PP, $S_{\mathcal{D}}, \epsilon$) to obtain an index set $\mathbb{K}_T \subseteq [\mathcal{K}]$.

Let $\mathbb{K}_{\mathcal{D}} = \{k_i | 1 \leq i \leq Q\}$ be the index set of keys corrupted by the adversary. We say that the adversary $\mathcal{A}$ wins the game if the following conditions hold:

1. $\Pr[\mathcal{D}(\text{Encrypt}(\text{PP}, M, S_{\mathcal{D}})) = M] \geq \epsilon$, where the probability is taken over the random choices of message M and the random coins of $\mathcal{D}$. A decryption blackbox satisfying this condition is said to be a *useful attributes-specific decryption blackbox*.
2. $\mathbb{K}_T = \emptyset$, or $\mathbb{K}_T \nsubseteq \mathbb{K}_{\mathcal{D}}$, or ($S_{\mathcal{D}}$ *does not satisfy* $\mathbb{A}_{k_t} \; \forall k_t \in \mathbb{K}_T$).

We denote by TRAdv$_{\mathcal{A}}$ the probability that adversary $\mathcal{A}$ wins this game.

Remark: For a useful attributes-specific decryption blackbox $\mathcal{D}$, the traced $\mathbb{K}_T$ must satisfy $(\mathbb{K}_T \neq \emptyset) \wedge (\mathbb{K}_T \subseteq \mathbb{K}_{\mathcal{D}}) \wedge (\exists k_t \in \mathbb{K}_T \; s.t. \; S_{\mathcal{D}} \; satisfies \; \mathbb{A}_{k_t})$ for traceability. (1) $(\mathbb{K}_T \neq \emptyset) \wedge (\mathbb{K}_T \subseteq \mathbb{K}_{\mathcal{D}})$ captures the preliminary traceability that the tracing algorithm can extract at least one malicious user and the coalition of malicious users cannot frame any innocent user. (2) $(\exists k_t \in \mathbb{K}_T \; s.t. \; S_{\mathcal{D}} \; satisfies \; \mathbb{A}_{k_t})$ captures *strong traceability* that the tracing algorithm can extract at least one malicious user whose private key enables $\mathcal{D}$ to have the decryption ability of decrypting ciphertexts generated under $S_{\mathcal{D}}$. We refer to [13, 19] on why strong traceability is desirable.

Definition 3. *A $\mathcal{K}$-user KP-ABE scheme is traceable if for all PPT adversaries $\mathcal{A}$ the advantage TRAdv$_{\mathcal{A}}$ is negligible in λ.*

We say that a $\mathcal{K}$-user KP-ABE scheme is *selectively traceable* if we add an **Init** stage before **Setup** where the adversary commits to the attribute set $S_{\mathcal{D}}$.

We emphasize that we are modelling public traceability, namely, the Trace algorithm does not need any secrets and anyone can perform the tracing from the public parameters only. Also note that we are modelling a stateless (resettable) decryption blackbox – the decryption blackbox is just an oracle and maintains no state between activations.

3   Augmented KP-ABE

Following the routes of [19] where CP-ABE's traceability is discussed, instead of constructing a traceable KP-ABE directly, we define a simpler primitive called

Augmented KP-ABE (or AugKP-ABE for short) and its security notions (message-hiding and index-hiding) first, then we show that an AugKP-ABE with message-hiding and index-hiding properties can be transformed to a secure KP-ABE with traceability. In Sect. 4, we propose an AugKP-ABE construction and prove its message-hiding and index-hiding properties in the standard model.

3.1 Definitions of Augmented KP-ABE

An Augmented KP-ABE (AugKP-ABE) has four algorithms: $\mathsf{Setup_A}$, $\mathsf{KeyGen_A}$, $\mathsf{Encrypt_A}$, and $\mathsf{Decrypt_A}$. The setup algorithm $\mathsf{Setup_A}$ and key generation algorithm $\mathsf{KeyGen_A}$ are the same as that of KP-ABE in Sect. 2.1. The encryption algorithm $\mathsf{Encrypt_A}$ takes one more parameter $\bar{k} \in [\mathcal{K}+1]$ as input, and is defined as follows.

$\mathsf{Encrypt_A}(\mathsf{PP}, M, S, \bar{k}) \to CT_S$. The algorithm takes as input PP, a message M, an attribute set $S \subseteq \mathcal{U}$, and an index $\bar{k} \in [\mathcal{K}+1]$, and outputs a ciphertext CT_S. S **is included in** CT_S, **but the value of** $\bar{k}$ **is not**.

The decryption algorithm $\mathsf{Decrypt_A}$ is also defined in the same as that of the KP-ABE in Sect. 2.1. However, the correctness definition is changed to the following.

Correctness. For all access structures $\mathbb{A} \subseteq 2^{\mathcal{U}} \setminus \{\emptyset\}$, $k \in [\mathcal{K}]$, $S \subseteq \mathcal{U}$, $\bar{k} \in [\mathcal{K}+1]$, and messages M: if $(\mathsf{PP}, \mathsf{MSK}) \leftarrow \mathsf{Setup_A}(\lambda, \mathcal{U}, \mathcal{K})$, $\mathsf{SK}_{k,\mathbb{A}} \leftarrow \mathsf{KeyGen_A}(\mathsf{PP}, \mathsf{MSK}, \mathbb{A})$, $CT_S \leftarrow \mathsf{Encrypt_A}(\mathsf{PP}, M, S, \bar{k})$, and $(S \ satisfies \ \mathbb{A}) \wedge (k \geq \bar{k})$, we have $\mathsf{Decrypt_A}(\mathsf{PP}, CT_S, \mathsf{SK}_{k,\mathbb{A}}) = M$.

Remark: Note that during decryption, as long as S satisfies $\mathbb{A}$, the decryption algorithm outputs a message, but only when $k \geq \bar{k}$, the output message is equal to the correct message, that is, if and only if $(S \ satisfies \ \mathbb{A}) \wedge (k \geq \bar{k})$, can $\mathsf{SK}_{k,\mathbb{A}}$ correctly decrypt a ciphertext under $(S, \bar{k})$. If we always set $\bar{k} = 1$, the functions of AugKP-ABE are identical to that of KP-ABE. In fact, the idea behind transforming an AugKP-ABE to a blackbox traceable KP-ABE, that we will show shortly, is to construct an AugKP-ABE with index-hiding property, and then always sets $\bar{k} = 1$ in normal encryption, while using $\bar{k} \in [N+1]$ to generate ciphertexts for tracing.

Security. We define the security of AugKP-ABE in the following three games.

$\mathsf{Game}^{\mathsf{A}}_{\mathsf{MH_1}}$. The first game, a message-hiding game denoted by $\mathsf{Game}^{\mathsf{A}}_{\mathsf{MH_1}}$, is similar to $\mathsf{Game_{MH}}$ except that the **Challenge** phase is

Challenge. $\mathcal{A}$ submits two equal-length messages M_0, M_1 and an attribute set S^*. The challenger flips a random coin $b \in \{0, 1\}$, and sends $CT_{S^*} \leftarrow \mathsf{Encrypt_A}(\mathsf{PP}, M_b, S^*, 1)$ to $\mathcal{A}$.

$\mathcal{A}$ wins the game if $b' = b$ under the **restriction** that S^* cannot satisfy any of the queried access structures $\mathbb{A}_{k_1}, \ldots, \mathbb{A}_{k_Q}$. The advantage of $\mathcal{A}$ is defined as $\mathsf{MH_1^A Adv}_{\mathcal{A}} = |\Pr[b' = b] - \frac{1}{2}|$.

$\mathsf{Game}^{\mathsf{A}}_{\mathsf{MH_{\mathcal{K}+1}}}$. The second game, a message-hiding game denoted by $\mathsf{Game}^{\mathsf{A}}_{\mathsf{MH_{\mathcal{K}+1}}}$, is similar to $\mathsf{Game_{MH}}$ except that the **Challenge** phase is

Challenge. $\mathcal{A}$ submits two equal-length messages M_0, M_1 and an attribute set S^*. The challenger flips a random coin $b \in \{0, 1\}$, and sends $CT_{S^*} \leftarrow$ $\mathsf{Encrypt}_\mathsf{A}(\mathsf{PP}, M_b, S^*, \mathcal{K} + 1)$ to $\mathcal{A}$.

$\mathcal{A}$ wins the game if $b' = b$. The advantage of $\mathcal{A}$ is defined as $\mathsf{MH}^\mathsf{A}_{\mathcal{K}+1}\mathsf{Adv}_\mathcal{A} = |\Pr[b' = b] - \frac{1}{2}|$.

Definition 4. *A $\mathcal{K}$-user Augmented KP-ABE scheme is message-hiding if for all PPT adversaries $\mathcal{A}$ the advantages $\mathsf{MH}^\mathsf{A}_1\mathsf{Adv}_\mathcal{A}$ and $\mathsf{MH}^\mathsf{A}_{\mathcal{K}+1}\mathsf{Adv}_\mathcal{A}$ are negligible functions of λ.*

Game$^\mathsf{A}_\mathsf{IH}$. The third game, called **index-hiding game**, requires that for any attribute set $S^* \subseteq \mathcal{U}$, no adversary can distinguish between an encryption using $(S^*, \bar{k})$ and one using $(S^*, \bar{k}+1)$ without a private key $\mathsf{SK}_{\bar{k}, \mathbb{A}_{\bar{k}}}$ where S^* satisfies $\mathbb{A}_{\bar{k}}$. The game takes as input a parameter $\bar{k} \in [\mathcal{K}]$ which is given to both the challenger and the adversary $\mathcal{A}$. The game proceeds as follows:

Setup. The challenger runs $\mathsf{Setup}_\mathsf{A}(\lambda, \mathcal{U}, \mathcal{K})$ and sends PP to $\mathcal{A}$.
Key Query. For $i = 1$ to Q, $\mathcal{A}$ adaptively submits (index, access structure) pair $(k_i, \mathbb{A}_{k_i})$ to the challenger, and the challenger responds with $\mathsf{SK}_{k_i, \mathbb{A}_{k_i}}$.
Challenge. $\mathcal{A}$ submits a message M and an attribute set S^*. The challenger flips a random coin $b \in \{0, 1\}$, and sends $CT_{S^*} \leftarrow \mathsf{Encrypt}_\mathsf{A}(\mathsf{PP}, M, S^*, \bar{k}+b)$ to $\mathcal{A}$.
Guess. $\mathcal{A}$ outputs a guess $b' \in \{0, 1\}$ for b.

$\mathcal{A}$ wins the game if $b' = b$ under the **restriction** that none of the queried pairs $\{(k_i, \mathbb{A}_{k_i})\}$ can satisfy $(k_i = \bar{k}) \wedge (S^* \ satisfies \ \mathbb{A}_{k_i})$. The advantage of $\mathcal{A}$ is defined as $\mathsf{IH}^\mathsf{A}\mathsf{Adv}_\mathcal{A}[\bar{k}] = |\Pr[b' = b] - \frac{1}{2}|$.

Definition 5. *A $\mathcal{K}$-user Augmented KP-ABE scheme is index-hiding if for all PPT adversaries $\mathcal{A}$ the advantages $\mathsf{IH}^\mathsf{A}\mathsf{Adv}_\mathcal{A}[\bar{k}]$ for $\bar{k} = 1, \ldots, \mathcal{K}$ are negligible functions of λ.*

We say that an Augmented KP-ABE scheme is *selectively index-hiding* if in Game$^\mathsf{A}_\mathsf{IH}$ we add an **Init** stage before **Setup** where the adversary commits to the challenge attribute set S^*.

3.2 Reducing Traceable KP-ABE to Augmented KP-ABE

Let $\Sigma_\mathsf{A} = (\mathsf{Setup}_\mathsf{A}, \mathsf{KeyGen}_\mathsf{A}, \mathsf{Encrypt}_\mathsf{A}, \mathsf{Decrypt}_\mathsf{A})$ be an AugKP-ABE, define $\mathsf{Encrypt}(\mathsf{PP}, M, S) = \mathsf{Encrypt}_\mathsf{A}(\mathsf{PP}, M, S, 1)$, then $\Sigma = (\mathsf{Setup}_\mathsf{A}, \mathsf{KeyGen}_\mathsf{A}, \mathsf{Encrypt}, \mathsf{Decrypt}_\mathsf{A})$ is a KP-ABE derived from Σ_A.

Theorem 1. *If Σ_A is message-hiding in Game$^\mathsf{A}_{\mathsf{MH}_1}$, then Σ is secure.*

Proof. Note that Σ is a special case of Σ_A where the encryption algorithm always sets $\bar{k} = 1$. Hence, Game$_\mathsf{MH}$ for Σ is identical to Game$^\mathsf{A}_{\mathsf{MH}_1}$ for Σ_A, which implies that $\mathsf{MHAdv}_\mathcal{A}$ for Σ in Game$_\mathsf{MH}$ is equal to $\mathsf{MH}^\mathsf{A}_1\mathsf{Adv}_\mathcal{A}$ for Σ_A in Game$^\mathsf{A}_{\mathsf{MH}_1}$, i.e., if Σ_A is message-hiding in Game$^\mathsf{A}_{\mathsf{MH}_1}$, then Σ is secure.

Now we construct a tracing algorithm Trace for Σ and show that if Σ_A is message-hiding in $\mathsf{Game}^A_{\mathsf{MH}_{\mathcal{K}+1}}$ and (selectively) index-hiding, then Σ (equipped with Trace) is (selectively) traceable against attributes-specific decryption blackbox.

$\mathsf{Trace}^{\mathcal{D}}(\mathsf{PP}, S_{\mathcal{D}}, \epsilon) \to \mathbb{K}_T \subseteq [\mathcal{K}]$: Given an attributes-specific decryption blackbox $\mathcal{D}$ associated with an attribute set $S_{\mathcal{D}}$ and probability $\epsilon > 0$, the tracing algorithm works as follows:[3]

1. For $k = 1$ to $\mathcal{K} + 1$, do the following:
   (a) The algorithm repeats the following $8\lambda(\mathcal{K}/\epsilon)^2$ times:
      i. Sample M from the message space at random.
      ii. Let $CT_{S_{\mathcal{D}}} \gets \mathsf{Encrypt}_A(\mathsf{PP}, M, S_{\mathcal{D}}, k)$.
      iii. Query oracle $\mathcal{D}$ on input $CT_{S_{\mathcal{D}}}$, and compare the output of $\mathcal{D}$ with M.
   (b) Let $\hat{p}_k$ be the fraction of times that $\mathcal{D}$ decrypted the ciphertexts correctly.
2. Let $\mathbb{K}_T$ be the set of all $k \in [\mathcal{K}]$ for which $\hat{p}_k - \hat{p}_{k+1} \geq \epsilon/(4\mathcal{K})$. Then output $\mathbb{K}_T$ as the index set of the private decryption keys of malicious users.

Note that the running time of Trace is cubic in $\mathcal{K}$. It can be made (almost) quadratic using binary search instead of a linear scan.

Theorem 2. *If Σ_A is message-hiding in $\mathsf{Game}^A_{\mathsf{MH}_{\mathcal{K}+1}}$ and index-hiding (resp. selectively index-hiding), then Σ is traceable (resp. selectively traceable).*

Proof. In the proof sketch below, we show that if the attributes-specific decryption blackbox output by the adversary is a useful one then the traced $\mathbb{K}_T$ will satisfy $(\mathbb{K}_T \neq \emptyset) \wedge (\mathbb{K}_T \subseteq \mathbb{K}_{\mathcal{D}}) \wedge (\exists k_t \in \mathbb{K}_T \; s.t. \; S_{\mathcal{D}} \; satisfies \; \mathbb{A}_{k_t})$ with overwhelming probability, which implies that the adversary can win the game $\mathsf{Game}_{\mathsf{TR}}$ only with negligible probability, i.e., $\mathsf{TRAdv}_{\mathcal{A}}$ is negligible. The selective case is similar.

Let $\mathcal{D}$ be the attributes-specific decryption blackbox output by the adversary, and $S_{\mathcal{D}}$ be the attribute set describing $\mathcal{D}$. Define

$$p_{\bar{k}} = \Pr[\mathcal{D}(\mathsf{Encrypt}_A(\mathsf{PP}, M, S_{\mathcal{D}}, \bar{k})) = M],$$

where the probability is taken over the random choice of message M and the random coins of $\mathcal{D}$. We have that $p_1 \geq \epsilon$ and $p_{\mathcal{K}+1}$ is negligible (for simplicity let $p_{\mathcal{K}+1} = 0$). The former follows from the fact that $\mathcal{D}$ is useful, and the latter is because Σ_A is message-hiding in $\mathsf{Game}^A_{\mathsf{MH}_{\mathcal{K}+1}}$. Then there must exist some $k \in [\mathcal{K}]$ such that $p_k - p_{k+1} \geq \epsilon/(2\mathcal{K})$. By the Chernoff bound it follows that with overwhelming probability, $\hat{p}_k - \hat{p}_{k+1} \geq \epsilon/(4\mathcal{K})$. Hence, we have $\mathbb{K}_T \neq \emptyset$.

For any $k \in \mathbb{K}_T$ (i.e., $\hat{p}_k - \hat{p}_{k+1} \geq \frac{\epsilon}{4\mathcal{K}}$), we know, by Chernoff, that with overwhelming probability $p_k - p_{k+1} \geq \epsilon/(8\mathcal{K})$. Clearly $(k \in \mathbb{K}_{\mathcal{D}}) \wedge (S_{\mathcal{D}} \; satisfies \; \mathbb{A}_k)$ since otherwise, $\mathcal{D}$ can be directly used to win the index-hiding game for Σ_A. Hence, we have $(\mathbb{K}_T \subseteq \mathbb{K}_{\mathcal{D}}) \wedge (S_{\mathcal{D}} \; satisfies \; \mathbb{A}_k \; \forall k \in \mathbb{K}_T)$.

---

[3] The tracing algorithm uses a technique based on that in broadcast encryption by [4,5,8].

4   An Efficient Augmented KP-ABE Scheme

We now propose an AugKP-ABE scheme which can be considered as combining the KP-ABE scheme of [14] and the traitor tracing scheme of [8]. We stress that this work is not a trivial combination of the two schemes, which may result in insecure or inefficient schemes, as discussed in [19]. The proposed AugKP-ABE scheme is highly expressive in supporting any monotonic access structures, and is efficient with ciphertext size $O(\sqrt{\mathcal{K}} + |S|)$, where $\mathcal{K}$ is the number of users in the system and S is the attribute set of the ciphertext. We prove that the scheme is adaptively message-hiding and selectively index-hiding in the standard model. Combining this AugKP-ABE scheme with the result in Sect. 3.2, we obtain a fully secure and highly expressive KP-ABE scheme which is simultaneously selectively traceable, and for a fully collusion-resistant blackbox traceable system the resulting KP-ABE scheme achieves the most efficient level to date, with overhead linear in $\sqrt{\mathcal{K}}$.

4.1   Preliminaries

Linear Secret-Sharing Schemes. As shown in [2], any monotonic access structure can be realized by a linear secret sharing scheme.

Definition 6 (Linear Secret-Sharing Schemes (LSSS)). *[27] A secret sharing scheme Π over attribute universe $\mathcal{U}$ is called linear (over $\mathbb{Z}_p$) if*

1. *The shares for each attribute form a vector over $\mathbb{Z}_p$.*
2. *There exists a matrix A called the share-generating matrix for Π. The matrix A has l rows and n columns. For $i = 1, \ldots, l$, the ith row A_i of A is labeled by an attribute $\rho(i)$ (ρ is a function from $\{1, \ldots, l\}$ to $\mathcal{U}$). When we consider the column vector $v = (s, r_2, \ldots, r_n)$, where $s \in \mathbb{Z}_p$ is the secret to be shared and $r_2, \ldots, r_n \in \mathbb{Z}_p$ are randomly chosen, then Av is the vector of l shares of the secret s according to Π. The share $\lambda_i = (Av)_i$, i.e., the inner product $A_i \cdot v$, belongs to attribute $\rho(i)$.*

Also shown in [2], every LSSS as defined above enjoys the linear reconstruction property, which is defined as follows: Suppose that Π is an LSSS for access structure $\mathbb{A}$. Let $S \in \mathbb{A}$ be an authorized set, and $I \subset \{1, \ldots, l\}$ be defined as $I = \{i : \rho(i) \in S\}$. There exist constants $\{\omega_i \in \mathbb{Z}_p\}_{i \in I}$ such that if $\{\lambda_i\}$ are valid shares of a secret s according to Π, $\sum_{i \in I} \omega_i \lambda_i = s$. Furthermore, these constants $\{\omega_i\}$ can be found in time polynomial in the size of the share-generating matrix A. For any unauthorized set, no such constants exist. In this paper, as of previous work, we use an LSSS matrix (A, ρ) to express an access structure associated to a private decryption key.

Composite Order Bilinear Groups [3]. Let $\mathcal{G}$ be a group generator algorithm, which takes a security parameter λ and outputs $(p_1, p_2, p_3, \mathbb{G}, \mathbb{G}_T, e)$ where p_1, p_2, p_3 are distinct primes, $\mathbb{G}$ and $\mathbb{G}_T$ are cyclic groups of order $N = p_1 p_2 p_3$, and $e : \mathbb{G} \times \mathbb{G} \to \mathbb{G}_T$ is a map such that: (1) (Bilinear) $\forall g, h \in \mathbb{G}, a, b \in$

$\mathbb{Z}_N$, $e(g^a, h^b) = e(g,h)^{ab}$, (2) (Non-Degenerate) $\exists g \in \mathbb{G}$ such that $e(g,g)$ has order N in $\mathbb{G}_T$. Assume that group operations in $\mathbb{G}$ and $\mathbb{G}_T$ as well as the bilinear map e are computable in polynomial time with respect to λ. Let $\mathbb{G}_{p_1}$, $\mathbb{G}_{p_2}$ and $\mathbb{G}_{p_3}$ be the subgroups of order p_1, p_2 and p_3 in $\mathbb{G}$ respectively. These subgroups are "orthogonal" to each other under the bilinear map e: if $h_i \in \mathbb{G}_{p_i}$ and $h_j \in \mathbb{G}_{p_j}$ for $i \neq j$, $e(h_i, h_j) = 1$ (the identity element in $\mathbb{G}_T$).

Complexity Assumptions. The message-hiding property of our AugKP-ABE scheme will be based on three assumptions (Assumptions 1, 2 and 3 in [15]) that are used by [14,15] to achieve full security of their ABE schemes, and the index-hiding property will be based on two assumptions (the Decision 3-Party Diffie-Hellman Assumption and the Decisional Linear Assumption) that are used by [8] to achieve traceability in the setting of broadcast encryption. We refer to [8,15] for the details of these assumptions.

Notations. Suppose the number of users $\mathcal{K}$ in the system equals m^2 for some m.[4] We arrange the users in an $m \times m$ matrix and uniquely assign a tuple (i,j) where $1 \leq i, j \leq m$, to each user. A user at position (i,j) of the matrix has index $k = (i-1) * m + j$. For simplicity, we directly use (i,j) as the index where $(i,j) \geq (\bar{i}, \bar{j})$ means that $((i > \bar{i}) \vee (i = \bar{i} \wedge j \geq \bar{j}))$. The use of pairwise notation (i,j) is purely a notational convenience, as $k = (i-1) * m + j$ defines a bijection between $\{(i,j) | 1 \leq i, j \leq m\}$ and $[\mathcal{K}]$. For a given vector $\boldsymbol{v} = (v_1, \dots, v_d)$, by $g^{\boldsymbol{v}}$ we mean the vector $(g^{v_1}, \dots, g^{v_d})$. Furthermore, for $g^{\boldsymbol{v}} = (g^{v_1}, \dots, g^{v_d})$ and $g^{\boldsymbol{w}} = (g^{w_1}, \dots, g^{w_d})$, by $g^{\boldsymbol{v}} \cdot g^{\boldsymbol{w}}$ we mean the vector $(g^{v_1 + w_1}, \dots, g^{v_d + w_d})$, i.e. $g^{\boldsymbol{v}} \cdot g^{\boldsymbol{w}} = g^{\boldsymbol{v}+\boldsymbol{w}}$, and by $e_d(g^{\boldsymbol{v}}, g^{\boldsymbol{w}})$ we mean $\prod_{k=1}^{d} e(g^{v_k}, g^{w_k})$, i.e. $e_d(g^{\boldsymbol{v}}, g^{\boldsymbol{w}}) = \prod_{k=1}^{d} e(g^{v_k}, g^{w_k}) = e(g,g)^{(\boldsymbol{v} \cdot \boldsymbol{w})}$ where $(\boldsymbol{v} \cdot \boldsymbol{w})$ is the inner product of $\boldsymbol{v}$ and $\boldsymbol{w}$. Given a bilinear group order N, one can randomly choose $r_x, r_y, r_z \in \mathbb{Z}_N$, and set $\boldsymbol{\chi}_1 = (r_x, 0, r_z)$, $\boldsymbol{\chi}_2 = (0, r_y, r_z)$, $\boldsymbol{\chi}_3 = \boldsymbol{\chi}_1 \times \boldsymbol{\chi}_2 = (-r_y r_z, -r_x r_z, r_x r_y)$. Let $span\{\boldsymbol{\chi}_1, \boldsymbol{\chi}_2\}$ be the subspace spanned by $\boldsymbol{\chi}_1$ and $\boldsymbol{\chi}_2$, i.e. $span\{\boldsymbol{\chi}_1, \boldsymbol{\chi}_2\} = \{\nu_1 \boldsymbol{\chi}_1 + \nu_2 \boldsymbol{\chi}_2 | \nu_1, \nu_2 \in \mathbb{Z}_N\}$. We can see that $\boldsymbol{\chi}_3$ is orthogonal to the subspace $span\{\boldsymbol{\chi}_1, \boldsymbol{\chi}_2\}$ and $\mathbb{Z}_N^3 = span\{\boldsymbol{\chi}_1, \boldsymbol{\chi}_2, \boldsymbol{\chi}_3\} = \{\nu_1 \boldsymbol{\chi}_1 + \nu_2 \boldsymbol{\chi}_2 + \nu_3 \boldsymbol{\chi}_3 | \nu_1, \nu_2, \nu_3 \in \mathbb{Z}_N\}$. For any $\boldsymbol{v} \in span\{\boldsymbol{\chi}_1, \boldsymbol{\chi}_2\}$, we have $(\boldsymbol{\chi}_3 \cdot \boldsymbol{v}) = 0$, and for random $\boldsymbol{v} \in \mathbb{Z}_N^3$, $(\boldsymbol{\chi}_3 \cdot \boldsymbol{v}) \neq 0$ happens with overwhelming probability.

4.2 AugKP-ABE Construction

$\mathsf{Setup}_A(\lambda, \mathcal{U}, \mathcal{K} = m^2) \to (\mathsf{PP}, \mathsf{MSK})$. Let $\mathbb{G}$ be a bilinear group of order $N = p_1 p_2 p_3$ (3 distinct primes, whose size is determined by λ), $\mathbb{G}_{p_i}$ the subgroup of order p_i in $\mathbb{G}$ (for $i = 1, 2, 3$), and $g, f \in \mathbb{G}_{p_1}$, $g_3 \in \mathbb{G}_{p_3}$ the generators of corresponding subgroups. The algorithm randomly chooses exponents $\alpha \in \mathbb{Z}_N$, $\{\alpha_i, r_i, z_i \in \mathbb{Z}_N\}_{i \in [m]}$, $\{c_j \in \mathbb{Z}_N\}_{j \in [m]}$, $\{a_x \in \mathbb{Z}_N\}_{x \in \mathcal{U}}$. The public parameter PP includes the description of the group and the following

---

[4] If the number of users is not a square, we add some "dummy" users to pad to the next square.

elements:

$$\Big(g, f, E = e(g,g)^\alpha, \{E_i = e(g,g)^{\alpha_i}, G_i = g^{r_i}, Z_i = g^{z_i}\}_{i\in[m]},$$

$$\{H_j = g^{c_j}\}_{j\in[m]}, \ \{U_x = g^{a_x}\}_{x\in\mathcal{U}} \Big).$$

The master secret key is set to $\mathsf{MSK} = (\alpha, \{\alpha_i, r_i\}_{i\in[m]}, \{c_j\}_{j\in[m]}, g_3)$. In addition, a counter $ctr = 0$ is included in MSK.

$\mathsf{KeyGen}_A(\mathsf{PP}, \mathsf{MSK}, (A, \rho)) \to \mathsf{SK}_{(i,j),(A,\rho)}$. A is an $l \times n$ LSSS matrix and ρ maps each row A_k of A to an attribute $\rho(k) \in \mathcal{U}$. It is required that ρ would not map two different rows to the same attribute[5]. The algorithm first sets $ctr = ctr + 1$ and computes the corresponding index in the form of (i, j) where $1 \le i, j \le m$ and $(i - 1) * m + j = ctr$. Then it randomly chooses $u = (\sigma_{i,j}, u_2, \ldots, u_n) \in \mathbb{Z}_N^n$, $w_2, \ldots, w_n \in \mathbb{Z}_N$, and $\{\xi_k \in \mathbb{Z}_N, R_{k,1}, R_{k,2} \in \mathbb{G}_{p_3}\}_{k\in[l]}$. Let $w = (\alpha, w_2, \ldots, w_n)$, the algorithm outputs a private key $\mathsf{SK}_{(i,j),(A,\rho)} = ((i,j), (A,\rho), K_{i,j}, K'_{i,j}, K''_{i,j}, \{K_{i,j,k,1}, K_{i,j,k,2}\}_{k\in[l]})$ where

$$K_{i,j} = g^{\alpha_i} g^{r_i c_j} f^{\sigma_{i,j}}, \ K'_{i,j} = g^{\sigma_{i,j}}, \ K''_{i,j} = Z_i^{\sigma_{i,j}},$$

$$\{K_{i,j,k,1} = f^{(A_k\cdot u)} g^{(A_k\cdot w)} U_{\rho(k)}^{\xi_k} R_{k,1}, \ K_{i,j,k,2} = g^{\xi_k} R_{k,2}\}_{k\in[l]}.$$

$\mathsf{Encrypt}_A(\mathsf{PP}, M, S, (\bar{i}, \bar{j})) \to CT_S$. The algorithm randomly chooses

$$\pi, \kappa, \tau, s_1, \ldots, s_m, t_1, \ldots, t_m \in \mathbb{Z}_N, \quad v_c, w_1, \ldots, w_m \in \mathbb{Z}_N^3.$$

In addition, the algorithm randomly chooses $r_x, r_y, r_z \in \mathbb{Z}_N$, and sets $\chi_1 = (r_x, 0, r_z)$, $\chi_2 = (0, r_y, r_z)$, $\chi_3 = \chi_1 \times \chi_2 = (-r_y r_z, -r_x r_z, r_x r_y)$. Then it randomly chooses $v_i \in \mathbb{Z}_N^3 \ \forall i \in \{1, \ldots, \bar{i}\}, v_i \in span\{\chi_1, \chi_2\} \ \forall i \in \{\bar{i}+1, \ldots, m\}$, and creates a ciphertext $\langle S, (R_i, R'_i, Q_i, Q'_i, Q''_i, T_i)_{i=1}^m, (C_j, C'_j)_{j=1}^m, P, \{P_x\}_{x\in S}\rangle$ as follows:

1. For each $i \in [m]$:

– if $i < \bar{i}$: it randomly chooses $\hat{s}_i \in \mathbb{Z}_N$, and sets

$$R_i = g^{v_i}, \ R'_i = g^{\kappa v_i},$$
$$Q_i = g^{s_i}, \ Q'_i = f^{s_i} Z_i^{t_i} f^\pi, \ Q''_i = g^{t_i}, \quad T_i = E_i^{\hat{s}_i} \cdot E^\pi.$$

– if $i \ge \bar{i}$: it sets

$$R_i = G_i^{s_i v_i}, \ R'_i = G_i^{\kappa s_i v_i},$$
$$Q_i = g^{\tau s_i(v_i\cdot v_c)}, \ Q'_i = f^{\tau s_i(v_i\cdot v_c)} Z_i^{t_i} f^\pi, \ Q''_i = g^{t_i},$$
$$T_i = M \cdot E_i^{\tau s_i(v_i\cdot v_c)} \cdot E^\pi.$$

---

[5] This restriction is inherited from the underlying KP-ABE scheme [14], and can be removed with the techniques in [14] similarly, with some loss of efficiency. The similar restriction in CP-ABE has been efficiently eliminated recently by Lewko and Waters in [16], but fully secure KP-ABE scheme without this restriction is not proposed yet.

2. For each $j \in [m]$:
   - if $j < \bar{j}$: it randomly chooses $\mu_j \in \mathbb{Z}_N$, and sets
     $C_j = H_j^{\tau(v_c + \mu_j \chi_3)} \cdot g^{\kappa w_j}, \quad C'_j = g^{w_j}$.
   - if $j \geq \bar{j}$: it sets $C_j = H_j^{\tau v_c} \cdot g^{\kappa w_j}, \quad C'_j = g^{w_j}$.

3. It sets $P = g^\pi, \{P_x = U_x^\pi\}_{x \in S}$.

$\mathsf{Decrypt}_\mathsf{A}(\mathsf{PP}, CT_S, \mathsf{SK}_{(i,j),(A,\rho)}) \to M$ or $\bot$. For a ciphertext $CT_S = \langle S, (R_i, R'_i, Q_i, Q'_i, Q''_i, T_i)_{i=1}^m, (C_j, C'_j)_{j=1}^m, P, \{P_x\}_{x \in S} \rangle$ and a private decryption key $\mathsf{SK}_{(i,j),(A,\rho)} = ((i,j), (A,\rho), K_{i,j}, K'_{i,j}, K''_{i,j}, \{K_{i,j,k,1}, K_{i,j,k,2}\}_{k \in [l]})$, if S does not satisfy (A,ρ), the algorithm outputs $\bot$, otherwise it

1. computes constants $\{\omega_k \in \mathbb{Z}_N\}$ such that $\sum_{\rho(k) \in S} \omega_k A_k = (1, 0, \ldots, 0)$, then computes

$$D_P = \prod_{\rho(k) \in S} \left(\frac{e(K_{i,j,k,1}, P)}{e(K_{i,j,k,2}, P_{\rho(k)})} \right)^{\omega_k} = \prod_{\rho(k) \in S} \left(e(f^{(A_k \cdot u)} g^{(A_k \cdot w)}, g^\pi) \right)^{\omega_k}$$
$$= e(f, g)^{\pi \sigma_{i,j}} e(g, g)^{\alpha \pi}.$$

2. computes

$$D_I = \frac{e(K_{i,j}, Q_i) \cdot e(K''_{i,j}, Q''_i)}{e(K'_{i,j}, Q'_i)} \cdot \frac{e_3(R'_i, C'_j)}{e_3(R_i, C_j)}.$$

3. computes $M = T_i/(D_P \cdot D_I)$ as the output message.

Correctness. Assume the ciphertext is generated from message M' and encryption index $(\bar{i}, \bar{j})$, we have

$$D_I = \begin{cases} E_i^{\tau s_i(v_i \cdot v_c)} / e(g, f)^{\pi \sigma_{i,j}}, & : (i > \bar{i}) \text{ or } (i = \bar{i} \wedge j \geq \bar{j}) \\ E_i^{\tau s_i(v_i \cdot v_c)} / (e(g, f)^{\pi \sigma_{i,j}} e(g, g)^{r_i s_i c_j \tau \mu_j(v_i \cdot \chi_3)}), & : (i = \bar{i} \wedge j < \bar{j}). \end{cases}$$

Thus, we have (1) if $(i > \bar{i}) \vee (i = \bar{i} \wedge j \geq \bar{j})$, then $M = M'$; (2) if $i = \bar{i} \wedge j < \bar{j}$, then $M = M' \cdot e(g, g)^{\tau s_i r_i c_j \mu_j (v_i \cdot \chi_3)}$; (3) if $i < \bar{i}$, then M has no relation with M'. The correctness details can be found in the full version [21].

4.3   AugKP-ABE Security

The following Theorems 3 and 4 show that our AugKP-ABE construction in Sect. 4.2 is message-hiding, and Theorem 5 shows that our construction is selec-tively index-hiding.

Theorem 3. *Suppose that Assumptions 1, 2, and 3 in [15] hold. Then no PPT adversary can win* $\mathsf{Game}_{\mathsf{MH}_1}^\mathsf{A}$ *with non-negligible advantage.*

Proof. The structure of the KP-ABE portion of our AugKP-ABE is similar to that of the KP-ABE in [14], the proof of Theorem 3 is also similar to that of [14]. Here we prove the theorem by reducing the message-hiding property of our AugKP-ABE scheme in $\mathsf{Game}^{\mathsf{A}}_{\mathsf{MH}_1}$ to the security of the KP-ABE in [14]. The proof details are provided in the full version [21].

Theorem 4. *No PPT adversary can win $\mathsf{Game}^{\mathsf{A}}_{\mathsf{MH}_{\mathcal{K}+1}}$ with non-negligible advantage.*

Proof. The argument for the message-hiding property in $\mathsf{Game}^{\mathsf{A}}_{\mathsf{MH}_{\mathcal{K}+1}}$ is very straightforward since an encryption to index $\mathcal{K} + 1 = (m + 1, 1)$ contains no information about the message. The simulator simply runs actual $\mathsf{Setup}_{\mathsf{A}}$ and $\mathsf{KeyGen}_{\mathsf{A}}$ algorithms and encrypts the message M_b by the challenge attribute set S^* and index $(m + 1, 1)$. Since for all $i = 1$ to m, the values of $T_i = E_i^{\hat{s}_i} \cdot E^{\pi}$ contains no information about the message, the bit b is perfectly hidden and $\mathsf{MH}^{\mathsf{A}}_{\mathcal{K}+1}\mathsf{Adv}_{\mathcal{A}} = 0$.

Theorem 5. *Suppose that the Decision 3-Party Diffie-Hellman Assumption and the Decisional Linear Assumption hold (referring to [8] for the details of the two assumptions). Then no PPT adversary can selectively win $\mathsf{Game}^{\mathsf{A}}_{\mathsf{IH}}$ with non-negligible advantage.*

Proof. Theorem 5 follows Lemmas 1 and 2 below.

Lemma 1. *If the Decision 3-Party Diffie Hellman Assumption holds, then for $\bar{j} < m$ no PPT adversary can selectively distinguish between an encryption to $(\bar{i}, \bar{j})$ and an encryption to $(\bar{i}, \bar{j} + 1)$ in $\mathsf{Game}^{\mathsf{A}}_{\mathsf{IH}}$ with non-negligible advantage.*

Proof. The proof of this lemma explores the techniques of combining a traitor tracing scheme and a KP-ABE scheme. When the adversary queries a private key with the index $(\bar{i}, \bar{j})$, the game restriction implies that the corresponding access structure must not be satisfied by the challenge attribute set S^*. In other words, we have to use a restriction on "attributes and access structure" to prove the index-hiding property on "index", which are very uncorrelated structures. The ciphertext components $Z_i^{t_i}$ (in Q_i') and $Q_i'' = g^{t_i}$ work like a "transmission gear" to intertwine the two structures, *securely* combining the tracing part ($f^{\tau s_i(v_i \cdot v_c)}$ for $i \geq \bar{i}$ and f^{s_i} for $i < \bar{i}$) and the KP-ABE part (f^{π}) together. The proof is given in Appendix A.

Lemma 2. *Suppose that the Decision 3-Party Diffie-Hellman Assumption and the Decisional Linear Assumption hold. Then for $1 \leq \bar{i} \leq m$ no PPT adversary can selectively distinguish between an encryption to $(\bar{i}, m)$ and one to $(\bar{i}+1, 1)$ in $\mathsf{Game}^{\mathsf{A}}_{\mathsf{IH}}$ with non-negligible advantage.*

Proof. Similar to the proof of Lemma 6.3 in [8], to prove this lemma we define the following hybrid experiments: H_1: Encrypt to $(\bar{i}, \bar{j} = m)$; H_2: Encrypt to $(\bar{i}, \bar{j} = m + 1)$; and H_3: Encrypt to $(\bar{i}+1, 1)$. Lemma 2 follows Claims 1 and 2 below.

Claim 1. *If the Decision 3-Party Diffie-Hellman Assumption holds, then no PPT adversary can selectively distinguish between experiment H_1 and H_2 with non-negligible advantage.*

Proof. The proof is identical to that of Lemma 1.

Claim 2. *Suppose that the Decision 3-Party Diffie-Hellman Assumption and the Decisional Linear Assumption hold. Then no PPT adversary can distinguish between experiment H_2 and H_3 with non-negligible advantage.*

Proof. The indistinguishability of H_2 and H_3 can be proven using the similar proof to that of Lemma 6.3 in [8], which was used to prove the indistinguishability of similar hybrid experiments for their Augmented Broadcast Encryption (AugBE) scheme. We will prove Claim 2 by a reduction from our AugKP-ABE scheme to the AugBE scheme in [8, Sect. 5.1]. The proof details are provided in the full version [21].

5   Conclusion

We proposed an expressive and efficient KP-ABE scheme that simultaneously supports fully collusion-resistant (and public) blackbox traceability and high expressivity (i.e. supporting any monotonic access structures). The scheme is proven fully secure in the standard model and selectively traceable in the standard model. Compared with the most efficient conventional (non-traceable) KP-ABE schemes in the literature with high expressivity and full security, our scheme adds fully collusion-resistant blackbox traceability with the price of adding only $O(\sqrt{\mathcal{K}})$ elements in the ciphertext and public key. Instead of directly building a traceable KP-ABE scheme, we constructed a simpler primitive called Augmented KP-ABE, and showed that an Augmented KP-ABE scheme with message-hiding and index-hiding properties is sufficient for constructing a secure KP-ABE scheme with fully collusion-resistant blackbox traceability.

Acknowledgment. The work described in this paper was supported in part by the Research Grants Council of the HKSAR, China, under Project CityU 123511, in part by the National Natural Science Foundation of China under Grant 61161140320, Grant 61371083, and Grant 61373154, and in part by the Prioritized Development Projects, Specialized Research Fund for the Doctoral Program of Higher Education of China, under Grant 20130073130004.

A   Proof of Lemma 1

Proof. Suppose there exists a PPT adversary $\mathcal{A}$ that can selectively break the index-hiding game for $(\bar{i}, \bar{j})$ with advantage ϵ. We build a PPT algorithm $\mathcal{B}$ to solve a Decision 3-Party Diffie-Hellman problem instance as follows.

$\mathcal{B}$ receives a Decision 3-Party Diffie-Hellman problem instance from the challenger as $(g, A = g^a, B = g^b, C = g^c, T)$. The problem instance will be given in the subgroup $\mathbb{G}_{p_1}$ of prime order p_1 in a composite order group $\mathbb{G}$ of order $N = p_1 p_2 p_3$, i.e., $g \in \mathbb{G}_{p_1}$, $a, b, c \in \mathbb{Z}_{p_1}$, $\mathcal{B}$ is given the factors p_1, p_2, and p_3, and its goal is to determine whether $T = g^{abc}$ or a random element from $\mathbb{G}_{p_1}$ [6].

Init. $\mathcal{A}$ gives $\mathcal{B}$ the challenge attribute set $S^* \subseteq \mathcal{U}$.

Setup. $\mathcal{B}$ chooses random exponents

$$\eta, \alpha \in \mathbb{Z}_N, \ \{\alpha_i \in \mathbb{Z}_N\}_{i \in [m]}, \ \{r_i, z_i' \in \mathbb{Z}_N\}_{i \in [m] \setminus \{\bar{i}\}}, \ \{c_j \in \mathbb{Z}_N\}_{j \in [m] \setminus \{\bar{j}\}},$$
$$r_{\bar{i}}', z_{\bar{i}}, c_{\bar{j}}' \in \mathbb{Z}_N, \ \{a_x \in \mathbb{Z}_N\}_{x \in S^*}, \ \{a_x' \in \mathbb{Z}_N\}_{x \in \mathcal{U} \setminus S^*}.$$

$\mathcal{B}$ gives $\mathcal{A}$ the following public parameter PP:

$$g, \ f = C^\eta, \ E = e(g,g)^\alpha, \ \{E_i = e(g,g)^{\alpha_i}\}_{i \in [m]},$$
$$\{G_i = g^{r_i}, \ Z_i = C^{z_i'}\}_{i \in [m] \setminus \{\bar{i}\}}, \{H_j = g^{c_j}\}_{j \in [m] \setminus \{\bar{j}\}}, \ G_{\bar{i}} = B^{r_{\bar{i}}'}, Z_{\bar{i}} = g^{z_{\bar{i}}}, H_{\bar{j}} = C^{c_{\bar{j}}'},$$
$$\{U_x = g^{a_x}\}_{x \in S^*}, \ \{U_x = C^{a_x'}\}_{x \in \mathcal{U} \setminus S^*}.$$

Note that $\mathcal{B}$ implicitly chooses $r_{\bar{i}}, z_i (i \in [m] \setminus \{\bar{i}\})$, $c_{\bar{j}}, a_x(x \in \mathcal{U} \setminus S^*) \in \mathbb{Z}_N$ such that

$$br_{\bar{i}}' \equiv r_{\bar{i}} \bmod p_1, \ cz_i' \equiv z_i \bmod p_1 \ \forall i \in [m] \setminus \{\bar{i}\}, \ cc_{\bar{j}}' \equiv c_{\bar{j}} \bmod p_1,$$
$$ca_x' \equiv a_x \bmod p_1 \ \forall x \in \mathcal{U} \setminus S^*.$$

Key Query. To respond to a query from $\mathcal{A}$ for $((i,j), (A, \rho))$ where A is an $l \times n$ matrix:

- If $(i,j) \neq (\bar{i}, \bar{j})$: $\mathcal{B}$ randomly chooses $u = (\sigma_{i,j}, u_2, \ldots, u_n) \in \mathbb{Z}_N^n$, $w_2, \ldots,$ $w_n \in \mathbb{Z}_N$ and $\{\xi_k \in \mathbb{Z}_N, R_{k,1}, R_{k,2} \in \mathbb{G}_{p_3}\}_{k=1}^l$. Let $w = (\alpha, w_2, \ldots, w_n)$, $\mathcal{B}$ creates a private key $\mathsf{SK}_{(i,j),(A,\rho)} = ((i,j), (A, \rho), K_{i,j}, K_{i,j}', K_{i,j}'', \{K_{i,j,k,1}, K_{i,j,k,2}\}_{k=1}^l)$ as

$$K_{i,j} = \begin{cases} g^{\alpha_i} g^{r_i c_j} f^{\sigma_{i,j}}, & : i \neq \bar{i}, j \neq \bar{j} \\ g^{\alpha_i} B^{r_{\bar{i}}' c_j} f^{\sigma_{i,j}}, & : i = \bar{i}, j \neq \bar{j} \\ g^{\alpha_i} C^{r_i c_{\bar{j}}'} f^{\sigma_{i,j}}, & : i \neq \bar{i}, j = \bar{j}. \end{cases}$$
$$K_{i,j}' = g^{\sigma_{i,j}}, \quad K_{i,j}'' = Z_i^{\sigma_{i,j}},$$
$$\{K_{i,j,k,1} = f^{(A_k \cdot u)} g^{(A_k \cdot w)} U_{\rho(k)}^{\xi_k} R_{k,1}, \quad K_{i,j,k,2} = g^{\xi_k} R_{k,2}\}_{k=1}^l.$$

- If $(i,j) = (\bar{i}, \bar{j})$: $\mathcal{B}$ randomly chooses $\sigma_{\bar{i},\bar{j}}', u_2', \ldots, u_n', w_2, \ldots, w_n \in \mathbb{Z}_N, \{\xi_k \in \mathbb{Z}_N\}_{k \in [l]}$ s.t. $\rho(k) \in S^*$, $\{\xi_k' \in \mathbb{Z}_N\}_{k \in [l]}$ s.t. $\rho(k) \notin S^*$, $\{R_{k,1}, R_{k,2} \in \mathbb{G}_{p_3}\}_{k=1}^l$. Let $u' = (0, u_2', \ldots, u_n'), w = (\alpha, w_2, \ldots, w_n)$. As (A, ρ) cannot be satisfied by S^*

---

[6] The situation is similar to that of the proof in [4,5] in the sense that the challenge is given in a subgroup of a composite order group and the factors are given to the simulator. Actually, Lewko and Waters [16] use this case explicitly as an assumption, i.e. the 3-Party Diffie-Hellman Assumption in a Subgroup.

(since $(i,j) = (\bar{i},\bar{j})$), $\mathcal{B}$ can efficiently find a vector $\boldsymbol{u}'' = (u_1'', u_2'', \ldots, u_n'') \in \mathbb{Z}_N^n$ such that $u_1'' = 1$ and $A_k \cdot \boldsymbol{u}'' = 0$ for all k such that $\rho(k) \in S^*$. Implicitly setting $\sigma_{\bar{i},\bar{j}} \in \mathbb{Z}_N$, $\boldsymbol{u} \in \mathbb{Z}_N^n$, $\{\xi_k \in \mathbb{Z}_N\}_{k \in [l] \ s.t. \ \rho(k) \notin S^*}$ as

$$\sigma_{\bar{i},\bar{j}}' - br_{\bar{i}}'c_{\bar{j}}'/\eta \equiv \sigma_{\bar{i},\bar{j}} \bmod p_1, \quad \boldsymbol{u} = \boldsymbol{u}' + \sigma_{\bar{i},\bar{j}}\boldsymbol{u}'',$$

$$\xi_k' + br_{\bar{i}}'c_{\bar{j}}'(A_k \cdot \boldsymbol{u}'')/a_{\rho(k)}' \equiv \xi_k \bmod p_1 \ \forall k \in [l] \ s.t. \ \rho(k) \notin S^*,$$

$\mathcal{B}$ creates a private key $\mathsf{SK}_{(\bar{i},\bar{j}),(A,\rho)} = ((\bar{i},\bar{j}), (A,\rho), \ K_{\bar{i},\bar{j}}, K_{\bar{i},\bar{j}}', K_{\bar{i},\bar{j}}'', \{K_{\bar{i},\bar{j},k,1}, K_{\bar{i},\bar{j},k,2}\}_{k=1}^l)$ as:

$$K_{\bar{i},\bar{j}} = g^{\alpha_{\bar{i}}}f^{\sigma_{\bar{i},\bar{j}}'}, \quad K_{\bar{i},\bar{j}}' = g^{\sigma_{\bar{i},\bar{j}}'}B^{-r_{\bar{i}}'c_{\bar{j}}'/\eta}, \quad K_{\bar{i},\bar{j}}'' = (g^{\sigma_{\bar{i},\bar{j}}'}B^{-r_{\bar{i}}'c_{\bar{j}}'/\eta})^{z_{\bar{i}}},$$

$$\{K_{i,j,k,1} = f^{(A_k \cdot \boldsymbol{u}')}g^{(A_k \cdot \boldsymbol{w})}U_{\rho(k)}^{\xi_k}R_{k,1}, \quad K_{i,j,k,2} = g^{\xi_k}R_{k,2}\}_{\rho(k) \in S^*},$$

$$\{K_{i,j,k,1} = f^{(A_k \cdot \boldsymbol{u}') + \sigma_{\bar{i},\bar{j}}'(A_k \cdot \boldsymbol{u}'')}g^{(A_k \cdot \boldsymbol{w})}U_{\rho(k)}^{\xi_k'}R_{k,1},$$

$$K_{i,j,k,2} = g^{\xi_k'}B^{r_{\bar{i}}'c_{\bar{j}}'(A_k \cdot \boldsymbol{u}'')/a_{\rho(k)}'}R_{k,2}\}_{\rho(k) \notin S^*}.$$

Challenge. $\mathcal{A}$ submits a message M. $\mathcal{B}$ randomly chooses

$$\pi', \ \tau', \ s_1, \ldots, s_{\bar{i}-1}, s_{\bar{i}}', s_{\bar{i}+1}, \ldots, s_m, \ t_1', \ldots, t_{\bar{i}-1}', t_{\bar{i}}, t_{\bar{i}+1}', \ldots, t_m' \ \in \mathbb{Z}_N,$$

$$\boldsymbol{w}_1, \ldots, \boldsymbol{w}_{\bar{j}-1}, \boldsymbol{w}_{\bar{j}}', \ldots, \boldsymbol{w}_m' \ \in \mathbb{Z}_N^3.$$

$\mathcal{B}$ randomly chooses $r_x, r_y, r_z \in \mathbb{Z}_N$, and sets $\boldsymbol{\chi}_1 = (r_x, 0, r_z)$, $\boldsymbol{\chi}_2 = (0, r_y, r_z)$, $\boldsymbol{\chi}_3 = \boldsymbol{\chi}_1 \times \boldsymbol{\chi}_2 = (-r_y r_z, -r_x r_z, r_x r_y)$. Then $\mathcal{B}$ randomly chooses

$$\boldsymbol{v}_i \in \mathbb{Z}_N^3 \ \forall i \in \{1, \ldots, \bar{i}\}, \quad \boldsymbol{v}_i \in span\{\boldsymbol{\chi}_1, \boldsymbol{\chi}_2\} \ \forall i \in \{\bar{i}+1, \ldots, m\},$$

$$\boldsymbol{v}_{c,p} \in span\{\boldsymbol{\chi}_1, \boldsymbol{\chi}_2\}, \quad \boldsymbol{v}_{c,q} = \nu_3 \boldsymbol{\chi}_3 \in span\{\boldsymbol{\chi}_3\}.$$

$\mathcal{B}$ sets the value of $\pi, \kappa, \tau, s_{\bar{i}}, t_i (i \in [m] \setminus \{\bar{i}\}) \in \mathbb{Z}_N$, $\boldsymbol{v}_c \in \mathbb{Z}_N^3$, $\{\boldsymbol{w}_j \in \mathbb{Z}_N^3\}_{j=\bar{j}}^m$ by implicitly setting

$$\boldsymbol{v}_c = a^{-1}\boldsymbol{v}_{c,p} + \boldsymbol{v}_{c,q}, \quad \pi' - a\tau' s_{\bar{i}}'(\boldsymbol{v}_{\bar{i}} \cdot \boldsymbol{v}_{c,q}) \equiv \pi \bmod p_1,$$

$$b \equiv \kappa \bmod p_1, \quad ab\tau' \equiv \tau \bmod p_1, \quad s_{\bar{i}}'/b \equiv s_{\bar{i}} \bmod p_1,$$

$$t_i' + \eta a\tau' s_{\bar{i}}'(\boldsymbol{v}_{\bar{i}} \cdot \boldsymbol{v}_{c,q})/z_i' \equiv t_i \bmod p_1 \ \forall i \in \{1, \ldots, \bar{i}-1\},$$

$$t_i' - \eta b\tau' s_i(\boldsymbol{v}_i \cdot \boldsymbol{v}_{c,p})/z_i' + \eta a\tau' s_{\bar{i}}'(\boldsymbol{v}_{\bar{i}} \cdot \boldsymbol{v}_{c,q})/z_i' \equiv t_i \bmod p_1 \ \forall i \in \{\bar{i}+1, \ldots, m\},$$

$$\boldsymbol{w}_{\bar{j}}' - cc_{\bar{j}}'\tau'\boldsymbol{v}_{c,p} \equiv \boldsymbol{w}_{\bar{j}} \bmod p_1,$$

$$\boldsymbol{w}_j' - ac_j\tau'\boldsymbol{v}_{c,q} \equiv \boldsymbol{w}_j \bmod p_1 \ \forall j \in \{\bar{j}+1, \ldots, m\}.$$

$\mathcal{B}$ creates a ciphertext $\langle S^*, (\boldsymbol{R}_i, \boldsymbol{R}_i', Q_i, Q_i', Q_i'', T_i)_{i=1}^m, (\boldsymbol{C}_j, \boldsymbol{C}_j')_{j=1}^m, P, \{P_x\}_{x \in S^*}\rangle$:

1. For each $i \in [m]$:

– if $i < \bar{i}$: it randomly chooses $\hat{s}_i \in \mathbb{Z}_N$, then sets

$$\boldsymbol{R}_i = g^{\boldsymbol{v}_i}, \quad \boldsymbol{R}_i' = B^{\boldsymbol{v}_i}, \quad Q_i = g^{s_i}, \quad Q_i' = f^{s_i}Z_i^{t_i'}f^{\pi'}, \quad Q_i'' = g^{t_i'}A^{\eta\tau' s_{\bar{i}}'(\boldsymbol{v}_{\bar{i}} \cdot \boldsymbol{v}_{c,q})/z_i'},$$

$$T_i = E_i^{\hat{s}_i} \cdot e(g^\alpha, g)^{\pi'} \cdot e(g^\alpha, A)^{-\tau' s_{\bar{i}}'(\boldsymbol{v}_{\bar{i}} \cdot \boldsymbol{v}_{c,q})}.$$

- if $i = \bar{i}$: it sets

$$R_i = g^{r_i' s_i' v_{\bar{i}}}, \quad R_i' = B^{r_i' s_i' v_{\bar{i}}},$$

$$Q_i = g^{\tau' s_i'(v_{\bar{i}} \cdot v_{c,p})} A^{\tau' s_i'(v_{\bar{i}} \cdot v_{c,q})}, \quad Q_i' = C^{\eta \tau' s_i'(v_{\bar{i}} \cdot v_{c,p})} Z_i^{t_i} f^{\pi'}, \quad Q_i'' = g^{t_{\bar{i}}},$$

$$T_i = M \cdot e(g^{\alpha_i}, Q_i) \cdot e(g^\alpha, g)^{\pi'} \cdot e(g^\alpha, A)^{-\tau' s_i'(v_{\bar{i}} \cdot v_{c,q})}.$$

- if $i > \bar{i}$: it sets

$$R_i = g^{r_i s_i v_i}, \quad R_i' = B^{r_i s_i v_i},$$

$$Q_i = B^{\tau' s_i(v_i \cdot v_{c,p})}, \quad Q_i' = Z_i^{t_i'} f^{\pi'}, \quad Q_i'' = g^{t_i'} B^{-\eta \tau' s_i(v_i \cdot v_{c,p})/z_i'} A^{\eta \tau' s_i'(v_{\bar{i}} \cdot v_{c,q})/z_i'},$$

$$T_i = M \cdot e(g^{\alpha_i}, Q_i) \cdot e(g^\alpha, g)^{\pi'} \cdot e(g^\alpha, A)^{-\tau' s_i'(v_{\bar{i}} \cdot v_{c,q})}.$$

2. For each $j \in [m]$:

- if $j < \bar{j}$: it randomly chooses $\mu_j' \in \mathbb{Z}_N$ and implicitly sets the value of μ_j such that $(ab)^{-1} \mu_j' \nu_3 - \nu_3 \equiv \mu_j \bmod p_1$, then sets
$C_j = B^{c_j \tau' v_{c,p}} \cdot g^{c_j \tau' \mu_j' v_{c,q}} \cdot B^{w_j}, \quad C_j' = g^{w_j}.$
- if $j = \bar{j}$: it sets $C_j = T^{c_j \tau' v_{c,q}} \cdot B^{w_j'}, \quad C_j' = g^{w_j'} \cdot C^{-c_j' \tau' v_{c,p}}.$
- if $j > \bar{j}$: it sets $C_j = B^{c_j \tau' v_{c,p}} \cdot B^{w_j'}, \quad C_j' = g^{w_j'} \cdot A^{-c_j \tau' v_{c,q}}.$

3. It sets $P = g^{\pi'} A^{-\tau' s_i'(v_{\bar{i}} \cdot v_{c,q})}, \quad P_x = (g^{\pi'} A^{-\tau' s_i'(v_{\bar{i}} \cdot v_{c,q})})^{a_x} \; \forall x \in S^*.$

If $T = g^{abc}$, then the ciphertext is a well-formed encryption to the index $(\bar{i}, \bar{j})$. If T is randomly chosen, say $T = g^r$ for some random $r \in \mathbb{Z}_{p_1}$, the ciphertext is a well-formed encryption to the index $(\bar{i}, \bar{j} + 1)$ with implicitly setting $\mu_{\bar{j}}$ such that $(\frac{r}{abc} - 1)\nu_3 \equiv \mu_{\bar{j}} \bmod p_1$.

Guess. $\mathcal{A}$ outputs a guess $b' \in \{0,1\}$ to $\mathcal{B}$, then $\mathcal{B}$ outputs this b' to the challenger as its answer to the Decision 3-Party Diffie-Hellman game.

Note that the distributions of the public parameter, private keys and challenge ciphertext are same as the real scheme, $\mathcal{B}$'s advantage in the Decision 3-Party Diffie-Hellman game will be exactly equal to $\mathcal{A}$'s advantage in selectively breaking the index-hiding game.

References

1. Attrapadung, N., Libert, B., de Panafieu, E.: Expressive key-policy attribute-based encryption with constant-size ciphertexts. In: Catalano, D., Fazio, N., Gennaro, R., Nicolosi, A. (eds.) PKC 2011. LNCS, vol. 6571, pp. 90–108. Springer, Heidelberg (2011)
2. Beimel, A.: Secure schemes for secret sharing and key distribution. Ph.D. thesis, Israel Institute of Technology, Technion, Haifa, Israel (1996)
3. Boneh, D., Goh, E.-J., Nissim, K.: Evaluating 2-DNF formulas on ciphertexts. In: Kilian, J. (ed.) TCC 2005. LNCS, vol. 3378, pp. 325–341. Springer, Heidelberg (2005)

4. Boneh, D., Sahai, A., Waters, B.: Fully collusion resistant traitor tracing with short ciphertexts and private keys. In: Vaudenay, S. (ed.) EUROCRYPT 2006. LNCS, vol. 4004, pp. 573–592. Springer, Heidelberg (2006)
5. Boneh, D., Waters, B.: A fully collusion resistant broadcast, trace, and revoke system. In: ACM Conference on Computer and Communications Security, pp. 211–220 (2006)
6. Cheung, L., Newport, C.C.: Provably secure ciphertext policy ABE. In: ACM Conference on Computer and Communications Security, pp. 456–465 (2007)
7. Garg, S., Gentry, C., Halevi, S., Sahai, A., Waters, B.: Attribute-based encryption for circuits from multilinear maps. In: Canetti, R., Garay, J.A. (eds.) CRYPTO 2013, Part II. LNCS, vol. 8043, pp. 479–499. Springer, Heidelberg (2013)
8. Garg, S., Kumarasubramanian, A., Sahai, A., Waters, B.: Building efficient fully collusion-resilient traitor tracing and revocation schemes. In: ACM Conference on Computer and Communications Security, pp. 121–130 (2010)
9. Goyal, V., Jain, A., Pandey, O., Sahai, A.: Bounded ciphertext policy attribute based encryption. In: Aceto, L., Damgård, I., Goldberg, L.A., Halldórsson, M.M., Ingólfsdóttir, A., Walukiewicz, I. (eds.) ICALP 2008, Part II. LNCS, vol. 5126, pp. 579–591. Springer, Heidelberg (2008)
10. Goyal, V., Pandey, O., Sahai, A., Waters, B.: Attribute-based encryption for fine-grained access control of encrypted data. In: ACM Conference on Computer and Communications Security, pp. 89–98 (2006)
11. Herranz, J., Laguillaumie, F., Ràfols, C.: Constant size ciphertexts in threshold attribute-based encryption. In: Nguyen, P.Q., Pointcheval, D. (eds.) PKC 2010. LNCS, vol. 6056, pp. 19–34. Springer, Heidelberg (2010)
12. Katz, J., Sahai, A., Waters, B.: Predicate encryption supporting disjunctions, polynomial equations, and inner products. In: Smart, N.P. (ed.) EUROCRYPT 2008. LNCS, vol. 4965, pp. 146–162. Springer, Heidelberg (2008)
13. Katz, J., Schröder, D.: Tracing insider attacks in the context of predicate encryption schemes. In: ACITA (2011). https://www.usukita.org/node/1779
14. Lewko, A.B., Okamoto, T., Sahai, A., Takashima, K., Waters, B.: Fully secure functional encryption: attribute-based encryption and (hierarchical) inner product encryption. IACR Cryptol. ePrint Arch. 2010, 110 (2010)
15. Lewko, A., Okamoto, T., Sahai, A., Takashima, K., Waters, B.: Fully secure functional encryption: attribute-based encryption and (Hierarchical) inner product encryption. In: Gilbert, H. (ed.) EUROCRYPT 2010. LNCS, vol. 6110, pp. 62–91. Springer, Heidelberg (2010)
16. Lewko, A., Waters, B.: New proof methods for attribute-based encryption: achieving full security through selective techniques. In: Safavi-Naini, R., Canetti, R. (eds.) CRYPTO 2012. LNCS, vol. 7417, pp. 180–198. Springer, Heidelberg (2012)
17. Li, J., Huang, Q., Chen, X., Chow, S.S.M., Wong, D.S., Xie, D.: Multi-authority ciphertext-policy attribute-based encryption with accountability. In: ASIACCS, pp. 386–390 (2011)
18. Li, J., Ren, K., Kim, K.: A2BE: accountable attribute-based encryption for abuse free access control. IACR Cryptol. ePrint Arch. 2009, 118 (2009)
19. Liu, Z., Cao, Z., Wong, D.S.: Blackbox traceable CP-ABE: how to catch people leaking their keys by selling decryption devices on ebay. In: ACM Conference on Computer and Communications Security, pp. 475–486 (2013)
20. Liu, Z., Cao, Z., Wong, D.S.: White-box traceable ciphertext-policy attribute-based encryption supporting any monotone access structures. IEEE Trans. Inf. Forensics Secur. 8(1), 76–88 (2013)

21. Liu, Z., Cao, Z., Wong, D.S.: Fully collusion-resistant traceable key-policy attribute-based encryption with sub-linear size ciphertexts. IACR Cryptol. ePrint Arch. **2014**, 676 (2014). http://eprint.iacr.org/2014/676
22. Okamoto, T., Takashima, K.: Fully secure functional encryption with general relations from the decisional linear assumption. In: Rabin, T. (ed.) CRYPTO 2010. LNCS, vol. 6223, pp. 191–208. Springer, Heidelberg (2010)
23. Ostrovsky, R., Sahai, A., Waters, B.: Attribute-based encryption with non-monotonic access structures. In: ACM Conference on Computer and Communications Security, pp. 195–203 (2007)
24. Rouselakis, Y., Waters, B.: Practical constructions and new proof methods for large universe attribute-based encryption. In: ACM Conference on Computer and Communications Security, pp. 463–474 (2013)
25. Sahai, A., Waters, B.: Fuzzy identity-based encryption. In: Cramer, R. (ed.) EUROCRYPT 2005. LNCS, vol. 3494, pp. 457–473. Springer, Heidelberg (2005)
26. Wang, Y.T., Chen, K.F., Chen, J.H.: Attribute-based traitor tracing. J. Inf. Sci. Eng. **27**(1), 181–195 (2011)
27. Waters, B.: Ciphertext-policy attribute-based encryption: an expressive, efficient, and provably secure realization. In: Catalano, D., Fazio, N., Gennaro, R., Nicolosi, A. (eds.) PKC 2011. LNCS, vol. 6571, pp. 53–70. Springer, Heidelberg (2011)
28. Waters, B.: Functional encryption for regular languages. In: Safavi-Naini, R., Canetti, R. (eds.) CRYPTO 2012. LNCS, vol. 7417, pp. 218–235. Springer, Heidelberg (2012)
29. Yu, S., Ren, K., Lou, W., Li, J.: Defending against key abuse attacks in KP-ABE enabled broadcast systems. In: Chen, Y., Dimitriou, T.D., Zhou, J. (eds.) SecureComm 2009. LNICST, vol. 19, pp. 311–329. Springer, Heidelberg (2009)

Integrating Ciphertext-Policy Attribute-Based Encryption with Identity-Based Ring Signature to Enhance Security and Privacy in Wireless Body Area Networks

Changji Wang[1,3]([✉]), Xilei Xu[2,3], Yuan Li[2,3], and Dongyuan Shi[2,3]

[1] National Pilot School of Software,
Yunnan University, Kunming 650500, China
wchangji@gmail.com
[2] School of Information Science and Technology,
Sun Yat-sen University, Guangzhou 510275, China
[3] Guangdong Key Laboratory of Information Security Technology,
Sun Yat-sen University, Guangzhou 510275, China

Abstract. The technology of wireless body area network (WBAN) has attracted intensive attention in recent years. For widespread deployment of WBANs, security and privacy issues must be addressed properly. Recently, Hu et al. proposed a fuzzy attribute-based signcryption scheme with the aim to provide security and privacy mechanisms in WBANs. In this paper, we first show Hu et al.'s scheme cannot achieve the claimed security properties. In particular, an adversary is capable of generating private keys for any set of attributes. Then we introduce a new cryptographic primitive named ciphertext-policy attribute-based ring signcryption (CP-ABRSC) by integrating the notion of ciphertext-policy attribute-based encryption with identity-based ring signature. We give formal syntax and security definitions for CP-ABRSC and present a provable secure CP-ABRSC scheme from bilinear pairings. Finally, we propose a novel access control framework for WBANs by exploiting CP-ABRSC scheme, which can not only provide semantic security, unforgeability and public authenticity, but also can provide participants privacy and fine-grained access control on encrypted health data.

Keywords: Fuzzy identity-based signcryption · Identity-based ring signature · Ciphertext-policy attribute-based encryption · Wireless body area network

1 Introduction

With the development of the modern society, the health care is attracting more and more attention, which triggers the introduction of novel technology-driven enhancements to current health care practices. Among them, a new type of network architecture, generally known as wireless body area network (WBAN) is considered as an

© Springer International Publishing Switzerland 2015
D. Lin et al. (Eds.): Inscrypt 2014, LNCS 8957, pp. 424–442, 2015.
DOI: 10.1007/978-3-319-16745-9_23

appropriate way for monitoring the body, which is made feasible by novel advances on lightweight, small-size, ultra-low-power, and intelligent monitoring wearable sensors [22].

WBAN can be utilized in diverse applications such as physiological and medical monitoring, human computer interaction, education and entertainment [7]. We illustrated a typical application of WBANs in the healthcare domain in Fig. 1, which allows inexpensive and continuous health monitoring with real-time updates of medical records through the Internet. The implanted intelligent physiological sensors in the human body will collect various vital signals (e.g., body temperature, blood pressure, heart rate, etc.) in order to monitor the patient's health status no matter their location, and these collected signals will be transmitted wirelessly to a controller (a mobile computing device like a PDA or smart phone). This device will transmit all information in real time or in non-real time to the third party remote server (e.g., health cloud server) to be stored, and these information will be shared by the patient's primary doctor, physicians and any other who needs to acquire the essential information for the patient's health. If an emergency is detected, the physicians will immediately inform the patient through the computer system by sending appropriate messages or alarms.

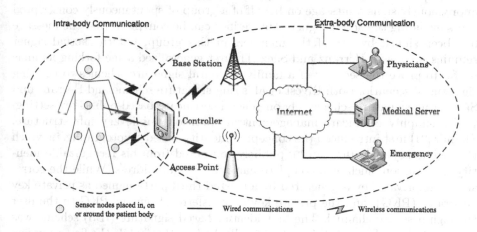

Fig. 1. Illustration of a typical WBAN for health care application

Unlike conventional sensor networks, WBANs deal with medical information, which has stringent requirements for security and privacy [14]. For widespread deployment of WBANs, security requirements such as fine-grained access control, participants privacy, data authenticity, confidentiality and non-repudiation must be addressed properly. Recently, Hu et al. [6] introduced a novel security mechanism in WBANs named fuzzy attribute-based signcryption (FABSC) without rigorous syntax and security definitions, and they presented a FABSC scheme by combining Sahai and Waters' fuzzy identity-based encryption scheme [18] with Yang et al.'s fuzzy identity-based signature scheme [25].

In this paper, we first point out there are several mistakes in Hu et al.'s scheme [6], and show that Hu et al.'s scheme does not hold the claimed security properties. In particular, an adversary can impersonate the key issuing authority and generate private keys for any set of attributes, thus totally break Hu et al.'s scheme. Then we introduce a new cryptographic primitive named ciphertext-policy attribute-based ring signcryption (CP-ABRSC) scheme by integrating the notion of ciphertext-policy attribute-based encryption scheme with identity-based ring signature scheme, and give formal syntax and security definitions for CP-ABRSC scheme. We propose a concrete CP-ABRSC scheme from bilinear pairings, and prove that the proposed construction is semantic security, existential unforgeability and strong anonymity. Finally, we present a secure, privacy-protected and fine-grained access control framework for WBANs by exploiting CP-ABRSC scheme, which can not only ensure data authenticity, confidentiality and non-repudiation, but also can provide participants privacy and fine-grained access control on encrypted health data.

1.1   Related Work

Identity-Based Ring Signature. To provide anonymity for signers, Rivest et al. [17] first introduced the concept of ring signature in 2001, where a user can anonymously signs a message on behalf of a group of spontaneously conscripted users including the actual signer. Any verifier can be convinced that the message has been signed by one of the members in this group, but the actual signer remains unknown. Herranz and Sáez [11] first generalized some forking lemmas useful to prove the security of a family of digital signature schemes to the ring signatures' scenario. Both Rivest et al.'s ring signature scheme and Herranz and Sáez's ring signature scheme rely on general certificate-based public-key setting.

To simplify certificate management in tradition public key infrastructure, Shamir [21] first introduced the concept of identity-based cryptography, in which the public key of a user can be publicly computed from his recognizable identity information, such as a complete name, an e-mail address. While the corresponding private key is generated by a trusted third party named as private key generator (PKG), and the private key is transferred from the PKG to the user through a secure channel. The first identity-based signature (IBS) scheme was constructed by Shamir [21] based on the RSA algorithm, while the first practical and secure identity-based encryption (IBE) scheme was proposed by Boneh and Franklin [4] from bilinear pairings. Since then, many IBE and IBS schemes based on the bilinear pairings were presented.

Zhang and Kim [26] first extended the concept of ring signature to the identity-based setting and proposed an ID-based ring signature (IBRS) scheme. Herranz and Sáez [12] improved Zhang and Kim's IBRS scheme and proved that the improved IBRS scheme is secure under the CDH assumption in the random oracle model. Chow et al. [5] proposed a more efficient IBRS scheme, which only takes two pairing operations for any group size, and the generation of the signature involves no pairing computations at all. Chow et al. also proved their IBRS scheme is secure under the CDH assumption in the random oracle model. The proof technique both in [5,12] is to apply the forking lemma for generic ring signature schemes [11].

Attribute-Based Encryption. Attribute-based encryption (ABE) was first introduced by Sahai and Waters [18] with the aim to provide an error-tolerant IBE that uses biometric identities. ABE can be viewed as an extension of the notion of IBE in which user identity is generalized to a set of descriptive attributes instead of a single string specifying the user identity. Compared with IBE, ABE has significant advantage as it achieves flexible one-to-many encryption instead of one-to-one, it is envisioned as a promising tool for addressing the problem of secure and fine-grained data sharing and decentralized access control [10].

According to access policy is associated with the ciphertext or private key, ABE can be divided into two categories: key-policy ABE (KP-ABE) and ciphertext-policy ABE (CP-ABE).

In a KP-ABE system, ciphertexts are labeled by the sender with a set of descriptive attributes, while users' private key are issued by the trusted attribute authority captures an policy that specifies which type of ciphertexts the key can decrypt. The first KP-ABE construction was provided by Goyal et al. [10], which was very expressive in that it allowed the access policies to be expressed by any monotonic formula over encrypted data. The system was proved selectively secure under the BDH assumption. Later, Ostrovsky et al. [16] proposed a KP-ABE scheme where private keys can represent any access formula over attributes, including non-monotone ones.

In a CP-ABE system, when a sender encrypts a message, they specify a specific access policy in terms of access policy over attributes in the ciphertext, stating what kind of receivers will be able to decrypt the ciphertext. Users possess sets of attributes and obtain corresponding attribute private keys from the attribute authority. Such a user can decrypt a ciphertext if his attributes satisfy the access policy associated with the ciphertext. The first CP-ABE scheme was proposed by Bethencourt et al. [3], but its security was proved in the generic group model. Waters [24] proposed a more expressive and efficient CP-ABE scheme, the size of a ciphertext depending linearly on the number of attributes involved in the specific policy for that ciphertext.

Signcryption. Encryption and signature are two basic cryptographic primitives to achieve confidentiality and authenticity. Zheng [27] first proposed the concept of signcryption, which can perform digital signature and public key encryption simultaneously in a single logical step with the cost in terms of both communication and computation significantly lower than *sign-then-encrypt* approach. Beak et al. [1] first gave the formal security notions for signcryption scheme via semantic security against adaptive chosen ciphertext attack and existential unforgeability against adaptive chosen message attack.

Malone-Lee [15] extended the concept of signcryption to the identity-based settings. Malone-Lee's work spurred a great deal of research on identity-based signcryption (IBSC), many IBSC schemes and IBSC schemes with additional properties have been proposed. In a conventional IBSC scheme, the message is hidden and thus the validity of the signcrypted ciphertext can be verified only after the unsigncryption process. Thus, a third party will not be able to verify whether the signcrypted ciphertext is valid or not. Selvi et al. [19] first proposed an IBSC scheme with public verifiability, which allows any one to verify the validity of signcrypted ciphertext without the knowledge of the message. To provide anonymity for the signcrypting party, Huang et al. [13] first introduced the concept of identity-based ring signcryption (IBRSC) scheme by combining the concept of IBRS and IBSC together. In an IBRSC scheme, a user can signcrypt a message along with the identities of a set of potential signcrypting parties (including the signcrypting party himself) without revealing which user in the set has actually produced the signcrypted ciphertext. IBRSC is very useful to protect privacy and authenticity of a collection of users who are connected through an ad hoc network.

Gagné et al. [9] first introduced the notion of attribute-based signcryption (ABSC), and proposed a threshold ABSC scheme where the access structure of user is limited in threshold structure and fixed when the user requests his attribute-based private key. Later, Emura et al. [8] proposed a dynamic threshold ABSC scheme, where access structures of the signcrypting party can be updated flexibly without re-issuing his attribute-based private key. Wang et al. [23] showed that both Gagné et al. threshold ABSC scheme and Emura et al. dynamic threshold ABSC scheme are not secure.

1.2 Paper Organization

The rest of the paper is organized as follows. We introduce some preliminaries in Sect. 2. We present security analysis of Hu et al.'s FABSC scheme in Sect. 3. We give formal syntax and security definitions of CP-ABRSC in Sect. 4, and describe our CP-ABRSC construction in Sect. 5. We present a secure, privacy-protected and fine-grained access control framework for WBANs by applying CP-ABRSC scheme in Sect. 6. Finally, we conclude the paper in Sect. 7.

2   Preliminaries

We denote by κ the system security parameter. If $\mathbf{S}$ is a set, we denote by $x \xleftarrow{\$} \mathbf{S}$ the operation of picking an element x uniformly at random from $\mathbf{S}$.

2.1   Bilinear Group Generator and Complexity Assumptions

Definition 1 (Bilinear Group Generator). *A bilinear group generator $\mathcal{G}$ is an algorithm that takes as input a security parameter κ and outputs a bilinear group $(p, \mathbf{G}_1, \mathbf{G}_2, \hat{e}, g)$, where $\mathbf{G}_1$ and $\mathbf{G}_2$ are cyclic groups of prime order p, g is a generator of $\mathbf{G}_1$, and $\hat{e} \colon \mathbf{G}_1 \times \mathbf{G}_1 \to \mathbf{G}_2$ is a bilinear map with the following properties:*

- *Bilinearity: For $g_1, g_2 \xleftarrow{\$} \mathbf{G}_1$ and $a, b \xleftarrow{\$} \mathbf{Z}_p^*$, we have $\hat{e}(g_1^a, g_2^b) = \hat{e}(g_1, g_2)^{ab}$.*
- *Non-degeneracy: There exists $g_1, g_2 \in \mathbf{G}_1$ such that $\hat{e}(g_1, g_2) \neq 1$.*
- *Computability: There is an efficient algorithm to compute $\hat{e}(g_1, g_2)$ for all $g_1, g_2 \in \mathbf{G}_1$.*

Definition 2 (CDH Assumption). *The computational Diffie-Hellman ass-umption in a prime p order group $\mathbf{G}$ states that, given (g, g^a, g^b), there is no prob-abilistic polynomial-time (PPT) adversary $\mathcal{A}$ can compute g^{ab} with non-negligible advantage, where $g \xleftarrow{\$} \mathbf{G}$ and $a, b \xleftarrow{\$} \mathbf{Z}_p^*$.*

Definition 3 (q-DBDHE Assumption). *The decisional q-parallel bilinear Diffie-Hellman exponent assumption in a prime order bilinear group $(p, \mathbf{G}_1, \mathbf{G}_2, \hat{e}, g)$ generated by $\mathcal{G}(1^\kappa)$ states that, given $X \xleftarrow{\$} \mathbf{G}_2$ and*

$$y = g, g^s, g^x, \dots, g^{(x^q)}, g^{(x^{q+2})}, \dots, g^{(x^{2q})}$$

$$\forall_{1 \le j \le q} \ g^{s \cdot b_j}, g^{x/b_j}, \dots, g^{(x^q/b_j)}, g^{(x^{q+2}/b_j)}, \dots, g^{(x^{2q}/b_j)}$$

$$\forall_{1 \le k \le q, k \ne j} \ g^{x \cdot s \cdot b_k/b_j}, \dots, g^{(x^q \cdot s \cdot b_k/b_j)},$$

there is no PPT adversary $\mathcal{A}$ can decide whether $X = \hat{e}(g, g)^{x^{q+1}s}$ with non-negligible advantage, where $x, s, b_1, \dots, b_q \xleftarrow{\$} \mathbf{Z}_p$.

2.2 Access Structure and Secret Sharing Schemes

Let $\mathbf{P} = \{\mathcal{P}_1, \mathcal{P}_2, \dots, \mathcal{P}_n\}$ be a set of parties. A collection $\mathbb{A} \subseteq 2^{\mathbf{P}}$ is monotone if for any set of parties $\mathbf{B}$ and $\mathbf{C}$, we have that if $\mathbf{B} \in \mathbb{A}$ and $\mathbf{B} \subseteq \mathbf{C}$ then $\mathbf{C} \in \mathbb{A}$. An access structure (respectively, monotone access structure) is a collec-tion (respectively, monotone collection) $\mathbb{A} \subseteq 2^{\mathbf{P}} \setminus \{\emptyset\}$. The sets in $\mathbb{A}$ are called the authorized sets, and the sets not in $\mathbb{A}$ are called the unauthorized sets [24].

In our context, the role of the parties is taken by the attributes. Thus, the access structure $\mathbb{A}$ will contain the authorized sets of attributes. We restrict our attention to monotone access structures. If a set of attributes ω satisfies an access structure $\mathbb{A}$, we denote it as $\mathbb{A}(\omega) = 1$.

A (t, n)-threshold scheme is a method of sharing a secret $s \in \mathbf{Z}_p$, which is chosen by the dealer (denoted by $\mathcal{D}$), among a set of n participants $\mathbf{P}$, in such a way that any t participants can compute the value of s, but no group of $t - 1$ participants can do so. Shamir [20] proposed a threshold secret sharing scheme by using polynomial interpolation, which is described as follows.

- $\mathcal{D}$ chooses randomly a polynomial $f(x) \in \mathbf{Z}_p[x]$ of degree $t - 1$ with $f(0) = s$, i.e. $f(x) = s + \sum_{j=1}^{t-1} a_j x^j \bmod p$, where $s \in \mathbf{Z}_p$ is the secret to be shared.
- $\mathcal{D}$ assigns every participant $\mathcal{P}_i$ with a unique random element $\alpha_i \in \mathbf{Z}_p^*$.
- $\mathcal{D}$ computes $s_i = f(\alpha_i)$ for $1 \le i \le n$ and gives the secret share s_i to $\mathcal{P}_i$ through a private channel.

Now a group $\mathbf{S} \subset \mathbf{P}$ of at least t participants, i.e. $|\mathbf{S}| \ge t$, can recover the secret s by using the following formula.

$$f(x) = \sum_{\mathcal{P}_i \in \mathbf{S}} \Delta_{\alpha_i, \mathbf{S}}(x) f(\alpha_i) = \sum_{\mathcal{P}_i \in \mathbf{S}} \Delta_{\alpha_i, \mathbf{S}}(x) s_i, \text{ where}$$

$$\Delta_{\alpha_i, \mathbf{S}}(x) = \prod_{\mathcal{P}_i \in \mathbf{S}, k \ne i} \frac{x - \alpha_k}{\alpha_i - \alpha_k} \bmod p.$$

On the other hand, it can be proved that if the subset $\mathbf{B} \subseteq \mathbf{P}$ such that $|\mathbf{B}| < t$ could not get any information about the polynomial $f(x)$.

Definition 4 (Linear Secret Sharing Scheme). *A SSS Π for an access structure $\mathbb{A}$ over a set of n participants $\mathbf{P}$ is called linear over $\mathbf{Z}_p$ if*

- *The shares for each participant form a vector over $\mathbf{Z}_p$.*
- *There exists a share-generating matrix $\mathbf{M}_{\ell \times n}$ for Π. For all $1 \leq i \leq \ell$, we let the function ρ defined the party labeling row i of $\mathbf{M}_{\ell \times n}$ as $\rho(i)$. When we consider the column vector $\boldsymbol{v} = (s, r_2, \ldots, r_n)^{\mathsf{T}}$, where $s \in \mathbf{Z}_p$ is the secret to be shared, and $r_2, \ldots, r_n \overset{\$}{\leftarrow} \mathbf{Z}_p$, then $\boldsymbol{\alpha} = \mathbf{M}_{\ell \times n} \boldsymbol{v}$ is the vector of ℓ shares of the secret s according to Π. The share $\alpha_i = (\mathbf{M}_{\ell \times n} \boldsymbol{v})_i$ belongs to party $\rho(i)$.*

Beimel [2] showed that every LSSS according to the above definition enjoys linear reconstruction property: Suppose that Π is a LSSS for the access structure $\mathbb{A}$. Let $\mathbf{S} \in \mathbb{A}$ be any authorized set. Define $\mathbf{I} = \{i | \rho(i) \in \mathbf{S}\} \subset \{1, 2, \ldots, \ell\}$. If $\{\lambda_i\}$ are valid shares of any secret s according to Π, then there exist constants $\{w_i \in \mathbf{Z}_p\}_{i \in \mathbf{I}}$ satisfying that

$$\sum_{i \in \mathbf{I}} w_i \lambda_i = s,$$

where these constants $\{w_i\}$ can be found in time polynomial in the size of $\mathbf{M}_{\ell \times n}$. For unauthorized sets, no such constants $\{w_i\}$ exist.

3   Security Analysis of Hu et al. FABSC Scheme

3.1   Review of Hu et al. FABSC Scheme

In Hu et al. FABSC scheme [6], a user's identity consists of n attributes, and the access structure of a user may be designated as 'd out of n attributes', which allows the user to obtain the data from the WBAN controller when the user has at least d attributes possessed by the data. For each identity, they specify an error-tolerance d. Hu et al. FABSC scheme is described as follows.

- **Setup:** The PKG performs as follows.
    1. Run $\mathcal{G}(1^{\kappa}) \rightarrow (p, \mathbf{G}_1, \mathbf{G}_2, \hat{e}, g)$, set the universe $\Omega = \{a_i\}_{i=1}^n$ where $a_i \in \mathbf{Z}_p^*$, and let $\mathbf{N}$ be the set $\{1, 2, \ldots, n+1\}$.
    2. Choose a cryptographic secure hash function $H : \mathbf{G}_2 \times \{0,1\}^{32} \rightarrow \mathbf{G}_1$. Note that the hash function H is not clearly defined in [6].
    3. Pick $y \overset{\$}{\leftarrow} \mathbf{Z}_p^*$ and $g_2 \overset{\$}{\leftarrow} \mathbf{G}_1$, compute $g_1 = g^y$ and $U = \hat{e}(g_1, g_2)$.
    4. Select $t_1, t_2, \ldots, t_{n+1} \overset{\$}{\leftarrow} \mathbf{G}_1$, and define a function $T(x)$ as

$$T(x) = g_2^{x^n} \prod_{i=1}^{n+1} t_i^{\Delta_{i,\mathbf{N}}(x)}.$$

5. Select $I' \xleftarrow{\$} \mathbf{Z}_p$ and $I_i \xleftarrow{\$} \mathbf{Z}_p$, and compute $v' = g^{I'}$ and $v_i = g^{I_i}$ for $1 \le i \le m$. Note that I' is mistakenly written as $I' \xleftarrow{\$} \mathbf{G}_1$ in [6].

6. Set the master key $msk = y$, and the public system parameters $mpk = (\Omega, p, \mathbf{G}_1, \mathbf{G}_2, \hat{e}, g, g_1, g_2, t_1, \dots, t_{n+1}, v', v_1, \dots, v_m, U, H)$.

- **KeyGen:** The PKG generates attribute-based private keys for a user with a set $\mathbf{Id} \subseteq \Omega$ of identity attributes as follows.

  1. Select a $d - 1$ degree polynomial $q(x) \xleftarrow{\$} \mathbf{Z}_p[x]$ such that $q(0) = y$.
  2. Pick $r_1, r_2, \dots, r_{|\mathbf{Id}|} \xleftarrow{\$} \mathbf{Z}_p$, and compute $D_i = g_2^{q(i)} T(i)^{r_i}$ and $d_i = g^{-r_i}$ for $a_i \in \mathbf{Id}$.
  3. Set the private key sets corresponding to the set $\mathbf{Id}$ of identity attributes as $K_{\mathbf{Id}} = \{D_i, d_i\}_{a_i \in \mathbf{Id}}$.

The private keys of a controller with identity attributes $\mathbf{Id}_C$ are denoted as

$$K_{\mathbf{Id}_C} = \{D_{C,i}, d_{C,i}\}_{a_i \in \mathbf{Id}_C} = \{g_2^{q_C(i)} T(i)^{r_{C,i}}, \ g^{-r_{C,i}}\}_{a_i \in \mathbf{Id}_C},$$

where $q_C(x) \xleftarrow{\$} \mathbf{Z}_p[x]$ is a $d - 1$ degree polynomial such that $q_C(0) = y$, and $r_{C,i} \xleftarrow{\$} \mathbf{Z}_p$ for all $a_i \in \mathbf{Id}_C$. Note that they are mistakenly written as $K_{\mathbf{Id}_C} = \{D_C, d_C\} = \{g_2^{q_C(i)} T(i)^{r_C}, g^{-r_C}\}$ in [6]. Similarly, the private keys of a doctor Victor with identity attributes $\mathbf{Id}_V$ are denoted as

$$K_{\mathbf{Id}_V} = \{D_{V,i}, d_{V,i}\}_{a_i \in \mathbf{Id}_V} = \{g_2^{q_V(i)} T(i)^{r_{V,i}}, \ g^{-r_{V,i}}\}_{a_i \in \mathbf{Id}_V},$$

where $q_V(x) \xleftarrow{\$} \mathbf{Z}_p[x]$ is a $d - 1$ degree polynomial such that $q_V(0) = y$, and $r_{V,i} \xleftarrow{\$} \mathbf{Z}_p$ for all $a_i \in \mathbf{Id}_V$. Note that they are mistakenly written as $K_{\mathbf{Id}_V} = \{D_V, d_V\} = \{g_2^{q_V(i)} T(i)^{r_V}, g^{-r_V}\}$ in [6].

- **Signcrypt:** The signcryption algorithm is run by the controller with identity attributes $\mathbf{Id}_C$ for a message M, where the message can be represented as an m-bit element in the group $\mathbf{G}_2$, i.e., $M = (\mu_1, \dots, \mu_m)$. The algorithm produces a signcrypted ciphertext E encrypted with identity attributes $\mathbf{Id}'$ and signed with identity attributes $\mathbf{Id}_C$. Denote by $tt \in \{0, 1\}^{32}$ a time stamp and by θ a predefined time limit for message decryption. The controller performs the following steps.

  1. Select $r_1, r_2, \dots, r_m \xleftarrow{\$} \mathbf{Z}_p$, and compute $t = \sum_{i=1}^m r_i$, $\widetilde{M} = H(M \| tt)$, $E_1 = M \cdot U^t$, $E_2 = g^{-t}$, $E_{3,i} = T(i)^t$ for $a_i \in \mathbf{Id}'$. Note that these $r_1, r_2, \dots, r_m$ will not be used in the rest of the signcrypt algorithm, so $t \leftarrow \sum_{i=1}^m r_i$ is equivalent to chooses $t \xleftarrow{\$} \mathbf{Z}_p$. And $E_{3,i}$ for $a_i \in \mathbf{Id}'$ are mistakenly written as $E_3 = \{T(i)^t\}$ in [6].

  2. For $a_i \in \mathbf{Id}_C$, compute $S_{1,i} = D_{C,i} \cdot (v' \prod_{j=1}^m v_j^{\mu_j} \tilde{M})^t = g_2^{q_C(i)} T(i)^{r_{C,i}} \cdot (v' \prod_{j=1}^m v_j^{\mu_j} \tilde{M})^t$ and $S_{2,i} = d_{C,i} = g^{-r_{C,i}}$. Note that $S_{1,i}$ and $S_{2,i}$ are mistakenly written as $S_1 = g_2^{q_C(i)} T(i)^{r_C} \cdot (v' \prod_{j=1}^m v_j^{\mu_j} \tilde{M})^t$ and $S_2 = g^{-r_C}$ in [6].

3. Set the signcrypted ciphertext

$$E = (E_1, E_2, \{E_{3,i}\}_{a_i \in \mathbf{Id}'}, tt, \mathbf{Id}', \{S_{1,i}\}_{a_i \in \mathbf{Id}_C}, \{S_{2,i}\}_{a_i \in \mathbf{Id}_C}, \mathbf{Id}_C).$$

Note that E is mistakenly written as $(E_1, E_2, E_3, S_1, S_2, tt, \mathbf{Id}')$ in [6].

- **Designcrypt:** The designcryption algorithm is run by a receiver Victor with identity attributes $\mathbf{Id}_V$, which is described as follows.
  1. Upon receiving signcrypted ciphertext E, Victor checks the current time $\overline{tt}$. If $|\overline{tt} - tt| \leq \theta$, Victor sets $\mathbf{S} = \mathbf{Id}' \cap \mathbf{Id}_V$, and computes

$$M' = E_1 \cdot \prod_{a_i \in \mathbf{S}} (\frac{\hat{e}(d_{V,i}, E_{3,i})}{\hat{e}(D_{V,i}, E_2)})^{\Delta_{i,\mathbf{s}}(0)}, \quad \widetilde{M}' = H(M'\|tt)$$

  2. Check whether the following equation holds or not.

$$U = \prod_{a_i \in \mathbf{Id}_C} [\hat{e}(S_{1,i}, g) \cdot \hat{e}(S_{2,i}, T(i)) \cdot \hat{e}(E_2, v' \prod_{j=1}^{m} v_j^{\mu_j} \widetilde{M}')]^{\Delta_{i,\mathbf{Id}_C}(0)}.$$

Note that the above equation is mistakenly written in [6] as

$$U = \prod_{i \in \mathbf{S}} [\hat{e}(S_1, g) \cdot \hat{e}(S_2, T(i)) \cdot \hat{e}(E_2, v' \prod_{j=1}^{m} v_j^{\mu_j} \widetilde{M}')]^{\Delta_{i,\mathbf{s}}(0)}.$$

  3. If it holds, it represents M' is valid and outputs $M = M'$. Otherwise, it represents M' is invalid and asks the controller to resend the message.

3.2 Security Analysis of Hu et al.'s FABSC Scheme

Theorem 1. *Hu et al.'s FABSC scheme can not resist private key forgery attack. i.e., an adversary can impersonate the trusted key server and generate private keys for any set of attributes, thus totally break Hu et al.'s FABSC scheme.*

Proof. Suppose an adversary $\mathcal{A}$ who acts as a signcrypting party with identity consists of 4 attributes, i.e., $\mathbf{ID}_C = \{a_1, a_2, a_3, a_4\}$ and the threshold is set as $d = 2$. The adversary can get the value $D_i = g_2^{q(i)} T(i)^{r_i}$ such that $q(0) = y$ for $a_i \in \mathbf{ID}_C$. Since $d = 2$, if $\mathcal{A}$ can cancel $T(i)^{r_i}$ from D_i, then $\mathcal{A}$ can use $\{D_1, D_2\}$, $\{D_1, D_3\}$, $\{D_1, D_4\}$, $\{D_2, D_3\}$ and $\{D_2, D_4\}$ to compute g_2^y.

For simplicity, let $T_i = T(i)^{r_i}$ and $\Delta_{ij} = \Delta_{a_i, \{a_i, a_j\}}(0)$. Note that Δ_{ij} is not necessarily equal to Δ_{ji}. The adversary can get:

$$
\begin{aligned}
X_1 &= g_2^y T_1^{r_1 \Delta_{12}} T_2^{r_2 \Delta_{21}}, & X_2 &= g_2^y T_1^{r_1 \Delta_{13}} T_3^{r_3 \Delta_{31}} \\
X_3 &= g_2^y T_1^{r_1 \Delta_{14}} T_4^{r_4 \Delta_{41}}, & X_4 &= g_2^y T_2^{r_2 \Delta_{23}} T_3^{r_3 \Delta_{32}} \\
X_5 &= g_2^y T_2^{r_2 \Delta_{24}} T_4^{r_4 \Delta_{42}}, & X_6 &= g_2^y T_3^{r_3 \Delta_{34}} T_4^{r_4 \Delta_{43}}
\end{aligned}
$$

Let $\Delta_{1,1} = \Delta_{12} - \Delta_{13}$, $\Delta_{1,2} = \Delta_{12} - \Delta_{14}$, $\Delta_{2,1} = \Delta_{21} - \Delta_{23}$ and $\Delta_{2,2} = \Delta_{21} - \Delta_{24}$. Then $\mathcal{A}$ can compute:

$$Y_1 = \frac{X_1}{X_2} = T_1^{\Delta_{1,1}} T_2^{\Delta_{21}} T_3^{-\Delta_{31}}, \quad Y_2 = \frac{X_1}{X_3} = T_1^{\Delta_{1,2}} T_2^{\Delta_{21}} T_4^{-\Delta_{41}}$$

$$Y_3 = \frac{X_1}{X_4} = T_1^{\Delta_{12}} T_2^{\Delta_{2,1}} T_3^{-\Delta_{32}}, \quad Y_4 = \frac{X_1}{X_5} = T_1^{\Delta_{12}} T_2^{\Delta_{2,2}} T_4^{-\Delta_{42}}$$

Furthermore, let $\Delta_{1,3} = \Delta_{1,1} - \Delta_{1,2}$, $\Delta_{1,4} = \Delta_{1,1}\Delta_{32} - \Delta_{12}\Delta_{31} \neq 0$, $\Delta_{1,5} = \Delta_{1,2}\Delta_{42} - \Delta_{12}\Delta_{41}$, $\Delta_{2,3} = \Delta_{21}\Delta_{32} - \Delta_{2,1}\Delta_{31}$, $\Delta_{2,3} = \Delta_{21}\Delta_{32} - \Delta_{2,1}\Delta_{31}$, $\Delta_{2,4} = \Delta_{21}\Delta_{42} - \Delta_{2,2}\Delta_{41}$ and $\Delta_{2,5} = \Delta_{2,2}\Delta_{1,5} - \Delta_{2,4}\Delta_{1,4} \neq 0$, then $\mathcal{A}$ can compute:

$$Z_1 = \frac{Y_1^{\Delta_{32}}}{Y_3^{\Delta_{31}}} = T_1^{r_1 \Delta_{1,4}} T_2^{r_2 \Delta_{2,3}}, \quad Z_2 = \frac{Y_2^{\Delta_{42}}}{Y_4^{\Delta_{41}}} = T_1^{r_1 \Delta_{1,5}} T_2^{r_2 \Delta_{2,4}}$$

$$Z_3 = \frac{Z_1^{\Delta_{1,5}}}{Z_2^{\Delta_{1,4}}} = T_2^{r_2 \Delta_{2,5}}, \quad Z_4 = Z_2^{\frac{1}{\Delta_{2,4}}} = T_1^{r_1 \Delta_{1,5}/\Delta_{2,4}} T_2^{r_2}$$

Next, $\mathcal{A}$ can find

$$T_2^{r_2} = Z_3^{\Delta_{2,5}^{-1}}, \quad T_1^{r_1} = [\frac{Z_2}{T_2^{r_2 \Delta_{2,4}}}]^{\Delta_{1,5}^{-1}} \Rightarrow g_2^y = \frac{X_1}{T_1^{r_1 \Delta_{12}} T_2^{r_2 \Delta_{21}}}$$

$\mathcal{A}$ can generate the private key sets $K_{\mathbf{Id}} = \{D_i, d_i\}_{a_i \in \mathbf{Id}}$ corresponding to identity attributes $\mathbf{Id} \subset \Omega$ as follows.

- Choose $\alpha_{d-1}, \ldots, \alpha_1 \overset{\$}{\leftarrow} \mathbb{Z}_q^*$ and $r_1, r_2, \ldots, r_{|\mathbf{Id}|} \overset{\$}{\leftarrow} \mathbb{Z}_p$.
- For all $a_i \in \mathbf{Id}$, compute

$$D_i = g_2^{\alpha_{d-1} i^{d-1}} g_2^{\alpha_{d-2} i^{d-2}} \cdots g_2^{\alpha_1 i} g_2^y T(i)^{r_i}$$
$$= g_2^{\alpha_{d-1} i^{d-1} + \alpha_{d-2} i^{d-2} \cdots + \alpha_i + y} T(i)^{r_i} = g_2^{q(i)} T(i)^{r_i}$$
$$d_i = g^{-r_i}$$

- Set the private key sets corresponding to identity attributes $\mathbf{Id}$ as $K_{\mathbf{Id}} = \{D_i, d_i\}_{a_i \in \mathbf{Id}}$.

Thus, $\mathcal{A}$ is capable of generating private keys for any set of attributes with the help of g_2^y, and the forged private key sets and the actual private key sets generated by the trusted party corresponding to identity attributes $\mathbf{Id}$ are computationally indistinguishable, i.e., they have the same probability distributions. Thus Hu et al. FABSC scheme is completely broken.

This completes the proof.

4   Syntax and Security Definitions of CP-ABRSC Scheme

A CP-ABRSC scheme can be defined by the following six PPT algorithms:

- **Setup:** The probabilistic setup algorithm is run by the trusted PKG. It takes as input a security parameter κ. It outputs the public system parameters mpk, and the master key msk which is known only to the PKG.
- **IBKeyGen:** The probabilistic identity-based private key generation algorithm is run by the PKG. It takes as input the public parameters mpk, the master key msk, a user identity ID submitted by a user $\mathcal{U}$. It outputs the corresponding identity-based private key sk_{ID}.
- **ABKeyGen:** The probabilistic attribute-based private key generation algorithm is run by the PKG. It takes as input the public parameters mpk, the master key msk, and a set ω of attributes owned by a user $\mathcal{U}$. It outputs an attribute private key dk_{ω} corresponding to the set ω of attributes.
- **Signcrypt:** The probabilistic signcrypt algorithm is run by a signcrypting party. It takes as input the public parameters mpk, a message msg, an ad-hoc group of ring members $\mathbf{U} = \{\mathcal{U}_1, \mathcal{U}_2, \ldots, \mathcal{U}_n\}$ with corresponding identities $\mathbf{ID} = \{\text{ID}_i\}_{i=1}^n$, the signcrypting party's identity-based private key sk_{ID_s} with $\text{ID}_s \in \mathbf{ID}$, and an access structure $\mathbb{A}$ over the universe of attributes. It outputs a signcrypted ciphertext C. Note that $\mathbb{A}$ and $\mathbf{ID}$ are contained in the signcrypted ciphertext.
- **PubVerify:** The deterministic public verifiability algorithm is run by any outside receivers. It takes as input the public parameters mpk, a signcrypted ciphertext C. It outputs a bit b which is 1 if the signcrypted ciphertext C is generated by a certain member in the group $\mathbf{U}$, or 0 if the signcrypted ciphertext C is not generated by any member in the group $\mathbf{U}$.
- **UnSigncrypt:** The deterministic unsigncryption algorithm is run by a receiver. It takes as input the public parameters mpk, a signcrypted ciphertext C, the receiver's attribute-based private key dk_{ω}. It outputs the message msg if $\mathbb{A}(\omega) = 1$ and $\text{ID}_s \in \mathbf{ID}$. Otherwise it outputs a reject symbol $\perp$.

The set of algorithms must satisfy the following consistency requirement:

$$\mathbf{Setup}(1^\kappa) \to (mpk, msk), msg \xleftarrow{\$} \{0,1\}^*, \text{ID}_s \xleftarrow{\$} \mathbf{ID},$$
$$\mathbf{IBKeyGen}(mpk, msk, \text{ID}_s) \to sk_{\text{ID}_s}, \mathbf{ABKeyGen}(mpk, msk, \omega) \to dk_{\omega},$$
$$\text{If } \mathbb{A}(\omega) = 1 \text{ and } \mathbf{SignCrypt}(mpk, \mathbf{ID}, sk_{\text{ID}_s}, \mathbb{A}, msg) \to C,$$
$$\text{Then } \mathbf{UnSignCrypt}(mpk, dk_{\omega}, \mathbf{ID}, \mathbb{A}, C) = msg \text{ holds.}$$

The property of indistinguishability under chosen plaintext attack (IND-CPA) is considered a basic requirement for provably secure public key encryption schemes. For CP-ABRSC, we define IND-CPA in the selective model by the following game between an adversary $\mathcal{A}$ and a challenger $\mathcal{C}$.

- **Init:** $\mathcal{A}$ declares the access structure $\mathbb{A}^*$ that he wishes to be challenged upon.
- **Setup:** $\mathcal{C}$ runs the setup algorithm on input a security parameter κ, gives public parameters mpk to $\mathcal{A}$, while keeps the master key msk secret.
- **Phase 1:** $\mathcal{A}$ is allowed to issue the following queries adaptively.

- Singing private key queries on identity ID_i. $\mathcal{C}$ runs **IBKeyGen**(mpk, msk, ID_i) and sends sk_{ID_i} back to $\mathcal{A}$.
- Decrypting private key queries on a set ω_i of attributes. If $\mathbb{A}^*(\omega_i) \neq 1$, then $\mathcal{C}$ runs **ABKeyGen**(mpk, msk, ω_i) and sends dk_{ω_i} back to $\mathcal{A}$. Otherwise, $\mathcal{C}$ rejects the request.

- Challenge: $\mathcal{A}$ submits two equal length messages msg_0 and msg_1, a set $\mathbf{ID}^* = \{\mathsf{ID}_i^*\}_{i=1}^n$ of identities to $\mathcal{C}$. The challenger then flips a random coin b and picks an identity $\mathsf{ID}_i^* \xleftarrow{\$} \mathbf{ID}^*$. Finally, $\mathcal{C}$ sends the corresponding signcrypted ciphertext C^* to $\mathcal{A}$ by running $sk_{\mathsf{ID}_i^*} \leftarrow$ **IBKeyGen**($mpk, msk, \mathsf{ID}_i^*$) and $C^* \leftarrow$ **Signcrypt**($mpk, \mathbf{ID}^*, sk_{\mathsf{ID}_i^*}, \mathbb{A}^*, msg_b$).
- Phase 2: Phase 1 is repeated.
- Guess: $\mathcal{A}$ outputs a guess b' of b.

The advantage of $\mathcal{A}$ is defined as $\mathrm{Adv}_{\mathcal{A}}(\kappa) = \Pr[b' = b] - \frac{1}{2}$.

Definition 5. *A CP-ABRSC scheme is said to be IND-CPA secure in the selective model if* $\mathrm{Adv}_{\mathcal{A}}(\kappa)$ *is negligible in the security parameter* κ.

Remark 1. The above security model deals with insider security, since the adversary is assumed to have access to the private key of a signcrypting party who belong to ring members $\mathbf{U}^*$ chosen for the challenge phase. This means that the confidentiality is preserved even if a signcrypting party's private key is compromised.

The property of existential unforgeability against adaptive chosen message and identity attack (EUF-CMIA) is considered a basic requirement for provably secure IBRS schemes. For CP-ABRSC, we define EUF-CMIA by the following game played between an adversary $\mathcal{A}$ and a challenger $\mathcal{C}$.

- Setup: Same as in the above IND-CPA game.
- Find: $\mathcal{A}$ is allowed to issue the following queries adaptively.
  - Singing private key queries on identity ID_i. $\mathcal{C}$ gets sk_{ID_i} by running **IBKeyGen**(mpk, msk, ID_i), and sends sk_{ID_i} back to $\mathcal{A}$.
  - Decrypting private key queries on set of attributes ω_i. $\mathcal{C}$ gets dk_{ω_i} by running **ABKeyGen**(mpk, msk, ω_i), and sends dk_{ω_i} back to $\mathcal{A}$.
  - Signcrypt queries on ($msg, \mathbf{ID}, \mathbb{A}$). $\mathcal{C}$ picks an identity $\mathsf{ID}_i \xleftarrow{\$} \mathbf{ID}$, gets signcrypted ciphertext C by running $sk_{\mathsf{ID}_i} =$ **IBKeyGen**(mpk, msk, ID_i) and $C =$ **Signcrypt**($mpk, \mathbf{ID}, sk_{\mathsf{ID}_i}, \mathbb{A}, msg$), and sends C back to $\mathcal{A}$.
- Forgery: Finally, $\mathcal{A}$ produces a new triple ($C^*, \mathbf{ID}^*, \mathbb{A}^*$). The only restriction is that ($\mathbf{ID}^*, \mathbb{A}^*$) does not appear in the set of previous signcryption queries during find stage and each of signing private keys in $\mathbf{ID}^*$ is never returned by any singing private key queries. $\mathcal{A}$ wins the game if **PubVerify** ($mpk, C^*, \mathbf{ID}^*, \mathbb{A}^*$) = 1.

The advantage of $\mathcal{A}$ is defined as the probability that it wins.

Definition 6. *A CP-ABRSC scheme is said to be EUF-CMIA secure if no polynomially bounded adversary $\mathcal{A}$ has non-negligible advantage in the above game.*

Remark 2. The above security model deals with insider security since the adversary is assumed to have access to the private key of the receiver used for generation of the signcrypted ciphertext C^*. This means that the unforgeability is preserved even if a receiver's private key is compromised.

Definition 7. *A CP-ABRSC scheme is publicly verifiable if given a signcrypted ciphertext C along with* **ID** *and* $\mathbb{A}$, *anyone can verify that C is a valid signcryption by some member with identity* $\mathsf{ID}_s \in \mathbf{ID}$ *to receivers specified by the access structure* $\mathbb{A}$, *without knowing any decryption private key dk_ω such that $\mathbb{A}(\omega) = 1$.*

Definition 8. *A CP-ABRSC scheme is strong anonymous if for any signcrypting party group* **U** *of n_s members with identities* **ID**, *any message msg and signcrypted ciphertext C, the probability to identify the actual signcrypting party is not better than a random guess, i.e., an adversary outputs the identity of actual signcrypting party with probability $1/n_s$ if he is not a member of* **U**, *and with probability $1/(n_s - 1)$ if he is a member of* **U**.

5   Our CP-ABRSC Construction

The proposed CP-ABRSC construction is described as follows.

- **Setup:** The PKG first defines the universe $\Omega = \{\mathsf{atr}_i\}_{i=1}^n$ of attributes, runs bilinear group generator $\mathcal{G}(1^\kappa) \rightarrow (p, \mathbf{G}_1, \mathbf{G}_2, \hat{e}, g)$, chooses $x \xleftarrow{\$} \mathbf{Z}_p^*$, $y \xleftarrow{\$} \mathbf{Z}_p^*$, and $h_i \xleftarrow{\$} \mathbf{G}_1$ for $1 \leq i \leq n$, computes $h = g^x$ and $Y = \hat{e}(g,g)^y$. The PKG also picks two cryptographic hash functions $H_1 : \{0,1\}^* \rightarrow \mathbf{G}_1$ and $H_2 : \{0,1\}^* \rightarrow \mathbf{Z}_p^*$, sets the master secret key $msk = (x, g^y)$, and publishes system parameters $mpk = (p, \mathbf{G}_1, \mathbf{G}_2, \hat{e}, g, \Omega, h_1, \ldots, h_n, h, Y, H_1, H_2)$.

- **IBKeyGen:** Given an identity ID_i, the PKG sets user's public key $g_{\mathsf{ID}_i} = H_1(\mathsf{ID}_i) \in \mathbf{G}_1$, computes the corresponding private key $sk_{\mathsf{ID}_i} = g_{\mathsf{ID}_i}^x$, then sends the signing private key sk_{ID_i} to the user via a secure channel.

- **ABKeyGen:** Given a set $\omega \subseteq \Omega$ of attributes owned by a user, the PKG first chooses $t \xleftarrow{\$} \mathbf{Z}_p^*$, computes $K = g^y g^{xt}$, $L = g^t$, and $K_i = h_i^t$ for all $\mathsf{atr}_i \in \omega$. The PKG then sets the corresponding attribute private key $dk_\omega = (K, L, \{K_i\}_{attr_i \in \omega})$, and sends dk_ω to the user via a secure channel.

- **Signcrypt:** Let **U** be an ad-hoc group of n_s members with identities **ID** $= \{\mathsf{ID}_i | 1 \leq i \leq n_s\}$ including the actual signcrypting party with identity ID_j where $1 \leq j \leq n_s$. To signcrypt a message $msg \in \mathbf{G}_2$ on behalf of the group **U** under a LSSS access structure $(\mathbf{M}_{\ell \times n}, \rho)$, the signcrypting party chooses $s, y_2, \ldots, y_n, r_1, \ldots, r_\ell \xleftarrow{\$} \mathbf{Z}_p^{(n+\ell)}$, sets $v = (s, y_2, \ldots, y_n)^\mathsf{T}$, computes $C' = msg \cdot \hat{e}(g,g)^{sy}$, $C'' = g^s$, $\lambda_i = \mathbf{M}_i \cdot v$, $C_i = g^{x\lambda_i} h_{\rho(i)}^{-r_i}$ and $D_i = g^{r_i}$ for $i = 1$ to ℓ, where $\mathbf{M}_i$ is the vector corresponding to the i-th row of the share-generating matrix $\mathbf{M}_{\ell \times n}$. For all $1 \leq i \neq j \leq n_s$, the signcrypting party chooses $g_i \xleftarrow{\$} \mathbf{G}_1$, compute $h_i = H_2(C' \| (\mathbf{M}_{\ell \times n}, \rho) \| \mathbf{ID} \| g_i)$, $g_j = g_{ID_j}^s / \prod_{i=1, i \neq j}^{n_s} g_i g_{ID_i}^{h_i}$, $h_j = H_2$

$(C'\|(\mathbf{M}_{\ell\times n},\rho)\|\mathbf{ID}\|g_j)$, $\sigma_1 = sk_{ID_j}^{h_j+s}$, $\tilde{g} = \prod_{i=1}^{n} g_i$ and $\sigma_2 = H_2(msg\|\tilde{g}\|Y^s)$. Finally, the signcrypting party outputs the following ring signcrypted ciphertext

$$C = (C', C'', \{C_i, D_i\}_{i=1}^{\ell}, \{g_i\}_{i=1}^{n_s}, \sigma_1, \sigma_2, \mathbf{ID}, (\mathbf{M}_{\ell\times n}, \rho)).$$

- **PubVerify:** Any receiver can check the validity of the signcrypted ciphertext C against a set $\mathbf{ID}$ of identities. For $1 \leq i \leq n_s$, any receiver can compute $h_i = H_2(C'\|(\mathbf{M}_{\ell\times n},\rho)\|\mathbf{ID}\|g_i)$, and check the equation $\hat{e}(h, \prod_{i=1}^{n}(g_i g_{ID_i}^{h_i})) = \hat{e}(g, \sigma_1)$. It outputs 1 if the equation holds, or 0 if the equation does not hold.

- **UnSigncrypt:** Upon receiving the signcrypted ciphertext C, the receiver uses his decryption private key $dk_{\boldsymbol{\omega}}$ corresponding to the set $\boldsymbol{\omega}$ of attributes to recover and verify the signcrypted ciphertext as follows.
  1. Determine whether the set $\boldsymbol{\omega}$ of attributes satisfy the access structure $\mathbb{A}$ described by $(\mathbf{M}_{\ell\times n},\rho)$. If not, the receiver rejects the signcrypted ciphertext C.
  2. For $1 \leq i \leq n_s$, compute $h_i = H_2(C'\|(\mathbf{M}_{\ell\times n},\rho)\|\mathbf{ID}\|g_i)$, and check

$$\hat{e}(h, \prod_{i=1}^{n} g_i g_{ID_i}^{h_i}) \overset{?}{=} \hat{e}(g, \sigma_1).$$

If the equation does not hold, reject the signcrypted ciphertext C.
  3. Define $\mathbf{I} = \{i|\rho(i) \in \boldsymbol{\omega}\} \subset \{1, 2, \dots, \ell\}$. Let $\{w_i \in \mathbf{Z}_p\}$ be a set of constants such that if $\{\lambda_i\}$ are valid shares of y according to $(\mathbf{M}_{\ell\times n},\rho)$, then $\sum_{i\in\mathbf{I}} w_i\lambda_i = y$. Note there could potentially be different ways of choosing the w_i values to satisfy this.
  4. The receiver computes

$$V = \frac{\hat{e}(C'', K)}{\prod_{i\in\mathbf{I}}(\hat{e}(C_i, L)\hat{e}(D_i, K_{\rho(i)}))^{w_i}}, \quad msg' = \frac{C'}{V}, \quad \tilde{g} = \prod_{i=1}^{n_s} g_i$$

  5. Check $\sigma_2 \overset{?}{=} H_2(msg'\|\tilde{g}\|V)$. If it holds, the receiver accepts and outputs the message msg. Otherwise, rejects and outputs error symbol $\perp$.

Theorem 2. *The proposed CP-ABRSC construction is correct.*

Proof. The correctness can be verified as follows.

$$\hat{e}(h, \prod_{i=1}^{n_s} g_i g_{ID_i}^{h_i}) = \hat{e}(g^x, \prod_{i=1,i\neq j}^{n_s} (g_i g_{ID_i}^{h_i} g_j g_{ID_j}^{h_j})) = \hat{e}(g^x, g_{ID_j}^{s+h_j})$$

$$= \hat{e}(g, g_{ID_j}^{x(s+h_j)}) = \hat{e}(g, sk_{ID_j}^{s+h_j}) = \hat{e}(g, \sigma_1)$$

$$V = \frac{\hat{e}(C'', K)}{\prod_{i\in\mathbf{I}}(\hat{e}(C_i, L)\hat{e}(D_i, K_{\rho(i)}))^{w_i}} = \frac{\hat{e}(g^s, g^{xt})\hat{e}(g^s, g^y)}{\prod_{i\in\mathbf{I}}(\hat{e}(g^{\lambda_i x}, g^t)\hat{e}(h_{\rho(i)}^{-r_i}, g^t)\hat{e}(g^{r_i}, h_{\rho(i)}^t))^{w_i}}$$

$$= \frac{\hat{e}(g^s, g^{xt})\hat{e}(g^s, g^y)}{\hat{e}(g^x, g^t)^{\sum_{i\in\mathbf{I}}\lambda_i w_i}} = \hat{e}(g, g)^{sy} = Y^s$$

$$msg' = \frac{C'}{V} = \frac{msg \cdot Y^s}{Y^s} = msg$$

Theorem 3. *The proposed CP-ABRSC scheme satisfies strong anonymity.*

Proof. Since $s \xleftarrow{\$} \mathbf{Z}_q^*$ and $g_i \xleftarrow{\$} \mathbf{G}_1$ for $1 \leq i \neq j \leq n$ are generated uniformly at random. All components of C except σ_1 do not contain any identity information bound to them. Thus, we only need to check whether $\sigma_1 = sk_{ID_j}^{h_j+s}$ will leak information about the actual signcrypting party. Anyone can compute $g_{ID_j}^s$ according to $g_{ID_j}^s = g_j \prod_{i \neq j} g_i g_{ID_i}^{h_i}$, and tries to determine whether a user with identity ID_k is the actual signcrypting party by verifying the following equation:

$$\hat{e}(g_k \prod_{i=1,\, i \neq k}^{n_s} g_i g_{ID_i}^{h_i}, h) \cdot \hat{e}(g_{ID_k}^{h_k}, h) \overset{?}{=} \hat{e}(\sigma_1, g)$$

The above equation holds for all $1 \leq k \leq n_s$ as

$$\hat{e}(g_k \prod_{i=1,\, i \neq k}^{n_s} g_i g_{ID_i}^{h_i}, h) \cdot \hat{e}(g_{ID_k}^{h_k}, h) = \hat{e}(g_{ID_k}^s, h) \cdot \hat{e}(g_{ID_k}^{h_k}, h)$$

$$= \hat{e}(g_{ID_k}^{s+h_k}, g^x) = \hat{e}(g_{ID_k}^{x(s+h_k)}, g) = \hat{e}(sk_{ID_k}^{s+h_k}, g) = \hat{e}(\sigma_1, g)$$

Thus, we conclude that even an adversary with unbounded computing power has no advantage in identifying the actual signcrypting party over random guessing.

Theorem 4. *The proposed CP-ABRSC construction is IND-CPA secure in the selective model under the q-DBDHE assumption.*

Proof. We omit the proof here due to page limitation and will be given in the full version of this paper.

Theorem 5. *The proposed CP-ABRSC construction is EUF-CMIA secure in the adaptive model under the CDH assumption.*

Proof. We omit the proof here due to page limitation and will be given in the full version of this paper.

6   Application of CP-ABRSC Scheme in WBAN

In this section, we present a secure, privacy-protected and fine-grained access control framework for WBANs by exploiting CP-ABRSC scheme. Figure 2 illustrates the proposed framework for WBANs, which involves five participants:

- One hospital authority (HA) who acts as the PKG. HA is responsible for generating system public parameters, issuing signing private keys for controllers based on their identities and decryption private keys for healthcare providers based on their attributes (credentials).
- Multiple wearable or implanted sensors, which can sense and process vital signs (heart rate, blood pressure, oxygen saturation, activity) or environmental parameters (location, temperature, humidity, light), and transfer the relevant data to the corresponding controller.

- Multiple controllers who aggregate information from sensors and ultimately convey the information about health status across existing networks to the medical server. Each controller can be uniquely identified by the registered patient's identity who owns the controller, and obtained its signing private key from the HA that bind it to the claimed identity.
- One central medical server (such as a cloud storage server maintained by a cloud service provider) who keeps personal health information of registered users and provides various services to the users and healthcare providers. We consider honest but curious medical server as those in [13,20]. That means the server will try to find out as much secret information in the stored personal health information as possible, but they will honestly follow the protocol in general. The server may also collude with a few malicious users in the system. On the other hand, some users will also try to access personal health information beyond their privileges. For example, a pharmacy may want to obtain the prescriptions of patients for marketing and boosting its profits. To do so, they may even collude with other users.
- Multiple healthcare providers (include doctors, nurses, researchers etc.) who may access the patients' health information and provide health services. Healthcare providers are identified by their attributes and obtain their decryption private keys that bind them to claimed attributes from the HA. For example, a physician would receive "Hospital A, Chief Physician, Master of Internal Medicine, Division Director, Cardiovascular Medicine" as her attributes from the HA.

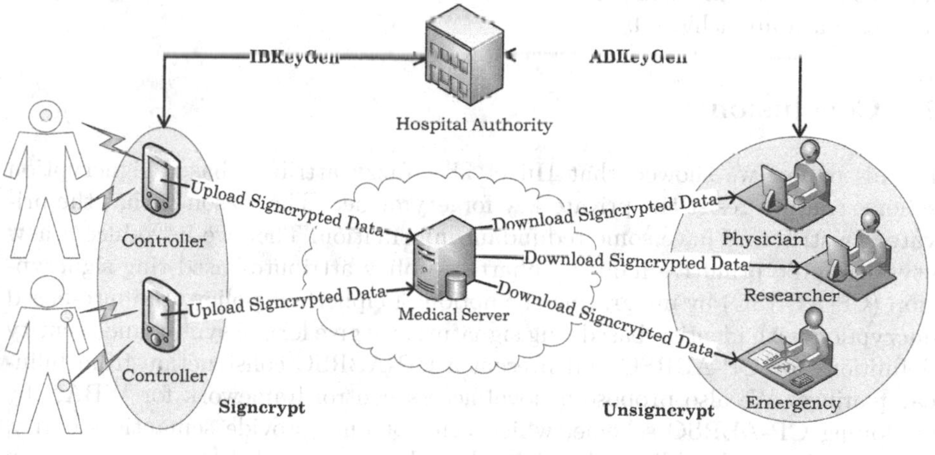

Fig. 2. Application of CP-ABRSC Scheme in BAN

Sensors in and around the body collect the vital signals of the patient continuously and transmit the collected signals to the corresponding controller regularly. The controller aggregates the received signals and signcrypts the aggregated information msg as follows.

- Choose a group of identities $\mathbf{ID} = \{\mathsf{ID}_i\}_{i=1}^{n_s}$ that includes the controller's own identity ID_j where $1 \leq j \leq n_s$.
- Generate an access policy $(\mathbf{M}_{\ell \times n}, \rho)$ based on attributes of authorized healthcare providers. For example, a policy may look like "(Organization = Hospital A $\vee$ Organization = Hospital B) $\wedge$ (Specialty = Internal Medicine) $\wedge$ (Profession = Physician)". Here controllers specify their own privacy policies to prevent the medical server and unauthorized users from learning the contents of corresponding patients' health data.
- Run the Signcrypt algorithm to get the signcrypted ciphertext C with the controller's private key, the aggregated health information msg, identities $\mathbf{ID}$ and access policy $(\mathbf{M}_{\ell \times n}, \rho)$ as input.

The controller uploads the signcrypted ciphertext C along with identities $\mathbf{ID}$ and access policy $(\mathbf{M}_{\ell \times n}, \rho)$ to the medical server. The medical server can verify the signcrypted ciphertext C by running the PubVerify algorithm. Since the signcrypted ciphertext C is actually signed by the controller on the CP-ABE ciphertext using Chow et al. IBRS scheme [5], data authenticity and unforgeability, anonymity for controller (includes untraceability and unlinkability) are achieved.

The healthcare providers can download the signcrypted health information C that contained the access policy $\mathbb{A}$ described as $(\mathbf{M}_{\ell \times n}, \rho)$ from the medical server, and they can open C only if they have suitable attribute-based private keys dk_ω where $\mathbb{A}(\omega) = 1$. Since the signcrypted ciphertext C is actually encrypted by the controller using Waters CP-ABE scheme [24], data confidentiality, anonymity for medical personnel and fine-grained access control on encrypted medical data are achieved.

7   Conclusion

In this paper, we showed that Hu et al.'s fuzzy attribute-based signcryption scheme can not resist the private key forgery attack. The reason is that the private key structure have some redundant information. Then we introduce a new cryptographic primitive named ciphertext-policy attribute-based ring signcryption (CP-ABRSC) by integrating the notion of ciphertext-policy attribute-based encryption with identity-based ring signature, we give formal syntax and security definitions for CP-ABRSC and present a CP-ABRSC construction from bilinear pairings. We also propose a novel access control framework for WBAN by exploiting CP-ABRSC scheme, which can not only provide semantic security, unforgeability and public authenticity, but also can provide participants privacy and fine-grained access control on encrypted health data.

Acknowledgment. This research is jointly funded by National Natural Science Foundation of China (Grant No. 61173189) and Innovative Research Team Project in Yunnan University.

References

1. Baek, J., Steinfeld, R., Zheng, Y.: Formal proofs for the security of signcryption. In: Naccache, D., Paillier, P. (eds.) PKC 2002. LNCS, vol. 2274, pp. 80–98. Springer, Heidelberg (2002). http://dx.doi.org/10.1007/3-540-45664-3_6

2. Beimel, A.: Secure schemes for secret sharing and key distribution. Ph.D. thesis, Israel Institute of Technology, Technion, Haifa, Israel (1996)

3. Bethencourt, J., Sahai, A., Waters, B.: Ciphertext-policy attribute-based encryption. In: IEEE Symposium on Security and Privacy, 2007, SP 2007, pp. 321–334, May 2007

4. Boneh, D., Franklin, M.: Identity-based encryption from the weil pairing. In: Kilian, J. (ed.) CRYPTO 2001. LNCS, vol. 2139, pp. 213–229. Springer, Heidelberg (2001). http://dx.doi.org/10.1007/3-540-44647-8_13

5. Chow, S.S.M., Yiu, S.-M., Hui, L.C.K.: Efficient identity based ring signature. In: Ioannidis, J., Keromytis, A.D., Yung, M. (eds.) ACNS 2005. LNCS, vol. 3531, pp. 499–512. Springer, Heidelberg (2005). http://dx.doi.org/10.1007/11496137_34

6. Chunqiang, H., Nan, Z., Hongjuan, L., Xiuzhen, C., Xiaofeng, L.: Body area network security: a fuzzy attribute-based signcryption scheme. IEEE J. Sel. Areas Commun. **31**(9), 37–46 (2013)

7. Cordeiro, C., Fantacci, R., Gupta, S., Paradiso, J., Smailagic, A., Srivastava, M.: Body area networking: technology and applications. IEEE J. Sel. Areas Commun. **27**(1), 1–4 (2009)

8. Emura, K., Miyaji, A., Rahman, M.S.: Toward dynamic attribute-based signcryption (poster). In: Parampalli, U., Hawkes, P. (eds.) ACISP 2011. LNCS, vol. 6812, pp. 439–443. Springer, Heidelberg (2011). http://dx.doi.org/10.1007/978-3-642-22497-3_32

9. Gagné, M., Narayan, S., Safavi-Naini, R.: Threshold attribute-based signcryption. In: Garay, J.A., De Prisco, R. (eds.) SCN 2010. LNCS, vol. 6280, pp. 154–171. Springer, Heidelberg (2010). http://dx.doi.org/10.1007/978-3-642-15317-4_11

10. Goyal, V., Pandey, O., Sahai, A., Waters, B.: Attribute-based encryption for fine-grained access control of encrypted data. In: Proceedings of the 13th ACM Conference on Computer and Communications Security, CCS 2006, pp. 89–98. ACM, New York (2006). http://doi.acm.org/10.1145/1180405.1180418

11. Herranz, J., Sáez, G.: Forking lemmas for ring signature schemes. In: Johansson, T., Maitra, S. (eds.) INDOCRYPT 2003. LNCS, vol. 2904, pp. 266–279. Springer, Heidelberg (2003). http://dx.doi.org/10.1007/978-3-540-24582-7_20

12. Herranz, J., Sáez, G.: New identity-based ring signature schemes. In: López, J., Qing, S., Okamoto, E. (eds.) ICICS 2004. LNCS, vol. 3269, pp. 27–39. Springer, Heidelberg (2004). http://dx.doi.org/10.1007/978-3-540-30191-2_3

13. Huang, X., Susilo, W., Mu, Y., Zhang, F.: Identity-based ring signcryption schemes: cryptographic primitives for preserving privacy and authenticity in the ubiquitous world. In: 19th International Conference on Advanced Information Networking and Applications, 2005, AINA 2005, vol. 2, pp. 649–654, March 2005

14. Li, M., Yu, S., Guttman, J.D., Lou, W., Ren, K.: Secure ad hoc trust initialization and key management in wireless body area networks. ACM Trans. Sen. Netw. **9**(2), 18:1–18:35 (2013). http://doi.acm.org/10.1145/2422966.2422975

15. Malone-Lee, J.: Identity-based signcryption. Cryptology ePrint Archive, Report 2002/098 (2002). http://eprint.iacr.org/

16. Ostrovsky, R., Sahai, A., Waters, B.: Attribute-based encryption with non-monotonic access structures. In: Proceedings of the 14th ACM Conference on

Computer and Communications Security, CCS 2007, pp. 195–203. ACM, New York (2007). http://doi.acm.org/10.1145/1315245.1315270

17. Rivest, R.L., Shamir, A., Tauman, Y.: How to leak a secret. In: Boyd, C. (ed.) ASIACRYPT 2001. LNCS, vol. 2248, pp. 552–565. Springer, Heidelberg (2001). http://dx.doi.org/10.1007/3-540-45682-1_32

18. Sahai, A., Waters, B.: Fuzzy identity-based encryption. In: Cramer, R. (ed.) EUROCRYPT 2005. LNCS, vol. 3494, pp. 457–473. Springer, Heidelberg (2005). http://dx.doi.org/10.1007/11426639_27

19. Selvi, S.S.D., Sree Vivek, S., Pandu Rangan, C.: Identity based public verifiable signcryption scheme. In: Heng, S.-H., Kurosawa, K. (eds.) ProvSec 2010. LNCS, vol. 6402, pp. 244–260. Springer, Heidelberg (2010). http://dx.doi.org/10.1007/978-3-642-16280-0_17

20. Shamir, A.: How to share a secret. Commun. ACM **22**(11), 612–613 (1979). http://doi.acm.org/10.1145/359168.359176

21. Shamir, A.: Identity-based cryptosystems and signature schemes. In: Blakely, G.R., Chaum, D. (eds.) CRYPTO 1984. LNCS, vol. 196, pp. 47–53. Springer, Heidelberg (1985). http://dx.doi.org/10.1007/3-540-39568-7_5

22. Ullah, S., Higgins, H., Braem, B., Latre, B., Blondia, C., Moerman, I., Saleem, S., Rahman, Z., Kwak, K.: A comprehensive survey of wireless body area networks. J. Med. Syst. **36**(3), 1065–1094 (2012). http://dx.doi.org/10.1007/s10916-010-9571-3

23. Wang, C.J., Huang, J.S., Lin, W.L., Lin, H.T.: Security analysis of Gagne et al'.s threshold attribute-based signcryption scheme. In: 2013 5th International Conference on Intelligent Networking and Collaborative Systems (INCoS), pp. 103–108, September 2013

24. Waters, B.: Ciphertext-policy attribute-based encryption: an expressive, efficient, and provably secure realization. In: Catalano, D., Fazio, N., Gennaro, R., Nicolosi, A. (eds.) PKC 2011. LNCS, vol. 6571, pp. 53–70. Springer, Heidelberg (2011). http://dx.doi.org/10.1007/978-3-642-19379-8_4

25. Yang, P., Cao, Z., Dong, X.: Fuzzy identity based signature with applications to biometric authentication. Comput. Electr. Eng. **37**(4), 532–540 (2011). http://www.sciencedirect.com/science/article/pii/S0045790611000589

26. Zhang, F., Kim, K.: ID-based blind signature and ring signature from pairings. In: Zheng, Y. (ed.) ASIACRYPT 2002. LNCS, vol. 2501, pp. 533–547. Springer, Heidelberg (2002). http://dx.doi.org/10.1007/3-540-36178-2_33

27. Zheng, Y.: Digital signcryption or how to achieve cost(signature & encryption) ≪ cost(signature) + cost(encryption). In: Kaliski Jr, B.S. (ed.) CRYPTO 1997. LNCS, vol. 1294, pp. 165–179. Springer, Heidelberg (1997). http://dx.doi.org/10.1007/BFb0052234

Elliptic Curve

Parallelized Software Implementation of Elliptic Curve Scalar Multiplication

Jean-Marc Robert[1,2](✉)

[1] Team DALI, Université de Perpignan, Perpignan, France
[2] LIRMM, UMR 5506, Université Montpellier 2 and CNRS, Montpellier, France
`jean-marc.robert@univ-perp.fr`

Abstract. Recent developments of multicore architectures over various platforms (desktop computers and servers as well as embedded systems) challenge the classical approaches of sequential computation algorithms, in particular elliptic curve cryptography protocols. In this work, we deploy different parallel software implementations of elliptic curve scalar multiplication of point, in order to improve the performances in comparison with the sequential counter parts, taking into account the multi-threading synchronization, scalar recoding and memory management issues. Two thread and four thread algorithms are tested on various curves over prime and binary fields, they provide improvement ratio of around 15 % in comparison with their sequential counterparts.

Keywords: Elliptic curve cryptography · Parallel algorithm · Efficient software implementation

1 Introduction

Elliptic curve cryptography (ECC) is widely used in a large number of protocols: secret key exchanges, asymmetric encryption-decryption, digital signatures. . . The main operation in these protocols is the scalar multiplication (ECSM) defined as $k \cdot P$ where P is a point of order r on an elliptic curve $E(\mathbb{F}_q)$ and $k \in [0, r[$ is an integer. The scalar multiplication is computed with *Double-and-add* approaches which consist of sequences of several hundreds of doublings and additions of curve points. It is thus a costly operation which might be implemented efficiently.

In this paper we consider parallel approaches for software implementation of scalar multiplication. There are two versions of the *Double-and-add* scalar multiplication: the left-to-right and the right-to-left depending on the way the bits of k are scanned. On the one hand, the left-to-right version cannot be parallelized due to the strong dependence of the consecutive doublings and additions. On the other hand, the right-to-left version is easier to parallelize: this was noticed by Moreno and Hasan in [15]. Indeed, in [15], the authors provide an algorithm consisting in one thread producing the points $2^i P$ through consecutive doublings, which are then consumed by a second thread performing all the necessary additions. They did not provide any implementation results of

© Springer International Publishing Switzerland 2015
D. Lin et al. (Eds.): Inscrypt 2014, LNCS 8957, pp. 445–462, 2015.
DOI: 10.1007/978-3-319-16745-9_24

their approach. In practice this can be challenging to implement efficiently the synchronizations between the two threads.

When the elliptic curve is defined over a binary field $\mathbb{F}_{2^m}$, a formula exists (cf. [5,12]) which computes efficiently the halving of a point, i.e., $\frac{1}{2}P$. This makes possible to perform the scalar multiplication through a sequence of halvings and additions of points. This can be used to parallelize the scalar multiplication into two totally independent threads: one thread performing a halve-and-add scalar multiplication and a second thread performing a double-and-add. This approach has been implemented by Taverne *et al.* in [20] showing a significant speed-up compared to non-parallelized versions.

In this paper we first explore the implementation of the two threads parallel approach of Moreno and Hasan [15]. Specifically, we analyze three different strategies to perform synchronization between both threads: using `signals`, `mutexes` or `busy-waiting` approaches, we propose a synchronization strategy based on this analysis. We also study the best approach for the coding of the integer k: this impacts the number of additions and post-computations, i.e., the work load of the thread performing the additions.

We then investigate a four thread parallelization of the scalar multiplication in $E(\mathbb{F}_{2^m})$. This approach combines the *Double/halve-and-add* algorithm of [20] with the approach of Moreno and Hasan.

We provide experimental results for two curves defined over a prime field $p = 2^{255} - 19$ and for the two binary elliptic curves B409 and B233 recommended by NIST in [18]. Our experimental results show that the parallelized scalar multiplication is up to 15 % faster than their non-parallelized counterparts (depending on the curve type and the field size).

The remaining of the paper is organized as follows: in Sect. 2 we review basic definitions of elliptic curve and scalar multiplication algorithms. In Sect. 3, we present our implementation approaches of scalar multiplication. We then provide in Sect. 4 the experimental results and comparisons with the state of the art. We end the paper in Sect. 5 with some concluding remarks.

2   Background on Elliptic Curve Scalar Multiplication

In this section, we briefly review basic results concerning elliptic curve and their use in cryptography. For further details on this matter we refer the reader to [9]. An elliptic curve over a finite field $E(\mathbb{F}_q)$ is the set of point $(x, y) \in \mathbb{F}_q^2$ satisfying a smooth curve equation of degree 3 in x and y plus a point at infinity $\mathcal{O}$. A group law can be defined using the so-called chord-and-tangent approach, providing formulas in terms of point coordinates which compute doubling $2P$ and addition $P + Q$ in the group. The element $\mathcal{O}$ is the neutral element of the group. Cryptographic protocols are based on the intractability of the discrete logarithm problem: given a generator of the group P and a point Q, compute k such that $Q = kP$. The most costly operation involved in most ECC protocols is the scalar multiplication: given $P \in E(\mathbb{F}_q)$ and an integer k, the scalar multiplication consists in computing $kP = P + P + \cdots + P$ (k times). The elliptic curves used in practice are defined either over prime field $\mathbb{F}_p$ with p prime or over binary field $\mathbb{F}_{2^m}$. In the remainder of this section, we briefly review explicit formulas and algorithms for scalar multiplication over these two fields.

2.1 Scalar Multiplication over Prime Field

Weierstrass Elliptic Curve. An elliptic curve E over a prime field $\mathbb{F}_p$ is generally defined by a short Weierstrass equation:

$$E : y^2 = x^3 + ax + b, (a, b) \in \mathbb{F}_p^2.$$

Then, in this case, addition and doubling on $E(\mathbb{F}_p)$ works as follows: let $P_1 = (x_1, y_1), P_2 = (x_2, y_2), P_3 = (x_3, y_3)$, be three points of E such that $P_3 = P_1 + P_2$, then we have:

$$\begin{cases} x_3 = \lambda^2 - x_1 - x_2, \\ y_3 = \lambda(x_1 - x_3) - y_1, \end{cases} \text{where} \begin{cases} \lambda = \frac{y_2 - y_1}{x_2 - x_1} \text{ if } P_1 \neq P_2, \\ \lambda = \frac{3x_1^2 + a}{2y_1} + x_1 \text{ if } P_1 = P_2. \end{cases}$$

Jacobi Quartic Curves over Prime Field. This curve was suggested by Billet *et al.* in [4]. The curve equation of E is:

$$y^2 = x^4 - \frac{3}{2}\theta x^2 + 1, \theta \in \mathbb{F}_p.$$

For such curve, the addition and doubling formulas are unified. Let $P_1 = (x_1, y_1)$, $P_2 = (x_2, y_2), P_3 = (x_3, y_3)$, be three points of E such that $P_3 = P_1 + P_2$, then we have:

$$\begin{cases} x_3 = (x_1 y_2 + y_1 x_2)/(1 - (x_1 x_2)^2), \\ y_3 = ((1 + (x_1 x_2)^2)(y_1 y_2 + 2ax_1 x_2) + 2x_1 x_2(x_1^2 + x_2^2))/(1 - (x_1 x_2)^2)^2. \end{cases}$$

The Jacobi Quartic curve is isomorphic to the following Weierstrass elliptic curve:

$$y^2 = x^3 + ax + b \text{ where } a = (-16 - 3\theta^2)/4 \text{ and } b = -\theta^3 - a\theta.$$

Elliptic Curve Point Operations. The most expensive field operation is the inversion which roughly requires several tens of field multiplications. In order to avoid such operation, additions and doublings utilize projective coordinate system. In our implementation, we consider two systems: the Jacobian coordinate where the point $(X : Y : Z)$ corresponds to the affine point $(X/Z^2, Y/Z^3)$ and the $XYZZ$ coordinate system where the point $(X : XX : Y : Z : ZZ)$ corresponds to the affine point $(X/Z, Y/ZZ)$ with $XX = X^2$ and $ZZ = Z^2$. Explicit formulas for addition and doubling in these systems can be found in [1]

The resulting complexities are shown in Table 1, which shows that the complexities of the Jacobi Quartic curve operations are better than for the Weierstrass equation case. Moreover, based on the elliptic curve formula database in [1], the Jacobi Quartic curves provide the most efficient point operation among all known curves and formulas. This is the reason why we used such curve and these formulas in our implementations.

Table 1. Weierstrass curve and Jacobi Quartic curve point operations, M = multiplications, S = squaring, R = field reduction.

Complexity comparison for: point operations	Weierstrass with Jacobian coord	Jacobi Quartic curve with $XXYZZ$ coord
Doubling	$4M + 4S + 8R$	$3M + 4S + 7R$
Mixed addition	$9M + 3S + 12R$	$6M + 3S + 9R$
Full projective addition	$13M + 2S + 15R$	$7M + 4S + 11R$

Scalar Multiplication Algorithm. The basic method to compute a scalar multiplication consists in scanning the bits k_i of $k = \sum_{i=0}^{t-1} k_i \cdot 2^i$ and performing a sequence of doubling followed by an addition when $k_i = 1$. This approach is described in Algorithm 1.

In order to reduce the number of additions, the non adjacent form (NAF) and the window non adjacent form (W-NAF) recoding of the scalar are well-known methods, which reduce the number of non zero digit representing the scalar. In the binary scalar representation, half of the digits are either zero or one on average. In the NAF representation, one uses three digits instead of two: $k = \sum_{i=0}^{t} k_i \cdot 2^i$ with $k_i \in \{-1, 0, 1\}$ and there are only $t/3$ non zero digits k_i on average.

The W-NAF representation extends this concept by using more digits: $k = \sum_{i=0}^{t} k_i \cdot 2^i$ with $k_i \in \{-(2^{w-1} - 1), \ldots, -5, -3 - 1, 0, 1, 3, 5, \ldots, (2^{w-1} - 1)\}$. The number of non zero digits is now $t/(w + 1)$ on average. Algorithm 1 can be adapted to use k recoded as NAF or W-NAF. The complexities of the resulting scalar multiplication are given in Table 2.

Table 2. Complexity comparison between binary, NAF and W-NAF scalar representation in terms of t the bit length of the scalar.

	Nb. of doublings	Nb. of additions
Double-and-add	$t - 1$	$t/2$
NAF Double-and-add	t	$t/3$
W-NAF Double-and-add	t	$t/(w + 1) + 2^{w-2} - 1$

The reader may refer to [8] for further details and algorithms to compute NAF and W-NAF representation.

2.2 Elliptic Curve Scalar Multiplication over Binary Field

An elliptic curve E over a binary field $\mathbb{F}_{2^m}$ is the set of points $P = (x, y) \in \mathbb{F}_{2^m}^2$ satisfying the following equation:

$$E : y^2 + xy = x^3 + ax^2 + b, (a, b) \in \mathbb{F}_{2^m}^2.$$

Algorithm 1. Left-to-Right Double-and-add	Algorithm 2. Right-to-left Halve-and-add
Require: $k = (k_{t-1}, \ldots, k_1, k_0), P \in E(\mathbb{F}_{2^m})$ **Ensure:** $Q = k \cdot P$ 1: $Q \leftarrow \mathcal{O}$ 2: **for** i from $t - 1$ downto 0 3: $Q \leftarrow 2 \cdot Q$ 4: **if** $k_i = 1$ **then** 5: $Q \leftarrow Q + P$ 6: endif 7: **endfor** 8: **return** (Q)	**Require:** $P \in E(\mathbb{F}_{2^m})$ of order r and k an integer in $[0, r[$ **Ensure:** $Q = k \cdot P$ 1: Compute $k' = 2^t \cdot k \mod r = \sum_{i=0}^{t} k_i' 2^i$ with $t = \lfloor \log_2(r) \rfloor + 1$ 2: $Q \leftarrow \mathcal{O}$ 3: **for** i from t downto 0 4: **if** $k_i = 1$ **then** 5: $Q \leftarrow Q + P$ 6: endif 7: $P \leftarrow P/2$ 8: **endfor** 9: **return** (Q)

Let $P_1 = (x_1, y_1), P_2 = (x_2, y_2), P_3 = (x_3, y_3)$, be three points of E such that $P_3 = P_1 + P_2$, then we have:

$$\begin{cases} x_3 = \lambda^2 + \lambda + x_1 + x_2 + a, \\ y_3 = (x_1 + x_3)\lambda + x_3 + y_1, \end{cases} \text{where} \begin{cases} \lambda = \frac{y_1 + y_2}{x_1 + x_2} \text{ if } P_1 \neq P_2, \\ \lambda = \frac{y_1}{x_1} + x_1 \text{ if } P_1 = P_2. \end{cases} \quad (1)$$

Elliptic Curve Scalar Multiplication with Halving. It was noticed by Knudsen in [12] that over a binary field, halving of points is possible in case of points of odd order since 2 admits an inverse modulo the order of the point. In other words, point *halving* is the reciprocal operation of point doubling: given $Q = (u, v) \in E(\mathbb{F}_{2^m})$, one looks for $P = (x, y) \in E(\mathbb{F}_{2^m}), P \neq -P$ such as $Q = 2 \cdot P$. Based on Eq. (1), we know that x, y, u and v satisfy the following relations:

$$\lambda = x + y/x \quad (2)$$
$$u = \lambda^2 + \lambda + a \quad (3)$$
$$v = x^2 + u(\lambda + 1) \quad (4)$$

Consequently, in order to compute P, we first have to solve Eq. (3) to get λ (which means solve $\lambda^2 + \lambda = u + a$), then, Eq. (4) gives $x = \sqrt{v + u(\lambda + 1)}$, and finally, Eq. (2) gives $y = \lambda x + x^2$. The reader may refer to Knudsen in [12] and Fong *et al.* in [5] for further details. In practice, this can be implemented efficiently and has roughly the same cost as two field multiplications (see [20]).

The *Double-and-add* method can be modified into a *Halve-and-add* scalar multiplication. Preliminary, we need to change the scalar. Assuming the point P to be multiplied is of odd order r, we compute $k' = 2^t \cdot k \mod r = \sum_{i=0}^{t} k_i' 2^i$ with $t = \lfloor \log_2(r) \rfloor + 1$. Then, we have $k \equiv k'/2^t \equiv \sum_{i=0}^{t} k_i' 2^{i-t} \mod r$ and the scalar multiplication can be computed as follows:

$$k \cdot P = (k_t' + k_{t-1}' \cdot 2^{-1} + \ldots + k_0' 2^{-t}) \cdot P.$$

This can be computed as a sequence of halvings and additions as shown in Algorithm 2.

Cost of Elliptic Curve Point Operations. Over a field of characteristic 2, and in order to avoid the inversions during the computation, which is the most expensive field operation again, one may use projective coordinate systems. The most interesting systems are the Lopez-Dahab ($\mathcal{LD}$, as shown in [8]) and the Kim-Kim ($\mathcal{KK}$, see [11]) projective coordinate systems. With such point representation, the addition and doubling operations do not include any inversion as shown in Table 3, and the whole scalar multiplication is computed with a significant speed-up. Table 3 shows that the complexities of $\mathcal{KK}$ are slightly better and then, when possible, we give the preference to the $\mathcal{KK}$ coordinate system.

Table 3. Elliptic curve point operations, M = multiplications, S = squaring, SR = square root, QS = quadratic solver, R = reduction.

Point operation	Coord. system	Cost
Doubling	$\mathcal{LD}$	$4M + 4S + 8R$
Mixed addition	$\mathcal{LD}$	$9M + 4S + 13R$
Projective addition	$\mathcal{LD}$	$13M + 4S + 17R$
Doubling	$\mathcal{KK}$	$4M + 5S + 7R$
Mixed addition	$\mathcal{KK}$	$8M + 4S + 9R$
Halving	Affine	$1M + 1SR + 1R + 1QS$

3   Strategies for Parallel Implementation of Scalar Multiplication

In this section, after a quick review of the implementation strategies used for the field operations, we expose how we elaborate the parallelized algorithm, taking into account all the constraints for such concurrent programming.

The platform used for the experimentations is an Optiplex 990 DELL®, with a Linux 12.04 operating system. The processor is an Intel core i7®-2600 Sandy Bridge 3.4 GHz. This processor owns four physical cores, which corresponds to the maximum thread number of our implementations. The code is written in C language and compiled with gcc version 4.6.3.

3.1   Field Implementation Strategies

Prime Field Implementation Strategies. We considered the prime field $\mathbb{F}_p$, with $p = 2^{255} - 19$, which was introduced by Bernstein in [2]. To compute the field operations, we reused the publicly available code of Adam Langley in [13]. Based on our experiments, the code of Langley is significantly more efficient compared to low level functions of the GMP library [6] for the considered field. In the code of Adam Langley a field element is stored in a table of five 64 bit words, each word containing only 51 bits. This allows a better management of carries in field addition and subtraction operations. The field multiplications and squarings are performed in two steps, which are multiprecision integer multiplication (respectively squaring) and a modular reduction of integer of 510 bit size (less than p^2) into field element of size 255 bits (reduction modulo p). The multiprecision integer multiplications and squarings are computed with the schoolbook method. The squaring operation is optimized with the usual trick which reduces the number of word multiplications. The reduction modulo $p = 2^{255} - 19$ of 510 bit size integer consists in multiplying by 19 the 255 most significant bits and adding the result to the 255 least significant bits. An inversion of a field element is computed using the Itoh-Tsujii method [10]: $a^{-1} \equiv a^{p-2} \mod p$, and the exponentiation to $p - 2$ is performed with a sequence of squarings and multiplications.

Binary Field Implementation Strategies. Our implementations deal with NIST recommended fields $\mathbb{F}_{2^{233}} = \mathbb{F}[x]/(x^{233} + x^{74} + 1)$ and $\mathbb{F}_{2^{409}} = \mathbb{F}[x]/(x^{409} + x^{87} + 1)$. Concerning the binary polynomial multiplication, we apply a small number of recursions of the Karatsuba algorithm. The Karatsuba algorithm breaks the m bit polynomial multiplication into several 64 bit polynomial multiplications. Such 64 bit multiplications are computed with the PCLMUL instruction, available on Intel Core i7 processors. Due to the special form of the irreducible polynomials, the reduction is done with a small number of shifts and bit-wise XORs. We compute the field inversion with the Itoh-Tsujii algorithm, that is a sequence of field multiplications and multisquarings performed with look-up table. For field squaring, square root and quadratic solver (needed in halvings), we also use a look-up table method, which is the fastest way according to our tests.

Remark 1. The use of Karatsuba for binary fields and schoolbook method for the prime field is due to the relative cost of word addition compared to word multiplication and to carry managements on our platform. Indeed, integer word additions and multiplications have roughly the same cost (1 vs 2 cycles). The use of Karatsuba algorithm for $\mathbb{F}_p$ decreases the number of word multiplications, but, in counter part, it increases the number of additions and carry managements. For binary field the relative cost of addition (bit-wise XOR) and multiplications (PCLMUL instruction) is more important: 1 cycle vs 10 cycles. In this case Karatsuba is efficient to decrease the timing of a field multiplication.

3.2   Parallelization

The left-to-right *Double-and-add* algorithm (see Algorithm 1 page 449) does not allow any parallelization of the computations, due to the read-after-write dependency inside each loop iteration, between step 5 (addition) and step 3 (doubling). It is necessary to use the right-to-left variant of this algorithm (see Algorithm 3) which allows the parallelization. Indeed Algorithm 3 can be parallelized into two threads as follows:

- A producer-thread performing the sequence of doublings generating the points $2^i P$.
- An addition-thread accumulating the points generated by the producer-thread.

In the sequential case, the left-to-right *Double-and-add* algorithm (Algorithm 1) is better, because the point addition in step 5 can use a mixed coordinate addition. This is faster than the projective addition used in the right-to-left version in step 4 (Algorithm 3). We will see that this penalty is overcome in most of the cases, thanks to the parallelization.

The *Halve-and-add* algorithm (Algorithm 2 page 449) can also be parallelized with two threads. Indeed, since the computation in step 7 of Algorithm 2 only depends on the same step in the previous loop iteration (read-after-write dependency), the sequence of halvings (step 7) can be performed in a separate thread (the producer-thread) and the addition in an addition-thread which accumulates the points generated by the producer-thread.

Algorithm 3. Right-to-left Double-and-add

Require: $k = (k_{t-1}, \ldots, k_1, k_0), P \in E(\mathbb{F}_q)$
Ensure: $Q = k \cdot P$
1: $Q \leftarrow \mathcal{O}$
2: **for** i from 0 to $t - 1$ **do**
3:     **if** $k_i = 1$ **then**
4:         $Q \leftarrow Q + P$
5:     **end if**
6:     $P \leftarrow 2 \cdot P$
7: **end for**
8: **return** (Q)

Synchronization Between Threads. Both parallelization (right-to-left *Double-and-add*, Algorithm 3 and *Halve-and-add*, Algorithm 2) are classical producer-consumer configurations.

The safest way to guarantee absolute correct computation is to use a strong synchronization device, processing the computation by small batches: the producer-thread computes and stores a small batch of point doublings/halvings, sends a signal in order to trigger the addition computation in the addition-thread only concerning the batch in shared memory. In parallel, the producer-thread goes on with the next batch and the addition-thread waits the end of each batch before processing the corresponding additions (in the way described by Mueller in [16] or by Tannenbaum in [19]). In our case, on the one hand, the batch size has to be small to compute the maximum of additions in parallel. But on the other hand, if the batches are too small, the synchronization cost would increase, due to the bigger number of synchronization signals to manage. This is especially true as the granularity of doublings/halvings and additions (several hundreds of processor clock cycles) is too small in comparison with the cost of synchronization barriers and signals.

The three following methods can be used to synchronize the two threads:

- mutex. A mutex is a mutual exclusion lock provided by the pthread library used to synchronize threads. When a thread holds a mutex, another thread, trying to take it, is locked, waiting for the releasing of the mutex from the first thread. Mutexes are generally used to protect critical sections of code. The cost of a lock or an unlock is about 150–200 processor clock cycles, which is almost negligible.
- signals: they are used in the inter-thread and inter-process communication. A thread waiting for a signal is put in a sleeping state until another thread sends the corresponding signal. Then, the thread wakes up and goes on running. The sleeping state allows savings of resources which are then available for another process. In our experience and on our platform, the cost to send a signal is about 2000 clock cycles.
- busy-waiting: this method consists in using a shared flag (in the global memory) and use it to keep the addition-thread in a *busy-waiting* loop while waiting for the producer-thread to output the next point and modify the flag. The main drawback of this method is to waste processor resources.

According to our experiments, signals are too costly compared to the two other techniques. The busy-waiting and mutex techniques almost give the same results in terms of performance, although the mutex method is slightly better in some cases. Thus we decided to use exclusively mutexes.

Proposed Synchronization Method. Our strategy was to avoid the use of mutex synchronization as much as possible. We chose to use only one single mutex: at the very beginning of the computation the mutex keeps the addition-thread in an inactive state while a first batch of doublings or halvings is computed by the producer-thread. At the end of the computation of this batch, the producer-thread releases the mutex and pursues the whole sequence of doubling without performing any further locking on the mutex. This approach is depicted in Fig. 1.

The correctness of the final result depends on the size of the first batch of points before the mutex releasing, which ensures that the writings of the point stored in shared memory by the doubling thread precedes the reading of the same point by the addition thread. If this batch is too small and in case of long sequence of zeros in the binary or NAF scalar representation, one can meet a violation of the read-after-write dependency, and the computation is not correct. To avoid this configuration, we carefully tuned this batch size in order to have the error rate as close as possible to zero. In our test results shown below, this error rate is limited to less than 1 %. This is a compromise chosen in order to limit the first batch of doublings/halvings size, and to get the best performances. But at this step, such an error rate remains unacceptable.

In order to eliminate these errors, we added a test on the addition-thread. In the producer-thread, we used a variable which is stored in global memory as the loop counter. This allows to check if the addition processed uses a point

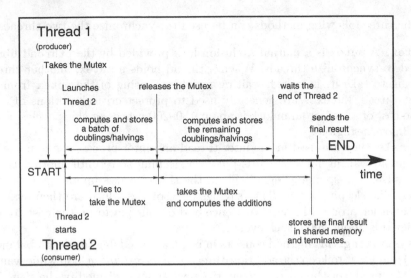

Fig. 1. Synchronization and thread processing for our ECSM implementation

which has already been computed by the producer-thread, i.e. the read-after-write dependency is ensured. The cost of this test is almost negligible, although the use of a global memory counter is not totally free. When an error is detected (that is to say a read-after-write dependency violation), we break the addition-thread loop, and launch a sequential computation of $k \cdot P$. Due to the small error rate, the cost of this rescue computation, which frequency is near zero, is negligible on average.

Algorithm 4 presents an algorithmic formulation in the case of Right-to-left *Double-and-add* scalar multiplication of this approach, including the elimination of the error computations due to a synchronization failure.

Impact of Scalar Recoding. In the sequential case, it is a useful technique to recode the scalar using NAF and W-NAF to speed-up the computation (as previously mentioned in Subsect. 2.1 page 447). In the parallel algorithms, the situation is different. Indeed, the NAF and W-NAF recodings reduce the number of additions performed by the addition-thread. This fact can be seen when analyzing the amount of computations performed by the two threads. We can evaluate this amount using the results given Table 2 and Table 1 in the case of curves over $E(\mathbb{F}_p)$, and using the results given Table 2 and Table 3 in the case of curves over $E(\mathbb{F}_{2^m})$. For simplicity we assumed that $S = 0.8M$ in $\mathbb{F}_p$ and that a squaring and square root are negligible in $\mathbb{F}_{2^m}$ and that the cost of a quadratic solver is roughly $1M$. The resulting complexities are given in Table 4.

Table 4, we remark that, generally, the amount of computation of the addition-thread is larger than the producer-thread for the binary coding. When using the NAF recoding the amount of computation of the two threads are roughly the same. Finally, the use of W-NAF makes the amount of computation of the

Table 4. Complexity of the two threads for a t-bit scalar coded in binary, NAF and W-NAF, in multiplication number.

Recoding	Double-and-add over $\mathbb{F}_p$			Double-and-add over $\mathbb{F}_{2^m}$			Halve-and-add over $\mathbb{F}_{2^m}$		
	Producer-thread	Addition-thread	Post-comp	Producer-thread	Addition-thread	Post-comp	Producer-thread	Addition-thread	Post-comp
binary	$6.2tM$	$5.1tM$	0	$4tM$	$6.5tM$	0	$2tM$	$4tM$	0
NAF	$6.2tM$	$3.4tM$	0	$4tM$	$4.33tM$	0	$2tM$	$2.66tM$	0
W-NAF ($w = 4$)	$6.2tM$	$2.04tM$	$33M$	$4tM$	$2.6tM$	$39M$	$2tM$	$1.6tM$	$39M$

addition-thread significantly smaller than the producer thread. This means that when using W-NAF recoding, the addition-thread progresses faster and even would have to wait for the producer-thread to output new points. But in any case, the addition-thread terminates after the producer-thread. Moreover in the W-NAF case, the post-computations delay the output of the results after the end of the producing process, since in the parallel algorithms, this final reconstruction cannot be done before the end of the parallelized additions.

These remarks are confirmed by the chronogram given in Fig. 2 which shows the different timings required by each thread related to the recoding used for the execution of the parallelized halve-and-add for scalar multiplication in $E(\mathbb{F}_{2^{233}})$. This fact leads us to opt for the NAF recoding for our implementations.

In addition to this choice, and in order to improve the performances, we implement a variable initial batch size of doublings/halvings in the producer-thread. Indeed, the number of additions performed by the addition-thread depends on the Hamming weight of the NAF representation of the scalar (i.e. the non-zero digits). As stated previously, a read-after-write dependency violation can appear if this batch is too small. But if the Hamming weight of the scalar is higher, the risk of this dependency violation is lower, and the batch size can be reduced in this case. This improvement applies only when the addition-thread has roughly the same running time as the producer thread. This concerns the *Double-and-add* approach over $\mathbb{F}_p$ and $\mathbb{F}_{2^m}$ but not the *Halve-and-add* approach (cf. Table 4).

When possible, we use the variable initial batch size in the producer-thread.

3.3 Four-Thread Parallel Version over Binary Elliptic Curve

Over binary field, the parallelization proposed by Taverne *et al.* in [20] splits the scalar multiplication into two independent threads. Specifically, they split the t-bit scalar $k = k_1 + k_2$ where k_1 and k_2 are as follows

$$k = \underbrace{(k'_t 2^{t-\ell} + \ldots + k'_\ell)}_{k_1} + \underbrace{(k'_{\ell-1} 2^{-1} + \ldots + k'_0 2^{-\ell})}_{k_2}. \tag{5}$$

In general ℓ is close to $t/2$ and represents the length of the *Halve-and-add* subkey. Then the computations can be parallelized into one thread computing $k_1 P$ with the *Double-and-add* algorithm and a second thread computing $k_2 P$ with the *Halve-and-add* algorithm.

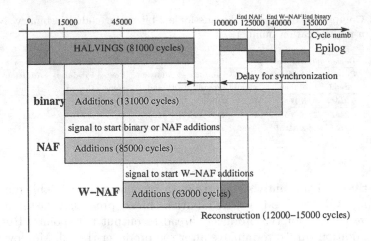

Fig. 2. Chronogram of the *halve-and-add* computation with binary, NAF and W-NAF scalar representation over B233

We propose to combine the approach of Taverne *et al.* with the parallelization approach discussed in Subsect. 3.2. This results in a four-thread algorithm: the partial scalar multiplication $k_1 P$ is computed with the parallel two-thread algorithm // *Double-and-add* and $k_1 P$ is computed with the parallel two-thread algorithm // *Halve-and-add*. This four-thread approach is shown in Fig. 3. This approach increases the level of parallelization, but it also requires additional thread launching and management. Therefore, this algorithm works better on large fields, as it will be shown in the next section.

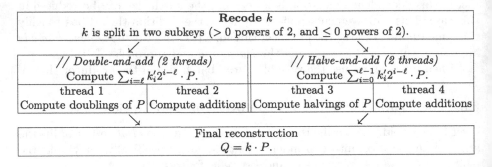

Fig. 3. Four-thread algorithm.

Our implementations of the four-thread of Fig. 3 use the following strategies for thread launching and synchronization:

Algorithm 4. Parallel *Double-and-add* Elliptic Curve Scalar Multiplication

Require: scalar k, $P \in \mathbb{F}_{2^m}$.
Ensure: kP.
   (Barrier)

Compute Doublings (producer-thread)

1: $D[0] \leftarrow P$

2: **for** $glblMmry.i = 1$ to $initBtchSze$ **do**
3:    //*Doubling* LD projective
      $D[glblMmry.i] \leftarrow D[glblMmry.i - 1] \times 2$
4: **end for**
5: signal to thread addition
6: **for** $glblMmry.i = initBtchSze + 1$
   to $M - 1$ **do**
7:    //*Doubling* LD projective
      $D[glblMmry.i] \leftarrow D[glblMmry.i - 1] \times 2$
8: **end for**

Compute Additions (addition-thread)

9: $Q \leftarrow \mathcal{O}$
10: Wait for signal from thread Doubling

11: **for** $i = 0$ to $M - 1$ **do**
12:    **if** $i > glblMmry.i - 1$ **then**
13:       launch rescue computation ($Q \leftarrow kP$)
14:       break
15:    **end if**
16:    **if** $k_i = 1$ **then**
17:       //Full LD projective addition
         $Q \leftarrow Q + D[i]$
18:    **end if**
19: **end for**

   (Barrier)
20: **return** Q

- The threads are launched in this order: (1) the *Halve-and-add* producer-thread, which launches (2) the *Double-and-add* producer-thread which launches (3) the *Halve-and-add* addition-thread which finally launches (4) the *Double-and-add* addition-thread.
- The recoding of the scalar is done by (3) the *Halve-and-add* addition-thread before launching (4) the *Double-and-add* addition-thread.
- Due to the delay of thread launching and key recoding computation, it is not necessary to use mutexes with initial batch size of halvings or doublings of points for each producer-thread.
- In order to eliminate computation errors due to synchronization failure, we use the same method as the one described Subsect. 3.2. Thus, we have two global memory counters, one for the doubling producer-thread and one for the halving producer-thread. Each addition-thread compare its own counter with the global counter of its corresponding, and launches a partial rescue computation if a synchronization dependency violation is detected.

4 Timings

The platform used for the experimentations is an Optiplex 990 DELL®, with a Linux 12.04 operating system. The processor is an Intel Core i7®-2600 Sandy Bridge 3.4 GHz, which owns four physical cores. The code is written in C language, compiled with gcc version 4.6.3. The *Hyperthreading*® BIOS and also the Turbo-boost® options have been deactivated on our platform in order to measure the performances as accurately as possible.

Since the operating system has the possibility to preempt the resources in order to launch another task, we avoid such difficulties by choosing to run our

Table 5. Timings (in clock cycles)

		Binary Field		Prime Field $\mathbb{F}_p$	
		B233	**B409**	**Weierstrass**	**Jacobi Quartic**
		References			
Sequential 1 thread	*Double-and-add*	159000	706000 (W-NAF, $w=4$)	256631 (NAF)	222558
	Halve-and-add	135000	534000 (W-NAF, $w=4$)	-	-
2 threads	*Dbl/Hlv-and-add*	98000	347000 (W-NAF, $w=4$)	-	-
		NAF // *Double-and-add*			
2 threads	mean	**154621**	**598491**	**218606**	**184048**
	Doublings	114713	505662	168958	134398
	Additions-D	120748	522869	125990	87415
		NAF // *Halve-and-add*			
2 threads	mean	**126639**	**430222**		
	Halvings	81630	300113		
	Additions-D	85107	373534		
		NAF // *Halve-Double-and-add*			
4 threads	mean	**133273** ($\ell = 151$)	**324395** ($\ell = 246$)		
	Doublings	44672	202393		
	Additions-D	39333	200660		
	Halvings	67076	199625		
	Additions-H	55615	217534		

codes in a recovery mode shell. But we noticed that the codes generally run well in normal operating system conditions too, although perturbations may be observed in a few cases.

We give the parameters of the curves used in the experimentation in Appendix:

- Appendix A.1 for the B233 and B409 binary field curves;
- Appendix A.2 for the Weierstrass prime field curve;
- Appendix A.3 for the Jacobi-Quartic prime field curve.

Table 5 shows the results of the proposed parallel strategies for scalar multiplication implementations. The above timings include the error detection and correction due to erroneous thread synchronization.

The performances are measured using one hundred batches of 2000 computations, each batch with a different random scalar. The minimum value of each batch is considered and the average value gives the performance. With this measurement process, we take into account the variations due to the different Hamming weights of the scalars.

For each case we provide the detailed duration of each thread. We notice that, generally, the overall computation finishes around several tens of thousands cycles after the producer-thread. These timings might correspond to the delayed start of the addition-thread (due to the initial batch size of points computed by the producer thread) and the synchronization and thread management time. For the four-thread versions, the given value ℓ corresponds to the scalar bit size of

Table 6. Performance comparison with the state of the art

	Scalar multiplication	Curve	Security	Processor	Method	Cycles
$E(\mathbb{F}_p)$	Hamburg [7]	Montgomery	126	Intel core i7 SB	Montgomery ladder	153000
	Langley [13]	Curve25519	128	Intel core i7 SB	Montgomery ladder	229000[a]
	Bernstein [2,3]	Curve25519	128	Intel core i7 SB	Montgomery ladder	194000
	Longa et al. [14]	jac256189	128	Intel core 2 Duo	WNAF D&A	337000
	Longa et al. [14]	ted256189	128	Intel core 2 Duo	WNAF D&A	281000
	This work	jac25519	128	Intel core i7 SB	//NAF D&A	184048
$E(\mathbb{F}_{2^m})$	Ngre et al. [17]	B233	112	Intel core i7 SB	WNAF D-H&A	98000
	Taverne et al. [20]	B233	112	Intel core i7 SB	WNAF D-H&A	102000
	This work	B233	112	Intel core i7 SB	//NAF H&A 2 th.	126639
	Ngre et al. [17]	B409	192	Intel core i7 SB	WNAF D-H&A	347000
	Taverne et al. [20]	B409	192	Intel core i7 SB	WNAF D-H&A	358000
	This work	B409	192	Intel core i7 SB	//NAF D-H&A 4 th.	324395

[a] compiled and run on our platform.

the *Halve-and-add* computation (cf. Eq. (5)). We have evaluated the overhead due to error management due to wrong error synchronization and it represents, in average, roughly 2–6% of the overall computation time.

Concerning the results over $\mathbb{F}_{2^{233}}$, we remark that the four-thread version is not competitive. This might be due to the synchronization and thread creation and management cost. Furthermore, the speed-ups with the two thread versions are not very important.

Over $\mathbb{F}_{2^{409}}$, the situation is different since the four-thread version is now better: it requires 324395 clock cycles whereas the two-thread parallel W-NAF *Double/halve-and-add* necessitates 347000 clock cycles (6.6 % improvement). The speed-ups provided by the two-thread versions is also more important: between 15 % (*Double-and-add* case) and 19.5 % (*Halve-and-add* case).

Concerning the results over $\mathbb{F}_p$, we first notice that a scalar multiplication over a Jacobi Quartic is faster than over a Weierstrass curve. This corroborates the complexities of the curve operations shown in Table 1. We also notice that the tested two-thread parallelization provides performance improvements of around 15 %–17 % compared to the NAF sequential *Double-and-add* approach.

Comparison. We give in Table 6 some published results in the literature. Over, $\mathbb{F}_p$, the work of Longa is on Intel Core 2 with $p = 2^{256} - 189$ and Hamburg is over a Sandy Bridge with $p = 2^{252} - 2^{232} - 1$ and has smaller key. The other works deal with the same processor and on the same fields as the one considered in this paper. We can see that, in the case of $E(\mathbb{F}_{2^{233}})$ our two-thread approach is not competitive with best know results. In the cases of $E(\mathbb{F}_{2^{409}})$, the proposed approach improves by 9.4 % the previous best known timings reported in [20]. Finally, the timing provided in [7] is better than the timing obtain by our method, but Hamburg uses a slightly smaller field and key size. On the other hand, we improve the best known results for curve defined for a 128 bit security level.

5 Conclusion

In this work, we have considered parallelized software implementations of scalar multiplication kP over $E(\mathbb{F}_{2^m})$ and $E(\mathbb{F}_p)$. We first have considered the parallelization suggested by Moreno et Hasan in [15] which splits the right-to-left scalar multiplication into two threads: one producer-thread computing $2^i P$ or $2^{-i}P$ for $i = 1, \ldots, t$ and one addition-thread which accumulates these points to compute kP. We have proposed a lightweight approach for thread synchronization. In addition, in order to avoid remaining computation errors due to dependency violation, we proposed a low cost checking method of the synchronization between threads with a rescue computation. We have also evaluated the best approach for the scalar recoding in this context. In the special case of $E(\mathbb{F}_{2^m})$ we have combined this approach to the parallelized Double/halve-and-add approach of [20]. The experimental results show that these parallelization techniques provide some speed-up on elliptic curve scalar multiplication computations compared to previously best known implementations. Indeed, over prime field and binary fields, in most cases the parallelization provides an improvement of roughly 15 % on the computation time.

Acknowledgement. We would like to thank Christophe Nègre for his valuable and helpful comments.

This work has been suported by a PHD grant from PAVOIS project (ANR 12 BS02 002 01).

A Appendix: Curve Parameters

A.1 Elliptic Curves over Binary Field

The curve equation is:

$$y^2 + xy = x^3 + x^2 + b \text{ where } b \in \mathbb{F}_{2^m}.$$

The parameters are for B233:

$a = 1$,
$h = 2$,
$f(x) = x^{233} + x^{74} + 1$,
$b = $ 0x00000066 647ede6c 332c7f8c 0923bb58 213b333b 20e9ce42 81fe115f 7d8f90ad,
$r = $ 0x00000100 00000000 00000000 00000000 0013e974 e72f8a69 22031d26 03cfe0d7.

where the order of the curve is $n \times h$. For B409 we have:

$a = 1$,
$h = 2$,
$f(x) = x^{409} + x^{87} + 1$,
$b = $ 0x0021a5c2 c8ee9feb 5c4b9a75 3b7b476b 7fd6422e f1f3dd67 4761fa99 d6ac27c8
a9a197b2 72822f6c d57a55aa 4f50ae31 7b13545f,
$r = $ 0x01000000 00000000 00000000 00000000 00000000 00000000 000001e2 aad6a612
f33307be 5fa47c3c 9e052f83 8164cd37 d9a21173.

A.2   Weierstrass Curve over Prime Field

The curve equation is:

$$y^2 = x^3 - 3x + b \text{ where } b \in \mathbb{F}_p.$$

The parameters are:

$p = 2^{255} - 19$
$b = $ 0x1d09bac9ffe9e7f8284aed0442552779bcdef2e62b9cb1d568513fa798b94003
$r = $ 0x8000000000000000000000000000000012c18945a05ad7f2edf026258ea5288ef

r is the prime order of P.

A.3   Jacobi Quartic Curve over Prime Field

The curve equation is:

$$y^2 = x^4 - \frac{3}{2}\theta x^2 + 1, \theta \in \mathbb{F}_p.$$

The parameters are:

θ = 0x1731beeea2156180446f9e5ab64af78d4ed3e0eb68d5070c10ef2468b910d5f7
number of points:
$h \times r$ = 0x800000000000000000000000000000002672bdbb41f31390c5527cab6e282744
= 4 · 0x20000000000000000000000000000000099caf6ed07cc4e431549f2adb8a09d1

The Jacobi Quartic curve is isomorphic to the following Weierstrass elliptic curve:

$$y^2 = x^3 + ax + b$$

where: $a = (-16 - 3\theta^2)/4$ and $b = -\theta^3 - a\theta$. Hence, in our case:

$a = $ 0xc500be2450246d16c114830a5d1aef9c2b80c567b4fd87562c69db659713ad2,
$b = $ 0xa38f53e5d27462dcdada9a78b9eac482ef06e855af92ca704060c551a9a5854.

References

1. Explicit formula database (2014). http://www.hyperelliptic.org/EFD/index.html
2. Bernstein, D.J.: Curve25519: new Diffie-Hellman speed records. In: Yung, M., Dodis, Y., Kiayias, A., Malkin, T. (eds.) PKC 2006. LNCS, vol. 3958, pp. 207–228. Springer, Heidelberg (2006)
3. Bernstein, D.J., Lange, T. (eds): eBACS: ECRYPT Benchmarking of Cryptograhic Systems (2012). http://bench.cr.yp.to/. Accessed 25 May 2014
4. Billet, O., Joye, M.: The Jacobi model of an elliptic curve and side-channel analysis. In: Fossorier, Marc P.C., Høholdt, Tom, Poli, Alain (eds.) AAECC 2003. LNCS, vol. 2643, pp. 34–42. Springer, Heidelberg (2003)

5. Fong, K., Hankerson, D., López, J., Menezes, A.: Field inversion and point halving revisited. IEEE Trans. Comput. **53**(8), 1047–1059 (2004)
6. Granlund, T., The GMP Development Team: GNU MP: The GNU Multiple Precision Arithmetic Library, 5.0.5 edition (2012). http://gmplib.org/
7. Hamburg, M.: Fast and compact elliptic-curve cryptography. Technical report, Cryptology ePrint Archive, Report 2012/309 (2012). http://eprint.iacr.org/
8. Hankerson, D., Hernandez, J.L., Menezes, A.: Software implementation of elliptic curve cryptography over binary fields. In: Paar, C., Koç, Ç.K. (eds.) CHES 2000. LNCS, vol. 1965, pp. 1–24. Springer, Heidelberg (2000)
9. Hankerson, D., Menezes, A., Vanstone, S.: Guide to Elliptic Curve Cryptography. Springer, New York (2004)
10. Itoh, T., Tsujii, S.: A fast algorithm for computing multiplicative inverses in $GF(2^m)$ using normal bases. Inf. Comput. **78**(3), 171–177 (1988)
11. Kim, K.H., Kim, S.I.: A new method for speeding up arithmetic on elliptic curves over binary fields. D.P.R. of Korea, Technical report, National Academy of Science, Pyongyang(2007)
12. Knudsen, E.W.: Elliptic scalar multiplication using point halving. In: Lam, K.-Y., Okamoto, E., Xing, C. (eds.) ASIACRYPT 1999. LNCS, vol. 1716, pp. 135–149. Springer, Heidelberg (1999)
13. Langley, A.: C25519 code (2008). http://code.google.com/p/curve25519-donna/
14. Longa, P., Gebotys, C.: Efficient techniques for high-speed elliptic curve cryptography. In: Mangard, S., Standaert, F.-X. (eds.) CHES 2010. LNCS, vol. 6225, pp. 80–94. Springer, Heidelberg (2010)
15. Moreno, C., Hasan, M.A.: SPA-resistant binary exponentiation with optimal execution time. J. Cryptographic Eng. **1**(2), 87–99 (2011)
16. Mueller, F.: A library implementation of POSIX threads under UNIX. In: USENIX Winter, pp. 29–42 (1993)
17. Nègre, C., Robert, J.-M.: Impact of optimized field operations AB, AC and AB + CD in scalar multiplication over binary elliptic curve. Technical report hal-00724785, HAL, July 2014
18. Gallagher, P., Furlani, C.: Digital Signature Standard (DSS). In: FIPS Publications, vol. FIPS 186-3, p. 93. NIST (2009)
19. Tannenbaum, A.S.: Modern Operating Systems (2009). http://www.freewebs.com/ictft/sisop/Tanenbaum_Chapter2.pdf
20. Taverne, J., Faz-Hernández, A., Aranha, D.F., Rodríguez-Henríquez, F., Hankerson, D., López, J.: Speeding scalar multiplication over binary elliptic curves using the new carry-less multiplication instruction. J. Cryptographic Eng. **1**(3), 187–199 (2011)

A Note on Diem's Proof

Song Tian[1,2,3]($\boxtimes$), Kunpeng Wang[1], Bao Li[1], and Wei Yu[1]($\boxtimes$)

[1] State Key Laboratory of Information Security,
Institute of Information Engineering,
Chinese Academy of Sciences, Beijing, China
{szts1987,yuwei_1_yw}@163.com
[2] Data Assurance and Communication Security Research Center,
Chinese Academy of Sciences, Beijing, China
[3] University of Chinese Academy of Sciences, Beijing, China

Abstract. Semaev's summation polynomials are suggested to be used to construct Index Calculus for elliptic curves over extension fields. The complexity of the proposed algorithm was first studied by Diem with the help of Weil restriction and intersection theory. The key ingredient is to analyse the probability that a uniformly distributed point has an isolated decomposition over the factor base. Following his tactics, we present his result in a simple manner by recourse to the theory of function fields and generic resultant.

Keywords: Summation polynomial · Elliptic curve · Discrete logarithm problem · Index Calculus

1 Introduction

Let G be a finite group and g an element of G. Given $h \in \langle g \rangle$, the smallest nonnegative integer r with $h = g^r$ is called the discrete logarithm of h with respect to the base g.

It is well known that Index Calculus yields algorithms to compute discrete logarithms in the multiplicative groups of finite prime fields and the Jacobians of hyperelliptic curves over finite fields which are faster than generic algorithms. This method can be divided into two stages: construct a linear system and solve this system. To obtain a linear system, we need first define a factor base which is a set of some special elements of the group, then check whether a randomly chosen element can be written as a sum of elements of the factor base, and if such a representation exists, it is added to the linear system. This is repeated until enough equations are collected. In 2004, Semaev [5] studied the possibility of a similar method for elliptic curves over finite prime fields. In his work, Semaev did not give an efficient algorithm because the probability of decomposing a randomly chosen element into a combination of elements of the factor base is

---

This work is supported in part by National Research Foundation of China under Grant No. 61272040, 61379137, and in part by National Basic Research Program of China (973) under Grant No. 2013CB338001.

© Springer International Publishing Switzerland 2015
D. Lin et al. (Eds.): Inscrypt 2014, LNCS 8957, pp. 463–471, 2015.
DOI: 10.1007/978-3-319-16745-9_25

464   S. Tian et al.

not good. However, summation polynomials he introduced in [5] become a useful tool for solving the discrete logarithm problem in elliptic curves over extension fields. With summation polynomials, Gaudry [3] and Diem [2] independently proposed more efficient algorithms. Gaudry's results depend crucially on some un-proven assumptions about the probability of success for a given point to be decomposed as a sum of elements in factor base. In [2], Diem proved these assumptions and gave an algorithm for solving the discrete logarithm problem in elliptic curves with proven complexity. The proofs of his results, which are hard to read, are mainly of theoretical interest.

The principal aim of this note is to present Diem's proof in the technical paper [2] in a simpler way. For simplicity, elliptic curves are defined throughout over finite fields of odd characteristic. The factor base defined in this note is essentially the same as that in [2], which is closely connected to some curve on the Weil restriction of the elliptic curve. The lower bound on the cardinality of the factor base is obtained by function field approach instead of properties of Weil restriction functor. The core of [2] is to give a lower bound on the probability that a uniformly distributed point has an isolated decomposition. This can be done by exploiting the trick of generic resultant twice. It is the second use of generic resultant that replaces all computations in Chow ring in [2].

This note is organized as follows. In Sect. 2, we give the definition of the factor base of Diem's attack and discuss its cardinality. In Sect. 3, we study the probability of obtaining a linear equation from summation polynomial.

2 Factor Base

In this section, we give another form of the factor base in [2] and make an estimate of its size via some facts from algebraic function field theory. For reader's convenience, these facts are listed in appendix.

Let $k = \mathbb{F}_q$ be a finite field of odd characteristic, K/k a field extension of odd degree n and σ the Frobenius automorphism of K/k. Let E/K be an elliptic curve given by Weierstrass equation

$$y^2 = f(x), f(x) \in K[x].$$

Assume that E/K satisfies the following condition(the Condition 2.7 in [2]): there exists a point $T \in \mathbb{P}^1(\bar{k})$ such that $T, \sigma(T), \cdots, \sigma^{n-1}(T)$ are all distinct and T is the only branch point of $\pi = x|_E : E \to \mathbb{P}^1$ in set $\{T, \sigma(T), \cdots, \sigma^{n-1}(T)\}$. By following lemma, we can choose $\lambda \in K$ uniformly at random until the elliptic curve $y^2 = f(x + \lambda)$ satisfies this condition.

Lemma 1 ([2, Lemma 2.10]). *Let $x_1, x_2, x_3 \in \bar{k}$. If λ follows the uniform distribution on K, then the probability that λ satisfies*

$$(x_1 - \lambda)^{q^i} \notin \{x_1 - \lambda, x_2 - \lambda, x_3 - \lambda\}(\forall i : 1 \leq i \leq n - 1)$$

is greater than $\frac{1}{3}$.

Proof. For $t = 1, 2, 3$, we see that $(x_1 - \lambda)^{q^i} = x_t - \lambda$ if and only if $\lambda^{q^i} - \lambda = x_1^{q^i} - x_t$. Since the map $\rho_i : K \to K, \lambda \mapsto \lambda^{q^i} - \lambda$ is k-linear, we have $\#\rho_i^{-1}(x_1^{q^i} - x_t) = \#\ker\rho_i \leq q^{\gcd(i,n)}$. Hence the number of the λ's that do not satisfy the condition is at most $3\sum_{i=1}^{n-1} q^{\gcd(i,n)} (\leq 3(n-1)q^{\frac{n}{2}})$. For $(q, n) = (3, 3), (3, 5)$, the probability in question is $\geq 1 - \frac{3\sum_{i=1}^{n-1} q^{\gcd(i,n)}}{q^n} > \frac{1}{3}$. For $(q, n) \neq (3, 3), (3, 5)$, the probability is

$$\geq 1 - \frac{3(n-1)}{q^{\frac{n}{2}}} \geq \min\{1 - \frac{3(3-1)}{5^{3/2}}, 1 - \frac{3(7-1)}{3^{7/2}}\} > \frac{1}{3}.$$

$\square$

Under above condition, the factor base of Diem's attack becomes

$$\mathfrak{B} = \{(x, y) \in E(K) : x \in k\} \cup \{\infty\}.$$

We need fix some notations before giving a lower bound of the size of $\mathfrak{B}$. Let $\bar{K}$ be an algebraic closure of K, $K(E)$ the function field of E. Fix a separable closure $K(x)^{sep}$ of $K(x)$ which contains $K(E)$ and $\bar{K}(x)$. Let $\sigma(x) = x$. Then the Galois group of $K(x)/k(x)$ is generated by σ, which can extend to an automorphism $\tilde{\sigma}$ of $K(x)^{sep}$. Let F' be the composition of fields $K(E), \tilde{\sigma}(K(E)), \cdots, \tilde{\sigma}^{n-1}(K(E))$. Then $[F' : K(x)] = [\bar{K}F' : \bar{K}(x)] = 2^n$, and the genus $g(F')$ of F' is at most $2^n(n-1) + 1$ by Abhyankar's Lemma and Hurwitz genus formula.

Choose $\tilde{\sigma}$ of order n(see [1] for its existence) and denote by F the fixed field of $\tilde{\sigma}$ in F'. Let y_0 be a root of $y^2 - f(x) \in K(x)[y]$ in $K(E)$. Then for $i = 1, 2, \cdots, n-1$, $y_i = \tilde{\sigma}(y_{i-1})$ is a root of $y^2 - \sigma^i(f) \in K(x)[y]$ in $\tilde{\sigma}^i(K(E))$. Since the order of $\tilde{\sigma}$ is n, $\tilde{\sigma}(y_{n-1}) = y_0$. For $j = (j_0, j_1, \cdots, j_{n-1}) \in \{0, 1\}^n$, let

$$z_j = z_j(y_0, y_1, \cdots, y_{n-1}) = (-1)^{j_0}y_0 + (-1)^{j_1}y_1 + \cdots + (-1)^{j_{n-1}}y_{n-1}.$$

Then the minimal polynomial of $y_0 + y_1 + \cdots + y_{n-1}$ over $K(x)$ is $g(t) = \prod_{j \in \{0,1\}^n}(t - z_j)$. Since $[F : k(x)] = 2^n$ and $y_0 + y_1 + \cdots + y_{n-1} \in F$, it follows that

$$F = k(x, y_0 + y_1 + \cdots + y_{n-1}).$$

If L/K' is a function field with constant field K', by Place$(L/K', 1)$ we mean the set of places of degree 1 of L/K'. Since F'/F is a constant field extension, any $D \in$ Place$(F/k, 1)$ has only one extension D' in F'/K, which is of degree 1 and denoted by $D'|D$. It is obvious that $D' \cap K(E) \in$ Place$(K(E)/K, 1)$, $D' \cap k(x) = D \cap k(x) \in$ Place$(k(x)/k, 1)$. Since the elements of Place$(K(E)/K, 1)$ correspond to the K-rational points of E, we have a map

$$\varphi : \text{Place}(F/k, 1) \to \mathfrak{B}, D \mapsto D' \cap K(E).$$

Proposition 2. *φ is injective.*

Proof. Let $\tau \in$ Gal$(F'/K(x))$ such that $\tau(y_i) = -y_i$ for $i = 0, 1, \cdots, n-1$. Then $\tau(F) = F$.

Let $D \in \mathrm{Place}(F/k, 1)$ and $D' = \mathrm{Con}_{F'/F}(D)$. It is sufficient to show

1. If $\tau(D) = D$, then $\{D_i \in \mathrm{Place}(F/k, 1) : D_i | D \cap k(x)\} = \{D\}$;
2. If $\tau(D) \neq D$, then $\{D_i \in \mathrm{Place}(F/k, 1) : D_i | D \cap k(x)\} = \{D, \tau(D)\}$ and $\varphi(D) \neq \varphi(\tau(D))$.

If $\tau(D) = D$, then $\tau(D') = D'$. Since $\#\{g \in \mathrm{Gal}(F'/K(x)) : g(D') = D'\} = e(D'|D' \cap K(x)) \leq 2$, we have $\{g \in \mathrm{Gal}(F'/K(x)) : g(D') = D'\} = \{\mathrm{Id}, \tau\}$. If there was a $D_0 \in \{D_i \in \mathrm{Place}(F/k, 1) : D_i | D \cap k(x)\}$ other than D, then there would be a $g \in \mathrm{Gal}(F'/K(x))$ such that $g(D') = D_0' = \mathrm{Con}_{F'/F}(D_0)$. For such a g, we would have $g^{-1}\tilde{\sigma}g\tilde{\sigma}^{-1}(D') = g^{-1}\tilde{\sigma}g(D') = g^{-1}\tilde{\sigma}(D_0') = g^{-1}(D_0') = D'$ and $g^{-1}\tilde{\sigma}g\tilde{\sigma}^{-1}|_{K(x)} = \mathrm{Id}$, so $g^{-1}\tilde{\sigma}g\tilde{\sigma}^{-1} = \mathrm{Id}$ or τ. Now let $g(y_i) = \epsilon_i y_i, \epsilon_i \in \{1, -1\}(0 \leq i \leq n - 1)$. If $g^{-1}\tilde{\sigma}g\tilde{\sigma}^{-1} = \mathrm{Id}$, then it would follow from $\tilde{\sigma}g(y_i) = g\tilde{\sigma}(y_i)$ that g is Id or τ, and we have a contradiction. If $g^{-1}\tilde{\sigma}g\tilde{\sigma}^{-1} = \tau$, then we would have from $\tilde{\sigma}g(y_i) = g\tau\tilde{\sigma}(y_i)$ that $\epsilon_i = -\epsilon_{i+1 \mod n}$. Again we have a contradiction.

Now assume that $\tau(D) \neq D$. Since $\tau \in \mathrm{Aut}(F/k(x))$, we have $\tau(D)|D \cap k(x)$ and $\deg(\tau(D)) = f(\tau(D)|D \cap k(x)) = f(D|D \cap k(x)) = 1$. Hence, $\mathrm{Con}_{F'/F}(\tau(D)) = \tau(D') \neq D'$.

We claim that $D' \cap \tilde{\sigma}^i(K(E)) \neq \tau(D') \cap \tilde{\sigma}^i(K(E))$ for $i = 0, 1, \cdots, n - 1$. If it did not hold for some i, then there would be a $g \in \mathrm{Gal}(F'/\tilde{\sigma}^i(K(E)))$ such that $g(\tau(D')) = D'$. A similar argument as above shows that $g\tilde{\sigma}g^{-1}\tilde{\sigma}^{-1}$ is different from Id and $g\tau$. Therefore we would have $\{\mathrm{Id}, g\tau, g\tilde{\sigma}g^{-1}\tilde{\sigma}^{-1}\} \subset \{h \in \mathrm{Gal}(F'/K(x)) : h(D') = D'\}$, which is contrary to the fact that $\#\{h \in \mathrm{Gal}(F'/K(x)) : h(D') = D'\} = e(D'|D' \cap K(x)) \leq 2$.

The above claim says that $D' \cap K(x)$ has two extensions in $\tilde{\sigma}^i(K(E))$ for each i, which implies that $e(D'|D' \cap k(x)) = 1$. So $e(D|D' \cap k(x)) = 1$. By the Theorem 3.8.3 of [7], the fixed field of $\{h \in \mathrm{Gal}(F'/k(x)) : h(D') = D'\}$ in F' is F, hence we have $\{h \in \mathrm{Gal}(F'/k(x)) : h(D') = D'\} = <\tilde{\sigma}>$.

If there was a $D_0 \in P(F/k, 1)$ other than $D, \tau(D)$ such that $D_0|D \cap k(x)$, then we would have a $g \in \mathrm{Gal}(F'/K(x))$ such that $g(D') = \mathrm{Con}_{F'/F}(D_0)$. Hence, $g\tilde{\sigma}g^{-1}(D') = D'$. Note that $g\tilde{\sigma}g^{-1}|_{K(x)} = \sigma$, so we have $g\tilde{\sigma}g^{-1} = \tilde{\sigma}$, and this implies that $g = \mathrm{Id}$ or τ, which is impossible.

Proposition 3. *Let* $\log_2 q \geq 5n$. *Then* $\#\mathfrak{B} \geq \frac{q+1}{2}$.

Proof. It follows from the above proposition that

$$\begin{aligned} \#\mathfrak{B} &\geq \#\mathrm{Place}(F/k, 1) \\ &\geq q + 1 - 2g(F) \cdot q^{\frac{1}{2}} \\ &\geq q + 1 - n \cdot 2^{n+2} \cdot q^{\frac{1}{2}}. \end{aligned}$$

Since $n \cdot 2^{n+2} \cdot q^{\frac{1}{2}} \leq 2^{\frac{n}{2}} \cdot 2^{n+2} \cdot q^{\frac{1}{2}} = 2^{\frac{3n+4}{2}} q^{\frac{1}{2}}$, we have $\#\mathfrak{B} \geq \frac{q+1}{2}$ provided that $\log_2 q \geq 5n$.

3   Isolated Decomposition

The method to generate linear equations is to solve systems of multivariate polynomial equations over $\mathbb{F}_q$ which are constructed from summation polynomials by restriction of scalars. If the algebraic set defined by the associated system is not finite, then the algorithm used to find a point of the algebraic set may fail. Hence the next step is to find out what can guarantee that the algebraic set is a finite set, or in other words, of dimension 0. To this end, we need the Bezout's theorem in multiprojective space $(\mathbb{P}^1)^m$ and the theory of generic resultant for multihomogeneous polynomials.

3.1   The Bezout's Theorem and the Generic Resultant

Let us recall that the closed algebraic subvariety of $(\mathbb{P}^1)^m$ is given by

$$Z(\{H_j(X_1 : Y_1; \cdots ; X_m : Y_m) : 1 \leq j \leq t\}),$$

where the polynomial H_j is homogeneous in each of the m sets $\{X_1, Y_1\}$, $\cdots$, $\{X_m, Y_m\}$ of variables. Then the Bezout's theorem in $(\mathbb{P}^1)^m$ (cf. [6, Chap.IV]) can be stated as follows.

Proposition 4. *Let* $Z(H_1, \cdots, H_m)$ *be a closed algebraic subvariety of* $(\mathbb{P}^1)^m$, *and* $\deg H_j = (d_{j,1}, \cdots, d_{j,m})$ *for* $j = 1, 2, \cdots, m$. *Then the intersection number of effective divisors* $H_1, \cdots, H_m$ *is*

$$H_1 \cdots H_m = \sum_{\tau \in \mathfrak{S}_m} \prod_{j=1}^{m} d_{j,\tau(j)},$$

where $\mathfrak{S}_m$ *is the symmetric group of degree* m.

Let $A = (a_1, \cdots, a_r) \in \mathbb{N}_+^r$. We set

$$M_A = \{X_1^{u_1} Y_1^{v_1} X_2^{u_2} Y_2^{v_2} \cdots X_r^{u_r} Y_r^{v_r} : u_i, v_i \in \mathbb{N}, u_i + v_i = a_i, \forall i : 1 \leq i \leq r\}.$$

For $i = 1, 2, \cdots, r + 1$, fix some $A_i = (a_{i,1}, \cdots, a_{i,r}) \in \mathbb{N}_+^r$ and let $C_{i,w}$ be an indeterminate for each $w \in M_{A_i}$. Then we have the ith generic multihomogeneous polynomial

$$G_i = \sum_{w \in M_{A_i}} C_{i,w} w.$$

It is of multidegree A_i over the polynomial ring $k[\{C_{i,w} : w \in M_{A_i}, i = 1, \cdots, r + 1\}]$. The properties of generic resultant we need are given in the following proposition(cf. [2, Proposition 4.8]).

Proposition 5

1. *There is an irreducible polynomial* $\mathrm{Res} \in k[\{C_{i,w} : w \in M_{A_i}, i = 1, \cdots, r+1\}]$ *with the following property: For all field extensions k'/k and homogeneous polynomials $H_1, \cdots, H_{r+1} \in k'[X_1, Y_1, \cdots, X_r, Y_r]$ of degrees $A_1, \cdots, A_{r+1}$, the zero set of $\{H_1, \cdots, H_{r+1}\}$ is not empty if and only if $\mathrm{Res}(H_1, \cdots, H_{r+1}) = 0$, where $\mathrm{Res}(H_1, \cdots, H_{r+1})$ is obtained by substituting the coefficients of $H_1, \cdots, H_{r+1}$ for the generic coefficients.*

2. *For each $i = 1, \cdots, r+1$, Res is homogeneous in the coefficients of G_i and of degree $(\mathrm{Perm}(B_1), \cdots, \mathrm{Perm}(B_{r+1}))$. Here $\mathrm{Perm}(B_i)$ is the permanent of the matrix $B_i = (a_{l,j})_{1 \le l, j \le r, l \ne i}$.*

3.2   Probability of Isolated Decomposition

We adopt Diem's definition of summation polynomial. Dehomogenization of the polynomial with respect to $Y_1, \cdots, Y_m$ is Semaev's summation polynomial. As indicated in [2], these polynomials can be calculated by interpolation.

Definition 6. *For each integer $m \ge 2$, the mth summation polynomial of E/K is an irreducible multihomogeneous polynomial $S_m(X_1 : Y_1; X_2 : Y_2; \cdots; X_m : Y_m) \in K[X_1, Y_1, X_2, Y_2, \cdots, X_m, Y_m]$, which has the property that for any $P_1, \cdots, P_m \in E(\bar{K})$, $S_m(\pi(P_1), \pi(P_2), \cdots, \pi(P_m)) = 0$ if and only if there exist $\epsilon_1, \cdots, \epsilon_m \in \{\pm 1\}$ such that $\epsilon_1 P_1 + \cdots + \epsilon_m P_m = \infty$. The multidegree of S_m is $\deg S_m = (2^{m-2}, 2^{m-2}, \cdots, 2^{m-2})$.*

Let $P \in E(K)$. Let $\{b_1, \cdots, b_n\}$ be a k-base of K. Then there are $S_{n+1}^{(1)}, \cdots, S_{n+1}^{(n)} \in k[X_1, Y_1, \cdots, X_n, Y_n]$ such that

$$S_{n+1}(X_1 : Y_1; \cdots; X_n : Y_n; \pi(P)) = \sum_{i=1}^{n} S_{n+1}^{(i)} b_i.$$

If the algebraic set $Z(S_{n+1}^{(1)}, \cdots, S_{n+1}^{(n)}) \subseteq (\mathbb{P}^1)^n$ is zero-dimensional, then Rojas's algorithm([4]) is suggested by Diem to find all k-rational points.

Definition 7. *An n-tuple $(P_1, \cdots, P_n) \in E(K)^n$ is called a decomposition of P if $P_1 + \cdots + P_n = P$ and $\pi(P_i) \in \mathbb{P}^1(k)$ for $i = 1, \cdots, n$. A decomposition is called isolated if $(\pi(P_1), \cdots, \pi(P_n))$ is an isolated point of the associated algebraic set $Z(S_{n+1}^{(1)}, \cdots, S_{n+1}^{(n)})$ in $(\mathbb{P}^1)^n$.*

Let $Q_1, \cdots, Q_n \in \mathbb{P}^1(k)$. If $S_{n+1}(Q_1, \cdots, Q_n, \pi(P)) = 0$, then for $i = 0, 1, \cdots, n-1$, we have $\sigma^i(S_{n+1})(Q_1, \cdots, Q_n, \sigma^i(\pi(P))) = 0$. Therefore the point $(Q_1, \cdots, Q_n, \pi(P), \sigma(\pi(P)), \cdots, \sigma^{n-1}(\pi(P)))$ is on the algebraic set

$$V = Z(H_1, \cdots, H_n) \subset (\mathbb{P}^1)^n \times (\mathbb{P}^1)^n,$$

where

$$H_j = \sigma^{j-1}(S_{n+1}(X_{1,1} : Y_{1,1}; \cdots; X_{1,n} : Y_{1,n}; X_{2,j} : Y_{2,j}))$$

is obtained from $S_{n+1}(X_{1,1} : Y_{1,1}; \cdots ; X_{1,n} : Y_{1,n}; X_{2,j} : Y_{2,j})$ by applying the σ^{j-1} to the coefficients for $1 \leq j \leq n$.

Let Pr_1, Pr_2, $Pr_{1,i}$ be the maps defined by

$$Pr_1 : V \longrightarrow (\mathbb{P}^1)^n$$
$$(X_{1,1} : Y_{1,1}; \cdots ; X_{2,n} : Y_{2,n}) \longmapsto (X_{1,1} : Y_{1,1}; \cdots ; X_{1,n} : Y_{1,n}),$$
$$Pr_2 : V \longrightarrow (\mathbb{P}^1)^n$$
$$(X_{1,1} : Y_{1,1}; \cdots ; X_{2,n} : Y_{2,n}) \longmapsto (X_{2,1} : Y_{2,1}; \cdots ; X_{2,n} : Y_{2,n}),$$
$$Pr_{1,i} : (\mathbb{P}^1)^n \longrightarrow \mathbb{P}^1$$
$$(X_{1,1} : Y_{1,1}; \cdots ; X_{1,n} : Y_{1,n}) \longmapsto (X_{1,i} : Y_{1,i}).$$

Denote by V_Q the preimage of point $Q = (Q_1, \cdots, Q_n)$ under Pr_2. Then for the point $Q = (\pi(P), \sigma(\pi(P)), \cdots, \sigma^{n-1}(\pi(P)))$, we have

$$V_Q \cong Z(\{H_j(X_{1,1} : Y_{1,1}; \cdots ; X_{1,n} : Y_{1,n}; \sigma^{j-1}(\pi(P))) : j = 1, \cdots, n\})$$
$$= Z(\{S_{n+1}^{(j)}(X_{1,1} : Y_{1,1}; \cdots ; X_{1,n} : Y_{1,n}) : j = 1, \cdots, n\}),$$

where the isomorphism is given by the restriction of Pr_1. From Proposition 4, we see that $\#V_Q \leq n! \cdot 2^{n(n-1)}$ if $\dim V_Q = 0$.

The next lemma, due to Diem, deals with the problem for which Q the dimension of V_Q is zero. We include his proof here.

Lemma 8. *There exists a multihomogeneous polynomial $C \in K[X_{2,1}, Y_{2,1}, \cdots, X_{2,n}, Y_{2,n}]$ of degree $(n! \cdot 2^{n(n-1)}, \cdots, n! \cdot 2^{n(n-1)})$ such that V_Q is not zero-dimensional only if Q is a point of $Z(C)$ in $(\mathbb{P}^1)^n$.*

Proof. For each $i = 1, \cdots, n$, regarding $H_1, \cdots, H_n$ as polynomials in $K[X_{1,i}, Y_{1,i}, X_{2,1}, Y_{2,1}, \cdots, X_{2,n}, Y_{2,n}][X_{1,1}, Y_{1,1}, \cdots, X_{1,i-1}, Y_{1,i-1}, X_{1,i+1}, Y_{1,i+1}, \cdots, X_{1,n}, Y_{1,n}]$, we have a resultant $R_i \in K[X_{1,i}, Y_{1,i}, X_{2,1}, Y_{2,1}, \cdots, X_{2,n}, Y_{2,n}]$. Since $\dim V = \dim(\mathbb{P}^1)^n = n$, R_i is not trivial and of degree $(n! \cdot 2^{n(n-1)}, (n-1)! \cdot 2^{n(n-1)}, \cdots, (n-1)! \cdot 2^{n(n-1)})$ by Proposition 5. Note that V_Q is non-zero dimensional if and only if $Pr_{1,i}(V_Q) = \mathbb{P}^1$ for some i, which means that $R_i(X_{1,i} : Y_{1,i}; Q) \equiv 0$. Fix some non-zero coefficient $C_i \in K[X_{2,1}, Y_{2,1}, \cdots, X_{2,n}, Y_{2,n}]$ of $R_i \in K[X_{2,1}, Y_{2,1}, \cdots, X_{2,n}, Y_{2,n}][X_{1,i}, Y_{1,i}]$ for each $i = 1, \cdots, n$, then the product of polynomials $C_1, \cdots, C_n$ gives a polynomial C in the statement.

We can use the trick of resultant a second time. Then the computations in Chow ring in [2] can be avoided.

Lemma 9. *Let $G \in K[X_{1,1}, Y_{1,1}, \cdots, X_{1,n}, Y_{1,n}]$ be the resultant of $H_1, \cdots, H_n$, C with respect to the variables $X_{2,1}, Y_{2,1}, \cdots, X_{2,n}, Y_{2,n}$. Then G has degree $(n \cdot n! \cdot 2^{2n(n-1)}, \cdots, n \cdot n! \cdot 2^{2n(n-1)})$.*

Let $H = G\sigma(G) \cdots \sigma^{n-1}(G) \in k[X_{1,1}, Y_{1,1}, \cdots, X_{1,n}, Y_{1,n}]$. Then the number of k-rational points of H is at most $n^3 \cdot n! \cdot 2^{2n(n-1)} \cdot (q+1)^{n-1}$, which can be proved by induction on n and the fact that a univariate polynomial of degree d over $\mathbb{F}_q$ has at most d roots in $\mathbb{F}_q$(cf. [2, Lemma 4.27]).

For point $(P_1, \cdots, P_n) \in \mathfrak{B}^n$, if $H(\pi(P_1), \cdots, \pi(P_n)) \neq 0$, then it follows from Proposition 5 that point $(\pi(P_1), \cdots, \pi(P_n), \sigma^0(\pi(-\sum_{i=1}^n P_i)), \cdots, \sigma^{n-1}(\pi(-\sum_{i=1}^n P_i)))$ is not a zero of C, which implies that V_Q is zero-dimensional for point $Q = (\sigma^0(\pi(-\sum_{i=1}^n P_i)), \cdots, \sigma^{n-1}(\pi(-\sum_{i=1}^n P_i)))$.

From above discussion, all points in the set $\{P \in E(K) : (\pi(P), \sigma(\pi(P)), \cdots, \sigma^{n-1}(\pi(P))) \in Pr_2(\alpha(\mathfrak{B}^n) - \alpha(\mathfrak{B}^n) \cap H(k))\}$ have isolated decompositions, where $\alpha(\mathfrak{B}^n)$ is the set formed by points $(\pi(P_1), \cdots, \pi(P_n), \sigma^0(\pi(-\sum_{i=1}^n P_i)), \cdots, \sigma^{n-1}(\pi(-\sum_{i=1}^n P_i)))$ with $(P_1, \cdots, P_n) \in \mathfrak{B}^n$, and H is regarded as a polynomial in $k[X_{1,1}, Y_{1,1}, \cdots, X_{1,n}, Y_{1,n}, X_{2,1}, Y_{2,1}, \cdots, X_{2,n}, Y_{2,n}]$. To get a lower bound on the number of these points, we have

$$\#(\alpha(\mathfrak{B}^n) - \alpha(\mathfrak{B}^n) \cap H(k)) \geq \frac{1}{2^n} \#\mathfrak{B}^n - \#\alpha(\mathfrak{B}^n) \cap H(k)$$

$$\geq \frac{1}{2^n} \cdot (\frac{q+1}{2})^n - n^3 \cdot n! \cdot 2^{2n(n-1)} \cdot (q+1)^{n-1}$$

$$(\text{if } \log_2 q \geq 5n),$$

hence

$$\#Pr_2(\alpha(\mathfrak{B}^n) - \alpha(\mathfrak{B}^n) \cap H(k)) \geq \frac{1}{n! \cdot 2^{n(n-1)}} \#(\alpha(\mathfrak{B}^n) - \alpha(\mathfrak{B}^n) \cap H(k))$$

$$\geq \frac{(q+1)^{n-1}}{n! \cdot 2^{n(n+1)}} \cdot (q+1 - n! \cdot n^3 \cdot 2^{2n^2}).$$

Let $\epsilon > 0$. Since $n! \cdot n^3 \cdot 2^{2n^2} < 2^{(2+\epsilon)n^2-1}$ holds for sufficiently large n, it follows that $\frac{(q+1)^{n-1}}{n! \cdot 2^{n(n+1)}} \cdot (q+1 - n! \cdot n^3 \cdot 2^{2n^2}) \geq \frac{q^n}{n! \cdot 2^{n^2+n+1}} \geq 2q^{n-\frac{1}{2}}$, provided that $\log_2 q \geq (2+\epsilon)n^2$ and n is large enough. This leads to the following result.

Proposition 10. *Let $\epsilon > 0$. For n large enough and $\log_2 q \geq (2+\epsilon)n^2$, if P follows the uniform distribution on $E(K)$, then the probability that P has an isolated decomposition is not less than $q^{-\frac{1}{2}}$.*

Appendix: Galois Extensions of Function Fields

Here, we give the notations and facts from algebraic function theory that are used in Sect. 2.

Definition 11. *Let L_2/K_2 be an algebraic extension of function field L_1/K_1, and let D_1, D_2 are places of L_1/K_1, L_2/K_2 respectively. If $D_1 \subseteq D_2$, then D_2 is called an extension of D_1 and denoted by $D_2|D_1$.*

If $D_2|D_1$, then $D_1 = D_2 \cap F_1$. Any place D_1 of L_1/K_1 has only finitely many extensions in L_2. If L_2/L_1 is a Galois extension and $D_{2,1}, D_{2,2}, \cdots, D_{2,r}$ are all extensions of D_1 in L_2, then the Galois group $\mathrm{Gal}(L_2/L_1)$ acts transitively on the set $\{D_{2,1}, D_{2,2}, \cdots, D_{2,r}\}$ and the ramification indices $e(D_{2,i}|D_1)$ of $D_{2,i}$ over D_1 are equal.

Definition 12. *Let L_2/K_2 be an algebraic extension of function field L_1/K_1, D_1 a place of L_1/K_1. We define*

$$Con_{L_2/L_1}(D_1) = \sum_{D_2|D_1} e(D_2|D_1)D_2.$$

If L_2/L_1 is a constant field extension, then L_2/L_1 is unramified and the degree of divisor is invariant under Con_{L_2/L_1}.

Lemma 13 (Abhyankar's Lemma). *Let L'/L be a finite separable extension of function fields. Let $L' = L_1 L_2$ be the composition of intermediate fields $L \subset L_1, L_2 \subset L'$, and let D be a place of L with extensions D_1, D_2, D' in L_1, L_2, L'. Assume that there exists i such that $\gcd(e(D_i|D), \text{Char } L) = 1$. Then*

$$e(D'|D) = \text{lcm}\{e(D_1|D), e(D_2|D)\}.$$

Lemma 14 *([7, Theorem3.8.3]). Let L'/L be a Galois extension of function fields, D a place of L with an extension D' in L'. Let M be an intermediate field of L'/L, $D_M = D' \cap M$. Denote by $Z(D'|D)$ the fixed field of $G_Z(D'|D) = \{g \in \text{Gal}(L'/L) : g(D') = D'\}$ in L'. Then*

(1) The order of $G_Z(D'|D)$ is $e(D'|D)f(D'|D)$.
(2) $M \subseteq Z(D'|D)$ if and only if $e(D_M|D) = f(D_M|D) = 1$.
(3) $M \supseteq Z(D'|D)$ if and only if D' is the only extension of D_M in L'.

References

1. Diem, C.: The GHS attack in odd characteristic. J. Ramanujan Math. Soc. **18**(1), 1–32 (2003)
2. Diem, C.: On the discrete logarithm problem in elliptic curves. Compos. Math. **147**(01), 75–104 (2011)
3. Gaudry, P.: Index calculus for abelian varieties of small dimension and the elliptic curve discrete logarithm problem. J. Symb. Comput. **44**(12), 1690–1702 (2009)
4. Rojas, J.M.: Solving degenerate sparse polynomial systems faster. J. Symb. Comput. **28**(1), 155–186 (1999)
5. Semaev, I.: Summation polynomials and the discrete logarithm problem on elliptic curves, February 2004). http://eprint.iacr.org/2004/031
6. Shafarevich, I.R.: Basic algebraic geometry. 1: Varieties in projective space (Transl. from the Russian by M. Reid.), 2nd, rev. and exp. edn. (1994)
7. Stichtenoth, H.: Algebraic Function Fields and Codes, vol. 254. Springer, New York (2009)

Cryptographic Primitive
and Application

Stand-by Attacks on E-ID Password Authentication

Lucjan Hanzlik$^{(\boxtimes)}$, Przemysław Kubiak, and Mirosław Kutyłowski

Faculty of Fundamental Problems of Technology,
Wrocław University of Technology, Wrocław, Poland
{lucjan.hanzlik,przemyslaw.kubiak,miroslaw.kutylowski}@pwr.edu.pl

Abstract. We show that despite the cryptographic strength of the password authentication, we cannot exclude an attack by an adversary that penetrates the reader device at some moment, but apart from this is passive and manipulates neither the reader nor the microcontroller of the identity document. So even the most careful examination and certification of the smart cards and the readers cannot prevent attacks of this kind. We present concrete attack scenarios for PACE-GM, PACE-IM and SPEKE protocols.

We show that the weaknesses can be easily and effectively eluded via changing a few implementation details on the side of the reader. Our second contribution is that immunity against the attacks can be tested by the operator of the reader, thus replacing costly and unreliable certification process of black box devices.

Our more general contribution is to draw attention on hidden penetration attacks and to show that effective countermeasures are possible.

Keywords: Massive surveillance · Temporary penetration attack · Privacy · Tracing · Wireless communication · Personal identity card · Password authentication · PACE · SPEKE · Verifiability · Certification

1 Introduction

The traditional approach to create secure systems is to design a strong cryptographic scheme, preferably with a formal security proof, and to implement it in accordance to a widely accepted methodology. While the cryptographic schemes are today relatively mature, the weak point of the final product is usually the implementation process. The end product is often regarded as secure, if the scheme is implemented exactly as described at the abstract level. However, the idealized high level abstraction may disregard many crucial issues. Moreover, we make strong assumptions concerning security of the system components and trustworthiness of some actors of the process. Widespread usage of black box

Partially supported by National Science Centre, HARMONIA 4 Programme, DEC-2013/08/M/ST6/00928.

© Springer International Publishing Switzerland 2015
D. Lin et al. (Eds.): Inscrypt 2014, LNCS 8957, pp. 475–495, 2015.
DOI: 10.1007/978-3-319-16745-9_26

solutions makes it even worse – in most cases we blindly believe declarations of the providers or, at best, certificates issued by the third parties.

Recently, it became evident that this approach is quite naïve. The cases such as alleged eavesdropping on Chancellor Merkel's cell phone by US agencies as well as revelations of Edward Snowden indicate that the scope of gathering data in a hidden way may be much larger than openly admitted. At the moment the arguments based on reputation of the system providers (who *"would not dare to play in this way fearing to loose their reputation"*) are not that convincing as before.

Almost all attacks on cryptographic security systems are not breaking cryptographic mechanisms. Instead, the attackers take advantage of ill-designed systems that make room for penetration despite the strength of cryptographic protocols. Unlike design of cryptographic algorithms, integration of a cryptographic system is usually an ad hoc process, where an independent review is quite limited and serves more marketing purposes than assuring the system's properties. In fact, it seems that there is no implementation methodology that would provide guarantees comparable with the security level of cryptographic algorithms. Moreover, cryptographic systems may be designed to guard against most attackers, but at the same time may deliberately enable penetration by certain parties. There are many possibilities to make it work [15], but the problems are downscaled by decision makers. Fortunately, it seems that the cryptographic community starts to be concerned about these critical security issues [4].

In the current situation it is a great challenge to design solutions that are arguably secure not only in an idealized world, but also in the practical sense. Among others, they should guard against manufacturers, system providers and other presumably trustworthy parties. Unfortunately, there are quite a few designs of this kind. Many lessons can be learnt from the state-of the-art of e-voting schemes. It took many years of painful mistakes until the concept of verifiable E2E (end-to-end) systems became widely accepted. Currently, no party participating in the voting process and no hardware or software component should be blindly trusted. Moreover, verification of the claimed security properties are not delegated to third parties.

This paper is devoted to these problems in the area of personal identity documents and illegal tracing personal traffic. We show that a lot of care is necessary during deployment of these systems and that security of the system is not guaranteed by the current standards. We also define threat model that should be used for such systems, especially if they fall into category of Internet of Things.

1.1 Personal Identity Documents and Password Authentication

Personal identity documents are frequently equipped with a cryptographic microcontroller. For the reasons such as chip durability and user's convenience a wireless radio channel is chosen for the communication with the microcontroller. For example, wireless communication is used for biometric passports standardized by the ICAO organization. The price paid is lack of a direct control over the microcontroller: it can be activated by any reader without the consent of the document's owner, who might be even unaware about the interaction. In countries like UK and USA this leads to social resistance against introduction of personal identity documents, despite of rapidly growing scale of identity theft criminality.

Password Authentication. Password authentication should prevent activation of an electronic identity document without the consent of its owner. More specifically, it should guarantee achieving the following goals:

- A session can be established only when the reader presents a proof to the microcontroller showing that the reader knows the owner's password.
- It must be infeasible to derive the password from a transcript of communication between the reader and the microcontroller.
- A reader that is unaware about the password should not be able to learn it more efficiently than by trying the passwords one by one in separate communication sessions with the microcontroller. (Of course, due to limited entropy of the passwords that can be memorized by a user, a malicious reader having an electronic identity document in its range for a long time can perform a brute force attack.)

Deployment of Password Authentication. There is a growing interest on password authentication for electronic identity documents as a replacement for PIN based authentication. A large number of password authentication protocols have been designed by the academic community and the industry. However, the most important step might be an amendment to Commission Decision C(2006) 2909 by the European Commission. According to it a password authentication protocol must be implemented by the EU countries by December 31, 2014 for biometric passports. Furthermore, PACE v2 protocol [11] is to be used.

PACE has been designed by the German federal authority BSI [6] and later adopted by ICAO (International Civil Aviation Organization) as a component of Supplemental Access Control (SAC), which is a de facto new standard for electronic travel documents. Probably, due to availability of software and hardware products, integration with other protocols necessary for electronic identity documents, privacy "by-design" features, and the status of de facto standard, PACE has a great advantage to become dominant on the market. The strong position of PACE can be also attributed to the support of German and French authorities and the industry.

Due to the reasons mentioned above, in this paper we focus our attention on PACE protocols (both Integrated Mapping and General Mapping versions). We also discuss the former protocol SPEKE [12] covered by an US patent. (Note that PACE has not been developed in order to improve some security features of SPEKE, the main target was to avoid the US patent.)

Security of Password Authentication. Just like PACE and SPEKE, many advanced password authentication protocols are integrated with key exchange protocols and known under the name *password-based authenticated key exchange* (PAKE). Defining a relevant security model for them is an uneasy task. The problem is to create a clear model that would capture all relevant attack scenarios and potential threats. In order to be useful, the model must be simple enough to enable a short and clear but still rigorously formal security proof. These two goals are in some sense contradictory and frequently lead to oversimplifications.

So far there is an abundance of security models for authentication protocols. They are based on abstract games describing protocol executions in which the adversary is challenged to return some protected information. The major effort was to formulate a game that would cover all or almost all potential attack scenarios. An example of this approach is the CK-model [8] which allows to create security proofs in a generic way. Unfortunately, it has turned out that many issues require model extensions and models such as eCK [14] and its variants have been introduced.

The CK-model framework has also been adapted to the PAKE protocols. Abdalla et al. [1] introduce a real-or-random game between the adversary and the challenger in which the adversary is allowed, among others, to ask oracles to reveal the secret password. The adversary is not allowed to ask for the internal state of the user (e.g. the used random numbers). However, such attack scenarios seem to be realizable in the real world. Recently, the cryptographic community becomes more aware about these threats. For example, the paper [4] from CRYPTO'2014 shows that we have to take into account malicious implementations that are indistinguishable from the honest ones. We go beyond that: we are talking about indistinguishability of honest and dishonest actors in a system.

Let us mention that some formal proofs concerning the PACE protocol have been published (see e.g. [5]). Interestingly, they appeared long after presenting the protocol and including them in a de facto German standard. Moreover, some features have not been covered by the initial proofs (see e.g. the privacy issues [10]). This might be an evidence that making decisions about cryptographic protocols for the real world applications is much more complicated than the theoretically oriented and abstract process proposed by the academic community.

1.2   Adversary Model

The traditional approach of cryptographers is to assume that some secrets are never stolen by the adversaries: protocol designers implicitly assume that the secrets are stored safely in secure hardware devices and/or are effectively protected by their owners. Failure to fulfill these assumptions is usually treated as a problem outside the scope of a cryptographic design. Consequently, it is expected that the problem will be solved by implementation engineers, legal rules, etc.

There are many reasons why this approach is likely to fail in practice. First, a secure hardware device is a myth. There is a race between the manufacturers designing protection mechanisms and the attackers developing technologies to break them. Both the protection mechanisms and the attack methods are confidential and we may only guess what is the current state-of-the-art, especially when talking about technology available for the most powerful players. Second, many actors in the system can be tempted to create trapdoors or simply use confidential knowledge in order to circumvent security mechanisms. The purpose is not necessarily evil, it can be motivated by the issues of the national security. The cases like installing the trapdoors by Swiss reputable company CRYPTO AG should learn us that these situations are not hypothetical. The recently revealed scope of massive surveillance (also against political allies) confirms that the opportunities created by over-optimistic cryptographic design can be exploited.

Even if knowledge about the possibilities to create cryptographic trapdoors is available to the public for over two decades (see e.g. [15]), only recently it has started to attract enough attention.

Hidden Penetration Attacks. We consider systems consisting of two kinds of components: *black boxes* and *white boxes*. The contents of a white box is visible to its owner, the black box components can only be used in the way described by the protocol. For example, typical black boxes are the components implementing secret keys and PRNGs.

In this paper we consider the following model of an adversary attacking a system Z:

- at some moment of the lifecycle of Z the adversary succeeds to penetrate the resources of Z,
- during the penetration phase the adversary learns the contents of both: the white boxes and black boxes of Z,
- the adversary leaves no trace of the attack in Z. So the adversary cannot change the contents of white boxes. The adversary can change the contents of black boxes provided that no change of their behavior can be later detected by inspection of Z – both by reading white boxes and observing Z during its runtime. (So the program code can be changed only if it is in the black box. Note that in this case the model reduces to the one from [4].)
- during the penetration phase the adversary does not start any irregular activity in Z (like e.g., running own code in a way observable for third parties),
- apart from the data gathered during the penetration phase, the adversary may passively gather all messages exchanged over the public channels (e.g. over a radio channel).

So we consider a hidden adversary – just as for a good surveillance system.

The goal of the adversary is to gain confidential data known to the system after or before the penetration. This includes in particular the passwords entered to a reader during execution of password authentication protocols.

The hidden penetration attack model covers a number of practical situations: e.g. breaking into a system and its "secure components", physical leakage that can be measured only occasionally, or installing a weak or known seed to a PRNG.

In some cases the penetration phase is not really an attack but merely misuse of obligations. For example, a manufacturer or a provider of cryptographic equipment may store copies of the private keys pre-installed in the devices. Theoretically, even the possibility of creating the copies should be excluded. In reality, even the legal rules admit such services as storing copies of secret signing keys of the customers [9, Annex II, point 4]!

Unfortunately, may be with a few exceptions real world systems are not designed to be immune against the adversary defined above. Many "black box" security arguments fail under this new view and many devastating attacks might emerge in practice.

Immunity Against Hidden Penetration Attacks. We feel that providing immunity against penetration attacks cannot be left to the non-cryptographic mechanisms for the sheer reason that there is no idea how to do it effectively. In contrast to the replacement attacks, no audit or certification procedure can help, as long as the adversary leaves really no traces in the system attacked.

In our opinion the right way to deal with this problem is to design cryptographic mechanisms in a way that eliminates hidden penetration attacks or at least makes them difficult in practice.

1.3   Problems with Randomness

Choosing elements at random is one of the standard components of cryptographic protocols. Theoretically, there are many sources of physical randomness that can be used as external sources of true random numbers. However, there are many problems concerning this approach. So far the industrial practice focuses on detection of sources that have stochastic properties far from uniform distribution – see e.g. the suite of NIST randomness tests. However, the test results never guarantee that the generator's output is really unpredictable by an external observer. Last not least, the generator may contain undetectable hardware Trojans [3].

A pseudorandom number generator (PRNG) may be a good solution to these problems. The output of the PRNG might be provably indistinguishable from an ideal random source, while the circuitry of the generator can be checked against its specification. In many cases hybrid solutions are proposed: a truly random source is replaced by an output of a PRNG which involves input from a random source with some number of *entropy bits*. In this way we hope to combine advantages of a good PRNG and of a source of physical randomness despite its limited quality.

The main advantages of PRNG are: a low cost, possibility to implement on low-end devices and verifiability. On the downside, a PRNG offers no security when the adversary learns its internal state. Since a PRNG is deterministic, the adversary can compute all future outputs delivered by the PRNG once its internal state becomes known.

Certification Framework and Password Authentication for Inspection Systems. The Common Criteria Framework can be used to analyze and define security properties necessary in concrete applications. Protection Profiles (PP) list these properties, their role is to provide sufficient conditions for security guarantees against realistic threats. However, if a PP neglects some issues, then a device fulfilling this PP might be insecure despite conformance with the Common Criteria requirements.

An exemplary Protection Profile (PP) [7] of the Inspection System (IS) defines security features of the terminals used for document inspection, that is, the readers interacting with biometric passports. Among others, [7] states the properties of randomness used. According to Sect. 6.1.2.4, *"The Target of Evaluation Security Functionality shall provide a mechanism to generate random numbers that meet the functionality class K4 (...) with at least 64 bit entropy for the seed"*. According to [13, page 8], the class AIS20 K4 is comparable with the class DRG.3, and on page 70 of [13] we see that the generator from the class DRG.3 does not have to contain re-seed/seed update/refresh functionalities. As refers to the entropy, the document [13] defines *entropy source* as *"A component, device or event that generates unpredictable output values which, when captured and processed in some way, yields discrete values (usually, a bit string) containing entropy (Examples: electronic circuits, radioactive decay, RAM data of a PC, API functions, user interactions)."*. In particular, a True Random Number Generator (TRNG) is not required as a necessary entropy source for the PRNG. Moreover, PP [7] does not require it either. Consequently, the source of randomness could be a deterministic PRNG with a random seed with sufficient entropy and uploaded by a trusted party. As this is admitted by the PP, we should not expect that the electronic identification documents are equipped with better solutions: the rules of public procurements effectively eliminate such solutions due to the unit price factor.

1.4 Overview of the Rest of the Paper

In the rest of the paper we discuss in detail the problem of hidden penetration attacks for real world examples of password authentication. Our main goal is to draw attention to the problems neglected so far and to show that slightly reshaping the protocols on the implementation level can significantly improve the situation. Perhaps, this is the last moment to discuss such details, since after large scale deployment it will be hard to revert the design decisions.

In Sect. 2 we present hidden penetration attacks against PACE-GM, PACE-IM and SPEKE. The target of the hidden penetration attack are the readers and not the personal identity documents. Afterwards a passive adversary may derive the passwords used by the documents executing PACE or SPEKE and consequently trace the document owners. Note that the passive adversary might use a tiny electronic bug attached to a (genuine) inspection terminal.[1]

---

[1] The cases of spying PIN numbers with tiny cameras attached to ATM machines should serve as a warning that this in not only a hypothetical situation.

In Sect. 4 we propose a simple but generic countermeasure based on additional exponentiation using a fixed secret. The method works for any Diffie-Hellman-like key agreement and mitigates many hidden penetration attacks against the readers. One of the main advantages is that the new component can be a plug-in device from a different provider and that its behavior can be easily tested (unlike a PRNG). The modified protocol does not intertwine the PRNG secret and the new secret in a way that knowing the first gives the second while observing the traffic.

2   Attacks

2.1   Attack on PACE-GM

Protocol Description. First let us recall a specification of the PACE protocol with the general mapping. We do not attempt to explain the rationale of the protocol and refer the reader to the original papers. Our primary goal is to show the problem and an effective solution (Table 1).

Attack Scenario. For the attack we make the following assumptions:

- the readers consists of white boxes, with the exception of the PRNG, which are implemented as black boxes,
- the attacker performs the hidden penetration attack on all readers produced by a given manufacturer, and get the seeds installed there,
- the attacker can install a passive device recording the (encrypted) communication at the place where authentication protocol is executed between the readers and the identification documents; we call this device a *shadow reader*,
- at some moment the shadow reader may deliver the recorded sessions to the adversary.

We do not assume that the attacker has any control over readers deployment - they are out of control once they are sold and delivered to the customer. On the other hand, installing a shadow reader might be quite realistic. It can be easily left in an automatic booth entered for control by hundreds of people.

Now we describe the offline analysis on the protocol transcript performed by an adversary.

Step 1: Finding ID of the reader:
    The first message sent by the reader is $Y_B = g^{y_B}$, where y_B comes, according to our assumptions, from the PRNG installed in the reader. In particular, it does not depend on the password.
    The offline analysis considers each seed installed in the readers and makes trial computations deriving the exponent y_B and checks against Y_B. For efficiency, these computations can be done once by the adversary. For each result u of the PRNG the adversary stores the seed and the number of steps in a hash table based dictionary, where the address of the entry is based on

Table 1. PACE protocol, GM version

Card	Reader
holds:	holds:
π - password	π - password entered by the document owner
$\mathcal{G}$ - parameters, with g as a generator of a group of prime order q	
$K_\pi := \text{Hash}(0\|\pi)$	$K_\pi := \text{Hash}(0\|\pi)$
choose $s \leftarrow \mathbb{Z}_q$ at random	
$z := \text{ENC}(K_\pi, s)$	
$\xrightarrow{\ \mathcal{G},z\ }$	abort if $\mathcal{G}$ incorrect
	$s := \text{DEC}(K_\pi, z)$
choose $y_A \leftarrow \mathbb{Z}_q^*$ at random	choose $y_B \leftarrow \mathbb{Z}_q^*$ at random
$Y_A := g^{y_A}$	$Y_B := g^{y_B}$
$\xleftarrow{\ Y_B\ }$	
abort if $Y_B \notin \langle g \rangle \setminus \{1\}$ $\xrightarrow{\ Y_A\ }$	abort if $Y_A \notin \langle g \rangle \setminus \{1\}$
$h := Y_B^{y_A}$	$h := Y_A^{y_B}$
$\hat{g} := h \cdot g^s$	$\hat{g} := h \cdot g^s$
choose $y'_A \leftarrow \mathbb{Z}_q^*$ at random	choose $y'_B \leftarrow \mathbb{Z}_q^*$ at random
$Y'_A := \hat{g}^{y'_A}$	$Y'_B := \hat{g}^{y'_B}$
$\xleftarrow{\ Y'_B\ }$	
abort if $Y'_B = Y_B$ $\xrightarrow{\ Y'_A\ }$	abort if $Y'_A = Y_A$
$K := Y'^{\,y'_A}_B$	$K := Y'^{\,y'_B}_A$
$K_{\text{ENC}} := H(1\|K)$	$K_{\text{ENC}} := H(1\|K)$
$K'_{\text{SC}} := H(2\|K)$	$K'_{\text{SC}} := H(2\|K)$
$K_{\text{MAC}} := H(3\|K)$	$K_{\text{MAC}} := H(3\|K)$
$K'_{\text{MAC}} := H(4\|K)$	$K'_{\text{MAC}} := H(4\|K)$
$T_A := \text{MAC}(K'_{\text{MAC}}, (Y'_B, \mathcal{G}))$	$T_B := \text{MAC}(K'_{\text{MAC}}, (Y'_A, \mathcal{G}))$
$\xleftarrow{\ T_B\ }$	
abort if $T_B \neq \text{MAC}(K'_{\text{MAC}}, (Y'_A, \mathcal{G}))$ $\xrightarrow{\ T_A\ }$	abort if $T_A \neq \text{MAC}(K'_{\text{MAC}}, (Y'_B, \mathcal{G}))$

the hash value computed for g^u. The database need not to be particularly large concerning the limited number of devices produced and a relatively low usage rate.

Note that once the ID of the reader is found, the adversary also learns what is the current step number of the PRNG. This enables to derive quickly its next outputs.

Step 2: Finding h:

As $h = Y_A^{y_B}$, it can be immediately derived from y_B found in the Step 1 and from Y_A known from the transcript of the communication between the reader and the ID document.

Step 3: Reconstruction of K:

Since the adversary can find the next output from the PRNG after yielding y_B, the adversary learns y_B'. So in particular the adversary can reconstruct K as $Y_A'^{y_B'}$. Then he follows the steps of the protocol in order to derive the session keys.

Step 4: Reconstruction of $\hat{g}$:

The adversary computes $\hat{g}$ as $(Y_B')^{(y_B')^{-1 \bmod q}}$. Furthermore, he computes $S :=$ $\hat{g}/h$. Note that according to the protocol, we have $S = g^s$.

Step 5: Deriving the password:

For each password π, the adversary computes $s_\pi := \mathrm{DEC}(\mathrm{Hash}(0\|\pi), z)$ and tests whether $S = g^{s_\pi}$. If the equality holds, then with a high probability the password has been guessed correctly.

Complexity of the offline analysis (apart from the precomputation) is quite low. Step 5 is a brute force search, but there is only a limited number of possibilities due to the low entropy of the passwords.

The attack complexity is reduced by the fact that deriving the session key and deriving the password are in some sense separated in PACE. This is an interesting situation, since such a separation makes the protocol stronger from another point of view (inability to test correctness of the password until the very end of the protocol execution).

2.2   Attack on PACE-IM

In the description above, Encoding is a function specified in the standard. It maps the elements to points of an elliptic curve (Table 2).

Attack Scenario. For the attack we make the same assumptions as in Sect. 2.1. The offline analysis performed by the adversary consists of the following steps:

Step 1: Finding ID of the reader:

The first message sent by the reader is β which is the output of the PRNG. In particular, it does not depend on the password and other parameters.

The offline analysis considers each seed installed in the readers and makes trial computations deriving β. As for the previous attack, a precomputation may decrease complexity of this phase.

Note that once the ID of the reader is found, the adversary also learns what is the current step number of the PRNG. Thereby, the next output of the PRNG can be derived.

Step 2: Finding Z:

Once the state of the PRNG is recovered, the adversary knows y and can compute $Z := y \cdot X$.

Table 2. PACE protocol, Integrated Mapping version

Card	Reader
holds:	holds:
π - password, *params* - parameters, with g a generator of a group $\mathcal{G}$ of prime order q	π - password entered by the document owner
choose $s \leftarrow \{0,1\}^{l_1}$ at random $z := \mathrm{ENC}(\pi, s)$	
$\xrightarrow{\ params, z\ }$	abort if $\mathcal{G}$ incorrect
	$s := \mathrm{DEC}(\pi, z)$ choose $\beta \leftarrow \{0,1\}^{l_2}$ at random
$\xleftarrow{\quad \beta \quad}$	
$\hat{G} = \mathrm{Encoding}(\mathrm{Hash}(s, \beta))$ choose $x \leftarrow \mathbb{Z}_q$ at random $X := x \cdot \hat{G}$	$\hat{G} = \mathrm{Encoding}(\mathrm{Hash}(s, \beta))$
$\xrightarrow{\quad X \quad}$	
	choose $y \leftarrow \mathbb{Z}_q$ at random $Y := y \cdot \hat{G}$
$\xleftarrow{\quad Y \quad}$	
$Z = x \cdot Y$ derive the session keys from Z	$Z = y \cdot X$ derive the session keys from Z
send MACs to confirm the keys	
exchange messages in a secure channel	

Step 3: Reconstruction of $\hat{G}$:
  The adversary computes $\hat{G} := y^{-1} \cdot Y$.

Step 4: Deriving the password:
  For each possible password π, the adversary computes $s_\pi := \mathrm{DEC}(\pi, z)$ and tests whether $\hat{G} = \mathrm{Encoding}(\mathrm{Hash}(s_\pi, \beta))$. If the equality holds, then with a high probability the password has been guessed correctly.

As in the previous attack, the complexity is low apart from the Step 1 which depends on the number of readers installed and the time with no supervision. However, now it is even easier to transmit some data in the outputs as β observed by the attacker is directly the output of the generator. In case of PACE-GM it was harder, as no output of the generator is presented directly.

2.3   Attack on SPEKE

Protocol Description. First we recall the design of SPEKE password authentication protocol (Table 3).

Table 3. SPEKE protocol

Card	Reader
holds:	holds:
π - password	π - password entered by the document owner
a safe prime p, $\quad p = 2q + 1$	
$g := (\mathrm{Hash}(\pi))^2 \bmod p$	$g := (\mathrm{Hash}(\pi))^2 \bmod p$
choose $y_A \leftarrow \mathbb{Z}_q^*$ at random	choose $y_B \leftarrow \mathbb{Z}_q^*$ at random
$Y_A := g^{y_A}$	$Y_B := g^{y_B}$
$\xleftarrow{\quad Y_B \quad}$	
abort if $Y_B \notin [2, p-2]$ $\quad\xrightarrow{\quad Y_A \quad}$	abort if $Y_A \notin [2, p-2]$
$K := Y_B^{y_A}$	$K := Y_A^{y_B}$

Attack Scenario. The assumptions concerning the attack scenario are the same as in Sect. 2.1.

Step 1: Finding ID of the reader:
In case of SPEKE finding identity of the reader by a passive eavesdropper is problematic. However, the adversary can easily find it in an active way. The adversary may appear just once with a legitimate ID document, enter the correct password and observe the message Y_B sent by the reader. As the correct password is entered by the adversary, he knows also the parameter g used by the reader. Afterwards, the adversary may analyze the transcript of the communication with his ID document in order to match the parameter Y_B from the protocol transcript with g^u, where u is the potential output of the PRNG generator.

Step 2: Reconstruction of K:
After Step 1, the identity of the reader is known, but not the step number of the reader at the moment when the communication session is recorded. For each step number the adversary computes the output u of the PRNG and computes a candidate for K as Y_A^u. Then the adversary decrypts the messages exchanged between the reader and the ID document using a candidate key K. When the plaintexts obtained make sense, the guess u for y_B might be regarded as correct.

Step 3: Reconstruction of g:
The adversary computes g as $(Y_B)^{(y_B)^{-1}} \bmod q$.

Step 4: Deriving the password:
For each password π, the adversary checks if $g = (\mathrm{Hash}(\pi))^2 \bmod p$. Thanks to the pseudorandomness of the hash function Hash, the correct password is likely to be the only one password fulfilling this equality.

2.4   Tracing Threats

Once the adversary learns the password of an ID document, he may deploy the readers that activate the ID documents in their range, trying the known passwords. Any successful attempt (i.e. an attempt with the correct password) results in detection of an ID document. Note that a failed attempt is not recorded by the ID document as they are passive and do not create transaction logs. A shadow reader can be hidden, for example, in a hotel elevator used by the guests.

Thereby we see that a mechanism used for privacy protection turns out to be a simple and effective tool for tracing the holders of ID documents - violating the main principle of privacy protection. Of course, the attack can be prevented by electromagnetic shielding the identity document. In case of biometric passports the cover pages may be used for this purpose (as in the US biometric passports), however this makes automatic border control slower (preferably, the traveller should lay down the passport unopened on the reader). For ID-1 ISO/IEC 7810 identification cards a protective case would be necessary – which is somewhat inconvenient for the document holders.

3   Trusted Exponentiation Strategy

Our goal is an implementation strategy for building readers so that we get a system with the following features:

1. the readers can still communicate with any document executing the original protocol in exactly the same way,
2. the attacks described in Sect. 2 do not work anymore,
3. a user can effectively check whether the reader is following the particular implementation strategy preventing the attacks from Sect. 2.

The proposed changes are on the low level of the protocol design and simple enough to be practical. Their main achievement is that there is no need to trust (unconditionally) the hardware manufacturers.

3.1   Naive Protection Strategies

In this section we present simple solutions that protect against the attacks presented in Sect. 2.

To start with, one could argue that we can reset the seed using new entropy. Our PRNG would, on demand, compute a new seed using the bits outputted by a true random number generator (TRNG). In the literature such a generator is called *hybrid random number generator*. However, we have no assurance that the generator really uses the new random bits. Instead, it may compute a deterministic function (unknown to an adversary) on the old seed to produce a new one. There is no way to test that this is not the case whenever the device follows the black box design principle.

The second solution is that one could enable a way to seed the generator post-production so that the manufacturer will not know the seed. However, if the generator uses some bits of a TRNG as input, then we cannot distinguish if our seed was really set. On the other hand, if the generator is a pure PRNG, then we cannot be certain that at some point someone will re-seed our PRNG (since, for security reasons we do not retain the seed, we cannot verify that).

The simplest solution is just to use a TRNG as our generator. Beside the fact that it is hard to distinguish if the output comes from such a generator or a PRNG, TRNG degrade over time and it is also hard to prove that under all conditions the output of the TRNG is still unpredictable.

In all cases the proposed solutions are simple, but as shown they do not cover point 3 of the above list of requirements. Namely, the participants of the protocol have no assurance that the protection was indeed used.

3.2   Deriving Random Exponents

In almost all cases the protocols presented above have used the random elements as exponents in the following steps of the Diffie-Hellman key exchange:

step 1. choose $r < q$ at random,
step 2. compute $U := V^r$,
step 3. present U to the other party involved in the protocol.

Furthermore, the number r will never be used again except for deriving the shared key via exponentiation:

step ... given an element W obtained from the other party compute $L := W^r$
after step ... use L as a shared value with the other party of the protocol.

We propose an architecture composed of two components: a PRNG (or a true RNG) and an Exponentiation Unit (ExpU) holding a private key. Each ExpU holds a secret key x installed separately. The idea is that instead of using the exponent r from the PRNG unit, the protocol uses $r \cdot x \bmod q$, where q is the prime order of the group used for the computations. Note that if r has uniform probability distribution in $\mathbb{Z}_q^*$, then the same applies for $r \cdot x \bmod q$. So we may rewrite the above steps as follows:

Implementation with Exponentiation Unit

step 1. obtain r from a PRNG,
step 2. compute $U' := V^r$,
step 2A. send U' to ExpU and get back $U := (U')^x$,
step 3. present U to the other party involved in the protocol.

For the additional steps of the protocol:

step ... given an element W obtained from the other party compute $L' := W^r$,
step ... A. send L' to ExpU and get back $L := (L')^x$,
after step ... A use L as a shared value with the other party of the protocol.

Note that the above procedure <u>never</u> produces the random element $r \cdot x \bmod q$ despite the fact that it is implicitly used by the protocol. So in particular there is no way to leak this ephemeral secret from the device. Of course, if V^x is known (as the reader's public key), then the Steps 2 and 2 A can be executed alternatively as $U := (V^x)^r$ (however, this is not possible for deriving L).

The ExpU is intended to be a separate hardware unit that holds a private key x and given an input M it outputs M^x. This could be a smart card with an interface restricted to a single operation of raising to power x. As the smart cards using asymmetric algorithms based on the Discrete Logarithm Problem must perform exponentiations, obtaining such a smart card is possible by restricting and adjusting the code executed by the smart card. Similarly, it could be a standard cryptographic co-processor, again with restricted functionalities.

3.3 Control Mechanism

While the protocols described so far should provide a good security level, the main problem is to make sure that they are really executed in the way described. So in particular, one can pretend that an implementation is based on an ExpU and in fact execute the protocol in the standard way. This is possible, since the outputs given to the other participant of the protocol have exactly the same probability distribution. However, we may include an additional procedure that enables to detect that ExpU is not used. On the side of the other protocol participant there are the following changes:

- One of two modes of execution must be used: the regular mode or the control mode. If the regular mode is used, then the execution is just as described by the authentication scheme.
- In the control mode:
  - after obtaining U from Step 3 request r from the reader,
  - check that $U = X^r$, where $X = V^x$ is the public key of the ExpU,
  - terminate the protocol execution after the check.

Remark 1. The control procedure is based on the fact that the reader must provide $U = X^r$ where r is known. Then the shared Diffie-Hellman value L equals W^{rx}, where W is given by the verifier. Therefore, the reader must be able to raise W^r to the power x. However, as W is a random challenge given after committing to r via U, the probability distribution of W^r is uniform as well. Therefore, to proceed with the protocol execution the reader has to be able to raise random elements to power x.

The above code can be executed if V is a fixed generator and therefore X can be declared in advance by the reader. If this is not the case, then the control mechanism is slightly different. In case of the protocol PACE it will be based on the following property of V:

$V = G^{x \cdot z \cdot a + b}$, where a, b are known to the verifier, z is known to the reader, and $X = G^x$ is the public key of ExpU.

The control procedure requires also knowledge of $\tilde{X} = G^{x^2}$. It runs as follows:

- after obtaining $U = V^{rx}$ from Step 3 request r and z from the reader,
- check whether $U = \tilde{X}^{rza} \cdot X^{rb}$,
- terminate the protocol execution after the check.

Obviously, if the protocol is correctly executed by the reader, then

$$\tilde{X}^{rza} \cdot X^{rb} = G^{x^2 rza + xrb} = (G^{xza+b})^{rx} = V^{rx} = U.$$

Remark 2. Let us note that it is possible to get $\tilde{X}$ from the ExpU unit. Namely, for the public key X we may take a random element k, compute X^k and feed it to the ExpU. (The ExpU cannot refuse X^k as it has uniform probability distribution.) The result returned is $S = X^{kx} = G^{x^2 k}$. So it suffices to compute $S^{k^{-1} \bmod q}$ to get $\tilde{X}$.

3.4 Key Update Mechanism

The mechanisms based on the secrets stored in the hardware components can be trusted as long as the adversary has no advantage over the user(s) of the device. The solution based on ExpU has the advantage that one can effectively control whether the device is really using the key x corresponding to the declared control public key $X = G^x$, but still the adversary may know x and thereby perform the attacks such as described in Sect. 2. This may occur in particular, if the key x is preinstalled in the device. In order to defer such problems we propose the following simple procedure:

Initialization: the manufacturer chooses a secret x at random and preinstalls it in the ExpU; the corresponding public key $X = G^x$ is retained for control purposes. The ExpU stores two keys: the permanent key x (it could be for instance hardwired by storing it in the ROM memory) and an update key y, which is initially set to 1. The key y is stored in a non-volatile memory.

Key update: a user can execute the update procedure at any time. He has to choose an update key k and upload it to ExpU. The ExpU should then perform the operation $y := y \cdot k$. At the same time, the public key used for the control purposes should be updated as follows: $X := X^k$.

Note that the control public key equals G^{xy} where x is set by the manufacturer and y is set by the party operating the device. In this way no party knows the exponent xy and it suffices that at least one party chooses its component at random.

Remark 3. Note that it seems to be impossible to propose a generic way to update a PRNG in a similar way. After updating the seed, in general there is no way to find any properties of the PRNG output that would convince the user that the update has been performed in the way intended by the user. Therefore, a malicious device can cheat the user about the update and perform it in a way convenient for the attacker.

3.5 Security Features

We may claim that a protocol where the exponent is derived in the way proposed is as secure as the original protocol with random exponents, provided that the random number generator is secure and the secret key is secure. However, we claim more: the protocol is secure if the random number generator is secure, but the key x gets exposed.

Proposition 1. *If there is an attack against a protocol with the implementation using ExpU, where the secret key x from the ExpU is leaked, then there is an attack against the original protocol.*

Proof (sketch). The proposition follows easily by the observation that there is a one-to-one relation between the protocol executions of the original scheme and the modified scheme based on the ExpU implementation. Namely, we have to replace each random exponent r from the original scheme by $r/x \bmod q$ and recompute all intermediate values.   □

One may ask what happens if the PRNG is attacked, as discussed in Sect. 2. Of course, the answer depends on the details of the scheme. However, an implementation with the ExpU holding a secret key x offers for instance the following advantages:

- If the device computes $(h^r)^x$ where h and r are known to the adversary (r is the output of the PRNG and h is the parameter obtained from the other protocol participant in clear text), then it is infeasible for the adversary to derive $(h^r)^x$, if the discrete logarithm of h is unknown. Thereby, the adversary cannot reconstruct $(h^r)^x$, while its counterpart in the regular implementation (i.e. h^r) can be easily derived.
- Given u, v, r' as well as h, r and h^{rx}, it is infeasible for the adversary to decide whether $v = u^{r'x}$. Thereby, it might be infeasible to decide whether two executions involve the same device.

In this extended abstract we skip a formal security proof for the ExpU implementations for protocols PACE-IM, PACE-GM and SPEKE presented in the next section.

4 Secure Password Authentication with Trusted Exponentiation

In this section we specify how to implement the proposed changes to password authentication protocols discussed in Sect. 2.

Modified PACE-GM. The solution proposed in Sect. 3 can be almost directly implemented for PACE with General Mapping. By $u := \mathtt{ExpU}(v)$ we will denote the result of a call with parameter v to the secure Exponentiation Unit of a device. That is, if the ExpU holds a secret key x, then returned value u equals v^x (Table 4).

Table 4. Modified PACE protocol, General Mapping version

Card	Reader
holds:	holds:
π - password	π - password entered by the document owner
$\mathcal{G}$ - parameters, with g a generator of a group of prime order q	
card's ExpU private key y	reader's ExpU private key x
card's ExpU public key $Y = g^y$	reader's ExpU public key $X = g^x$
$K_\pi := \mathrm{Hash}(0\|\pi)$	$K_\pi := \mathrm{Hash}(0\|\pi)$
choose $s_0 \leftarrow \mathbb{Z}_q$ at random	
$s := \mathrm{Hash}(\mathrm{ExpU}(s_0))$	
$z := \mathrm{ENC}(K_\pi, s)$	
$\xrightarrow{\quad \mathcal{G},z \quad}$	abort if $\mathcal{G}$ incorrect
	$s := \mathrm{DEC}(K_\pi, z)$
choose $y_A \leftarrow \mathbb{Z}_q^*$ at random	choose $y_B \leftarrow \mathbb{Z}_q^*$ at random
$Y_A := Y^{y_A}$	$Y_B := X^{y_B}$
$\xleftarrow{\quad Y_B \quad}$	
abort if $Y_B \notin \langle g \rangle \backslash \{1\}$ $\xrightarrow{\quad Y_A \quad}$	abort if $Y_A \notin \langle g \rangle \backslash \{1\}$
$\underline{h_1} := Y_B^{y_A}$	$\underline{h_2} := Y_A^{y_B}$
$h := \mathrm{ExpU}(\underline{h_1})$	$h := \mathrm{ExpU}(\underline{h_2})$
$\hat{g} := h \cdot g^s$	$\hat{g} := h \cdot g^s$
choose $y_A' \leftarrow \mathbb{Z}_q^*$ at random	choose $y_B' \leftarrow \mathbb{Z}_q^*$ at random
$\underline{Y_A'} := \hat{g}^{y_A'}$	$\underline{Y_B'} := \hat{g}^{y_B'}$
$Y_A' := \mathrm{ExpU}(\underline{Y_A'})$	$Y_B' := \mathrm{ExpU}(\underline{Y_B'})$
$\xleftarrow{\quad Y_B' \quad}$	
abort if $Y_B' = Y_B$ $\xrightarrow{\quad Y_A' \quad}$	abort if $Y_A' = Y_A$
$\underline{K_1} := Y_B'^{y_A'}$	$\underline{K_2} := Y_A'^{y_B'}$
$K := \mathrm{ExpU}(\underline{K_1})$	$K := \mathrm{ExpU}(\underline{K_2})$
proceed as for the regular PACE-GM	

Note that the modified protocol executes three more exponentiations on the side of the reader and four more exponentiations on the side of the card. The average time of point multiplication (elliptic curve equivalence of exponentiation) on modern Java Cards is about 100 ms. This concerns 256-bit curves which provide a decent security level of 128 bits (according to NIST [2]). Thus, the execution time on the side of the smart card may increase by about half of a second.

Table 5. Modified PACE protocol, Integrated Mapping version

Card	Reader
holds:	holds:
π - password	π - password entered by the document owner
params - parameters, with g a generator of a group $\mathcal{G}$ of prime order q	
card's ExpU private key a	reader's ExpU private key b
choose $s_0 \leftarrow \mathbb{Z}_q$ at random	
$s := \mathrm{Hash}(\mathrm{ExpU}(s_0))$	
$z := \mathrm{ENC}(\pi, s)$ $\xrightarrow{\quad params, z \quad}$	abort if $\mathcal{G}$ incorrect
	$s := \mathrm{DEC}(\pi, z)$
	choose β_0 at random
	$\beta := l_2$ leading bits of $\mathrm{Hash}(\mathrm{ExpU}(\beta_0))$
$\xleftarrow{\quad \beta \quad}$	
$\hat{G} = \mathrm{Encoding}(\mathrm{Hash}(s, \beta))$	$\hat{G} = \mathrm{Encoding}(\mathrm{Hash}(s, \beta))$
choose $x \leftarrow \mathbb{Z}_q$ at random	
$\underline{X} := x \cdot \hat{G}$	
$X = \mathrm{ExpU}(\underline{X})$	
$\xrightarrow{\quad X \quad}$	
	choose $y \leftarrow \mathbb{Z}_q$ at random
	$\underline{Y} := y \cdot \hat{G}$
$\xleftarrow{\quad Y \quad}$	$Y = \mathrm{ExpU}(\underline{Y})$
$\underline{Z_1} = x \cdot Y$	$\underline{Z_2} = y \cdot X$
$Z := \mathrm{ExpU}(\underline{Z_1})$	$Z := \mathrm{ExpU}(\underline{Z_2})$
derive the session keys from Z	derive the session keys from Z
send MACs to confirm the keys	
exchange messages in a secure channel	

Modified PACE-IM. In case of PACE-IM the ExpU operates on the elliptic curves and performs multiplications with the secret scalar (Table 5).
Note that the modification increases the number of multiplications with a scalar by three on each side.

Modified SPEKE. In case of SPEKE modification is based directly on the method from Sect. 3.
Note that the modification increases the number of exponentiations by two on each side (Table 6).

Table 6. Modified SPEKE protocol

Card	Reader
holds:	holds:
π - password	π - password entered by the document owner
a safe prime p, $p = 2q + 1$	
card's $\mathtt{ExpU}$ private key y	reader's $\mathtt{ExpU}$ private key x
$g := (\mathrm{Hash}(\pi))^2 \bmod p$	$g := (\mathrm{Hash}(\pi))^2 \bmod p$
choose $y_A \leftarrow \mathbb{Z}_q^*$ at random	choose $y_B \leftarrow \mathbb{Z}_q^*$ at random
$\underline{Y_A} := g^{y_A}$	$\underline{Y_B} := g^{y_B}$
$Y_A := \mathtt{ExpU}(\underline{Y_A})$	$Y_B := \mathtt{ExpU}(\underline{Y_B})$
	$\xleftarrow{\quad Y_B \quad}$
$\xrightarrow{\quad Y_A \quad}$	
abort if $Y_B \notin [2, p-2]$	abort if $Y_A \notin [2, p-2]$
$\underline{K_1} := Y_B^{y_A}$	$\underline{K_2} := Y_A^{y_B}$
$K := \mathtt{ExpU}(\underline{K_1})$	$K := \mathtt{ExpU}(\underline{K_2})$

5   Final Remarks

The attacks described in this paper are really simple and easy to execute unless strict (and expensive) organizational countermeasures are implemented. The lesson learnt from this particular example should be that we can talk about really secure cryptographic protocols when we consider them as a complete solution, with all its (potentially dishonest) actors and life-cycle of the system components. On the other hand, we have tried to convince that sometimes an improvement can be achieved with simple cryptographic means without substantial changes in the high level description of the cryptographic protocols. This is yet another evidence that for security products the devil is in the detail.

Acknowledgment. We would like to thank an anonymous reviewer for remarks on security model for real world computations. Some valuable comments are incorporated almost literally. We also thank INSCRYPT PC chairs for many helpful suggestions.

References

1. Abdalla, M., Fouque, P.-A., Pointcheval, D.: Password-based authenticated key exchange in the three-party setting. In: Vaudenay, S. (ed.) PKC 2005. LNCS, vol. 3386, pp. 65–84. Springer, Heidelberg (2005)
2. Barker, E.B., Barker, W.C., Burr, W.E., Polk, W.T., Smid, M.E.: Sp 800-57. Recommendation for Key Management, part 1: General (revised). Technical report, Gaithersburg, MD, United States (2007)

3. Becker, G.T., Regazzoni, F., Paar, C., Burleson, W.P.: Stealthy dopant-level hardware Trojans: extended version. J. Crypt. Eng. 4(1), 19–31 (2014)
4. Bellare, M., Paterson, K.G., Rogaway, P.: Security of symmetric encryption against mass surveillance. In: Garay, J.A., Gennaro, R. (eds.) CRYPTO 2014, Part I. LNCS, vol. 8616, pp. 1–19. Springer, Heidelberg (2014). Also IACR Cryptology ePrint Archive 2014, 438 (2014)
5. Bender, J., Fischlin, M., Kügler, D.: Security analysis of the PACE key-agreement protocol. In: Samarati, P., Yung, M., Martinelli, F., Ardagna, C.A. (eds.) ISC 2009. LNCS, vol. 5735, pp. 33–48. Springer, Heidelberg (2009)
6. BSI: Advanced Security Mechanisms for Machine Readable Travel Documents 2.11. Technische Richtlinie TR-03110-3 (2013)
7. Bundesamt für Sicherheit in der Informationstechnik: Common Criteria Protection Profile for Inspection Systems (IS), BSI-CC-PP-0064 (2010). https://www.commoncriteriaportal.org/files/ppfiles/pp0064b_pdf.pdf
8. Canetti, R., Krawczyk, H.: Analysis of key-exchange protocols and their use for building secure channels. In: Pfitzmann, B. (ed.) EUROCRYPT 2001. LNCS,Canetti, R., Krawczyk vol. 2045, pp. 453–474. Springer, Heidelberg (2001)
9. European Parliament and the Council: Regulation (EU) no 910/2014 on Electronic Identification and Trust Services for Electronic Transactions in the Internal Market and Repealing Directive 1999/93/EC (2014). http://eur-lex.europa.eu/legal-content/EN/TXT/HTML/?uri=CELEX:32014R0910&from=EN
10. Hanzlik, L., Krzywiecki, Ł., Kutyłowski, M.: Simplified PACE|AA protocol. In: Deng, R.H., Feng, T. (eds.) ISPEC 2013. LNCS, vol. 7863, pp. 218–232. Springer, Heidelberg (2013)
11. ISO/IEC JTC1 SC17 WG3/TF5 for the International Civil Aviation Organization: Supplemental Access Control for Machine Readable Travel Documents. Technical report version 1.02, March 08 (2011)
12. Jablon, D.P.: Extended password key exchange protocols immune to dictionary attacks. In: WETICE, pp. 248–255. IEEE Computer Society (1997)
13. Killmann, W., Schindler, W.: A proposal for functionality classes for random number generators (2011). https://www.bsi.bund.de/SharedDocs/Downloads/DE/BSI/Zertifizierung/Interpretationen/AIS_20_Functionality_classes_for_random_number_generators_e.pdf
14. LaMacchia, B.A., Lauter, K., Mityagin, A.: Stronger security of authenticated key exchange. IACR Cryptology ePrint Archive 2006, 73 (2006)
15. Young, A.L., Yung, M.: Malicious Cryptography - Exposing Cryptovirology. Wiley, New York (2004)

Stegomalware: Playing Hide and Seek with Malicious Components in Smartphone Apps

Guillermo Suarez-Tangil[✉], Juan E. Tapiador, and Pedro Peris-Lopez

Department of Computer Science, Universidad Carlos III de Madrid,
Avda. Universidad 30, 28911 Leganes, Madrid, Spain
guillermo.suarez.tangil@uc3m.es,
{jestevez,pperis}@inf.uc3m.es

Abstract. We discuss a class of smartphone malware that uses stegano-graphic techniques to hide malicious executable components within their assets, such as documents, databases, or multimedia files. In contrast with existing obfuscation techniques, many existing information hiding algorithms are demonstrably secure, which would make such *stegomalware* virtually undetectable by static analysis techniques. We introduce various types of stegomalware attending to the location of the hidden payload and the components required to extract it. We demonstrate its feasibility with a prototype implementation of a stegomalware app that has remained undetected in Google Play so far. We also address the question of whether steganographic capabilities are already being used for malicious purposes. To do this, we introduce a detection system for stegomalware and use it to analyze around 55 K apps retrieved from both malware sources and alternative app markets. Our preliminary results are not conclusive, but reveal that many apps do incorporate steganographic code and that there is a substantial amount of hidden content embedded in app assets.

Keywords: Smartphone security · Malware · Steganography · Obfuscation

1 Introduction

Malware for smartphones has rocketed over the last few years. Such a phenomenon is intimately related to the popularity of smartphone platforms and the substantial rise in the number of apps available for download in online markets. While these two facts have contributed to create new business models and reshape the way we communicate, malware writers have taken advantage of the possibilities offered by smartphones for spying on the user's activities, stealing his identity, or committing fraud, among other malicious activities [30].

Thwarting malware attacks in smartphones is still a formidable challenge. On the one hand, battery-powered smartphones do not possess enough computing

© Springer International Publishing Switzerland 2015
D. Lin et al. (Eds.): Inscrypt 2014, LNCS 8957, pp. 496–515, 2015.
DOI: 10.1007/978-3-319-16745-9_27

capabilities to constantly check for attempts of executing malicious operations. Furthermore, distinguishing what is malware from what is not is far from being easy. In any case, the most common distribution strategy for smartphone malware is still the use of both official and unofficial markets [25]. Attackers simply upload malicious apps to the market, sometimes using a stolen identity, and users get infected by just downloading and installing the app.

In the case of official markets, operators are generally concerned about the security of the software they distribute. To address malware attacks, most markets implement a revision process that presumably includes various security checkings [15]. Malware writers are constantly seeking ways of evading detection. For instance, in the so-called update attacks, the app just contains a "hook" that, once installed in the user's device, downloads and executes a malicious payload from a external server pretending to be a required update of the app.

Smartphone malware is becoming increasingly stealthy and recent specimens are relying on advanced code obfuscation techniques to evade detection by security analysts [24]. For example, *DroidKungFu* has been one of the major Android malware outbreaks. It started on June 2011 and has already spanned over at least six variants. *DroidKungFu* has been mostly distributed through official or alternative markets by piggybacking the malicious payload into a variety of legitimate applications. The malicious payload with *DroidKungFu*'s actual capabilities is encrypted into the app's assets folder and decrypted at runtime using a key placed within a local variable belonging to a specific class module. *GingerMaster* is another representative example of smartphone malware that deliberately tries to hide itself. In this case, the main payload is stored as PNG and JPEG pictures in the asset folder. Such files are loaded as regular pictures by a special hook within the app and then interpreted as code.

Examples such as the two described above do not abound yet but are becoming increasingly common. In the case of traditional platforms such as PCs, malware writers have made use of obfuscation techniques for decades, ranging from simple packing algorithms to using more sophisticated polymorphic and metamorphic engines (see [16] for an excellent overview). Such techniques transform malicious code to make it difficult to understand, analyze, and detect by static analysis.

1.1 Contributions

Motivated by the increasingly creative ways used by smartphone malware to hide malicious components and evade detection, we pose the question of how information hiding techniques could be used by malware writers to achieve the same purpose. Contrarily to the ad hoc—and, in many cases, sloppy—measures seen today to obfuscate malicious components, modern information hiding techniques could provide a simple yet theoretically robust way of hiding executable pieces in assets such as pictures, databases, multimedia files, etc. To the best of our knowledge, this is the first paper that looks into this issue.

In summary, in this paper we make the following contributions:

1. We describe a class of smartphone malware that uses steganographic techniques to hide malicious executable components within an app's assets. We call this "stegomalware" and argue that steganographic algorithms provide malware writers with a mechanism for hiding malicious payloads more secure than obfuscation techniques currently in use.
2. We discuss various architectures for stegomalware depending on the location of the asset with hidden capabilities and the algorithm required to extract it.
3. We show that current app markets may be vulnerable to stegomalware. In particular, we describe a prototype implementation of a stegomalware sample for Android platforms that is available for download in Google Play and has remained undetected so far.
4. We introduce a detection system for stegomalware that combines steganalysis techniques with the detection of steganographic algorithms in the app code.
5. Using an implementation of our stegomalware detection system, we address the question of whether steganographic capabilities are being already used for malicious purposes. We analyze around 55 K apps retrieved from both malware sources and alternative app markets. Even though our preliminary results are not conclusive, we found that many apps do incorporate steganographic code and that there is a substantial amount of hidden content embedded in app assets.

2   Information Hiding Techniques and Related Work

In this section, we provide a brief background on information hiding techniques, including some formal definitions, common stegosystems, and techniques to detect the presence of hidden information. Because of their popularity, we will focus our discussion on stegosystems that hide information in pictures. We will revisit this point later and discuss alternative stegosystems for the type of malware discussed in this paper. Interested readers can find good introductions to this discipline in [13,17,19]. Finally, we briefly review the literature related to malware in smartphones.

2.1   Stegosystems

Modern steganography studies techniques to hide the presence of information by embedding secret messages within other, seemingly harmless digital objects. According to the standard terminology of information hiding [18], the original object used to hide information is called the *covertext*, whereas the same object after embedding the secret message is called the *stegotext*. The embedding process depends on a key and the adversary (known as the *warden*) is generally assumed to know everything but the key.

Formally, a symmetric[1] stegosystem is a triple of probabilistic polynomial-time algorithms $(\mathsf{SK}, \mathsf{SE}, \mathsf{SD})$ with the following properties [3]:

---

[1] This definition can be naturally extended to public-key stegosystems [3].

- The key generation algorithm SK takes a security parameter n as input and returns a stegokey sk.
- The steganographic encoding algorithm SE takes as input the security parameter n, the stegokey sk, and a message $m \in \{0,1\}^l$ and outputs a stegotext c belonging to the covertext space C. The algorithm may access the distribution of C if needed.
- The steganographic decoding algorithm SD takes as input the security parameter n, the stegokey sk, and an element $c \in C$, and outputs either a message $m \in \{0,1\}^l$ or a special symbol $\perp$ denoting that no message is embedded in c.

A stegosystem must be *reliable*, i.e., the probability that

$$SD(1^n, sk, SE(1^n, sk, m)) \neq m \tag{1}$$

must be negligible in n for all messages m and all stegokeys sk. Informally speaking, the *security* of a stegosystem is related to the probability that an adversary detects the presence of an embedded message. Thus, a secure stegosystem is one in which covertexts and stegotexts are indistinguishable. This notion can be formally established as a distinguishability experiment similar to those common in the field of provable security. In turn, this gives rise to various notions for secure stegosystems, including perfect security, statistical security, and computational security (see [3] for further details).

Apart from their security, there are other aspects of practical relevance for stegosystems. One is their *capacity*, defined as the maximum number of secret bits that can be securely embedded. In most practical stegosystems there is a trade-off between security and capacity, so the longer the embedded message, the more distinguishable the stegotext becomes. Another relevant property is the *robustness* of the stegosystem. Informally speaking, robustness measures the amount of distortions that a stegotext can endure until recovering the embedded message becomes impossible. This is a key property for applications such as watermarking and fingerprinting, where the focus may not be on hiding the presence of some embedded mark, but on preventing an adversary from removing it without degrading the quality of the data object.

2.2 Common Steganographic Algorithms

A variety of stegosystems have been proposed for embedding data in all sorts of digital sources, including text file formats, compiled code, images, audio, and video. There are, in fact, few digital formats where some opportunity for steganography has not been identified. Stegosystems for multimedia objects— and, in particular, for images—are among the most popular schemes because of the high embedding capacity offered by digital pictures and the proliferation of images in Internet. Thus, most papers in this field have concentrated on JPEG images, although the underlying ideas and algorithms are generally applicable to other formats as well.

The JPEG image format is based on taking the discrete cosine transform (DCT) of 8×8-pixel blocks of the image, producing 64 DCT coefficients. Once quantized, the least-significant bits (LSB) of each coefficient are modified to embed hidden messages. Note that the modification of a single DCT coefficient affects all 64 pixels in the block. We next describe four popular JPEG stegosystems that use some form of LSB embedding in the frequency domain (see [4] for a recent survey).

Jsteg Proposed by Derek Upham, this is one of the earliest stegosystems for JPEG images [27]. Jsteg replaces the LSB of the DCT coefficients by the secret message bits, skipping those coefficients with the values 0 or 1. The image is scanned sequentially and the algorithm does not support random bit selection. The key, if any, is used to encrypt the message before embedding. Jsteg-shell is a popular Windows front-end for Jsteg that encrypts the message with RC4.

JPHide This is a stegosystem proposed by Allan Latham that supports compression of the secret message and encryption with Blowfish. The algorithm is also based on replacing the LSB of the DCT coefficients but does not do it sequentially. Instead, it uses a fixed table to determine which coefficient will be changed next. Furthermore, a pseudorandom number generator (PRNG) is used to skip some of them, where the probability of skipping changes depending on how many bits have been embedded already and how many are left. JPHide can also use the second LSB in some cases.

OutGuess This is yet another JPEG stegosystem using LSB encoding in the DCT coefficients. Contrarily to the two previous algorithms, OutGuess chooses the coefficients randomly using a PRNG initialized with a user-provided password. The content is also encrypted using RC4 with the same password used for the PRNG.

F5 Developed by Andreas Westfeld, F5 can be seen as an evolution of the stegosystems described above. It introduces a number of novel ideas, including the use of a matrix encoding to reduce the number of necessary changes and a permutative straddling to uniformly spread out the modifications over the whole covertext. F5 reduces the propagation of steganographic information over the carrier medium. This feature makes F5 robust against certain distortions such as resizes or rotations. Full details are available in [29].

2.3 Steganalysis

Steganographic encoding algorithms leave traces on the stegotexts as a consequence of the alterations required to embed the message. Such traces are instrumental in facilitating detection, i.e., distinguishing whether an object has or has not embedded information. This is the main goal of a *passive warden*, i.e., an adversary who can read objects and must determine if a secret communication is taking place. Note that a correct detection defeats the main purpose of steganography, which is hiding the very presence of a communication. In general, exposing the content of such a secret communication is another problem entirely.

Contrarily to passive wardens, an *active warden* is not concerned with detecting secret communications, but with destroying them. Active warden techniques introduce deliberate modifications in all digital objects in the hope that any potential hidden content would be rendered unusable. In some domains, this is just too costly and some form of sampling must be performed.

In what follows, by *steganalysis* we will refer to the process of distinguishing whether an object has or has not hidden information. A steganalytic technique is often presented in the form of *distinguisher*, this being a test that returns some measure of the likelihood of the input sample having embedded information. Steganalytic techniques can be classified according to various criteria. A *targeted* distinguisher focuses on the artifacts produced by one specific algorithm and, therefore, can only detect if that algorithm has been used. Contrarily, a *blind* (or *universal*) distinguisher identifies statistical alterations caused by any steganographic encoding algorithm [7]. An example of a blind steganalysis is the χ^2-attack, based on applying a χ^2 test to compare the distributions of adjacent DCT coefficients, which is similar in images with hidden data embedded [20]. Fridrich et al. provide in [9,10] an overview of feature-based steganalysis for JPEG images and its implications for future designs of stegosystems.

There are a number of freely available implementations of the main steganalytic techniques proposed so far, including:

Stegdetect [20] is a popular and free steganalytic tool. It includes a number of distinguishers to detect the presence of hidden data in images and is able to identify the method used during embedding process. **Stegdetect** is the *de facto* tool used by security and forensic practitioners due to its excellent capabilities and its free and open nature. A recent study by Khalind et al. [14] has revisited its features and warned about the implications of its false positive ratio.

VSL (Virtual Steganographic Laboratory) [8] is a suite of steganalytic techniques that includes some of the most popular techniques, including RS-Analysis (a distinguishing algorithm for LSB methods) and a blind steganalysis technique based on Support Vector Machines.

SSS (Simple Steganalysis Suite) [2] is another publicly available implementation of various image steganalytic techniques, including χ^2-attack and various histogram-based tests.

2.4 Thwarting Malware in Smartphones

A substantial amount of recent work has addressed the problem of analyzing malware in smartphones using a variety of techniques [6,25]. Static analysis techniques are well known in traditional malware detection and have recently gained popularity as efficient mechanisms for market protection [1,26]. However, current static techniques fail to identify malicious components when they are obfuscated or embedded separately from the code (e.g., hidden into an image) [12,22]. Approaches based on dynamic code analysis [11] are promising, but current works [6,21,23] only provide an holistic understanding of the behavior of an app. This feature challenges the identification of malware using steganography.

More recent approaches focus on detecting hidden functionality within components of an app [24]. Although this technique has shown to be promising, it requires a non-negligible overhead derived from the dynamic execution of every app analyzed. As regards the various ways to hide or locate hidden code in apps using steganography and steganalysis, to the best of our knowledge this is the first work addressing this issue in smartphones.

3   Stegomalware

This section introduces the idea of an app that uses a stegosystem to hide a malicious component within its assets and then extracts and executes it dynamically. We then discuss various architectures for such a stegomalware and describe a prototype implementation for Android platforms that is available for download in Google Play and has remained undetected so far.

3.1   Hiding Malicious Code in App Assets

Malware developers can use steganographic capabilities to hide malicious components within an app resources. Such resources depend on the particular app and may include images, audio, video, databases, and text files in a variety of formats (e.g., plain text, XML, and HTML). Practical stegosystems for all these digital objects have been proposed, some of them with a reasonable security level. Moreover, there is a variety of freely available implementations of such stegosystems that are exceedingly simple to use within an app.

The main goal pursued by an attacker who uses a stegosystem to securely embed a piece of malicious code into an app resource is to evade detection, particularly static analysis based approaches implemented by market operators. Hiding malicious components may also difficult malware analysis, as the payload will be located in places that security analysts could overlook. More importantly, the piece of malicious code is not accessible to analysis. This is a key difference—and a substantial advantage for malware writers—with respect to traditional malware obfuscation techniques: after obfuscation, malicious code may be difficult to recognize, but it is still somewhere in the app. In contrast, in a stegomalware specimen the malicious component is revealed at execution time only. Thus, it will not match any signature even if the search includes the asset where it is hidden.

A stegomalware contains the following three basic components:

- A *stegotext* R, this being one of the app assets. R is the result of embedding a malicious payload p with a steganographic encoding algorithm SE using some stegokey.
- A *stegokey* sk required to extract p from R. In case of using a symmetric stegosystem, sk is the stegokey that was used to embed p in R.
- A *steganographic decoding algorithm* SD needed to recover p from R using sk.

These three elements (R, sk, SD) can be packaged together and distributed with the app, or dynamically retrieved at runtime from an external server. Based on this, we next describe three architectural choices for stegomalware. This list does not intend to be exhaustive and more complex variants are possible.

3.2   Type 0: Autonomous Stegomalware

One simple choice is to have all the stego material (R, sk, SD) distributed with the app. The asset R is the less problematic of the three, as it can just be put in the asset folder with the remaining resources. The algorithm SD must be part of the code assets so the app can invoke it to retrieve p. The stegokey sk must be part of the app too, either hardcoded as a variable somewhere in the code, or else distributed in some other asset (e.g., a text file, a database, etc.). Note that sk is not necessarily a random string, as many stegosystems accept alphanumeric passwords from which subsequent keying material is derived.

This type of stegomalware is fully autonomous and does not depend on a remote infrastructure to achieve its goals (see Table 1). This, however, comes at a price: the presence of R, sk, and SD may facilitate detection. We will discuss this issue in more detail later in Sect. 4.

Table 1. Three variants of stegomalware and their activation procedures.

Type	Locally Available	Remotely Available	Activation
Type 0	R, sk, and SD	nothing	1. Get sk and R from the app resources
			2. Recover the payload: $p = \mathsf{SD}(sk, R)$
			3. Execute p
Type I	l and SD	R and sk	1. Get URL l from the app resources
			2. Connect to l and retrieve (R, sk)
			3. Recover the payload: $p = \mathsf{SD}(sk, R)$
			4. Execute p
Type II	l and SD	R and sk	1. Get URL l from the app resources
			2. Connect to l and retrieve (R, sk, SD)
			3. Recover the payload: $p = \mathsf{SD}(sk, R)$
			4. Execute p

3.3   Type I: Stegoupdate Attacks

A more flexible alternative to Type 0 stegomalware consists of retrieving the stegotext R remotely during activation. This allows for using different malicious payloads depending on the attacker's goals, the particular target, etc. The app must necessarily contain the decoding algorithm SD and possibly sk, although it is generally more convenient to have sk associated with the particular R and, therefore, also dynamically downloaded. This variant introduces the need for incorporating into the app the location (e.g., an URL) of the external server from where (R, sk)

will be fetched. The activation procedure is quite simple (see Table 1) and involves fetching R and sk, extracting p locally, and then executing it.

This type of stegomalware can be seen as a variant of the classical update attacks [30] in which the malicious payload is downloaded embedded into an innocuous-looking object. This would evade detection schemes based on monitoring update traffic and preventing the execution of downloaded code.

3.4   Type II: Agnostic Stegomalware

In a more radical setting, the app could be totally agnostic of the stegosystem used to embed the payload. Thus, after activation the app would connect to a remote site and download R, sk, and SD. The decoding algorithm would be then used to extract the malicious payload from R.

The key idea here is not to distribute SD within the app, so even a detailed code analysis would not raise any suspicion. This idea admits some minor variations. For example, R and sk might actually be part of the app, so the update engine just downloads SD and uses it to extract the malicious payload. Similarly, in a collusion scenario, the app could pass R and sk on to another app in the same device that implements SD. This second app would first extract the payload and pass it back to the original app (e.g., by putting it in shared space) or just execute it.

3.5   A Proof-of-Concept Implementation

We implemented a prototype of a simple autonomous stegomalware (Type 0) for Android platforms. The app is named LikeImage and contains a malicious payload embedded into an image that is part of the app resources (see Fig. 1). In execution time, the payload is extracted from the image and executed dynamically. We used an open source Java implementation[2] of F5 for JPEG images as the underlying stegosystem. Figure 2 shows the original image and the one that is actually distributed with the app after embedding the payload. As expected, both images look exactly the same.

The malicious payload is a Java JAR library compiled in Dalvik Executable (DEX) Format. Figure 3 shows a fragment of the main class. The payload contains an update engine that requests instructions from a remote Command &Control (C&C) server in the form of a second payload that might change depending on the interests of the attacker and the target device. This second payload is then retrieved, dynamically loaded, and executed in the user device. Initially, the update component used in our first demonstration was programmed to exfiltrate the IMEI of the device. The app was submitted to Google Play in early June 2014 and passed all security controls. At the time of this writing, it is still available in the market[3], although our C&C server has been instructed to always return an innocuous payload and the app was modified to leak nothing from the device.

---

[2] https://code.google.com/p/f5-steganography.
[3] https://play.google.com/store/apps/details?id=es.uc3m.cosec.likeimage.

Fig. 1. Our stegomalware app (LikeImage).

(a) Original image.   (b) Image with embedded payload.

Fig. 2. Asset used by our stegomalware proof-of-concept.

```
package es.uc3m.cosec;
public class Malware{
    String getPayload(){
        String payload;
        ...
        return payload;
    }
}
```

```
DEX version '035'
Class #0    –
    Class  : 'LMalware;'
    Methods    –
    #0           : in LMalware;
    name         : 'getPayload'
    type         : 'String;'
    source_file: Malware.java
```

(a) Java code.   (b) DEX dump.

Fig. 3. Code snippet of the malicous payload.

4   Searching for Stegomalware in the Wild

After the ideas introduced in previous sections, we next describe our efforts so far to find out if malware with steganographic capabilities is already in the wild. Our focus has been on types 0 and I, and we have only searched for apps using image or audio (mp3) stegosystems. With this in mind, we built a detector whose operating principle revolves around three main ideas:

1. Detect the presence of capabilities to execute dynamically loaded code. This is essential to transfer control to any downloaded or extracted payload.
2. Identify assets that are suspect of containing embedded messages (steganalysis).
3. Identify steganographic decoding algorithms in the app code.

In the remaining of this section we describe the experimental setting used in our study and discuss the main results obtained so far.

4.1   Experimental Setting

Our study is based on a dataset composed of around 54 K apps retrieved from two different sources (see Table 2). On the one hand, we downloaded around 31 K apps from Aptoide[4] (AP), a distributed marketplace for Android apps. Contrarily to popular markets such as Google Play, in Aptoide there is not a centralized repository for apps, but each user manages their own "store". On the other hand, we retrieved around 22 K Android malware samples from a popular virus repository: VirusShare[5] (VS).

We implemented a detection framework for Android apps in Java and Python. The detector makes use of several open source tools that allows us to extract static information from Android apps and unpackage their resources [5]. It also relies on existing open-source implementations of various steganalytics tools, such as those described in Sect. 2.3. Finally, we also integrated some parts

---

[4] http://www.aptoide.com/.
[5] http://www.virusshare.com/.

Table 2. App sources used in our experimentation.

Source	#Apps	Type
Aptoide (AP)	31,935	Presumably goodware
VirusShare (VS)	22,707	Known malware
Total	54,642	

of the implementation of Anagram [28], an anomaly-based intrusion detection system for application-layer traffic based on n-gram analysis and Bloom filters. As explained in detail later, this is used as the basis for a detector of steganographic code.

4.2   Step 1: Selecting Apps with Payload Execution Capability

The first component in our detection framework identifies apps containing the capability to execute payloads. For instance, Android provides a runtime environment that allows apps to dynamically load libraries. Detecting such capabilities may help to discard those apps that make use of a stegosystem for purposes other than hiding a malicious payload (e.g., to check a watermark).

In our current implementation, we pay special attention to apps with one or more of the following features:

– *Native code*, i.e., apps that contain components using native-code languages such as C and C++.
– *Dynamic code*, i.e. apps that contain functions to dynamically load executable code or libraries.
– *Reflection code*, i.e., apps that attempt to dynamically inspect other fragments of code at runtime.

We assign a "code execution score" S_C to each app $\mathcal{A}$ based on this. Specifically:

$$S_C(\mathcal{A}) = s_{\text{native}} + s_{\text{dynamic}} + s_{\text{reflection}} \tag{2}$$

where s_x is a factor measuring the risk of having code of type x. We experimentally determined such factors to be $s_{\text{native}} = 0.1$, $s_{\text{dynamic}} = 0.2$, and $s_{\text{reflection}} = 0.5$.

We computed the score defined in Eq. (2) for all apps in our dataset. The results are shown in the Venn diagrams provided in Fig. 4. We discovered that 3,605 apps in the AP dataset (around 11 %) contain some form of advanced code. Similarly, our analysis over the VS dataset reported 1,855 samples (around 8 %) with advanced code. It can also be observed that reflection code is the most common operation, as it appears in 4,239 and 1,834 apps in AP and VS, respectively.

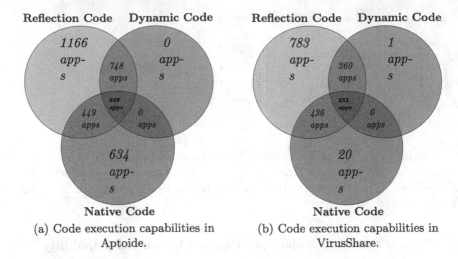

Reflection Code   Dynamic Code

1166 apps

748 apps

0 apps

805 apps

449 apps

0 apps

634 apps

Native Code

(a) Code execution capabilities in Aptoide.

Reflection Code   Dynamic Code

783 apps

360 apps

1 apps

255 apps

436 apps

0 apps

20 apps

Native Code

(b) Code execution capabilities in VirusShare.

Fig. 4. Number of apps containing native, dynamic, and reflection code in our dataset.

4.3 Step 2: Flagging Suspicious Assets

Our second contributing factor to the overall detection score is related to the presence of hidden information embedded in the app assets. When analyzing the distribution of potential stegotexts in our dataset, we observed that many apps contain a large number of images. For instance, several apps contained over 5 K images each. Additionally, we also found that audio files are less common. In our subsequent analysis, we limited the search for stegotexts to JPEG images and MP3 audio files.

We use `Stegdetect` to determine if a given candidate is likely to contain a hidden payload within any of its image or audio files. `Stegdetect` admits as input a sensitivity threshold ranging between 0.1 and 10 that is used by the underlying distinguishers. The higher this threshold is, the more sensitive the test. Furthermore, the output provides a confidence level in the detection of hidden messages that takes 3 possible values: low confidence (*), medium confidence (**), and high confidence (***).

As in the first step above, we computed a numerical score with each app summarizing the likelihood of having embedded information in its contents. This "steganalysis score" S_H is computed as:

$$S_H(\mathcal{A}) = \max \left\{ 1, \sum_{c \in R(\mathcal{A})} \frac{\text{conf}(c)}{3} \right\} \tag{3}$$

where $R(\mathcal{A})$ is the set of potential stegotexts in the app (i.e., its resources/assets) and $\text{conf}(c) \in \{1, 2, 3\}$ is the confidence level returned by `Stegdetect`, or 0 in the case that no distinguisher returns true.

Table 3 summarizes the experimental results obtained and shows the number of images matching each steganographic method for a relatively low sensitivity value (1.0). There is an extremely high number of matches. Specifically, `Stegdetect` reports matches for more than 20 K images distributed in approximately 3 K apps in the case of AP, and almost 10 K images over 2 K apps for VS. In terms of embedding algorithms, `JPHide` seems to be the most popular steganographic tool used with about 85 % the matches for both AP and VS apps. These results will be further discussed later when analyzing individual apps.

Table 3. Number of matches per steganographic tool/method (`JPHide`, `OutGuess`, `JSteg`, `F5`, `appended`, `camouflage`, and `alpha-channel`) and confidence (*, **, and ***) for all samples retrieved from Aptoide (AP) and VirusShare (VS).

	JPHide			OutGuess			JSteg			F5			Other		
	(*)	(**)	(***)	(*)	(**)	(***)	(*)	(**)	(***)	(*)	(**)	(***)	appended	camouflage	alpha-channel
AP	9,774	2,219	8,421	517	441	2,203	10	5	0	0	0	3	257	1	5
	20,414			3,161			15			3			263		
	23,804 matches in 3,364 apps														
VS	3,924	3,924	2,570	137	143	1,260	6	0	0	0	0	0	213	0	0
	10,418			1,540			6			0			213		
	12,177 matches in 2,009 apps														

4.4 Step 3: Identifying Steganographic Code

A common factor between Type 0 and Type I stegomalware is the presence of a steganographic decoding algorithm SD in the app code required to extract the malicious payload from the stegotext (asset). Determining if a piece of compiled code includes some unknown function SD is not straightforward. We approached this problem statistically using N-gram analysis as follows:

1. For each candidate steganographic decoding algorithm SD that we wish to detect, we create a simple app with just one class that uses SD.
2. We extract from the compiled app all N-grams. To do this, the app is treated as a stream of bytes and a sliding window of length N is passed through it. To efficiently store all N-grams, we relied on the Bloom filter implementation used in Anagram [28] for a content-based anomaly detector.
3. In detection mode, we extract all N-grams from a candidate app and check how many of them are found in the content filter associated with SD. The more the number of hits, the more likely the app contains an implementation of SD identical to that used to generate the content filter.

We generated content filters for various Java open-source steganographic tools, including all stegosystems enumerated in Sect. 2. We experimented with different values of N and noticed that for $N < 10$ it generates too many false positives, while too high values (i.e., $N > 50$) results in a very ineffective detector. Thus, for each candidate algorithm SD we created 5 different models, one for each $N = \{10, 20, 30, 40, 50\}$.

As with the two previous factors, we associate with each app a score quantifying the likelihood of the app containing one of the steganographic decoding algorithms SD modeled in our content filters. The score is computed in two steps. First, for each algorithm SD and each value of N we compute a normality score by counting the fraction of N-grams of the app $\mathcal{A}$ that are found in the content filter:

$$S_S(\mathcal{A}, \text{SD}, N) = \frac{\#Hits}{|\mathcal{A}| - N + 1} \tag{4}$$

The score $S_A(\mathcal{A}, \text{SD})$ for algorithm SD is then computed as the average score for the five values of N. Finally, the overall score for the app $S_S(\mathcal{A})$ is just the average score for all algorithms SD. In our experimentation, we determined that a good practice is filtering out those algorithms whose score $S_S(\mathcal{A}, \text{SD})$ is lower than a given threshold (around 0.6 for our datasets). This is the approach followed in the results reported next.

Figure 5 shows the score distribution computed for those apps that had one or more of the code execution features described in Sect. 4.2. The score is extremely low for most of them, suggesting that they do not contain traces of steganographic algorithms—at least, not the ones for which we have a content filter. There are, however, a reduced number of apps that seem to contain steganographic-like operations, with scores ranging from 0.2 to almost 0.6. The majority of these apps come from the AP dataset.

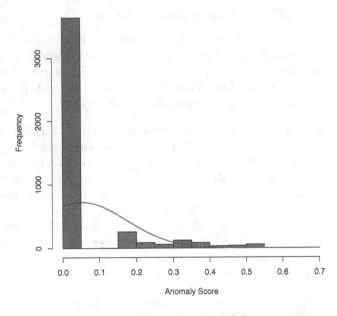

Fig. 5. Distribution of $S_S(\mathcal{A})$.

Note that we have not considered the scenario where an attacker uses multiple SD algorithms at the same time. Although our implementation reports the average anomaly score per SD, we could easily extend it to evaluate all SD together.

4.5 Putting It All Together

The three separate scores introduced above can be combined into a single value to measure the likelihood of an app containing stegomalware. We propose using the following score:

$$S(\mathcal{A}) = S_S(\mathcal{A})^{1-S_H(\mathcal{A})\cdot S_C(\mathcal{A})} \qquad (5)$$

For instance, an app containing dynamic and reflection code, for which the SD detection score is 0.6, and with one picture with hidden content with confidence 100 % (***) is assigned a score of $0.6^{1-(1.0\cdot0.7)} = 0.858$.

The rationale behind expression (5) is to have a power-law score where the basis is determined by the likelihood of detecting a steganographic decoding algorithm in the app. If one is reasonably sure of the fact that the app has steganographic capabilities, then the base score is raised to a factor that considers the existence of code execution capabilities and the output of steganalysis over the app's assets. In doing so, we pursue to reduce the effects of the high false positive ratio of current steganalysis techniques.

4.6 Main Findings

Table 4 shows the top 10 specimens with the highest score in the AP and VS datasets. In VS there is only three samples with a score higher than 0.5, whereas in Aptoide more than 200 samples present such higher scores. For a reference value, our stegomalware demo app LikeImage has an associated score of $0.64^{1-(1\cdot0.7)} = 0.738$.

Table 4. Top 10 scores for specimens from Aptoide and VirusShare. The score of our stegomalware PoC app Likeimage is shown as a reference.

Rank	Aptoide	VirusShare
1	0.767	0.746
2	0.734	0.731
3	0.725	0.584
4	0.724	0.068
5	0.701	0.068
6	0.699	0.068
7	0.699	0.067
8	0.697	0.067
9	0.695	0.067
10	0.695	0.066
Our PoC Likeimage: 0.738		

We carried out some manual inspection of the top scored apps from both Aptoide and VirusShare. We next summarize our main findings:

512   G. Suarez-Tangil et al.

```
VS  sample  a8869e0ec4fb1acf12ae01d5294c0c2b.apk
    91138622720e0cf3b119c1c50846f21fbf09aae7.jpg  —  jphide(*)
    6391e903918fa0ecc9951f32249759ee3d6ddb04.jpg  —  jphide(*)
    f76575600c3387444c079942530fd9f9d62aa08b.jpg  —  jphide(*)
    e8112b2ac65c10384e23dff0b0119313b07e8967.jpg  —  jphide(**)
    ae8267310a55b3192befac4041a98226cffc17b5.jpg  —  jphide(**)
    5d212aa85edf8db1664c98520b23dd54564e7463.jpg  —  jphide(***)
    9304c888d43f87943d766a12d01b0ef41ad53ac5.jpg  —  jphide(***)
    5af4d7ea15ce36d3a21d75ff3bf33a87e850b1a1.jpg  —  jphide(***)
    f35ea0096b63f624f13782ce8644ebf81b4ca352.jpg  —  jphide(***)
    90cebeec08fa513d9a1b0ce03f6d55fbb2fbd918.jpg  —  jphide(***)
    dc854fda81cb39db12584466d2160924ab183022.jpg  —  jphide(***)
    834344afa40f4bfb7391e165014f78f0f736187a.jpg  —  jphide(***)
    78701455b319ebc4a4ff8aa18026cffc1f1716c7.jpg  —  jphide(***)
    ...
```

Fig. 6. Fragment of the Stegdetect report on various images extracted from a VS sample. The app produced over 70 hits.

- *Most of the inspected apps are definitely not stegomalware as defined in this paper, but many of them behave very much like it.*

We found various cases of apps manipulating images through operations very similar to those used by some LSB encoding algorithms. For instance, several apps use some algorithms contained in the Apache Commons Imaging library[6], including Huffman coding for constructing minimum-redundancy codes, which is used by some steganographic tools.

- *A huge amount of apps contain images that clearly have embedded information.*

For example, Fig. 6 shows a fragment of the report on a VS sample with more than 70 (***) hits. However, to the best of our understanding, such hidden messages are never extracted during the app execution. We did not try to break the password by brute force and recover the embedded messages, but the headers used by the encoding algorithms are recognizable. Interestingly, many of these apps use cryptographic functions either to obfuscate payloads piggybacked together with the app (in the case of VS apps), or else as part of the legitimate function of the app (in some AP apps).

Such a large amount of images with hidden content suggest two different conclusions. Firstly, many apps might contain stegotexts for purposes other than stegomalware as defined in this paper. Some form of watermarking or other copyright protection techniques is a natural explanation, but they might have other uses. Secondly, even though Provos and Honeyman [19] report a very low percentage of false positives for Stegdetect, a recent study by Khalind et al. [14] suggests that it depends on the chosen sensitivity and that it can be quite high in some cases. Thus, it may be the case that a fraction of those images do not actually have any embedded information.

- *We found many cases of apps that use naïve methods to hide their malicious components.*

---

[6] http://commons.apache.org/proper/commons-imaging/.

```
public long a(byte[] param){
    return
    (0xFF & param[3]) +
    ((0xFF & param[4]) << 16) +
    ((0xFF & param[2]) << 40) +
    ((0xFF & param[0]) << 48) +
    (param[1] << 56);
}
```

```
private static String a(){
    return new StringBuilder()
        .append(i.a(200))
        .append(i.a(194))
        .append(i.a(216))
        .append(i.a(92))
        .toString();
}
```

(a) Sample 224058dbe82d71248cf93c5 transforms an array of bytes and returns the hidden value.

(b) Sample a8869e0ec4017d5294c0c2b plays with the way strings are declared in order to hide them.

Fig. 7. Examples of methods found in various apps to hide malicious components.

For instance, the sample with identifier

f0f65bd7287cf 83bfabcd22cbf6a0c8c

from VS simply stores the payload in a file called *data.png*. At runtime it uses several methods, including one with the suspicious name:

cn.bighead.utils.Encoding.covert2Url(covert)

to extract the required components. Figure 7 shows additional examples of suspicious operations used to conceal malicious information.

5   Conclusions and Future Work

In this paper, we have introduced and discussed the notion of stegomalware, a class of malware attacks using information hiding techniques to evade detection and hinder the task of security analysts. We have argued that this is a far more powerful capability than traditional obfuscation techniques currently observed in most smartphonne malware samples, such as disguising the malicious payload as an image or audio file by simply faking the file extension, or putting it at the end of another asset. We have introduced different architectural choices for smartphone stegomalware and demonstrated its viability with a proof-of-concept app that has remained undetected in the Google Play market for nearly 4 months so far.

In an attempt to check whether something similar to our notion of stegomalware is actually being used, we have proposed a detection framework that combines evidences gathered from the use of code execution capabilities, the presence of hidden messages detected by steganalysis, and the identification of steganographic decoding algorithms in an app's code. We have applied our detector to a dataset of around 55 K apps comprising both known malware and apps gathered from an alternative market.

Unfortunately, the preliminary results of our search for stegomalware in the wild are not conclusive. However, our study so far has produced a few interesting conclusions. For example, we have found that a substantial amount of apps do have assets with hidden information. Furthermore, many goodware apps such as games, productivity, and e-health tools incorporate steganographic decoding algorithms. We do not claim that such apps are stegomalware as defined in this paper, since there are many legitimate uses of steganography (e.g., intellectual property protection). Nevertheless, we believe this issue deserves more attention. In particular, *we recommend market operators to include steganalysis and other forms of stegomalware detection among their analysis techniques.*

Our work can be extended in a number of ways. For instance, the detection framework implemented so far only includes image and audio steganography. There are stegosystems for many other digital objects that are often found among an app's resources, such as XML and HTML documents, databases, and other multimedia files. We will incorporate appropriate distinguishers and content filters to our prototype to check such assets.

Acknowledgements. We are very grateful to the anonymous reviewers for constructive feedback and insightful suggestions that helped to improve the quality of the original manuscript. This work was supported by the MINECO grant TIN2013-46469-R (SPINY: Security and Privacy in the Internet of You).

References

1. Arp, D., Spreitzenbarth, M., Hübner, M., Gascon, H., Rieck, K.: Drebin: effective and explainable detection of android malware in your pocket. In: Proceedings of Network and Distributed System Security Symposium (NDSS), February 2014
2. Bastien, F.: Sss - simple steganalysis suite (Visited 2014). https://code.google.com/p/simple-steganalysis-suite/
3. Cachin, C.: Digital steganography. In: van Tilborg, H.C.A. (ed.) Encyclopedia of Cryptography and Security, pp. 159–164. Springer, US (2005)
4. Cheddad, A., Condell, J., Curran, K., Mc Kevitt, P.: Digital image steganography: survey and analysis of current methods. Signal Process. **90**(3), 727–752 (2010)
5. Desnos, A., et al.: Androguard: Reverse engineering, malware and goodware analysis of android applications (Visited December 2013), https://code.google.com/p/androguard
6. Egele, M., Scholte, T., Kirda, E., Kruegel, C.: A survey on automated dynamic malware-analysis techniques and tools. ACM Comp. Surv. **44**(2), 1–42 (2012)
7. Farid, H., Siwei, L.: Detecting hidden messages using higher-order statistics and support vector machines. In: Petitcolas, F.A.P. (ed.) IH 2002. LNCS, vol. 2578, pp. 340–354. Springer, Heidelberg (2002)
8. Forczmanski, P., Wegrzyn, M.: Open virtual steganographic laboratory. In: International Conference on Advanced Computer Systems (ACS-AISBIS) (2009). http://vsl.sourceforge.net/
9. Fridrich, J.: Feature-based steganalysis for JPEG images and its implications for future design of steganographic schemes. In: Fridrich, J. (ed.) IH 2004. LNCS, vol. 3200, pp. 67–81. Springer, Heidelberg (2004)

10. Fridrich, J., Goljan, M., Hogea, D.: New methodology for breaking steganographic techniques for JPEGs. In: International Society for Optics and Photonics Electronic Imaging 2003, pp. 143–155 (2003)
11. Gao, J., Bai, X., Tsai, W.T., Uehara, T.: Mobile application testing: a tutorial. Computer 47(2), 46–55 (2014)
12. Huang, H., Zhu, S., Liu, P., Wu, D.: A framework for evaluating mobile app repackaging detection algorithms. In: Huth, M., Asokan, N., Čapkun, S., Flechais, I., Coles-Kemp, L. (eds.) TRUST 2013. LNCS, vol. 7904, pp. 169–186. Springer, Heidelberg (2013)
13. Johnson, N.F., Jajodia, S.: Exploring steganography: seeing the unseen. Computer 31(2), 26–34 (1998)
14. Khalind, O.S., Hernandez-Castro, J.C., Aziz, B.: A study on the false positive rate of Stegdetect. Digit. Invest. 9(3), 235–245 (2013)
15. Oberheide, J., Miller, C.: Dissecting the android bouncer. In: SummerCon (2012)
16. O'Kane, P., Sezer, S., McLaughlin, K.: Obfuscation: the hidden malware. IEEE Secur. Priv. 9(5), 41–47 (2011)
17. Petitcolas, F.A., Anderson, R.J., Kuhn, M.G.: Information hiding-a survey. Proc. IEEE 87(7), 1062–1078 (1999)
18. Pfitzmann, B.: Information hiding terminology. In: Anderson, R. (ed.) IH 1996. LNCS, vol. 1174, pp. 347–350. Springer, Heidelberg (1996)
19. Provos, N., Honeyman, P.: Hide and seek: an introduction to steganography. IEEE Secur. Priv. 1(3), 32–44 (2003)
20. Provos, N., Honeyman, P.: Detecting steganographic content on the internet. Technical report, Center for Information Technology Integration University of Michigan (2001)
21. Rastogi, V., Chen, Y., Enck, W.: AppsPlayground: automatic security analysis of smartphone applications. In: Proceedings of the Third ACM Conference on Data and Application Security and Privacy CODASPY '13, pp. 209–220. ACM, New York (2013)
22. Rastogi, V., Chen, Y., Jiang, X.: Droidchameleon: evaluating android anti-malware against transformation attacks. In: Proceedings of the 8th ACM SIGSAC Symposium on Information, Computer and Communications Security ASIA CCS '13, pp. 329–334. ACM, New York (2013)
23. Shabtai, A., Tenenboim-Chekina, L., Mimran, D., Rokach, L., Shapira, B., Elovici, Y.: Mobile malware detection through analysis of deviations in application network behavior. Comput. Secur. 43, 1–18 (2014)
24. Suarez-Tangil, G., Tapiador, J.E., Lombardi, F., Pietro, R.D.: Thwarting Obfuscated malware via differential fault analysis. IEEE Comput. 47(6), 24–31 (2014)
25. Suarez-Tangil, G., Tapiador, J.E., Peris, P., Ribagorda, A.: Evolution, detection and analysis of malware for smart devices. IEEE Commun. Surv. Tutorials 16(2), 961–987 (2014)
26. Suarez-Tangil, G., Tapiador, J.E., Peris-Lopez, P., Blasco, J.: Dendroid: a text mining approach to analyzing and classifying code structures in android malware families. Expert Syst. Appl. 41(1), 1104–1117 (2014)
27. Upham, D.: Jsteg (1997). http://www.tiac.net/users/korejwa/jsteg.htm
28. Wang, K., Parekh, J.J., Stolfo, S.J.: Anagram: a content anomaly detector resistant to mimicry attack. In: Advances in Intrusion Detection. pp. 226–248 (2006)
29. Westfeld, A.: F5-A steganographic algorithm. In: Moskowitz, I.S. (ed.) IH 2001. LNCS, vol. 2137, p. 289. Springer, Heidelberg (2001)
30. Zhou, Y., Jiang, X.: Dissecting android malware: characterization and evolution. In: IEEE Symposium on Security and Privacy. pp. 95–109 (2012)

A Lightweight Security Isolation Approach for Virtual Machines Deployment

Hongliang Liang[(✉)], Changyao Han, Daijie Zhang,
and Dongyang Wu

Beijing University of Posts and Telecommunications, Beijing, China
{hliang, changyao}@bupt.edu.cn

Abstract. Cloud computing has changed the way of IT services; virtualization technology is the foundation of it, which directly affects the security and reliability of the cloud computing platform. From the point of virtualization technology security, we study to integrate mandatory access control mechanism into virtual machines deployment to control resources available for virtual machines, design and implement a lightweight MAC-based strong isolation and migration approach between virtual machines. Experiments show that our method is effective in isolation and migration, and with less performance overload.

Keywords: Virtualization · Hypervisor · Isolation · Migration · Mandatory access control

1 Introduction

Cloud computing cause a profound impact on the way of IT resources management and utilization. Providers can easily adjust the allocation of resources and services architecture, such as Amazon AWS [1], Microsoft Azure [2] and Google Cloud Platform [3]. So customers can require resources according to their own need. With the rapid development of cloud computing services, Cloud computing customers are increasingly concerned about their services security. However, traditional well-known protection methods are not well suited. Physical isolation achieves security protection through spatial separation between networks or computer equipment. Firewalls establish security gateway between Internet and Intranet to audit network traffic packets to implement security isolation. Intrusion detection systems can be deployed to monitor the health of the network and systems, attempting to detect a variety of attacks, so new security mechanisms are needed for layering, clustering and adjusting cloud computing infrastructure [4].

In this paper, we study to utilize the Linux kernel's Mandatory Access Control (MAC) [5] security mechanism to secure virtual machines deployments, provide strong isolation between guest virtual machines (VM), thus securing cloud computing platform at the infrastructure level. We make the following contributions in this paper:

First, we design and implement a lightweight secure isolation method for virtual machines deployment. We integrate mandatory access control mechanism into libvirt to control resources available to guests VM, implementing strong isolation between guests VM.

D. Lin et al. (Eds.): Inscrypt 2014, LNCS 8957, pp. 516–529, 2015.
DOI: 10.1007/978-3-319-16745-9_28

Second, we design and implement a trusted channel mechanism to secure virtual machines migration. We introduce security label in socket communication to ensure VM migration secure.

The remainder of this paper is organized as follows. Section 2 describes our motivation. Section 3 presents the design method. Implementation is detailed in Sect. 4. We evaluate the system in Sect. 5, including isolation, migration, and performance. Section 6 describes related work. We conclude in Sect. 7.

2 Virtualization Security Analysis

According to different demand, Virtualization technology can be divided into processes virtualization, server virtualization, network virtualization and storage virtualization. We focus on server virtualization in the paper. As the basis for cloud computing, virtualization technology enables customers to deploy multiple virtual machines on a single hardware, which are managed by the virtual machine monitor (hypervisor) [7].

Virtual machine monitor can be divided into two types:

Type I: Monitor runs directly on the host physical hardware, providing customers with virtual machines runtime environment, such as Xen [8], VMWare ESXi [9].

Type II: Monitor runs on the host operating system as an application program, such as QEMU/KVM [10], VirtualBox [11].

Virtual machine monitor is a privileged program, which deploys multiple guest virtual machines on the same hardware, and isolates each guest virtual machines. Isolation protection reduces the security threats from attackers and malicious programs, however, virtualization environment faces new attacks, for example, the virtualization layer between host system/hardware and virtual machines is becoming the new attack target. An attacker could exploit the vulnerability of hypervisor to attack virtualization platform, host and other virtual machines on the same platform, causing serious threats, as shown in Fig. 1.

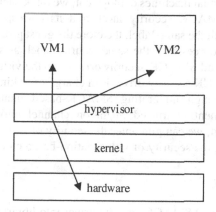

Fig. 1. Virtualization platform security threats

Therefore, the security isolation is growing the focus of virtualization security researchers and attackers. Attackers use virtual machine escape and other attack methods to threat the reliability of virtualization isolation [12]. At the same time, some research projects, such as HyperSentry [13], HyperSafe [14], CloudVisor [15], aim to reinforce the hypervisor, which undoubtedly increase the complexity of hypervisor software, thus cause potential security threat. Another research projects (NOVA [16], NoHype [17], etc.) split the privileged function of virtualization layer and reduce the risk of threats, while involve big change to existing virtualization solutions. sHype [18], sVirt [19] integrate MAC mechanism to enhance hypervisor security but are complex. Our major idea is inspired by sHype and sVirt.

In this paper, we will use the host system's security mechanisms to enhance the security of isolation in type II virtualization environment. Our method directly uses Linux kernel existing MAC security mechanism to reinforce isolation, without making any change to hypervisor software. we utilize Linux kernel's SMACK [6] module to implement lightweight MAC-based security isolation for virtual machines deployment.

3   Design Overview

Mandatory access control is a kind of system security mechanism, and is implemented in Linux kernel based on Linux Security Module (LSM) [20] framework. Typically, we can use MAC mechanism to control subject (such as process) access to the object (such as file or other process). Access control to processes in host system and isolation to virtual machines deployed in cloud platform is similar. For example, different guest virtual machines run as the processes of host system, and need to be controlled to access image files and other resources on host. Therefore, we use host system's MAC mechanism to reinforce isolation protection between virtual machine processes. When deploying and running virtual machines, different virtual machine processes will be controlled to only access their own corresponding image files and related resources, thus achieving strong isolation.

Our system is built on QEMU/KVM hypervisor in Linux, to use Linux's SMACK module to reinforce the isolation of guest virtual machines, controlling resources available to guests. In virtual machines deployment, we mark each guest virtual machine process with unique SMACK security label, and its corresponding image file and resources are marked with the same label, therefore the guest process can only access the guest image file and resources with the same security label, as shown in Fig. 2.

We implement and add SMACK security driver into the virtualization management library libvirt [21]. SMACK security driver is in charge of linking libvirt with SMACK module in kernel. Through integrating SMACK-based mandatory access control mechanism and implementing trusted migration channel into the virtual machines deployment management, we can guarantee the security of isolation and easy to deploy features, thus improving the security of virtualization-based cloud computing platform.

4   Implementation

In this section, first, we add SMACK security driver into libvirt, described in Sect. 4.1. Second, we present how to integrate SMACK mandatory access control mechanism into virtual machines deployment in Sect. 4.2. Finally, we describe the implementation of SMACK-labelled trusted migration channel in Sect. 4.3.

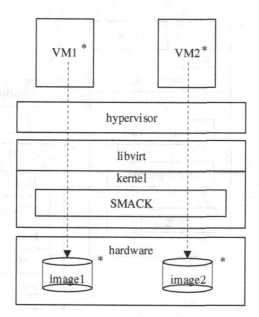

Fig. 2. MAC-based security isolation (* represents SMACK security label)

4.1   SMACK Security Driver Module for Libvirt

Libvirt is a virtualization management library, which supports a variety of hypervisors by integrating back-end drivers, including QEMU/KVM, Xen, etc. Libvirt provides a set of API and users can build tools to manage virtual machines, such as virsh, virt-manager.

Libvirt consists of various driver modules. In order to integrate SMACK mandatory access control into virtual machines deployment, we design to add SMACK security driver module into libvirt. In virtual machines deployment, SMACK security driver module will work with other driver modules such as hypervisor driver, storage driver to manage SMACK label, as shown Fig. 3.

We implement SMACK security driver for libvirt to support SMACK kernel module. Below we describe the main functions implementation of SMACK security driver, including security label generation, process label management, image files label management and socket label management.

Security label generation. When using libvirt to deploy virtual machines, different guest virtual machines are assigned with different security labels, so we construct SMACK label (of ASCII strings type) based on the UUID of each virtual machine. In libvirt, a guest virtual machine's configuration is stored in the structure _virDomain-Def, whose member uuid [VIR_UUID_BUFLEN] saves the UUID of the virtual machine, we constructed virtual machine SMACK label through adding UUID with SMACK_PREFIX, namely SMACK_PREFIX_UUID. SmackGenSecurityLabel() implements the function.

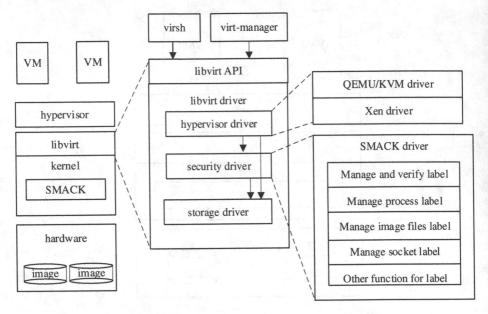

Fig. 3. Libvirt driver modules

Process label management. Linux system provides /proc/PID/attr interface to get/set process security attributes. The content of /proc/PID/attr/current interface is the security label of process PID and the process can set its own security label via writing security label to the interface. In SMACK security driver, we implement SmackSet-SecurityProcessLabel() and SmackSetSecurityChildProcessLabel() function to set process security labels. SmackSetSecurityProcessLabel() calls smack_set_label_for_self() function to directly write security label to /proc/PID/attr/current interface. When a process creates a child process, the security label of parent process and child process are consistent by default, SmackSetSecurityChildProcessLabel() function is used to set the security label of child process different with the parent process.

Image files label management. The SMACK security label of files is stored in the extended attribute (security.SMACK64) of file system. To mark image files with security label, we need to write the security label information to the attribute security. SMACK64 of image files. In SMACK security driver, we implement SmackSetSe-curityImageLabel() and SmackRestoreSecurityImageLabel() to manage image file's security label, the former is responsible to mark image files' security label when virtual machines start, the latter to reset image files' security label to default values when virtual machines close. In order to write security label to the extended attribute of image file, the key is to locate the image position, libvirt provides structure _virDo-mainDiskDef to store virtual machine configuration information, whose member field src points to the location of the image file. SmackSetSecurityImageLabel() use setattr() to write security label to the security.SMACK64 attribute of image file. SmackRestoreSecurityImageLabel() uses the same method to locate image file and reset its security label to the default.

Socket label management. Because Linux system does not provide socket security attribute interface for SMACK in /proc/PID/attr, we modify the Linux kernel and add two interfaces (/proc/PID/attr/sockoutcreate and /proc/PID/attr/sockincreate) to set socket SMACK security label. After a process writes SMACK label to /proc/PID/attr/sockoutcreate, the process create socket, which sends packets carrying the security label. For SMACK label written in /proc/PID/attr/sockincreate interface, SMACK will compare it with the label stored in received packets and determine whether to accept them according to the access rules. If sockoutcreate and sockincreate interfaces are empty, then packets will carry default security label and SMACK will also use the default label to make access control decision. Function SmackSetSecuritySocketLabel() and SmackSet- SecurityDaemonSocketLabel() are added to write SMACK label to /proc/PID/attr/sockoutcreate and /proc/PID/attr/sockincreate interfaces. We will discuss the details of sockoutcreate and sockincreate in Sect. 4.3.

4.2   Integrating SMACK Mechanism to Libvirt

SMACK is designed to provide a lightweight mandatory access control mechanism. Subject (such as process) and Object (such as file) need to be marked with security labels. The default SMACK security policy restricts that only when the subject and object have the same label, the subject can access read/write/execute/append (r/w/x/a) the object.

In Sect. 4.1, we implement SMACK security driver for libvirt and it will be registered and initialized by daemon libvirtd on startup. In order to integrate SMACK mandatory access control to virtual machines deployment, virtual machine process, image files and other resources should be marked with SMACK security label. Libvirt uses hook functions to call SMACK security driver when virtual machines are started by libvirt. At runtime, virtual machine process and image file will have the same SMACK security label, so SMACK mandatory access control mechanism will take effect, as shown in Fig. 4.

Hook functions to set security label. In libvirt, function qemuProcessStart() is responsible to start the virtual machine. During startup, virtual machine boot process and image files will be set SMACK security label by different hook functions, which are called during two phases, the initialization phase and the setup phase.

In initialization phase, hook function virSecurityManagerGenLabel() initializes security label, it actually call SMACK security driver function SmackGenSecurityLabel() to generate and save security labels for process, image files and other resources. Function virSecurityManagerSetChildProcessLabel() will initialize security label for child process, it actually call SMACK security driver function SmackSetSecurityChildProcessLabel() to save specific security labels for the child process.

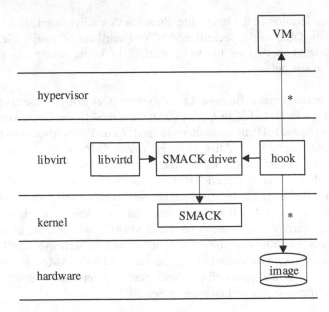

Fig. 4. Integrating SMACK mechanism with Libvirt (* represents SMACK security label)

In setup phase, libvirt main process first creates a child process as virtual machine boot process, and then enters wait state, the child process will call function smack_-set_label_for_self() to write security label to /proc/PID/attr/current. Then, the virtual machine boot process will possess security label, then notifies the parent process and enters wait state. The parent process will wake up and call function virSecurityMan-agerSetAllLabel() to mark image files and other resources, which actually calls SMACK security driver SmackSetSecurityImageLabel() and other functions. Finally, parent process notifies child process, the child process (boot process) will access image file to start virtual machine. As shown in Fig. 5.

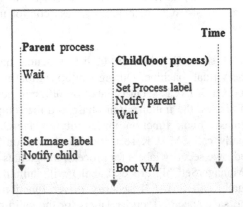

Fig. 5. The process of starting VM in libvirt

4.3 Migrating Virtual Machines Under Trusted Access Control

Libvirt supports hypervisor (QEMU/KVM) for VM's non-shared storage live migration [22]. To provide secure migration between source host and destination host, we use SMACK security driver to control the migration. As shown in Fig. 6, guest virtual machine running on the source host has security label. When migration happens, a socket link will be established between the source and destination hosts. At source host, SMACK driver uses CIPSO [23] to carry security label in packets. At destination host, SMACK driver extracts security label from received packets, judge whether the label meet SMACK security policies, if meet, it will receive the packets to prepare virtual machine, if not, it will discard the packets. When the migration is completed, the VM on the source host stops running, the migrated VM will continue running on the destination host, with the same running state and same security label as running on the source host.

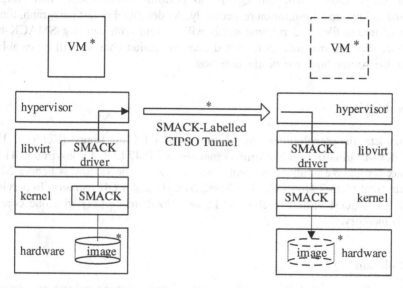

Fig. 6. VM migration under trusted access control (* represents SMACK security label)

Supporting socket label in SMACK. In Sect. 4.1, we mention to add Linux kernel with /proc/PID/attr/sockoutcreate and/proc/PID/attr/sockincreate interfaces for setting socket security. We will discuss it in detail. SMACK module is implemented based on LSM framework, process-related SMACK label is saved in structure task_smack, whose member field smk_task corresponding to/proc/PID/attr/current, we add two members smk_sk_in and smk_sk_out corresponding to/proc/PID/attr/sockincreate and/ proc/PID/attr/sockoutcreate, modified structure task_smack is shown in Fig. 7. Moreover, we modify function smack_sk_alloc_security() to examine smk_sk_in and smk_sk_out before setting socket label. Other functions such as smack_getprocattr() and smack_setprocattr() are also modified to add support for sockincreate and sockoutcreate.

H. Liang et al.

```
struct task_smack {
struct smack_known * smk_task;   // label for access control
struct smack_known * smk_forked; // label when the fork
struct smack_known * smk_sk_in; // label for socket in
struct smack_known * smk_sk_out; // label for socket out
struct list_head smk_rules;
struct mutex smk_rules_lock;
).
```

Fig. 7. Structure task_smack in SMACK

Trusted migration channel. Non-shared storage live migration is supported through ensuring both virtual machine image files and runtime status to be migrated between different hosts. Libvirt will call QEMU to perform "drive-mirror" and "migrate" command to implement migration respectively. As described in last paragraph, the data packets of image files and runtime status will be sent with carrying SMACK-based CIPSO security information, so a trusted communication channel will be established between the source host and destination host.

5 Evaluation

We complete the development based on libvirt-1.1.4 on Linux (Kernel 3.11.10) QEMU/KVM virtualization platform, comprising of 1847 LOC for libvirt and 41 LOC for Linux kernel. We build experiment environment as follows, host is Fedora 20 with hardware Intel Core 2 Duo CPU T6670@2.20 GHz and 4 GB memory, hypervisor is QEMU 1.4.2, guest virtual machine is Fedora cloud image [24] with one core plus 512 MB memory.

5.1 Isolation

First, we will verify that each running guest virtual machine process and its corresponding image file have the same SMACK security label and different virtual machines with different security labels. We start guest virtual machine fedora-01 and fedora-02, as shown in Fig. 8, two virtual machines are running simultaneously.

```
[root@src fedora]# virsh list
 Id    Name                         State
----------------------------------------------------
 2     fedora20-01                  running
 3     fedora20-02                  running
```

Fig. 8. Virtual Machines fedora-01 and fedora-02 run simultaneously

Since we employ QEMU as hypervisor, each virtual machine process is QEMU process. Different virtual machine processes have different SMACK Security labels, as shown in Fig. 9.

```
[root@src fedora]# ps -eZ | grep qemu
smack-a8a2ceb9-94f2-4d9c-8a8b-6f39a2e77aaa 4748 ? 00:00:19 qemu-system-x86
smack-7ce54237-811e-4079-9b61-5978012582ac 4762 ? 00:00:18 qemu-system-x86
```

Fig. 9. The SMACK labels of running VM fedora-01 and fedora-02

Figure 10 shows the security labels of virtual machine image files. Comparing with Fig. 9, the security label of each virtual machine process is consistent with the corresponding image file. SMACK will ensure virtual machine process only access the image files with same label, thus achieving strong isolation between virtual machines in deployment.

```
[root@src fedora]# chsmack /var/lib/libvirt/images/Fedora-i386-20-01.qcow2
/var/lib/libvirt/images/Fedora-i386-20-01.qcow2 access="smack-a8a2ceb9-94f2-4d9c
-8a8b-6f39a2e77aaa"
[root@src fedora]# chsmack /var/lib/libvirt/images/Fedora-i386-20-02.qcow2
/var/lib/libvirt/images/Fedora-i386-20-02.qcow2 access="smack-7ce54237-811e-4079
-9b61-5978012582ac"
```

Fig. 10. The SMACK labels of VM fedora-01 and fedora-02 image files

To test mandatory access control for virtual machines isolation deployment, we use libvirt to start virtual machine and intentionally mark image file with different security labels from virtual machine boot process. In such a case, we simulate that a guest process attacks other guest images. When virtual machine boot process try accessing image file to start virtual machine, SMACK mandatory access control mechanism will prevent boot process access image file, resulting in virtual machine failing to start (Permission denied), as shown in Fig. 11.

```
[root@src fedora]# virsh start fedora20-01
error: Failed to start domain fedora20-01
error: internal error: process exited while connecting to monitor: Warning: opti
on deprecated, use lost_tick_policy property of kvm-pit instead.
qemu-system-x86_64: -drive file=/var/lib/libvirt/images/Fedora-i386-20-01.qcow2,
if=none,id=drive-virtio-disk0,format=qcow2: could not open disk image /var/lib/l
ibvirt/images/Fedora-i386-20-01.qcow2: Permission denied

[root@src fedora]# virsh start fedora20-02
error: Failed to start domain fedora20-02
error: internal error: process exited while connecting to monitor: Warning: opti
on deprecated, use lost_tick_policy property of kvm-pit instead.
qemu-system-x86_64: -drive file=/var/lib/libvirt/images/Fedora-i386-20-02.qcow2,
if=none,id=drive-virtio-disk0,format=qcow2: could not open disk image /var/lib/l
ibvirt/images/Fedora-i386-20-02.qcow2: Permission denied
```

Fig. 11. VM processes fail to access image files

Above experiments show that SMACK provides a lightweight mandatory access control mechanism, which simplifies security policy configuration. Security policy requires subject (virtual machine process) to access object (resources) when both have the same security label. Security policies are customized and extended by modifying the configuration files.

5.2 Migration

QEMU supports non-shared storage live migration. Because we integrate SMACK security driver into libvirt, source host and destination host establish SMACK-labelled CIPSO tunnel to migrate image files and runtime state. After migration, virtual machine running on destination host has the same security label as the one on source host.

```
[root@target fedora]# ps -eZ | grep qemu
smack-7ce54237-811e-4079-9b61-5978012582ac 6693 ? 00:00:50 qemu-system-x86
[root@target fedora]# chsmack /var/lib/libvirt/images/Fedora-i386-20-02.qcow2
/var/lib/libvirt/images/Fedora-i386-20-02.qcow2 access="smack-7ce54237-811e-4079
-9b61-5978012582ac"
```

Fig. 12. The SMACK label of VM fedora-02 after migration

When migration is completed, fedora-02 VM stops on source host and runs on destination host. Figure 12 shows the security label of fedora-02 process and image file on destination host. Comparing with Figs. 9 and 10, it is consistent with the security labels when running on source host.

5.3 Performance

We measure the overhead introduced by MAC-based security isolation protection for performance evaluation. For a throughout evaluation, we consider the start-up time of guest OS, application benchmarks, and micro-benchmark as evaluation factors.

Guest OS start-up time. We test virtual machines start-up time in three scenarios: without security driver, with SELinux security driver [19] and with SMACK security driver.

Table 1. Comparison of VM start-up time

Security driver	Start-up time (Overhead ratio[a])
Without	13.4 s (0 %)
With SELinux	13.65 s (1.8 %)
With SMACK	13.46 s (0.4 %)

[a] Overhead ratio = (without/with security driver time - without security driver time) ÷ without security driver time

As shown in Table 1, integrating security driver to libvirt causes extra time overhead in starting VM, due to the need to mark virtual machines boot process, image files and other resources with security label during VM start-up. Overhead with SMACK is less than the case with SELinux since of its lightweight design.

Application benchmarks. We perform application-level tests to measure the impacts on guest OS due to host MAC-based security mechanism's. We compile the Linux 3.11.10 kernel to evaluate performance. Figure 13 shows the normalized performance results of the case with SELinux protection and the case with SMACK protection using pure kernel (without any MAC protection) as baseline. Overall, MAC-based security mechanism introduces performance less 1 % overhead. And, SMACK has less overhead than SELinux.

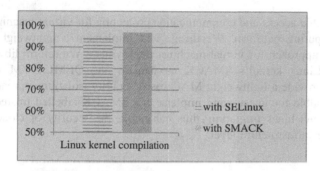

Fig. 13. The performance comparison of application benchmarks

Microbenchmark. We use UnixBench [26] to test the guest VM performance. The final scores for the cases without MAC, with SELinux, and with SMACK, are respectively 460, 458.2(0.4 % downgrade), and 459.7(0.06 % downgrade). The results is consistent with the previous application-level test, SMACK has better performance.

In conclusion, MAC-based security isolation mechanism results in performance overhead, Comparing SELinux and SMACK, SMACK introduces less overhead and provides a lightweight method for security isolation.

6 Related Work

Aiming at the security of hypervisor, HyperSentry [13] proposed a monitor program about the integrity of the virtual machine monitor. It developed a specific framework to evaluate the integrity of privileged part of virtualization layer. HyperSafe [14] proposed a lightweight protection solution for the type I virtual machine monitor, making the monitor has a self-protection ability to maintain own integrity. CloudVisor [15] used nested virtualization technology to guard the virtual machine monitor, achieving to protect the confidentiality and integrity of virtual machines in a cloud environment.

NOVA [16] established a microkernel-based virtual machine monitor, to minimize the size of the privileged program using user space virtualization technology. It changed the traditional monolithic structure into component-based and enforced the least privilege principle to different components. NoHype [17] removed the virtualization layer to reduce the security threats of the virtual machine monitor. The method utilized hardware virtualization to establish virtualization execution environment, each virtual machine was locked to a unique processor unit, using hardware to fixe memory allocation. As no needing monitor to control the virtual device, it reduced the threat surface.

There were studies to use mandatory access control mechanism to enhance hypervisor security. sHype [18] implemented mandatory access control based on the Xen hypervisor, sVirt [19] integrated SELinux into virtual machines deployment in QEMU/KVM hypervisor to achieve mandatory access control.

7 Conclusions

Providing suitable access and communication protection for virtual machines is crucial in cloud computing platform. We design and implement a lightweight mandatory access control approach for virtual machines deployment. We reform libvirt library to integrate the Linux kernel's SMACK mechanism into QEMU/KVM virtualization environment, provide a lightweight MAC-based security isolation approach to control resources available for guests, and implement a SMACK-labelled trusted channel to secure virtual machines migration, thus enhancing the security of cloud computing platform at the infrastructure level.

References

1. Amazon Web Services. http://aws.amazon.com/cn/
2. Microsoft Azure. http://azure.microsoft.com/zh-cn/
3. Google Cloud. https://developers.google.com/cloud/
4. Grobauer, B., Walloschek, T., Stocker, E.: Understanding cloud computing vulnerabilities. IEEE Secur. Priv. **9**, 50–57 (2011)
5. http://en.wikipedia.org/wiki/Mandatory_access_control
6. SMACK Project. http://schaufler-ca.com/
7. http://en.wikipedia.org/wiki/Hypervisor
8. Xen project. http://www.xenproject.org/
9. https://www.vmware.com/cn/products/vsphere/features/esxi-hypervisor.html
10. KVM project. http://www.linux-kvm.org/
11. Vitrualbox project. https://www.virtualbox.org/
12. Elhage, N.: Virtunoid: Breaking out of KVM, Black Hat USA (2011)
13. Azab, A.M., Ning, P., Wang, Z., Jiang, X., Zhang, X., Skalsky, N.C.: HyperSentry: enabling stealthy in-context measurement of hypervisor integrity. In: Proceedings of the 17th ACM Conference on Computer and Communications Security, pp. 38–49 (2010)
14. Wang, Z., Jiang, X.: HyperSafe: a lightweight approach to provide lifetime hypervisor control-flow integrity. In: Proceedings of the 2010 IEEE Symposium on Security and Privacy, pp. 380–395 (2010)
15. Zhang, F., Chen, J., Chen, H., Zang, B.: Cloudvisor: retrofitting protection of virtual machines in multi-tenant cloud with nested virtualization. In: Proceedings of the 23th ACM Symposium on Operating Systems Principles, pp. 203–216 (2011)

16. Steinberg, U., Kauer, B.: NOVA: a microhypervisor-based secure virtualization architecture. In: Proceedings of the 5th European Conference on Computer Systems, pp. 209–222 (2010)
17. Keller, E., Szefer, J., Rexford, J., Lee, R.B.: NoHype: virtualized cloud infrastructure without the virtualization. ACM SIGARCH Comput. Archit. News **38**, 350–361 (2010)
18. Sailer, R., Jaeger, T., Valdez, E., Caceres, R., Perez, R., Berger, S., Griffin, J.L., van Doorn, L.: Building a MAC-based security architecture for the Xen open-source hypervisor. In: Proceedings of the 21st Annual Computer Security Applications Conference, pp. 276–285 (2005)
19. sVirt project. http://selinuxproject.org/page/SVirt/
20. Morris, J., Smalley, S., Kroah-Hartman, G.: Linux security modules: general security support for the linux kernel. In: USENIX Security Symposium (2002)
21. libvirt project. http://libvirt.org/
22. http://wiki.libvirt.org/page/NBD_storage_migration
23. https://tools.ietf.org/id/draft-ietf-cipso-ipsecurity-01.txt
24. http://fedoraproject.org/zh_CN/get-fedora#clouds
25. SELinuxproject. http://selinuxproject.org/page/Main_Page
26. http://code.google.com/p/byte-unixbench/

A Novel Approach to True Random Number Generation in Wearable Computing Environments Using MEMS Sensors

Neel Bedekar[1(✉)] and Chiranjit Shee[2]

[1] Institute for Interdisciplinary Information Sciences, Tsinghua University,
FIT Building, 1-208, Beijing 100084, China
neel.bedekar@gmail.com
[2] Airbus Industrie, Xylem Building, Mahadevapura, Bangalore 560038, India

Abstract. Micro Electro Mechanical Systems (MEMS) sensors (accelerometer, gyroscope, and compass) offer a practical approach for true random number generation. Entropy values of 0.99 close to theoretical value of 1, and large Kullback-Leibler distances were obtained in this study [1]. The main contribution of this work was the generation of high quality random number strings, when the MEMS sensor was at complete rest, a configuration in which these sensors were heretofore considered to be inadequate. This was accomplished by using the initial noise in the sensing mechanisms for the MEMS sensors. The compass output stream passed the highest number of NIST Tests; 11/15 and 14/15 under stationary and complete motion, respectively [2–4]. Short burst (<1 s) strings passed 13 out of 15 NIST tests, and applying the Barak-Impagliazzo-Wigderson recursive extractor led to successful results in all 15 NIST tests. Interleaving MEMS output with audio resulted in a string that passed 14 out of 15 NIST tests.

Keywords: MEMS · Invensense · Nasiri · Accelerometer · Gyroscope · Compass · Random number · Entropy · NIST · Kullback-Leibler · Extractor · Von neumann · Barak-Impagliazzo-Wigderson · Audio

1 Introduction

Random numbers play a key role in information security. They are used in the generation of encryption keys and inputs to cryptographic functions. The most common use of random numbers is in generating the encryption keys for symmetric protocols like DES (Data Encryption Standard) and the asymmetric standards like ECC (Elliptical Curve Cryptography) [5]. Both encryption techniques rely on the strength of the random number generation, and any weakness or bias can be exploited to attack the system.

Wearable computing is generally considered as the next frontier of technology innovation. The Apple watch, Samsung Gear and Moto 360 are first versions of products that are expected to see widespread application. The total worldwide market for computers was 296 million units in 2013, while cell phone shipments reached 1.8 billion units [6]. The wearable computing market is expected to reach 19 million units in 2014. In the next 5–10 years due to the vast increases in features, the total market for

© Springer International Publishing Switzerland 2015
D. Lin et al. (Eds.): Inscrypt 2014, LNCS 8957, pp. 530–546, 2015.
DOI: 10.1007/978-3-319-16745-9_29

wearable devices and mobile phones is expected to cross 3 billion units [7]. It is vitally important to have a secure source of random number generation in wearable computing environments. The key challenges are small form factor, battery life constraints, and the lack of the traditional sources of random number generation due to the limitations in electronics that can be incorporated into the wearable device.

Micro Electro Mechanical System (MEMS) based sensors are ubiquitous in cell phones, gaming devices, and wearable electronics. For example, MEMS sensors are used to understand a phone's position and rotate the screen appropriately. MEMS based motion sensors (accelerometers, gyroscopes and compass) are excellent candidates for true non-deterministic random number generation in these environments. iSuppli estimates that the number of accelerometers shipped has grown from 700 million in 2009 to 2.3 billion in 2013 [8]. Every wearable computing device (phones, watches, medical sensors) is expected to have MEMS based motion sensors given their low cost (<\$ 1), small form factor and the ability to integrate with existing semiconductor electronic circuitry. MEMS sensors are thus ideal high entropy sources to generate random numbers [9].

2 Random Number Generation

Pseudo Random Number Generators (PRNG) rely on algorithms and are generated by specific computational formulae. PRNGs are deterministic and can be generated in a fast manner. However, they provide an easier hurdle for an adversary in that once the computation formulae are compromised, an attacker can gain access to the random number output [10].

True Random Number Generators (TRNGs) rely on a physical source of randomness that exists in nature or on the circuitry to generate the output. By its very nature, TRNGs are much more secure, since the hurdle for an attacker is substantially higher. Even if an adversary gains insight into the process of generating TRNGs, it is very difficult to model the output [11].

TRNGs have been obtained from numerous physical phenomena; white noise, keystrokes, radioactive decay measurements, and audio input to name a few. Until the early 2000s, it was unfeasible to use motion sensors, since the underlying technology (piezoresistive effect and capacitive accelerometers) was expensive and could not fit into a small form factor [12, 13].

2.1 Background Research

Silicon MEMS are a relatively new invention, and only in the last seven years have they begun to see widespread adoption with the smart phone revolution. Voris, Saxena and Halevi in 2011 were the first to show true random number generation in RFID tags using accelerometers [9]. They achieved speeds up to 85 bits/s. Lauradoux, Ponge and Roeck developed an entropy estimation algorithm for the iPhone accelerometer, gyroscope and compass sensors in 2011 [14]. Their work focused mainly on the entropy estimation, and they did not evaluate the properties of the random number strings generated. There is significant literature in using the accelerometers as a source of entropy in providing key exchange protocols for interactions between mobile devices. Groza and Mayrhofer developed several protocols for secure key exchange [15]. Studer, Passaro and Bauer developed the "Shake on it" protocol for secure exchange between mobile devices [16]. Chagnaadorj and Tanaka developed accelerometer based gesture input encryption schemes to facilitate secure exchange between mobile devices [17]. Several researchers have used physical noise to study random number generation. Zhou et al., studied the generation of true random numbers using mouse movement [18]. Chen I-Te studied the random number generation from audio and video sources and found that 98 % of the data passed all the NIST tests [13].

2.2 Research Goals

The first iPhone launched in 2007 only featured a 3 axis accelerometer. Compass sensors were incorporated into iPhone 3G in June 2009, and gyroscopes saw adoption in iPhone 4 (Sept. 2010). Due to the newness of the high sensitivity MEMS sensors, their use in true random number generation has not been studied in literature. Given this gap in research, the goal of this study was to evaluate all three sensors, and benchmark the NIST Test Suite performance of the resulting random number strings. An additional goal of this work was to evaluate the combination of uncorrelated entropy sources such as the microphone with the MEMS sensor output.

2.3 Silicon Micro Electro Mechanical Systems Sensors (MEMS)

The advent of silicon MEMS has radically expanded the market for motion sensing. MEMS sensors are typically built on top of CMOS circuitry (Complementary Metal Oxide Semiconductor), and are thus able to leverage the cost advantages, die shrinkage and computing power due to all the CMOS developments of the last forty years [19]. This study used MEMS sensors from InvenSense Inc., which use the patented Nasiri fabrication process, a state of the art vertical integration of MEMS with CMOS circuitry that allows for a very small form factor package, industry leading accuracy, and low cost [20] (Fig. 1).

MEMS accelerometers are based on the motion of a proof mass attached to a mechanical anchor in a bulk substrate. If the proof mass sees an acceleration it experiences a motion that can be measured due to capacitance change between the proof mass and reference electrode [22, 23].

MEMS gyroscopes are the most complicated sensors with a much higher manufacturing requirement than accelerometers. A proof mass within the sensor is driven into oscillation by an actuator. When the sensor is rotated it experiences a Coriolis force, which depends on the orientation of the angular velocity with the velocity of the proof mass. The Coriolis force, velocity of the proof mass and the angular motion are perpendicular to each other. A sensing of this Coriolis force leads to the measurement of angular velocity [24].

MEMS magnetic field sensors are based on the Hall effect, which is the voltage gradient produced in an electrical conductor when a magnetic field is applied in a direction perpendicular to the electric field. The output voltage is directly proportional to the magnetic field [25] (Fig. 2).

CMOS-MEMS Structure

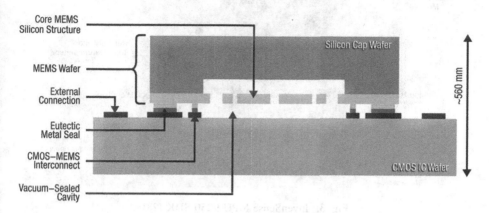

Fig. 1. CMOS MEMS Structure(21)

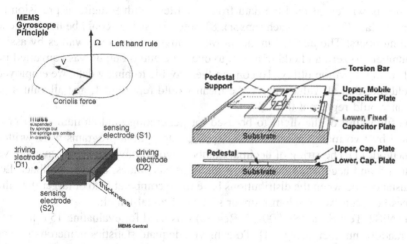

Fig. 2. MEMS Sensors – Principles of operation [26, 27]

3 Experimental Methods

The MPU 9250 outputs three axes accelerometer, gyroscope and compass data with 9 degrees of precision after the decimal point via a Bluetooth interface. The MPU 9250 allows a Bluetooth access with a resolution of 16 bits and gyroscope, accelerometer and compass ranges of ±2000 dps, ±16 g and ±4800 µT The python utility supplied with the Software Development Kit (SDK) was used for data collection. For the audio input, we utilized the microphone of an iPhone 5 and "I said what" app from Tapparatus to collect the input at 500 % sensitivity (Fig. 3).

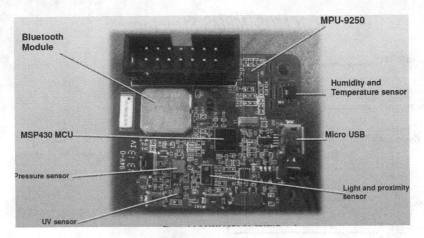

Fig. 3. InvenSense MPU 9250 SDK [28]

The data was initially collected as csv files (comma separated values) and a python program was written to read the data from csv files. With 9 digits of precision and 9 columns of data (3 axes for each sensor), 81 potential strings could be used to generate random numbers. The decimal numbers were converted to binary values by assigning even numbers to zero and odd numbers to one. The audio output was converted into a. wav file using a python utility. To convert the.wav file to binary data, we employed the threshold rule that all values above the mean would represent 1, and all values below the mean would represent 0.

The suitability of the digits to be used for generating random number strings was evaluated by two main approaches and appropriate python programs were written for each method; (1) The string of decimal numbers would have to display average values close to 4.5 and acceptable mean, skew and kurtosis values, and (2) Kullback-Leibler (KL) distances between the distributions have to be comparable to or larger than the KL distances between two random number strings of equal length.

The NIST Test Suite (SP 800-22, Rev 1a) was used for evaluating 15 different tests of the random number strings [4]. To achieve adequate statistics numerous successive runs were concatenated together to produce outputs strings of sufficient lengths for the NIST tests.

This study analyzed the MEMS sensor output for two boundary conditions; complete rest in an enclosure that minimized outside vibrations (most challenging environment) and continuous motion (board strapped on the wrist of a player playing ping pong). The major focus of this study was the complete rest condition, and two main data collection methods were employed. In the first case, ten measurements of 2 min each were concatenated to yield a random number string for NIST testing.

When any measurement process is undertaken, there is an initial noise inherent in the capacitive sensing and Hall effect voltage measurement mechanisms for the accelerometer, gyroscope and compass sensors. After 1–2 s, the noise dies down, and the underlying values of the sensors are displayed. MEMS sensors have traditionally been thought of as being severely challenged in the complete rest condition; since there is no motion, there is no fluctuation in the measured values, and hence no entropy for generating random number strings. To overcome this problem, this study focused on utilizing the noise inherent in the measurement process. Rapid measurements (<1 s) were performed and 100 such measurements were concatenated together to yield strings for NIST Test Suite measurements. Rapid (<1 s) measurements would only capture the noise, and could provide adequate entropy even when the MEMS sensor was at complete rest.

4 Results

4.1 Data Output – MEMS Sensor Output

Figure 4 shows that the gyroscope sensor output was 10 %–20 % more than the accelerometer, and the compass sensor output was 70 %–80 % lower than the accelerometer. Unlike an accelerometer, which relies on only one proof mass, a gyroscope works on the analysis of the Coriolis force on a mass set in motion. This requires two MEMS devices to achieve the required output. Gyroscopes also have to exhibit high sensitivity because of the smaller magnitude of the Coriolis forces that are being detected. Higher sensitivities imply more accurate capacitive sensing mechanisms and hence higher frequency of data output. As a result, 10 %–20 % more harvestable data was generated from gyroscopes than from accelerometers.

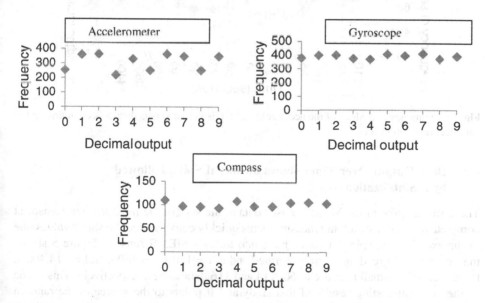

Fig. 4. Frequency of the decimal output of the 8th digit from the accelerometer, gyroscope, and compass x axis (MEMS sensor in complete motion)

The compass on the other hand works on the Hall effect principle in which a voltage is generated due to the Lorentz force of an orthogonal magnetic field on a conductor carrying electrical current. The Hall effect sensing mechanism inherently generated less random number data (but of a higher quality as later part of this study will show) than the capacitive sensing mechanism of the accelerometer and gyroscope.

4.2   Speed of Data Extraction

The InvenSense MPU 9250 outputs 43 unique characters in the first five seconds after the measurement was triggered. The preceding sections have shown that Digits 3 to Digit 9 past the decimal point can be used to harvest the data. With 3 axes measurement of the accelerometer, gyroscope and compass values, this gave us an effective output of 500 bits per second even in the most challenging case of the sensor at complete rest. One of the key disadvantages of true random number generators is the slow output which renders them difficult to use in statistical or cryptography applications. With an output of 500 bits/sec, the InvenSense MEMS sensors overcame one of the key disadvantages of TRNGs.

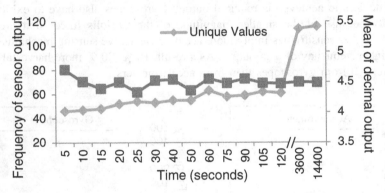

Fig. 5. Frequency and Mean of the accelerometer x axis output versus data collection time in the complete rest condition

4.3   Data Output Over Time Shows an Initial Spike Followed by a Stabilization

There was a spike in the MEMS sensor data at the start of the measurement (sensor at complete rest) after which the measurements quickly converged to the stable values due to the excellent damping circuitry built into today's MEMS sensors. Figure 5 shows that at t = 5 s, 43 readings were generated and at t = 120 s, t = 3600, and t = 14,400 s, there was only a small increase to 60, 114 and 116 data values, respectively. This is one of the most interesting results of this study and it points to the source of the random number generation. When the sensor detection process is initiated, there is a significant spike in readings and 43 unique readings were obtained in the first five seconds. In the time from 5 s to 120 s, there was only a small increase of 17 unique data values. Due to the excellent damping circuitry, only a small increase of 54 data values was obtained in the time period from t = 120 s (2 min) to t = 3600 (1 h). Increasing the data collection time to 14,400 s (4 h) led to a negligible increase of 2 data values.

When the MEMS sensor is at complete rest, there are two sources of fluctuation in values; (1) electrical noise inherent in the sensing process for the accelerometer, gyroscope, and compass sensors and (2) fluctuation in the measured values due to external vibrations, electrical, audio or other external effects.

The initial spike in the number of data values was most likely due to the noise in the sensing process of the accelerometer, gyroscope, and compass values. The subsequent small increase observed over time was due to the excellent damping circuitry built into the MEMS sensors. Before undertaking this study, it was thought that the random number generation would most likely be due to fluctuations in the measured values from the sensors, but this study proved that it was the noise in the measurement process that caused the initial accumulation of random numbers. State of the art MEMS sensors used in today's wearable computing devices have excellent damping circuitry, and it is for-tuitous that the sensing mechanism was responsible for the bulk of random number generation. Mean values converged to 4.5 at time intervals of t = 10 s, and stayed fairly constant after that. Since decimal output is being analyzed, the expected mean is 4.5, and these results indicate a uniform distribution (Fig. 6).

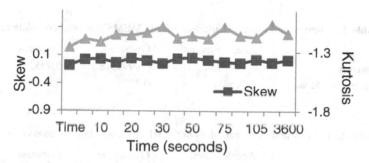

Fig. 6. Skew and Kurtosis for the accelerometer x axis versus data collection time in the complete rest condition

$$Skewness = \frac{\sum_{i=1}^{N}(Y_i - Y_m)}{(N-1)s^3} \quad (1)$$

Y_m = mean, s = standard deviation, N = number of data points

$$Kurtosis = \frac{\sum_{i=1}^{N}(Y_i - Y)^4}{(N-1)s^4} - 3 \quad (2)$$

Skewness is a measure of the symmetry of the distribution whereas Kurtosis is a measure of how peaked is the distribution compared with a normal distribution. Low kurtosis values indicate a flattish profile. [4]

The skew and kurtosis parameters stayed fairly constant for all data collection times. A perfectly random distribution of decimal characters would show skewness and kurtosis values of 0 and ~ -1.2. Even with time spans as small as 5 s, the observed values were very close to the ideal indicating that the MEMS sensor output was in-line with that of an unbiased sensor.

The results of this section showed that the decimal outputs from the MEMS sensor were an excellent source of physical noise to create random number strings.

4.4 Entropy Calculations

Entropy is a measure of the uncertainty of the measured value. In the case of binary data since we have two states, the theoretical maximum entropy is 1 (2 states) (Tables 1, 2).

Entropy H (x) is [29]

$$H(x) = -\sum p(x) \log p(x) \tag{3}$$

Table 1. Entropy and mean calculations for the MEMS sensor at complete rest

	Accelerometer	Gyroscope	Compass
Entropy	0.9992954	0.9998548	0.995252
Arithmetic Mean	0.5156	0.4929	0.5405

Table 2. Entropy and mean calculations for MEMS sensor in continuous motion

	Accelerometer	Gyroscope	Compass
Entropy	0.999995	0.999909	0.999887
Arithmetic Mean	0.5012	0.5055	0.5063

In this experiment we converted the decimal output from the MEMS sensor into binary values. Entropy values close to the perfect value of 1 were obtained even when the device was at complete rest. This is a very important conclusion as it shows the suitability of using any of three MEMS sensors for generating the random number output. Arithmetic mean was very close to 0.5 indicating a uniform binary distribution.

4.5 Kullback – Leibler Distance

The MEMS sensor outputs three axis data for three sensors, yielding nine potential values. Each reading is outputted with 9 digits of arithmetic precision. A total of 81 potential decimal outputs can thus be used to generate random numbers, and it is important to ascertain if these are independent distributions.

The Kullback–Leibler distance is proposed as a way of evaluating the suitability of combining the data. Kullback-Leibler divergence was first proposed in 1951 as a way to quantify the divergence between two distributions [1]. The KL divergence of probability distribution B from A is defined as

$$D_{KL}(A||B) = \sum_i \ln\left(\frac{A(i)}{B(i)}\right) x A(i) \tag{4}$$

D is always greater than zero and D_{KL} (A||B) is not equal to D_{KL} (B||A)

In this study, KL distance was used as a key metric to determine if the distributions obtained from each digit (out of the 9 digits after the decimal point) from the three axes of the accelerometer, gyroscope and compass could be combined. Higher KL values would indicate distributions that are far apart, and hence suitable for combination in generating large random number strings.

To evaluate the absolute value of the KL distance, the Pseudo Random Number Generation (PRNG) utility in python was used to create two additional distributions, Random 1 and 2, of the same length as the test string under evaluation. If the KL distance between two sensor data streams was higher than that between the two random number streams, then combining the sensor data was a viable technique of generating large strings of random numbers.

Tables 3 and 4 show the KL distance values for different test conditions. These results clearly show that (1) More than 60 %–70 % of the KL distance values for the MEMS sensors were greater than those observed between two random distributions. Combining data from various streams from the MEMS sensor is hence an excellent way of generating random number strings, (2) Successive runs yielded varying KL distances indicating an inherent non-deterministic nature of the MEMS sensor and (3) KL distances increased more than 50x when the sensor was in motion indicating a very robust performance for scenarios where the MEMS sensor was in motion. Combining data streams from digits 3 to 9 of the nine sensor axes was thus an excellent approach of generating long random number strings.

Table 3. KL distance between 9th digit of the accelerometer z axis for four consecutive runs (Random 1 to Random 2 KL distance was 1296). Shaded regions indicate KL distances higher than those obtained between two random number strings of equivalent length.

	29-1	29-2	29-3	29-4
29-1	0	268	2361	4163
29-2	99	0	2401	3867
29-3	3220	2968	0	2634
29-4	682	928	3052	0

4.6   Sensor in Motion

When the sensor was in full motion, more than 70 % of the KL distances were higher than those generated between two pseudo random number sequences in python. The KL

Table 4. KL distance between 9th digit of the z axis of the accelerometer (2), gyroscope (5) and compass (8) for four consecutive runs (Random 2 to Random 1 KL distance was 310 in the first table and 624 for the 2nd table).

	29-1	29-2	59-1	59-2	89-1	89-2
29-1	0	252	825	680	612	371
29-2	98	0	961	825	666	425
59-1	1160	1283	0	737	263	438
59-2	1448	1580	880	0	585	343
89-1	1546	1376	848	1074	0	362
89-2	1620	1535	1057	776	350	0

	29-3	29-4	59-3	59-4	89-3	89-4
29-3	0	1490	1270	1392	243	525
29-4	1950	0	790	1314	591	344
59-3	1607	1177	0	1158	587	759
59-4	1907	1798	1931	0	498	204
89-3	2065	2903	2125	2060	0	400
89-4	2673	2538	2317	1778	381	0

distances increased by 50x versus the sensor at rest configuration. Thus, a high relative entropy was obtained between the different distributions, when the sensor was in motion.

4.7   NIST Data Analysis

NIST Test Suite (SP 800-22, Rev 1a) was used for testing [30]. The data files were generated by using the 3rd to 9th digit of the accelerometer, gyroscope and compass axes outputs. To achieve sufficient long bit streams more than 50–100 test files were concatenated to generate the test file. If 7 out of 10 streams passed a test, then the sensor was considered to have passed the corresponding test. The Frequency Test measures the proportion of zeros and ones for the entire sequence, and measures whether they fall in the expected range of a random sequence. This is the most basic test and the remaining tests depend on passing this test. A majority of the test configurations in this study passed this test.

4.8   Compass Sensor Output Passed the Maximum NIST Tests

The most interesting conclusion was the substantial improvement in the NIST test results observed for the compass sensor versus the accelerometer and gyroscope, in spite of data counts that are typically 70 %–80 % lower (Table 3). For the stationary configuration, the compass sensor passed 11 out of 15 tests, and for continuous motion, successful results were obtained on 14 out of 15 tests.

The compass sensor is based on the Hall effect sensor wherein the output voltage changes due to the magnetic field. The superior compass sensor results were most likely due to the higher quality randomness inherent in the sensing of the Hall effect voltage.

4.9   Von Neumann and Barak-Impagliazzo-Wigderson Extractor

In 1951, Von Neumann proposed an extractor to generate a random binary sequence from a biased stream [2]. If a string is composed of 01001110, then the extractor will pair up this distribution into pairs (01), (00), (11), (10). Each pair which has the same digits is discarded and in the remaining pairs the second digit is discarded to yield us a string of 01 from the preceding stream. The Von Neumann extractor caused a slight degradation in the result with the resulting bit stream passing 6 out of 15 tests. The biggest contribution of the Von Neumann extractor is to yield an equal frequency of 0 s and 1 s. However, the entropy and mean calculations from previous sections have shown that the MEMS sensor output is exceptionally uniform. Hence the Von Neumann extractor did not cause any improvements in these tests. Since the Over Lapping Template tests the occurrence of predetermined strings, and the Linear Complexity test measures the length of Linear Feedback Shift Registers, the forced alteration due to Von Neumann extractor caused these tests to fail [30] (Table 6).

Table 5. NIST Test Suite results for MEMS sensor in complete motion (Pass threshold 7 out of 10)

Parameter	Accelerometer only	Gyroscope only	Compass only
Frequency	10/10	10/10	7/10
BlockFrequency	10/10	10/10	8/10
CumulativeSums	10/10	10/10	7/10
CumulativeSums	10/10	10/10	7/10
Runs	3/10	1/10	8/10
LongestRun	0/10	0/10	7/10
Rank	9/10	10/10	10/10
FFT	1/10	4/10	8/10
Nonoverlapping Template	10/10	10/10	10/10
OverlappingTemplate	10/10	8/10	9/10
Universal	10/10	10/10	10/10
ApproximateEntropy	0/10	0/10	0/10
Serial	0/10	2/10	7/10
Serial	0/10	5/10	8/10
LinearComplexity	9/10	10/10	10/10

The Barak-Impagliazzo-Wigderson two source extractor works on the principle of a recursive sum of a · b + c and the resulting bit stream passed 10 out of the total 15 NIST tests (Table 6) (Boaz Barak, 2006). The two big areas of improvement were in the Longest Run and FFT test. The Longest Run test checks the length of the 0 s and 1 s with that expected for a random distribution. Even though the frequency data showed excellent results indicating an even distribution of 0 s and 1 s, the native data from InvenSense MEMS sensor typically failed the Runs Test indicating a problem in the length of the runs. The Barak-Impagliazzo-Wigderson two source extractor addressed this problem, and a higher $\chi 2$ was obtained, which led to a passing of the Longest Run test. The FFT test detects periodic features that would indicate a non random distribution, and the recursive extractor overcame the limitations observed with the raw MEMS sensor data.

Table 6. NIST Test results with extractors MEMS sensor at rest (Pass threshold 7 out of 10)

Parameter	Stationary Sensor	Von Neumann Extractor	Barak-Impagliazzo-Wigderson
Frequency	10/10	10/10	8/10
BlockFrequency	6/10	10/10	0/10
CumulativeSums	10/10	10/10	8/10
CumulativeSums	10/10	10/10	8/10
Runs	0/10	0/10	1/10
LongestRun	0/10	0/10	7/10
Rank	10/10	5/10	9/10
FFT	0/10	1/10	10/10
NonOverlapping Template	10/10	10/10	10/10
OverlappingTemplate	7/10	0/10	10/10
Universal	10/10	10/10	10/10
ApproximateEntropy	0/10	0/10	0/10
Serial	0/10	0/10	0/10
Serial	0/10	0/10	0/10
LinearComplexity	9/10	4/10	8/10

4.10   Short Duration Measurements and Barak-Impagliazzo-Wigderson Extractors Led to a Dramatic Improvement in Results

This study tested the stationary MEMS sensor output in two data acquisition scenarios; (1) 2 min data collection time, and (2) 1 s data collection time. NIST test results for the short data collection time (1 s) showed a dramatic improvement in the test results, and random number strings passed 13 out 15 NIST tests as compared with 8 out of 15 for the 2 min data collection case. The 1 s data collection times predominantly captured the measurement noise, thus leading to better NIST test results (Table 7).

Table 7. NIST test results for different data collection times and extractor application (Pass threshold 7 out of 10)

Parameter	Stationary Sensor (2 min)	After Barak-Impagliazzo-Wigderson	Stationary Sensor (1 sec)	After Barak-Impagliazzo-Wigderson
Frequency	10/10	8/10	7/10	10/10
BlockFrequency	6/10	0/10	10/10	10/10
CumulativeSums	10/10	8/10	8/10	10/10
CumulativeSums	10/10	8/10	7/10	10/10
Runs	0/10	1/10	7/10	8/10
LongestRun	0/10	7/10	6/10	10/10
Rank	10/10	9/10	9/10	10/10
FFT	0/10	10/10	9/10	10/10
Non Overlapping Template	10/10	10/10	10/10	10/10
OverlappingTemplate	7/10	10/10	9/10	10/10
Universal	10/10	10/10	10/10	10/10
ApproximateEntropy	0/10	0/10	10/10	10/10
Serial	0/10	0/10	5/10	9/10
Serial	0/10	0/10	8/10	10/10
LinearComplexity	9/10	8/10	10/10	10/10

4.11 Adding an Uncorrelated Audio Output Led to a Dramatic Improvement in NIST Test Results

While the raw audio random number data stream only passed 6 out of the 15 NIST tests, interleaving this data with the MEMS sensors yielded a string that passed 14 out of the 15 tests (Table 8). Interleaving also effectively doubled the size of the output data.

4.12 Future Directions

The Apple watch uses visible and infrared LEDs in addition to photo sensors to detect heart rate, and several other health parameters. Other wearable devices are expected to include numerous unique sensors. In the future in addition to MEMS sensors, we will be able to use numerous other sensors to generate a random number string. Combining the output of the sensors that measure non-correlated parameters will most likely give rise to a stronger random number string.

5 Conclusions

This study proved that MEMS sensors offered excellent sources of entropy for true random number generation. While the original thesis that the MEMS sensors would serve as an excellent random number generation source did turn out to be true, the source

Table 8. NIST test results for MEMS + Interleaved audio random number strings (Pass threshold 7 out of 10)

Parameter	MEMS	Audio	Interleaved MEMS +Audio
Frequency	7/10	10/10	8/10
BlockFrequency	10/10	0/10	10/10
CumulativeSums	8/10	10/10	8/10
CumulativeSums	7/10	10/10	8/10
Runs	7/10	0/10	10/10
LongestRun	6/10	0/10	10/10
Rank	9/10	8/10	10/10
FFT	9/10	1/10	10/10
NonOverlapping Template	10/10	0/10	10/10
OverlappingTemplate	9/10	0/10	10/10
Universal	10/10	10/10	10/10
ApproximateEntropy	10/10	0/10	6/10
Serial	5/10	0/10	9/10
Serial	8/10	0/10	9/10
LinearComplexity	10/10	10/10	8/10

of entropy was a new discovery in this study. When the MEMS sensor was evaluated in the complete rest condition, this work showed that the bulk of the random number generation happened in the first five seconds of data collection. The source of the entropy was the noise in the sensing mechanism, as opposed to the fluctuations in the measured value. Hence, it was not a surprise that the MEMS sensor yielded an adequate random number string in the stationary configuration. This is the most unique contribution of this study in that it shines light on the origin of the random number generation.

While the overall MEMS sensor passed 8 out of 15 NIST tests, compass outputs offered superior NIST test result pass rates of 11/15 and 14/15 for the stationary and continuous motion cases, respectively. The Hall effect voltage sensing mechanism generated higher quality random number streams than the capacitive sensing mechanism used for accelerometers and gyroscopes. Concatenating short duration (<1 s) measurements yielded strings that passed 13 out of the 15 NIST tests as compared with 8 out of 15 NIST tests for the 2 min measurement case. Applying the Barak-Impagliazzo-Wigderson two source recursive extractor yielded a random number string that passed all 15 NIST tests.

Even in the most challenging situation of the MEMS sensor at complete rest, this work has shown high random number generation output speeds of 500 bits/sec. The Kullback-Leibler (KL) distances for successive runs, different axes data, and varying configurations (stationary versus continuous motion) showed significant relative entropy. High entropy values close to the theoretical value of 1, and very little skew point to high quality random number data streams obtained in this study. The addition of an uncorrelated output such as audio to MEMS sensors led to a dramatic improvement in the results. The resulting random number string passed 14 out of the 15 NIST tests, and offered a data output twice as large as the input. In conclusion, this study showed that even under the most challenging condition of the sensor being at complete rest, the MEMS sensors offered excellent entropy, and high data output speeds of 500–1000 bits per second, thus overcoming the traditional challenge of true random number generators.

Acknowledgments. The authors would like to thank Professor John Steinberger for helpful discussions and support through the summer internship. This work was performed at Tsinghua University, Institute for Interdisciplinary Information Sciences, Beijing, China and their support is greatly appreciated. The authors also appreciate the support of Cameron Ballingall and InvenSense Inc., which provided the MEMS sensors used in this study. The authors would also like to thank the reviewers for their comments and suggestions.

References

1. Leibler, R.A., Kullback, S.: On Information and Sufficiency. Ann. Math. Stat. **22**, 79–86 (1951)
2. Von Neumann, J.: Various techniques used in connection with random digits. In: Forsythe, G.E., Germond, H.H., Householder, A.S. (eds.) Monte Carlo Method. Government Printing Office, Washington (1951)
3. Barak, B., Impagliazzo, R., Wigderson, A.: Extracting randomness using few independent sources. SIAM J. Comput. **36**, 1095–1118 (2006). 4, Philadelphia USA: Society for Industrial and Applied Mathematics
4. NIST. Engineering and Statistics Handbook. s.l.: NIST (2012)
5. Whitfield, D., Hellman, M.: New directions in cryptography. IEEE Trans. Inf. Theory **IT-22** (6), 644–654 (1976)
6. Gartner: Forecast: PCs, ultramobiles and mobile phones 2014. Gartner Group, Egham (2014)
7. Llamas, Ramon T.: Worldwide Wearable Computing Device 2014–2018 Forecast and Analysis. IDC, Boston (2014)
8. Invensense Annual Report, Form 10-K. San Jose, CA, USA: s.n., March 2014
9. Voris, J., Saxena, N., Halevi, T.: Accelerometers and Randomness: Perfect Together. ACM, Hamburg, Germany (2011). WiSec
10. Dodis, Y., Pointcheval, D., Ruhault, S., Vergnaud, D., Wichs, D.: Security analysis of pseudo-random number generators with input. In: ACM Conference on Computer and Communication Security. ACM, Berlin, Germany (2013)
11. Barak, Boaz, Shaltiel, Ronen, Tromer, Eran: True random number generators secure in a changing environment. In: Walter, Colin D., Koç, Çetin Kaya, Paar, Christof (eds.) CHES 2003. LNCS, vol. 2779, pp. 166–180. Springer, Heidelberg (2003)
12. Wang, X., Xue, Q., Lin, T.: A novel true random number generator based on mouse movement and a one dimensional chaotic map. s.l. Math. Prob. Eng. **2012**, 12 (2012). Hindawi Publishing
13. Chen, I.T.: Random numbers generated from audio and video sources s.l. Math. Prob. Eng. **2013**, 7 (2013). Hindawi Publishing
14. Lauradoux, C., Ponge, J., Roeck, A.: Online Entropy Estimation for Non-Binary Sources and Applications on iPhone. Institut National de Recherche en Informatique et en Automatique, pp. 19–42 (2011)

15. Mayrhofer, R., Groza, B. (eds.): Simple accelerometer based wireless pairing with heuristic trees. In: Proceedings of the 10th International Conference on Advances in Mobile Computing and Multimedia. ACM (2012)
16. Studer, A., Timothy, P., Bauer, L.: Don't bump, shake on it; the expoitation of a popular accelerometer-based smart phone exchange and its secure replacement. In: Proceedings of the 27th Annual Computer Security Applications Conference, pp. 333–342. ACM, New York (2011)
17. Chagnaadorj, O., Tanaka, J.: MimicGesture: Secure Device Pairing with Accelerometer-Based Gesture Input. In: Han, Y.-E., Park, D.-S., Jia, W., Yeo, S.-S. (eds.) Ubiquitous Information Technologies and Applications. Lecture Notes in Electrical Engineering, vol. 214, pp. 59–67. Springer, Heidelberg (2013)
18. Zhou, Q., Liao, X., Wong, K., Hu, Y., Xiao, D.: True random number generator based on mouse movement and chaotic hash function s.l. Inf. Sci. **179**, 3442–3450 (2009). Elsevier
19. Nasiri, Steven: A Critical Review of MEMS Gyroscope Technology and Commercialization Status. Invensense Inc, San Jose (2006)
20. Nasiri, S.: Wafer-Scale Packaging and Integration Are Credited for New Generation of Low-Cost MEMS Motion Sensor Products. Invensense, San Jose (2006)
21. Daneman, M., Lim, M., Assaderaghi, F.: Evolution of MEMS towards a semiconductor model. Hearst Business Coummunications, New York (2012)
22. Nasiri, S., Seeger, J., Yaralioglu, G.: *20080314147* Saratoga, California (2008)
23. Qiu, J., Seeger, J., Castro, A., Tchertkov, I., Li, R.: *20120125104* Sunnyvale, California (2012)
24. Seeger, J., Nasiri, S., Castro, A: *2010132460* Menlo Park, USA (2010)
25. Dixon-Warren, J.: MEMS J (February 2011). www.memsjournal.com
26. MEMS Sensors; How it works technically. MEMS Central (2014). http://memscentral.com/Secondlayer/mems_motion_sensor_perspectives-2.htm. Accessed 20 July 2014
27. Silicon Designs. Silicon Designs (2014). http://www.silicondesigns.com/tech.html. Accessed 20 July 2014
28. Invensense MPU 9250 SDK. Invensense Inc, San Jose, California (2013)
29. Cover, T.M., Thomas, J.A.: Entropy, Relative Entropy and Mutual Information. Elements of Information Theory, pp. 12–25. John Wiley and Sons, New York (1991)
30. A Statistical Test Suite For Random and Pseudorandom Number Generators For Cryptographic Applications. National Institute of Standards and Technology, Gathiersburg, Maryland, USA (2010)

Author Index

Printed in the United States
By Bookmasters